1000
Solved Problems in Fluid Mechanics
(Includes Hydraulic Machines)

AF860173

About the Author

K Subramanya retired as Professor of Civil Engineering from the Indian Institute of Technology, Kanpur. He obtained his Bachelor's degree in Civil Engineering from Mysore University and Master's degree from the University of Madras. He obtained another Master's degree and Ph. D degree from the University of Alberta, Edmonton, Canada. He has over 35 years of teaching experience and has made significant contributions in the area of Water Resources Engineering. He is the author of several successful books which include *Flow in Open Channels, 2e* (TMH 1992) and *Engineering Hydrology, 2e* (TMH 1994).

Dr. Subramanya has published over eighty technical papers in various national and international journals and conferences. He is a fellow of the Institution of Engineers (India); and Member of the Indian Society of Technical Education, Indian Society for Hydraulics and Indian Water Resources Association.

Currently, he is a practicing consultant in Water Resources Engineering at Bangalore.

1000 Solved Problems in Fluid Mechanics

(Includes Hydraulic Machines)

K Subramanya
Formerly Professor
Department of Civil Engineering
Indian Institute of Technology
Kanpur

Tata McGraw-Hill Publishing Company Limited
NEW DELHI

McGraw-Hill Offices
New Delhi New York St Louis San Francisco Auckland Bogotá Caracas
Kuala Lumpur Lisbon London Madrid Mexico City Milan Montreal
San Juan Santiago Singapore Sydney Tokyo Toronto

Tata McGraw-Hill

Copyright © 2005, by Tata McGraw-Hill Publishing Company Limited.

No part of this publication may be reproduced or distributed in any form or by any means, electronic, mechanical, photocopying, recording, or otherwise or stored in a database or retrieval system without the prior written permission of the publishers. The program listings (if any) may be entered, stored and executed in a computer system, but they may not be reproduced for publication.

This edition can be exported from India only by the publishers,
Tata McGraw-Hill Publishing Company Limited.

ISBN(13 digit) 978-0-07-014476-7
ISBN(10 digit) 0-07-014476-1

Published by the Tata McGraw-Hill Publishing Company Limited,
7 West Patel Nagar, New Delhi 110 008, typeset in Times New Roman at
Tej Composers, WZ-391, Madipur Village, New Delhi 110 063

Delhi 110 032

RCXCRDLIDRYLA

The McGraw-Hill Companies

Contents

Chapter 8 Boundary Layer Concepts 175

Chapter 9 Drag and Lift on Immersed Bodies 194

Chapter 10 Turbulent Pipe Flow 210

Chapter 11 Pipe Flow Systems 232

Chapter 12 Flow in Open Channels 271

Chapter 13 Flow Measurement 314

Chapter 14 Unsteady Flow 349

Chapter 15 Compressible Flow 370

Preface

Fluid mechanics is an important constituent of the undergraduate syllabi of many engineering disciplines. The subject is given considerable importance in Civil, Mechanical and Chemical engineering programmes at the core as well as at professional levels. While problem-solving is an important aspect of learning fluid mechanics, a typical textbook in this subject does not provide adequate space for this aspect due to various constraints. Therefore, the students who are required to solve problems in various tests and examinations, find the subject of fluid mechanics difficult to cope with in their studies.

The book, primarily based on my earlier book, *Theory and Applications in Fluid Mechanics* (TMH, 1993) is designed as an essential supplement to any textbook in fluid mechanics with the needs of undergraduate engineering students in mind. The typical undergraduate syllabus in Fluid Mechanics is covered in sixteen chapters. In each of the chapters, an outline of the basic theory is given and is followed by a large number of carefully chosen worked examples covering all the sub-areas of the chapter theme. A set of objective questions, with answers provided at the end of the book, will help the students in quick review of the chapter and also in thorough understanding of the concepts. Further, a set of tested problems with answers is provided in each chapter to help the students in their practice. The contents of each chapter are designed to help the students in learning all aspects of the subject matter, viz. theory, application, information manipulation and processing. The set of multiple-choice objective questions provided at the end of the book in Appendix A cuts across various chapters and will be specially be useful to those preparing for competitive examinations. Every attempt has been made to make the book user-friendly.

SI units are used throughout the book and a logical sequence of topics is followed. In sixteen chapters, the book covers the syllabus of Fluid Mechanics and Hydraulic Machinery for the undergraduate programmes in Civil, Mechanical and Chemical engineering disciplines. The topics covered include Fluid Properties; Fluid Statics; Kinematics and Inviscid Flows; Energy and Momentum Equations and Their Applications; Dimensional Analysis and Similitude; Laminar and Turbulent Flows; Boundary Layer Concepts; Drag and Lift on Submerged Bodies; Pipe Flow Systems; Open Channel Flows; Flow Measurements; Unsteady Flows; Compressible Flows; and Hydraulic Machines comprising Turbines, Rotodynamic and Reciprocating Pumps. With this coverage, the book will be extremely useful to the engineering and AMIE students. Further, students preparing for competitive examinations such as UPSC examinations and GATE will find the book of invaluable help. Teachers will find the book beneficial as source material for designing question papers for various examinations.

Suggestions for improvement of the book would be gratefully appreciated.

K Subramanya

1

Properties of Fluids

CONCEPT REVIEW

1.1 INTRODUCTION

Fluid mechanics is the science of the mechanics of fluids. The terms fluid is used to denote matter in liquid or gaseous phase. As opposed to a solid, a fluid will undergo continuous deformation when subjected to shearing forces and the deformation will continue so long as the shearing force continues to act on the fluid body.

1.1.1 Continuum

In engineering problems dealing with fluids, one generally deals with dimensions that are very large compared to molecular sizes. The space between the molecules is not considered and the fluid properties are considered to vary continuously in space. The density of fluid is thus a point function. This method of considering fluid as a continuous mass is called as *continuum principle*. Except in dealing with rarified gases, all normal fluid mechanics analysis deals with fluid as a continuum.

1.1.2 Units

In fluid mechanics four fundamental dimensions namely mass, length, time and temperature are involved. These days the SI units are adopted in describing the various parameters of fluid flow. In this system the fundamental units are:

Mass [M]	kilogram	kg
Length [L]	metre	m
Time [T]	second	s
Temperature [θ]	Kelvin (for thermodynamic calculations) or °Celsius	K °C

Based upon these fundamental units a number of derived units are developed. The most commonly used derived units is the unit of force which is *newton* (N). A *newton* of force corresponds to an acceleration of 1 m/s^2 of a mass of 1 kg. The commonly used derived terms and their units are listed in Table 1.1.

Table 1.1 *Commonly Used Derived Terms in Fluid Mechanics*

Derived term	*Dimension*	*SI unit*	*Abbreviation*
Area	(L^2)	m^2	
Volume	(L^3)	m^3	
Velocity	(LT^{-1})	m/s	
Acceleration	(LT^{-2})	m/s^2	
Force	(MLT^{-2})	N	
Pressure (or stress)	($ML^{-1}T^{-2}$)	N/m^2	Pascal Pa = N/m^2
Energy (or work)	(ML^2T^{-2})	N.m	Joule J = N.M.
Power	(ML^2T^{-3})	J/s	Watt W = J/s
Density	(ML^{-3})	kg/m^3	
Viscosity	($ML^{-1}T^{-1}$)	kg/m.s (= N.s/m^2)	Pa.s
Surface tension	(MT^{-2})	N/m	

1.2 DENSITY AND SPECIFIC WEIGHT

The *density* ρ of a fluid is its mass per unit volume. The units are kg/m^3. In general the density of a fluid depends upon temperature and pressure. For incompressible fluids (liquids) the variation of density with pressure is however small.

The *specific weight* γ of a fluid is its weight per unit volume. Thus,

$$\boxed{\gamma = \rho g \text{ (in units of N/m}^3\text{)}}$$

The standard value of acceleration due to gravity is 9.806 m/s^2 and is usually taken as 9.81 m/s^2. At standard temperature (20°C) and pressure (1 atmospheric pressure or 760 mm of mercury), the density of water is 998 kg/m^3. Thus, the specific weight of water (at 20°C and 1 atmospheric pressure) is $\gamma = 998 \times 9.81 = 9790$ N/m^3 = 9.79 kN/m^3.

Relative Density (RD) of a fluid is the ratio of its density to that of pure water at standard conditions. Hence it is dimensionless. Thus,

$$\text{RD} = \frac{\text{Density of substance}}{\text{Density of pure water at standard temperature and pressure}}$$

$$= \frac{\text{Density of the substance (kg/m}^3\text{)}}{998\text{ (kg/m}^3\text{)}} \text{ at 20°C}$$

The term *specific gravity* used in CGS and FPS units is the same as relative density.

If the relative density of a liquid is 0.850, it means that its density is 0.850 × 998 = 848.3 kg/m^3 at 20°C.

Commonly used values of approximate relative densities in fluid flow calculations are 1.0 and 13.6 for water and mercury respectively. When no other information is available, the following standard values (corresponding to 20°C and standard atmospheric pressure) are used:

Item	*water*	*Air*
Density ρ	998 kg/m^3	1.205 kg/m^3
Relative density (RD)	1.00	1.204×10^{-3}
Specific weight γ	9790 N/m^3 (= 9.79 kN/m^3)	11.82 N/m^3

(Unless otherwise stated the above standard values for ρ and γ will be used in this book).

[**Note:** For approximate/quick calculations for water, ρ = 1000 kg/m^3 and γ = 9.8 kN/m^3 can be used].

1.3 PRESSURE

Pressure is the compressive stress on the fluid and is given by

$$\text{Pressure } p = \frac{\text{Force } F}{\text{Area } A} \text{ for uniform pressure}$$

$$= \frac{dF}{dA} \text{ for variable pressure}$$

The units are N/m^2 = Pa.

Other commonly used units are

kPa = kilo pascals = 1000 N/m^2

bar = 100 kPa = 10^5 N/m^2

Sometimes the pressure is expressed in terms of the height h of an equivalent column of fluid of density ρ. Thus,

$$\boxed{p = \rho g h = \gamma h} \qquad (1.1)$$

$$\text{and } h \text{ (meteres of fluid)} = \frac{p}{\rho g} = \frac{p\ (\text{N/m}^2)}{\gamma\ (\text{N/m}^3)}$$

In such cases h is called the pressure head.

For example: (i) A pressure head of 5 m of water is equivalent to a pressure of

$$5.0 \times 9790 = 48950 \text{ Pa} = 48.95 \text{ kPa}$$

(ii) A pressure head of 3 cm of mercury = $h\gamma$

$$= \frac{3}{100} \times (13.6 \times 9790) \text{ Pa}$$
$$= 3994 \text{ Pa} = 3.994 \text{ kPa}$$

1.4 SHEAR STRESS AND VISCOSITY

While the pressure, a normal stress, is encountered in both fluid static and dynamic situations, the shear stress (τ) is encountered in the motion of all real fluids. The unit of shear stress is N/m^2 and is designated in Pa or kPa depending on the magnitude.

1.4.1 Viscosity

Viscosity is the resisting property of a fluid to shearing force. The shear stress τ is related to the deformation rate by the *Newton's law of viscosity*, as

$$\boxed{\tau = \mu \frac{du}{dy}} \qquad (1.2)$$

where $\frac{du}{dy}$ = velocity gradient in the y direction and

μ = coefficient of viscosity.

The fluids which obey Newton's law of viscosity are known as *Newtonian fluids*.

The coefficient of viscosity μ (also known variously as coefficient of dynamic viscosity, absolute viscosity or simply viscosity) has the units

$$\mu = \frac{\tau}{(du/dy)} = \frac{\text{N/m}^2}{\left(\frac{\text{m}}{\text{s}}\cdot\frac{1}{\text{m}}\right)} = \text{N.s/m}^2 = \text{Pa.s}$$

The ratio of the coefficient of dynamic viscosity to the density of the fluid is designated by Greek letter (nu) ν. Hence,

$$\nu = \frac{\mu}{\rho} = \frac{\text{N.s/m}^2}{\text{kg/m}^3} = \frac{\text{kg.m}^{-1}\text{s}^{-1}}{\text{kg.m}^{-3}} = \text{m}^2/\text{s}$$

This term ν has the dimensions of $[L^2/T]$ and is known as *kinematic viscosity* of the fluid.

Sometimes the coefficient of dynamic viscosity μ is designated by *poise* (abbreviated as P) or *centipoise* (abbreviated as CP) where

$$1 \text{ poise} = 1 \frac{\text{gm}}{\text{cm.s}} = \frac{1 \text{ dyne. second}}{\text{cm}^2}$$

$$= \frac{10^{-5}}{(10^{-2})^2}\frac{\text{N.s}}{\text{m}^2} = \frac{1}{10} \text{ Pa.s}$$

$$1 \text{ centipoise} = \frac{1}{100} \text{ poise} = \frac{1}{1000} \text{ Pa.s}$$

The kinematic viscosity ν is sometimes expressed in *stoke* or *centi stoke* where

$$1 \text{ stoke} = 1 \frac{\text{cm}^2}{\text{s}} = 1\,(10^{-2})^2 \frac{\text{m}^2}{\text{s}} = 10^{-4} \text{ m}^2/\text{s}$$

$$1 \text{ centistoke} = \frac{1}{100} \text{ stoke} = 10^{-6} \text{ m}^2/\text{s}$$

The coefficient of viscosity μ depends upon the temperature. Generally, for liquids the value of μ decreases with an increase in the temperature, and for gases the value of μ increases with an increase in the temperature.

Table 1.2 gives the viscosities of some commonly used fluids at 20°C and standard presssure.

1.5 SURFACE TENSION

A liquid forms an interface with a second liquid or gas. The surface energy per unit area of interface is known as *surface tension* or *coefficient of surface tension* σ. The most common interfaces and values of σ, for clean surface at 20°C, are:

$\sigma = 0.073$ N/m for air-water interface.

and $\sigma = 0.480$ N/m for air-mercury interface.

Note that the surface tension σ has the dimension of force/unit length (N/m).

When a liquid interface interacts with a solid surface, a contact angle θ is formed. For water-clean glass surface $\theta \approx 0°$ and for mercury-clean glass $\theta \approx 130°$.

Due to surface tension, pressure changes occur across a curved interface. The pressure difference between inside and outside of a curved surface Δp is related to the radius of curvature R and surface tension σ as:

Table 1.2 *Properties of Some Common Fluids at 20°C and 1 atm Pressure*

Fluids	*Density* $\rho(kg/m^3)$	*Dynamic viscosity* $\mu(Ns/m^2)$	*Kinematic viscosity* $\nu(m^2/s)$	*Surface tension* $\sigma(N/m)$	*Bulk modulus* $K\ (N/m^2)$
(a) Liquids					
Water	998	1.00×10^{-3}	1.00×10^{-6}	7.28×10^{-2}	2.19×10^{9}
Sea water	1025	1.07×10^{-3}	1.04×10^{-6}	7.28×10^{-2}	2.28×10^{9}
Petrol	680	2.92×10^{-4}	4.29×10^{-7}	2.16×10^{-2}	9.58×10^{8}
Kerosene	804	1.92×10^{-3}	2.39×10^{-4}	2.80×10^{-2}	1.43×10^{9}
Glycerine	1260	1.49	1.18×10^{-3}	6.33×10^{-2}	4.34×10^{9}
Mercury	13550	1.56×10^{-3}	1.15×10^{-7}	4.84×10^{-1}	2.55×10^{10}
SAE 10 oil	917	1.04×10^{-1}	1.13×10^{-4}	3.60×10^{-2}	1.31×10^{9}
SAE 30 oil	917	2.90×10^{-1}	3.16×10^{-4}	3.50×10^{-2}	1.38×10^{9}
Castor oil	960	9.80×10^{-1}	1.02×10^{-3}	3.92×10^{-2}	1.44×10^{9}
	$\rho\,(kg/m^3)$		$\mu\,(Ns/m^2)$	$\nu\,(m^2/s)$	Specific heat ratio, $k = C_p/C_v$
(b) Gases					
Air	1.205		1.80×10^{-5}	1.494×10^{-5}	1.40
Carbon dioxide	1.840		1.48×10^{-5}	0.804×10^{-5}	1.28
Hydrogen	0.084		0.90×10^{-5}	10.714×10^{-5}	1.40
Nitrogen	1.160		1.76×10^{-5}	1.517×10^{-5}	1.40
Methane	0.668		1.34×10^{-5}	2.000×10^{-5}	1.30
Oxygen	1.330		2.00×10^{-5}	1.504×10^{-5}	1.40
Water vapour	0.747		1.01×10^{-5}	1.352×10^{-5}	1.33

(1) For the interior of a liquid cylinder

$$\boxed{\Delta p = \frac{\sigma}{R}} \qquad (1.3)$$

(2) For a spherical droplet

$$\boxed{\Delta p = \frac{2\sigma}{R}} \qquad (1.4)$$

(3) A soap bubble has two surfaces and the pressure difference is given by

$$\boxed{\Delta p = \frac{4\sigma}{R}} \qquad (1.5)$$

Thus the pressure inside a droplet or a soap bubble will be higher than the surrounding atmosphere. The pressure inside will be higher, the smaller the size of the droplet or bubble.

1.5.1 Capillarity

Liquids have both cohesion and adhesion, which are forms of molecular attraction. Capillarity, the rise (or fall) of liquid in small-diameter tubes is due to this attraction. Liquids, such as water, which wet a surface cause capillary rise. In nonwetting liquids (e.g. mercury) capillary depression is caused.

For a cylindrical glass tube the capillary rise (or depression) h (Fig. 1.1) is given by

$$\boxed{h = \frac{2\sigma\cos\theta}{\gamma R}} \qquad (1.6)$$

where θ = contact angle,
γ = unit weight of the liquid (= ρg),
R = radius of curvature of the glass tube
σ = coefficient of surface tension.

(*Note:* Capillary rise is usually measured to the bottom of the meniscus)

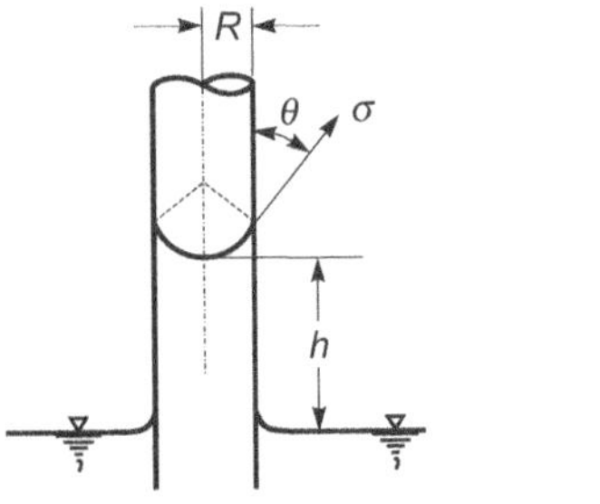

Fig. 1.1 *Capillary Rise*

For clean glass and water θ can be assumed to be zero. For clean mercury–air–glass interface, θ = 130°.

1.6 COMPRESSIBILITY

Gasses are highly compressible and their relaionship between pressure, temperature and volume is expressed by perfect gas equation

$$\boxed{\frac{p}{\rho} = pv = RT} \qquad (1.7)$$

where p = absolute pressure
ρ = mass density
v = specific volume = $1/\rho$
T = absolute temperature in Kelvin
= (273 + °C)

R = characteristic gas constant = $\left(\frac{m^2}{s^2 K}\right)$

For any gas $$R = \frac{8312}{M} \qquad (1.8)$$

where M = molecular weight of gas.
For air M = 28.97, giving R_{air} = 287 $m^2/(s^2K)$.
Another fundamental equation of a perfect gas between two state points is

$$\boxed{\frac{p_1}{\rho_1^n} = \frac{p_2}{\rho_2^n}} \qquad (1.9)$$

where p is absolute pressure.

(i) If the process is *isothermal* (i.e. at constant temperature),

$$n = 1.0$$

(ii) If the process is *adiabatic* (i.e. without heat transfer) and without friction (*isentropic*)

$$n = k$$

where $k = C_p/C_v$ = ratio of specific heat at constant pressure (C_p) and that at constant volume (C_v). For air and diatomic gases $k \approx 1.4$. Values of k for various gases are given in Table 1.2 (b).

Combining Eq. 1.7 and Eq. 1.9

$$\boxed{\frac{T_2}{T_1} = \left(\frac{p_2}{p_1}\right)^{(k-1)/k}} \qquad (1.10)$$

1.6.1 Bulk Modulus of Elasticity

The compressibility of a fluid is expressed by the *bulk modulus of elasticity K* as

$$\boxed{K = \frac{dp}{(-d\forall/\forall)}} \qquad (1.11)$$

where dp = change in pressure
and $d\forall$ = change in the volume $\forall$.

Since $\left(-\frac{d\forall}{\forall}\right)$ is dimensionless, the dimension of K is the same as that of p, viz. N/m^2 = (Pa).

For a perfect gas

(i) $K = p$ for isothermal process, and
(ii) $K = kp$ for isentropic process.

By definition, the compressibility of a fluid is *inversely* proportional to its value of K.

1.6.2 Velocity of Propagation of Sound (C)

Sound is propagated in a fluid due to compressibility of the medium, and the speed of sound C is given by

$$\boxed{C = \sqrt{K/\rho}} \tag{1.12}$$

where K = bulk modulus of elasticity of the medium and ρ = mass density of the fluid.

1.7 VAPOUR PRESSURE

Vapour Pressure is the pressure at which a liquid boils and is in equilibrium with its own vapor. In many liquid flow situations, low pressures approaching vapor pressure of the liquid may occur. When this happens, the liquid flashes into vapour, forming rapidly expanding cavity. This phenomenon, known as *Cavitation,* has serious implications on the operating performance of hydraulic machines and passages of high-speed flows, (see Chapter 16, Sec. 16.3.6 for further details).

Vapour Pressure of a liquid depends upon temperature and increases with it. At 20°C, water has a vapour pressure (P_v) of 2,34 kPa (i.e. vapour pressure head (P_v/γ) = 0.24 m).

WORKED EXAMPLES

1.1 A mass of liquid weighs 500 N when exposed to standard earth's gravity $g = 9.806$ m/s^2.

(a) What is its mass?
(b) What will be its weight in a planet with acceleration due to gravity of (i) 3.5 m/s^2 and (ii) 18.0 m/s^2?

Solution: Let W = weight of the liquid and m = its mass.

(a) $W = mg$

$500 = m \times 9.806$

$m = \frac{500}{9.806}$ = **50.99 kg**

(b) The mass of the fluid remains constant regardless of its location. Hence m = 50.99 kg at all locations.

(i) If $g_1 = 3.5$ m/s^2

$W_1 = mg_1 = 50.99 \times 3.5$ = **178.46 N**

(ii) If $g_2 = 18.0$ m/s^2

$W_2 = mg_2 = 50.99 \times 18.0 = \boxed{917.8 \text{ N}}$

1.2 A body has a weight of 400 N at standard earth's gravity of 9.806 m/s^2. When a force of 800 N is applied to the body, what is the value of the acceleration (i) on earth and (ii) on moon where the gravity is 1.6 m/s^2?

Solution: The mass of the body,

$$m = \frac{\text{weight}}{\text{gravity}} = \frac{400}{9.806} = 40.791 \text{ kg}$$

This mass is constant and does not change with location. Hence, when a force F is applied, by Newton's second law,

$$F = ma$$

or acceleration $\quad a = F/m$

which is independent of g. Hence, both on the earth, as well as on moon,

$a = \frac{800}{40.791}$ = **19.612 m/s^2**

1.3 A liquid at 20°C has a relative density of 0.80 and a kinematic viscosity of 2.3 centistoke. Determine its (i) unit weight and (ii) dynamic viscosity in Pa.s.

Solution:

(i) Unit weight $\gamma = \rho g = (\rho_{water} \times \text{RD}) \times g$

Taking ρ_{water} at 20° as 998 kg/m^3,

$\gamma = 998 \times 0.8 \times 9.81$

= 7832.3 N/m^3 = **7.832 kN/m^3**

(ii) Dynamic viscosity $\mu = \nu\rho$

ν = 2.3 centistoke = 2.3×10^{-6} m^2/s

$\rho = 998 \times 0.8 = 798.4$ kg/m^3

$\mu = 2.3 \times 10^{-6} \times 798.4$ = **1.836×10^{-3} Pa.s.**

1.4 The space between two parallel plates kept 3 mm apart is filled with an oil of dynamic viscosity 0.2 Pa.s. What is the shear stress on the lower fixed plate, if the upper one is moved with a velocity of 1.50 m/s? (Refer Fig 1.2).

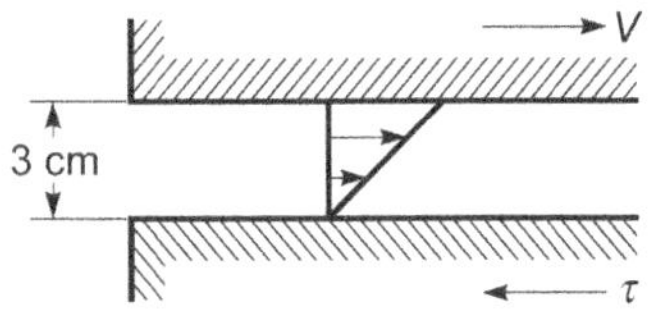

Fig. 1.2

Solution: Since the gap between the plates is very small, a linear variation of velocity can be assumed.

$$\frac{du}{dy} = \frac{V}{h} = \frac{1.50}{3 \times 10^{-3}} = 500(\text{s}^{-1})$$

τ = Shear stress on the bottom plate

$$= \mu \frac{du}{dy} = 0.2 \times 500 = 100 \text{ N/m}^2$$

1.5 The velocity distribution in a viscous flow over a plate is given by

$$u = 4y - y^2 \text{ for } y \le 2 \text{ m}$$

where u = velocity in m/s at a point distant y from the plate. If the coefficient of dynamic viscosity is 1.5 Pa.s determine the shear stress at $y = 0$ and at $y = 2.0$ m.

Solution: Given

$$u = 4y - y^2$$

Therefore $\frac{du}{dy} = 4 - 2y$

Shear stress $\tau = \frac{du}{dy} = \mu\,(4 - 2y)$

At $y = 0$, $\tau_0 = 4\mu = 4 \times 1.5 =$ **6.0 Pa.s**
At $y = 2.0$ m, $\tau_2 = \mu(4 - 4) =$ **0**

1.6 The velocity distribution near the solid wall at a section in a laminar flow is given by

$$u = 5.0 \sin (5\pi y)$$

for $y \le 0.10$ m. Compute the shear stress at a section at (i) $y = 0$, (ii) $y = 0.05$ m and (iii) $y = 0.10$ m. The dynamic viscosity of the fluid is 5 poise.

Solution: Given,

$$\mu = 5 \text{ poise} = \frac{5}{10} \text{ Pa.s}$$

Since $u = 5.0 \sin (5\pi y)$

$$\frac{du}{dy} = 5.0 \times 5\pi \cos (5\pi y)$$

Shear stress $\tau = \mu \frac{du}{dy} = \frac{5}{10} \times 25\pi \cos (5\pi y) = 12.5\pi \cos (5\pi y)$

(i) At $y = 0$, $\tau = 12.5\pi \cos (0) = 12.5\pi =$ **39.27 N/m²**

(ii) At $y = 0.05$ m, $\tau = 12.5\pi \cos (5\pi \times 0.05)$
$= 12.5\pi \times 0.707 =$ **27.76 N/m²**

(iii) At $y = 0.10$ m, $\tau = 12.5\pi \cos (5\pi \times 0.1) =$ **0**

1.7 A 90 N rectangular solid block slides down a 30° inclined plane. The plane is lubricated by a 3 mm thick film of oil of relative density 0.90 and viscosity 8.0 poise. If the contact area is 0.3 m², estimate the terminal velocity of the block (Refer. Fig. 1.3).

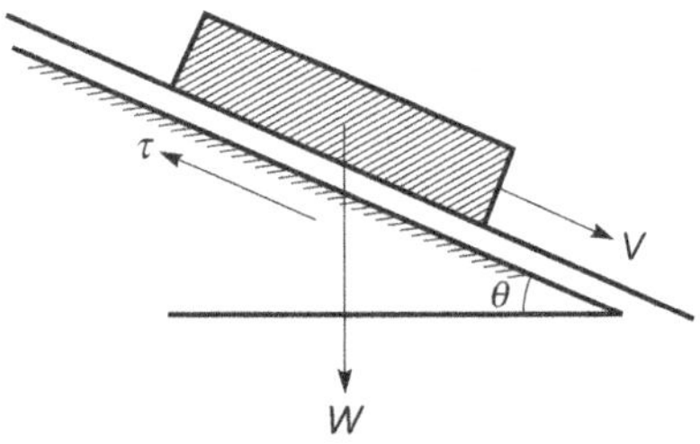

Fig. 1.3

Solution: Given

$$W = 90 \text{ N}$$
$$V = \text{terminal velocity}$$
$$\theta = 30°$$

At the terminal velocity, the sum of the forces acting on the block in the direction of its motion is zero. Hence

$$W \sin \theta - \tau A = 0$$

where τ = shear stress on the block and
A = area of the block.

$$\tau = \mu \frac{du}{dy} = \mu \frac{V}{h} \text{ where } h = \text{thickness of oil film}$$

$$\mu = 8 \text{ poise} = \frac{8}{10} \text{ Pa.s} = 0.8 \text{ Pa.s}$$
$$h = 3 \text{ mm} = 3 \times 10^{-3} \text{ m}$$
$$A = 0.3 \text{ m}^2$$

Substituting the various values in the above equation,

$$90 \sin 30° - \frac{(0.8\,V)}{3 \times 10^{-3}} \times (0.3) = 0$$

$\therefore$ $V = \frac{45}{80} =$ **0.5625 m/s**

1.8 A thin plate is placed between two flat surfaces h cm apart such that the viscosity of liquids on the top and bottom of the plate are μ_1 and μ_2 respectively. Determine the position of the thin plate such that the viscous resistance to uniform motion of the thin plate is minimum. [Assume h to be very small].

Solution: Let y be the distance of the thin flat plate from the top flat surface (Fig. 1.4) and V = velocity of the thin plate.

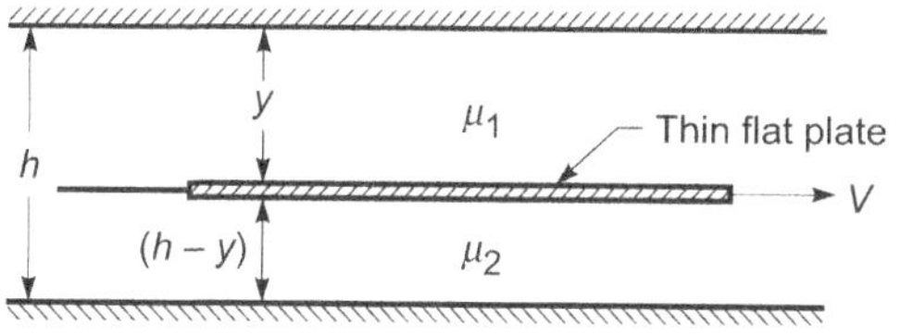

Fig. 1.4

Shear stress on the top portion $\tau_1 = \mu_1 \dfrac{du}{dy} = \mu_1 \dfrac{V}{y}$

Since the gap in the bottom portion = $h - y$

Shear stress, $\tau_2 = \mu_2 \dfrac{V}{(h-y)}$

Force on both sides of the plate

$$F = A(\tau_1 + \tau_2) = VA\left[\frac{\mu_1}{y} + \frac{\mu_2}{h-y}\right]$$

where A = area of the thin plate.

For F to be minimum $\dfrac{dF}{dy} = 0$

$$-\frac{\mu_1}{y^2} + \frac{\mu_2}{(h-y)^2} = 0$$

$$\frac{y^2}{(h-y)^2} = \left(\frac{\mu_1}{\mu_2}\right)$$

or

$$\frac{y}{(h-y)} = \sqrt{\mu_1/\mu_2}$$

1.9 A cylindrical shaft of 90 mm diameter rotates about a vertical axis inside a fixed cylindrical tube of length 50 cm and 95 mm internal diameter. If the space between the tube and the shaft is filled by a lubricant of dynamic viscosity 2.0 poise, determine the power required to overcome viscous resistance when the shaft is rotated at a speed of 240 rpm (Refer. Fig. 1.5).

Solution:

V = Circumferential velocity of the shaft

$$= \omega r = \frac{2\pi N}{60} \times r$$

$$= \frac{2\pi \times 240}{60} \times \left(\frac{0.090}{2}\right) = 1.131 \text{ m/s}$$

Clearance $h = \dfrac{95-90}{2} = 2.5$ mm

Assuming linear variation of velocity across the gap,

Velocity gradient $\dfrac{du}{dr} = \dfrac{V}{h}$

$$= \frac{1.131}{2.5 \times 10^{-3}} = 452.4 \text{ s}^{-1}$$

μ = 2.0 poise = 0.2 Pa.s

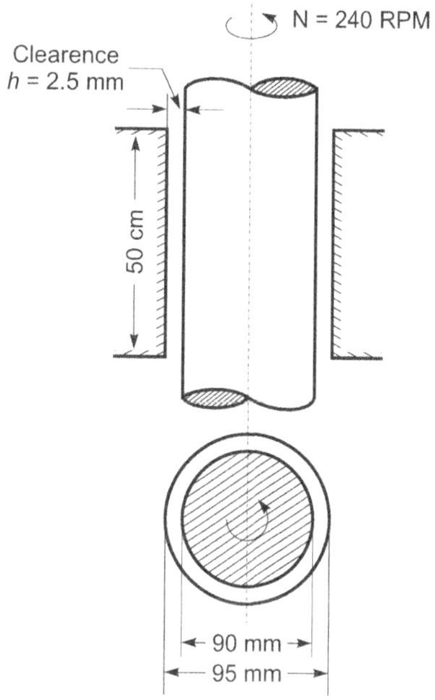

Fig. 1.5

Shear stress on the shaft $\tau = \mu \dfrac{du}{dr}$

$= 0.2 \times 452.4 = 90.48$ Pa

Shear force $F_s = \tau \times 2\pi r \times L$

$$= 90.48 \times \left(2\pi \times \frac{0.09}{2}\right) \times 0.50$$

$= 12.791$ N

Torque $T = F_s r$

$$= 12.791 \times \frac{0.09}{2} = 0.6756 \text{ N.m}$$

Power required $= P = \dfrac{2\pi N}{60} \cdot T$

$$= \frac{2\pi \times 240}{60} \times 0.6756 = \boxed{14.5 \text{ W}}$$

1.10 A sleeve 10 cm long encases a vertical metal rod 3.0 cm in diameter with a radial clearance of 0.02 mm. If when immersed in an oil of viscosity 6.0 poise, the effective weight of the sleeve is 7.5 N, will the sleeve slide down the rod and if so at what velocity? (Refer Fig 1.6)

Solution: Given

μ = 6 poise = 0.6 Pa.s

$h = 0.02$ mm $= 0.02 \times 10^{-3}$ m

Let V = Velocity of the sleeve when sliding down.

Shear stress $\tau = \mu \dfrac{du}{dr} = \mu \dfrac{V}{h}$

$$= (0.6) \times \frac{V}{2 \times 10^{-5}} = 3 \times 10^4 \text{ V}$$

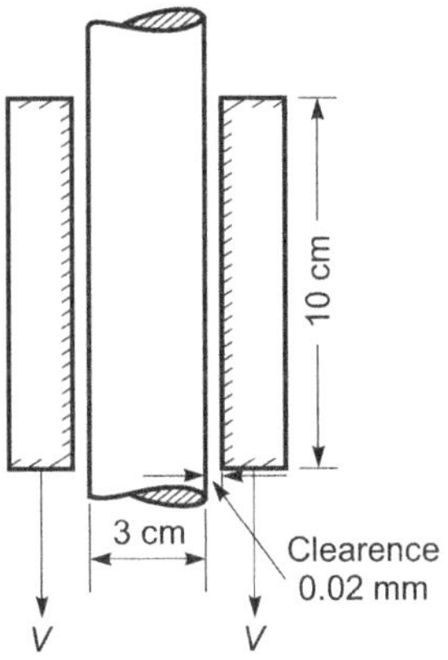

Fig. 1.6

At terminal velocity

Shear force = Submerged weight of the sleeve.

$\therefore \quad (2\pi rL) \times \tau = W_s$

$$\left(2\pi \times \frac{0.03}{2} \times \frac{10}{100} \times 3 \times 10^4\right) V = 7.5$$

$$282.74\ V = 7.5$$

$$V = 0.02653 \text{ m/s} = \mathbf{2.66 \text{ cm/s}}$$

1.11 A hydraulic lift used for lifting automobiles has a 25 cm diameter ram which slides in a 25.018 cm diameter cylinder, the annular space being filled with oil having a kinematic viscosity of 3.7 cm^2/s and relative density of 0.85. If the rate of travel of the ram is 15 cm/s find the frictional resistance when 3.3 m of ram is engaged in the cylinder.

Solution:

$$\mu = \nu\rho = 3.7 \times 10^{-4} \times 0.85 \times 998$$
$$= 0.3139 \text{ Pa.s}$$

Shear stress $\tau = \mu \dfrac{du}{dy} = \mu \dfrac{V}{h}$

$$= 0.3139 \times \frac{0.15}{(25.018 - 25.00)/(2 \times 10^2)}$$
$$= 523.1 \text{ N/m}^2$$

Frictional resistance $F_s = A\tau$

$$F_s = (\pi DL) \times \tau$$
$$= \pi \times \frac{25}{100} \times 3.3 \times 523.1$$
$$= 1356 \text{ N} = \mathbf{1.356 \text{ kN}}$$

1.12 A circular disc of radius R is kept at a small height above a fixed bed by means of a layer of oil of viscosity μ. If the disc is rotated at an angular velocity ω, obtain an expression for viscous torque on the disc (Refer Fig 1.7).

Solution: Consider an element of disc of width dr at a radial distance r.

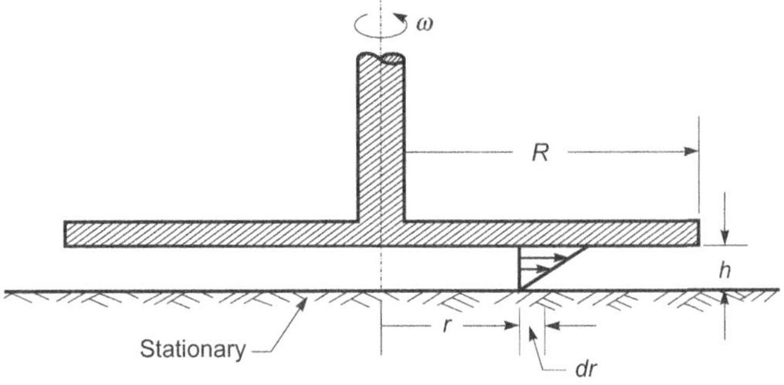

Fig. 1.7

Velocity at this radius = $V = r\omega$.

Assuming linear variation of velocity with depth in the gap h,

Shear stress $\tau = \mu \dfrac{V}{h} = \dfrac{\mu}{h} \omega r$

Viscous torque on the element

$$dT = \frac{\mu}{h} \omega r\, (2\pi r dr) r$$
$$= \frac{\mu\omega}{h} \cdot 2\pi r^3 dr$$

Total torque $\quad T = \displaystyle\int_0^R dT = \int_0^R \frac{\mu\omega}{h}\, 2\pi r^3 dr$

$$= \left[\frac{\mu\omega}{h} \cdot 2\pi \frac{r^4}{4}\right]_0^R$$

or $\quad \mathbf{T} = \dfrac{\mathbf{1}}{\mathbf{2}} \cdot \dfrac{\pi\mu\omega}{\mathbf{h}} \cdot \mathbf{R^4}$

1.13 Two coaxial cylinders 10 cm and 9.75 cm in diameter and 2.5 cm high have both their ends open and have a viscous liquid filled in between. A torque of 1.2 N.m is produced on the inner cylinder when the outer one rotates at 90 rpm. Determine the coefficient of viscosity of the liquid (Refer Fig 1.8).

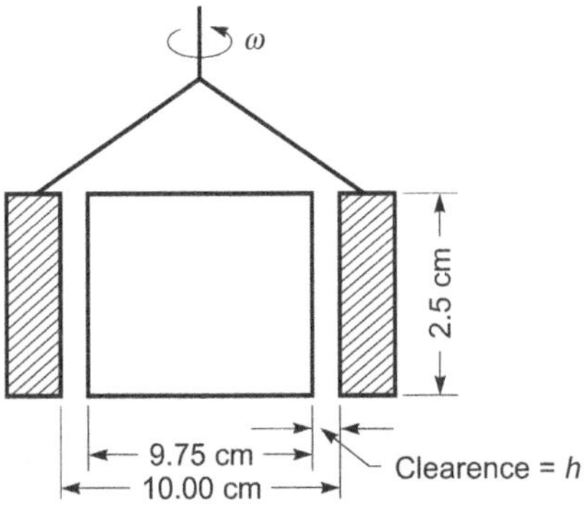

Fig. 1.8

Solution: Tangential velocity

$$V = \frac{\pi DN}{60} = \frac{\pi \times 0.10 \times 90}{60}$$
$$= 0.4712 \text{ m/s}$$

Radial clearance $h = \frac{(10.00 - 9.75)}{2}$ cm

$= 0.125$ cm or 0.00125 m

Velocity gradient $\frac{du}{dr} = \frac{V}{h} = \frac{0.4712}{0.00125} = 377$

Shear stress $\tau = \mu \frac{du}{dr} = 377\mu$

Shear force $F_s = \tau \times$ area

$$= 377\mu \left(2\pi \times \frac{0.1}{2} \times \frac{2.5}{100}\right)$$
$$= 2.961\mu$$

Torque $= F_s \times$ lever arm $= F_s\, r$

$$T = 2.961\, \mu \times \frac{(0.1)}{2} = 0.14805\, \mu$$

But Torque $T = 1.2$ N.m

Therefore $0.14805\, \mu = 1.2$

$\therefore$ $\mu = \frac{120}{0.14805}$ = **8.106 Pa.s**

1.14 A solid cone of radius r_0, and vertex angle 2θ is to rotate at an angular velocity ω. An oil of viscosity μ and thickness h fills the gap between the cone and the housing. Determine the torque T required to rotate the cone (Refer Fig. 1.9).

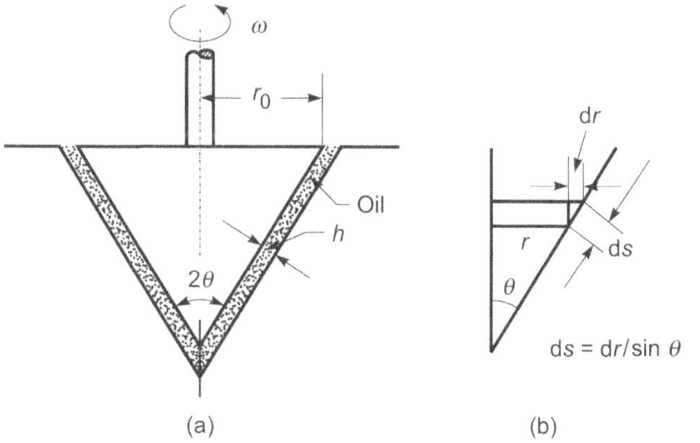

Fig. 1.9

Solution: At any radius $r \le r_0$,

$$u = \omega r$$

Shear stress on the inclined wall $\tau = \mu \frac{du}{dy}$

$$= \mu \frac{V}{h} = \mu \frac{\omega r}{h}$$

Considering an elemental area $(2\pi r\, ds)$

$$= 2\pi r \cdot \frac{dr}{\sin\theta}$$

d (torque) $= dT = r\, d(\text{force})$

$$= r\tau 2\pi r \cdot \frac{dr}{\sin\theta}$$
$$= r\,\mu \frac{\omega r}{h}\, 2\pi r \frac{dr}{\sin\theta}$$
$$= \mu \frac{2\pi\omega}{h} \frac{1}{\sin\theta} \cdot r^3\, dr$$

$$\text{Torque } T = \int_0^{r_0} dT = \frac{2\pi\omega\mu}{h\sin\theta} \int_0^{r_0} r^3\, dr$$

$$\boldsymbol{T = \frac{\pi\omega\mu}{2h\sin\theta} r_0^4}$$

1.15 If the surface tension at air-water interface is 0.073 N/m, what is the pressure difference between inside and outside of an air bubble of diameter 0.01 mm?

Solution: An air bubble has only one surface. Hence

$$\Delta p = \frac{2\sigma}{R}$$
$$= \frac{2 \times 0.073}{\left(\frac{0.01}{2}\right) \times 10^{-3}} = 29200 \text{ N/m}^2$$
$$= \textbf{29.2 kPa}$$

1.16 If the surface tension at the soap-air interface is 0.088 N/m, calculate the internal pressure in a soap bubble of 3 cm diameter.

Solution: In a soap bubble, there are two interfaces. Hence

$$\Delta p = \frac{4\sigma}{R} = \frac{4 \times 0.088}{\left(\frac{3}{2} \times 10^{-2}\right)}$$

= 23.47 N/m² above atmospheric pressure

1.17 A U-tube has two limbs of internal diameter 6 mm and 16 mm respectively and contains some water. Calculate the difference in water levels in the two limbs. Surface tension of water = 0.073 N/m.

Solution: Let h = difference in water levels in the two limbs. By assuming the angle of contact $\theta = 0°$,

$$\Delta p = \rho g h = \left(\frac{2\sigma}{R_1} - \frac{2\sigma}{R_2}\right)$$

$$h \times 998 \times 9.81 = 2 \times 0.073 \left(\frac{1}{3 \times 10^{-3}} - \frac{1}{8 \times 10^{-3}}\right)$$

$$h = 3.1 \times 10^{-3} \text{ m} = \textbf{3.1 mm}$$

1.18 A clean tube of internal diameter 3 mm is immersed in a liquid with a coefficient of surface tension = 0.48 N/m. The angle of contact of the liquid with the glass can be assumed to be 130°. The density of the liquid is 13,600 kg/m^3. What would be the level of the liquid in the tube relative to the free surface of the liquid outside the tube.

Solution: The liquid in the tube rises (or falls) due to capillarity. The capillary rise (or fall)

$$h = \frac{2\sigma \cos\theta}{\gamma R}$$

Here $R = \frac{3}{2}$ mm $= 1.5 \times 10^{-3}$ m

$\theta = 130°$, $\sigma = 0.48$ N/m

$\gamma = \rho g = 13.6 \times 10^3 \times 9.81$

$$\therefore \quad h = \frac{2 \times 0.48 \times \cos 130°}{(13.6 \times 10^3 \times 9.81) \times (1.5 \times 10^{-3})}$$

$= -3.08 \times 10^{-3}$ m $=$ **–3.08 mm**

Therefore, there is a capillary depression of 3.08 mm.

1.19 Assuming that the interstices in a clay are of size equal to one tenth the mean diameter of the grain, estimate the height to which water will rise in a clay soil of average grain diameter 0.05 mm. Assume surface tension at air-water interface as 0.073 N/m.

Solution:

Diameter of pores $= 2R = 0.005$ mm

$R = 0.00025$ mm $= 2.5 \times 10^{-6}$ m

$$\Delta h = \frac{2\sigma}{\gamma R} \text{ by assuming } \theta = 0°$$

$$= \frac{2 \times 0.073}{(9.81 \times 998) \times (2.5 \times 10^{-6})}$$

$=$ **5.95 m**

1.20 Calculate the work done in blowing a soap bubble of diameter 12 cm. Assume the surface tension of soap solution = 0.040 N/m.

Solution: The soap bubble has two interfaces.

Work done = Surface tension × total surface area

$$= 0.040 \times 4\pi\left(\frac{12}{2} \times 10^{-2}\right)^2 \times 2$$

$=$ **36.2 × 10^{-4} N.m**

1.21 What force is necessary to lift a thin platinum wire ring of 3.0 cm diameter from a water surface. Assume the surface tension of water = 0.0728 N/m and neglect the weight of the wire.

Solution: Two water surfaces at the ring resist the lifting force. Referring to Fig. 1.10, by assuming $d << D$

$$F = 2[\pi D \sigma]$$

$$= 2 \times \pi \times \frac{3.0}{100} \times 0.0728 = \mathbf{0.0137\ N}$$

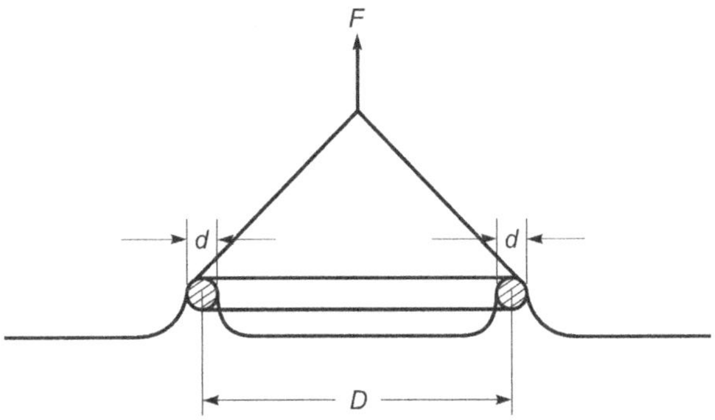

Fig. 1.10

1.22 Using perfect gas law, estimate the density of (i) air and (ii) CO_2, at 60°C and 120 kPa (abs) pressure.

Solution:

(i) For air M = Molecular weight = 28.97

Gas constant $R = \frac{8312}{28.97} = 287$

$p = 120 \times 10^3$ Pa. (abs)

$T = 273 + 60 = 333$ K

Density $\rho = \frac{p}{RT} = \frac{120 \times 10^3}{287 \times 333}$

(ii) For CO_2, $M = 44$

Gas constant $R = \frac{8312}{44} = 189$

Density $\rho = \frac{120 \times 10^3}{(189)(333)} = \mathbf{1.907\ kg/m^3}$

1.23 Air at 20°C and 200 kPa (abs) contained in a cylinder is compressed to half its volume. Find the pressure and temperature inside the cylinder if the process is (a) Isothermal and (b) Isentropic with $k = 1.4$.

Solution:

(a) In *isothermal process* $p_1 v_1 = p_2 v_2$

$$\therefore \quad p_2 = p_1 \frac{v_1}{v_2} = 200 \times \left(\frac{1}{0.5}\right)$$

$=$ **400 kPa (abs)**

The temperature will remain constant and hence

$$T_2 = T_1 = 20°\text{C}$$

(b) In *isentropic process*:

$$pv^k = \text{constant}$$

$$p_2 = p_1 \left(\frac{v_1}{v_2}\right)^k = 200 \left(\frac{1}{0.5}\right)^{1.4}$$

$=$ **527.8 kPa (abs)**

For temperature,

$$pv = RT \text{ and } pv^k = \text{constant}$$

$$RTv^{k-1} = \text{constant}$$

i.e. $$\frac{T_2}{T_1} = \left(\frac{v_1}{v_2}\right)^{k-1}$$

$$T_1 = 20°\text{C} = 273 + 20 = 293 \text{ K}$$

$$\frac{v_1}{v_2} = 2.0 \text{ and } (k - 1) = 0.4$$

$$T_2 = 293\,(2)^{0.4} = 386.6 \text{ K}$$

$$= (386.6 - 273) = \mathbf{113.6°C}$$

1.24 Gas A at 100 kPa (abs) is compressed isothermally and Gas B at 80 kPa (abs) is compressed isentropically ($k = 1.4$). What is the bulk modulus of elasticity of each gas and which is more compressible?

Solution:

Gas A: In isothermal change $K = p$

Hence $K_A = p =$ **100 kPa**

Gas B: In adiabatic change $K = kp$

Hence $K_B = kp = 1.4 \times 80 =$ **112 kPa**

Since $K_A < K_B$, gas A is more compressible than gas B, in the notified situation.

1.25 What change in pressure is required to compress a gas under isothermal conditions to quarter of its original volume?

Solution: If V is the original volume

$$\Delta V = (V_2 - V_1) = \left(\frac{1}{4}V - V\right) = -\frac{3}{4}V$$

$$dp = \left(-\frac{dV}{V}\right)K$$

where K is the bulk modulus of elasticity.

At isothermal conditions $K = p$. Hence change in pressure

$$\Delta p = \left(\frac{3}{4}\right)p$$

where p = initial pressure of the gas.

1.26 At 20°C, the bulk modulus of elasticity of kerosene is 1.43×10^9 Pa and its relative density 0.804. Calculate the velocity of propagation of sound in kerosene at this temperature.

Solution:

Density of kerosene $\rho = 0.804 \times 998$
$= 802.4$ kg/m^3

Velocity of sound $C = \sqrt{K/\rho}$

$$= \sqrt{\frac{1.43 \times 10^9}{802.4}} = \mathbf{1335 \text{ m/s}}$$

1.27 Determine the pressure increase required to reduce the volume of water by 1.5%, if its bulk modulus of elasticity is 2.2×10^9 Pa.

Solution: Let V = Volume of water

Change in volume $= dV = -\frac{1.5}{100}V = -0.015\,V$

$$-\frac{dV}{V} = 0.015$$

Increase in pressure $= \Delta p = \left(-\frac{dV}{V}\right)K$

$= 2.2 \times 10^9 \times 0.015$

$= 3.3 \times 10^7$ Pa

$=$ **3.3 × 10⁴ kPa**

1.28 The velocity of propagation of sound in air is calculated by assuming the process to be isentropic. What is the velocity of sound at 80°C?

Solution:

For air, gas constant $R = 287$

$$T = 80°\text{C} = 273 + 80 = 353 \text{ K}$$

Since the process is isentropic, $K = kp$.

Assuming the air as perfect gas

$$p = \rho R T$$

Hence the speed of sound

$$C = \sqrt{K/\rho} = \sqrt{\frac{k\rho RT}{\rho}} = \sqrt{kRT}$$

For air $k = 1.4$ and hence

$$C = \sqrt{1.4 \times 287 \times 353} = \mathbf{376.6 \text{ m/s}}$$

OBJECTIVE QUESTIONS

1.1 A perfect fluid (also known as an ideal fluid) is
(a) a real fluid
(b) the one which obeys perfect gas laws
(c) compressive and gaseous
(d) incompressible and frictionless

1.2 The concept of continuum in fluid flow assumes that the characteristics length of the flow is
(a) smaller than the mean free path of the molecules
(b) larger than the mean free path of the molecules
(c) larger than the dimensions of the suspended particles
(d) larger than the wavelength of sound in the medium

1.3 When a shear stress is applied to a substance it is found to resist it by static deformation. The substance is a
(a) liquid (b) solid
(c) gas (d) fluid

1.4 When subjected to shear force, a fluid
(a) deforms continuously no matter how small the shear stress may be
(b) deforms constinuously only for large shear forces
(c) undergoes static deformation
(d) deforms continuously only for small shear stresses

1.5 An object with a mass of 5 kg as determined on earth is placed in a planet which has an acceleration due to gravity of 3 m/s^2. Its mass and weight in this planet are respectively
(a) 5 kg and 15 N
(b) 1.53 kg and 5 N
(c) 15 kg and 15 N
(d) 15 kg and 5 N

1.6 The condition of "No Slip" at rigid boundaries is applicable to
(a) flow of Newtonian fluids only
(b) flow of ideal fluids only
(c) flow of all real fluids
(d) flow of non-Newtonian fluids only

1.7 Newton's law of viscosity for a fluid states that the shear stress is
(a) proportional to angular deformation
(b) proportional to rate of angular deformation
(c) inversely proportional to angular deformation
(d) inversely proportional to rate of angular deformation

1.8 The viscosity of
(a) liquids increases with temperature
(b) gases increases with temperature
(c) fluids decreases with temperature
(d) fluids increases with temperature

1.9 For a fluid, the shear stress was found to be directly proportional to the rate of angular deformation. The fluid is classified as
(a) Newtonian (b) Non-Newtonian
(c) Dilatant fluid (d) Thixotropic

1.10 If the relationship between the shear stress τ and the rate of shear strain du/dy is expressed as

$$\tau = k\left[\frac{du}{dy}\right]^n$$

the fluid with the exponent $n < 1$ is known as
(a) Pseudoplastic fluid
(b) Bingham fluid
(c) Dilatant fluid
(d) Newtonian plastic

1.11 A real fluid is any fluid which
(a) has surface tension and is incompressible
(b) has zero shear stress
(c) has constant viscosity and density
(d) has viscosity

1.12 A fluid indicated the following shear stress and deformation rates:

du/dy (units)	0	1	2	4
τ (units)	10	15	20	30

This fluid is classified as
(a) Newtonian (b) Bingham plastic
(c) Dilatant (d) Pseudoplastic

1.13 If the shear stress τ and shear rate (du/dy) relationship of a material is plotted with τ on the Y-axis and du/dy on the X-axis, the behaviour of an ideal fluid is exhibited by
(a) a straight line passing through the origin and inclined to the X-axis
(b) the positive X-axis

(c) the positive Y-axis
(d) a curved line passing through the origin

1.14 In the following Fig. 1.11 the line A describes the rheological behaviour of a fluid. The fluid can be classified as

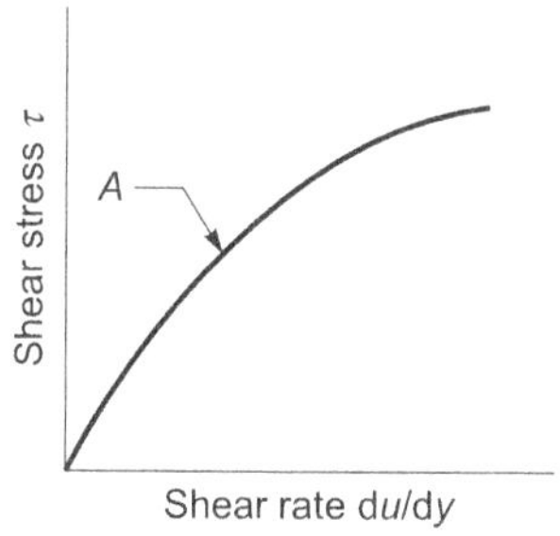

Fig. 1.11

(a) Newtonian (b) Bingham plastic
(c) Ideal (d) Non-Newtonian

1.15 Shear stress for a general fluid motion is respresented by

$$\tau = \mu \left(\frac{du}{dy}\right)^n + B$$

where, n and B are constants. A Newtonian fluid is given by
(a) n > 1 and B = 0
(b) n = 1 and B = 0
(c) n > 1 and B ≠ 0
(d) n < 1 and B = 0

1.16 The following shear stress-shear rate relationship was obtained for a fluid:

du/dy (units)	0	1	3	5
τ (units)	0	6	18	30

The fluid is classified as
(a) Bingham plastic (b) Dilatant
(c) Newtonian (d) Ideal

1.17 The dimensions of the coefficient of dynamic viscosity in [M, L, T] notation system are
(a) $M L^{-1} T$ (b) $M L^{-1} T^{-1}$
(c) $M^{-1} L T$ (d) $M L T^{-1}$

1.18 The dimensions of the coefficient of dynamic viscosity in [F, L, T] notation system are
(a) $F T L^{-2}$ (b) $F L^{-1} T^{-1}$
(c) $F L^2 T^{-1}$ (d) $F T^{-2} L$

1.19 A Newtonian fluid fills the clearance between a shaft and a sleeve. When a force of 800 N is applied to the shaft, parallel to the sleeve, the shaft attains a speed of 1.5 cm/s. If a force of 2.4 kN is applied instead, the shaft would move with a speed of
(a) 1.5 cm/s (b) 13.5 cm/s
(c) 0.5 cm/s (d) 4.5 cm/s

1.20 Typical example of a non-Newtonian fluid of pseudoplastic veriety is
(a) Water (b) Air
(c) Blood (d) Printing ink

1.21 The kinematic viscosity ν is related to the dynamic viscosity μ and density ρ as $\nu =$
(a) μ/ρ (b) $\mu\rho$
(c) ρ/μ (d) $\mu/\rho g$

1.22 A flow of a viscous fluid with $\mu = 1.0$ Ns/m^2 has a velocity distribution given by $u = 0.90\ y - y^2$. The shear stress at $y = 0.45$ m is
(a) 0.90 N/m^2 (b) ∞
(c) zero (d) – 0.90 N/m^2

1.23 A perfect gas
(a) is same as an ideal fluid
(b) satisfies pv^k = constant
(c) has zero velocity
(d) satisfies $p/\rho = RT$ relation

1.24 A perfect gas
(a) is a perfect fluid
(b) does not have viscosity
(c) is incompressible
(d) does not really exist

1.25 In an isentropic process
(a) pv = constant (b) p/T = constant
(c) p/v^k = constant (d) pv^k = constant

1.26 The bulk modulus of elasticity for a liquid, K
(a) is a function of both temperature and pressure
(b) at any given temperature decreases continuously with pressure
(c) at any pressure increases continuously with temperature
(d) is a constant

1.27 In a sample of water an increase of pressure by 18 MN/m^2 caused 1% reduction in the volume. The bulk modulus of elasticity of this sample, in MN/m^2, is
(a) 1.80 (b) 180
(c) 1800 (d) 0.18

1.28 Broadly speaking water is
(a) 10 times more compressible than steel
(b) 80 times more compressible than steel
(c) 80 times less compressible than steel
(d) 800 times less compressible than steel

1.29 The bulk modulus of elasticity K for a gas at constant temperature is

(a) p/ρ (b) ρT
(c) ρRT (d) p

1.30 The bulk modulus of elasticity for a gas undergoing adiabatic process pv^k = constant is,

(a) p/ρ (b) p
(c) kp (d) constant

1.31 Kerosene is known to have a bulk modulus of elasticity $K = 1.43 \times 10^9$ N/m^2 and a relative density of 0.806. The speed of sound in kerosene, in m/s, is

(a) 1333 (b) 1075
(c) 1197 (d) 184

1.32 The bulk modulus of elasticity K of

(a) a gas is larger than that of a solid
(b) a liquid is smaller than that of a gas
(c) a liquid is larger than that of a solid
(d) a solid is larger than that of a liquid

1.33 If the capillary rise of water in a 2 mm diameter tube is 1.5 cm, the height of capillary rise in a 0.5 mm diameter tube, in cm, will be

(a) 10.0 (b) 1.5
(c) 6.0 (d) 24.0

1.34 If the surface tension of water-air interface is 0.073 N/m, the gauge pressure inside a rain drop of 1 mm diameter is,

(a) 146.0 N/m^2 (b) 0.146 N/m^2
(c) 73.0 N/m^2 (d) 292.0 N/m^2

1.35 The excess pressure (above atmospheric) inside a soap bubble of diameter 1 cm, by assuming the surface tension of soap solution to be 0.04 N/m is,

(a) 32.0 N/m^2 (b) 16.0 N/m^2
(c) 160.0 N/m^2 (d) 0.32 N/m^2

1.36 The capillary rise in a 3 mm tube immersed in a liquid is 15 mm. If another tube of diameter 4 mm is immersed in the same liquid the capillary rise would be

(a) 11.25 mm (b) 20.00 mm
(b) 8.44 mm (d) 26.67 mm

1.37 An apparatus produces water droplets of size 70 μm. If the coefficient of surface tension of water in air is 0.07 N/m, the excess pressure in these droplets, in kPa, is,

(a) 5.6 (b) 4.0
(c) 8.0 (d) 13.2

1.38 If the coefficient of surface tension of water in air is 0.07 N/m, the diameter of a tube that can be used to keep the capillary height between 1.80 cm to 2.00 cm is,

(a) 1.65 mm (b) 3.33 cm
(c) 1.65 cm (d) 1.40 cm

1.39 At a liquid-air-solid interface the contact angle θ measured in the liquid is less than 90°. The liquid is,

(a) Wetting
(b) Non-wetting
(c) Ideal
(d) Does not form a stable bubble

1.40 If σ_{Lg} is surface tension at liquid-gas interface, σ_{gs} is surface tension at gas-solid interface, and σ_{Ls} is surface tension at liquid-solid interface, a small drop of liquid when dropped on to a solid surface will remain in equilibrium without spreading if

(a) $|\sigma_{Lg} - \sigma_{Ls}| > \sigma_{gs}$ (b) $|\sigma_{gs} - \sigma_{Ls}| > \sigma_{Lg}$
(c) $|\sigma_{Lg} - \sigma_{Ls}| < \sigma_{gs}$ (d) $|\sigma_{gs} - \sigma_{Ls}| < \sigma_{Lg}$

1.41 The predominant fluid property associated with cavitation phenomenon is

(a) Surface tension
(b) Vapour pressure
(c) Mass density
(d) Bulk modulus of elasticity

1.42 At 20°C, pure water will have a vapour pressure, in kPa, of about

(a) 0.5 (b) 2.34
(c) 101.3 (d) 8.67

1.43 At 100°C, at sea level, pure water will have a vapour pressure, in kPa, of about

(a) 0.50 (b) 2.3
(c) 10.1 (d) 101.3

PROBLEMS

1.1 The relative density of a fluid is 1.26 and its dynamic viscosity is 1.50 Pa.s. Calculate its (i) specific weight and (ii) kinematic viscosity.

(*Ans.* γ = 12.336 kN/m^3, ν = 1.193 × 10^{-3} m^2/s)

1.2 From a table of properties of liquids it was found that, at 20°C, carbon tetrachloride has a dynamic viscosity of 9.67 × 10^{-4} Pa.s and a kinematic viscosity of 6.08 × 10^{-7} m^2/s. Calculate its (i) relative density and (ii) specific weight.

(*Ans. RD* = 1.5936, γ = 15.602 kN/m^3)

1.3 A flat plate 60 cm × 120 cm slides on SAE 10 oil (μ = 1.04 × 10^{-1} Pa.s) over a large plane surface. What force is required to drag the plate at 3.5 m/s if the oil film is 3 mm thick?

(*Ans. F* = 87.36 N)

1.4 Glycerin has a density of 1260 kg/m^3 and a kinematic viscosity of 0.00183 m^2/s. What shear stress is required to deform this fluid at a strain rate of 10^4 s^{-1}?

(*Ans.* τ = 14.9 kPa)

1.5 The velocity distribution in the flow of a thin film of oil down an inclined channel is given by

$$u = \frac{\gamma}{2\mu}(d^2 - y^2) \sin \alpha$$

where d = depth of flow, α = angle of inclination of the channel to the horizontal, u = velocity at a depth y below the free surface, γ = unit weight of oil and μ = dynamic viscosity of oil. Calculate the shear stress: (i) on the bottom of the channel, (ii) at mid-depth and (iii) at the free surface.

(*Ans.* At $y = d$, $\tau_0 = -\gamma d \sin \alpha$;
at $y = d/2$, $\tau_{1/2} = -\gamma (d/2) \sin \alpha$;
at $y = 0$, $\tau_t = 0$)

1.6 Two large plane surfaces are 20 mm apart and the gap contains oil of dynamic viscosity 0.60 Pa.s. A thin plate of cross sectional area 0.50 m^2 is to be pulled through the gap at a constant velocity of 0.60 m/s. The location of the plate will have to be such that it is 8 mm from one of the surfaces. Neglecting edge effects, estimate the force required for pulling the plate as above.

(*Ans. F* = 37.5 N)

1.7 A flat plate 30 cm × 50 cm slides on oil (μ = 0.75 Pa.s) over a large plane surface. What force is required to drag the plate at a uniform velocity of 1.6 m/s, if the separating oil film is 0.2 mm thick?

(*Ans. F* = 900 N)

1.8 A square plate 50 cm × 50 cm weighing 200 N slides down an inclined plane of slope 1 vertical: 2.5 horizontal with a uniform velocity of 0.40 m/s. If a thin layer of oil of thickness 0.5 cm fills the space between the plate and the inclined plane, determine the coefficient of viscosity of the oil.

(*Ans.* μ = 1.4 Pa.s)

1.9 A flat plate of area 0.15 m^2 is to be towed up an inclined plane of slope 1 vertical : 3 horizontal on a thin layer of oil (μ = 0.75 Pa.s.) of thickness 0.2 mm. If the weight of the plate is 250 N, estimate the force required to pull the plate at 1.6 m/s velocity.

(*Ans. F* = 979 N)

1.10 A block with a base of 15 cm × 20 cm and weighing 20 N is allowed to slide down a long inclined plane of slope 1 vertical : 5 horizontal. A thin film of oil (μ = 1.5 poise) of thickness 0.3 mm exists on the surface of the inclined plane. Estimate the terminal speed of the block.

(*Ans. V* = 26.15 cm/s)

1.11 A piston of 7.95 cm diameter and 30 cm long works in a cylinder of 8.0 cm diameter. The annular space of the piston is filled with an oil of viscosity 2 poise. If an axial load of 10 N is applied to the piston, calculate the speed of movement of the piston.

(*Ans. V* = 16.68 cm/s)

1.12 Two discs of 20 cm diameter are placed 1 mm apart and the gap is filled with an oil of viscosity 0.8 kg/m.s. Determine the power required to rotate the upper disc at 600 rpm while holding the lower one stationary.

(*Ans. P* = 496 W)

1.13 A metal plate of size 0.60 m × 0.60 m and 1 mm thick and weighing 25 N is to be lifted up edgewise with a uniform velocity of 0.2 m/s in the gap between two flat surfaces. The plate is in the middle of the gap of width 2 mm and the gap contains oil of relative density 0.85 and viscosity 1.6 poise. Calculate the vertical force required for this job.

(*Ans.* F_v = 88.08 N)

1.14 A 90 mm diameter shaft rotates at 1200 rpm in a 100 mm long journal bearing of 90.5 mm internal diameter. The annular space in the bearing is filled with oil having a dynamic viscosity of 0.12 Pa.s. Estimate the power dissipated as heat.

(*Ans.* P = 434 W)

1.15 A vertical shaft has a hemispherical bottom of radius R which rotates inside a bearing of identical shape at its end. An oil film of thickness h and viscosity μ is maintained in the bearing. Estimate the viscous torque in the shaft when it rotates with an angluar velocity ω.

$$\left(\text{Ans. } T = \frac{4\pi\omega\mu R^4}{3h}\right)$$

1.16 A liquid drop of radius R breaks up into 64 smaller drops, all of equal size. Calculate the work done.

(*Ans.* $W = 12\ \pi\sigma R^2$)

1.17 A tube of internal diameter 2 mm is dipped vertically into a vessel containing mercury. The lower end of the tube is 2 cm below the mercury surface. Estimate the pressure of air inside the tube required to blow a hemispherical bubble at the lower end. Assume surface tension of mercury = 0.40 N/m.

(*Ans.* Δp = 3.468 kPa)

1.18 Obtain an expression for the height of capillary rise for a fluid of surface tension σ and contact angle θ between two parallel vertical plates at a distance B apart.

$$\left(\text{Ans. } h = \frac{2\sigma\cos\theta}{\gamma B}\right)$$

1.19 If the surface tension of water in contact with air is 0.075 N/m, what correction need be applied towards capillary rise in the manometric readings in tubes of 3 mm diameter?

(*Ans.* Correction: –10.2 mm to the readings).

1.20 Air is introduced through a nozzle into a tank of water to form a stream of bubbles of size 2 mm. Calculate by how much the pressure in the nozzle must exceed that of the surrounding water? Assume σ = 0.073 N/m.

(*Ans.* Δp = 143.4 pa)

1.21 Estimate the pressure inside a water droplet of size 0.3 mm. Assume σ = 0.0728 N/m.

(*Ans.* Δp = 0.97 kPa)

1.22 In measuring the unit surface energy of an oil of relative density (RD) = 0.85 by the bubble method, a tube with an internal diameter of 1.5 mm is immersed to a depth of 1.25 cm in the oil. Air is forced through the tube forming a bubble at the lower end. What magnitude of unit surface energy is indicated by a maximum bubble pressure of 1470 kPa?

(*Ans.* σ = 0.01613 N/m)

1.23 A perfect gas has its pressure doubled and its specific volume decreased by one third. If the initial temperature was 30°C what is the final temperature?

(*Ans.* T_2 = 131°C)

1.24 Chlorine at 100 kPa (abs) and 15°C is isentropically compressed to one fifth of its volume. Assuming k = 1.4, estimate the final temperature and pressure.

(*Ans.* T_2 = 275.2°C,
p_2 = 951.8 kPa)

1.25 Oxygen at 50°C and 90 kPa (abs) is compressed to 40% of its volume. What is the temperature and pressure of the gas if the process is (a) adiabatic (k = 1.4) and (b) isothermal.

(*Ans.* (a) p_2 = 324.6 kPa (abs), T_2 = 193°C
(b) p_2 = 225 kPa (abs), T_2 = 50°C)

1.26 (i) If the pressure on a sample of water is increased by 10^3 kPa above atmospheric pressure what percentage reduction in the volume of water is observed? Assume $K_{water} = 2.2 \times 10^9$ Pa.
(ii) To achieve the same reduction in volume as above in a sample of air by isothermal process what increase in the pressure is needed?

(*Ans.* (i) 0.0455% (ii) 0.0461 kPa)

1.27 If $K_{mercury} = 2.62 \times 10^{10}$ Pa, determine the velocity of propagation of sound in mercury.

(*Ans.* C = 1388 m/s)

1.28 The velocity of propagation of sound in air (C) is calculated by assuming the process to be isentropic. What error in C is involved if the sound propagation is assumed to occur isothermally ?

(*Ans.* 15.5% less)

1.29 Air at 60°C and standard atmospheric pressure has a density of 1.060 kg/m^3. What is the velocity of propagation of sound in this medium? Assume the process to be isentropic and k = 1.4.

(*Ans.* C = 365.8 m/s)

1.30 A liquid with a volume of 0.2 m^3 at 300 kPa is subjected to a pressure of 3000 kPa and its volume is found to decrease by 0.2%. Calculate the bulk modulus of elasticity of the liquid.

(*Ans.* $K = 1.35 \times 10^9$ Pa)

2

Fluid Statics

CONCEPT REVIEW

2.1 MEASUREMENT OF PRESSURE

Different aspects of pressure distribution in incompressible fluids and compressible fluid when at rest are described in this section. Use of manometers to measure pressure differences are highlighted.

2.1.1 Pressure in Incompressible Fluids

Fluid pressure at a point is the same in all directions. In a fluid at rest the pressure intensities in a horizontal plane are equal. For incompressible fluids (i.e. for liquids and gases where compressibility effects can be ignored) the variation of pressure in vertical direction (Fig. 2.1) is given by

$$\boxed{\frac{dp}{dz} = -\gamma} \qquad (2.1)$$

$$(p_2 - p_1) = \gamma(Z_1 - Z_2) \qquad (2.1\text{-a})$$

$$\boxed{(p + \gamma Z) = (p_1 + \gamma Z_1) = (p_2 + \gamma Z_2) = \text{Constant}} \qquad (2.2)$$

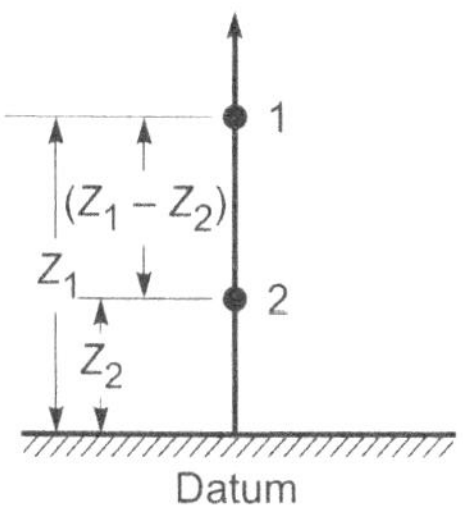

Fig. 2.1

where γ = Specific weight of the fluid

and Z = Vertical distance measured from a datum (positive upward).

At a free surface the pressure is atmospheric. If h is the depth below the free surface of a point M, the absolute pressure at M (Fig. 2.2) is

$$\boxed{p_m(\text{abs}) = \gamma h + p_{\text{atm}}}$$

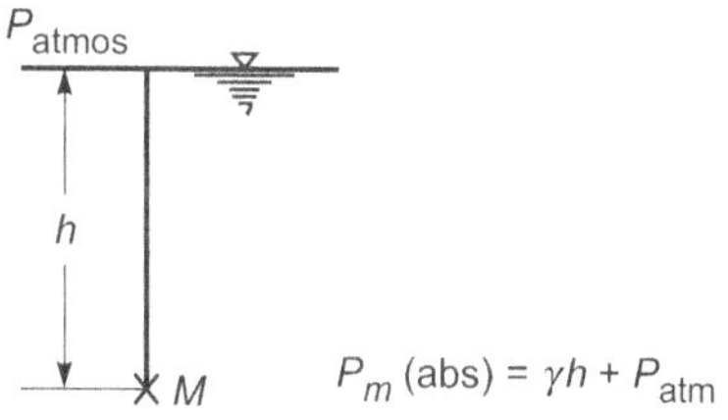

Fig. 2.2

If the pressure *in excess of atmosphere* is recorded then

$$\boxed{p_m(\text{abs}) - p_{\text{atm}} = p_{\text{m}} = \gamma h} \qquad (2.3)$$

(Note that h is measured positive downwards from the liquid surface).

The pressure p_{m} is then called *gauge pressure*.

The linear variation of pressure with depth below the free surface is known as *hydrostatic pressure distribution*.

The variation of gauge pressure in a liquid below the free surface is shown in Fig. 2.3. From this $p_1 = \gamma h_1$ and $p_2 = \gamma h_2$, or

$$\boxed{(p_2 - p_1) = \gamma(h_2 - h_1)}$$

Note that in the above the atmospheric pressure was assumed as the datum, i.e. reference with a zero value.

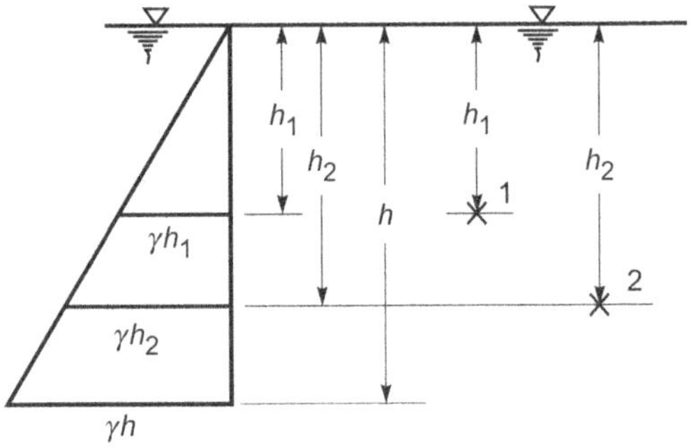

Fig. 2.3

Different references can be taken and depending upon the reference pressures we have:

Absolute pressure is the pressure measured above the absolute zero, a thermodynamic concept. Absolute pressures are designated with (abs) following the symbol or numeral to distinguish from other forms. Thus, for example, 12.0 kPa(abs); p_m (abs) are absolute pressures. Absolute pressures cannot be negative.

Gauge pressure is the pressure measured with respect to local *atmospheric pressure*. Gauge pressures are extensively used in engineering practice and as such are indicated with a symbol or a numeral without any other explanatory notation, e.g. 14.0 kPa, – 3.2 kPa, p_m are gauge pressures. Note that gauge pressures can be positive or negative. Negative gauge pressures are also called *vacuum pressures*.
It is seen that

Absolute pressure = (Local atmospheric pressure) + (gauge pressure)

Pressure has the dimension of [Force/Area]= [F L^{-2}] and is usually expressed in *pascals* Pa (= N/m^2); kilo pascals kPa (= 10^3 N/m^2); height h of a column of a fluid of specific weight γ, in *bars* (= 10^5 Pa) or *atmospheres* (= number of standard atmospheric pressure value). The pressures are commonly indicated as gauge pressures and unless a pressure is specifically marked absolute the pressure is treated as gauge pressure. The *atmosphere*, however, is an exception and is an absolute pressure unit. Gauge pressures are commonly measured by a *Bourdon gauge*. Differences in pressures are measured by manometers.
Local atmospheric pressure (i.e. the absolute pressure of the atmosphere at a place) is measured by a mercury *barometer*. The local atmospheric pressure varies with the elevation above mean sea level and local meteorological conditions. For engineering application, a standard atmospheric pressure at mean sea level at 15°C is often used. The value of this standard atmospheric pressure (called 1 atmosphere) is

1 atm = 10.3 m of water
= 760 mm of mercury
= 101.3 kPa
= 10130 mbar

Aneroid barometer is another instrument commonly used to measure local atmospheric pressure.

2.1.2 Pressure in Compressible Fluids

For a compressible fluid the density changes with change in pressure and temperature. For a perfect gas we have

$$\boxed{p = \rho R T} \tag{2.4}$$

where p = absolute pressure
T = temperature in Kelvin
R = gas constant
ρ = mass density

Since $\dfrac{dp}{dz} = -\rho g$

$$\boxed{\int dp = -g \int \rho\, dz} \tag{2.5}$$

Depending upon the process involved, i.e. isothermal or adiabatic, the corresponding variation of pressure can be evaluated.

Isothermal Process In an isothermal process $T = T_0$ = constant
Hence

$$\frac{dp}{dz} = -\frac{p}{RT_0} g$$

$$\int_1^2 \frac{dp}{p} = -\frac{g}{R}\int_{Z_1}^{Z_2} \frac{dz}{T_0}$$

$$\boxed{\frac{p_2}{p_1} = \exp\left[\frac{-g(Z_2 - Z_1)}{RT_0}\right]} \tag{2.6}$$

Adiabatic Process For the case of adiabatic process (zero heat transfer), if there is no friction (isentropic)

$$\frac{p}{\rho^k} = \text{Constant} = C_s \tag{2.7}$$

where k = adiabatic constant for the gas. Combining with perfect gas law (Eq. 2.4) we get

$$\frac{T}{\rho^{k-1}} = \text{Constant}; \quad \frac{T}{p^{(k-1)/k}} = \text{Constant}$$

Substituting in Eq. 2.5 and simplifying yields

$$\frac{p_2}{p_1} = \left[1 - \frac{(k-1)}{k} g(Z_2 - Z_1)\frac{\rho_1}{p_1}\right]^{k/(k-1)} \quad (2.8)$$

or

$$\boxed{\frac{p_2}{p_1} = \left[1 - \frac{(k-1)}{k} g\frac{(Z_2 - Z_1)}{RT_1}\right]^{k/(k-1)}} \quad (2.9)$$

The variation of the temperature with Z is given by

$$\boxed{\frac{T_2}{T_1} = \left[1 - \frac{(k-1)}{k} \cdot \frac{g(Z_2 - Z_1)}{RT_1}\right]} \quad (2.10)$$

The rate of variation of the temperature with elevation (dT/dZ) is known as *lapse rate* and is given by

$$\boxed{L = \frac{dT}{dZ} = -\frac{g}{R}\left(\frac{k-1}{k}\right)} \quad (2.11)$$

2.1.3 Manometers

For a fluid at rest it is seen from Eq. 2.1(a) that $(p_2 - p_1) = \gamma(Z_2 - Z_1)$. The pressure difference between two points can thus be measured by a static column of a liquid. A manometer is a device to determine the pressure in a fluid by balancing it against a column of liquid. Figure 2.4 shows some commonly used forms of manometers. In a manometer the basic point to remember is that, in a continuous mass of the same static fluid, the pressure at the points in any horizontal plane will be the same.

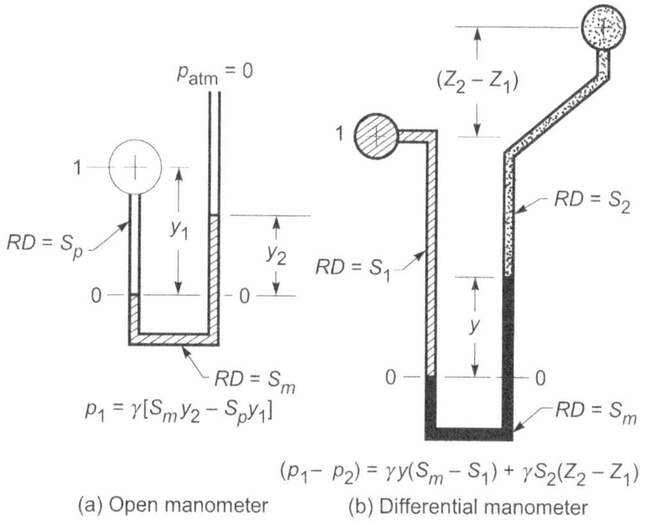

Fig. 2.4 *Manometers*

2.2 FORCES ON PLANE SURFACES

An important problem in the design of hydraulic structures and other structures which interact with fluids is the computation of hydrostatic forces on plane surfaces. Computations of magnitude and point of application of hydrostatic forces on plane surfaces are described.

2.2.1 Magnitude of Force on a Plane

When a plane area is immersed in a static liquid with its plane making an angle θ with the free liquid surface (Fig. 2.5) the total hydrostatic force on *one side* of the area is

$$\boxed{F = \gamma \bar{h} A} \quad (2.12)$$

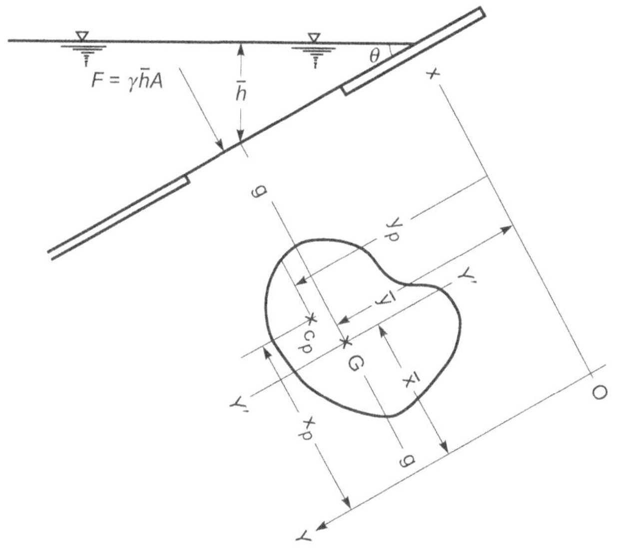

Fig. 2.5 *Centre of Pressure*

where γ = specific weight of the liquid
$\bar{h}$ = depth of the centre of gravity of the area below the free surface
A = area of the immersed plane.

It may be noted that the force F is independent of the angle of inclination θ so long as the depth of the centroid $\bar{h}$ is unchanged.

2.2.2 Centre of Pressure

The point of application of the force F on the submerged area is called the *centre of pressure*. Considering the line of intersection of the plane area with the liquid surface (Line OX) as the reference axis, the centre of pressure is located along the plane at

$$\boxed{y_p = \bar{y} + \frac{I_{gg}}{A\bar{y}}} \quad (2.13)$$

where I_{gg} = moment of inertia about an axis parallel to OX and passing through the centre of gravity of the area
$\bar{y}$ = location of the centre of gravity with respect to the axis OX
A = area of the plane area

Note that the distances y are measured *along the plane* from the axis OX.

The lateral position of the centre of pressure with respect to any axis OY perpendicular to OX and lying in the plane of the lamina is

$$x_p = \bar{x} + \frac{I_{xy}}{A\bar{y}} \tag{2.14}$$

where I_{xy} = product of inertia $\left(= \int_A xy\, dA\right)$ of the area about axis GY′, passing through the centre of gravity of the area and parallel to OY and OX.

When either of the centroidal axes $x = \bar{x}$ or $y = \bar{y}$ is an axis of symmetry, $I_{xy} = 0$ and $x_p = \bar{x}$.

2.3 FORCES ON CURVED SURFACES

When the fluid static force on a curved submerged surface is desired, it is convenient to consider the horizontal and vertical components of the force separately.

2.3.1 Horizontal Component

The horizontal component of hydrostatic force in any chosen direction on any area (plane or curved) is equal to the projection of the area on a vertical plane normal to the chosen direction. The horizontal force acts through the centre of pressure of the vertical projection.

2.3.2 Vertical Component

The vertical component of the hydrostatic force on any surface (plane or curved) is equal to the weight of volume of liquid extending above the surface of the object to the level of the free surface. This vertical component passes through the centre of gravity of the volume considered. The volume and the free surface can be real or imaginary.

2.3.3 Tensile Stress in a Pipe or Shell

In a circular pipe subjected to high pressure, the pressure centre can be taken to be at the centre of the pipe. The tensile circumferential stress (*hoop stress*) in a pipe wall subjected to an internal pressure of p (Fig. 2.6) is

$$\text{hoop stress } \sigma_h = \frac{pD}{2t} \tag{2.15}$$

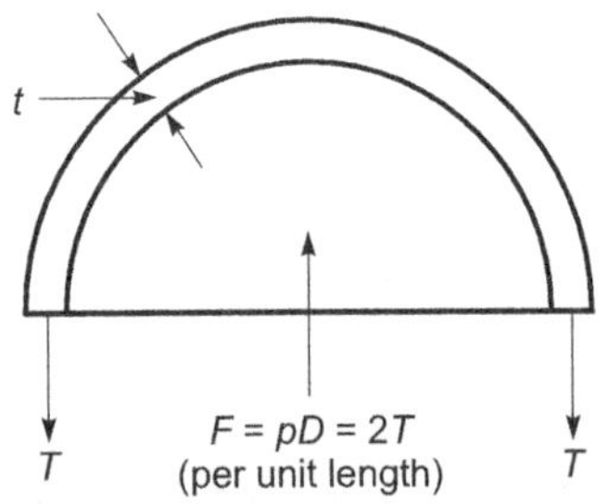

Fig. 2.6

where D = diameter of the pipe
t = thickness of pipe.

This formula assumes $t/D < 0.1$ and hence is based on thin cylinder theory.

If the ends of a cylinder are closed and the cylinder has a fluid under pressure, a longitudinal stress σ_L is produced in the cylinder. This stress is given by

$$\sigma_L = \frac{1}{2}\sigma_h = \frac{pD}{4t} \tag{2.16}$$

For thin spherical shells the tensile stress is

$$\sigma_s = \frac{pD}{4t} \tag{2.17}$$

2.4 BUOYANCY

When a body is submerged or floating in a static fluid the resultant force exerted on it by the fluid is called *buoyancy force*. This buoyancy force is always vertically upwards, and has the following characteristics.

1. The buoyancy force is equal to the weight of the fluid displaced by the solid body.
2. The buoyancy force acts through the centre of gravity of the displaced volume, called the *centre of buoyancy*.
3. A floating body displaces a volume of fluid whose weight is equal to the weight of the body.

2.4.1 Stability

A submerged body is stable if the centre of gravity of the body lies below the centre of buoyancy.

For a floating body the stability depends upon the type of couple that is formed for small angular displacements. For a body shown in Fig. 2.7(a) the centre of gravity is G and the centre of buoyancy is B. Initially it is stable with G above B. Figure 2.7 (b) shows the same body with a small displacement. If B′ is the new centre of buoyancy a vertical from B′ intersects the line of symmetry through

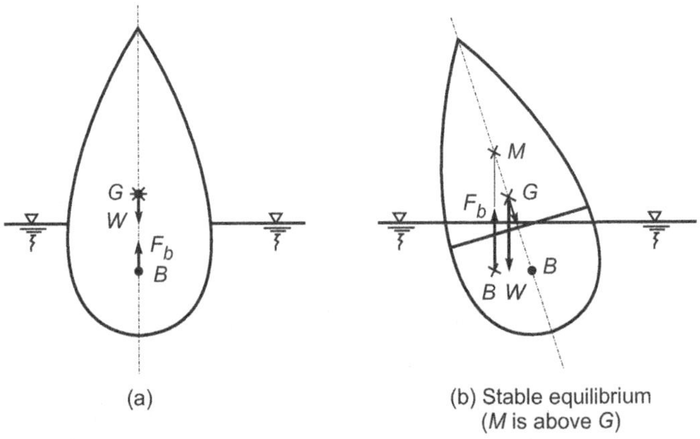

Fig. 2.7

G at M. M is known as the *meta centre*. If M is above G, then MG the metacentric height is positive and the equilibrium is *stable*. If M is below G, MG is negative and equilibrium is *unstable*. The *metacentric height* MG is independent of magnitude of angular rotation (so long as it is small) and is given by

$$MG = \frac{I}{\forall} - BG \quad (2.18)$$

In this equation

I = Moment of inertia of the *water line area* about an axis through the centre of the area and perpendicular to the axis of tilt (longitudinal axis).

BG = Vertical distance between the centre of gravity and centre of buoyancy.

$\forall$ = Volume of the fluid displaced by the body.

If M coincides with G, MG is zero, the body is said to be in *neutral equilibrium*.

2.5 RIGID BODY MOTION

When a fluid mass in a container is subjected to a motion such that there is no relative motion between the particles, such a motion is known as *rigid body motion*. The motion can be either translation or rotation at constant acceleration or a combination of both. As there is no relative motion there is no shear stress in such a motion and the pressure distribution is similar to that in fluids at rest, of course modified by the combined action of gravity and fluid acceleration.

2.5.1 Translation

If a container with a fluid is given a translation (a linear motion) with a uniform acceleration the piezometric head will have a gradient in the direction of motion.

If the motion is in the x-direction with a constant acceleration a_x then

$$-\frac{dh}{dx} = \tan\theta = \frac{a_x}{g} \quad (2.19)$$

where $h = (p/\gamma + z)$ = piezometric head above datum

θ = Inclination of hydraulic grade line.

= Inclination of water surface, measured clockwise with respect to the x-direction.

Thus, if a vessel containing a liquid is given an acceleration a_x in x-direction (Fig. 2.8.) the surface will back up against the farthest side, i.e. it will have increasing depth in (−x) direction.

If a closed tank without a free surface is involved, an imaginary free surface equivalent to the piezometric head line can be considered. This piezometric head line will be inclined to the x-direction such that

$$\boxed{\tan\theta = a_x/g}$$

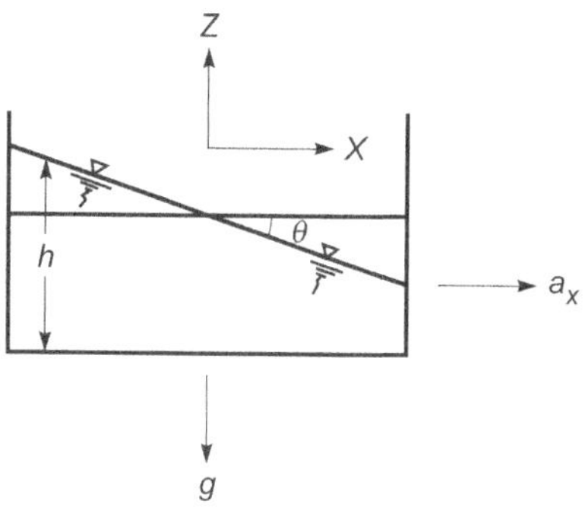

Fig. 2.8

It follows from the above that if acceleration is solely in the vertical direction (+ z direction) then $a_x = 0$ and $\tan\theta = 0$. This means that the liquid surface will remain horizontal. However, the pressure p_h at any depth h below the free surface will now be

$$\boxed{p_h = \gamma h\left(1 + \frac{a_z}{g}\right)} \quad (2.20)$$

In this a_z = vertical acceleration in + z direction (if the acceleration is vertically downwards, a_z is taken as negative).

In vertical acceleration the liquid suffers an apparent gravity equal to $(g + a_z)$.

If the acceleration is a_s in any direction s, then the components a_x and a_z in x- and z-directions are considered. The fluid surface will now have an inclination $\tan\theta$ given by

$$\boxed{-\frac{dh}{dx} = \tan\theta = \frac{a_x}{(g + a_z)}} \quad (2.21)$$

2.5.2 Rigid Body Rotation

When a vessel containing a liquid with a free surface is rotated about an axis, the free surface will be a paraboloid of revolution given by

$$\boxed{y = \frac{\omega^2 r^2}{2g}} \quad (2.22)$$

where ω = angular velocity

y = height of the free surface above the vertex at a radial distance r from the axis.

At any two points r_1 and r_2 from the axis

$$(y_2 - y_1) = \frac{\omega^2}{2g}(r_2^2 - r_1^2)$$

Since $\omega r = V$ = tangential velocity.

$(y_2 - y_1) = \Delta y$ = difference in the liquid surface elevation between the points 2 and 1 (Fig. 2.9)

$$= \frac{V_2^2}{2g} - \frac{V_1^2}{2g} = \Delta\,(V^2/2g)$$

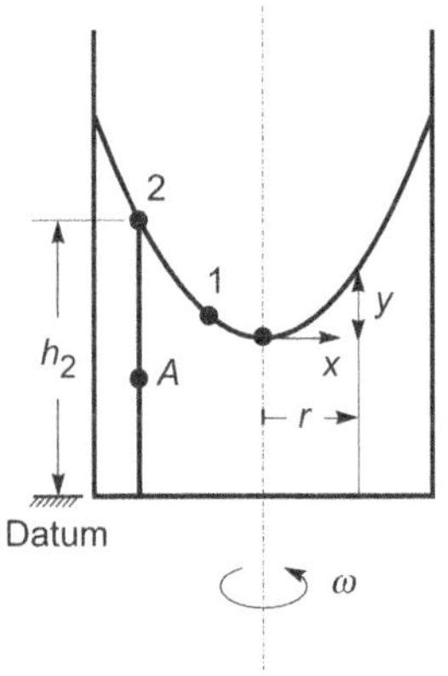

Fig. 2.9

= difference in the velocity head at these two points

The pressure distribution in any vertical line at a radial distance r will, however, remain hydrostatic. At point 2,

$$h_2 = \frac{p_A}{\gamma} + z_A$$ for all values of A on this vertical line.

If the free surface does not exist, the piezometric head will follow the relation for y (Eq. 2.22) as:

$$(h - h_0) = \frac{\omega^2 r^2}{2g}$$

where h = piezometric head above a datum at any radial distance r from the axis

h_0 = value of h at $r = 0$, i.e. on the axis

ω = angular velocity.

The piezometric head $h = \left(\frac{p}{\gamma} + z\right)$ will vary with r as a paraboloid of revolution and this surface can be considered as an imaginary liquid surface.

The volume of a paraboloid of revolution is one half the volume of the circumscribing cylinder.

WORKED EXAMPLES

A. Measurement of Pressure

2.1 Find the pressure represented by a column of

(i) 10 cm of water.

(ii) 5 cm of oil of relative density 0.75.

(iii) 2 cm of mercury.

Solution:

Pressure $p = \gamma h$

(i) For water column: $\gamma = 9790$ N/m³

$$p = 9790 \times \frac{10}{100} = \mathbf{979\ N/m^2}$$

(ii) For the oil column: $\gamma = 0.75 \times 9790$
$= 7342.5$ N/m³

$$p = 7342.5 \times \frac{5}{100} = \mathbf{367.13\ N/m^2}$$

(iii) For mercury column: $\gamma = 13.6 \times 9790$
$= 133144$ N/m³

$$p = 133144 \times \frac{2}{100} = \mathbf{2662.9\ N/m^2}$$

2.2 Find the absolute pressure at a depth of 5 m below the surface of a liquid of relative density 0.85. The barometer reading on the surface is 750 mm of mercury.

Solution:

For the liquid: $\gamma = 0.85 \times 9790 = 8321.5$ N/m³

Atmospheric pressure $= 0.750 \times 13.6 \times 9790$
$= 99858$ N/m²

Absolute pressure in the liquid at 5.0 m depth

= pressure due to 5 m of liquid + Atmospheric pressure

$= (5 \times 8321.5) + 99858$

$= 141465.5$ N/m² = **141.466 kPa**

2.3 A 6 m deep tank contains 4 m of water and 2 m of oil of relative density 0.88. Determine the pressure at bottom of the tank (Refer Fig 2.10).

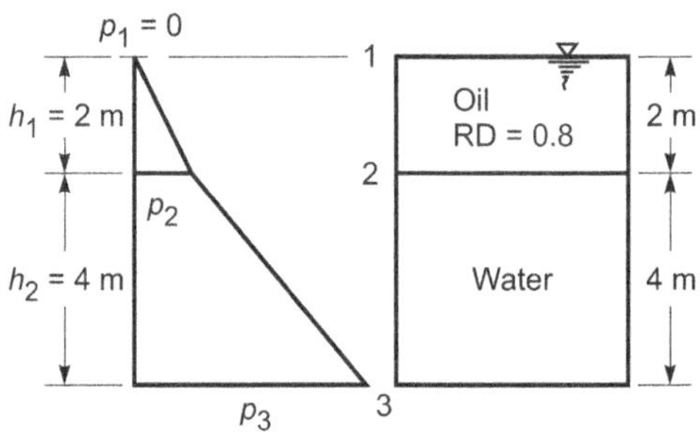

Fig. 2.10

Solution:

First determine the pressure at the oil-water interface.

$p_2 = p_1$ + pressure due to 2 m of oil

$= p_1 + (\gamma_0 \times 2)$

$p_1 = 0$, $\gamma_0 = 0.88 \times 9790 = 8615.2$ N/m³

$\therefore \quad p_2 = 8615.2\ (2) = 17230.4\ \text{N/m}^2$

For water $\gamma_w = 9790\ \text{N/m}^3$

$$p_3 = p_2 + \text{pressure due to 4 m of water}$$
$$= 17230.4 + (9790 \times 4)$$
$$= 56390.4\ \text{N/m}^2 = \mathbf{56.390\ kPa}$$

2.4 Assuming an isothermal process, find the pressure at an elevation of 2500 m above sea level. The mass density and atmospheric pressure at an elevation of 500 m above sea level are known to be 1.1677 kg/m^3 and 95480 Pa. What is the density of air at that level?

Solution: For an isothermal atmosphere, the pressure distribution is given by

$$\frac{p_2}{p_1} = \exp\left[-\frac{g(Z_2 - Z_1)}{RT_0}\right]$$

Since $\dfrac{p_1}{\rho_1} = RT_0$

$$\frac{p_2}{p_1} = \exp\left[-\frac{\rho_1 g(Z_2 - Z_1)}{p_1}\right]$$

$$p_2 = 95480 \exp\left[-\frac{(1.1677 \times 9.81) \times 2000}{95480}\right]$$
$$= 75111\ \text{Pa}$$

$$\frac{p_2}{\rho_2} = \frac{p_1}{\rho_1}$$

$$\therefore \quad \rho_2 = \frac{p_2}{p_1}\rho_1$$

$$= \frac{75111}{95480} \times 1.1677 = \mathbf{0.9186\ kg/m^3}$$

2.5. Calculate the atmospheric pressure and density at an elevation of 3500 m above sea level, given the density and atmospheric pressure at an elevation of 1000 m above sea level to be 1.1120 kg/m^3, and 89,890 Pa respectively. Assume the process to be adiabatic.

Solution: For an adiabatic atmosphere

$$\frac{p_2}{p_1} = \left[1 - \left(\frac{k-1}{k}\right)\frac{g(Z_2 - Z_1)}{RT_1}\right]^{\frac{k}{(k-1)}}$$

since $\dfrac{p_1}{\rho_1} = RT_1$

$$\frac{p_2}{p_1} = \left[1 - \left(\frac{k-1}{k}\right)\frac{(g\rho_1)(Z_2 - Z_1)}{p_1}\right]^{(k/k-1)}$$

k is assumed as equal to 1.4. Hence

$$\frac{k}{(k-1)} = \frac{1.4}{0.4} = 3.5$$

$$p_2 = 89{,}890\left[1 - \left(\frac{1.4-1}{1.4}\right)\frac{(9.81 \times 1.1120)\ (3500 - 1000)}{89{,}890}\right]^{3.5}$$
$$= 89{,}890\ [\ 1 - 0.086683]^{3.5}$$
$$= 65446\ \text{pa} = 65.446\ \text{kPa}$$

$$\frac{p_1}{\rho_1^k} = \frac{p_2}{\rho_2^k}$$

$$\therefore \quad \rho_2 = \left(\frac{p_2}{p_1}\right)^{1/k}\rho_1$$

$$= \left(\frac{65446}{89890}\right)^{1/1.4}(1.1120) = \mathbf{0.88665\ kg/m^3}$$

2.6 Determine the pressure difference between points A and B, for the set up shown in Fig. 2.11

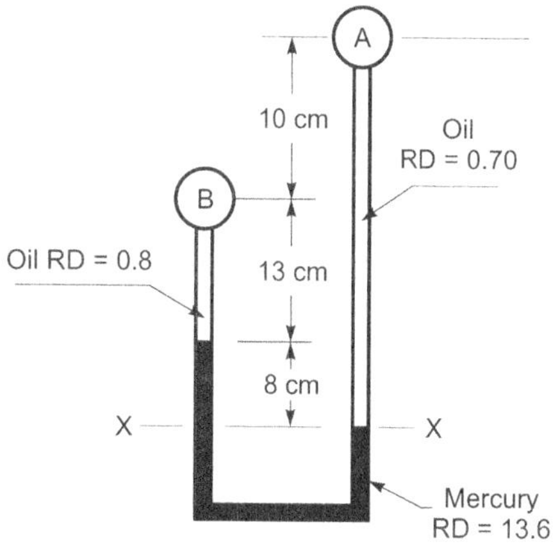

Fig. 2.11

Solution: Equating the pressures on both the limbs at the horizontal plane XX (Unit weight of water is taken as 9790 N/m^3):

$$p_A + [0.10 + 0.13 + 0.08] \times \gamma_A$$
$$= p_B + 0.13\ \gamma_B + 0.08\ \gamma_m$$
$$\gamma_A = 0.70 \times 9790 = 6853\ \text{N/m}^3$$
$$\gamma_B = 0.80 \times 9790 = 7832\ \text{N/m}^3$$
$$\gamma_m = 13.6 \times 9790 = 133144\ \text{N/m}^3$$
$$(p_A - p_B) = -0.31 \times 6853 + 0.13 \times 7832 + 0.08 \times 133144$$
$$= 9545\ \text{Pa} = \mathbf{9.545\ kPa}$$

2.7 For the manometer shown in Fig. 2.12 calculate the pressure difference between points M and N.

Solution: Equating the pressures at both the limbs along the horizontal plane XX

$$p_m - \gamma_w \cdot (0.06 + 0.035) = p_N - \gamma_w\ (0.12 + 0.06) - \gamma_0(0.035)$$
$$\gamma_w = \text{unit wt. of water}$$
$$= 9790\ \text{N/m}^3\ \text{(assumed)}$$

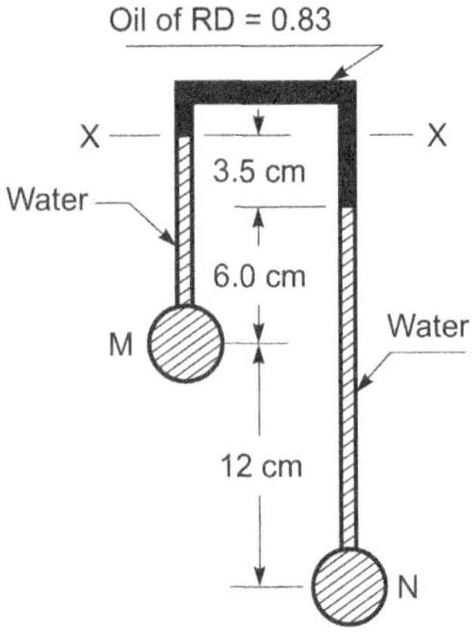

Fig. 2.12

$$\gamma_0 = \text{unit wt. of oil} = 0.83 \times 9790$$
$$= 8125.7 \text{ N/m}^3$$
$$\therefore \quad (p_M - p_N) = 9790 \times 0.095 - 9790 \times 0.18 - 8125.7 \times 0.035$$
$$= -1116.5 \text{ N/m}^2 = -1.1165 \text{ kPa}$$

Pressure at N is larger than at M by **1.117 kPa**

2.8 Two pressure tanks are built one inside the other (Fig. 2.13). A bourdon gauge M connected to the inner tank reads 20 kPa. Another bourdon gauge N connected to the outer tank reads 35 kPa. An aneroid barometer reads 750 mm of mercury. Calculate the absolute pressure recorded at M and N in of mercury.

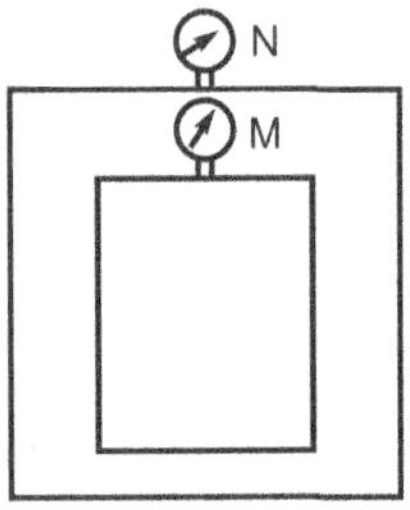

Fig. 2.13

Solution: A bourdon gauge records the gauge pressure relative to the pressure of the medium surrounding the tube.

Local atmospheric pressure is measured by the aneroid barometer.

In the present case local atmospheric pressure outside the gauge N = 750 mm.

Hence absolute pressure at N =

$$p_N(\text{abs}) = 750 + \left(\frac{35}{13.6 \times 9.79} \times 1000\right)$$

= 1012.9 mm of mercury (abs)

The gauge M reads relative to its surrounding pressure of 1012.9 mm of mercury (abs).

Hence, $p_M(\text{abs}) = 1012.9 + \left(\dfrac{20 \times 1000}{13.6 \times 9.79}\right)$

= 1163.1 mm of mercury (abs)

2.9 The tank in Fig. 2.14 is closed at top and contains air at a pressure p_A. Calculate the value of p_A for the manometer readings shown.

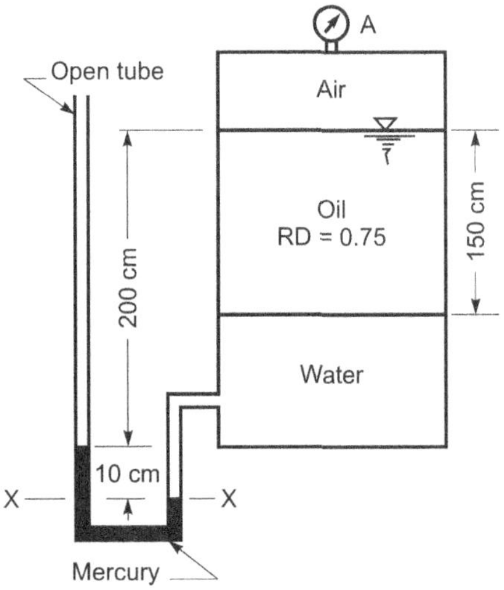

Fig. 2.14

Solution: Considering the pressure at the horizontal plane X–X:

$$p_A + 1.5 \times \gamma_0 + (2.0 - 1.5)\, \gamma_w + 0.10\, \gamma_w = 0.10\, \gamma_m$$
$$\gamma_o = \text{specific wt. of oil} = 0.75 \times 9790$$
$$= 7342.5 \text{ N/m}^3$$
$$\gamma_w = \text{specific wt. of water} = 9790 \text{ N/m}^3$$
$$\gamma_m = \text{specific wt. of mercury} = 13.6 \times 9790$$
$$= 133144 \text{ N/m}^3$$
$$\therefore \quad p_A = (0.10 \times 133144) - 0.50 \times 9790 - 1.5 \times 7342.5 - 0.1 \times 9790$$
$$= -3573.4 \text{ Pa} = \mathbf{-3.573 \text{ kPa}}$$

2.10 Determine the pressure difference $(p_M - p_N)$ when the manometer indicates as shown in Fig. 2.15.

Solution: Equating the pressure across the horizontal plane X-X:

$$p_M + 0.20\, \gamma_o + 0.16\, \gamma_o = p_N + 0.15\, \gamma_o + 0.15\, \gamma_m - 0.12\, \gamma_o + 0.12\, \gamma_m + 0.04\, \gamma_m$$
$$\gamma_o = \text{specific weight of oil} = 0.8 \times 9790$$
$$= 7832 \text{ N/m}^3$$
$$\gamma_m = \text{specific weight of mercury}$$
$$= 13.6 \times 9790 = 133144 \text{ N/m}^3$$
$$(p_M - p_N) = -0.36 \times 7832 + 0.15 \times 7832 - 0.12 \times 0.7832 + (0.15 + 0.12 + 0.04) \times 133144$$
$$= -2584.6 + 41274.6 = 38690 \text{ Pa}$$
$$= \mathbf{38.690 \text{ kPa}}$$

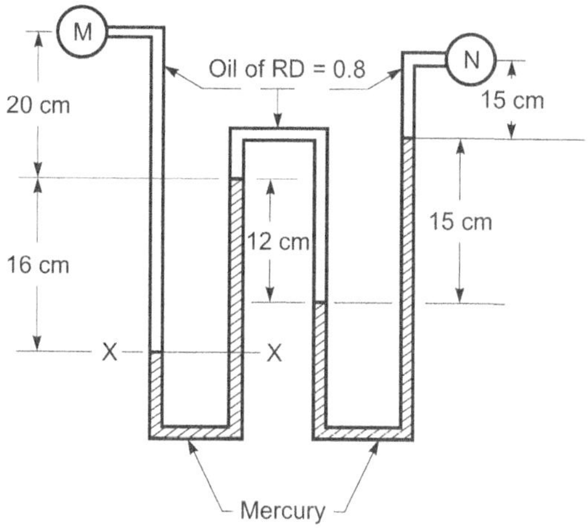

Fig. 2.15

2.11 A single leg mercury manometer (Fig. 2.16) is often used to measure differential pressure through reading of a single column only. A constant zero of the scale is adopted for all pressures.

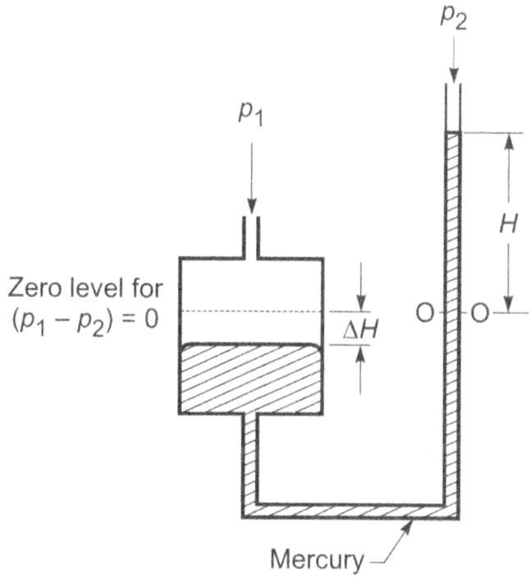

Fig. 2.16

(a) What should be the diameter of the cistern if the indicated mercury column H is to be within 0.5% the true height corresponding to a pressure differential of $(p_2 - p_1)$.

(b) If the ratio of the areas of cistern to that of the tube is 1/500, what percentage error is involved in the difference in pressure by reading only limb of height H?

Solution: The column height is measured with reference to zero level corresponding to $(p_1 - p_2) = 0$. When a pressure differential is applied the mercury in the cistern will go down by ΔH and the true column height is $(\Delta H + H)$.

(a) It is required that $H + \Delta H \le 1.005\, H$

$$\Delta H \le 0.005\, H$$

$$\frac{\Delta H}{H} \le 0.005$$

Since (Area of cistern) × ΔH = (area of tube) × H

For minimum area of cistern A_c; $\Delta H\, A_c = aH$

and $$\frac{\Delta H}{H} = 0.005 = \frac{a}{A_c} = \left(\frac{d}{D_c}\right)^2$$

Hence D_c = Minimum diameter of cistern

$$= d/\sqrt{0.005} = \mathbf{14.142\, d}$$

(b) If $\dfrac{a}{A_c} = 1/500$

then $$\frac{\Delta H}{H} = \frac{1}{500}$$

Percentage error in

$$H = \frac{\Delta H}{H} \times 100 = \frac{1}{500} \times 100 = \mathbf{0.2\%}$$

2.12 A glass tube of uniform bore is bent into the form of a square of sides a and filled with equal amounts of three immiscible liquids of densities ρ_1, ρ_2 and ρ_3. It is known that $\rho_1 < \rho_2 < \rho_3$. If the tube arrangement is placed in a vertical plane (i.e. two sides vertical) and if one of the vertical sides is completely filled with the liquid of density ρ_2.

(a) Show that

$$\frac{1}{3}(2\rho_3 + \rho_1) > \rho_2 > \frac{1}{3}(\rho_3 + 2\rho_1)$$

(b) If the relative densities of the first and third liquids are 1.0 and 1.2 respectively, find the range of the relative densities of the second liquid which makes the above arrangement possible.

Solution: Referring to Fig. 2.17, let E, F and G be the interfaces.

Let EA = x.

Then DE = DA – EA = $(a - x)$

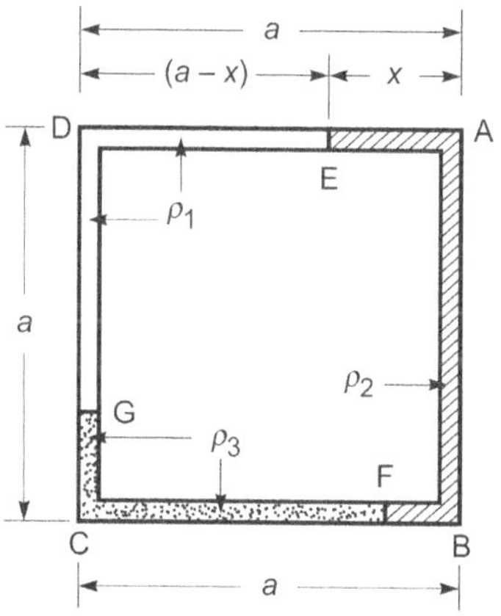

Fig. 2.17

Total tubing length = $4a$

Length of each liquid = $\frac{4}{3}a$

Hence for liquid 1:

$$EG = \frac{4}{3}a$$

$$DG = \frac{4}{3}a - (a - x) = \frac{1}{3}a + x$$

For liquid 3:

$$GC = a - \left(\frac{1}{3}a + x\right) = \frac{2}{3}a - x$$

$$FB = a - \left(\frac{2}{3}a + x\right) = \frac{1}{3}a - x$$

$$\left(\text{Check: FB + BA + AE} = \frac{1}{3}a - x + a + x = \frac{4}{3}a\right)$$

At the Interface F: The pressure balance is Pressures of (Column DG + Column GC) = Pressure of column AB

$$\rho_1 g\left(\frac{1}{3}a + x\right) + \rho_3 g\left(\frac{2}{3}a - x\right) = \rho_2 g a$$

$$x(\rho_3 - \rho_1) = \frac{1}{3}a\,(2\rho_3 + \rho_1 - 3\rho_2)$$

$$x = \frac{1}{3}a\,(2\rho_3 + \rho_1 - 3\rho_2)/(\rho_3 - \rho_1)$$

It is known that $x > 0$ and also $x < \frac{1}{3}a$

Hence $0 < x < \frac{a}{3}$

$\therefore \quad 0 < \frac{2\rho_3 + \rho_1 - 3\rho_2}{\rho_3 - \rho_1} < 1$

Also since $\rho_1 < \rho_2 < \rho_3$ the denominator $(\rho_3 - \rho_1)$ is positive. Hence the numerator is

$$0 < (2\rho_3 + \rho_1 - 3\rho_2) < 1$$

or $(2\rho_3 + \rho_1) > 3\rho_2$

or $\rho_2 < \frac{1}{3}(2\rho_3 + \rho_1)$ (i)

Also since $\frac{2\rho_3 + \rho_1 - 3\rho_2}{\rho_3 - \rho_1} < 1$

$$2\rho_3 + \rho_1 - 3\rho_2 < (\rho_3 - \rho_1)$$

or $3\rho_2 < \rho_3 + 2\rho_1$

$\rho_2 > \frac{1}{3}(\rho_3 + 2\rho_1)$ (ii)

Hence from inequalities (i) and (ii)

$$\mathbf{\frac{1}{3}(2\rho_3 + \rho_1) > \rho_2 > \frac{1}{3}(\rho_3 + 2\rho_1)}$$

(b) $\rho_1 = 1.0$, $\rho_3 = 1.2$

Hence $\rho_2 > \frac{1}{3}(\rho_3 + 2\rho_1)$

i.e. $> \frac{1}{3}(1.2 + 2 \times 10)$

> 1.0667

Also $\rho_2 < \frac{1}{3}(2\rho_3 + \rho_1)$

$< \frac{1}{3}(2 \times 1.2 + 1.0)$

< 1.1333

Hence $\mathbf{1.0667 < \rho_2 < 1.1333}$

B. Forces On Plane Surfaces

2.13 A rectangular plate 0.6 m wide and 1.2 m deep lies within a water body such that its plane is inclined at 45° to the horizontal and the top edge is 0.70 m below the water surface. Determine the total pressure force on one side of the plate and the location of the centre of pressure (Refer Fig. 2.18).

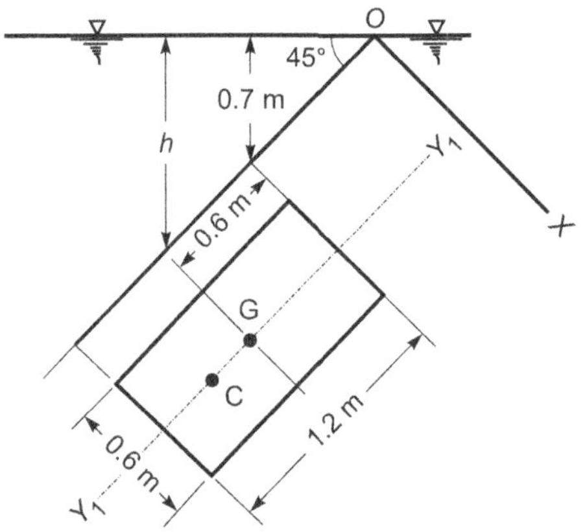

Fig. 2.18

Solution: Referring to Fig. 2.18

$\bar{h}$ = depth of CG of the plate = 0.7 + 0.6 sin 45° = 1.1243 m

Total pressure force $F = \gamma_A \bar{h}$

$= 9790 \times (0.6 \times 1.2) \times 1.1243$

$= 7924.9$ N = **7.925 kN**

Centre of pressure:

Because of symmetry $x_p = \bar{x}$, i.e. C lies on the axis Y_1Y_1 passing through the CG of the area.

$$y_p = \bar{y} + \frac{I_{gg}}{A\bar{y}}$$

$$\bar{y} = \frac{\bar{h}}{\sin 45°} = \frac{1.1243}{1/\sqrt{2}} = 1.59 \text{ m}$$

$$I_{gg} = \frac{bd^3}{12} = \frac{0.6 \times (1.2)^3}{12} = 0.0864$$

$$y_p = 1.59 + \frac{0.0864}{(0.6 \times 1.2) \times (1.59)}$$

$$= \mathbf{1.665\ m}$$

[y_p is measured along the plane of the area from the axis OX]

2.14 A plane area in the form of a right angle triangle of height h is immersed vertically in water with its vertex at the water surface. Calculate the total force on one side of the triangular plane and the location of the centre of pressure.

Solution:

$$\text{Total force } F = \gamma A \bar{h}$$

$$= \gamma\left(\frac{1}{2}bh\right) \times \frac{2h}{3} = \frac{\gamma}{3}bh^2$$

To determine the centre of pressure, consider the axis OX and OY as shown in Fig. 2.19. For an element of width x and height dy at a depth y,

$$\frac{b}{h} = \frac{x}{y} \quad \text{or} \quad y = \frac{h}{b}x$$

$$dy = \frac{h}{b}dx$$

Taking moments of force on the element about OY and intergrating

$$F \cdot x_p = \int_0^h \gamma y (x\,dx) \frac{x}{2}$$

$$\frac{\gamma}{3}bh^2 x_p = \int_0^b \gamma \frac{h}{b} x x \frac{h}{b} dx \frac{x}{2}$$

$$= \int_0^b \frac{\gamma h^2}{2b^2} x^3 dx$$

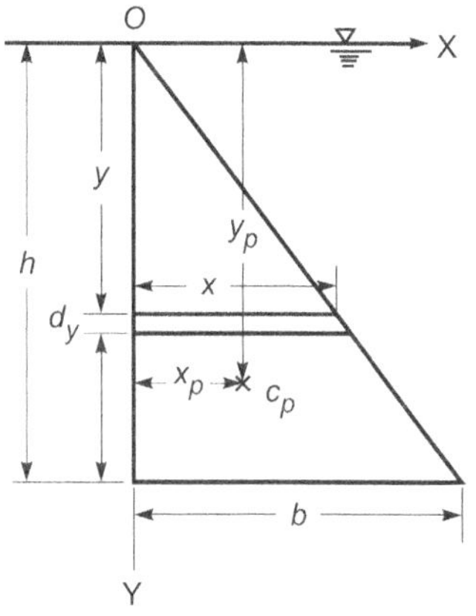

Fig. 2.19

$$= \frac{\gamma h^2 b^2}{8}$$

$$\mathbf{x_p = \frac{3}{8}b}$$

Similarly for y_p:

$$\frac{\gamma}{3}bh^2 y_p = \int_0^h \gamma y (x\,dy) y = \int \gamma y^2 \left(\frac{by}{h}\right) dy$$

$$= \frac{\gamma b}{h}\int y^3 dy = \frac{\gamma b}{h}\frac{h^4}{4}$$

$$\mathbf{y_p = \frac{3}{4}h}$$

2.15 A circular plate of diameter 0.75 m is immersed in a liquid of relative density 0.80 with its plane making an angle of 30° with the horizontal. The centre of the plate is at a depth of 1.50 m below the free surface. Calculate the total force on one side of the plate and the location of the centre of pressure.

Solution:

Referring to Fig. 2.20,

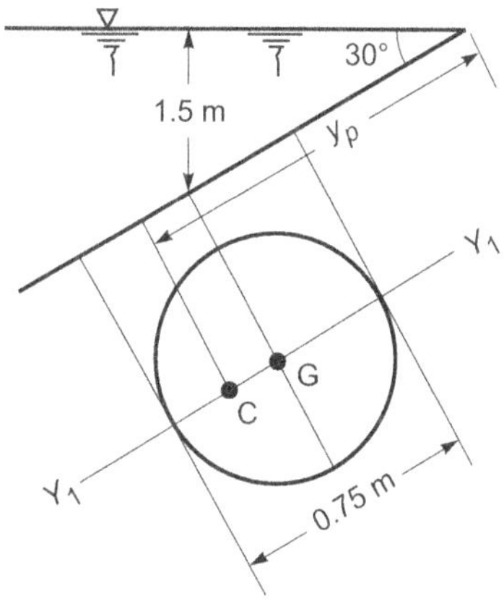

Fig. 2.20

$$\gamma = 0.80 \times 9790 = 7832 \text{ N/m}^3$$

$$\bar{h} = 1.5 \text{ m}$$

F = Force on one side of the plate

$$= \gamma A \bar{h}$$

$$= 7832 \times \left[\frac{\pi}{4}\cdot(0.75)^2\right] \times 1.5$$

$$= \mathbf{5190\ N}$$

By symmetry $x_p = \bar{x}$, i.e. the centre of pressure lies on the Y_1Y_1 axis through the CG of the area. To find y_p:

$$y_p = \bar{y} + \frac{I_{gg}}{A\bar{y}}$$

$$\bar{y} = 1.5/\sin 30° = 3.0 \text{ m}$$

$$I_{gg} = \frac{\pi}{64} D^4 = \frac{\pi}{64} \times (0.75)^4$$

$$y_p = 3.00 + \frac{\frac{\pi}{64}(0.75)^4}{\frac{\pi}{4} \times (0.75)^2 \times (3.0)}$$

$$= \mathbf{3.0117\ m}$$

2.16 A vertical isosceles triangular gate with its vertex up has a base width of 2 m and a height of 1.5 m. If the vertex of the gate is 1 m below the free water surface, find the total pressure force and the position of the centre of pressure on one side of the plate.

Solution:

Total pressure force $F = \gamma A \bar{h}$

$$= 9790 \times \left(\frac{1}{2} \times 2 \times 1.5\right)\left(1 + \frac{2}{3} \times 1.5\right)$$

$$= \mathbf{29370\ N}$$

Referring to Fig. 2.21,

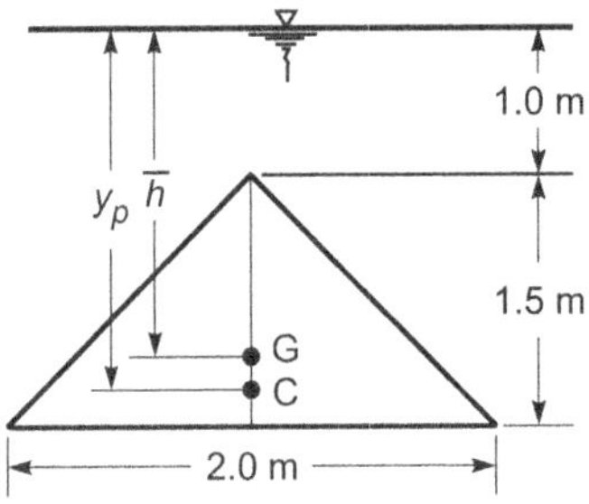

Fig. 2.21

By symmetry $x_p = \bar{x}$, i.e. the centre of pressure lies on the vertical axis passing through the CG of the plate.

$$y_p = \bar{y} + \frac{I_{GG}}{A\bar{y}}$$

Since the plane is vertical, $\bar{y} = \bar{h} = 2.0$ m

$$A = \frac{1}{2} \times 2 \times 1.5 = 1.5\ \text{m}^2$$

$$I_{GG} = \frac{1}{36} \times bh^3$$

$$= \frac{1}{36} \times 2.0\,(1.5)^3$$

$$= 0.1875\ \text{m}^4$$

$$y_p = 2.0 + \frac{0.1875}{1.5 \times 2.0}$$

$$= \mathbf{2.0625\ m}$$

2.17 A cubical tank has sides of 1.5 m. It contains water in the lower 0.6 m depth. The upper remaining part is filled with oil of relative density 0.9. Calculate for one vertical side of the tank (i) the pressure force and (ii) position of the centre of pressure.

Solution:

$\gamma_{oil} = 0.9 \times 9790 = 8811$ N

$\gamma_{water} = 9790$ N

p_0 = pressure at the top = 0

p_i = pressure at the interface = 8811 × 0.9 = 7930 N/m^2

p_b = pressure on the bottom = 7930 + 9790 × 0.6 = 13804 N/m^2

F_1 = Force on the top 0.9 m of a side.

$$= \frac{1}{2} \times 7930 \times (0.9 \times 1.5) = 5353\ \text{N}$$

acting at $\left(\frac{2}{3} \times 0.9\right)$ = 0.6 m below the surface.

F_2 = Part of force on the bottom 0.6 m of a side = 7930 × 0.6 × 1.5 = 7137 N

acting at $\left(0.9 + \frac{0.6}{2}\right)$ = 1.2 m below the surface.

F_3 = remaining part of force on the bottom 0.6 m of a side

$$= \frac{1}{2} \times (13804 - 7930) \times 0.6 \times 1.5$$

$$= \frac{1}{2} \times 5874 \times 0.6 \times 1.5 = 2643\ \text{N}$$

acting at $\left(0.9 + \frac{2}{3} \times 0.6\right)$ = 1.3 m below the surface.

Total force = $F_1 + F_2 + F_3$ = 15133 N
= **15.133 kN**

Centre of pressure = y_p

$$= \frac{(5353 \times 0.6) + (7137 \times 1.2) + (2643 \times 1.3)}{15133}$$

$$= \mathbf{1.005\ m}$$

By symmetry $\bar{x} = x_p$, i.e. the centre of pressure acts on the vertical passing through the CG of the side.

2.18 A trapezoidal plate of parallel sides a and $2a$ and height h is immersed vertically in water with its side of length a horizontal and topmost. The top edge is at a depth h below the water surface. Determine the total force on one side of the plate and the location of the centre of pressure.

Solution: The trapezium can be considered to be made of a rectangle (width a × height h) and a triangle (base a × height h) as in Fig. 2.22.

1. Pressure force F_1 on rectangular part

$$F_1 = \gamma(ah)\left(h + \frac{h}{2}\right) = \frac{3}{2}\gamma a h^2$$

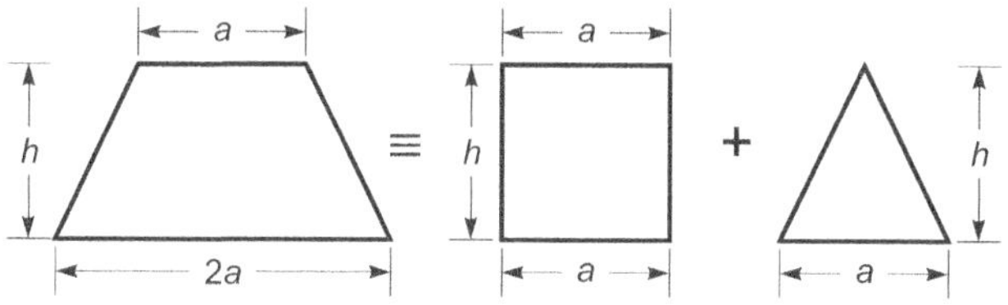

Fig. 2.22

Centre of pressure of force F_1

$$h_{c1} = \bar{y} + \frac{I_{gg}}{A\bar{y}} = \left(h + \frac{h}{2}\right) + \frac{ah^3/12}{(ah)\left(h + \frac{h}{2}\right)}$$

$$= \frac{3h}{2} + \frac{h}{12} \times \frac{2}{3} = \frac{14}{9}h$$

2. Pressure force on the triangular part:

$$F_2 = \gamma\left(\frac{ah}{2}\right)\left(h + \frac{2}{3}h\right) = \gamma a h^2\left(\frac{5}{6}\right)$$

Centre of pressure of F_2

$$h_{c2} = \left(h + \frac{2}{3}h\right) + \frac{ah^3/36}{\left(\frac{ah}{2}\right)\left(h + \frac{2}{3}h\right)}$$

$$= \frac{50+1}{30}h = \frac{51}{30}h$$

Total thrust $F = F_1 + F_2$

$$F = \gamma ah\left[\left(h + \frac{h}{2}\right) + \frac{1}{2}\left(h + \frac{2}{3}h\right)\right]$$

$$= \gamma ah^2\left(\frac{7}{3}\right) = \mathbf{\frac{7}{3}\gamma ah^2}$$

Centre of pressure of total force F

$$h_c = \frac{F_1 h_{c1} + F_2 h_{c2}}{F}$$

$$= \frac{\left(\frac{3}{2}\gamma ah^2 \times \frac{14}{9}h\right) + \left(\gamma ah^2 \times \frac{5}{6}\right)\left(\frac{51}{30}h\right)}{\frac{7}{3}\gamma ah^2}$$

$$= \frac{\frac{7}{3}h + \frac{17}{12}h}{7/3}$$

$$\mathbf{h_c = \frac{45}{28}h}$$

2.19 Figure 2.23 shows a rectangular flash board AB which is 4.5 m high and is pivoted at C. What must be the maximum height of C above B so that the flash board will be on the verge of tipping when the water surface is at A?

Solution: For critical stability, the centre of pressure must be at C.

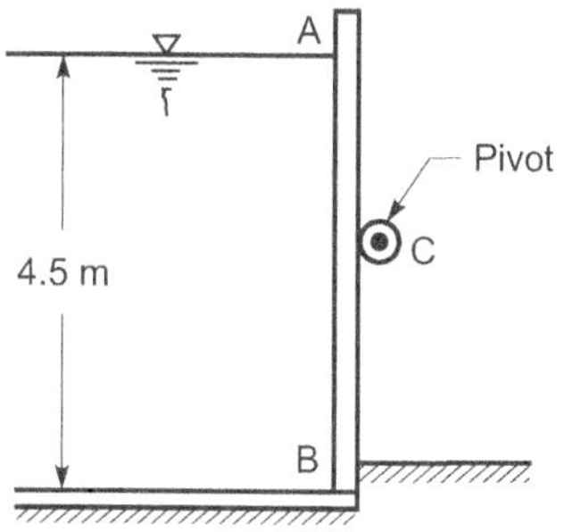

Fig. 2.23

Thus h = height of C above B.

$$= \frac{4.5}{3} = \mathbf{1.5\ m}$$

2.20 If the pivot of the flash board in Example 2.19 is at a height h = 1.5 m, determine the reactions at B and C when the water surface is 4.0 m above B.

Solution: Consider unit width of the flash board

$$\text{Total force } F = 9.790 \times \frac{4}{2} \times (4 \times 1)$$

$$= 78.32 \text{ kN/m}$$

acting at 4/3 m above B.

Let R_B and R_C be the reactions at B and C respectively. Taking moments of forces about C,

$$78.32 \times \left(1.5 - \frac{4}{3}\right) - R_B \times 1.5 = 0$$

$$\mathbf{R_B = 8.7\ kN/m}$$

$$R_C = F_1 - R_B = 78.32 - 8.70$$

$$= \mathbf{69.62\ kN/m}$$

2.21 A closed tank as shown in Fig. 2.24 contains water which is pressurized with a pressure of 10 kN on the top. A rectangular gate MN of width 0.6 m and height 1.5 m is provided on the inclined face. Calculate the total force on the gate and the location of the centre of pressure.

Solution:

From the Fig. 2.24

h_1 = depth of M below free surface = 2.0 m

h_2 = depth of N below free surface
= 2.0 + 1.5 sin 45° = 3.06 m.

$\bar{h}_{CG}$ = 2.0 + 0.75 sin 45° = 2.53 m

$\bar{y}$ = 2.53/sin 45° = 3.578 m

The force on the gate MN can be considered to be made up of two parts [see Fig. 2.24(b)].

F_1 = force due to uniform pressure p_a acting uniformly over the surface.
$= p_a A$ = (10 × 1.5 × 0.6) = 9 kN

acting at CG of the area h_{cg} = 2.53 m

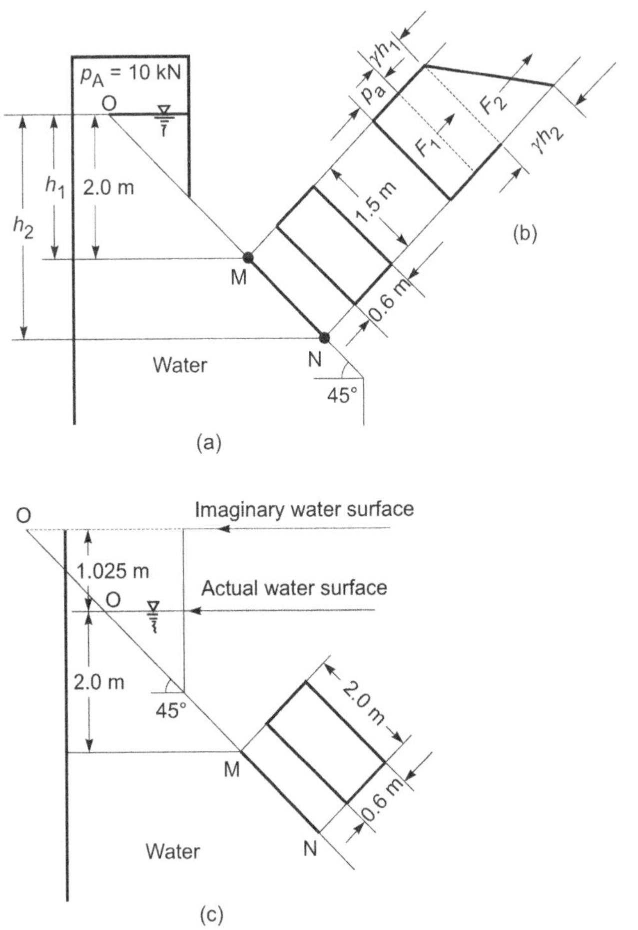

Fig. 2.24 *Example 2.21*

i.e $\quad \bar{y} = 2.53/\sin 45° = 3.578$ m along OM axis.

F_2 = Force due to hydrostatic pressure of liquid

$= \gamma A \bar{h}$

$= 9.79 \times (1.5 \times 0.6) \times 2.53 = 22.29$ kN

This force F_2 acts at y_{c1}, where

$$y_{c1} = \bar{y} + \frac{I_{gg}}{A\bar{y}}$$

$$= 3.578 + \frac{0.6 \times (1.5)^3/12}{(0.6 \times 1.5)\,(3.578)}$$

$= 3.6308$ m

h_c = vertical depth of this y_{c1}

$= 3.6308 \times \sin 45° = 2.5674$ m

Total force $F = F_1 + F_2$

$= 9 + 22.29 = \mathbf{31.29\ kN}$

This force acts at y_c.

Taking moments about CG

$$F \times (y_c - \bar{y}) = F_2 \times (y_{c1} - \bar{y})$$

$$y_c - \bar{y} = \frac{22.29}{31.29} \times (3.6308 - 3.578) = 0.0376 \text{ m}$$

$$y_c = 3.578 + 0.0376 = 3.616 \text{ m}$$

Depth of centre of pressure from the water surface = h_p = 3.616 sin 45° = **2.557 m**

Alternative method:

Consider the equivalent column of water corresponding to a pressure of 10 kN

$$h_e = \frac{10.0}{9.79} = 1.0215 \text{ m}$$

Now an imaginary water surface at a height h_e above the actual water surface be considered as reference surface. Hence force on the gate

$$F = \gamma A \bar{h}$$

$$= 9.79 \times (0.6 \times 1.5) \times \left(3.0215 + \frac{15}{2}\sin 45°\right)$$

$= 31.29$ kN

y'_p = location of centre of pressure on the O′M axis

$$= \bar{y}' + \frac{I_{GG}}{A\bar{y}'}$$

$$\bar{y}' = \left(\frac{3.0215}{\sin 45°} + 0.75\right) = 5.023$$

$$\frac{I_{GG}}{A\bar{y}'} = \frac{(0.6 \times (1.5)^3)/12}{(0.6 \times 1.5)\,(5.023)} = 0.0373$$

$$y'_p = 5.023 + 0.0373 = 5.060 \text{ m}$$

y_p = distance of centre of pressure from actual water surface

$= 5.060 - (1.0215/\sin 45°) = 3.616$ m

Depth of centre of pressure from actual water surface = 3.616 sin 45°

= **2.557 m**

2.22 For the system shown in Fig. 2.25 calculate the height H of oil at which the rectangular hinged gate will just begin to rotate counterclockwise.

Solution:

Force due to oil $F_1 = \gamma_0 A \bar{h}$

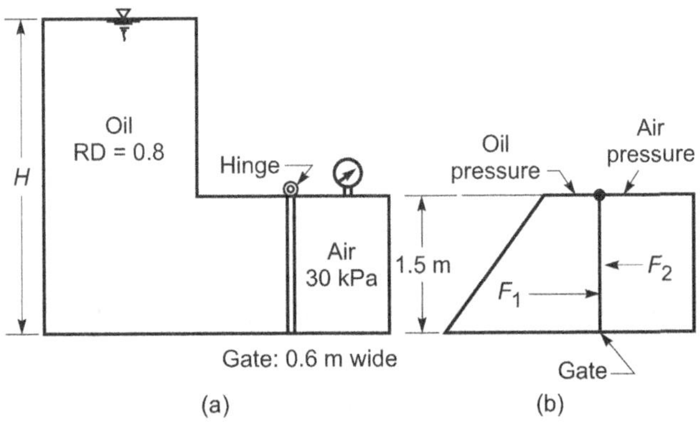

Fig. 2.25

$\gamma_{oil} = 0.8 \times 9.79 = 7.832$ kN/m^3
$A = (0.6 \times 1.5) = 0.9$ m^2

$$\bar{h} = (H - 1.5) + \left(\frac{1.5}{2}\right)$$

$$F_1 = 7.832 \times 0.9 \times \{(H - 1.5) + 0.75\}$$
$$= 7.0488\,H - 5.2866$$

Centre of pressure of force F_1

$$h_{c1} = \bar{h} + \frac{I_{gg}}{A\bar{h}}$$

$$= (H - 0.75) + \frac{0.6 \times (1.5)^3/12}{(0.6 \times 1.5)(H - 0.75)}$$

$$= (H - 0.75) + \frac{0.1875}{(H - 0.75)}$$

(2) Force due to air pressure

$$F_2 = pA$$
$$= 30 \times (0.6 \times 1.5) = 27 \text{ kN}$$

Centre of pressure of this force (below the oil surface)

$$h_{c2} = \bar{h} = (H - 0.75)$$

Taking the moments about the hinge:

$$F_1 \times [h_{c1} - (H - 1.5)] = F_2 \times [h_{c2} - (H - 1.5)]$$
$$\times \{7.0488\,H - 5.2866\}$$

$$\times \left\{(H - 0.75) - (H - 1.5) + \frac{0.1875}{(H - 0.75)}\right\}$$
$$= 27 \times \{(H - 0.75) - (H - 1.5)]\}$$

$$\{7.0488\,H - 5.2866\}\left\{0.75 + \frac{0.1875}{(H - 0.75)}\right\}$$
$$= 27 \times 0.75$$

On solving by trial and error

$$\mathbf{H = 4.33\ m}$$

2.23 A semicircular gate having a radius of 1.2 m and weighing 12 kN can swing about its diameter which is kept horizontal. The water surface on one side is 5.0 m above its diameter. Determine the minimum force required to keep the gate closed. The gate lies in a plane making an angle of 30° with the water surface.

Solution: In the Fig. 2.26, O_1O_2 B is the semicircular gate.

$$\bar{y} - y_0 = OG = \frac{4R}{3\pi} = \frac{4 \times 1.2}{3\pi} = 0.5093 \text{ m}$$

Since $\quad y_0 = \dfrac{5.0}{\sin 30°} = 10.0$ m, $\bar{y} = 10.5093$ m

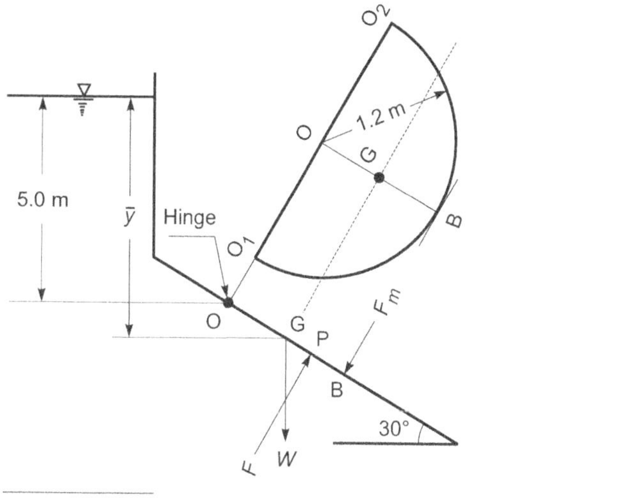

Fig. 2.26

$\bar{h}$ = depth of CG below water surface
$= 5.0 + \text{OG} \sin 30° = 5.0 + 0.5093 \sin 30°$
$= 5.2546$ m

Force on the gate (normal to the plane of the gate)

$$= F = \gamma A\bar{h} = 9.79\,\frac{(\pi \times 1.2^2)}{2} \times 5.2546$$
$$= 116.36 \text{ kN}$$

Location of centre of pressure on OB

$$y_p = \bar{y} + \frac{I_{gg}}{A\bar{y}} = \bar{y} + \frac{I_{0_10_2}}{A\bar{y}} - \frac{(\text{OG})^2}{\bar{y}}$$

$$y_p = 10.5093 + \left(\frac{\pi D^4}{128}\right)\frac{1}{\left(\frac{\pi D^2}{8}\right)} \times \frac{1}{10.5093} - \frac{(0.5093)^2}{10.5093}$$

Substituting $D = 2.5$ m, $y_p = 10.5218$ m

$$\text{OP} = y_p - y_o = 0.5218 \text{ m}$$

Let F_m = minimum force required to lift the gate.
By taking moments about O,

$$= W \times \text{OG} \cos 30° = F \times \text{OP} - F_m \times \text{OB}$$
$$12 \times 0.5093 \times 0.866 = 116.36 \times 0.5218 - 1.2\,F_m$$
$$5.293 = 60.717 - 1.2\,F_m$$
$$\mathbf{F_m = 46.19\ kN}$$

2.24 A sheet piling holds fresh water and salt water (relative density = 1.035) on either sides of it as shown in Fig. 2.27. Find the moment about the base M of the resultant force per unit length of piling.

Solution:
Consider 1 m length of piling

F_1 = force due to salt water
$= \gamma_1 A_1 \bar{h}_1$

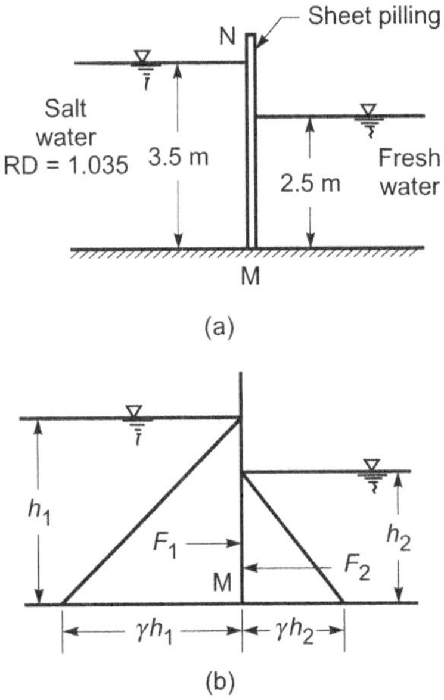

Fig. 2.27

$$= (1.035 \times 9.79)\,(3.5 \times 1.0)\left(\frac{3.5}{2}\right)$$

$= 62.062$ kN

h_{c1} = Centre of pressure of this force F_1

$$= \bar{h}_1 + \frac{I_{gg}}{A_1 \bar{h}_1} = \frac{3.5}{2} + \frac{1.0 \times (3.5)^3/12}{(3.5 \times 1.0)\left(\frac{3.5}{2}\right)}$$

$= 1.75 + 0.5833 = 2.333$ m.

Lever arm $a_1 = 3.5 - 2.333 = 1.167$ m

F_2 = Force due to fresh water = $\gamma_2 A_2 \bar{h}_2$

$$= (9.79)\,(2.5 \times 1)\left(\frac{2.5}{2}\right) = 30.594 \text{ kN}$$

$$h_{c2} = \bar{h}_2 + \frac{I_{gg}}{A_2 \bar{h}_2} = \frac{2.5}{2} + \frac{[1.0 \times (2.5)^3]/12}{(2.5 \times 1.0)(2.5/2)}$$

$= 1.667$ m. Lever arm $a_2 = 2.5 - 1.667 = 0.833$ m

Net moment about $M = M_1 = F_1 a_1 - F_2 a_2$

(Taking clockwise as positive)

$M = (62.062 \times 1.167) - (30.594 \times 0.833)$

$= 72.43 - 25.48 =$ **46.95 kN.m (clockwise)**

2.25 A tank of 2 m length and of cross section shown in Fig. 2.28 contains water. The tank is made of 8 mm steel plates. What is the force on the bottom due to water? What are the longitudinal tensile stresses in the side walls (AB) if (a) the tank is suspended from the top and (b) it is supported at the bottom ?

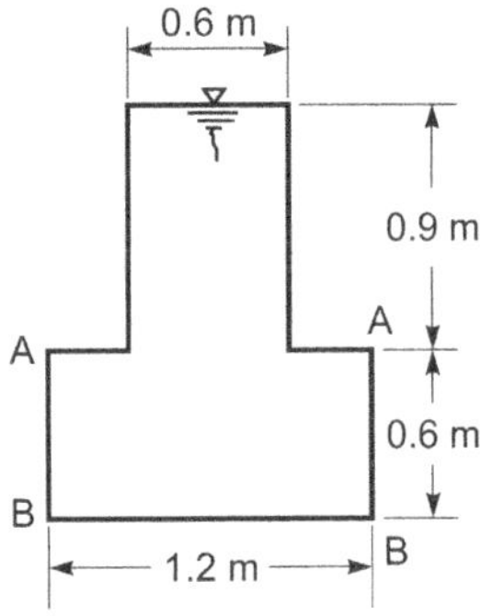

Fig. 2.28

Solution: Force on the bottom due to water

$F_1 = \gamma A \bar{h}$

$= 9.79 \times (1.2 \times 2.0) \times 1.5 = 35.24$ kN

Force on the surface AA $= 9.79 \times (0.6 \times 2.0) \times 0.9$

$= 10.57$ kN

When suspended from top the stress on the side walls

$$\sigma = \frac{35.24}{(1.2 + 1.2 + 2.0 + 2.0) \times 8/1000}$$

= **688.3 kPa**

When supported from bottom the stress on the side walls

$$\sigma = \frac{10.57}{(1.2 + 1.2 + 2.0 + 2.0) \times 8/1000}$$

= **206.4 kPa**

2.26 A square lamina is immersed in water with one of the diagonals vertical as shown in Fig. 2.29(a). Calculate the force on one side and the location of the centre of pressure.

Solution:

Area $A = a^2 = 0.36 \text{ m}^2$

$$\bar{h} = 0.6 + \frac{a}{\sqrt{2}} = 0.6 + \frac{0.6}{\sqrt{2}}$$

$= 1.0243$ m

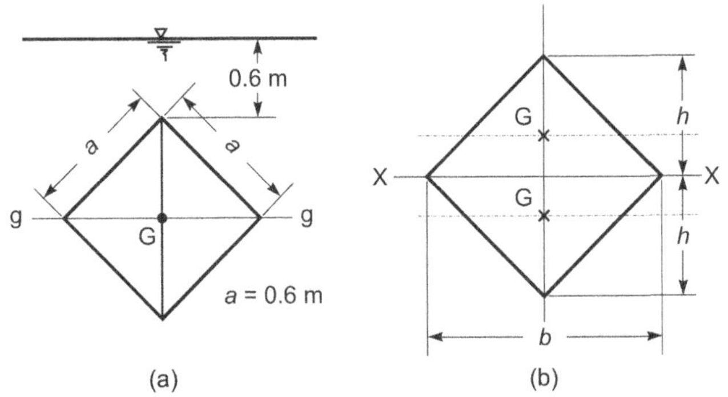

Fig. 2.29

Consider two triangular laminas of base b and height h, as shown in Fig. 2.29(b). Considering the axis XX

$$I_{xx} = \left\{\frac{1}{36} bh^3 + \left(\frac{bh}{2}\right)\frac{h^2}{9}\right\} \times 2$$

$$= \frac{1}{12} bh^3 \times 2 = \frac{bh^3}{6}$$

For the present case $I_{xx} = I_{gg}$

$$b = 2a/\sqrt{2} \text{ and } h = a/\sqrt{2}$$

$$I_{gg} = \frac{1}{6} \cdot \frac{2a}{\sqrt{2}} \frac{a^3}{2\sqrt{2}} = \frac{1}{12} a^4$$

$$= \frac{1}{12}(0.6)^4 = 0.0108$$

Force on one side of the lamina

$$F = \gamma A\bar{h} = 9.79 \times (0.36)(1.0243)$$

$$= \mathbf{3.61 \text{ kN}}$$

Centre of pressure:

By symmetry $x_p = 0$

$$h_p = y_p = \bar{h} + \frac{I_{gg}}{A\bar{h}}$$

$$= 1.0243 + \frac{0.0108}{(0.36)(1.0243)}$$

$$= \mathbf{1.0536 \text{ m}}$$

C. Forces on Curved Surfaces

2.27 A horizontal cylindrical barrier of 2.5 m diameter retains water up to its top on one side. Calculate the horizontal and vertical forces due to water per metre width of this barrier.

Solution:

Referring to Fig. 2.30 and considering 1 m width,

$$F_H = \text{horizontal force}$$

$$= \gamma\,(\text{Projected area of the cylinder MSN on a vertical M'N'}) \times \bar{h}$$

$$= 9.79 \times (2.5 \times 1) \cdot \left(\frac{2.5}{2}\right) = 30.59 \text{ kN}$$

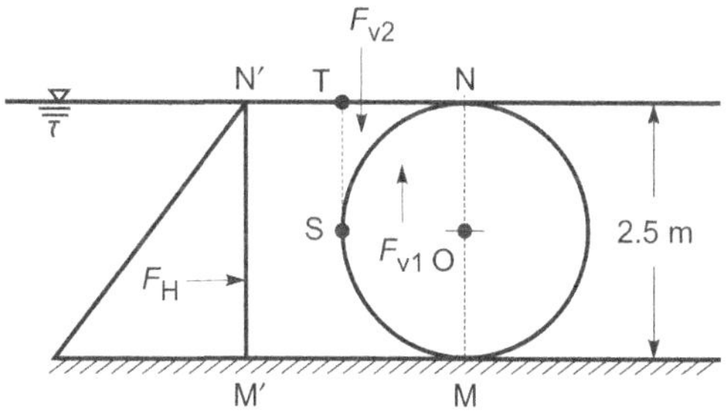

Fig 2.30

Vertical force

$$F_V = F_{V1} \text{ on MS} - F_{V2} \text{ on SN}$$

$$= \gamma\,(\text{volume NMST} - \text{volume NST})$$

$$= \gamma\left\{\left(\frac{1}{4} \times \frac{\pi \times (2.5)^2}{4}\right) + \left(\frac{2.5}{2} \times \frac{2.5}{2}\right)\right\} \times 1$$

$$- \gamma\left\{\left(\frac{2.5}{2} \times \frac{2.5}{2}\right) - \left(\frac{\pi \times (2.5)^2}{4 \times 4}\right)\right\} \times 1$$

$$= \gamma \times \frac{1}{2} \times \frac{\pi}{4}(2.5)^2 \times 1 = 24.03 \text{ kN}$$

Hence $\quad \mathbf{F_H = 30.59 \text{ kN}}$

$\mathbf{F_V = 24.03 \text{ kN}}$

2.28 A door in a tank is in the form of a quadrant of a cylinder of 1.5 m radius and 1.8 m wide (Fig. 2.31). Calculate the resultant force on the door and its location on the gate.

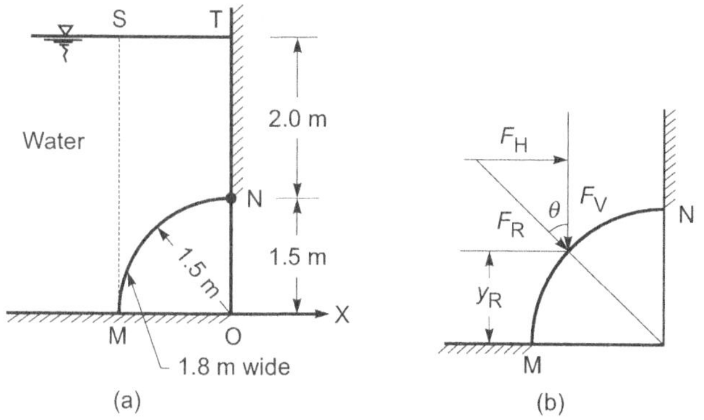

Fig. 2.31

Solution: Horizontal component of the force

$$F_H = \gamma\,(\text{Projected area}) \times \bar{h}$$

$$= 9.79 \times (1.5 \times 1.8) \times \left(2.0 + \frac{1.5}{2}\right)$$

$$= 72.69 \text{ kN}$$

Vertical component of the force

$$F_V = \gamma\,(\text{volume TNMS})$$

$$= 9.79\left[(1.5 \times 1.8) \times (2.0 + 1.5) - \frac{1}{4} \times \pi \times (1.5)^2 \times 1.8\right]$$

$$= 61.37 \text{ kN}$$

Resultant force

$$F_R = \sqrt{F_H^2 + F_V^2}$$

$$= \sqrt{(72.69)^2 + (61.37)^2} = \mathbf{95.13 \text{ kN}}$$

If θ is the inclination of F_R to vertical

$$\tan\theta = \frac{F_H}{F_V} = \frac{95.13}{61.37} = 1.55$$

$$\theta = 57.173°$$

The resultant passes through O.

y_R = vertical height of resultant on the gate.
$= R\cos\theta = 1.5\cos(57.173°) = \mathbf{0.813\ m}$

2.29 The lower corner of a water tank has the shape of a quadrant of a circle of radius 1.2 m. The water surface is 2.4 m above the centre of curvature. The water tank is 3.0, long. Find the magnitude, direction and location of the total force exerted by the water surface on this curved surface.

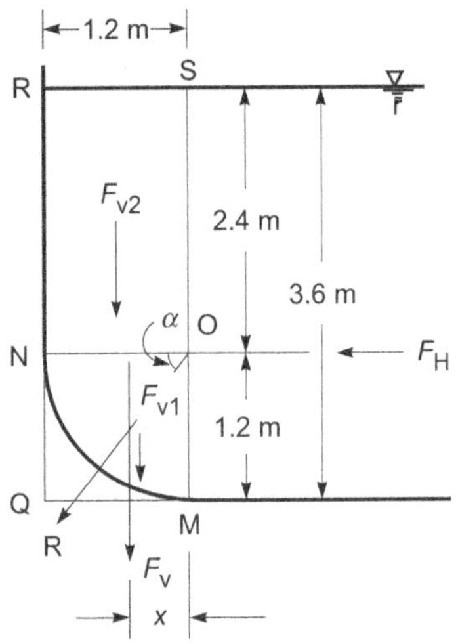

Fig. 2.32

Solution:
Vertical component:

F_V = volume contained in (OMN + ONRS)
$= F_{V1} + F_{V2}$

$$F_V = \left[\frac{\pi}{4}\times(1.2)^2\times 3.0\times 9.79\right] + (2.4\times 1.2 \times 3.0\times 9.79)$$

$= 33.22 + 84.59 = 117.81$ kN

F_{V1} acts at $\left(\frac{4}{3\pi}\times 1.2\right) = 0.5093$ m from OM

F_{V2} acts at 0.6 m from OM.
By taking moments of vertical forces about O,

$$F_V\cdot x = (33.22\times 0.5093) + (84.59)\times 0.6$$

$$x = \frac{67.67}{117.81} = 0.574\text{ m}$$

Horizontal component F_H = Force on the projection of curved face MN on a vertical plane

$= \gamma A\bar{h}$

$= 9.79\times(1.2\times 3)(2.4 + 0.6)$
$= 105.73$ kN

Depth of centre of pressure of $F_H = h_c = \bar{h} + \frac{I_{gg}}{A\bar{h}}$

$$= 3.0 + \frac{\frac{1}{12}\times[3\times(1.2)^3]}{(3\times 1.2)\times 3}$$

$= 3.04$ m

Resultant force on the curved face MN

$$R = \sqrt{F_V^2 + F_H^2} = \sqrt{(117.81)^2 + (105.73)^2}$$

$= \mathbf{158.3\ kN}$

If α is the inclination of R to the horizontal

$$\tan\alpha = \frac{F_V}{F_H} = \frac{117.81}{105.73} = 1.114$$

$\boldsymbol{\alpha = 48°\ 5'\ 36''}$

The resultant passes through the centre of curvature O.

2.30 A hemispherical projection of diameter 1.2 m exists on one of the vertical sides of a tank. If the tank contains water to an elevation of 3.0 m above the centre of the hemisphere, calculate the horizontal and vertical forces acting on the projection.

Solution: Referring to Fig. 2.33
Vertical force $F_V = F_{V1} - F_{V2}$
= weight of volume of water NSTQ – weight of volume of water MSTQ.
= weight of water contained by the hemisphere MSN

$$= \gamma\cdot\frac{1}{2}\cdot\left(\frac{4}{3}\pi R^3\right)$$

$$= 9.79\times\frac{1}{2}\times\frac{4}{3}\times\pi\,(1.2)^3 = 35.44\text{ kN}$$

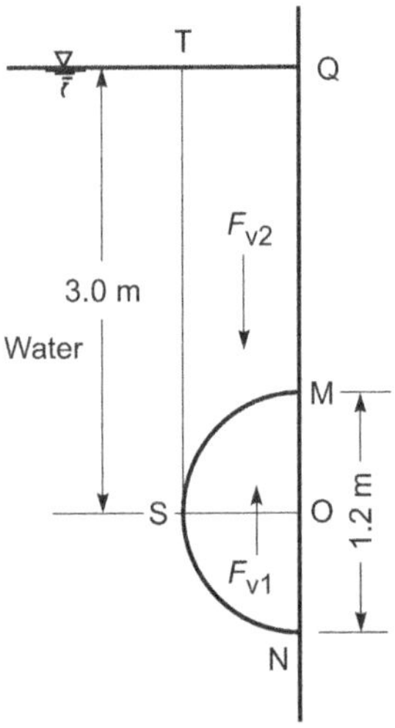

Fig. 2.33

Horizontal force $F_H = \gamma A \bar{h}$

where A = projected area of the hemisphere = πR^2

$$F_H = 9.79 \times \pi \times (1.2)^2 \times 3.0$$
$$= \mathbf{132.87\ kN}$$

2.31 A hemispherical bulge of diameter 1.2 m is provided in the bottom of a tank. If the depth of water above the horizontal floor of the tank is 3.0 m, calculate the magnitude and direction of the resultant force on the hemisphere.

Solution: Referring to Fig. 2.34, by symmetry the net horizontal force = 0

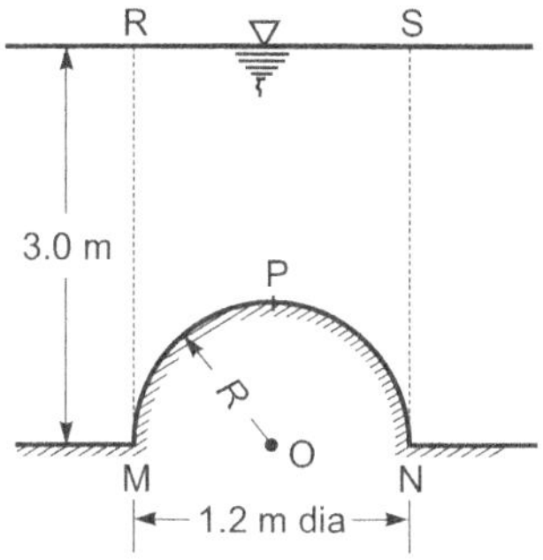

Fig. 2.34

Vertical force F_V = weight of fluid above the hemisphere MPN

$$= \gamma\left[\pi R^2 H - \frac{1}{2}\cdot\frac{4}{3}\pi R^3\right]$$

$$= 9.79\left[\pi \times (0.6)^2 \times 3.0 - \frac{2}{3}\pi(0.6)^3\right]$$

$$= \mathbf{28.79\ kN}$$

Resultant force is the same as the vertical force F_V = 28.79 kN acting vertically at the centre of the hemisphere.

2.32 A sector gate in the form of circular arc of radius 5 m retains water to a height of 4 m above its sill as in Fig. 2.35. Calculate the magnitude and direction of the resultant force per unit length of the gate.

Solution: Consider a gate width of 1 m

$\sin\theta = 2/5$, $\theta = 23.578°$, $2\theta = 47.156°$

F_H = horizontal component of force on the gate

$= \gamma$ (projected area) $\times \bar{h}$

$$= 9.79 \times (4 \times 1) \times \frac{4}{2} = 78.32\ \text{kN}$$

F_V = Vertical component of the force on the gate
= weight of water displaced by the gate
= γ [sector OMSN – triangle OMN]

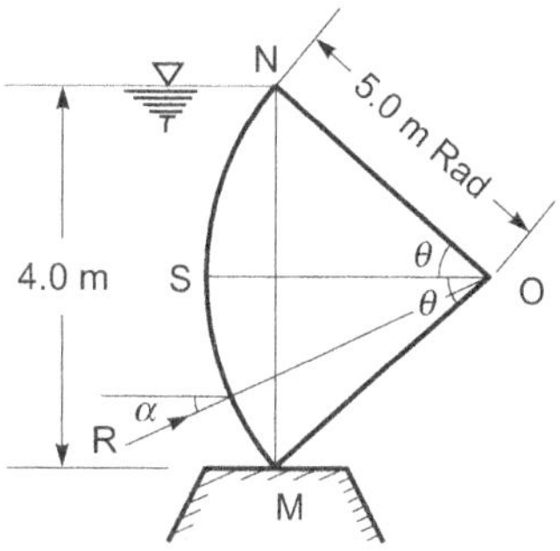

Fig. 2.35

$$= 9.79\left[\frac{47.156}{360}\times\pi\times5^2 - \frac{1}{2}\times4\times5\cos 23.578°\right]$$

$$= 10.99\ \text{kN}$$

Resultant force on the gate = $R = \sqrt{F_H^2 + F_V^2}$

$$R = \sqrt{(78.32)^2 + (10.99)^2} = \mathbf{79.08\ kN}$$

If R is inclined at an angle α to the horizontal

$$\tan\alpha = \frac{F_V}{F_H} = \frac{10.99}{78.32} = 0.14032$$

$$\boldsymbol{\alpha = 7.988°}$$

Since the surface of the gate is cylindrical, the resultant R passes through the centre of curvature O.

2.33 A cylindrical gate of diameter 2.5 m retains two liquids on either side of it as shown in Fig. 2.36. Estimate the resultant fluid force acting on unit length of the gate.

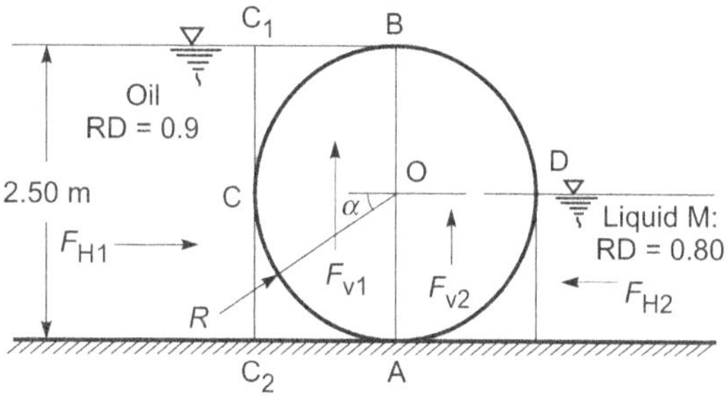

Fig. 2.36

Solution: Consider 1 m length of gate

Force due to oil:

Horizontal force $F_{H1} = \gamma_o$ (Projected area of ACB) $\times \bar{h}$

$$= 0.9 \times 9.79 \times (2.5 \times 1) \times \left(\frac{2.5}{2}\right)$$

$$= 27.53\ \text{kN}$$

This force acts at a height of $\frac{2.5}{3}$ = 0.833 m above level of A.

Vertical force F_{V1} = [weight of volume of oil ACC_1B – weight of volume of oil BCC_1]
= weight of volume ACB

$$= \gamma_o \times \frac{1}{2} \times \frac{\pi D^2}{4} \times 1.0$$

$$= 0.9 \times 9.79 \times \frac{1}{2} \times \frac{\pi}{4} \times (2.5)^2 \times 1.0$$

= 21.63 kN

This force acts at a distance of $\frac{4(D/2)}{3\pi}$

$$= \frac{4 \times 2.5/2)}{3\pi}$$

= 0.531 m from vertical at A.

Force due to liquid M:
Horizontal force F_{H2} (Right to left)

$$= \gamma_L \text{ (projected area)} \times \bar{h}$$

$$= 0.80 \times 9.79 \times (1.25 \times 1) \times \frac{1.25}{2}$$

= 6.12 kN

This force acts at $\frac{1.25}{3}$ = 0.417 m above level of A.

Vertical force F_{V2}

$$= \gamma_o \times [\text{volume AOD}]$$

$$= 0.8 \times 9.79 \times \left[\frac{\pi}{4} \times (2.5)^2 \times \frac{1}{4}\right] \times 1.0$$

= 9.61 kN

This force acts at a distance of $\frac{4}{3\pi}\left(\frac{D}{2}\right) = \frac{4 \times 2.5}{3\pi \times 2}$

= 0.531 m from the vertical at A.

Resultant force:

R_H = horizontal component = $F_{H1} - F_{H2}$
= 27.53 – 6.12
= 21.41 kN

R_V = vertical component = $F_{V1} + F_{V2}$
= 21.63 + 9.61
= 31.24 kN

Resultant force $F = \sqrt{R_H^2 + R_V^2}$

$$= \sqrt{(21.41)^2 + (31.24)^2}$$

= 37.87 kN

Let α = inclination of the resultant to the horizontal. Then

$$\tan \alpha = \frac{R_V}{R_H} = \frac{31.24}{21.41} = 1.459$$

$$\boldsymbol{\alpha = 55° \, 34' 32''}$$

The resultant acts normal to the cylindrical surface and passes through the centre of curvature O of the cylindrical surface.

2.34 A 1.6 m wide × 3.0 m long × 2.4 m high rectangular tank has a semi-cylindrical cover of 1.6 m dia × 3.0 m long, as shown in Fig. 2.37(a). The tank contains oil of relative density 0.90 under pressure. If gauge pressure at the bottom of the tank reads 60 kPa find the net force on the curved roof.

Solution: First convert the base pressure of 60 kPa into equivalent column of oil (h_o)

$$60 = \gamma_o \times h_o$$

$$h_o = \frac{60}{(9.79 \times 0.9)} = 6.81 \text{ m}$$

An imaginary free surface of oil at an elevation of 6.81 m above the base AB can be imagined [see Fig. 2.37 (b)].

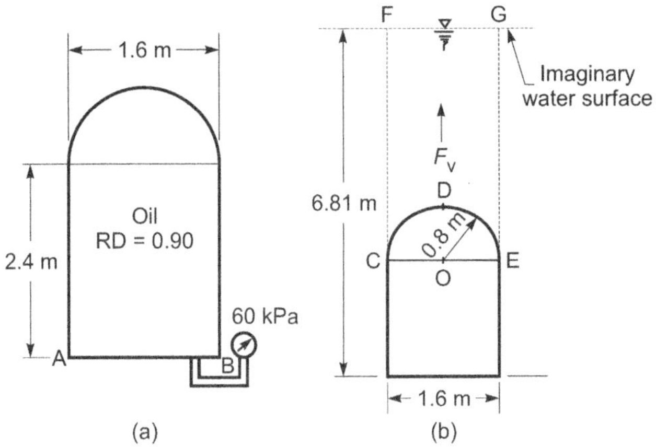

Fig. 2.37

By symmetry, horizontal force on the cylindrical cover is zero. The vertical force F_V

= γ_o [volume of prism COEGF – curved volume CDE]

= (0.9 × 9.79) ×

$$\left[(1.6 \times 3.0)(6.81 - 2.4) - \left(\frac{\pi}{2} \times \frac{(1.6)^2}{4} \times 3\right)\right]$$

= 8.811[21.168 – 3.016] = **159.9 kN**

This force acts vertically at the centre of the tank along OD.

2.35 A cylindrical tank of 3 m diameter and height 1.5 m has a hemispherical dome. The tank contains oil of relative density 0.83 [Fig. 2.38(a)]. The dome is joined to the cylindrical portion by 6 equally spaced bolts. If the pressure gauge at a point A, 0.6 m above the base of the tank, reads 50 kPa determine the force on each bolt.

Solution: First consider equivalent of pressure P_A in terms of oil column.

$$P_A = \gamma_o h_A$$
$$50 = (0.83 \times 9.79) \cdot h_A$$
$$h_A = 6.153 \text{ m of oil}$$

The imaginary oil surface at an elevation of $h_A = 6.153$ m is now considered. [Fig. 2.38(b)].

Above the base plane MN of the dome, the elevation of the imaginary oil surface is

$$= 6.153 - 0.9 = 5.253 \text{ m}$$

By symmetry, there is no horizontal force on the dome.

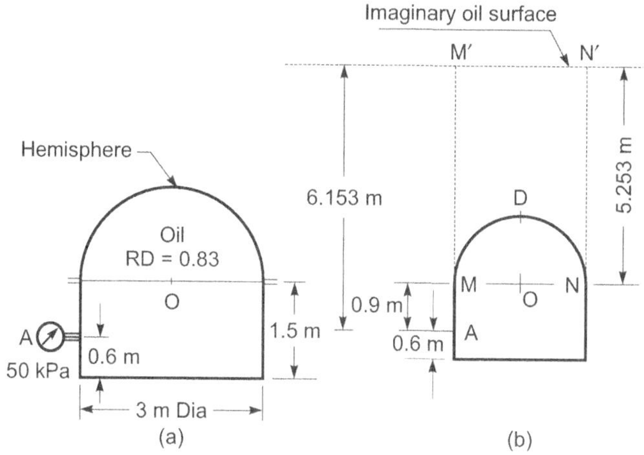

Fig. 2.38 *Example 2.35*

The vertical force F_V = weight of oil above the dome surface up to the imaginary oil surface = weight of volume MDN N′M′.

$$= (0.83 \times 9.79)\left[\left(\frac{\pi(3.0)^2}{4} \times 5.253\right) - \left(\frac{1}{2} \times \frac{4}{3}\pi\left(\frac{3.0}{2}\right)^3\right)\right]$$

$$= 8.126\ (37.131 - 7.069) = 244.28 \text{ kN}$$

This force is shared by six bolts.

$$\text{Tensile force on each bolt} = \frac{244.28}{6} \text{ kN}$$
$$= \mathbf{40.71\ kN}$$

2.36 A masonry weir is of trapezoidal cross section with a top width of 2.0 and of height 5 m with upstream slope of 1 vertical in 0.1 horizontal and a downstream slope of 1 vertical in 0.75 horizontal. If the weir has water stored up to its crest on the upstream side and has a tailwater of 2 m depth on the downstream, calculate, per unit length of the weir,

(i) the resultant force on the base of the weir and
(ii) the minimum and maximum stresses on the base of the weir.

(Assume specific weight of masonry as 22 kN/m^3 and neglect uplift forces).

Solution: Refer to Fig. 2.39 consider 1 m length of the weir.

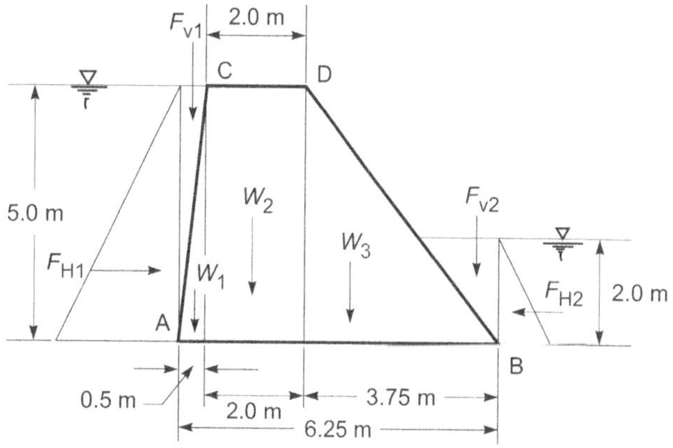

Fig. 2.39

The force acting on the weir are:

(i) weight of masonry $W = W_1 + W_2 + W_3$
(ii) vertical force due to water on the upstream slope, F_{V1}.
(iii) vertical force due to tailwater on the downstream slope, F_{V2}.
(iv) horizontal water force on the upstream side, F_{H1}.
(v) horizontal water force on the downstream side, F_{H2}.

The magnitudes of these forces, their distance from the toe of the weir (edge B) and the moments of these forces about B are tabulated as in Table 2.1.

$$\Sigma V = \text{Sum of vertical forces} = 480.7 \text{ kN}$$
$$\Sigma H = \text{Sum of horizontal forces} = 102.8 \text{ kN}$$

$$R = \text{Resultant} = \sqrt{\Sigma V^2 + \Sigma H^2}$$
$$= \sqrt{(480.7)^2 + (102.8)^2}$$
$$= \mathbf{491.6\ kN}$$

If θ is the inclination of the resultant to horizontal

$$\tan\theta = \frac{\Sigma V}{\Sigma H} = \frac{480.7}{102.8} = 4.676, \text{ and } \theta = \mathbf{77.93°}$$

$$\Sigma M = 1818.2 - 204 = 1614.2 \text{ kN.m}$$

x = distance of point of action of the resultant

$$\text{from } B = \frac{\Sigma M}{\Sigma V} = \frac{1614.2}{480.7} = 3.358$$

As b = base width = 6.25 m, the Eccentricity $e = x - \dfrac{b}{2}$

$$= 3.358 - \frac{6.25}{2} = 0.233 \text{ m}$$

Maximum and minimum stresses

Table 2.1 *Magnitude of the forces and moments and of the lever arm about B-Example 2.36*

Force	*Description*	*Horiz. force (kN)*	*Vert. force (kN)*	*Lever arm about B (m)*	*Moment (Clockwise) (kN.m)*	*Moment (Anticlockwise) (kN.m)*
W_1	$(0.5 \times 5) \times \frac{1}{2} \times 1 \times 22$		27.5	5.917		162.7
W_2	$(2.0 \times 5.0) \times 1.0 \times 22$		220.0	4.75		1045.0
W_3	$(3.75 \times 5) \times \frac{1}{2} \times 1 \times 22$		206.25	2.50		515.6
F_{V1}	$(0.5 \times 5) \times \frac{1}{2} \times 1 \times 9.79$		12.24	6.083		74.5
F_{V2}	$(1.5 \times 2.0) \times \frac{1}{2} \times 1 \times 9.79$		14.69	0.50		7.3
F_{H1}	$(5 \times 1) \times \frac{5}{2} \times 9.79$	122.38		1.667	204.0	
F_{H2}	$(2 \times 1) \times \frac{2}{2} \times 9.79$	– 19.58		0.667		13.1
	Sum	**102.8**	**480.7**		**204.0**	**1818.2**

$$\sigma_{1,2} = \frac{\Sigma V}{b}\left(1 \pm \frac{6e}{b}\right)$$

$$= \frac{480.7}{6.25}\left(1 \pm \frac{6 \times 0.233}{6.25}\right)$$

σ_1 = **maximum stress = 94.1 kPa**

σ_2 = **minimum stress = 59.7 kPa**

2.37 A 2.5 m diameter pipe carries oil (relative density = 0.85) at a pressure of 1000 kPA. If the maximum allowable stress of the pipe material is 120 MPa, what thickness of the pipe wall is required?

Solution: Hoop stress (circumferential tensile stress) in a cylinder

$$\sigma = \frac{pD}{2t}$$

If $\sigma = f_a$ = allowable stress in the material.
Minimum thickness t required is

$$t = \frac{pD}{2 f_a}$$

$$= \frac{1000 \times 2.5}{2 \times (120 \times 1000)} = 0.01042 \text{ m}$$

= **10.42 mm**

A thickness of 10.5 mm can therefore be used.

[***Note:*** The relative density of the fluid does not have any role when once the pressure inside the pipe is stipulated.]

2.38 The gates of a 8 m wide lock include an angle of 120° in the closed position. Each gate is held on by two hinges one placed at 0.8 m and another at 6.0 m from the bottom of the lock. If the water levels are 9.0 m and 3.0 m on the upstream and downstream respectively, determine (i) the resultant force due to water pressure and (ii) magnitudes of reaction at the hinges.

Solution:

Let F = resultant force due to water on each gate acting at the centre of each gate, point O

R = resultant reaction at the hinges

P = reaction between two gates at the common contact area, acting perpendicular to contact area, in this case normal to the centre line

The reaction R must pass through C (see Fig. 2.40), the point of intersection of F and P, due to equilibrium of the system. In triangle ACB, because CO is the perpendicular bisector of AB,

$$\angle CBA = \angle CAB = \theta$$

$$\therefore \quad R = P$$

Also $$P = R = \frac{F}{2 \sin \theta}$$

[***Note*** that it is assumed that F, P and R are in the same plane.]

$$B = \text{width of each gate} = AB$$

$$= \frac{4}{\cos 30°} = 4.62 \text{ m}$$

On the upstream:

$$F_1 = 9.79 \times (4.62 \times 9) \times \frac{9}{2} = 1831.8 \text{ kN}$$

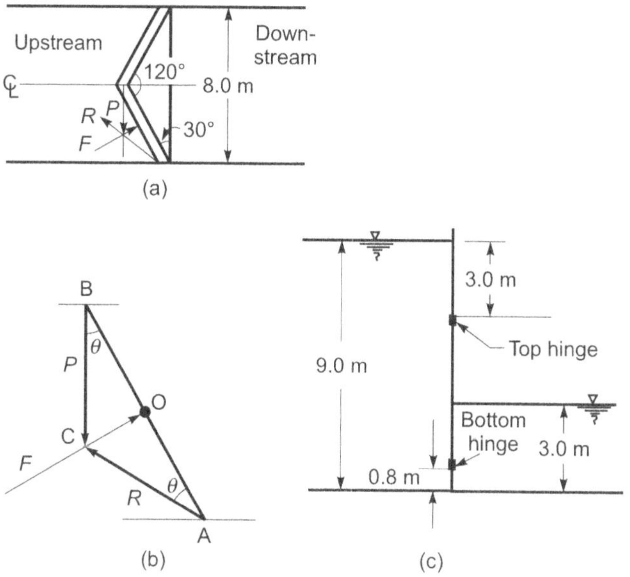

Fig. 2.40 *Lock Gates*

acts at $\frac{H_1}{3}$ = 3.0 m from bottom of the gate on the downstream.

$$F_2 = 9.79(4.62 \times 3) \times \frac{3}{2} = 203.5 \text{ kN}$$

acts at 1.0 m from the bottom.

(i) Net water force $F = F_1 - F_2 = 1831.8 - 203.5$ $= \mathbf{1628.3\ kN}$

acts at h from base where

$$F \times h = F_1 h_1 - F_2 h_2$$

$$1628.3 \times h = (1831.8 \times 3) - (203.5 \times 1)$$

$$h = \mathbf{3.25\ m\ from\ the\ bottom}$$

(ii) $P = R = \dfrac{F}{2 \sin 30°} = \dfrac{1628.3}{2 \sin 30°} = 1628.3$ kN

Let R_T = reaction at the top hinge.

R_B = reaction at the bottom hinge.

$$R_T + R_B = 1628.3$$

Also by taking moments about the bottom hinge

$$R_T(6.0 - 0.8) = R(3.25 - 0.8)$$

$$5.2\, R_T = 1628.3 \times 2.45$$

$$R_T = \mathbf{767.2\ kN}$$

$$R_B = (R - R_T) = \mathbf{861.1\ kN}$$

2.39 A spherical object of 1.5 m diameter is completely immersed in a water reservoir and chained to the bottom. If the chain has a tension of 5.30 kN find the weight of the object when it is taken out of the reservoir into the air.

Solution: Refer to Fig. 2.41. The tension in the chain indicates that the buoyant force is larger than weight of the object in air.

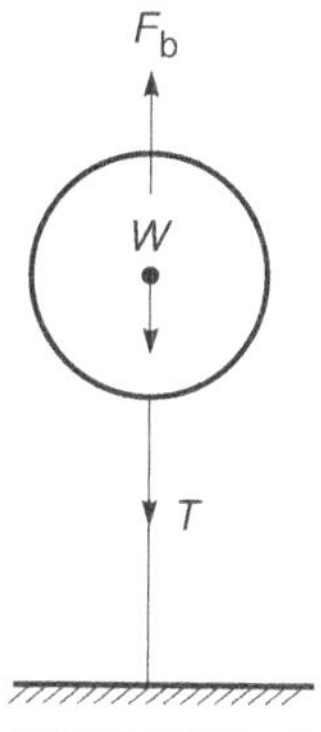

Fig. 2.41

Hence

$$T = B - W$$

or

$$W = B - T$$

$$= \frac{4}{3}\pi \times \left(\frac{1.5}{2}\right)^3 \times 9.79 - 5.30$$

$$= 17.30 - 5.30 = \mathbf{12.0\ kN}$$

2.40 A body weighs 20 N and 10 N when weighed under submerged conditions in liquids of relative densities 0.8 and 1.2 respectively. Determine its volume and weight in air.

Solution:

Let weight in air = W and volume = $V\!\!\!\!-$

Weight in a liquid of RD = S ; $W_S = W - \gamma S V\!\!\!\!-$

$\therefore$ $W_{S1} = W - \gamma S_1 V\!\!\!\!-$

$$W_{S2} = W - \gamma S_2 V\!\!\!\!-$$

$$\gamma V\!\!\!\!- (S_2 - S_1) = W_{S2} - W_{S1}$$

$$V\!\!\!\!- = \frac{W_{S2} - W_{S1}}{\gamma (S_2 - S_1)} = \frac{20 - 10}{9.79 \times 10^3 \times (1.2 - 0.8)}$$

$$= 2.5536 \times 10^{-3} \text{ m}^3 = \mathbf{2.5536\ L}$$

Weight in air $W_S = W_{S1} + \gamma S_1 V\!\!\!\!-$

$$= 20 + 9790 \times 0.8 \times 2.5536 \times 10^{-3}$$

$$= \mathbf{40\ N}$$

2.41 A hydrometer has a stem of diameter D and a weight W. Find the distance on the stem h between markings corresponding to relative densities of 1.0 and S.

Solution: Refer to Fig. 2.42

Let $V\!\!\!\!-$ = submerged volume when immersed in water (RD = 1.0).

$$W = \gamma V\!\!\!\!-$$

When immersed in a liquid with RD = S where $S < 1.0$

$$W = \left(V\!\!\!\!- + \frac{\pi D^2}{4} \times h\right) \times \gamma \cdot S$$

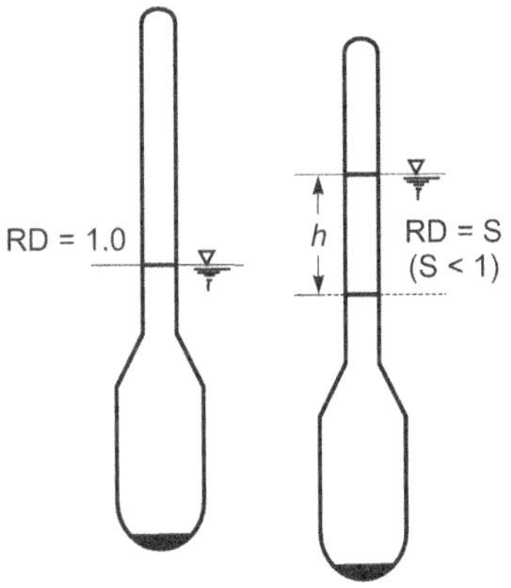

Fig. 2.42

$$\therefore \quad \gamma V = \gamma V S + \frac{\pi D^2}{4} h \gamma \cdot S$$

or
$$h = \frac{V}{(\pi D^2/4)}\left(\frac{1}{S} - 1\right) = \frac{W}{\gamma(\pi D^2/4)}\left(\frac{1}{S} - 1\right)$$

Note that when $S > 1$, h will be negative, i.e. the markings will be below the mark corresponding to RD = 1.0.

2.42 If a hydrometer weighs 0.2 N and has a stem diameter of 5 mm, find the distance between markings corresponding to relative densities of (i) 1.0 and 0.95 and (ii) 1.0 and 1.05.

Solution:

(i) h_1 = distance between marking of RD of 1.0 and 0.95.

$$h_1 = \frac{0.2}{9790\left(\frac{\pi}{4} \times \left(\frac{5}{1000}\right)^2\right)}\left(\frac{1}{0.95} - 1\right)$$

= 0.05476 m = **54.76 mm**

(ii) h_2 = distance between markings of RD of 1.0 and 1.05.

$$= \frac{0.2}{9790\left(\frac{\pi}{4} \times \left(\frac{5}{1000}\right)^2\right)} \times \left(\frac{1}{1.05} - 1\right)$$

= – 0.04954 = **– 49.54 mm**

h_2 will be below the marking corresponding to relative density of 1.0.

2.43 A 10 cm side cube weighing 5 N is immersed in a liquid of relative density 0.8 contained in a rectangular tank of cross sectional area 15 cm × 15 cm. If the tank contained liquid to a height of 8 cm before the immersion determine the levels of the bottom of the cube and the liquid surface.

Solution: Refer to Fig. 2.43

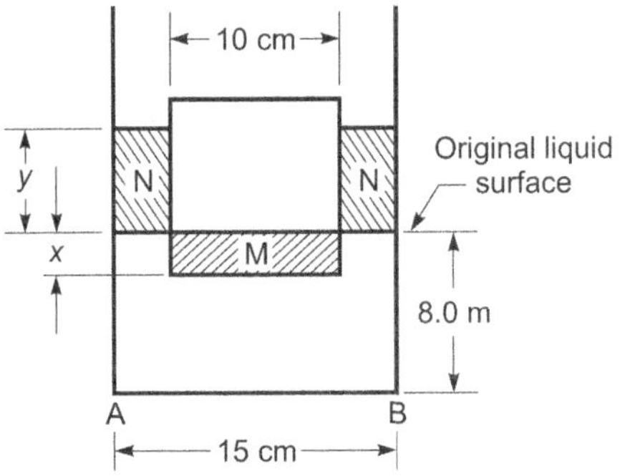

Fig. 2.43

Let x = depth to which the bottom of the cube falls below original liquid surface (cm)

y = height of the rise of liquid above the original liquid surface (cm)

$x + y$ = depth of submergence of the cube (cm)

Volume M = Volume N

$$(10 \times 10 \cdot x) = (15^2 - 10^2) \cdot y$$
$$x = 1.25\ y$$
$$W = \text{weight of cube} = 5\text{ N}$$
$$W = \text{buoyant force}$$
$$= \frac{(10 \times 10) \cdot (x + y)}{10^6} \times 0.8 \times 9790$$
$$5 = 0.7832\ (x + y)$$
$$2.25\ y = \frac{5}{0.7832}$$
$$y = 2.8374\text{ cm}$$
$$x = 3.5467\text{ cm}$$

Elevation of:

bottom of cube above plane AB = (8.00 – 3.547)
= **4.453 cm**

liquid surface above plane AB = (8.00 + 2.837)
= **10.837 cm**

2.44 A cylindrical can open at one end is 20 cm in diameter and 40 cm long. The can weighs 6 N. The can is carefully put in a large tank of water vertically with its open end down so that the air in it is trapped (see Fig. 2.44). If the air in it can be assumed to undergo isothermal compression, calculate the elevations of the water surface inside the can and the top of the can relative to the water surface in the tank. The atmospheric pressure is 100 kPa and the temperature is 20°C.

Solution:

Buoyant weight = weight of can

$$9790 \times \frac{\pi}{4} \times (0.2)^2 \times H_b = 6$$

$$\therefore \quad H_b = 0.0195\text{ m} = 1.95\text{ cm}$$

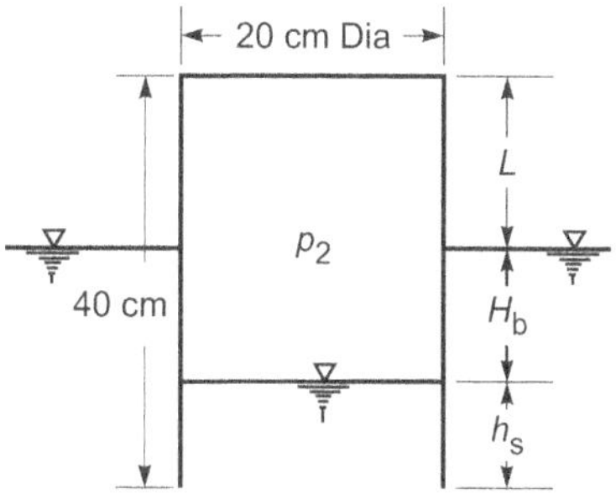

Fig. 2.44

Also if p_2 = pressure inside the can (abs)
and p_a = atmospheric pressure (abs)

$(p_2 - p_a) \times$ Area of can = weight of the can

$$p_2 - p_a = \frac{6.0}{\left(\frac{\pi}{4} \times (0.2)^2\right)} = 191 \text{ Pa}$$

$$p_2 = 191 + p_a$$

From isothermal relationship

$$p_a V_a = p_2 V_2 = \text{constant}$$

$$V_2 = \frac{p_a}{p_2} V_a = \frac{100000}{(100000 + 191)} \times (A \times 0.40)$$

$$h_s A = V_a - V_2 = (A \times 0.40)\left[1 - \frac{100000}{100191}\right]$$

$$h_s = 0.40 \times 1.9064 \times 10^{-3} = 7.626 \times 10^{-4} \text{ m}$$
$$= 0.0763 \text{ cm}$$
$$L = 40 - 1.950 - 0.076 = 37.974 \text{ cm}$$

1. The top of the can is L = **37.974 cm above the water surface** in the tank.
2. The water surface in the can is **1.95 cm below the water surface** in the tank.

2.45 A rectangular container having base area of 1.5 sq. m and length 0.9 m is floating in water with the open end downwards as shown in Fig. 2.45. If the difference in water levels inside and outside the container is 10 cm; (1) determine the mass of the container; (2) What force will be required to depress the bottom of the container to a depth of 10 m in water if the trapped air has constant temperature?

[Atmospheric pressure = 100 kPa]

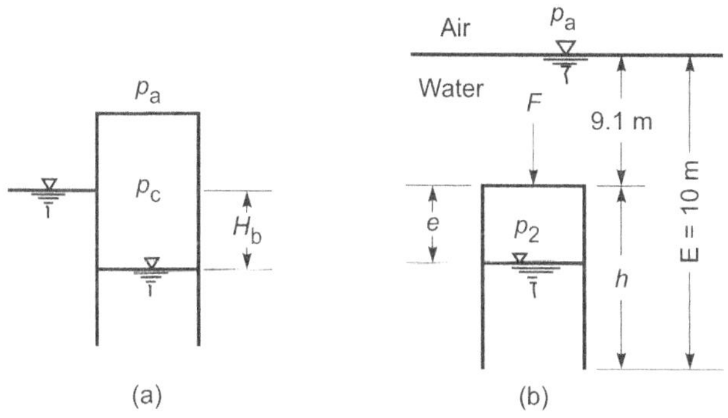

Fig. 2.45

Solution:

Case 1:

Let p_a = atmospheric pressure
p_c = pressure inside the container

$(p_c - p_a) \times$ area of the container = weight of the container.

But $(p_c - p_a) = \gamma H_b = 9790 \times 0.10 = 979 \text{ N/m}^2$.

Hence W = weight of the container
$= 979 \times 1.5 = 1468.5$ N
$= 1.469$ kN

$$\text{Mass of container} = \frac{1468.5}{9.81}$$
$$= \mathbf{149.7 \text{ kg}}$$

Case 2: When the container is completely immersed to a depth E below the water surface.

e = height of air column
h = total height of the container

From isothermal consideration

$$p_a V_a = p_2 V_2$$

where p = pressure and V = volume.

$$p_a \cdot (\text{Area}) \times h = p_2 \times (\text{area}) \times e$$

$$\therefore \quad \frac{e}{h} = \frac{p_a}{p_2}$$

Also from pressure consideration $p_2 = p_a + \gamma(E - h + e)$

Here $p_a = 100$ kPa, $h = 0.9$ m
$E = 10$ m and $\gamma = 9.79$ kN

$$\text{Hence} \quad \frac{100}{p_2} = \frac{e}{0.9}$$

$$\text{or} \quad p_2 e = 90 \quad \text{or} \quad p_2 = \frac{90}{e}$$

$$\frac{90}{e} = p_2 = 100 + 9.79(10 - 0.9 + e)$$
$$90 = [100 + 89.09 + 9.79\, e]\, e$$
$$9.79\, e^2 + 189.09\, e - 90 = 0$$

Solving, $e = 0.4648$ m

Buoyant force $F_b = \gamma e \text{ (Area)}$
$= 9.79 \times 0.4648 \times 1.5$
$= 6.8258$ kN

For equilibrium F_b = weight + external force
$= W + F$

External force $F = F_b - W$
$= 6.825 - 1.469$
$=$ **5.356 kN**

2.46 A wooden block in the form of a rectangular prism floats with its shortest axis vertical. The block is 40 cm long, 20 cm wide and 15 cm deep with a depth of immersion of 12 cm. Calculate the position of the metacentre and comment on the stability of the block.

Solution: Referring to Fig. 2.46.

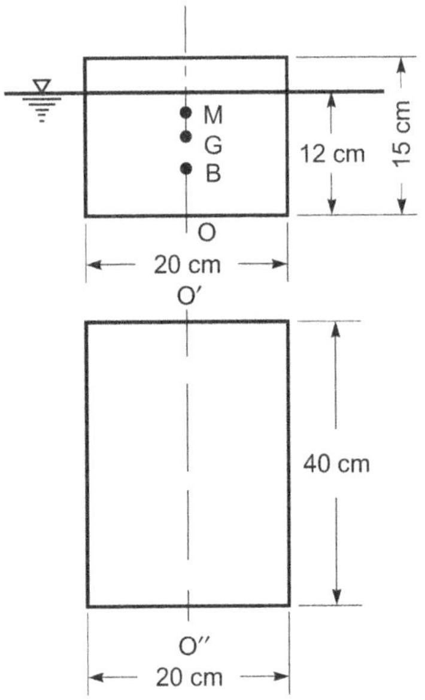

Fig. 2.46

Weight of body = weight of displaced volume of water
$= (12 \times 20 \times 40) \times 9.79 \times 10^{-3}$ N
(1 cm^3 of water weighs 9.79×10^{-3} N)
= 93.98 N

OB = height of centre of buoyancy above the base of the block

$$= \frac{12}{2} = 6 \text{ cm}$$

OG = height of centre of gravity of the block above O

$$= \frac{15}{2} = 7.5 \text{ cm}$$

If M is the metacentre

$$BM = \frac{I}{V\!\!\!/}$$

I = moment of inertia of the water line area about O′O″

$$= \frac{40 \times (20)^3}{12}$$

$V\!\!\!/$ = volume of fluid displaced by the body = (20 × 12 × 40)

$$\therefore \quad BM = \frac{40 \times (20)^3/12}{(40 \times 20 \times 12)} = \mathbf{2.778\ cm}$$

$$MG = BM - BG = BM - (OG - OB)$$
$$= 2.778 - (7.5 - 6.0) = 1.278 \text{ cm}$$

Since M is above G, the body is in **stable equilibrium.**

2.47 A block of wood of rectangular cross section, of sides a and b, and of length L has a relative density of S. If the block is to float in water with its longest axis horizontal and the length a vertical (Fig. 2.47), find the ratio b/a to have a stable equilibrium.

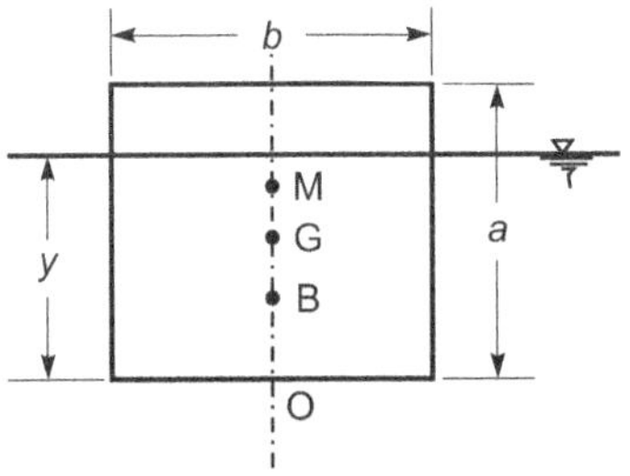

Fig. 2.47

Solution:

OG = height of centre of gravity above the base.
$= a/2$

Weight of the block $= S\gamma(Lba)$
= buoyant force $= \gamma bLy$

where y = depth of immersion.
$y = Sa$

OB = height of centre of buoyancy above the base.

$$= \frac{y}{2} = \frac{Sa}{2}$$

If M is the metacentre

$$BM = \frac{I}{V\!\!\!/} = \frac{\frac{1}{12}(Lb^3)}{(bL)(Sa)} = \frac{b^2}{12\,Sa}$$

$$\therefore \quad OM = OB + BM = \frac{Sa}{2} + \frac{b^2}{12\,Sa}$$

MG = metacentric height = OM – OG

$$= \frac{Sa}{2} + \frac{b^2}{12\,Sa} - \frac{a}{2}$$

$$= \frac{b^2}{12\,Sa} - \frac{a}{2}(1 - S)$$

For stability MG > 0

$$\text{Hence} \quad \frac{b^2}{12\,Sa} > \frac{a}{2}(1 - S)$$

$$\frac{b^2}{a^2} > 6\,S(1 - S)$$

$$\text{or} \quad \boldsymbol{\frac{b}{a} > \sqrt{6S(1 - S)}}$$

2.48 A conical object of base diameter D and height H floats in a liquid of relative density S_L. The apex is downwards and the axis vertical (Fig. 2.48). If the relative density of the cone material is S_c show that for stable equilibrium

$$H^2 < \frac{1}{4}\left[\frac{D^2 (S_c/S_L)^{1/3}}{1-(S_c/S_L)^{1/3}}\right]$$

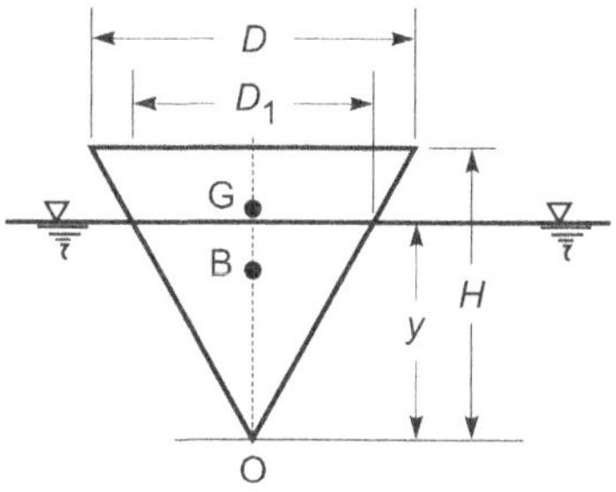

Fig. 2.48

Solution: Let the height of the portion under the liquid be y. Diameter of the cone at the water surface

$$= D_1 = \frac{D}{H}y.$$

$$W_c = \text{Weight of cone} = \frac{1}{3}\frac{\pi D^2}{4} H S_c \gamma$$

where γ = unit weight of water.

$$W_c = \frac{\pi D^2 H}{12}\gamma S_c$$

W_L = weight of liquid displaced

$$= \frac{\pi D_1^2 y}{12}\gamma S_L = \frac{\pi}{12}\frac{D^2}{H^2} y^3 \gamma S_L$$

But weight of cone = weight of liquid displaced, i.e.

$$W_c = W_L$$

$$\frac{\pi D^2 H \gamma S_c}{12} = \frac{\pi D^2}{12 H^2}\cdot y^3 \gamma S_L$$

$$y^3 = H^3\left(\frac{S_c}{S_L}\right) \text{ or } y = H\left(\frac{S_c}{S_L}\right)^{1/3}$$

Let B be the centre of buoyancy.

$$OB = \frac{3}{4}y = \frac{3}{4}H\left(\frac{S_c}{S_L}\right)^{1/3}$$

$$BM = \frac{I}{V} = \frac{\pi D_1^4/64}{\frac{1}{3}\pi\frac{D_1^2}{4}\cdot y} = \frac{3}{16}\frac{D_1^2}{y}$$

$$= \frac{3}{16}\frac{D^2}{H^2}y = \frac{3}{16}\frac{D^2}{H}\left(\frac{S_c}{S_L}\right)^{1/3}$$

Now $\quad OM = OB + BM$

$$= \frac{3}{4}H\left(\frac{S_c}{S_L}\right)^{1/3} + \frac{3}{16}\frac{D^2}{H}\left(\frac{S_c}{S_L}\right)^{1/3}$$

Further $\quad OG = \frac{3}{4}H$

For stable equilibrium

$$OG < OM$$

$$\frac{3}{4}H < \frac{3}{4}H\left(\frac{S_c}{S_L}\right)^{1/3} + \frac{3}{16}\frac{D^2}{H}\left(\frac{S_c}{S_L}\right)^{1/3}$$

i.e. $$\frac{3}{4}H\left[1-\left(\frac{S_c}{S_L}\right)^{1/3}\right] < \frac{3}{16}\frac{D^2}{H}\left(\frac{S_c}{S_L}\right)^{1/3}$$

$$\boldsymbol{H^2 < \frac{1}{4}\left[\frac{D^2(S_c/S_L)^{1/3}}{1-(S_c/S_L)^{1/3}}\right]}$$

2.49 A solid cylinder of diameter D and height H having a relative density of S floats in water with its axis vertical (Fig. 2.49). Obtain an expression for the maximum value of H/D for which the cylinder will float in stable equilibrium with its axis vertical.

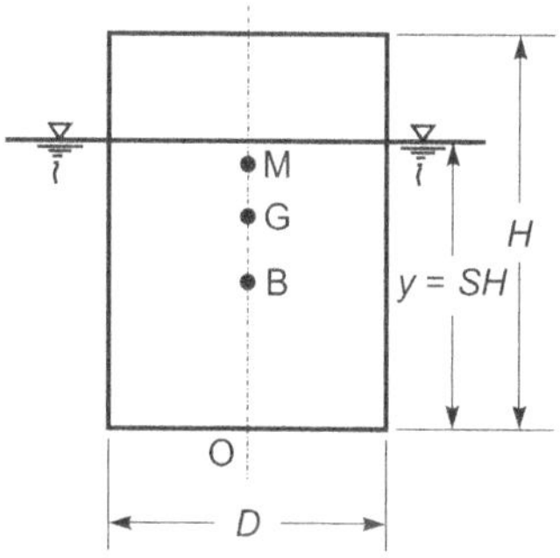

Fig. 2.49

Solution:

Let γ = specific weight of water.

Weight of cylinder = weight of water displaced.

If A = Cross sectional area of cylinder, $A = \pi D^2/4$

y = depth of immersion of the cylinder.

Weight of cylinder = $AH\gamma S$

= weight of displaced volume of water

= γyA

Hence $\quad y = SH$

OB = height of the centre of buoyancy = $\frac{SH}{2}$

OG = height of the centre of gravity of the cylinder = $\frac{H}{2}$

If M is the metacentre,

$$BM = \frac{I}{V} = \frac{\pi D^4/64}{\frac{\pi D^2}{4}\times SH} = \frac{1}{16}\times\frac{D^2}{SH}$$

$$OM = OB + BM = \frac{SH}{2} + \frac{1}{16} \times \frac{D^2}{SH}$$

For stable equilibrium with the axis vertical

$$OM > OG$$

$$\frac{SH}{2} + \frac{1}{16} \times \frac{D^2}{SH} > \frac{H}{2}$$

or $$\frac{H}{2}(1 - S) < \frac{D^2}{16SH}$$

$$\left(\frac{H}{D}\right)^2 < \frac{1}{8S(1-S)}$$

$$\boldsymbol{\frac{H}{D} < \frac{1}{\sqrt{8S(1-S)}}}$$

2.50 A solid cone of diameter 24 cm and height 20 cm floats with its vertex downwards in water as shown in Fig. 2.50. Is this cone in stable equilibrium?

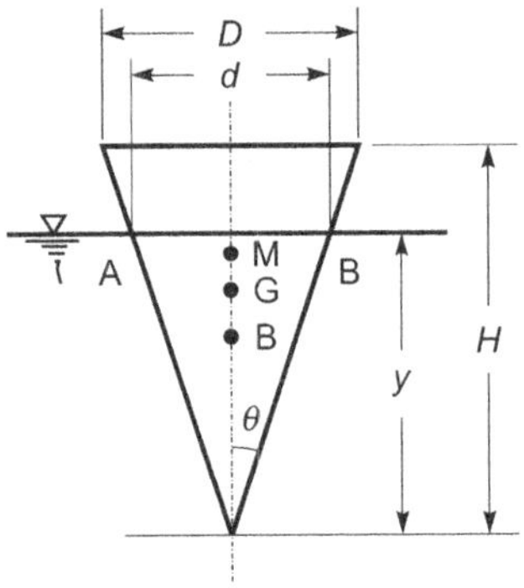

Fig. 2.50

Solution: For the cone

$$D = 24 \text{ cm}$$
$$H = 20 \text{ cm}$$
$$S = 0.8$$

Let θ = Semi-vertex angle

$$\tan \theta = \frac{12}{20} = 0.6, \; \theta = 30.96°$$

Diameter of cone at water surface

$$d = 2y \tan \theta$$

$$\text{weight of cone} = \frac{1}{3} \times \frac{\pi D^2}{4} \times H \times \gamma S$$
$$= \text{weight of water displaced}$$
$$= \frac{1}{3}\frac{\pi d^2}{4} y\gamma$$

$$\therefore \quad D^2HS = d^2y$$

$$4y^3\left(\frac{D}{24}\right)^2 = D^2HS = 4y^3 \tan^2 \theta = 4y^3\left(\frac{D}{2H}\right)^2$$

$$y^3 = H^3S \text{ or } y = H(S)^{1/3} = HS^{1/3}$$

$$y = 20 \times (0.8)^{1/3} = 18.566 \text{ cm}$$

If B is the centre of buoyancy

$$OB = \frac{3}{4}y = 0.75 \times 18.566 = 13.925 \text{ cm}$$

$$OG = \frac{3}{4}H = 0.75 \times 20 = 15.000$$

d = diameter at water surface
$= 2y \tan \theta = 22.280$ cm

$$BM = \frac{I}{V} = \frac{\pi d^4/64}{\frac{1}{3}\left(\frac{\pi d^2}{4}\right) \cdot y}$$

$$= \frac{3}{16}\left(\frac{d^2}{y}\right) = \frac{3}{4}y \tan^2 \theta$$

$$= 5.014 \text{ cm}$$

$$OM = OB + BM$$
$$= 13.925 + 5.014 = 18.939 \text{ cm}$$
$$OG = 15.000$$
$$MG = (18.939 - 15.000) = 3.939 \text{ cm}$$

(i.e. M is above G by 3.939 cm)

Hence the cone is under ***stable equilibrium***.

2.51 An open tank 5 m long and 2.0 m deep and 3.0 m wide contains oil of relative density 0.9 to a depth of 0.9 m (Fig. 2.51). If the tank is accelerated along its length on a horizontal track at a constant value of 3.0 m/s^2, determine (a) the new position of the oil surface and (b) pressures at the bottom of the tank at the front and rear edges.

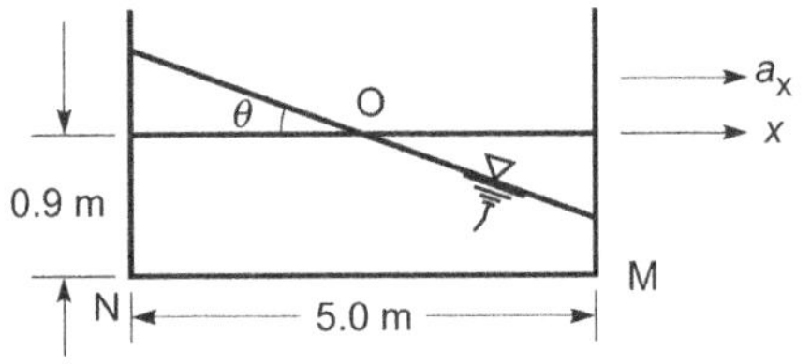

Fig. 2.51

Solution: The new oil surface will be inclined at an angle θ given by

$$\tan \theta = \frac{a_x}{g} = \frac{3.0}{9.81}$$
$$= 0.3058$$
$$\theta = 17°$$

(a) The depth of oil at the front edge M is

$$h_M = 0.9 - \frac{5}{2} \times \tan \theta = \mathbf{0.135 \text{ m}}$$

The depth of oil at the rear edge N is

$$h_N = 0.9 + \frac{5}{2} \tan \theta = \mathbf{1.665 \text{ m}}$$

(b) Pressure at M = $p_M = \gamma_s h_M = \gamma_{water} \times S \times h_M$

$= 9.79 \times 0.9 \times 0.135$

$= \mathbf{1.189\ kN/m^2}$

Pressure at N = $p_N = \gamma_s h_N$

$= 9.79 \times 0.9 \times 1.665$

$= \mathbf{14.67\ kN/m^2}$

2.52 The tank in Problem 2.51 is given a horizontal acceleration of 4.5 m/s^2. Calculate the amount of spill.

Solution: Refer to Fig. 2.52(a).

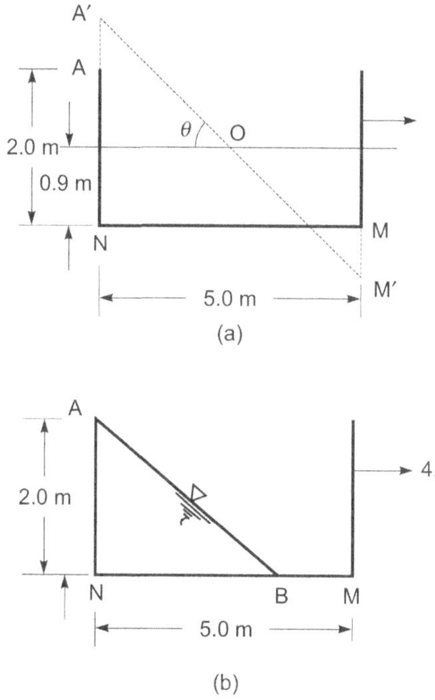

Fig. 2.52

Inclination of the oil surface

$$\tan\theta = \frac{a_x}{g} = \frac{4.5}{9.81} = 0.4587$$

If there were no spill, the oil surface would swing about an axis at O. Piezometric head at

$$N = h'_N = 0.9 + \frac{5}{2} \times \tan\theta = 2.0468 \text{ m}$$

Since this is larger than the depth of the tank there will be a spill of the oil. The new oil surface will have a depth h_N = depth of tank = 2.0 at N and a slope of θ.

X intercept of the surface at the bottom

$$\frac{2.0}{\tan\theta} = \frac{2.0}{0.4587} = 4.36$$

AB is the new oil surface [Fig. 2.52(b)].

Volume of oil = ΔANB × width of tank

$$= \frac{1}{2} \times 2.0 \times 4.36 \times 3.0 = 13.08 \text{ m}^3$$

Original volume of oil = $0.9 \times 5.0 \times 3.0 = 13.50$ m^3

Spill of oil = $13.50 - 13.08 = \mathbf{0.42\ m^3}$

2.53 A spherical tank of 1.2 m radius is half filled with oil of relative density 0.8. If the tank is given a horizontal acceleration of 10 m/s, calculate the inclination of the oil.

Solution: The oil surface will be inclined at an angle θ to the horizontal where

$$\tan\theta = \frac{a_x}{g}$$

$$= \frac{10.0}{9.81} = 1.0194$$

In Fig. 2.53, ROS is the original water surface and MON is the new surface at an acceleration of a_x. The surface tilts around O.

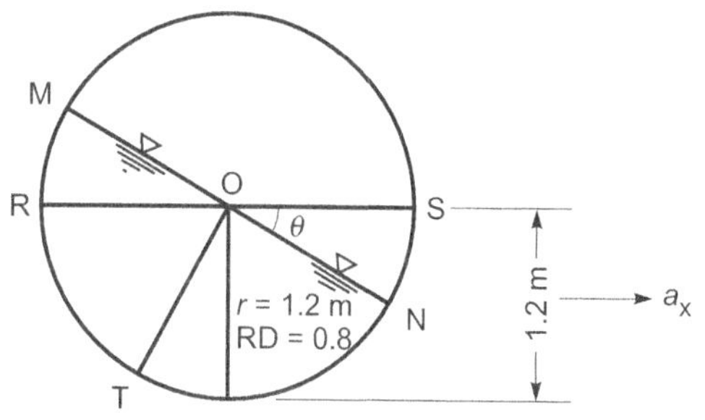

Fig. 2.53

The maximum pressure acts on the boundary point where the depth (measured normal to the free surface) is maximum.

In this case the maximum depth is OT = radius = 1.2 m

Hence $\left(\frac{p}{\gamma}\right)_m = 1.2$

$$p_m = 9.79 \times 0.8 \times 1.2 = \mathbf{9.4\ kN}$$

2.54 A closed tank 6 m long, 2 m wide and 1.8 m deep initially contains water to a depth of 1.2 m. The top has an opening in the front part to have air space at atmospheric pressure. If the tank is given a horizontal acceleration at a constant value of 2.4 m/s^2 along its length, calculate the total pressure force on the top of the tank.

Solution: In Fig. 2.54, RS is the original water surface. After the acceleration of a_x = 2.4 m/s^2, the water surface slope is

$$\tan\theta = \frac{a_x}{g} = \frac{2.4}{9.81} = 0.2446$$

$$\theta = 13.75°$$

As there is no spill of water, the air space will remain same as at start.

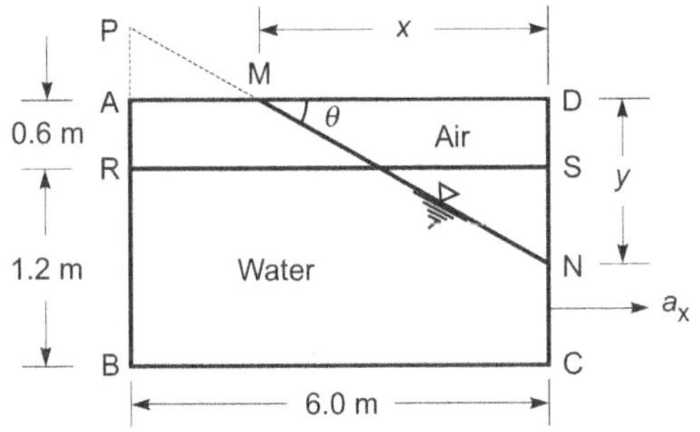

Fig. 2.54

Air space volume, $\forall = 0.6 \times 6.0 \times 2 = 7.2\ m^3$

Let MN be the new water surface at an inclination of θ to the horizontal. If MD = x and DN = y, B = breadth of the tank.

$$y = x \tan \theta$$

and

$$V = \frac{1}{2} xyB = \frac{1}{2} x^2 B \tan \theta$$

$$7.2 = \frac{1}{2} x^2 \times 2 \times 0.2446$$

$$x^2 = \frac{7.2}{0.2446} = 29.436$$

$$x = 5.4255 \text{ m} \quad \text{and} \quad y = 1.327 \text{ m}$$

Hence CN = depth of water in the front
= 1.80 – 1.327
= 0.473 m

AM = 6.0 – 5.4255 = 0.5745 m

AP = AM tan θ = 0.5745 × 0.2446
= 0.1405 m

The pressure profile on the top is represented by the triangle APM extending over the width. Pressure force on the top

$$F_t = \left(\frac{1}{2} \times \text{AP} \times \text{AM} \times \text{Breadth}\right)\gamma$$

$$= \frac{1}{2} \times 0.1405 \times 0.5745 \times 2.0 \times 9.79$$

$$F_t = 0.790 \text{ kN}$$

The force acts vertically upwards at $\frac{\text{AM}}{3} = 0.1915$ m from A at the mid-width section.

[***Note*** that in this case the free surface does not tilt at the mid length. As there is no spill that volume of water and air volume are conserved.]

2.55 A closed tank 10 m long, 3 m high and 2 m wide contains oil of relative density 0.81 and is given a horizontal acceleration of 0.25 g to the right in the direction of 10 m side. Estimate the pressure difference between (a) a point on the top rear edge and a point on the top front edge, (b) a point on the bottom front edge and a point on the top front edge, (c) Sketch the lines of equal pressure.

Solution: Refer to Fig. 2.55.

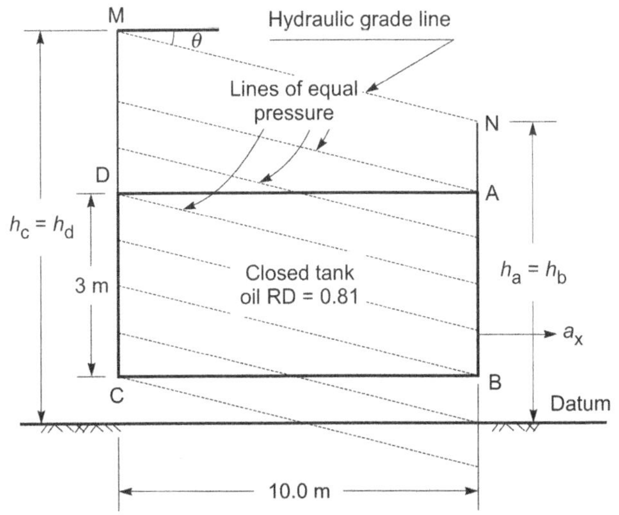

Fig. 2.55

It is required to find:

(a) $p_a - p_d$ and

(b) $p_b - p_a$ and

(c) Lines of equal pressure

At an acceleration of a_x let MN be the hydraulic grade line. Its inclination

$$\tan \theta = \frac{h_d - h_a}{L} = \frac{a_x}{g} = 0.25$$

(a) $(h_d - h_a) = 0.25 \times 10 = 2.5$ m

But $$(h_d - h_a) = \left(\frac{p_d}{\gamma_s} + Z_d\right) - \left(\frac{p_a}{\gamma_s} + Z_a\right)$$

and $$Z_d = Z_a$$

Hence $$(p_d - p_a) = \gamma_s \times 2.5$$

$$= 9.79 \times 0.81 \times 2.5 = \mathbf{19.83\ kPa}$$

(b) Along BA the hydraulic grade line is constant.

Hence $$h_b = h_a$$

$$\left(\frac{p_b}{\gamma_s} + Z_d\right) - \left(\frac{p_a}{\gamma_s} + Z_a\right) = 0$$

$$(p_b - p_a) = \gamma_s (Z_a - Z_b) = 9.79 \times 0.81 \times 3.0$$

$$= \mathbf{23.79\ kPa}$$

(c) Since the pressure distribution is hydrostatic in any vertical direction and the hydraulic grade line is inclined at θ to horizontal (line MN) the lines of equal pressure will be parallel to MN, as shown in Fig. 2.55.

2.56 A tank ABCDEF shown in Fig. 2.56(a) is filled with water. A small opening at E keeps the pressure at E atmospheric.

(a) Calculate the acceleration a_x required to cause onset of cavitation at A.

(b) What will be the pressures at that acceleration at points B, C, D and F?

[Assume local atmospheric pressure head = 10.0 m of water and vapour pressure head = 0.5 m (abs) of water].

Solution: At the acceleration a_x the hydraulic grade line will be inclined at θ, given by

$$\tan\theta = \frac{a_x}{g}$$

Since $\quad p_e$ = pressure at E = atmospheric pressure

the hydraulic grade line will pass through E as shown in Fig. 2.56(b) by the line MEN.

Then above an arbitrary datum:

$$h_e = h_f \quad \text{and} \quad h_a = h_b$$

Also $\quad (h_e - h_a) = AF \cdot \tan\theta = 5\tan\theta$

At the onset of cavitation at A, $p_a = p_v$ = vapour pressure. Considering absolute pressures,

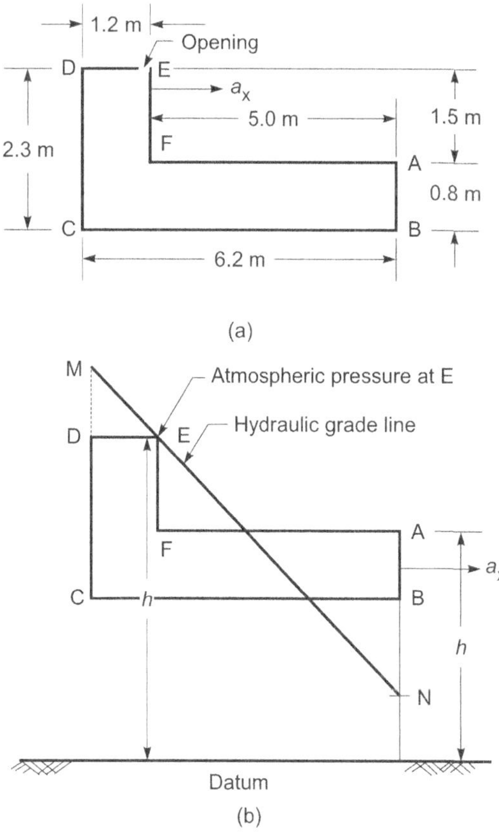

Fig. 2.56

$$(h_e - h_a) = \left(\frac{p_e}{\gamma} + Z_e\right) - \left(\frac{p_a}{\gamma} + Z_a\right)$$

$$= \left(\frac{p_e}{\gamma} - \frac{p_a}{\gamma}\right) + (Z_e - Z_a)$$

$$= (10.0 - 0.5) + (1.5) = 11.0 \text{ m}$$

But $(h_e - h_a) = 5\tan\theta$

Hence $\quad \tan\theta = \dfrac{a_x}{g} = \dfrac{11.0}{5} = 2.2$

or $\quad a_x = 2.2\,g = 2.2 \times 9.81 = \mathbf{21.585\ m/s^2}$

(b) Pressure at B:

$$(h_b - h_a) = \left(\frac{p_b}{\gamma} + Z_b\right) - \left(\frac{p_a}{\gamma} + Z_a\right) = 0$$

$$\frac{p_b}{\gamma} = \frac{p_a}{\gamma} + (Z_a - Z_b) = 0.5 + 0.8 = 1.3 \text{ m}$$

$$p_b = 1.3 \times 9.79 = \mathbf{12.727\ kPa\ (abs)}$$

Pressure at F:

$$(h_e - h_f) = \left(\frac{p_e}{\gamma} + Z_e\right) - \left(\frac{p_f}{\gamma} + Z_f\right) = 0$$

$$\frac{p_f}{\gamma} = \frac{p_e}{\gamma} + (Z_e - Z_f) = 10.0 + 1.5 = 11.5 \text{ m}$$

$$p_f = 11.5 \times 9.79 = \mathbf{112.585\ kPa\ (abs)}$$

Pressure at D:

$$(h_d - h_e) = 1.2\tan\theta = 1.2 \times 2.2 = 2.64 \text{ m}$$

Also $\quad (h_d - h_e) = \left(\dfrac{p_d}{\gamma} + Z_d\right) - \left(\dfrac{p_e}{\gamma} + Z_e\right)$

$$\therefore \quad \frac{p_d}{\gamma} = \frac{p_e}{\gamma} + (Z_e - Z_d) + 2.64$$

$$= 10.0 + 0 + 2.64 = 12.64 \text{ m}$$

$$p_d = 9.79 \times 12.64 = \mathbf{123.75\ kPa\ (abs)}$$

Pressure at C:

$$(h_d - h_c) = \left(\frac{p_d}{\gamma} + Z_d\right) - \left(\frac{p_c}{\gamma} - Z_c\right) = 0$$

$$\frac{p_c}{\gamma} = \frac{p_d}{\gamma} + (Z_d - Z_c)$$

$$= 12.64 + 2.03 = 14.94 \text{ m}$$

$$p_c = 9.79 \times 14.94 = \mathbf{146.26\ kPa\ (abs)}$$

2.57 An open tank 3.0 long, 1.5 wide and 2.0 m deep contains 1.2 m depth of water and moves down an inclined plane of 30° inclination to horizontal with an acceleration of 2.0 m/s^2. Determine the free surface slope and pressures at the bottom of the tank, in the front and rear ends of the tank.

Solution: Refer to Fig. 2.57

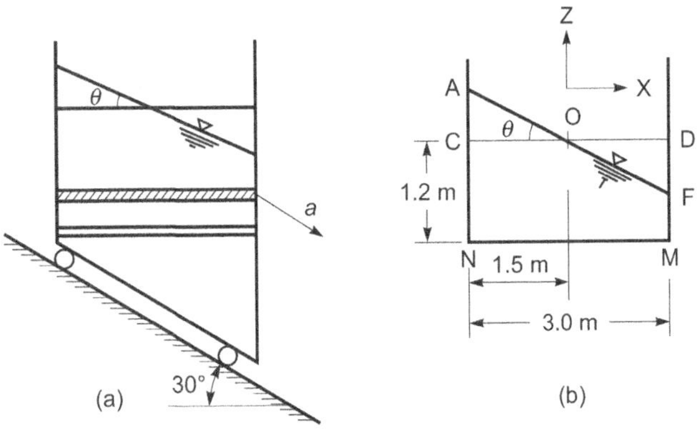

Fig. 2.57

Resolving the acceleration a_s into x- and z-components:

$$a_x = 2 \cos 30° = 1.732 \text{ m/s}^2$$
$$a_z = -2 \sin 30° = -1.00 \text{ m/s}^2$$

Water surface slope (θ measured clockwise from x-direction)

$$\tan \theta = \frac{a_x}{(a_z + g)} = \frac{1.732}{(-1+9.81)} = 0.1966$$

or $\theta = \mathbf{11.12°}$

Piezometric head at the rear

$$h_N = 1.2 + \frac{3.0}{2} \times \tan \theta$$
$$= 1.2 + 0.2949 = 1.495 \text{ m}$$

Piezometric head at the front

$$h_M = 1.2 - \frac{3.0}{2} \tan \theta = 0.905 \text{ m}$$

Pressures at N and M:

$$p_N = 1.495 \times 9.79 = \mathbf{14.64 \text{ kPa}}$$
$$p_M = 0.905 \times 9.79 = \mathbf{8.86 \text{ kPa}}$$

2.58 An open tank contains water to a depth of 1.5 m. The tank is put in an elevator which accelerates at 2.5 m/s^2. Calculate the pressure force per metre length on one of the sides of the tank when the (a) acceleration is upwards and (b) acceleration is vertically downwards.

Solution:

(a) When the acceleration is upwards

$$a_z = +2.5 \text{ m/s}^2$$

Pressure at any depth h below the free surface

$$p = \gamma h \left(1 + \frac{a_z}{g}\right)$$
$$= \gamma h \left(1 + \frac{2.5}{9.81}\right) = 1.2548 \gamma h$$

Pressure force on a side wall per metre width

$$F_H = 1.2548 \times 9.79 \times 1.5 \times \frac{1.5}{2}$$
$$= \mathbf{13.82 \text{ kN}}$$

(b) When the acceleration is downwards

$$a_z = -2.5 \text{ m/s}^2$$

Pressure at any depth h below free surface

$$p = \gamma h \left(1 + \frac{a_z}{g}\right) = 0.7452 \gamma h$$

Pressure force on a side wall per metre width

$$F_H = 0.7452 \times 9.79 \times 1.5 \times \frac{1.5}{2}$$
$$= \mathbf{8.207 \text{ kN}}$$

2.59 An open cylindrical tank 1 m in diameter and 2 m high contains water up to 1.5 m depth. If the cylinder rotates about its vertical axis what maximum angular velocity can be attained without spilling any water?

Solution: At the maximum permissible angular velocity the water surface will just touch the top rim of the cylinder.

Referring to Fig. 2.58, let C represent the original water level at the axis. As the water surface is a paraboloid of revolution, rise of water surface at the edge above C

= distance OC = $y/2$

where y = water surface elevation at the outer edge above the vertex 0.

Hence $y = 2(2.0 - 1.5) = 1.0$ m

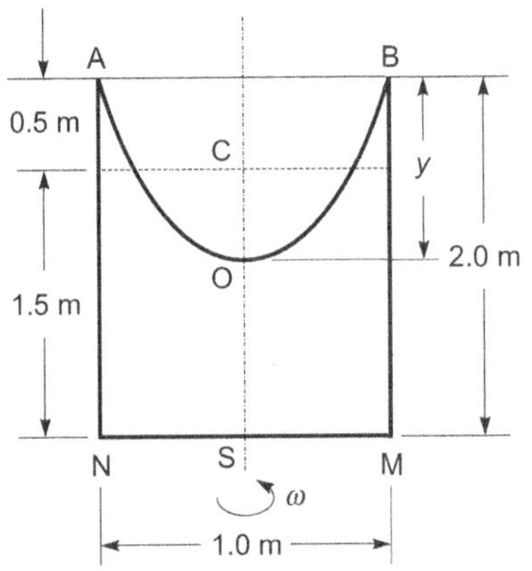

Fig. 2.58

In the present case

$$y = \frac{\omega^2 r^2}{2g}$$

$$1.0 = \frac{\omega^2 \times (0.5)^2}{2 \times 9.81}$$

$$\omega^2 = \frac{2 \times 9.81}{(0.5)^2} = 78.48$$

$$\omega = 8.858$$

But $\omega = \dfrac{2\pi N}{60}$ when N = rotations/min

$$N = \frac{60 \times 8.858}{2\pi} = \mathbf{84.6\ rpm}$$

2.60 An open cylindrical vessel of 20 cm diameter and 50 cm high contains water to a height of 30 cm (a) If the cylinder is rotated at 180 rpm about its axis, calculate the pressures on the bottom at the centre and at the edge of the vessel; (b) What speed of rotation would cause the water surface to just touch the top rim of the vessel?

Solution: Referring to Fig. 2.59

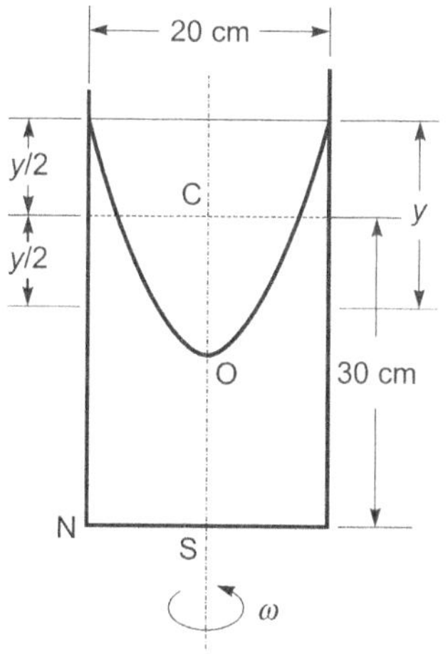

Fig. 2.59

(a) If N = rotations/min

Angular velocity $\omega = \dfrac{2\pi N}{60}$

In the present case

$$\omega = \frac{2\pi \times 180}{60} = 18.85 \text{ rad./s}$$

If y = water surface elevation above the vertex at a radial distance r, at the radial distance r = 10 cm

$$y = \frac{\omega^2 r^2}{2g} = \frac{(18.85)^2 (0.10)^2}{2 \times 9.81}$$

$= 0.181$ m $= 18.1$ cm

Rise at ends = $y/2$ above original level C.

h_N = piezometric head at a radial distance 10 cm on the base, i.e. point N.

= original depth + $y/2$

$$= 30.00 + \frac{18.10}{2} = 39.05 \text{ cm}$$

h_s = piezometric head at the centre of the base

$= h_N - y = 39.05 - 18.10 = 20.95$ cm

Pressure at N $= p_N = \gamma h_N$

$= 9.79 \times 0.3905 = \mathbf{3.823\ kPa}$

Pressure at S $= p_s = \gamma h_s$

$= 9.79 \times 0.2095 = \mathbf{2.051\ kPa}$

(b) In this case

$$\frac{y}{2} = 20 \text{ cm, or } y = 40 \text{ cm}$$

$h_s = 10$ cm and $h_N = 50$ cm

$$y = \frac{\omega^2 r^2}{2g}$$

or
$$\omega = \frac{\sqrt{2gy}}{r} = \frac{\sqrt{2 \times 9.81 \times 0.4}}{0.10}$$

$= 28.014$ rad./s

$$N = \text{revolutions/min} = \frac{60\omega}{2\pi}$$

$$= \frac{60 \times 28.014}{2\pi}$$

$= \mathbf{267.5\ rpm}$

2.61 An open cylinder 30 cm in diameter and 50 cm high is filled with water and rotated about its axis. Calculate the amount of water spilled when the speed of rotation is (a) 180 rpm and (b) 240 rpm.

Solution:

(a) When N = 180 rpm

$$\omega = \frac{2\pi \times 180}{60} = 18.85 \text{ rad./s}$$

At the wall of the cylinder, r = 15 cm = 0.15 m

$$y_{max} = \frac{\omega^2 r^2}{2g} = \frac{(18.85)^2}{2 \times 9.81} \times (0.15)^2$$

$= 0.4075$ m $= 4.075$ cm

The position of the water surface in the cylinder will be as shown in Fig. 2.60 (a). The water surface will start from the rim of the cylinder and extend downwards as a paraboloid to its vertex at O.

OS = The depth at the vertex

$= h_s - 50 - 40.75$

$= 9.25$ cm

Volume of the paraboloid AOB of height 40.75 cm

$= \dfrac{1}{2} \times$ (volume of enclosing cylinder)

= Volume of cylinder of height 20.375 cm

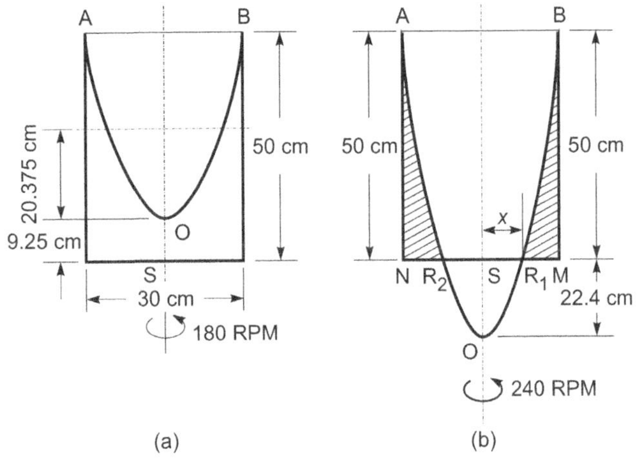

Fig. 2.60

Hence total volume of water inside the cylinder

= Volume corresponding to a depth of (9.25 + 20.375)

= 29.625 cm

$$= \frac{\pi}{4} \times (0.3)^2 \times 0.29625$$

$= 20.9 \times 10^{-3}\ m^3 = 20.9$ L

$$\text{Original volume} = \frac{\pi}{4} \times (0.3)^2 \times 0.50$$

$= 0.0353\ m^3 = 35.3$ L

Volume of water spilled = 35.3 – 20.9 = **14.4 L**

(b) When $N = 240$ rpm

$$\omega = \frac{2\pi \times 240}{60} = 25.13 \text{ rad./s}$$

At $r = 0.15$ m,

$$y_{max} = \frac{\omega^2 r^2}{2g} = \frac{(25.13 \times 0.15)^2}{2 \times 9.81}$$

= 0.724 m = 72.4 cm

As this value is larger than 50 cm it means that the theoretical paraboloid will extend below the base thereby leaving a part of the bottom uncovered by water. The paraboloid will start from the rim and extend up to its vertex O which is 72.4 cm below it as shown in Fig. 2.60(b).

Let x = radius of the exposed portion of the bottom of the cylinder

$$x = SR_1 = SR_2$$

$$\frac{y_{max}}{OS} = \frac{r^2}{x^2}$$

$$\therefore \quad x = r\sqrt{OS / y_{max}}$$

$$= 15 \times \sqrt{\frac{22.4}{72.4}}$$

= 8.34 cm

Volume of water spilled = Volume of the paraboloid AOB – Volume of the paraboloid R_1OR_2

$$= \frac{\pi r^2}{2} \times y_{max} - \frac{\pi x^2}{2} \times (OS)$$

$$= \left[\frac{\pi}{2} \times (0.15)^2 \times 0.724 - \frac{\pi}{2} \times (0.0834)^2 \times (0.224)\right]$$

$$= \frac{1}{2}[0.05118 - 0.004895]$$

$= 0.0232\ m^3 =$ **23.2 L**

2.62 A 90° conical vessel of height 2 m carries water whose volume is one half of the volume of the cone. Find the speed at which the vessel may be rotated to make the water just spill out.

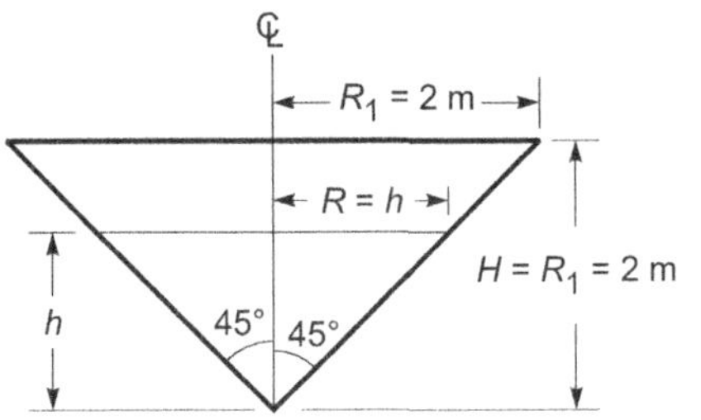

Fig. 2.61

Solution: Refer to Fig. 2.61,

$$\text{Volume of cone} = \frac{1}{3}\pi r^2 H = \frac{1}{3}\pi R_1^2$$

If h = height of water before rotation

$$\frac{1}{3}\pi R_1^3 (1/2) = \frac{1}{3}\pi h^3$$

$\therefore$ $h = R/(2^{1/3}) = 0.7937$

Since $R = 2.0$ m, $h = 1.5874$ m.

On rotation, $(\omega^2 R_1^2)/2g = (H - h) = (2.0 - 1.5874) = 0.4126$ m

$$\omega = \frac{2 \times 9.81 \times 0.4126}{(2.0)^2} = 2.0238$$

and $$\omega = \frac{2\pi N}{60}$$

Hence, $$\text{RPM} = N = \frac{60 \times 2.0238}{2\pi} = \mathbf{19.326\ rpm}$$

2.63 A closed cylinder of radius 10 cm height 30 cm is filled with water. If the cylinder is rotated about its vertical axis at a speed of 240 rpm, calculate the force exerted at the top and bottom covers of the cylinder.

Solution: Referring to Fig. 2.62, let, at the top, centre point be O.

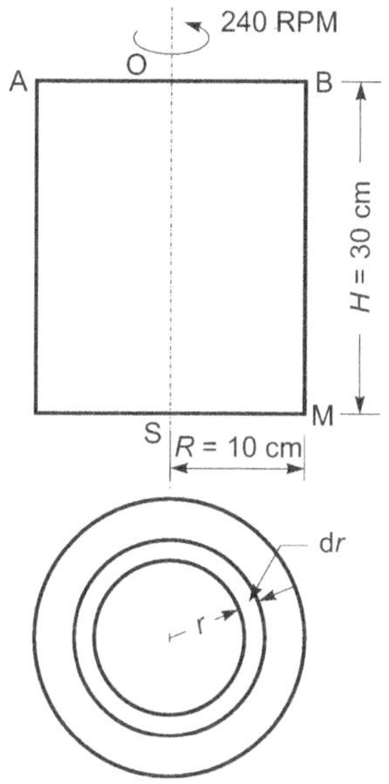

Fig. 2.62

Excess pressure due to rotation at the centre point = 0
Hence in the plane AOB pressure head at any radial distance r is

$$\frac{p}{\gamma} = \frac{\omega^2 r^2}{2g}$$

Considering a thin annular ring of radius r and thickness dr,
Pressure force on the top lid

$$F_T = \int_0^R p \cdot 2\pi r \cdot dr = \int \frac{\gamma \omega^2 r^2}{2g} 2\pi r \, dr$$

$$= \frac{\gamma \omega^2 \pi}{4g} R^4$$

In the present case,

$$\omega = \frac{2\pi N}{60} = \frac{2\pi \times 240}{60} = 25.13 \text{ rad./s}$$

$$R = 0.10 \text{ m}$$

$$\therefore \quad F_T = \frac{9.79 \times (25.13)^2 \times \pi}{4 \times 9.81} \times (0.10)^4$$

$= 0.0495$ kN = **49.5 N**

Force on the bottom = $F_T + \gamma H$ (Area of the base)
$= 49.5 + 9790 \times 0.30 \times \pi (0.10)^2$
= **141.8 N**

2.64 A closed cylindrical container of gas rotates about its axis in vertical position at a constant angular velocity ω. Neglecting gravity and assuming the gas temperature as constant and using the equation of state $p = \rho RT$, show that

$$\frac{p}{p_0} = \exp\left(\frac{\omega^2 r^2}{2RT}\right)$$

where p_0 = pressure on the axis.

Solution: The Euler equation relating the pressure gradient in normal direction is

$$\frac{dp}{dn} = \frac{dp}{dr} = \rho a$$

But $a = \omega^2 r$ and hence

$$\frac{dp}{dr} = \rho \omega^2 r$$

From equation of state $p = \rho RT$

$$dp = RT \, d\rho$$

$$\therefore \quad \frac{d\rho}{dr} = \rho\left(\frac{\omega^2 r}{RT}\right)$$

$$\frac{d\rho}{\rho} = \left(\frac{\omega^2 r}{RT}\right) dr$$

On integration $\ln \rho = \frac{1}{2}\frac{\omega^2 r^2}{RT} + C$

when $r = r_0$, $\rho = \rho_0$ and $p = p_0$

$$\therefore \quad \frac{\rho}{\rho_0} = \exp\left(\frac{1}{2}\frac{\omega^2 r^2}{RT}\right)$$

But $\frac{\rho}{\rho_0} = \frac{p}{p_0}$ and hence

$$\frac{p}{p_0} = \exp\left(\frac{\omega^2 r^2}{2RT}\right)$$

2.65 A U-tube contains a liquid of relative density 1.25 to a height of 25 cm in both the columns. If it is rotated about a symmetrical vertical axis which is 15 cm from both the limbs calculate the pressures at the bottom horizontal connecting tube at points M, N & S (Fig. 2.63) when the speed of rotation is (a) 120 rpm and (b) 240 rpm. Assume the diameter of the tube to be very small.

Solution:
Case (a):

$$\omega = \frac{2\pi N}{60} = \frac{2\pi \times 120}{60} = 12.57 \text{ rad./s}$$

In this case because of the symmetry the limbs AN and BM will be having a liquid column of 25 cm as at start. The points A and B being free surfaces will have atmospheric pressure. An imaginary free liquid surface will extend from A and B as a paraboloid with the vertex at 0. Thus the pressures on the limb MN will decrease

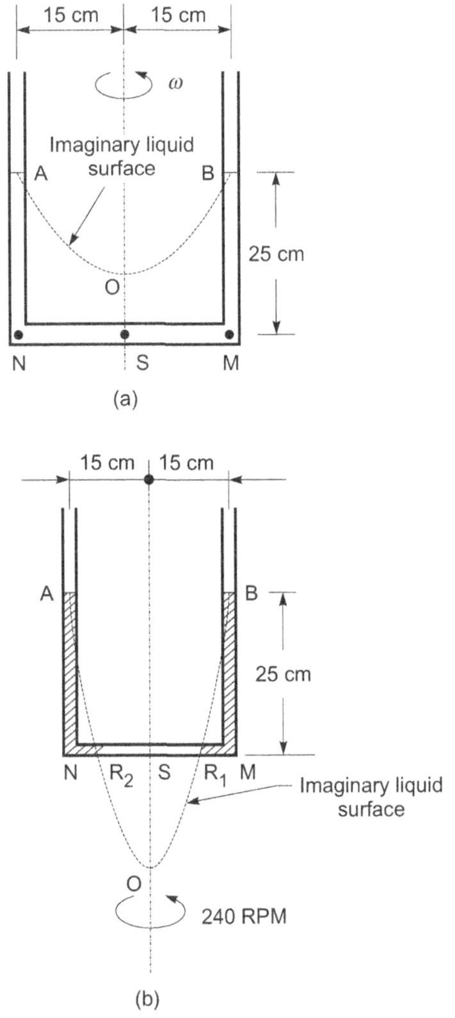

Fig. 2.63 *Example 2.65*

towards the centre to have a least value at S, the midpoint of the bottom limb.

$\therefore$ $BM = h_M = 25.0$ cm

$AN = h_N = 25.0$ cm

$$(BM - OS) = y = \frac{\omega^2 r^2}{2g} = \frac{(12.57)^2 \times (0.15)^2}{2 \times 9.81}$$

$= 0.181$ m $= 18.1$ cm

Hence $h_s = (25.0 - 18.1) = 6.9$ cm

Pressures at M, N and S are:

$p_M = \gamma h_M = 9.79 \times 1.25 \times 0.25 =$ **3.06 kPa**

$p_N = \gamma h_N = 9.79 \times 1.25 \times 0.25 =$ **3.06 kPa**

$p_S = \gamma h_S = 9.79 \times 1.25 \times 0.069 =$ **0.844 kPa**

Case (b):

When $N = 240$ rpm

$$\omega = \frac{2\pi \times 240}{60} = 25.13 \text{ rad./s}$$

$$(BM - OS) = y = \frac{\omega^2 r^2}{2g} = \frac{(25.13)^2 \times (0.15)^2}{2 \times 9.81}$$

$= 0.724$ m $= 72.4$ cm

Since BM = 25 cm, OS = – 72.4 + 25 = – 47.4 cm (i.e. O is below the point S at a depth of 47.4 cm) This signifies negative pressure at S. The free surface will be as in Fig. 2.63(b). As before, the pressure at M = pressure at N.

= pressure due to a column of 25 cm of liquid

= 3.06 kPa

Then it decreases to atmospheric pressure at R_1 and R_2, the point of intersection of the imaginary free surface with the limb MN.

At S, pressure $p_s = -\,47.4$ cm of liquid

$= -\,0.474 \times (1.25 \times 9.79)$

$=$ **– 5.8 kPa (gauge)**

[*Note:* The location of R_1 can be calculated by consi-dering points O and R_1 as

$$y = 0.474 \text{ m} = \frac{\omega^2 x^2}{2g}]$$

2.66 A U-tube shown in Fig. 2.64(a) is filled with a liquid of relative density 1.25 to a height of 15 cm in both the limbs. It is rotated about a vertical axis 15 cm from one limb and 30 cm from the other. If the speed of rotation is 60 rpm, find the difference in the liquid levels in two limbs. Also find the pressure at points M and N at the base of the U-tube.

Solution:
$$\omega = \frac{2\pi N}{60} = \frac{2\pi \times 60}{60}$$

$= 6.283$ rad./s

The liquid levels in both the limbs will be part of a paraboloid with vertex at point 0 [see Fig. 2.64(b)]

$$y = \frac{\omega^2 r^2}{2g}$$

$$\therefore \quad (y_1 - y_2) = \frac{\omega^2}{2g}(r_1^2 - r_2^2)$$

$$= \frac{(6.283)^2}{2 \times 9.81}[(0.30)^2 - (0.15)^2]$$

$= 0.136$ m $=$ **13.6 cm**

$$y_1 = \frac{\omega^2 r_1^2}{2g} = \frac{(6.283)^2}{2 \times 9.81} \times (0.30)^2$$

$= 0.181$ m $= 18.1$ cm

$y_2 = 18.1 - 13.6 = 4.5$ cm

There is another relationship relating to total volume of liquid in the tube. The horizontal portions MN is

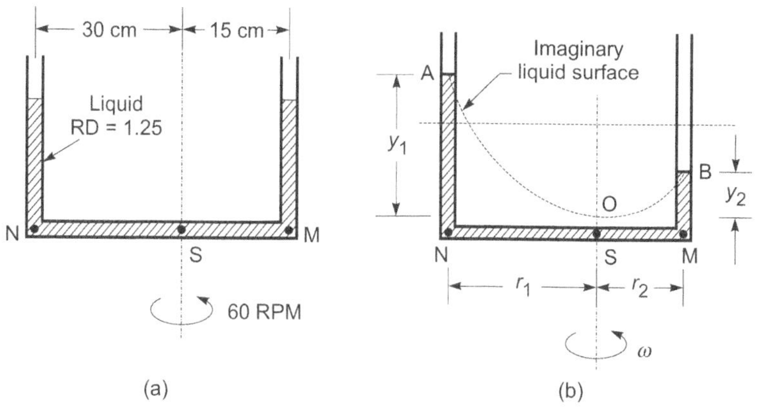

Fig. 2.64

common. The sum of two vertical columns of liquid before rotation

$$= (15 + 15) = 30 \text{ cm}$$

Also $\quad y_1 + y_2 = 18.1 + 4.5 = 22.6$ cm

The difference (30.0 – 22.6) = 7.4 cm is equally distributed in the two limbs to get the column heights in the right and left limbs as:

$$\text{Pressure head } h_S = \frac{7.4}{2} = 3.7 \text{ cm}$$

$$\text{Column NA} = h_N = 18.1 + \frac{7.4}{2} = 21.8 \text{ cm}$$

$$\text{Column MB} = h_M = 4.5 + \frac{7.4}{2} = 8.2 \text{ cm}$$

p_N = pressure at N = $1.25 \times 9790 \times 0.218$ = **2668 Pa**

p_M = pressure at M = $1.25 \times 9790 \times 0.082$ = **1003 Pa**

OBJECTIVE QUESTIONS

2.1 For a fluid at rest
(a) the shear stress depends upon the coefficient of viscosity
(b) the shear stress is maximum on a plane inclined at 45° to horizontal
(c) the shear stress is zero
(d) the shear stress is zero only on horizontal planes

2.2 If a Mohr circle is drawn for a fluid element inside a fluid body at rest, it would be
(a) a circle not touching the origin
(b) a circle touching the origin
(c) a point on the normal stress axis
(d) a point on the shear stress axis

2.3 Indicate the *incorrect* answer:
Hydrostatic pressure variation implies that
(a) the pressure varies linearly with depth
(b) the piezometric head $[p/\gamma + Z]$ in constant
(c) the density of the fluid in constant
(d) pressure varies linearly with distance

2.4 Normal stresses are of the same magnitude in all directions at a point in a fluid
(a) only when the fluid is frictionless
(b) only when the fluid is at rest
(c) only when there is no shear stress
(d) in all cases of fluid motion

2.5 The basic differential equation for the variation of pressure p in a static fluid with vertical distance y (measured upwards) is
(a) $dp = -\gamma dy$ (b) $dy = -\gamma dp$
(c) $dp = -\rho dy$ (d) $dp = -dy$

2.6 In an isothermal atmosphere the pressure
(a) decreases linearly with elevation
(b) decreases exponentially with elevation
(c) increases logarithmically with elevation
(d) varies inversely as the density

2.7 The piezometric head in a static liquid
(a) remains constant only on a horizontal plane
(b) increases linearly with depth below a free surface
(c) remains constant at all points in the fluid
(d) decreases linearly with depth below the free surface

2.8 Local atmospheric pressure is measured by
(a) Hydrometer (b) Barometer
(c) Hygrometer (d) Altimeter

2.9 Identify the correct statement:
(a) Local atmospheric pressure is always less than the standard atmospheric pressure
(b) Local atmospheric pressure depends only on the elevation of the place
(c) A barometer reads the difference between the local and standard atmospheric pressure
(d) Standard atmospheric pressure is 760 mm of mercury

2.10 Aneroid barometer measures
(a) local atmospheric pressure
(b) standard atmospheric pressure
(c) gauge pressure
(d) difference between the standard and local atmospheric pressures

2.11 Bourdon gauge measures
(a) absolute pressure
(b) gauge pressure
(c) local atmospheric pressure
(d) standard atmospheric pressure

2.12 In a barometer mercury is preferred over water because
(a) it is a better conductor of heat
(b) its surface is easier to read
(c) it has a higher density and lower vapour pressure
(d) it has a higher vapour pressure and bulk modulus

2.13 A barometer at a given location
(a) shows the local atmospheric pressure which is invariant with time
(b) always shows the local atmospheric pressure which may change with time
(c) shows the standard atmospheric pressure, if it is of aneroid type
(d) shows the local temperature if it is of mercury column type

2.14 In a mercury column-type barometer the correct local atmospheric pressure is obtained by adding correction due to vapour pressure of mercury as follows: H_a =
(a) $H_0 - h_v$ (b) $H_0 + h_v$
(c) H_0/h_v (d) $h_v - H_0$
where H_a = correct local pressure in mm of mercury, H_0 = observed barometer reading in mm of mercury and h_v = vapour pressure of mercury in mm.

2.15 When the barometer reads 740.0 mm of mercury, a pressure of 10 kPa suction at that location is equivalent to
(a) 10.02 m of water (abs)
(b) 9.87 m of water (abs)
(c) 88.53 kPa (abs)
(d) 0.043 kPa (abs)

2.16 The standard sea-level atmospheric pressure is equivalent to
(a) 10.0 m of fresh water of ρ = 998 kg/m^3
(b) 10.1 m of salt water of ρ = 1025 kg/m^3
(c) 12.5 m of kerosene of ρ = 800 kg/m^3
(d) 6.4 m of carbon tetrachloride of ρ = 1590 kg/m^3

2.17 The standard atmospheric pressure is 760 mm of mercury. At a certain location the barometer reads 710 mm of mercury. At this place an absolute pressure of 360 mm of mercury corresponds to a gauge pressure in mm of mercury.
(a) 400 mm of vacuum
(b) 350 mm of vacuum
(c) 360 mm of vacuum
(d) 710 mm

2.18 The standard atmospheric pressure is 101.32 kPa. The local atmospheric pressure at a location was 91.52 kPa. If a pressure is recorded as 22.48 kPa (gauge), it is equivalent to
(a) 123.80 kPa (abs) (b) 88.84 kPa (abs)
(c) 114.00 kPa (abs) (d) 69.04 kPa (abs)

2.19 A U-tube manometer measures
(a) absolute pressure at a point
(b) local atmospheric pressure
(c) difference in total energy between two points
(d) difference in pressure between two points

2.20 In the setup shown in Fig. 2.65 assuming the specific weight of water as 10 kN/m^3, the pressure difference between the two points A and B will be
(a) 10 N/m^3 (b) –10 N/m^3
(c) 20 N/m^3 (d) –20 N/m^3

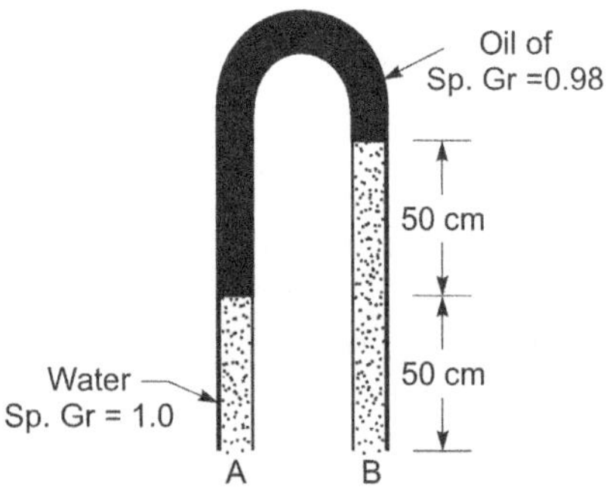

Fig. 2.65

2.21 An inclined manometer contains a liquid of relative density 0.8 and has an inclination of 30° to the horizontal. For a certain pressure the column length was 10 cm. If there is an uncertainty of 1° in the measurement of the angle of inclination, the calculated pressure would have an uncertainty of

(a) 1% (b) 0.28%
(c) 1.75% (d) 3.33%

2.22 A U-tube open at both ends and made of 8 mm diameter glass tube has mercury in the bottom to a height of 10 cm above the horizontal limb. If 19 cc of water is added to one of the limbs, the difference in mercury levels at equilibrium is

(a) 3.0 cm (b) 2.8 cm
(c) 1.0 cm (d) zero

2.23 For a submerged plane in a liquid, the resultant hydrostatic force F on one side of the plane is related to area A, centroidal depth $\bar{h}$, depth of the centre of pressure h_{cp} and depth of bottom edge h_b as F =

(a) $\gamma A h_{cp}$ (b) $\gamma A \bar{h}$
(c) $\gamma A h_b$ (d) $A \bar{h} / \gamma$

2.24 For an inclined plane submerged in a liquid the centre of fluid pressure on one side of the plane will be

(a) above the top edge of the area
(b) vertically below the centre of gravity
(c) below the centre of gravity
(d) in the same horizontal plane as the centre of gravity

2.25 An equilateral triangular plate is immersed in water as shown in Fig. 2.66. The centre of pressure below the water surface is at a depth of

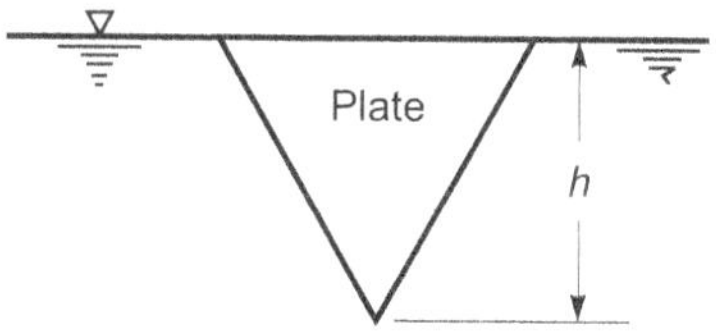

Fig. 2.66

(a) $3h/4$ (b) $h/3$
(c) $2h/3$ (d) $h/2$

2.26 When the water surface coincides with the top edge of a rectangular gate 2.5 m wide × 3 m deep, the depth of centre of pressure is

(a) 1.0 m (b) 1.5 m
(c) 2.0 m (d) 2.5 m

2.27 A circular plate of diameter D is submerged in water vertically so that the top most point is just at the water surface. The centre of pressure of the plate will be below the water surface at a depth of

(a) 5D/8 (b) 11D/16
(c) 2D/3 (d) 3D/4

2.28 The tank in Fig. 2.67 discharges water at constant rate for all water levels above the air inlet R. The height above datum to which water would rise in the manometer tubes M and N respectively, are

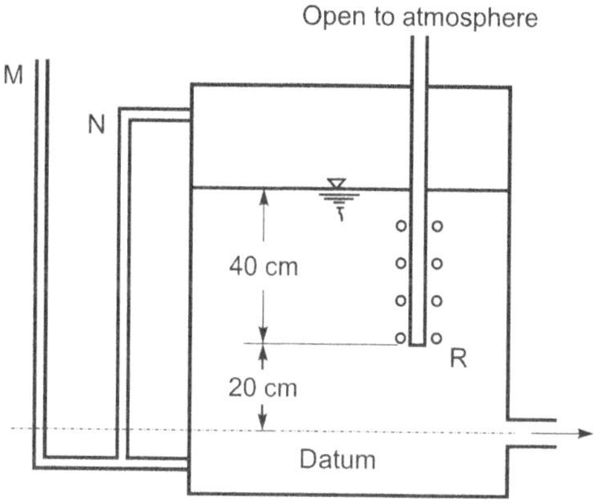

Fig. 2.67

(a) (60 cm, 20 cm) (b) (40 cm, 40 cm)
(c) (20 cm, 20 cm) (d) (20 cm, 60 cm)

2.29 A cylindrical tank of 2 m diameter is laid with its axis horizontal and is filled with water just to its top. The force on one of its end plate in kN, is

(a) 123.0 (b) 61.51
(c) 30.76 (d) 19.58

2.30 A rectangular plate 0.75 m × 2.4 m is immersed in a liquid of relative density 0.85 with its 0.75 m side horizontal and just at the water surface. If the plane of the plate makes an angle of 60° with the horizontal, the pressure force on one side of the plate, in kN, is

(a) 15.6 (b) 7.8
(c) 24.0 (d) 18.0

2.31 A circular annular plate bounded by two concentric circles of diameter 1.2 m and 0.8 m is immersed in water with its plane making an angle of 45° with the horizontal. If the centre of the circles is 1.625 m below the free surface, the total pressure force on one side of the plate in kN, is

(a) 7.07 (b) 10.0
(c) 14.14 (d) 18.0

2.32 A rectangular plate 30 cm × 50 cm is immersed vertically, in water with its longer side vertical. The total force on one side of the plate is estimated as 17.6 kN. If the plate is turned in the vertical plane at its centre of gravity by 90° and if all other

factors remain the same, the total force on one side of the plate would now be
(a) 8.8 kN (b) 15.6 kN
(c) 17.6 kN (d) 19.6 kN

2.33 A stationary liquid is so stratified that its density is $\rho_0(1 + h)$ at a depth h below the free surface. At a depth h in this liquid the pressure p in excess of $\rho_0 gh$ is
(a) $\rho_0 gh$ (b) $\rho_0 gh^2$
(c) $\rho_0 gh^3/3$ (d) $\rho_0 gh^2/2$

2.34 A curved surface is submerged in a static liquid. The horizontal component of pressure force on it is equal to
(a) the pressure force on a horizontal projection of the surface
(b) product of the surface are and pressure at the centre of gravity
(c) pressure force on a vertical projection of the surface
(d) weight of the liquid contained between the curved surface and the liquid surface

2.35 A hollow hemispherical object of diameter D was immersed in water with its plane surface coinciding with the free surface. The vertical component of force on the curved surface is given by F_v =
(a) $\frac{3}{8}\gamma\pi D^3$ (b) $\frac{1}{12}\gamma\pi D^3$
(c) $\frac{1}{24}\gamma\pi D^3$ (d) zero

2.36 A 2.0 m diameter penstock pipe carries water under a pressure head of 100 m. If the wall thickness is 7.5 mm, the tensile stress in the pipe wall, in MPa, is
(a) 65.3 (b) 130.5
(c) 231.0 (d) 1305.0

2.37 The centre of buoyancy of a submerged body
(a) coincides with the centre of gravity of the body
(b) coincides with the centroid of the displaced volume of the fluid
(c) is always below the centre of gravity of the body
(d) is always above the centroid of the displaced volume of liquid

2.38 An object weighing 100 N in air was found to weigh 75 N when fully submerged in water. The relative density of the object is
(a) 4.0 (b) 4.5
(c) 2.5 (d) 1.25

2.39 An iceberg has 12% of its volume projecting above the surface of the sea. If the density of sea water is 1025 kg/m^3, the density of the iceberg is
(a) 878 kg/m^3 (b) 1000 kg/m^3
(c) 1148 kg/m^3 (d) 902 kg/m^3

2.40 An object weighs 50 N in water and 80 N in an oil of relative density 0.80. Its volume in litres is
(a) 15.3 (b) 60.0
(c) 30.6 (d) 50.0

2.41 A floating body displaces
(a) a volume of liquid equal in magnitude to its own volume
(b) a volume of liquid equal to its own submerged weight
(c) a weight of liquid equal in magnitude to its own weight
(d) a weight of liquid which depends upon the volume of the container

2.42 A metal block is thrown into a deep lake. As it sinks deeper in water, the buoyant force acting on it
(a) increases
(b) remains the same
(c) decreases
(d) first increases and then decreases

2.43 When a block of ice floating on water in a container melts, the level of water in the container
(a) rises
(b) first falls and then rises
(c) remains the same
(d) falls

2.44 When a ship enters sea from a river one can expect it
(a) to rise a little
(b) to sink a little
(c) to remain at the same level of draft
(d) to rise or fall depending on whether it is of wood or steel

2.45 Buoyant force is the
(a) lateral force acting on a submerged body
(b) resultant force acting on a submerged body
(c) resultant hydrostatic force on a body due to fluid surrounding it
(d) the resultant force due to water on a body

2.46 A right circular wooden cone (Specific gravity = 0.8) with a base diameter of 0.6 m and hight of 0.8 m floats in water such that its axis is vertical and apex is downward. The immersed depth of cone is
(a) 0.480 m (b) 0.533 m
(c) 0.600 m (d) 0.743 m

2.47 A 12 cm steel cube (RD = 7.6) is submerged in a two layered fluid system. The bottom layer is mercury (rel. den = 13.6) and the top layer is kerosene. The height of the top surface of the cube above the interface of the two liquids is
(a) 4.76 cm (b) 2.95 cm
(c) 6.0 cm (d) zero

2.48 Two cubes of size 1.0 m sides, one of relative density = 0.60 and another of relative density = 1.15, are connected by a weightless wire and placed in a large tank of water. Under equilibrium the lighter cube will project above the water surface to a height of
(a) 10 cm (b) zero
(c) 50 cm (d) 25 cm

2.49 If B = centre of buoyancy, G = is the centre of gravity and M = metacentre, of a floating body, the body will be in stable equilibrium if
(a) MG = 0 (b) M is below G
(c) BG = 0 (d) M is above G

2.50 In a floating body I = moment of inertia of water line area about the longitudinal axis, $\forall$ = volume of displaced fluid, B = centre of pressure, G = centre of gravity and M = metacentre. For stable equilibrium of this body
(a) $BG = \frac{I}{\forall} + MG$ (b) $MG = \frac{I}{\forall} + BG$
(c) $MG = \left(\frac{I}{\forall}\right)/BG$ (d) $BG + MG = \frac{I}{\forall}$

2.51 A body is floating in a liquid as shown in Fig. 2.68. The centre of buoyancy, centre of gravity and metacentre are labeled as B, G and M respectively. The body is

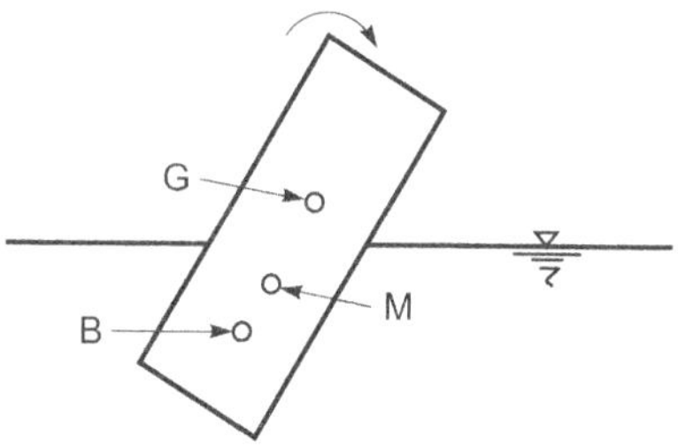

Fig. 2.68

(a) vertically stable
(b) vertically unstable
(c) rotationally stable
(d) rotationally unstable

2.52 An open box of base 2 m × 2 m contains a liquid of specific gravity 0.80 to a height of 2.5 m. When it is given an acceleration vertically upwards of 4.9 m/s^2, the pressure on the base of the tank, in kPa, is
(a) 9.8 (b) 36.8
(c) 19.6 (d) 29.4

2.53 A tank, open at top, contains a liquid with a relative density of 0.85 to a depth of 1.2 m. The acceleration which should be given to the tank to make the pressure at the bottom atmospheric is
(a) 9.8 m/s^2 vertically upward
(b) 9.8 m/s^2 vertically downward
(c) 8.33 m/s^2 vertically upward
(d) 8.33 m/s^2 vertically downward

2.54 An open tank contains a liquid and is made to slide down an inclined plane with uniform velocity. The free surface of the liquid
(a) will be parallel to the plane of the inclined plane
(b) will be horizontal
(c) will be inclined to the horizontal at an angle of $\tan^{-1} 1/g$
(d) will be inclined to the horizontal at an angle which depends upon the slope of the inclined plane

2.55 An open tank containing a liquid slides down a frictionless inclined plane. The free surface of the liquid will be
(a) inclined to the horizontal at an angle which depends upon the slope of the inclined plane
(b) horizontal
(c) parallel to the plane of the inclined plane
(d) inclined at an angle of ($\tan^{-1} 1/g$) to the horizontal

2.56 A U-tube shown in Fig. 2.69 is closed at A and open at B. It is filled with water.
The acceleration to the right needed to make the pressure at C atmospheric is

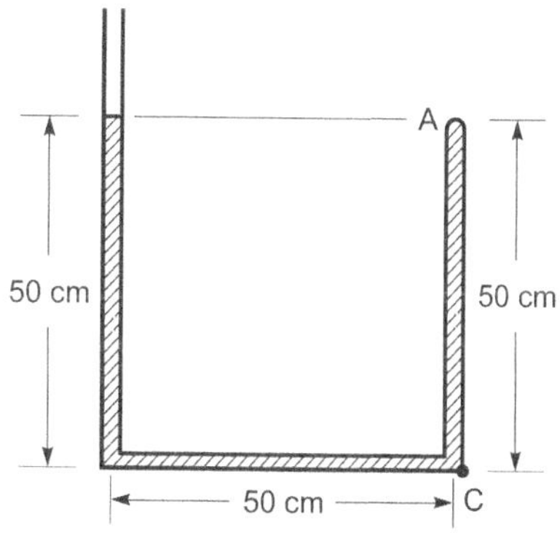

Fig. 2.69

(a) $\frac{1}{2}g$ (b) g

(c) 2g (d) $\frac{3}{2}g$

2.57 An open cubical tank was initially filled with water. When the tank was accelerated on a horizontal plane along one of its sides it was found that one third of the volume of water spilled out. The acceleration was

(a) $\frac{1}{3}g$ (b) $\frac{1}{2}g$

(c) $\frac{2}{3}g$ (d) g

2.58 An open cylindrical vessel filled with a liquid is falling freely with an acceleration *g*. The absolute pressure at any point in the liquid is

(a) zero
(b) above atmospheric pressure
(c) below atmospheric pressure
(d) equal to atmospheric pressure

2.59 Oil of density 800 kg/m^3 stands within 0.8 m from the top of an open tank 2.2 m high, 3.0 m wide and 8 m long. The greatest acceleration that this tank can have without any spilling over is

(a) 9.81 m/s^2 (b) 0.98 m/s^2
(c) 1.96 m/s^2 (d) 5.23 m/s^2

2.60 A liquid in a circular container is given a rigid body rotation about the axis of the cylinder. The piezometric line in a cross section is

(a) a horizontal line (b) a circular arc
(c) a parabola (d) a vertical line

2.61 An open cylindrical tank with its axis vertical is 0.8 m high and is 0.8 m in diameter. It is filled with an oil of density 800 kg/m^3 and is rotated at 120 rpm about the axis of the cylinder. The gauge pressure at the centre of the bottom of the tank is

(a) –0.49 m of oil (b) zero
(c) 1.29 m of oil (d) 6.27 kPa

2.62 A liquid in an open right circular cylinder is given rigid body rotation about the axis of the cylinder. The pressure distribution

(a) in any horizontal plane is uniform
(b) in any vertical plane is uniform
(c) on the bottom of the tank is uniform
(d) in any vertical plane is hydrostatic

2.63 An open circular cylinder 1.2 m high is filled with a liquid to its top. The liquid is given a rigid body rotation about the axis of the cylinder and the pressure at the centre of the bottom is found to be 0.3 m of liquid. The ratio of the volume of liquid spilled out of the cylinder to the original volume is

(a) 3/8 (b) 3/4
(c) 1/2 (d) 1/4

2.64 A liquid undergoing a rigid body rotation in a container is said to have

(a) circulatory flow
(b) circulation
(c) forced vortex motion
(d) free vortex motion

2.65 A 20 cm diameter open cylindrical container contains kerosene (relative density = 0.80) to a height of 20 cm. It is rotated about a vertical axis coinciding with the axis of the cylinder. If the bottom of the cylinder at the axis is just exposed, the speed of rotation is

(a) 94.6 rpm (b) 267.5 rpm
(c) 133.8 rpm (d) 535.0 rpm

2.66 If a cylinder containing a liquid is rotated about a vertical axis coinciding with the axis of the cylinder, the pressure in a vertical

(a) decreases as depth
(b) increases as depth
(c) decreases as square root of depth
(d) increases as square root of depth

2.67 A right circular cylinder is open at the top and filled with a liquid to its top level. It is rotated about its vertical axis at such a speed that half of the liquid spills out. The pressure at the point of intersection of the axis and the bottom is

(a) same as before rotation
(b) 1/2 the value before rotation
(c) 1/4 the value before rotation
(d) equal to the atmospheric pressure

2.68 For a U-tube containing water and rotating about an axis as in Fig. 2.70, the difference in water surface elevations ($h_a - h_b$) is

(a) 6.04 cm (b) 1.51 cm
(c) 3.02 cm (d) 24.15 cm

2.69 The U-tube shown in Fig. 2.71 contains water and is rotated at 120 rpm, about a vertical axis passing through the midpoint O of the horizontal limb. The pressure head at O is

(a) 8.05 cm (b) 11.95 cm
(c) 20.0 cm (d) 6.35 cm

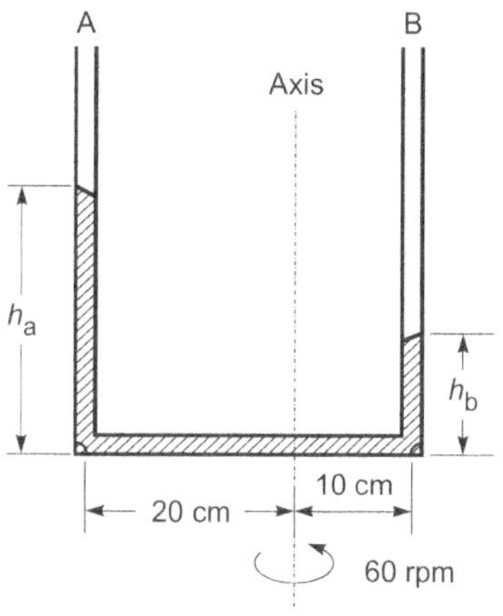

Fig. 2.70 *Question 2.68*

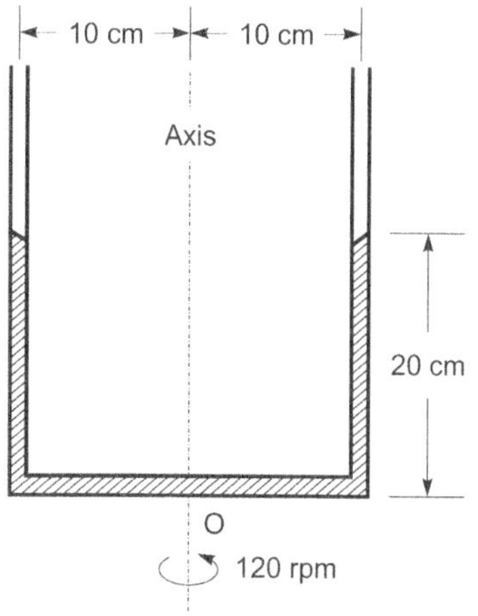

Fig. 2.71 *Question 2.69*

2.70 A 25 cm long pipe closed at one end (A) is filled with water and then the other end (B) is capped. It is then placed in a horizontal position and rotated at 30 rad./s about a vertical axis passing through the capped end, (Fig. 2.72). The difference of pressure between the two ends is
(a) 28.1 kPa
(b) 5.3 kPa
(c) 65.1 kPa
(d) 113.2 kPa

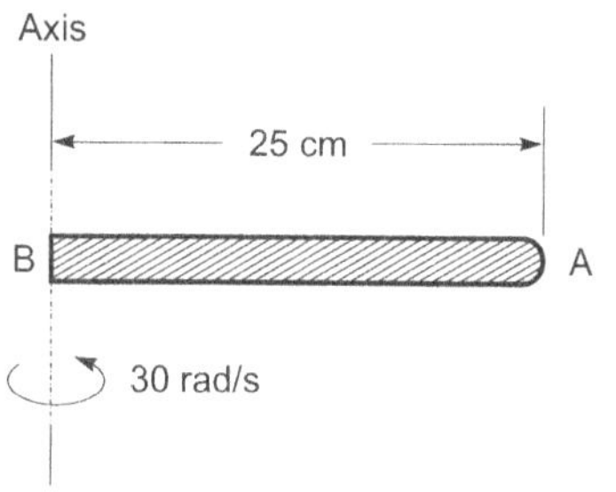

Fig. 2.72 *Question 2.70*

PROBLEMS

A. Measurement of Pressure

2.1 Calculate the pressure in pascals corresponding to
(a) 8 cm column of a liquid of relative density 0.80.
(b) 6 cm column of mercury.
(c) 2.0 m column of water.
(Ans. (a) 626.6 Pa, (b) 7988.6 Pa, (c) 19.58 kPa)

2.2 A bourdon gauge connected to a closed tank indicates 35 kPa at a place where the atmospheric pressure is 95.480 kPa. If the tank is now carried to a higher elevation where the atmospheric pressure is 70.11 kPa, what reading will be indicated by the gauge?
(Ans. p (gauge) = 60.37 kPa)

2.3 A hydraulic press has a ram of 150 mm and a plunger of 20 mm diameter. Find the force required on the plunger to lift a weight of 40 kN. If the plunger has a stroke of 0.40 m and makes 30 strokes per minute, determine the rate at which the weight is lifted per minute and the power required by the plunger.
(Ans. F = 0.711 kN, rate = 0.2133 m/min, **P** = 142 W)

2.4 An open tank contains water to a depth of 2.5 m and an oil of relative density 1.25 to a depth of 1.5 m. Determine the pressure at (a) the water surface (b) at the oil-water interface (c) at a depth of 3.5 m below the free surface and (d) at the bottom of the tank.
(Ans. (a) $p_1 = 0$, (b) p_2 = 24.475 kPa (c) p_3 = 36.713 kPa, (d) p_4 = 41.831 kPa)

2.5 Calculate the pressure and density of air at an elevation of 4500 m above sea level if the atmospheric pressure and density at sea level are 101.35 kPa and 1.2255 kg/m^3. Assume isothermal process.
(Ans. p_2 = 59.429 kPa, ρ_2 = 0.7186 kg/m^3)

2.6 Assuming adiabatic process exists, determine the pressure and density of air at an elevation of 500 m above sea level given that the atmospheric pressure and density at an elevation of 3000 m are 70.107 kPa and 0.9092 kg/m^3 respectively. Assume k = 1.4.

(Ans. p_2 = 95.054 kPa, ρ_2 = 1.1187 kg/m^3)

2.7 For the system shown in Fig. 2.73 calculate the air pressure p_A to make the pressure at N one third of that at M.

(Ans. p_A = 0.2937 kPa)

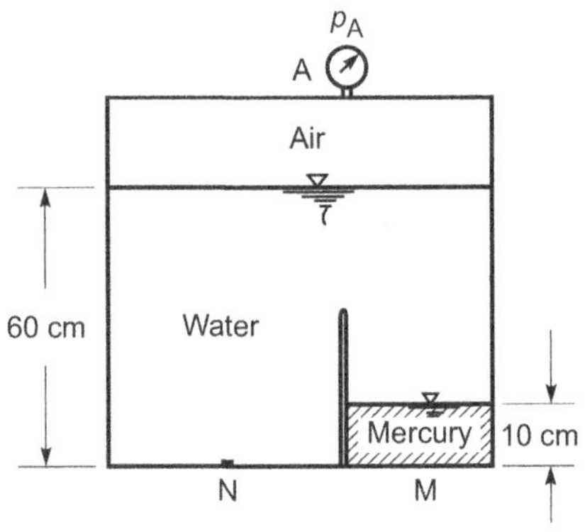

Fig. 2.73

2.8 For the system shown in Fig. 2.74 determine the pressures at A and B.

(Ans. p_A = 626.6 Pa (gauge), p_B = – 4072.6 Pa (gauge))

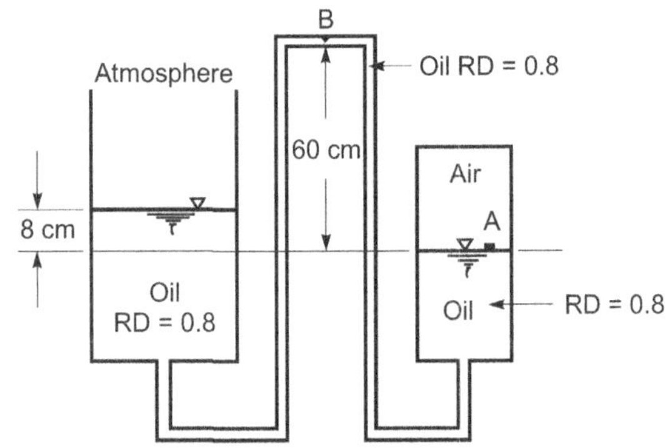

Fig. 2.74 *Problem 2.8*

2.9 If the pipe in Fig. 2.75 contains water and there is no flow, calculate the value of the manometer reading h.

(Ans. h = 0)

2.10 For the manometer arrangement of Problem 2.9 (Fig. 2.75) there is a flow of water from A towards B. If the manometer reading is h = 5 cm, calculate the pressure difference ($p_A - p_B$).

(Ans. ($p_A - p_B$) = 30.643 kPa))

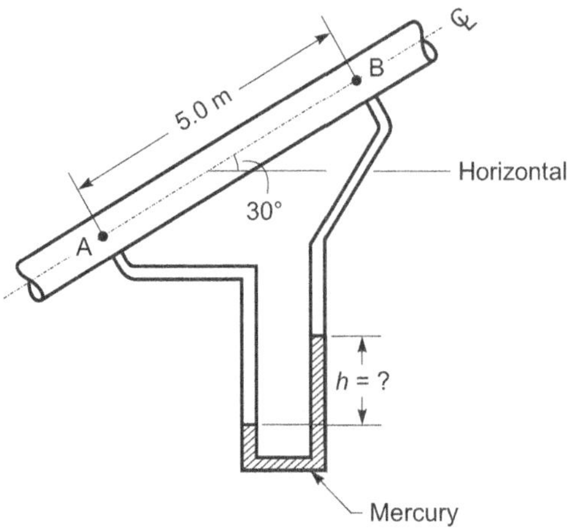

Fig. 2.75 *Problem 2.10*

2.11 Two points M and N in a pipe system (Fig. 2.76) are connected a mercury differential manometer. The connecting tube is filled with oil of relative density 0.85. The high pressure point M is 0.80 m above point N. If the mercury column reading is 8 cm (as shown in the figure), what is the pressure difference between M and N in (i) kPa and (ii) metres of water.

(Ans. (i) ($p_M - p_N$) = 3.33 kPa; (ii) h_w = 0.34 m))

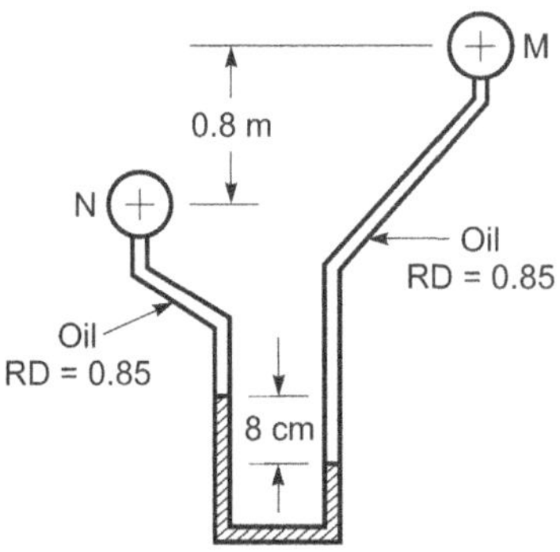

Fig. 2.76

2.12 A tube filled with mercury is placed in a bowl of mercury as in Fig. 2.77. The

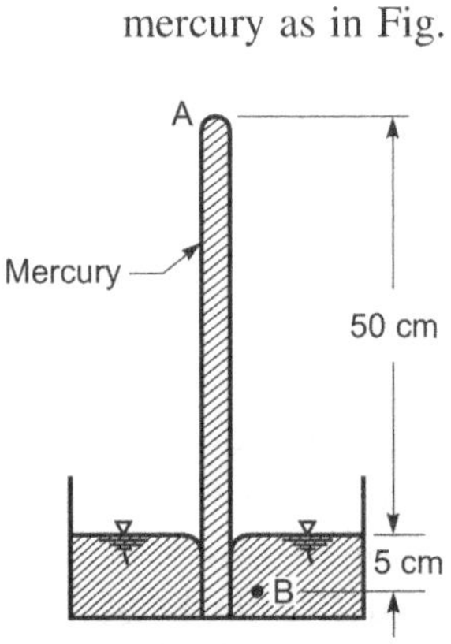

Fig. 2.77

tube is closed at point A and the other end is open. Calculate the absolute pressures at points A and B if the atmospheric pressure is 95.48 kPa.

(Ans. p_A (abs) = 28.91 kPa (abs), p_B (abs) = 102.137 kPa (abs))

2.13 The barometric pressure at sea level is 760 mm of mercury while on a mountain top it is found to be 685 mm. If the specific weight of air is assumed constant at 11.8 N/m^3, calculate the height of the mountain.

(Ans. H = 846.3 m)

2.14 Calculate the pressure and density of air at an elevation of 5000 m above mean sea level. The atmospheric pressure and temperature at sea level are 101.35 kPa and 15°C respectively. The temperature lapse rate is 0.007 K/m. The density of air at sea level is 1.2255 kg/m^3.

(Ans. p_2 = 52.854 kPa (abs), ρ_2 = 0.7413 kg/m^3)

B. Forces of Plane Surfaces

2.15 A vertical rectangular gate 2.0 m wide and 2.5 m high is subjected to water pressure on one side, the water surface being at the top of the gate. The gate is hinged at the bottom and is held by a horizontal chain at the top. Calculate the tension in the chain.

(Ans. F = 20.39 kN)

2.16 An annular ring cut in a sheet metal has 1.5 m outer diameter and 1.0 m inner diameter. It is inserted vertically in a liquid of relative density 0.90 with its centre 1.75 m below the surface. Calculate the total force on one side of this ring and the location of the centre of pressure.

(Ans. F = 15.138 kN; h_p = 1.866 m)

2.17 A stone masonry wall 6 m high and 2.5 m wide retains water to a height of 5 m on one side. If the specific weight of masonry is 25 kN/m^3, find the direction and magnitude of the resultant force per unit width of wall. Also find the point of action of the resultant force on the base of the wall. What are the maximum and minimum stresses at the base of the wall?

(Ans. R = 394.5 kN, θ = 71.9° to the horizontal, x = 1.794 m from the upstream, σ_{max} = 3.46 Pa, σ_{min} = – 46 kPa)

2.18 A right angled triangular plate of base b and height h is immersed in water vertically with its base horizontal and coincident with the water surface. Calculate the total pressure force on one side of the plate and the coordinates of the centre of pressure.

$$\left(Ans.\ F = \frac{\gamma b h^2}{6},\ x_p = \frac{b}{4},\ y_p = \frac{h}{2}\right)$$

2.19 A vertical gate of width 2.0 m and height 2.5 m controls a sluice opening in a dam. The top of the gate is 10 m below the water surface. If the gate weights 80 kN, find the vertical force required to raise the gate. The coefficient of friction between, the gate and the guides can be assumed to be 0.25. (Neglect buoyancy effect on the gate)

(Ans. F = 218 kN)

2.20 Some regular geometrical laminae are shown in Fig. 2.78. These laminae are immersed vertically with their upper edges or top most points on the water surface as indicated in the figures. Derive the expression for the centre of pressure as indicated against each lamina in Table 2.1.

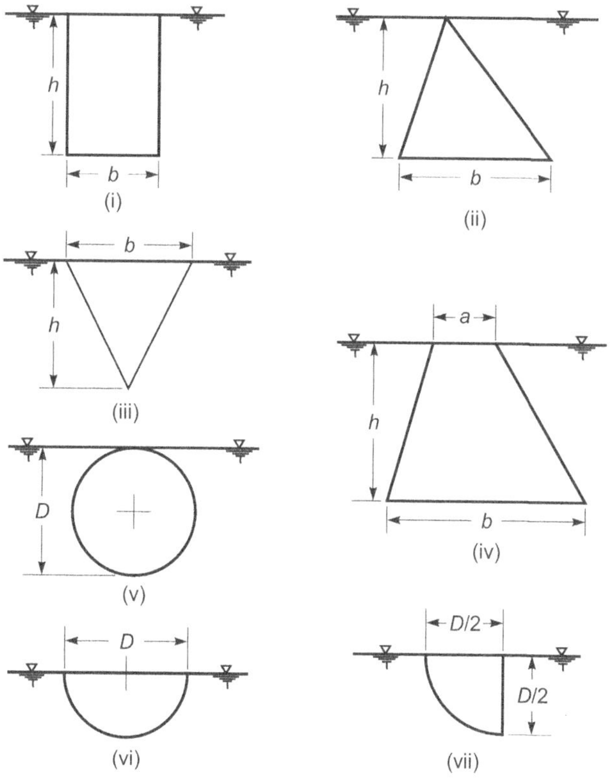

Fig. 2.78 *Problem 2.20*

2.21 A thin plate in the form of a rhombus is immersed in water with a vertex on the free surface and the diagonal through that vertex vertical. Show that the centre of pressure is located at a depth of $h_p = \frac{5}{7}h$ where h = length of the vertical diagonal.

Table 2.1 Location of centre of pressure — Problem 2.20

Lamina	*Description*	*Depth of centre of pressure h_p*
Fig. 2.78(i)	Rectangle	$\frac{2}{3}h$
Fig. 2.78(ii)	Trapezoid	$\frac{h}{2}\frac{(a+3b)}{(a+2b)}$
Fig. 2.78(iii)	Triangle with base on the free surface	$\frac{h}{2}$
Fig. 2.78(iv)	Triangle with vertex on the surface; base is horizontal	$\frac{3}{4}h$
Fig. 2.78(v)	Circle	$\frac{5}{8}D$
Fig. 2.78(vi)	Semicircle with diameter on the free surface	$\frac{3}{32}\pi D$
Fig. 2.78(vii)	Quadrant of circle with one of the bounding radii on the free surface	$\frac{3}{32}\pi D$

2.22 Calculate the force F required to hold the hinged door in Fig. 2.79 in closed position. The door is a 0.5 m square. An air pressure of 30 kPa acts over the water surface.

(Ans. F = 7.217 kN)

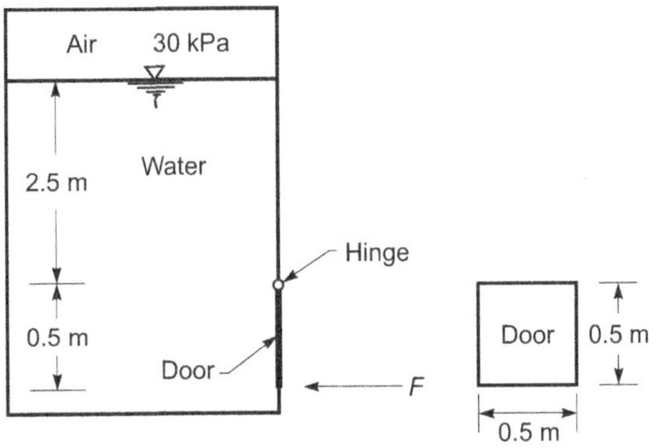

Fig. 2.79

2.23 A vertical gate 5 m × 2.5 m in size and weighing 0.5 tonnes slides along guides fitted on the side walls of an overflow spillway at its crest. The coefficient of friction between the gate and the guide is 0.25. What force will be exerted at the hoisting mechanism to lift the gate when the head of water over the crest is 2.0 m ?

(Ans: F = 3 tonnes)

2.24 The counterweight pivot gate shown in Fig. 2.80 controls the flow from a tank. The gate is rectangular and is 3 m × 2 m. Determine the value of the counterweight W such that the upstream water can be 1.5 m deep.

(Ans. W = 84.8 kN)

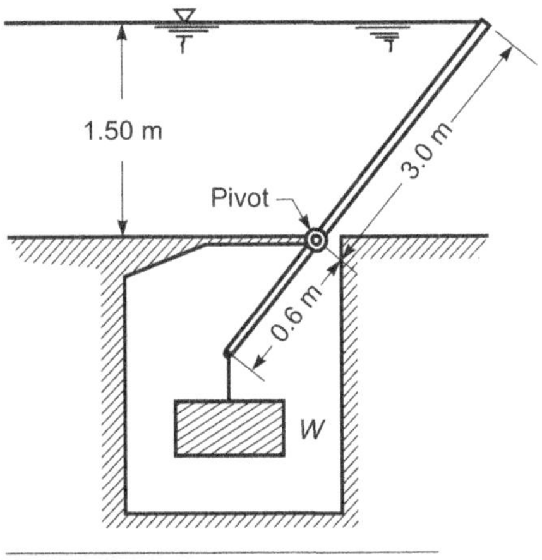

Fig. 2.80 *Problem 2.24*

2.25 The gate shown in Fig. 2.81 weighs 10 kN per metre width. Its centre of gravity is 0.3 m from the vertical face and 0.6 m above the lower horizontal face. Find the height H at which the gate will just topple about the hinge.

(Ans. H = 3.38 m)

2.26 A tank shown in Fig. 2.82 has an isosceles triangular gate hinged at the edge M. Find the horizontal force at M required to keep the gate closed. The weight of the gate is too small and can be neglected.

(Ans. F = 13.06 kN)

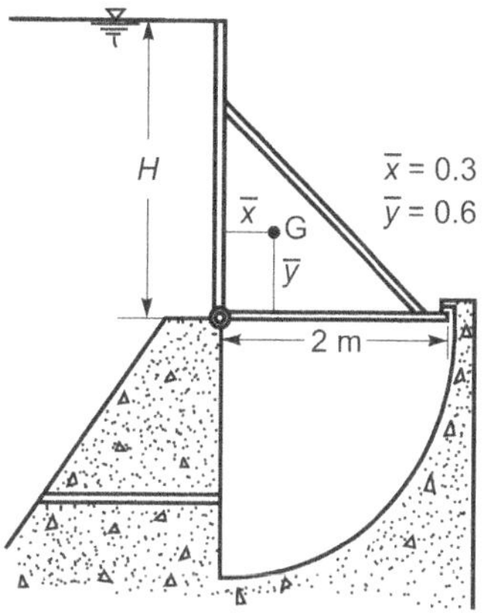

Fig. 2.81 *Problem 2.25*

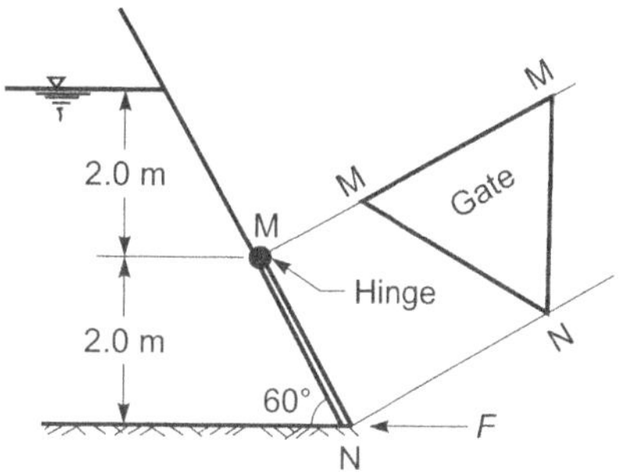

Fig. 2.82 *Problem 2.26*

2.27 The rectangular gate AB in Fig. 2.83 is 2.0 m high and 1.5 m wide. Calculate the net hydrostatic force on the gate.

(Ans. $F_1 - F_2 = 37.83$ kN)

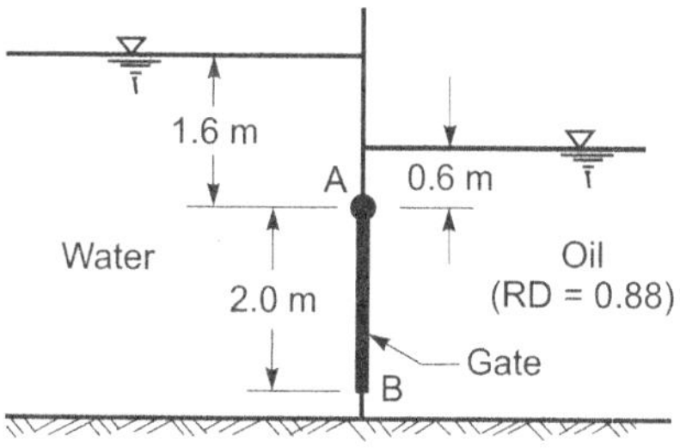

Fig. 2.83

2.28 A circular plate 3 m in diameter is submerged in water so that the greatest and least depths below the free surface are 2.0 m and 1.0 m respectively. Find (a) the total pressure force on one side of the plate and (b) the position of the centre of pressure.

(Ans. $F = 103.77$ kN, $h_p = 1.5416$ m)

2.29 For the flash board shown in Fig. 2.84 find the depth of water H and the compressive force on the strut per metre length of the flash board when the board is about to trip.

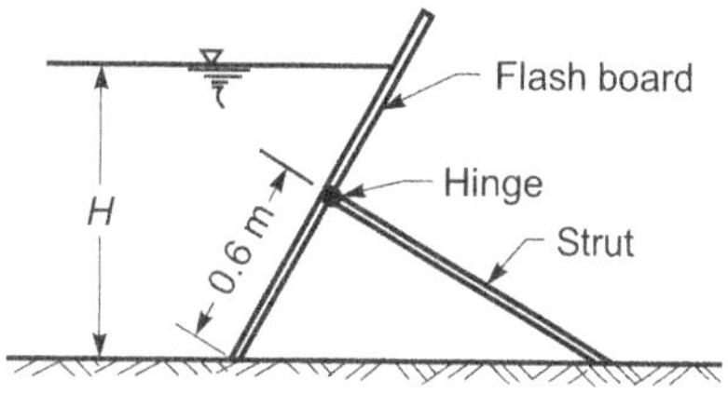

Fig. 2.84

(***Hint:*** At the time of tipping the centre of pressure coincides with the hinge.)

(Ans. $H = 1.559$ m; $F = 13.73$ kN)

2.30 A square aperture in the vertical side of a tank has one diagonal vertical and is completely covered by a plane plate hinged along one of the upper sides of the aperture. The diagonals of the aperture are 2 m long and the tank contains a liquid of relative density 1.15. The centre of aperture is 1.5 m below the free surface. Calculate the thrust exerted on the plate by the liquid and the position of the centre of pressure.

(Ans. $F = 33.78$ kN, $h_p = 1.611$ m)

2.31 A concrete gravity dam is in the form of a right angled triangle in section with the vertical side facing the water. The height of the dam is 10 m and the base width is 8 m. Assuming a specific weight of 25 kN/m^3 for the concrete, calculate the resultant force per unit length of dam and its point of action on the floor of the dam when it retains 10 m of water. What are the minimum and maximum vertical stresses on the base of the dam?
(Ans. $R = 1113.3$ kN, $\theta = 26.08°$, $\sigma_1 = 153.1$ kPa, $\sigma_2 = 96.9$ kPa)

2.32 A tank contains a suspension to a height H. The suspension can be considered to be a liquid whose density increases linearly, from ρ_0 at the surface, towards the bottom. Show that the pressure at any depth y is given by

$$p = \rho_0 g y + \frac{K}{2} g y^2 + p_0$$

where p_0 = pressure at the free surface and K = rate of increase of the density with depth. Show that the force per unit width on a vertical side of the tank is given by

$$F = \left[\frac{\rho_0 g H^2}{2} + \frac{K}{6} g H^3 + p_0 H\right]$$

C. Forces On Curved Surfaces

2.33 A circular cylinder of 1.8 m diameter and 2.0 m long is used for water level control in a tank. If at

a given instance it retains water as shown in Fig. 2.85 determine the reaction at the joint A.
(Ans. R_H = 31.72 kN, R = 61.96 kN R_v = 53.23 kN, tan θ = 0.596 where θ = inclination of R to the vertical. M_A = 36.0 kN.m (clockwise))

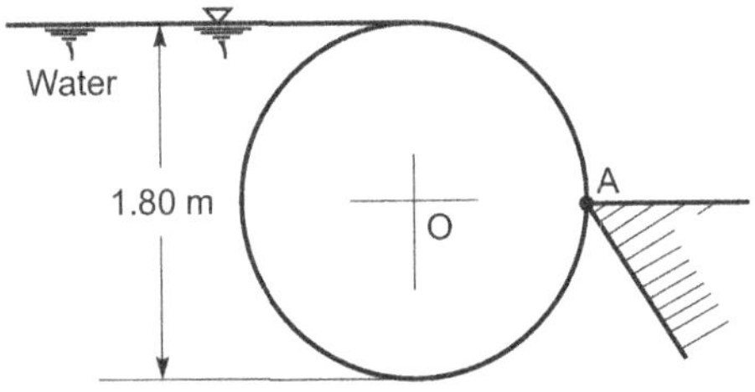

Fig. 2.85 *Problem 2.33*

2.34 A cylinder of diameter 0.6 m is located in water as shown in Fig. 2.86. The cylinder and the wall are smooth. If the length of the cylinder is 1.5 m, find (i) its weight, (ii) the resultant force exerted by the wall on the cylinder and (iii) the resultant moment about the centre of the cylinder due to water forces on the cylinder.
(Ans. 4410 kN, 661.5 kN, zero)

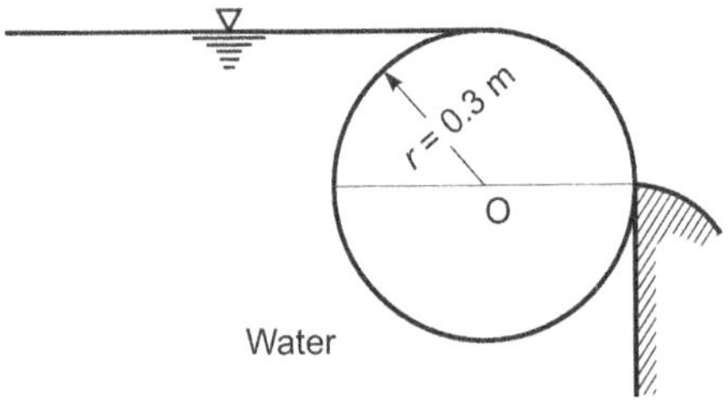

Fig. 2.86 *Problem 2.34*

2.35 A radial gate retains 5 m of water above the crest of a dam as shown in Fig. 2.87. Find the resultant water force on the gate per metre length.
(Ans. R = 124.4 kN, α = inclination of R to the horizontal = 10.235°. The resultant passes through the centre O.)

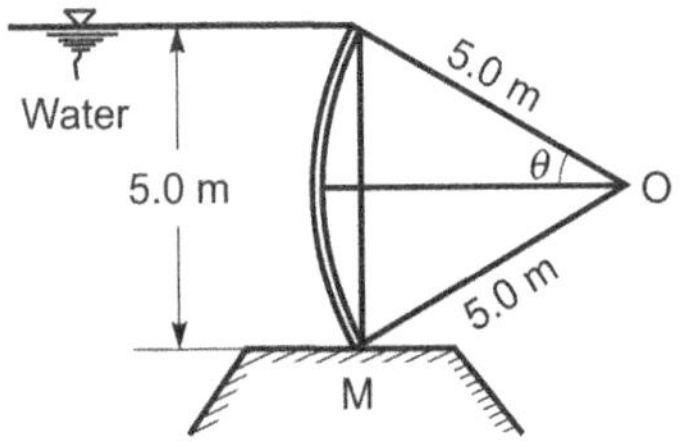

Fig. 2.87

2.36 For a cylindrical gate 4 m long shown in Fig. 2.88 calculate the resultant force due to fluid pressure.
(Ans. R = 219.64 kN, α = inclination of R to the horizontal = 40.34°, the resultant passes through the centre of the gate.)

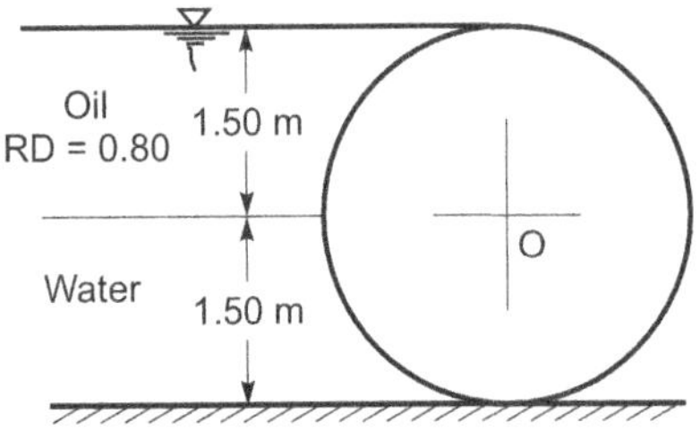

Fig. 2.88 *Problem 2.36*

2.37 For the container shown in Fig. 2.89 estimate the resultant force on the hemispherical bottom.
(Ans. R_V = 52.46 kN, R_H = 0 (due to symmetry))

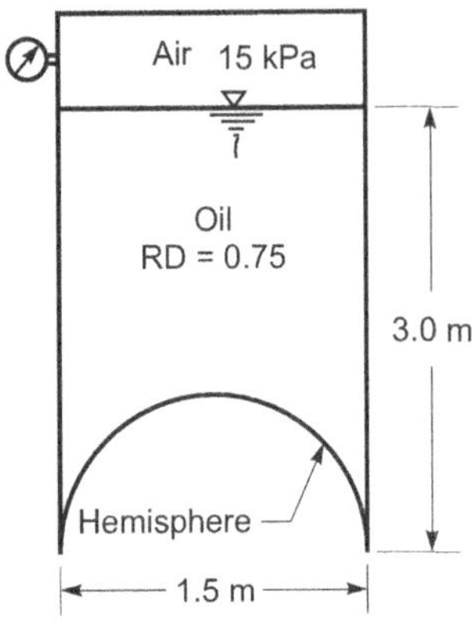

Fig. 2.89

2.38 A cylindrical barrier of diameter 2.0 m and width 3.0 m retains water on both sides with water surface elevations of 1.5 m and 0.5 m above the lowest point of the barrier, respectively. Calculate the resultant hydrostatic force on the barrier.
(Ans. R = 54.69 kN; α = inclinations of R to the horizontal = 57.5°)

2.39 A 1.8 m diameter cylindrical tank is laid with its axis horizontal on a level ground. Each of its ends are closed by a hemispherical dome. The tank contains oil of relative density 0.9 under pressure. If a pressure gauge on the top of the tank reads 22 kPa, calculate the resultant force on the hemispherical end.
(Ans. R = 54.69; α = inclination of R to the horizontal = 10°, the resultant passes through the centre of the hemisphere)

2.40 A cylindrical tank of diameter 2.5 m is founded with its axis horizontal. The tank contains water in the bottom half and an oil of relative density 0.82 on the top half with a common interface. The fluids are under pressure and a pressure gauge on the top of the cylinder reads 15 kPa. Calculate the force per metre length on the upper half of the cylindrical tank.

(Ans. F_V = 42.9 kN/m length (upward))

2.41 A conical valve of weight 2.5 kN deeps water from flowing out of a tank as shown in Fig. 2.90. It is held in position by a counter weight of 10 kN connected by a string passing over frictionless pullies as in Fig. 2.90. Find the maximum height *H* of water in the tank, which will make the device to function without any leak.

(Ans. *H* = 1.064 m)

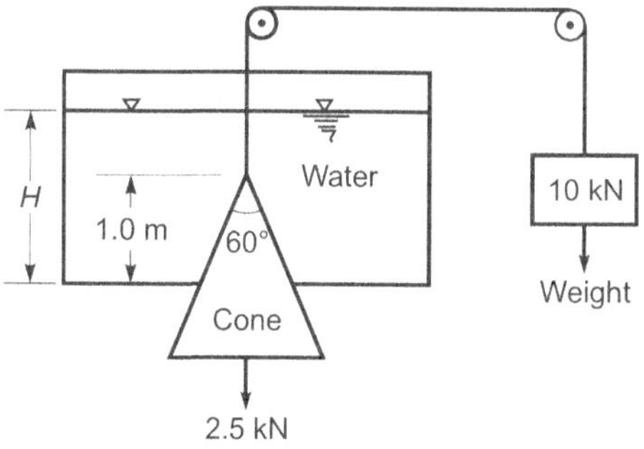

Fig. 2.90

2.42 A gravity dam has a cross section as shown in Fig. 2.91. It retains water to a depth of 20 m on the upstream. Calculate (i) the resultant force at the base per unit length of the dam, (ii) maximum and minimum stresses at the base. Assume specific weight of dam material to be 25 kN/m^3 and neglect all uplift pressures.

(Ans. (i) *R* = 5817 kN, α = inclination of *R* to the horizontal = 70.33°, (ii) σ_1 = 295.6 kPa on the upstream end of the base, σ_2 = 273.6 kPa on the downstream end of the base)

2.43 A rectangular water tank has its vertical sides joined to the horizontal bottom through a smooth curve which can be approximated to a quadrant of a circle of radius 1.6 m. If the tank contains 2.0 m of water and 2.5 m of oil relative density 0.8 over it, calculate the resultant force per metre length of the curved surface.

(Ans. *R* = 76.1 kN; α = inclination of *R* to the horizontal = 48.8°, the resultant passes through the centre of the arc.)

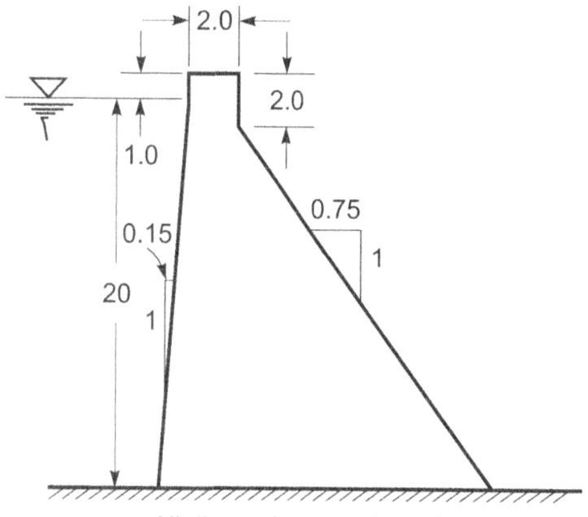

Fig. 2.91 *Problem 2.42*

2.44 A cylinder of 0.8 m diameter is made of 5 mm thick plates. If the maximum permissible stress in the plates is 100 MPa, calculate the maximum permissible pressure inside the cylinder.

(Ans. σ = 1.25 MPa)

2.45 A pressure vessel in the form of cylinder of 1.5 m diameter is to have a fluid under a pressure of 1.2 MPa. If the allowable stress in the material of the walls is 140 MPa find the minimum thickness of the walls.

(Ans. *t* = 6.5 mm)

2.46 A spherical vessel is made up of a material of permissible stress of 150 MPa. The diameter of the vessel is 15 m and the thickness of the sides is 5 mm. Find the maximum permissible pressure in this vessel.

(Ans. *p* = 2.0 MPa)

2.47 A pair of lock gates each of 3 m width make an angle of 120° when closed. The gates are supported by two hinges at 0.5 m and 4.5 m above the bottom. Determine the reaction force on each hinge when the lock has 4.8 m and 1.5 m of water on upstream and downstream respectively.

(Ans. R_{top} = 93.0 kN, R_{bottom} = 212.3 kN)

D. Buoyancy

2.48 A closed cylindrical tank of diameter 2.0 m, height 1.2 m and weighing 20 kN is floating with its axis vertical in sea water (relative density = 1.025).

(a) Find the depth of the cylinder below the water surface.

(b) What would be the depth of immersion if an additional load of 5.0 kN is added at the top?

(Ans. (a) h_1 = 0.6344 m, (b) h_2 = 0.793 m)

2.49 Two spheres each of 1.5 m diameter weigh 4.0 and 20.0 kN respectively. They are connected by a short rope and placed in a water reservoir. Find the tension in the rope and the percentage of volume of the lighter sphere that will be above water.

(Ans. T = 2.7 kN; percentage volume = 61.27%)

2.50 A metal sphere of volume V_m = 1 m^3, relative density S_m = 2 and fully immersed in water is attached by a flexible wire to a buoy of volume V_b = 1 m^3 and relative density S_b = 0.1 [See Fig. 2.92]. Calculate the tension T in the wire and volume of the buoy that is submerged.

(Ans. T = 1.96 kN, V_b = 0.2 m^3)

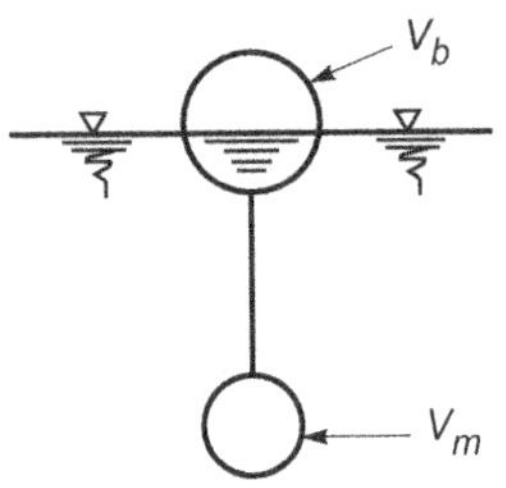

Fig. 2.92

2.51 An iceberg has a specific weight of 9.0 kN/m^3 and floats in sea water of specific weight 10.05 kN/m^3. What percentage of the total volume of the iceberg will be above the sea water surface?

(Ans. 10.5%)

2.52 A hydrometer is to be so built that the mark corresponding to relative density of 1.0 and 1.5 are 8 cm apart on a 5 mm diameter stem. How far from the mark of 1.0 will the relative density mark of 1.25 be located?

(Ans. h = 4.8 cm)

2.53 A hydrometer has a 6 mm diameter stem. The distance between markings of relative density 1.0 and 0.90 is 10 cm. Determine the weight of the hydrometer.

(Ans. W = 0.249 N)

2.54 A 10 cm cube of steel (relative density = 7.85) is to float on mercury in a container of square cross section, with a clearance of 1 cm all round. Find the weight of mercury (relative density = 13.6) required.

(Ans. W = 53 N)

2.55 An open cylindrical bucket 30 cm diameter and 50 cm long whose wall thickness and weight can be considered negligible is forced open end first into water until its lower edge is 10 m below the water surface. What force will be required to maintain this position? Assume that the trapped air undergoes any change under isothermal conditions. Atmospheric pressure = 100 kPa, temperature of water = 20°C.

(Ans. F = 177 N)

2.56 A pontoon of rectangular cross sectional area is 7.0 m long, 3.0 m wide and 1.5 m high. The depth of submergence of the pontoon is 0.9 m and its centre of gravity is 0.7 m above its bottom. Determine its metacentric height.

(Ans. MG = 0.583 m)

2.57 A solid cylinder of diameter 30 cm and height 1.20 m is to float in water with its axis vertical. Determine its stability if the relative density of the cylinder is 0.5.

(Ans. MG = – 0.2906 m; unstable)

2.58 A solid cylinder of diameter 30 cm and height 15 cm floats with its axis vertical in sea water (rel. den. = 1.03). If the relative density of the cylinder material is 0.9, examine the stability of the cylinder.

(Ans. MG = 3.34 cm; stable)

2.59 A cube of side a floats with one of its axes vertical in a liquid of relative density S_L. If the relative density of the cube material is S_c, find the condition for the metacentric height to be zero.

$$\left(Ans. \frac{S_L}{S_c} = 1.268 \text{ or } 4.732\right)$$

2.60 A solid cube of sides 0.5 m is made of a material of relative density 0.5. The cube floats in a liquid of relative density 0.95 with two of its faces horizontal. Examine its stability.

(Ans. MG = – 0.0393 m; unstable)

2.61 A rectangular barge of width b and depth of submergence H has its centre of gravity at the waterline. Find the metacentric height and the value of the ratio b/H for which the barge is stable.

$$\left(Ans. \text{MG} = \frac{b^2}{12H} - \frac{H}{2}, \text{ for stability } b/H \geq \sqrt{6}\right)$$

2.62 A solid cone with an apex angle of 60° and relative density S_c is floating in mercury (rel. den. = 13.6) with its vertex downwards and axis vertical. Determine the range of S_c over which the cone is in stable equilibrium.

(Ans. S_c > 5.738)

2.63 A cone of relative density 0.8 is to float in water with its axis vertical and vertex downwards. Find the least apex angle of the cone for stable equilibrium.

(Ans. $\theta = 31° \ 03' \ 33''$)

2.64 Show that, when a solid hemisphere of density ρ_s and radius r floats with its plane base immersed in a liquid of density ρ_f, the equilibrium is stable and the metacentric height is $\frac{3}{8} r \left(\frac{\rho_f}{\rho_s} - 1 \right)$.

E. Rigid Body Motion

2.65 An open tank is 7 m long, 2 m wide and 1.5 m deep. It contains oil of relative density 0.8 to a depth of 1.0 m. If the tank is given a horizontal acceleration at a constant value of 2.5 m/s^2 along its length, calculate the amount of oil spill. What are the pressures on the bottom of the tank at its front and rear end?

(Ans. Spill = 5.171 m^3, p_{front} = 0 (atmospheric), p_{rear} = 14.685 kPa)

2.66 An open tank 5 m long, 3 m wide and 2 m deep contains 1.5 m water. What is the maximum horizontal acceleration that can be given to the tank without causing spill over, and what is the alignment of the tank with respect to the movement?

(Ans. a_x = 3.27 along the width)

2.67 An open tank 3.0 m long, 2.0 m wide and 2.0 m deep contains water to a depth of 0.9 m. What minimum horizontal acceleration along the length should be given to have zero depth of the water along the front edge of the tank?

(Ans. a_x = 5.886 m/s^2)

2.68 A 2.5 m long open tank is mounted on a carriage which moves up a plane inclined at 35° to the horizontal at an acceleration of 2.0 m/s^2. What is the slope of the water surface? If the tank is 1.2 m deep and initially contains water to a height of 0.6 m, what would be the depths at the forward and rear edges of the tank?

(Ans. θ = 8.5° with lower depth in the front, y_{front} = 0.413 m, y_{rear} = 0.787 m)

2.69 A 30 cm diameter cylinder contains oil of specific weight 7.5 kN/m^3 to a height of 120 cm. Calculate the force on the bottom of the tank when the tank undergoes an acceleration of 3.5 m/s^2 (a) vertically downwards and (b) vertically upwards.

(Ans. (a) F_V = 0.863 kN, (b) F_V = 0.409 kN)

2.70 An open tank containing water slides down an inclined plane without friction. Show that the water surface will be (a) parallel to the plane if the acceleration is equal to the component of g along the inclined plane, (b) horizontal if the velocity of slide is constant.

2.71 A closed tank shown in Fig. 2.93 is given an acceleration of 3.6 m/s^2, the accelerations vector being inclined at an angle of 35° to the horizontal. Calculate the pressures at point M, N and R.

Ans. (P_M = 10.406 kPa, P_N = 25.73 kPa; P_R = 3.703 kPa)

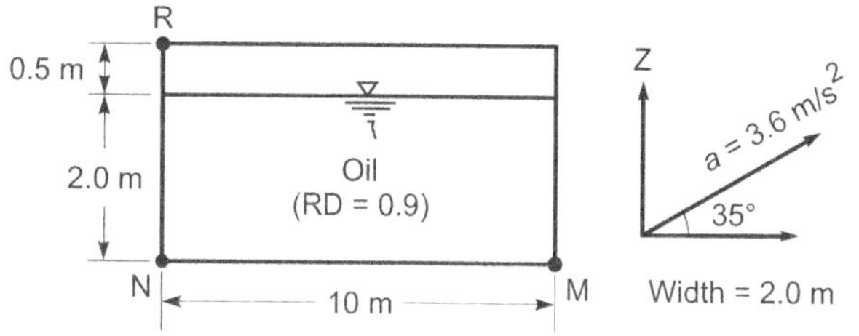

Fig. 2.93

2.72 An open cylindrical tank of 30 cm diameter is 40 cm high. The tank is filled with water to a depth of 30 cm. If the tank is rotated about the vertical axis of the cylinder, find the maximum speed of rotation that does not cause any spill of the liquid.

(Ans. N = 126 rpm)

2.73 A 40 cm diameter cylindrical tank is 35 cm high and is open at the top. Initially it contains water to a depth of 20 cm. If the tank is rotated about its vertical axis at 120 rpm, calculate the amount of liquid spilled out.

(Ans. V_s = 0.34 L)

2.74 If the tank in Problem 2.73 is rotated at 150 rpm, calculate the volume of water spilled out.

(Ans. V_s = 7.01 L)

2.75 A closed cylinder 40 cm in diameter and 40 cm in height is filled with oil of relative density 0.80. If the cylinder is rotated about its vertical axis at a speed of 200 rpm, calculate the thrust of oil on top as well as bottom of the cylinder.

(Ans. F_T = 440 N, F_B = 834 N)

2.76 A hemispherical bowl of radius 1.0 m is full of water and is to be rotated about its vertical axis at 30 rpm. Estimate the amount of water that will overflow.

(Ans. 0.790 m^3/s)

2.77 A U-tube has a liquid of relative density 0.85 in its limbs to a height of 50 cm above the horizontal limb of 30 cm length (Fig. 2.94). What will be the difference in elevation of the two free surfaces when the tube is rotated about a vertical axis 10 cm from one leg and 20 cm from the other, at 180 rpm?

(Ans. $(y_2 - y_1) = 54.3$ cm)

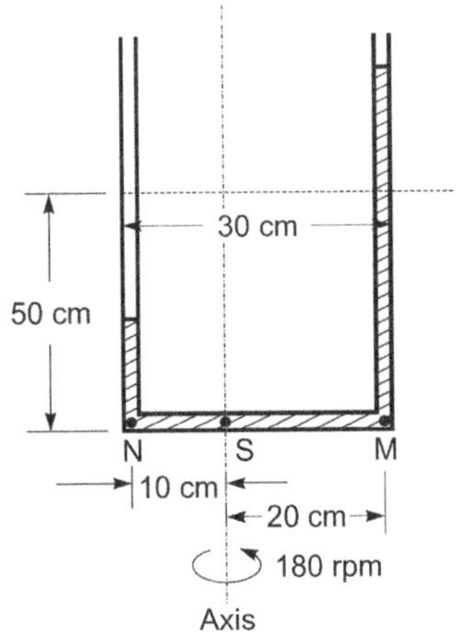

Fig. 2.94

2.78 For the U-tube of Problem 2.77, under the given conditions, determine the pressures at points N, S and M.

(Ans. $p_N = 1.79$ kPa, $p_S = 0.372$ kPa, $p_M = 6.042$ kPa)

2.79 A 20 cm high cylinder open at one end contains 10 cm of water and its top is connected to a 1.0 m lever arm. If the cylinder is rotated in the vertical plane at an angular velocity of 5 rad./s, calculate the pressure at the bottom of the cylinder when (a) it is at its highest point and (b) at its lowest point in the rotation. (Use an average radial distance of 1.15 m in calculating accelerations).

(Ans. (a) $p_b = 1890$ Pa, (b) $p_b = 3848$ Pa)

2.80 At what angular velocity must a U-tube, of 30 cm horizontal limb length and filled with water to a depth of 30 cm in the vertical limbs, be rotated about a vertical axis at mid distance from the vertical limbs, to cause cavitation at the point of intersection of the axis with the horizontal limb? The vapour pressure of water can be taken as 2.50 kPa (abs) and the atmospheric pressure as 95.5 kPa (abs). The density of water is 998 kg/m^3.

(Ans. $N = 869$ rpm)

(***Hint:*** Cavitation in water occurs when the local pressure reaches the vapour pressure)

2.81 For the U-tube containing mercury, shown in Fig. 2.95 what speed of rotation causes the differences in limb heights as indicated?

(Ans. $N = 48.8$ rpm)

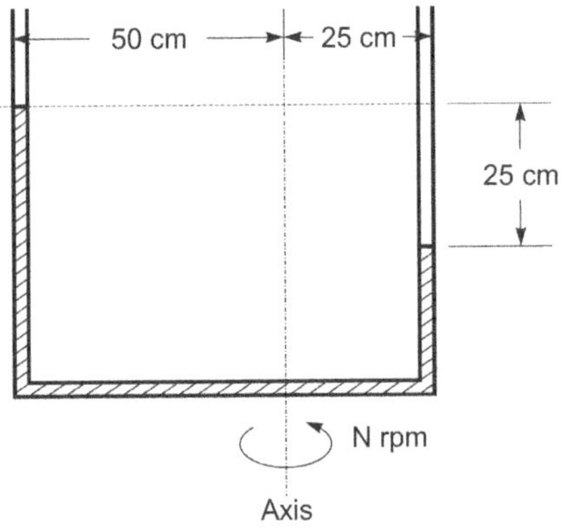

Fig. 2.95

3

Fluid Flow Kinematics

CONCEPT REVIEW

3.1 INTRODUCTION

The kinematics of fluid flow deals with the velocity, acceleration and other related aspects of space-time relations without specifically considering the associated forces. Thus the basic definitions and classification, aspects dealing with velocities and accelerations, continuity equation, stream function and potential functions to describe the fluid flow are dealt with in this chapter. The ideas of kinematics are essential in the successful application of energy and momentum equations (described in subsequent chapters) to solve various fluid flow problems. It is to be remembered that velocity and acceleration are vectors and hence have magnitude as well as direction.

To visualize the flow of fluids the following patterns generated by lines are used:

- A *stream line* as a line tangent to the velocity vector at every point in a given instant.
- A *path line* is the actual path traversed by a given fluid particle
- A *streak line* is a locus of particles which had earlier passed through a chosen point
- A *time line* is a set of fluid particles that form a line at a given instant.

Only stream lines are convenient for mathematical calculations. Other lines are used for flow visualizations.

3.2 CLASSIFICATION OF FLOW

(A) *Steady flow:* Fluid flow conditions at any point do not change with time. For example

$$\boxed{\frac{\partial \mathbf{V}}{\partial t} = 0, \frac{\partial p}{\partial t} = 0, \frac{\partial \rho}{\partial t} = 0}$$

In a steady flow steam line, path line and streak line are identical.

Unsteady flow: Flow parameters at any point change with time, e.g. $\frac{\partial \mathbf{V}}{\partial t} \neq 0$.

(B) *Uniform flow*: The velocity vector **V** is identically same at all points at a given instant. Thus

$$\boxed{\frac{\partial \mathbf{V}}{\partial t} = 0}$$

Non-uniform flow: The velocity vector **V** at any instant varies from point to point.

3.3 STREAMLINE

In a fluid flow, a continuous line so drawn that it is tangential to the velocity vector at every point is known as a streamline. If the velocity vector $\mathbf{V} = \mathbf{i}u + \mathbf{j}v + \mathbf{k}w$ then the differential equation of a streamline is given by

$$\boxed{\frac{dx}{u} = \frac{dy}{v} = \frac{dz}{w}} \tag{3.1}$$

3.4 ACCELERATION

Acceleration is a vector.

(i) In the natural co-ordinate system, viz. along and across a streamline (Fig. 3.1).

$$a = \frac{dV}{dt} \quad \text{and} \quad a = \sqrt{a_s^2 + a_n^2} \tag{3.2}$$

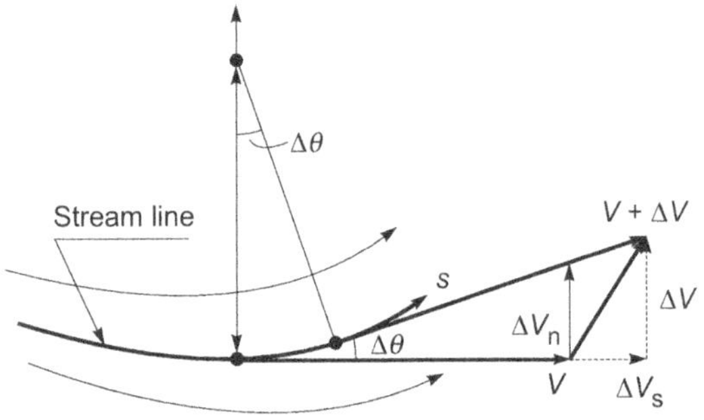

Fig. 3.1

In the tangential direction:

$$a_s = \frac{\partial V_s}{\partial t} + V\frac{\partial V_s}{\partial s} \tag{3.3}$$

In the normal direction

$$a_n = \frac{\partial V_n}{\partial t} + \frac{V_s^2}{r} \tag{3.4}$$

where r = radius of curvature of the streamline at the point, V_s = tangential component of the velocity V and V_n = normal component of velocity generated due to change in direction. The terms $\partial V_s/\partial t$ and $\partial V_n/\partial t$ are called *local accelerations.* Also $V(\partial V_s/\partial_s)$ = *tangential convective acceleration* and V^2/r = *normal convective acceleration.*

(ii) In Cartesian co-ordinates:

$$\mathbf{V} = \mathbf{i}u + \mathbf{j}v + \mathbf{k}w$$

Acceleration a_x, a_y and a_z in the x, y, z directions respectively are:

$$a_x = \frac{\partial u}{\partial t} + u\frac{\partial u}{\partial x} + v\frac{\partial u}{\partial y} + w\frac{\partial u}{\partial z} \tag{3.5}$$

$$a_y = \frac{\partial v}{\partial t} + u\frac{\partial v}{\partial x} + v\frac{\partial v}{\partial y} + w\frac{\partial v}{\partial z} \tag{3.6}$$

$$a_z = \frac{\partial w}{\partial t} + u\frac{\partial w}{\partial x} + v\frac{\partial w}{\partial y} + w\frac{\partial w}{\partial z} \tag{3.7}$$

(iii) In two-dimensional polar co-ordinates (Fig. 3.2)

$$a_r = \frac{\partial v_r}{\partial t} + v_r\frac{\partial v_r}{\partial r} + \frac{v_\theta}{r}\frac{\partial v_r}{\partial \theta} - \frac{v^2}{r} \tag{3.8}$$

$$a_\theta = \frac{\partial v_\theta}{\partial t} + v_r\frac{\partial v_\theta}{\partial r} + \frac{v_\theta}{r}\frac{\partial v_\theta}{\partial \theta} - \frac{v_r v_\theta}{r} \tag{3.9}$$

3.5 CONTINUITY EQUATION

3.5.1 In One-dimensional Analysis

In steady flow, mass rate of flow into stream tube is equal to mass rate of flow out of the tube

$$\rho_1 A_1 V_1 = \rho_2 A_2 V_2 \tag{3.10}$$

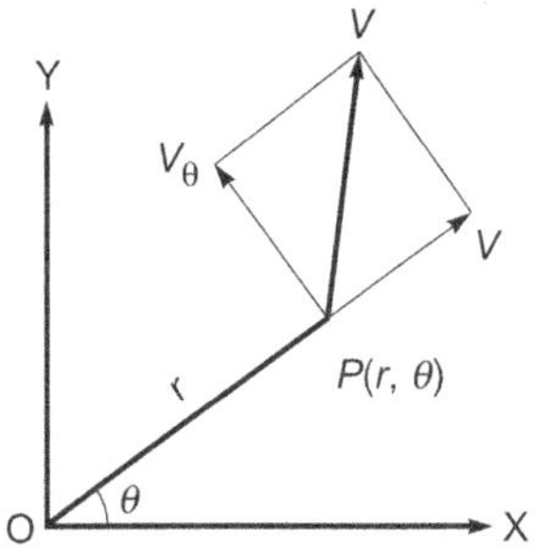

Fig. 3.2

(i) For incompressible fluid, under steady flow (Fig. 3.3).

$$\boxed{A_1V_1 = A_2V_2} \tag{3.11}$$

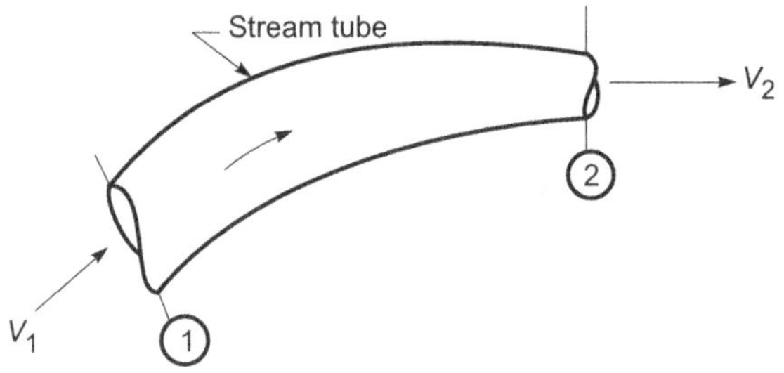

Fig. 3.3

(ii) When there is a variation of velocity across the cross section of a tube, for an incompressible fluid discharge, (Fig. 3.4)

$$\boxed{Q = \int_{A_1} v\,dA = \int_{A_2} v\,dA} \tag{3.12}$$

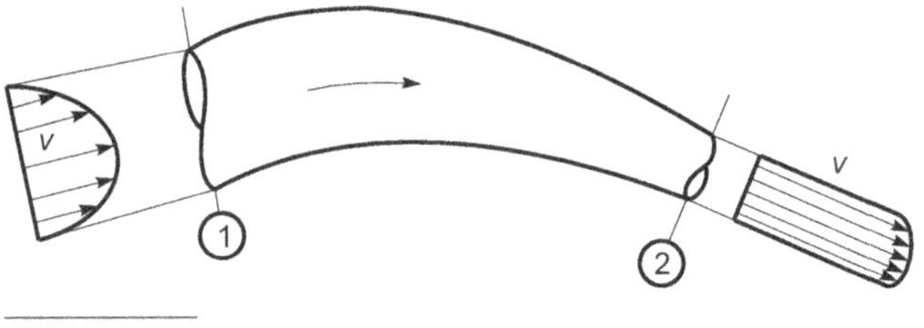

Fig. 3.4

3.5.2 In Differential Form

Cartesian co-ordinates:

$$\frac{\partial \rho}{\partial t} + \frac{\partial(\rho u)}{\partial x} + \frac{\partial(\rho v)}{\partial y} + \frac{\partial(\rho w)}{\partial z} = 0 \tag{3.13}$$

For incompressible fluid $(d\rho/dt) = 0$) and hence Eq. 3.13 is simplified as

$$\boxed{\frac{\partial u}{\partial x} + \frac{\partial v}{\partial y} + \frac{\partial w}{\partial z} = 0} \tag{3.14}$$

3.6 ROTATIONAL AND IRROTATIONAL MOTION

Consider a rectangular fluid element of sides dx and dy [(Fig. 3.5(a)]. Under the action of velocities acting on it let it undergo deformation as shown in Fig. 3.5(b) in a time dt.

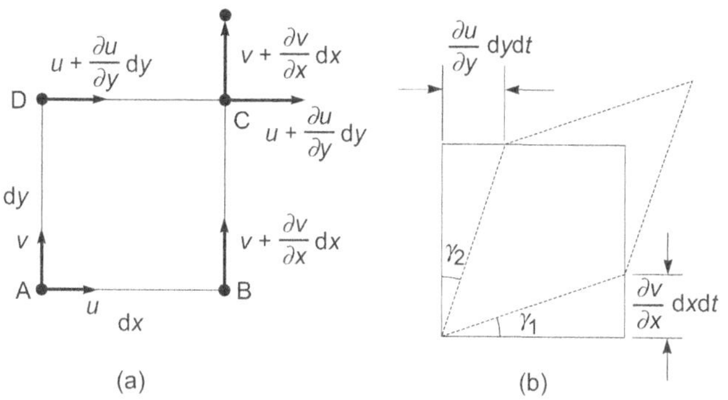

Fig. 3.5

$$\gamma_1 = \text{angular velocity of element AB} = \frac{\partial v}{\partial x}$$

$$\gamma_2 = \text{angular velocity of element AD} = \frac{\partial u}{\partial y}$$

Considering the anticlockwise rotation as positive, the average of angular velocities of two mutually perpendicular elements is defined as the rate of rotation.

Thus rotation about z-axis

$$\omega_z = \frac{1}{2}\left(\frac{\partial v}{\partial x} - \frac{\partial u}{\partial y}\right) \quad (3.15)$$

Thus for a three-dimensional fluid element, three rotational components as given in the following are possible:

$$\left.\begin{aligned} \text{About z axis, } \omega_z &= \frac{1}{2}\left(\frac{\partial v}{\partial x} - \frac{\partial u}{\partial y}\right) \\ \text{About y axis, } \omega_y &= \frac{1}{2}\left(\frac{\partial u}{\partial z} - \frac{\partial w}{\partial x}\right) \\ \text{About x axis, } \omega_x &= \frac{1}{2}\left(\frac{\partial w}{\partial y} - \frac{\partial v}{\partial z}\right) \end{aligned}\right\} \quad (3.16)$$

Fluid motion with one or more of the terms ω_x, ω_y or ω_2 different from zero is termed *rotational motion*.

Twice the value of rotation about any axis is called as *vorticity* along that axis. Thus the equation for vorticity along z-axis is $\zeta_z = 2w_z = \left(\frac{\partial v}{\partial x} - \frac{\partial u}{\partial y}\right)$.

A flow is said to be *irrotational* if all the components of rotation are zero,

viz. $\qquad \omega_x = \omega_y = \omega_z = 0,$

i.e. $\qquad \frac{\partial v}{\partial x} - \frac{\partial u}{\partial y} = 0, \; \frac{\partial u}{\partial z} - \frac{\partial w}{\partial x} = 0;$

and $\qquad \frac{\partial w}{\partial y} - \frac{\partial v}{\partial z} = 0$

Thus for a two-dimensional irrotational flow

$$\omega_z = \frac{1}{2}\left(\frac{\partial v}{\partial x} - \frac{\partial u}{\partial y}\right) = 0$$

or

$$\left(\frac{\partial v}{\partial x} - \frac{\partial u}{\partial y}\right) = 0 \quad (3.17)$$

3.6.1 Circulation

In rotational fluid motion, circulation is very useful concept. Circulation is defined as the line integral of the tangential component of the velocity taken around a closed contour (Fig. 3.6). The limiting value of circulation divided by the area of the closed contour, as the area tends to zero, is the vorticity along an axis normal to the area.

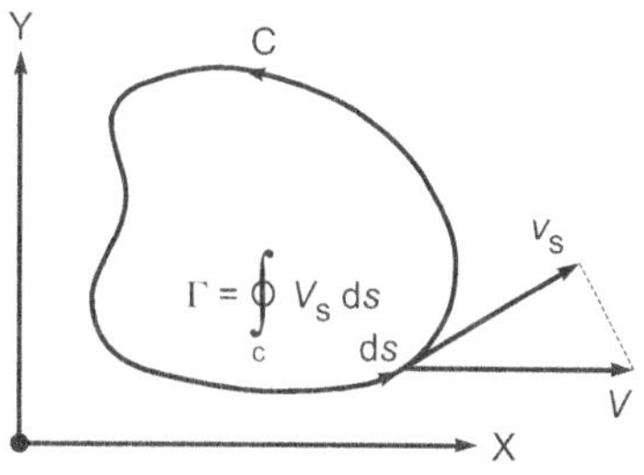

Fig. 3.6 *Circulation Concept*

3.7 STREAM FUNCTION

In a two-dimensional flow consider two streamlines S_1 and S_2. The flow rate (per unit depth) of an incompressible fluid across the two streamlines is constant and is independent of the path, (path *a* or path *b* from A to B in Fig. 3.7). A stream function ψ is so defined that it is constant along a streamline and the difference of ψs for the two streamlines is equal to the flow rate between them. Thus $\psi_B - \psi_A$ = flow rate between S_1 and S_2. The flow left to right is taken as positive, in the sign convention. The velocities u and v in x and y directions are given by

$$u = \frac{\partial \psi}{\partial y} \quad \text{and} \quad v = -\frac{\partial \psi}{\partial x} \quad (3.18)$$

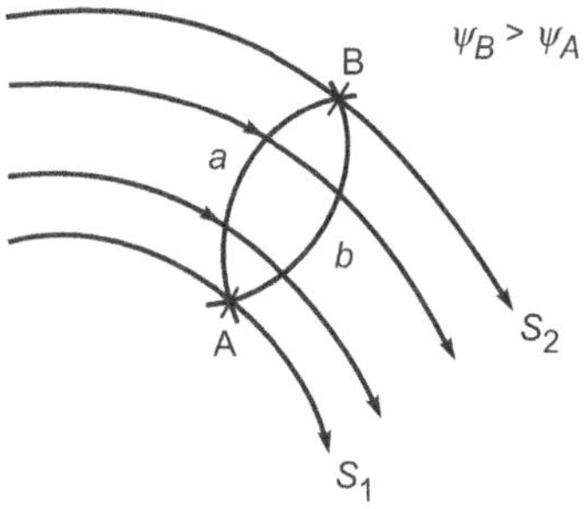

Fig. 3.7

The stream function ψ is defined as above for two dimensional flows only.

For an irrotational flow, $\frac{\partial v}{\partial x} - \frac{\partial u}{\partial y} = 0$

and hence, $-\frac{\partial^2 \psi}{\partial x^2} - \frac{\partial^2 \psi}{\partial y^2} = 0$

That is, the Laplace equation $\frac{\partial^2 \psi}{\partial x^2} + \frac{\partial^2 \psi}{\partial y^2} = 0$ (3.19)

is satisfied by the stream function in irrotational flow. Conversely, if ψ does not satisfy $\nabla^2 \psi = 0$, then the flow is rotational.

In polar coordinates $v_r = \frac{1}{r}\frac{\partial \psi}{\partial \theta}$ and $v_\theta = -\frac{\partial \psi}{\partial r}$

3.8 POTENTIAL FUNCTION

In irrotational flows, the velocity can be written as a gradient of a scalar function ϕ called velocity potential.

$$u = \frac{\partial \phi}{\partial x},\ v = \frac{\partial \phi}{\partial y} \text{ and } w = \frac{\partial \phi}{\partial z} \tag{3.20}$$

Considering the equation of continuity (Eq. 3.14) for an incompressible fluid,

$$\frac{\partial u}{\partial x} + \frac{\partial v}{\partial y} + \frac{\partial w}{\partial z} = 0$$

and substituting the expressions for u, v and w in terms of ϕ,

$$\nabla^2 \phi = \frac{\partial^2 \phi}{\partial x^2} + \frac{\partial^2 \phi}{\partial y^2} + \frac{\partial^2 \phi}{\partial z^2} = 0 \tag{3.21}$$

Thus the velocity potential satisfies the Laplace equation. Conversely, any function ϕ which satisfies the Laplace equation is a possible irrotational fluid flow case.
Lines of constant ϕ are called *equipotential lines* and it can be shown that these lines will form orthogonal grids with ψ = constant lines. This fact is used in the construction of flow nets for fluid flow analysis.

[***Note:*** Some authors define ϕ such that

$$u = -\frac{\partial \phi}{\partial x},\ v = -\frac{\partial \phi}{\partial y} \text{ and } w = -\frac{\partial \phi}{\partial z}$$]

3.9 RELATION BETWEEN Ψ AND ϕ FOR 2-DIMENSIONAL FLOW

- ϕ exists for irrotational flow only.
- $u = \frac{\partial \phi}{\partial x} = \frac{\partial \psi}{\partial y}$ (3.22)
- $v = \frac{\partial \phi}{\partial y} = -\frac{\partial \psi}{\partial x}$ (3.23)
- By continuity equation,
- $\frac{\partial^2 \phi}{\partial x^2} + \frac{\partial^2 \phi}{\partial y^2} = 0$ (3.21(a))
- By irrotational flow condition,
- $\frac{\partial^2 \psi}{\partial x^2} + \frac{\partial^2 \psi}{\partial y^2} = 0$ (3.19)
- ψ = constant along a streamline.
- ϕ = constant along an equipotential line which is normal to streamlines.

3.10 SOME COMMON FORMULAE IN CYLINDRICAL CO-ORDINATES

1. Equation of continuity:

$$\frac{1}{r}(\rho V_r) + \frac{\partial}{\partial r}(\rho V_r) + \frac{1}{r}\frac{\partial}{\partial \theta}(\rho V_\theta) = 0 \tag{3.24}$$

For incompressible fluid flow:

$$\frac{V_r}{r} + \frac{\partial V_r}{\partial r} + \frac{1}{r}\frac{\partial V_\theta}{\partial \theta} = 0 \tag{3.25}$$

2. Stream function ψ:

$$V_r = \frac{1}{r}\frac{\partial \psi}{\partial \theta} \tag{3.26}$$

$$V_\theta = -\frac{\partial \psi}{\partial r} \tag{3.27}$$

3. Potential function ϕ:

$$V_r = \frac{\partial \phi}{\partial r} \tag{3.28}$$

$$V_\theta = \frac{1}{r}\frac{\partial \phi}{\partial \theta} \tag{3.29}$$

4. Laplace equation $\frac{1}{r}\frac{\partial \phi}{\partial r} + \frac{\partial^2 \phi}{\partial r^2} + \frac{1}{r^2}\frac{\partial^2 \phi}{\partial \theta^2} = 0$ (3.30)

3.11 ELEMENTARY INVISCID PLANE FLOWS

Since the Laplace equation is linear, several interesting potential flow situations can be constructed by using elementary solutions and method of superposition. The basic flow types are Uniform flow, Source, Sink and Vortex. These are briefly described below.

3.11.1 Uniform Flow

A stream of constant velocity U in x-direction is shown in Fig. 3.8 and has

$$\psi = Uy \quad \text{and} \quad \phi = Ux \tag{3.31}$$

In polar coordinates $\psi = U\, r \sin\theta$ and $\phi = U\, r \cos\theta$ (3.32)

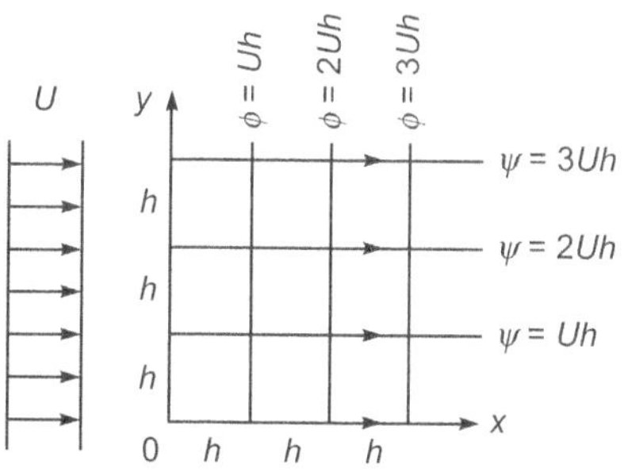

Fig. 3.8 *Uniform Flow*

3.11.2 Line Source and Sink

A two-dimensional flow emanating from a point in the x-y plane and imagined to flow uniformly in all directions is called a *source*. Since the two-dimensional source is a line in the z-direction, it is known as a *line source*.

The total flow per unit time per unit length of the line source is called the *strength m* of the source. The velocity at a radial distance r from the source is

$$v_r = \frac{m}{r} \tag{3.33}$$

The stream function ψ and the potential function ϕ for such line source is given by

$$\psi = m\theta \quad \text{and} \quad \phi = m \ln r \tag{3.34}$$

Figure 3.9 shows a source flow.

A sink is a negative source. It is a line in z-direction into which fluid flows radially in x-y plane (Fig. 3.10).

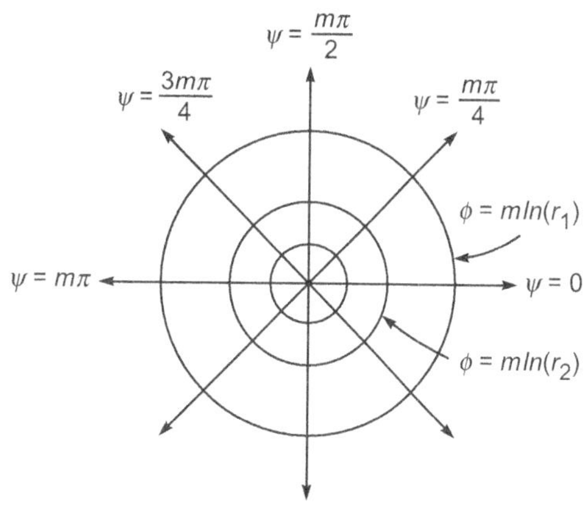

Fig. 3.9 *Line Source*

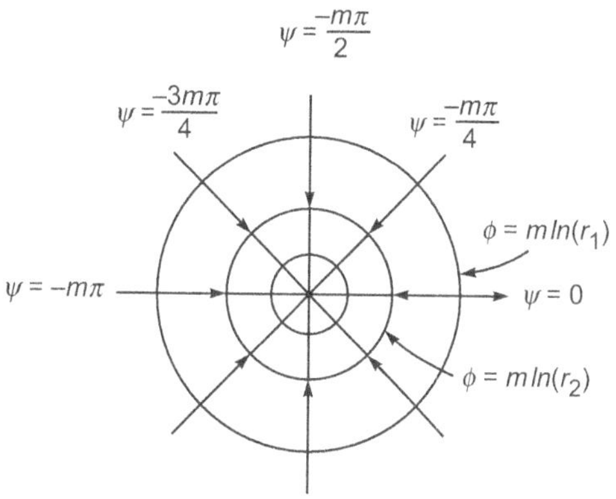

Fig. 3.10 *Line Sink*

3.11.3 Line Vortex

Suppose we reverse the role of ψ and ϕ in Fig. 3.9 yielding

$$\psi = -K \ln r \quad \text{and} \quad \phi = K\theta \tag{3.35}$$

from which we get $v_r = 0$ and $v_0 = K/r$ representing a circulating flow (Fig. 3.11). Such a flow is known as line *vortex* and K in Eq. 3.35 is known as *Vortex strength*. The centre of the vortex is a singular point and the circulation Γ of the vortex around a circular path about the centre is given by

$$\Gamma = 2\pi K \tag{3.36}$$

3.11.4 Two Dimensional Doublet

The limiting case of a line source approaching a line sink of equal strength while keeping constant the product of their strength and the distance between them (λ) is known as a two dimensional *doublet*. For a doublet.

$$\psi_{\text{doublet}} - \frac{(\lambda y)}{(x^2 + y^2)} = -\frac{\lambda \sin\theta}{r} \tag{3.37}$$

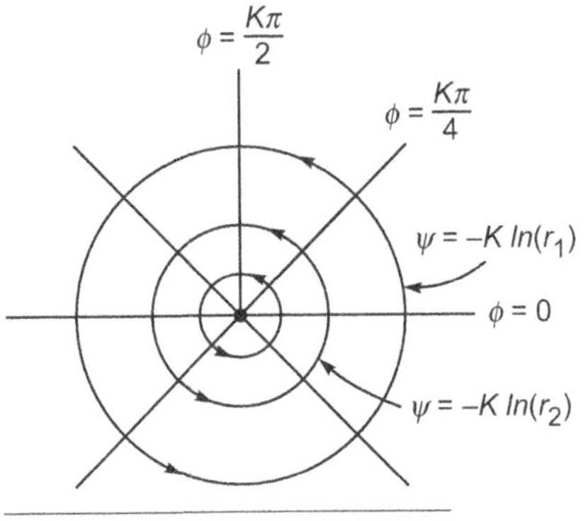

Fig. 3.11 *Line Vortex*

$$\phi_{doublet} = \frac{\lambda x}{(x^2 + y^2)} = \frac{\lambda \cos\theta}{r} \tag{3.38}$$

Figure 3.12 shows the streamlines and equipotential lines in a doublet.

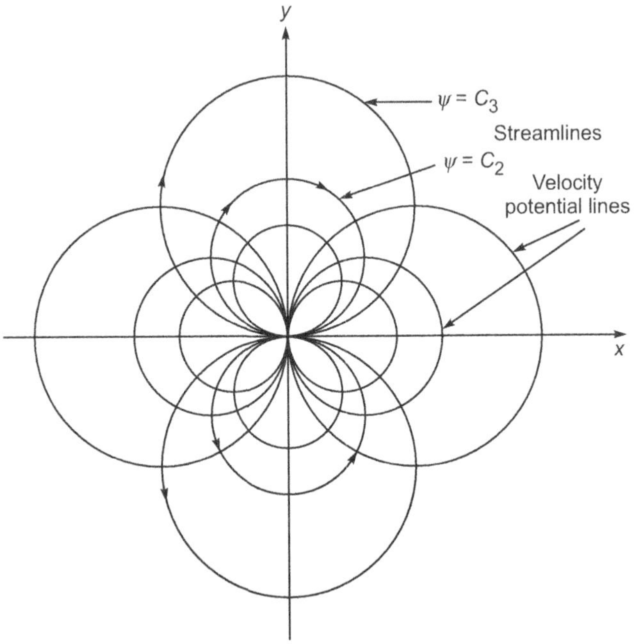

Fig. 3.12 *Doublet*

3.11.5 Other Inviscid Flows

Using the basic flow elements described above various flow situations can be created by the method of superposition. A few examples are given below in Table 3.1.

Table 3.1 *Some Ideal Fluid Flow Simulations*

Sl. No.	*Name*	*Combination (and Flow description)*	*Equation of Stream function*
1.	Rankine Half body	Source + uniform flow [curved, roughly elliptical half body]	$\psi = Ur \sin\theta + m\theta$
2.	Rankine oval	Source + sink + Uniform flow [cylindrical oval shaped body]	$\psi = Ur \sin\theta + m(\theta_1 - \theta_2)$
3.	Circular cylinder	Uniform flow + doublet [circular cylinder]	$\psi = Ur \sin\theta - \frac{\lambda \sin\theta}{r}$
4.	Rotating circular cylinder	Uniform flow + doublet + vortex [rotating circular cylinder]	$\psi = Ur \sin\theta - \frac{\lambda \sin\theta}{r} - K \ln r$

WORKED EXAMPLES

3.1 A circular pipe 10 cm in diameter has a 2 m length which is porous. In this porous section the velocity of exit is known to be constant (Fig. 3.13). If the velocities at inlet and outlet of the porous section are 2.0 m/s and 1.2 m/s respectively, estimate (i) the discharge emitted out through the walls of the porous pipe and (ii) the average velocity of this emitted discharge.

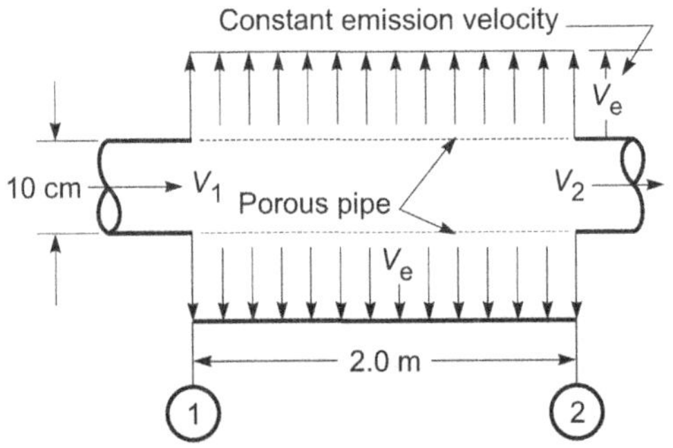

Fig. 3.13

Solution: A = area of pipe cross section

$$= \frac{\pi}{4} \times (0.1)^2 = 7.854 \times 10^{-3} \text{ m}^2$$

Q_1 = inlet discharge = $V_1 A$

$$= 2.0 \times 7.854 \times 10^{-3} = 0.01571 \text{ m}^3\text{/s}$$

Q_2 = outlet discharge = $V_2 A$

$$= 1.2 \times 7.854 \times 10^{-3} = 0.0094248 \text{ m}^3\text{/s}$$

(i) Q_e = discharge emitted through walls of the porous pipe

$$= Q_1 - Q_2 = 0.01571 - 0.0094248$$

$$= \mathbf{0.0052852\ m^3/s}$$

(ii) Surface area of emission $A_e = \pi DL$

$$= \pi \times 0.1 \times 2.0 = 0.6283 \text{ m}^2$$

$$V_e = \text{Velocity of emission} = \frac{Q_e}{A_e} = \frac{6.283 \times 10^{-3}}{0.6283}$$

$$= \mathbf{0.01\ m/s}$$

3.2 The axial velocity profile in a pipe is given by

$$u = u_m \left(1 - \frac{r}{R}\right)^n$$

where R = radius of the pipe and u_m = maximum velocity at $r = R$. Calculate the average velocity for $n = 1/5$ and $n = 1/2$.

Solution: Refer to Fig. 3.14

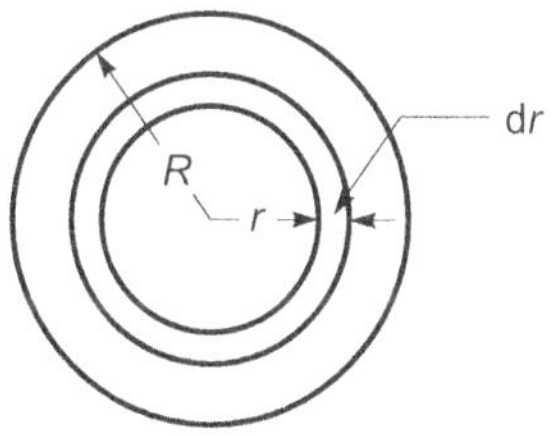

Fig. 3.14

If average velocity = V

Then

$$V.\pi R^2 = Q = \int_0^R u.2\pi r \cdot dr$$

$$V = \int_0^R \frac{u_m \cdot 2\pi r}{\pi R^2}\left(1 - \frac{r}{R}\right)^n dr$$

$$= 2u_m \int_{r/R=0}^{r/R=1} \frac{r}{R}\left(1 - \frac{r}{R}\right)^n d\left(\frac{r}{R}\right) = 2u_m$$

$$\left[-\frac{1}{(n+1)(n+2)}\left(1 - \frac{r}{R}\right)^{n+2} - \frac{1}{(n+1)}\frac{r}{R}\left(1 - \frac{r}{R}\right)^{n+1}\right]_0^1$$

$$= 2u_m \left[\frac{1}{(n+1)} - \frac{1}{(n+2)}\right]$$

when $n = 1/5$, $V = 2u_m \left(\frac{1}{1/5 + 1} - \frac{1}{1/5 + 2}\right)$

$$= 2u_m \times \frac{25}{66} = \mathbf{\frac{25}{33}\, u_m}$$

when $n = 1/2$, $V = 2u_m \left(\frac{1}{1/2 + 1} - \frac{1}{1/2 + 2}\right)$

$$= 2u_m \times 4/15 = \mathbf{8/15\, u_m}$$

3.3 A centrifugal pump has an axial inlet of 12 cm diameter and an impeller of 30 cm diameter. The width of the impeller at the outlet is 25 mm (Fig. 3.15). 5% of the outlet area can be assumed to be occupied by the blades. For a flow of 80 L/s, estimate the radial component of velocity at the outlet of the impeller. What is the axial velocity in the inlet pipe?

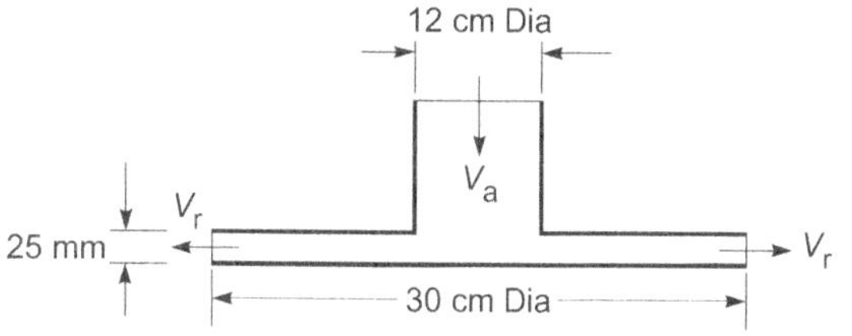

Fig. 3.15

Solution:

Discharge $Q = 0.08 \text{ m}^3\text{/s}$

At the outlet

Area $A = \pi \times 0.30 \times 0.025 \times 0.95$

$$= 0.02238 \text{ m}^2\text{/s}$$

Radial velocity at the edge of the impeller

$$= Q/A = 0.08/0.02238 = \mathbf{3.575\ m/s}$$

At the inlet, $V_a \cdot A_a = 0.08$

where V_a is the axial velocity, at the inlet pipe and A_a is the area of the inlet.

$$V_a = \frac{0.08}{\frac{\pi}{4}(0.12)^2} = \mathbf{7.074\ m/s}$$

3.4 Two large circular plates are kept at a distance y_0 apart and contain an incompressible flow in between. If the bottom plate is fixed and the top plate is moved downwards at a constant rate of V_0, (Fig. 3.16) estimate the velocity at which the fluid moves at a radial distance of r. What is the corresponding acceleration?

Solution: In time Δt, at a radial distance r, volume displaced by the upper plate

$$= V_0 \times \pi r^2$$

$$= \text{Radial out flow} = V_r \times 2\pi r \times y_0$$

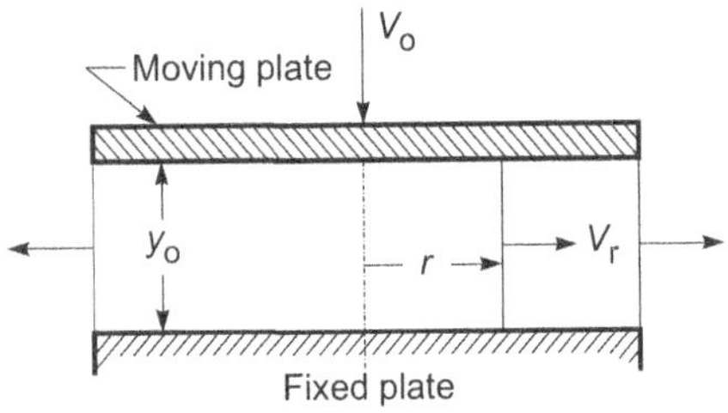

Fig. 3.16 *Example 3.4*

where V_r is the fluid velocity at a radial distance r.

$$V_r = \frac{V_0 \pi r^2}{2\pi r y_0} = \frac{V_0 r}{2 y_0}$$

As the flow is radially outwards, acceleration

$$a_r = V_r \frac{\partial V_r}{\partial r} = \frac{V_0 r}{2y_0}\left(\frac{V_0}{2y_0}\right) = \frac{V_0^2 r}{4y_0^2}$$

3.5 In a steady laminar boundary layer over a flat plate the velocity distribution is given by $\frac{u}{U_0} = 2\eta - 2\eta^3 + \eta^4$ where $\eta = y/\delta$ and δ = thickness of the boundary layer at a distance x from the leading edge of the plate (Fig. 3.17). Obtain an expression for the mass rate of outflow per unit width across an imaginary plane AB shown in the figure.

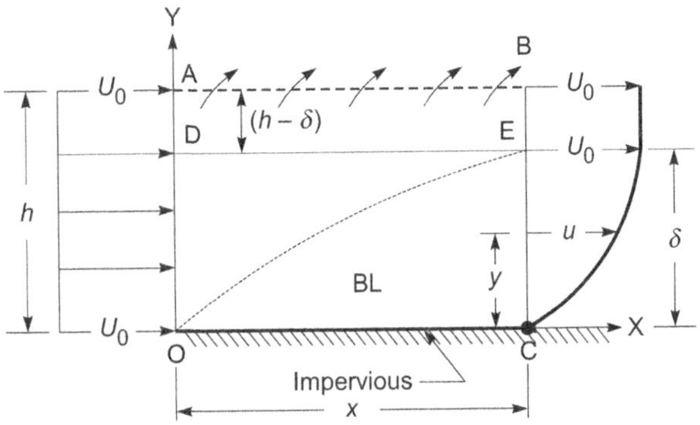

Fig. 3.17

Solution: By continuity equation for a volume ABCO: Mass rate of flow across AO – Mass rate across BC = Mass rate of flow across AB.

Since across section BC, $u = U_0$ for $y \geq \delta$,

the mass rate across AD = $\rho U_0(h - \delta)$ = mass rate across BE = $\rho U_0(h - \delta)$.

Hence, per unit width

$$\rho U_0 \delta - \int_0^\delta \rho u\, dy = \rho Q_{AB} = \text{mass rate across AB.}$$

$$\rho Q_{AB} = \rho U_0 \delta - \rho \int_0^\delta U_0 \left(2\frac{y}{\delta} - 2\left(\frac{y}{\delta}\right)^3 + (y/\delta)^4\right) dy$$

$$= \rho U_0 \left[\delta - \frac{2}{\delta}\frac{\delta^2}{2} + \frac{2}{\delta^3}\frac{\delta^4}{4} - \frac{1}{5}\frac{\delta^5}{\delta^4}\right]$$

$$= \rho U_0 \delta \left(1 - 1 + \frac{2}{4} - \frac{1}{5}\right) = \frac{\mathbf{3}}{\mathbf{10}} \boldsymbol{\rho U_0 \delta}$$

= Mass rate of outflow across AB per unit width.

3.6 Show that in an unsteady flow in a prismatic open channel the continuity equation is given by

$$\frac{\partial Q}{\partial x} + T\frac{\partial y}{\partial t} = 0$$

where T = top width of the channel section and Q = discharge at any instant t.

Solution: Refer to Fig. 3.18.

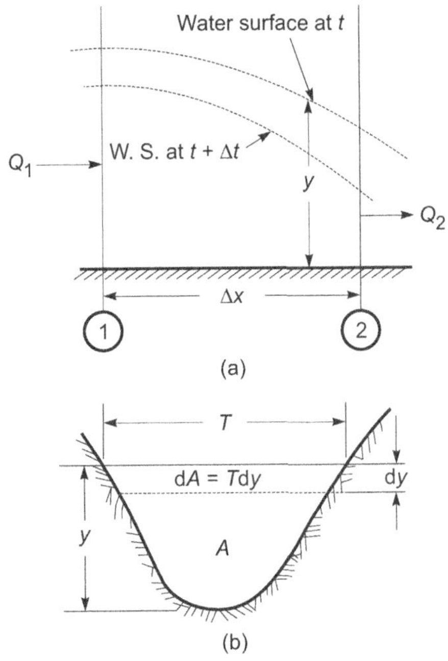

Fig. 3.18

Consider section 1 and 2 Δx apart. Let $Q_2 > Q_1$ at any instant t. Then

$$Q_2 - Q_1 = \frac{\partial Q}{\partial x} \cdot \Delta x$$

In a time interval Δt, volume rate of excess outflow over the inflow = $\frac{\partial Q}{\partial x} \Delta x \Delta t$. If the top width of the channel is T at a depth y, the water surface will drop by $\Delta y = \frac{\partial y}{\partial t} \Delta t$ and the decrease in storage between sections 1 and 2 is

$$-\Delta S = -\Delta A \times \Delta x = -T\Delta y\, \Delta x = -T\frac{\partial y}{\partial t} \Delta t\, \Delta x$$

By continuity consideration, $\frac{\partial Q}{\partial x} \Delta x \, \Delta t = - T \frac{\partial y}{\partial t} \Delta t \, \Delta x$

or
$$\frac{\partial Q}{\partial x} + T \frac{\partial y}{\partial t} = 0$$

3.7 Figure 3.19(a) shows a pipe network with junctions (nodes) at A, B, C, D, E and F. The numerals in the figure indicate the discharges at the nodes or in the pipes as the case is and the arrows indicate the directions of flows. By continuity equation determine the missing discharge values and their direction in pipes AB, BC, CD, BE and EF and at the node F.

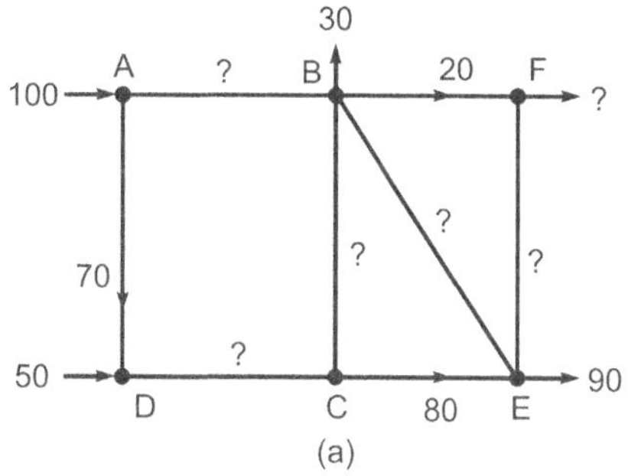

Fig. 3.19(a) *Example 3.7*

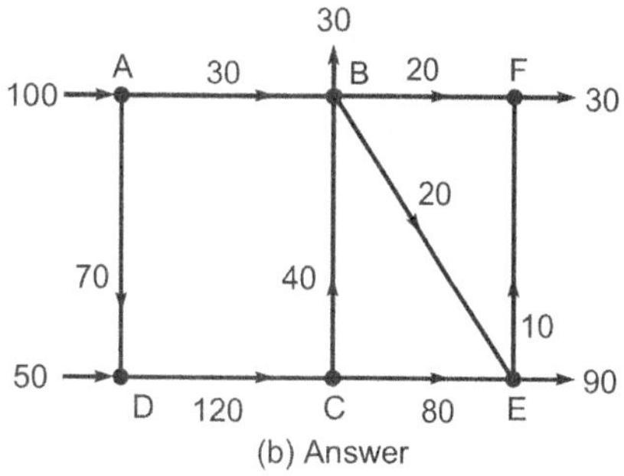

Fig. 3.19(b) *Example 3.7–Answer*

Solution: By continuity criterion the flow entering into a node must be equal to the flow going out of the node. Thus by considering flow into a node as positive, the algebraic sum of discharges at a node is zero.

Thus at node A:

$$100 - 70 - Q_{AB} = 0$$

or $\quad$ Q_{AB} **= 30 and Q_{AB} is from A to B.**

At node D: $70 + 50 - Q_{DC} = 0$

Q_{DC} **= 120 and Q_{DC} is from D to C.**

At node C: $120 - 80 - Q_{CB} = 0$

Q_{CB} **= 40 and Q_{CB} from C to B.**

At node B: $30 + 40 - 30 - Q_{BE} - 20 = 0$

Q_{BE} **= 20 and Q_{BE} is from B to E.**

At node E: $80 + 20 - Q_{EF} - 90 = 0$

Q_{EF} **= 10 and Q_{EF} is from E to F.**

At node F: $20 + 10 - Q_F = 0$

Q_F **= discharge out of node F = 30.**

The distribution of discharges are as in Fig. 3.19(b).

It can be seen now that at each node the continuity equation is satisfied.

3.8 For the following flows find the equation of the streamline passing through (1, 1).

(i) $\mathbf{V} = 3x\mathbf{i} - 3y\mathbf{j}$

(ii) $\mathbf{V} = - y^2\mathbf{i} - 6x\mathbf{j}$

Solution:

(i) $u = 3x$ and $v = - 3y$

The equation of a streamline in two-dimensional flow is

$$\frac{dx}{u} = \frac{dy}{v}$$

Here $\frac{dx}{3x} = - \frac{dy}{3y}$. On integration

$$\frac{1}{3} \ln x = - \frac{1}{3} \ln y + \frac{1}{3} \ln c$$

where $\quad$ c = a constant

$$\ln xy = \ln c \text{ or } xy = c$$

For the streamline passing through (1, 1),

c = 1 and hence the required streamline equation is

$$\boldsymbol{xy = 1.}$$

(ii) $u = - y^2$ and $v = - 6x$

$- \frac{dx}{y^2} = - \frac{dy}{6x}$ is the differential equation of the streamline.

$$\int 6x \, dx = \int y^2 \, dy + c$$

$$3x^2 - \frac{y^3}{3} = c. \qquad \text{Putting } x = 1, y = 1$$

$$c = 3 - \frac{1}{3} = 2\frac{2}{3}$$

Hence the equation of the required streamline is

$$3x^2 - \frac{y^3}{3} = \frac{8}{3}$$

or
$$\boldsymbol{9x^2 - y^3 = 8}$$

3.9 In a flow the velocity vector is given by $\mathbf{V} = 3x\mathbf{i} + 4y\mathbf{j} - 7z\mathbf{k}$. Determine the equation of the streamline passing through a point M = (1, 4, 5).

Solution: The equation of the streamline is

$$\frac{dx}{u} = \frac{dy}{v} = \frac{dz}{w}$$

Here $\quad$ $u = 3x$, $v = 4y$ and $w = - 7z$

Hence $\dfrac{dx}{3x} = \dfrac{dy}{4y} = -\dfrac{dz}{7z}$

Considering equations involving x and y, on integration

$$\frac{1}{3}\ln x = \frac{1}{4}\ln y + \ln C_1'$$

where C_1' = a constant

or $y = C_1 x^{4/3}$ (1)

where C_1 is another constant.

Similarly, by considering equations with x and z and on integration

$$\frac{1}{3}\ln x = -\frac{1}{7}\ln z + \ln C_2'$$

where C_2' = a constant

$$z = \frac{C_2}{x^{7/3}} \quad (2)$$

where C_2 is another constant.

Putting the coordinates of the point M (1, 4, 5)

from Eq. (1) $C_1 = \dfrac{4}{(1)^{4/3}} = 4$

from Eq. (2) $C_2 = 5 \times 1^{7/3}$

The streamline passing through M is given by

$$\mathbf{y = 4x^{4/3} \quad and \quad z = 5/x^{7/3}}$$

3.10 For the following velocity vectors determine the magnitude of velocity at A ($x = 2$, $y = -3$, $z = 1$, $t = 2$).

(i) $\mathbf{V} = (10t + xy)\mathbf{i} + (-yz - 10t)\mathbf{j} + (-yz + z^2/2)\mathbf{k}$

(ii) $\mathbf{V} = 4x\mathbf{i} + (-4y + 3t)\mathbf{j}$

Solution:

(i) $\mathbf{V} = u\mathbf{i} + v\mathbf{j} + w\mathbf{k}$ and A ($x = 2, y = -3, z = 1, t = 2$)

Here $u = 10t + xy \Rightarrow u_a = (10 \times 2) + (2 \times -3) = 14$

$v = -yz - 10t \Rightarrow v_a = -(-3 \times 1) - (10 \times 2) = -17$

$w = -yz + z^2/2 \Rightarrow w_a = -(-3 \times 1) + (1)^2/2 = 3.5$

Magnitude of velocity at A =

$$V_A = \sqrt{u_a^2 + v_a^2 + w_a^2}$$

$$= \sqrt{(14)^2 + (-17)^2 + (3.5)^2}$$

$$= \mathbf{22.3\ units}$$

(ii) $\mathbf{V} = 4x\mathbf{i} + (-4y + 3t)\mathbf{j} + (0)\mathbf{k}$

At A ($x = 2, y = -3, z = 1, t = 2$)

$u = 4x \Rightarrow u_a = 4 \times 2 = 8$

$v = -4y + 3t \Rightarrow v_a = (-4 \times -3) + 3 \times 2 = 18$

$w = 0 \Rightarrow w_a = 0$

Magnitude of velocity at A =

$$V_A = \sqrt{u_a^2 + v_a^2 + w_a^2}$$

$$= \sqrt{8^2 + 18^2} = \mathbf{19.7\ units}$$

3.11 The velocity along the centreline of a nozzle of length L is given by

$$V = 2t\left(1 - \frac{x}{2L}\right)^2$$

where V = velocity in m/s, t = time in seconds from commencement of flow, x = distance from inlet to the nozzle. Find the convective acceleration, local acceleration and the total acceleration when $t = 3$ s, $x = 0.5$ m and $L = 0.8$ m.

Solution:

(i) Local acceleration $= \dfrac{\partial V}{\partial t} = 2\left(1 - \dfrac{x}{2L}\right)^2$

at $t = 3$ s and $x = 0.5$ m, $\dfrac{\partial V}{\partial t} = 2\left(1 - \dfrac{0.5}{2 \times 0.8}\right)^2$

$= \mathbf{0.945\ m/s^2}$

(ii) Convective acceleration $= V\dfrac{\partial V}{\partial x}$

$$= 2t\left(1 - \frac{x}{2L}\right)^2 \cdot 2t \cdot 2\left(1 - \frac{x}{2L}\right)\left(-\frac{1}{2L}\right)$$

$$= -\frac{4t^2}{L}\left(1 - \frac{x}{2L}\right)^3$$

At $t = 3$ s and $x = 0.5$ m

Convective acceleration

$$= -\frac{4 \times 3^2}{0.8}\left(1 - \frac{0.5}{2 \times 0.8}\right)^3$$

$$= \mathbf{-14.623\ m/s^2}$$

(iii) Total acceleration = (local + convective) acceleration = 0.945 − 14.623 = **− 13.68 m/s²**

3.12 A 2.0 m long conical diffuser 20 cm in diameter at the upstream end has 80 cm diameter at the downstream end. At a certain instant the discharge through the diffuser is observed to be 200 L/s of water and is found to increase uniformly at a rate of 50 L/s per second. Estimate the local, convective and total acceleration at a section 1.5 m from the upstream end.

Solution: Refer to Fig. 3.20.

Diameter at section XX

$$D_X = \left(\frac{D_2 - D_1}{L}\right) \cdot x + D_1$$

$$= \left(\frac{0.8 - 0.2}{2}\right) \cdot x + 0.20$$

$$= 0.3x + 0.20$$

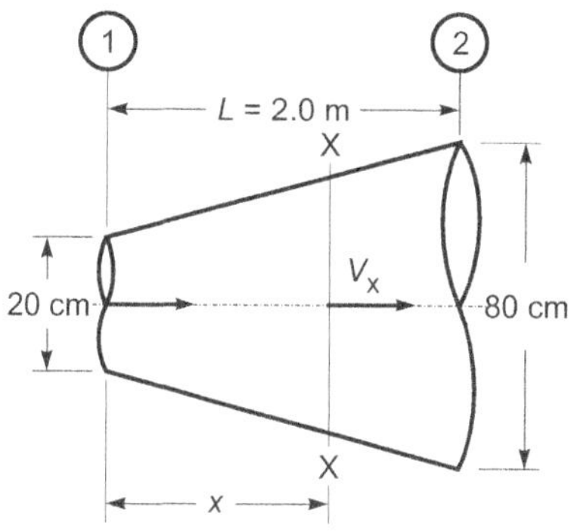

Fig. 3.20

At $x = 1.5$ m, $D = 0.65$ m

$$A = \text{area at section XX}$$

$$= \frac{\pi}{4} \times (0.65)^2 = 0.3318 \text{ m}^2$$

$$V = \text{Velocity} = \frac{Q}{A} = \frac{0.200}{0.3318} = 0.6027 \text{ m/s}$$

(i) $\frac{\partial V}{\partial t}$ = rate of increase of velocity

$$= \frac{\partial(Q/A)}{\partial t} = \frac{1}{A_x}\frac{\partial Q}{\partial t}$$

$$= \frac{1}{0.3318}(0.050) = \mathbf{0.1507 \text{ m/s}^2}$$

= local acceleration at XX

(ii) Convective acceleration $a_s = V_x \frac{\partial V_x}{\partial x}$

$$V_x = \frac{Q}{A_x} = \frac{Q}{\frac{\pi}{4}(0.3x + 0.2)^2} = \frac{0.20}{\frac{\pi}{4}(0.3x + 0.2)^2}$$

$$= 0.2546/(0.3x + 0.20)^2$$

$$\frac{\partial V_x}{\partial x} = (0.2546)(-2)(0.3x + 0.20)^{-3}(0.3)$$

$$= -0.15276/(0.3x + 0.20)^3$$

Hence convective acceleration

$$a_s = V_x \frac{\partial V_x}{\partial x} = \frac{0.2546}{(0.3x + 0.2)^2} \times \left(\frac{-0.15276}{(0.3x + 0.20)^3}\right)$$

$$= -\frac{0.03889}{(0.3x + 0.2)^5}$$

At $x = 1.5$ m, convective acceleration

$$a_s = -\frac{0.03889}{(0.3 \times 1.5 + 0.2)^5} = \mathbf{-0.3352 \text{ m/s}^2}$$

(iii) Total acceleration = local acceleration + convective acceleration

$$= \frac{\partial V_x}{\partial t} + V_x \frac{\partial V_x}{\partial t}$$

At $x = 1.5$ m, total acceleration $= 0.1507 - 0.3352$
$= \mathbf{-0.1845 \text{ m/s}^2}$

3.13 The velocity vector in an incompressible flow is given by

$$\mathbf{V} = (6xt + yz^2)\mathbf{i} + (3t + xy^2)\mathbf{j} + (xy - 2xyz - 6tz)\mathbf{k}$$

(i) Verify whether the continuity equation is satisfied.

(ii) Determine the acceleration vector at point A (1, 1, 1) at $t = 1.0$.

Solution:

(i) $\mathbf{V} = (6xt + yz^2)\mathbf{i} + (3t + xy^2)\mathbf{j} + (xy - 2xyz - 6tz)\mathbf{k}$
$= u\mathbf{i} + v\mathbf{j} + w\mathbf{k}$

$$u = 6xt + yz^2; \quad \frac{\partial u}{\partial x} = 6t$$

$$v = 3t + xy^2; \quad \frac{\partial v}{\partial y} = 2xy$$

$$w = xy - 2xyz - 6tz;$$

$$\frac{\partial w}{\partial z} = -2xy - 6t$$

$$\left(\frac{\partial u}{\partial x} + \frac{\partial v}{\partial y} + \frac{\partial w}{\partial z}\right) = 6t + 2xy - 2xy - 6t = 0$$

Hence the continuity equation is satisfied.

(ii) Acceleration $\mathbf{a} = a_x\mathbf{i} + a_y\mathbf{j} + a_z\mathbf{k}$

$$a_x = \frac{\partial u}{\partial t} + u\frac{\partial u}{\partial x} + v\frac{\partial u}{\partial y} + w\frac{\partial u}{\partial z}$$

$$= 6x + (6xt + yz^2)(6t) + (3t + xy^2)(z^2) + (xy - 2xyz - 6tz)(2yz)$$

At point A (1, 1, 1) and at $t = 1$,

$$a_x = 6 + (6 + 1)(6) + (3 + 1)(1) + (1 - 2 - 6)(2)$$
$$= 38 \text{ units}$$

$$a_y = \frac{\partial v}{\partial t} + u\frac{\partial v}{\partial x} + v\frac{\partial v}{\partial y} + w\frac{\partial v}{\partial z}$$

$$= 3 + (6xt + yz^2)(y^2) + (3t + xy^2)(2xy) + (xy - 2xyz - 6tz)(0)$$

At point A (1, 1, 1) and at $t = 1$

$$a_y = 3 + (6 + 1)(1) + (3 + 1)(2) = 18 \text{ units}$$

Similarly

$$a_z = \frac{\partial w}{\partial t} + u\frac{\partial w}{\partial x} + v\frac{\partial w}{\partial y} + w\frac{\partial w}{\partial z}$$

$$= -6z + (6xt + yz^2)(y - 2yz) + (3t + xy^2)(x - 2xz) + (xy - 2xyz - 6tz)(-2xy - 6t)$$

At point A (1, 1, 1) and at $t = 1$

$$a_z = -6 + (6 + 1)(1 - 2) + (3 + 1)(1 - 2) + (1 - 2 - 6)(-2 - 6)$$

$= -6 - 7 - 4 + 56 = 39$ units

Hence at A (1, 1, 1) and at $t = 1$,

$$\mathbf{a} = 38\mathbf{i} + 18\mathbf{j} + 39\mathbf{k}$$

3.14 For steady, incompressible flow, verify whether the following values of u and v are possible:

(i) $u = 4xy + y^2$, $v = 6xy + 3x$

(ii) $u = 2x^2 + y^2$, $v = -4xy$ and

(iii) $u = -x/(x^2 + y^2)$, $v = -y/(x^2 + y^2)$

Solution: For steady, incompressible flow the following continuity equation must be satisfied:

$$\frac{\partial u}{\partial x} + \frac{\partial v}{\partial y} = 0$$

(i) $u = 4xy + y^2$; $v = 6xy + 3x$

$$\frac{\partial u}{\partial x} = 4y\ ; \quad \frac{\partial v}{\partial y} = 6x$$

$$\therefore \frac{\partial u}{\partial x} + \frac{\partial v}{\partial y} = 4y + 6x \neq 0$$

Hence the flow is not possible.

(ii) $u = 2x^2 + y^2$; $v = -4xy$

$$\frac{\partial u}{\partial x} = 4x\ ; \quad \frac{\partial v}{\partial y} = -4x$$

$$\frac{\partial u}{\partial x} + \frac{\partial v}{\partial y} = 4x - 4x = 0$$

Hence the flow is possible.

(iii) $u = -x/(x^2 + y^2)$; $v = -y/(x^2 + y^2)$

$$\frac{\partial u}{\partial x} = \frac{-(x^2 + y^2) + 2x^2}{(x^2 + y^2)^2}\ ;$$

$$\frac{\partial v}{\partial y} = \frac{-(x^2 + y^2) + 2y^2}{(x^2 + y^2)^2}$$

$$\frac{\partial u}{\partial x} = \frac{(x^2 - y^2)}{(x^2 + y^2)^2}\ ; \quad \frac{\partial v}{\partial y} = \frac{(-x^2 + y^2)}{(x^2 + y^2)^2}$$

$$\therefore \frac{\partial u}{\partial x} + \frac{\partial v}{\partial y} = \frac{x^2 - y^2 - x^2 + y^2}{(x^2 + y^2)^2} = 0$$

Hence, the flow is possible.

3.15 Determine which of the velocity component sets given below satisfy the equation of continuity:

(a) $u = cx$ (b) $u = -cx/y$
$v = -cy$ $v = c \ln xy$

(c) $u = A \sin xy$ (d) $u = x + y$
$v = -A \sin xy$ $v = x - y$

(e) $u = 2x^2 + zy$
$v = -2xy + 3y^3 + 3zy$
$w = -\frac{3}{2} z^2 - 2xy - 6yz$

Solution:

(a) $u = cx$; $v = -cy$

$$\frac{\partial u}{\partial x} = c\ ; \quad \frac{\partial v}{\partial y} = -c$$

$$\frac{\partial u}{\partial x} + \frac{\partial v}{\partial y} = c - c = 0$$

Hence, the continuity equation is satisfied.

(b) $u = -cx/y$; $v = c \ln xy$

$$\frac{\partial u}{\partial x} = -c/y; \quad \frac{\partial v}{\partial y} = \frac{c}{xy} \cdot x = c/y$$

$$\frac{\partial u}{\partial x} + \frac{\partial v}{\partial y} = -c/y + c/y = 0$$

The continuity equation is satisfied.

(c) $u = A \sin xy$; $v = -A \sin xy$

$$\frac{\partial u}{\partial x} = Ay \cos xy; \quad \frac{\partial v}{\partial y} = -Ax \cos xy$$

$$\frac{\partial u}{\partial x} + \frac{\partial v}{\partial y} \neq 0$$

Hence the continuity equation is not satisfied.

(d) $u = x + y$; $v = x - y$

$$\frac{\partial u}{\partial x} = 1; \quad \frac{\partial v}{\partial y} = -1$$

$$\frac{\partial u}{\partial x} + \frac{\partial v}{\partial y} = 1 - 1 = 0$$

Hence the continuity equation is satisfied.

(e) $u = 2x^2 + zy$; $\frac{\partial u}{\partial x} = 4x$

$v = -2xy + 3y^2 + 3zy$;

$$\frac{\partial v}{\partial y} = -2x + 9y^2 + 3z$$

$$w = -\frac{3}{2} z^2 - 2xz - 6yz;$$

$$\frac{\partial w}{\partial z} = -3z - 2x - 6y$$

Hence $\frac{\partial u}{\partial x} + \frac{\partial v}{\partial y} + \frac{\partial w}{\partial z}$

$$= 4x - 2x + 9y^2 + 3z - 3z - 2x - 6y$$

$$\neq 0$$

The continuity equation is thus not satisfied.

3.16 Calculate the unknown velocity component in the following, so that the equation of continuity is satisfied.

(a) $u = A(x^2 + y^2)$ (b) $u = Ae^x$
$v = ?$ $v = ?$

(e) $u = -A \ln \frac{x}{L}$ (d) $u = ?$
$v = ?$ $v = Axy$

Solution:

(a) $u = A(x^2 + y^2)$

$$\frac{\partial u}{\partial x} = 2Ax = -\frac{\partial v}{\partial y}$$

$$v = \int -2Ax\,dy = \mathbf{-2A\,xy + f(x)}$$

(b) $u = Ae^x$

$$\frac{\partial u}{\partial x} = Ae^x = -\frac{\partial v}{\partial y}$$

$$v = \int -Ae^x\,dy = \mathbf{-Ae^x\,y + f(x)}$$

(c) $u = -A \ln (x/L)$

$$\frac{\partial u}{\partial x} = -\frac{A}{(x/L)}\frac{1}{L} = -\frac{A}{x} = -\frac{\partial v}{\partial y}$$

$$v = \int \frac{A}{x}\,dy = \mathbf{Ay/x + f(x)}$$

(d) $v = Axy$

$$\frac{\partial v}{\partial y} = Ax = -\frac{\partial u}{\partial x}$$

$$u = \int -Ax\,dx = \mathbf{-\frac{Ax^2}{2} + f(y)}$$

3.17 For the following flows, determine the components of rotation about the various axes.

(i) $u = xy^3z,\ v = -y^2z^2,\ w = yz^2 - \frac{y^3z^2}{2}$

(ii) $u = 3xy,\ v = \frac{3}{2}x^2 - \frac{3}{2}y^2$

(iii) $u = y^2,\ v = -3x$

Solution: The components of rotation about the various axes are:

$$\omega_z = \frac{1}{2}\left(\frac{\partial v}{\partial x} - \frac{\partial u}{\partial y}\right)$$

$$\omega_x = \frac{1}{2}\left(\frac{\partial w}{\partial y} - \frac{\partial v}{\partial z}\right)$$

$$\omega_y = \frac{1}{2}\left(\frac{\partial y}{\partial z} - \frac{\partial w}{\partial x}\right)$$

(i) $u = xy^3z;\ v = -y^2z^2;\ w = yz^2 - \frac{y^3z^2}{2}$

$$\omega_z = \frac{1}{2}\left(\frac{\partial v}{\partial x} - \frac{\partial u}{\partial y}\right) = \frac{1}{2}(0 - 3xy^2z) = -\frac{3}{2}xy^2z$$

$$\omega_x = \frac{1}{2}\left(\frac{\partial w}{\partial y} - \frac{\partial v}{\partial z}\right) = \frac{1}{2}\left(z^2 - \frac{3y^2z^2}{2} + 2y^2z\right)$$

$$\omega_y = \frac{1}{2}\left(\frac{\partial u}{\partial z} - \frac{\partial w}{\partial x}\right) = \frac{1}{2}(xy^3 - 0) = \frac{1}{2}xy^3$$

(ii) $u = 3xy\ ;\ v = \frac{3}{2}x^2 - \frac{3}{2}y^2$

$$\boldsymbol{\omega_z} = \frac{1}{2}\left(\frac{\partial v}{\partial x} - \frac{\partial u}{\partial y}\right) = \frac{1}{2}(3x - 3x) = \mathbf{0}$$

As the flow is two-dimensional in the *x–y* plane, $\boldsymbol{\omega_x = w_y = 0}$

(iii) $u = y^2;\ v = -3x$

$$\boldsymbol{\omega_z} = \frac{1}{2}\left(\frac{\partial v}{\partial x} - \frac{\partial u}{\partial y}\right) = \mathbf{\frac{1}{2}(-3 - 2y)}$$

As the flow is two-dimensional in *x–y* plane, $\boldsymbol{\omega_x = \omega_y = 0}$.

3.18 The velocity u in a laminar boundary layer over a flat plate is given by $\frac{u}{U_0} = \left[2\frac{y}{\delta} - \left(\frac{y}{\delta}\right)^2\right]$ where U_0 = constant and δ = boundary layer thickness and is given by $\delta = k\sqrt{x}$. Obtain an expression for the velocity component v across the boundary layer.

Solution: The continuity equation for two-dimensional flow is $\frac{\partial u}{\partial x} + \frac{\partial v}{\partial y} = 0$.

Here, $$u = U_0\left[2\frac{y}{\delta} - \left(\frac{y}{\delta}\right)^2\right]$$

$\therefore$ $$\frac{\partial u}{\partial x} = U_0\left[-2\frac{y}{\delta^2} + \frac{2y^2}{\delta^3}\right]\left[\frac{k}{2}\frac{1}{\sqrt{x}}\right]$$

Hence $$\frac{\partial v}{\partial y} = -\frac{\partial u}{\partial x} = kU_0\left[\frac{y}{\delta^2} - \frac{y^2}{\delta^3}\right]\frac{1}{\sqrt{x}}$$

$$v = \frac{kU_0}{\sqrt{x}}\left[\frac{y^2}{2\delta^2} - \frac{y^3}{3\delta^3}\right] + c$$

At $y = 0$, $v = 0$ and hence constant $c = 0$

$\therefore$ $$\mathbf{v = \frac{kU_0}{\sqrt{x}}\left[\frac{y^2}{2\delta^2} - \frac{y^3}{3\delta^3}\right]}$$

3.19 For the velocity components in a fluid flow given by $u = 2xy$ and $v = a^2 + x^2 - y^2$, show that the flow is possible. Obtain the relevant stream function.

Solution: $$u = 2xy\ ;\ v = a^2 + x^2 - y^2$$

$$\frac{\partial u}{\partial x} = 2y\ ;\ \frac{\partial v}{\partial y} = -2y$$

$$\frac{\partial u}{\partial x} + \frac{\partial v}{\partial y} = 2y - 2y = 0$$

The continuity equation for steady, incompressible flow is satisfied. Hence the flow is possible. The stream function ψ is related to u and v as

$$u = \frac{\partial \psi}{\partial y} = 2xy$$

$$\psi = \int 2xy\, dy = xy^2 + f(x)$$

$$-\frac{\partial \psi}{\partial x} = -y^2 - f'(x) = v = a^2 + x^2 - y^2$$

Hence $f'(x) = -(a^2 + x^2)$

$$f(x) = -a^2x - \frac{x^3}{3} + \text{constant}$$

Thus the relevant stream function is $\boldsymbol{\psi = xy^2 - a^2x - \frac{x^3}{3}}$ **+ constant.**

3.20 Obtain a relevant stream function to each of the following sets of velocity components of steady, incompressible flow:

(i) $u = -cx/y$, $v = c \ln xy$ (ii) $u = x + y$, $v = x - y$

(iii) $u = 2\,cx$, $v = -2cy$

Solution:

(i) $u = -cx/y$; $v = c \ln xy$

$$\frac{\partial u}{\partial x} = -c/y \; ; \; \frac{\partial v}{\partial y} = c/y$$

$$\frac{\partial u}{\partial x} + \frac{\partial v}{\partial y} = -c/y + c/y = 0$$

The flow is possible and ψ exists.

$$u = \frac{\partial \psi}{\partial y} = -cx/y$$

$$\psi = -cx \ln y + f(x)$$

$$-\frac{\partial \psi}{\partial x} = c \ln y - f'(x) = v = c \ln xy$$

$$= c \ln x + c \ln y$$

Hence $f'(x) = -c \ln x$

$$f(x) = -\int c \ln x \cdot dx = -c(x \ln x - x) + c_2$$

where c_2 = constant. Hence the stream function representing this flow is

$$\psi = -cx \ln y - cx \ln x + cx + c_2$$

$$\boldsymbol{\psi = -cx \ln xy + cx + c_2}$$

(ii) $u = x + y$; $v = x - y$

Therefore $\frac{\partial u}{\partial x} = 1$ and $\frac{\partial v}{\partial y} = -1$

$$\frac{\partial u}{\partial x} + \frac{\partial v}{\partial y} = 1 - 1 = 0.$$

Hence the flow is possible and ψ exists.

$$u = \frac{\partial \psi}{\partial y} = x + y$$

$$\psi = xy + \frac{y^2}{2} + f(x)$$

$$-\frac{\partial \psi}{\partial x} = -y - f'(x) = v = x - y$$

∴ $f'(x) = -x$

and $f(x) = \int -x\, dx = -\frac{x^2}{2} + c$

where c = a constant.

Hence

$$\psi = xy + \frac{y^2}{2} - \frac{x^2}{2} + c$$

∴ $\boldsymbol{\psi = \frac{1}{2}(y^2 - x^2) + xy + c}$

(iii) $u = 2cx$, $v = -2cy$

$$\frac{\partial u}{\partial x} = 2c \text{ and } \frac{\partial v}{\partial y} = -2c$$

$$\frac{\partial u}{\partial x} + \frac{\partial v}{\partial y} = 2c - 2c = 0.$$

Hene the flow is possible and ψ exists.

$$u = \frac{\partial \psi}{\partial y} = 2cx$$

$$\psi = 2cxy + f(x)$$

$$-\frac{\partial \psi}{\partial x} = -2cy - f'(x) = v = -2cy$$

Hence $f'(x) = 0$ and $f(x) = c_1$ = a constant

$$\boldsymbol{\psi = 2cxy + c_1}$$

3.21 For the following stream functions calculate the velocity at a point (2, 3).

(i) $\psi = 2xy$ (ii) $\psi = 3x^2y - y^3$

Solution:

(i) $u = \frac{\partial \psi}{\partial y} = 2x$

$$v = -\frac{\partial \psi}{\partial x} = -2y$$

At (2, 3), $u = 2 \times 2 = 4$

$v = -2 \times 3 = -6$

$V = \sqrt{u^2 + v^2} = \sqrt{4^2 + (-6)^2} = \sqrt{52}$ **units**

(ii) $u = \frac{\partial \psi}{\partial y} = 3x^2 - 3y^2$

$$v = -\frac{\partial \psi}{\partial x} = -6xy$$

At (2, 3), $u = 3 \times (2)^2 - 3(3)^2 = -15$

$v = -6 \times 2 \times 3 = -36$

$V = \sqrt{(-15)^2 + (-36)^2} =$ **39 units**

3.22 A flow is described by the stream function $\psi = 2\sqrt{3}xy$. Locate the point at which the velocity vector has a magnitude of 4 units and makes an angle of 150° with the x-axis.

Solution: The velocity vector is diagrammatically shown in Fig. 3.21

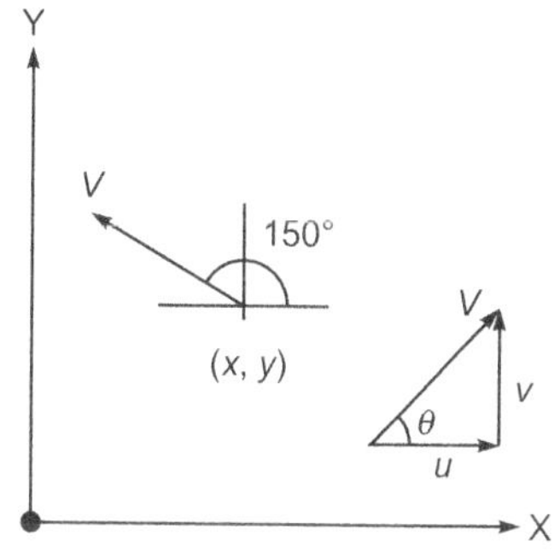

Fig. 3.21

$$\psi = 2\sqrt{3}\,xy$$

$$u = \frac{\partial \psi}{\partial y} = 2\sqrt{3}\,x$$

$$v = -\frac{\partial \psi}{\partial x} = -2\sqrt{3}\,y$$

$$V = \sqrt{u^2 + v^2}$$

$$4 = \sqrt{4(3)(x^2 + y^2)} \quad (1)$$

Also, $\tan\theta = \dfrac{v}{u}$

$$\tan 150° = -\frac{2\sqrt{3}}{2\sqrt{3}} \cdot \frac{y}{x}$$

$$-0.5774 = -y/x$$

$$y = 0.5774\,x \quad (2)$$

Substituting in Eq. (1)

$$4.0 = \sqrt{4 \times 3 \times [(x^2 + (0.5774)^2 x^2)]}$$

$$= \sqrt{16x^2} = 4x$$

$$x = 1.00$$

$$y = 0.5774$$

The required point (x, y) is **(1.00, 0.5774).**

3.23 The stream function ψ for uniform flow past a circular cylinder of radius a in polar co-ordinates is given by $\psi = U_\infty \left(1 - \dfrac{a^2}{r^2}\right) r \sin\theta$ where U_∞ is the free stream velocity parallel to the horizontal axis. Determine the velocity on the cylinder at $\theta = 90°$.

Solution:

$$V_r = \frac{1}{r}\frac{\partial \psi}{\partial \theta} = \frac{1}{r} U_\infty \left(1 - \frac{a^2}{r^2}\right) r \cdot \cos\theta$$

$$= U_\infty \left(1 - \frac{a^2}{r^2}\right) \cos\theta$$

At $r = a$ and $\theta = 90°$, $\mathbf{V_r = 0}$

$$V_\theta = -\frac{\partial \psi}{\partial r} = -U_\infty \sin\theta \left(1 - a^2\left(-\frac{1}{r^2}\right)\right)$$

$$= -U_\infty \sin\theta \left(1 + \frac{a^2}{r^2}\right)$$

At $r = a$ and $\theta = 90°$, $\mathbf{V_\theta = -2\,U_\infty}$

3.24 Values of ϕ for various flows are given. Determine the corresponding values of ψ.

(i) $\phi = 3xy$ (ii) $\phi = 4(x^2 - y^2)$

(iii) $\phi = x + y + 3$

Solution:

(i) $\phi = 3xy$

$$u = \frac{\partial \phi}{\partial x} = 3y = \frac{\partial \psi}{\partial y}$$

Hence $\psi = \dfrac{3}{2} y^2 + f(x)$

$$v = \frac{\partial \phi}{\partial y} = 3x = -\frac{\partial \psi}{\partial x}$$

$$= -f'(x)$$

$f'(x) = -3x$ and hence $f(x) = -3/2x^2 + c$

∴ $\boldsymbol{\psi = -3/2(x^2 - y^2) + c}$ **where c is a constant**

(ii) $\phi = 4(x^2 - y^2)$

$$u = \frac{\partial \phi}{\partial x} = 8x = \frac{\partial \psi}{\partial y}$$

Hence $\psi = 8xy + f(x)$

$$v = \frac{\partial \phi}{\partial y} = -8y = -\frac{\partial \psi}{\partial x}$$

$$= -8y - f'(x)$$

Hence $f'(x) = 0$ and $f(x) = \text{constant} = c$

∴ $\boldsymbol{\psi = 8xy + c}$

(iii) $\phi = x + y + 3$

$$u = \frac{\partial \phi}{\partial x} = 1 = \frac{\partial \psi}{\partial y}$$

Hence $\psi = y + f(x)$

$$v = \frac{\partial \phi}{\partial y} = 1 = -\frac{\partial \psi}{\partial x}$$

$$= -f'(x)$$

$f'(x) = -1$ and hence $f(x) = -x$

Hence $\boldsymbol{\psi = -x + y + c}$ **where c is a constant**

3.25 Verify whether the following functions are valid potential functions:

(i) $\phi = A\,xy$ (ii) $\phi = m \ln x$

(iii) $\phi = A(x^2 - y^2)$ (iv) $\phi = A \cos x$

Solution: A valid potential function satisfies the Laplace equation.

(i) $\phi = A\,xy$

$$\frac{\partial \phi}{\partial x} = Ay \; ; \; \frac{\partial \phi}{\partial y} = Ax$$

$$\frac{\partial^2 \phi}{\partial x^2} + \frac{\partial^2 \phi}{\partial y^2} = 0 + 0 = 0$$

Hence $\phi = Axy$ is a valid potential function.

(ii) $\phi = m \ln x$

$$\frac{\partial \phi}{\partial x} = \frac{m}{x} \; ; \; \frac{\partial \phi}{\partial y} = 0$$

$$\frac{\partial^2 \phi}{\partial x^2} = -\frac{m}{x^2} \; ; \; \frac{\partial^2 \phi}{\partial y^2} = 0$$

$$\frac{\partial^2 \phi}{\partial x^2} + \frac{\partial^2 \phi}{\partial x^2} = -m/x^2 \neq 0$$

Hence $\phi = m \ln x$ is not a valid potential function.

(iii) $\phi = A(x^2 - y^2)$

$$\frac{\partial \phi}{\partial x} = 2Ax \; ; \; \frac{\partial \phi}{\partial y} = -2Ay$$

$$\frac{\partial^2 \phi}{\partial x^2} = 2A \; ; \; \frac{\partial^2 \phi}{\partial y^2} = -2A$$

$$\frac{\partial^2 \phi}{\partial x^2} + \frac{\partial^2 \phi}{\partial y^2} = 2A - 2A = 0$$

Hence $\phi = A(x^2 - y^2)$ is a valid potential function.

(iv) $\phi = A \cos x$

$$\frac{\partial \phi}{\partial x} = -A \sin x \; ; \; \frac{\partial \phi}{\partial y} = 0$$

$$\frac{\partial^2 \phi}{\partial x^2} + \frac{\partial^2 \phi}{\partial y^2} = -A \cos x \neq 0$$

Hence $\phi = A \cos x$ is not a valid potential function.

3.26 Which of the following functions represent possible irrotational flow:

(i) $\psi = xy$ (iii) $\phi = Ur \cos \theta + \frac{U}{r} \cos \theta$

(ii) $\psi = A(x^2 - y^2)$ (iv) $\phi = \left(r - \frac{2}{r}\right) \sin \theta$

Solution: For an irrotational fluid flow phenomenon, ϕ as well as ψ satisfy Laplace equation.

(i) $\psi = xy$

$$\frac{\partial \psi}{\partial x} = y \; ; \; \frac{\partial \psi}{\partial y} = x$$

$$\frac{\partial^2 \psi}{\partial x^2} = 0 \; ; \; \frac{\partial^2 \psi}{\partial y^2} = 0$$

$$\frac{\partial^2 \psi}{\partial x^2} + \frac{\partial^2 \psi}{\partial x^2} = 0$$

Hence $\psi = xy$ represents a possible irrotational flow.

(ii) $\psi = A(x^2 - y^2)$

$$\frac{\partial \psi}{\partial x} = 2Ax \; ; \; \frac{\partial \psi}{\partial y} = -2Ay$$

$$\frac{\partial^2 \psi}{\partial x^2} + \frac{\partial^2 \psi}{\partial y^2} = 2A - 2A = 0$$

Hence $\psi = A(x^2 - y^2)$ represents a possible irrotational flow field.

(iii) $\phi = Ur \cos \theta + \frac{U}{r} \cos \theta$

Laplace equation is radial co-ordinates (r, θ)

$$\frac{1}{r}\frac{\partial \phi}{\partial r} + \frac{\partial^2 \phi}{\partial r^2} + \frac{1}{r^2}\frac{\partial^2 \phi}{\partial \theta^2} = 0$$

$$\frac{\partial \phi}{\partial r} = U \cos \theta - \frac{U}{r^2} \cos \theta$$

$$\frac{\partial^2 \phi}{\partial r^2} = +\frac{2U}{r^3} \cos \theta$$

$$\frac{\partial \phi}{\partial \theta} = -Ur \sin \theta - \frac{U}{r} \sin \theta$$

$$\frac{\partial^2 \phi}{\partial \theta^2} = -Ur \cos \theta - \frac{U}{r} \cos \theta$$

L.H.S. of Laplace equation is

$$\frac{1}{r}\left(U \cos \theta - \frac{U}{r^2} \cos \theta\right) + \left(\frac{2U}{r^3} \cos \theta\right) + \frac{1}{r^2}\left(-Ur \cos \theta - \frac{U}{r^2} \cos \theta\right)$$

$$= U \cos \theta \left(\frac{1}{r} - \frac{1}{r^3} + \frac{2}{r^3} - \frac{1}{r} - \frac{1}{r^3}\right)$$

$$= 0$$

The Laplace equation is satisfied and hence **the given function ϕ represents a possible irrotational flow.**

(iv) $\phi = \left(r - \frac{2}{r}\right) \sin \theta$

$$\frac{\partial \phi}{\partial r} = \left(1 + \frac{2}{r^2}\right) \sin \theta$$

$$\frac{\partial^2 \phi}{\partial r^2} = -\frac{2}{r^3} \sin \theta$$

$$\frac{\partial \phi}{\partial \theta} = \left(r - \frac{2}{r}\right) \cos \theta$$

$$\frac{\partial^2\phi}{\partial\theta^2} = -\left(r - \frac{2}{r}\right)\sin\theta$$

The Laplace equation, in radial coordinates (r, θ) is

$$\frac{1}{r}\frac{\partial\phi}{\partial r} + \frac{\partial^2\phi}{\partial r^2} + \frac{1}{r^2}\frac{\partial^2\phi}{\partial\theta^2} = 0$$

Substituting for L.H.S. terms

$$= \sin\theta\left[\frac{1}{r} + \frac{2}{r^3} - \frac{2}{r^3} - \frac{1}{r} + \frac{2}{r^3}\right]$$

$$= \sin\theta\left(\frac{2}{r^3}\right) \neq 0$$

Hence, **the given function does not represent any possible irrotational flow.**

3.26 Values of ψ for two flows are given. Determine the Corresponding values of potential function φ.

(i) $\psi = A(x^2 - y^2)$ (ii) $\psi = y^3 - 3x^2y$

Solution:

(i) $\psi = A(x^2 - y^2)$

$u = \partial\psi/\partial y = -2Ay = \partial\phi/\partial x$

Hence $\phi = -2Axy + f(y)$

$v = -\partial\psi/\partial x = -2Ax$ (i)

But $v = \partial\phi/\partial y = -2Ax + f'(y)$ (ii)

Comparing (i) and (ii), $f'(y) = 0$ and hence $f(y)$ = constant

$\therefore$ $\boldsymbol{\phi = -2Axy}$ **+ constant**

(ii) $\psi = y^3 - 3x^2y$

$u = \partial\psi/\partial y = 3y^2 - 3x^2 = \partial\phi/\partial x$

Hence $\phi = 3y^2x - x^3 + f(y)$

$v = -\partial\psi/\partial x = 6xy$ (i)

Also $v = \partial\phi/\partial y = 6xy + f'(y)$ (ii)

Comparing (i) and (ii), $f'(y) = 0$ and hence $f(y)$ = constant

Hence, $\boldsymbol{\phi = 3y^2x - x^3}$ **+ constant**

3.27 For the following functions in two-dimensional irrotational flow, determine the conjugate functions

(i) $\psi = m \ln r$ (ii) $\phi = \dfrac{m\cos\theta}{r}$

Solution: In radial co-ordinates,

$$v_r = \frac{\partial\phi}{\partial r} = \frac{1}{r}\frac{\partial\psi}{\partial\theta}$$

$$v_\theta = \frac{1}{r}\cdot\frac{\partial\phi}{\partial\theta} = -\frac{\partial\psi}{\partial r}$$

(i) For $\psi = m \ln r$

$$v_r = \frac{1}{r}\frac{\partial\psi}{\partial\theta} = 0 = \frac{\partial\phi}{\partial r}$$

$$\phi = f(\theta)$$

$$v_\theta = \frac{1}{r}\frac{\partial\phi}{\partial\theta} = \frac{1}{r}f'(\theta) = -\frac{\partial\psi}{\partial r} = -\frac{m}{r}$$

$f'(\theta) = -m$ and hence $f(\theta) = -m\theta + c$ where c = a constant. Hence, $\boldsymbol{\phi = -m\theta + c}$.

(ii) For $\phi = \dfrac{m\cos\theta}{r}$

$$v_r = \frac{\partial\phi}{\partial r} = \frac{m\cos\theta}{r^2} = \frac{1}{r}\frac{\partial\psi}{\partial\theta}$$

$$\frac{\partial\psi}{\partial\theta} = \frac{m}{r}\cos\theta$$

and $$\psi = \frac{m}{r}\sin\theta + f(r)$$

$$v_\theta = -\frac{\partial\psi}{\partial r} = -\frac{m}{r^2}\sin\theta - f'(r) = \frac{1}{r}\frac{\partial\phi}{\partial\theta}$$

$$= \frac{1}{r}\left(-\frac{m}{r}\sin\theta\right) = -\frac{m}{r^2}\sin\theta$$

Thus

$$f'(r) = -\frac{2m}{r^2}\sin\theta \text{ and}$$

$$f(r) = -\frac{2m}{r}\sin\theta + c$$

$$\therefore \quad \psi = \frac{m}{r}\sin\theta - \frac{2m}{r}\sin\theta + c$$

$$\boldsymbol{\psi = -\frac{m}{r}\sin\theta + c}$$

OBJECTIVE QUESTIONS

3.1 The flow field represented by the velocity vector

$$V = axi + by^2j + czt^2k$$

where a, b and c are constants, is

(a) three-dimensional and unsteady
(b) two-dimensional and steady
(c) three-dimensional and steady
(d) two-dimensional and unsteady

3.2 The flow of a liquid at constant rate in a conically tapered pipe is classified as

(a) steady, uniform flow
(b) steady, non-uniform flow

(c) unsteady, uniform flow
(d) unsteady, non-uniform flow

3.3 A *pathline* is the
(a) mean direction of a number of particles at the same instant of time.
(b) instantaneous picture of positions of all particles in the flow which passed a given point.
(c) trace made by a single particle over a period of time
(d) path traced by continuously injected tracer at a point

3.4 A *streamline* is a line
(a) which is normal to the velocity vector at every point
(b) which represents lines of constant velocity potential
(c) which is normal to the lines of constant stream function
(d) which is tangential to the velocity vector everywhere at a given instant

3.5 In a steady flow
(a) streamlines and pathlines are identical but are different from streaklines
(b) streakline and pathlines are identical but are different from streamlines
(c) streamline, streakline and pathline can all be different from each other
(d) none of the above

3.6 In two-dimensional flow the equation of a streamline is given as

(a) $\frac{dy}{u} = \frac{dx}{v}$ (b) $\frac{dx}{u} = \frac{dy}{v}$

(c) $\frac{dx}{dt} = u, \frac{dy}{dt} = v$ (d) $\frac{u}{dx} = \frac{dy}{v}$

3.7 The shape of the stream line passing through the origin in a flow field $u = \cos\theta$; $v = \sin\theta$ for a constant θ, is determined by
(a) $y = x^3$ (b) $y = x \cot^2\theta$
(c) $y = x \tan\theta$ (d) $y = \sin\theta$

38. A velocity vector **V** in two-dimensional flow is inclined at an angle θ to the X-axis. The resulting acceleration vector **a**
(a) will be always normal to **V**
(b) will be always parallel to **V**
(c) will have an inclination of $(90 - \theta)$ to the y-axis
(d) will have an inclination α to the X-axis which depends on the components of the acceleration.

3.9 A streamline is defined in terms of stream function ψ and potential function ϕ as

(a) ϕ = constant (b) $\frac{\partial\phi}{\partial s}$ = constant

(c) $\frac{\partial\psi}{\partial s}$ = constant (d) ψ = constant.

3.10 If $\psi = 2xy$, the magnitude of the velocity vector at (2, – 2) is
(a) $4\sqrt{2}$ (b) 4
(c) – 8 (d) $\sqrt{2}$

3.11 A continuity equation for steady two-dimensional compressible flow is

(a) $\rho\frac{\partial u}{\partial x} + \rho\frac{\partial v}{\partial y} = 0$

(b) $\frac{\partial u}{\partial x} + \frac{\partial v}{\partial y} = 0$

(c) $\frac{\partial(\rho u)}{\partial x} + \frac{\partial(\rho v)}{\partial y} = 0$

(d) $u\frac{\partial\rho}{\partial x} + v\frac{\partial\rho}{\partial y} = 0$

3.12 The continuity equation

$$\frac{\partial u}{\partial x} + \frac{\partial v}{\partial y} + \frac{\partial w}{\partial z} = 0$$

(a) is not valid for unsteady, incompressible fluids.
(b) is valid for incompressible fluids whether the flow is steady or unsteady
(c) is valid for steady flow, whether compressible or incompressible.
(d) is valid for ideal fluid flow only.

3.13 If u and v, the components of velocity in x and y directions respectively are given by

$$u = ax + by \quad \text{and} \quad v = cx + dy$$

then the condition to the satisfied is
(a) a + c = 0
(b) b + d = 0
(c) a + b + c + d = 0
(d) a + d = 0

3.14 Which of the following can be a set of velocity components in a two-dimensional flow?
(a) $u = x + y$; $v = x^2 + y^2$
(b) $u = x + y$; $v = x - y$
(c) $u = x\,y$; $v = x/y$
(d) $u = x^2 + y^2$; $v = x^2 - y^2$

3.15 A flow has diverging straight streamlines. If the flow is steady, the flow
(a) is a uniform flow with local acceleration
(b) has convective normal acceleration

(c) has convective tangential acceleration
(d) has convective normal as well as tangential accelerations.

3.16 A flow has parallel curved streamlines and is steady. This flow has
(a) tangential convective acceleration
(b) local acceleration
(e) normal convective as well as local acceleration
(d) normal convective acceleration

3.17 In a 2 m long tapered duct the area decreases as $A = (0.4 - 0.1x)$ where x is distance in metres. At a given instant a discharge of 0.48 m^3/s was flowing in the duct and it was found to increase at a rate of 0.12 m^3/s. The local acceleration at $x = 0$ in m/s^2 is
(a) 0.3 (b) 3.6
(c) 3.9 (d) – 0.30

3.18 In a two-dimensional flow acceleration component in the X-direction is given by a_x =
(a) $\frac{\partial u}{\partial t} + u\frac{\partial u}{\partial x} + v\frac{\partial v}{\partial y}$
(b) $u\frac{\partial u}{\partial x} + v\frac{\partial u}{\partial y}$
(c) $\frac{\partial u}{\partial t} + u\frac{\partial u}{\partial x} + v\frac{\partial u}{\partial y}$
(d) $u\frac{\partial u}{\partial t} + v\frac{\partial u}{\partial x} + u\frac{\partial v}{\partial y}$

3.19 In a natural coordinate system the acceleration a_n in the normal direction when local and convective terms are present is given by a_n =
(a) $\frac{\partial v}{\partial t} + v\frac{\partial v}{\partial t}$ (b) $\frac{v^2}{r} + v\frac{\partial v}{\partial r}$
(c) $\frac{d(v^2/r)}{dt}$ (d) $\frac{\partial v_n}{\partial t} + \frac{v^2}{r}$

3.20 The velocity of an incompressible fluid is given by

$$V = (Px - Q)\,i + Ry\,j + St\,k \text{ m/s}$$

where $P = 3\ s^{-1}$, $Q = 4$ m/s, $R = -3\ s^{-1}$. x and y are in m and t in s. The local and convective acceleration components at x = 1 m, y = 2 m and t = 5 s are respectively,
(a) $5k$ and $(-3i + 18j)$ m/s
(b) zero and $(-3i + 18j)$ m/s
(c) $5k$ and $(18i - 3j)$ m/s
(d) $-5k$ and $(-3i + 18j)$ m/s

3.21 A flow is said to be rotational when
(a) the streamlines are curved
(b) a velocity gradient in the normal direction to flow exists.
(c) every fluid element has finite angular velocity about its mass centre
(d) every fluid element has an angular velocity about a common axis

3.22 In three-dimensional motion of a fluid the component of rotation about the X-axis is ω_x =
(a) $\frac{1}{2}\left(\frac{\partial w}{\partial y} - \frac{\partial v}{\partial z}\right)$ (b) $\frac{1}{2}\left(\frac{\partial u}{\partial z} - \frac{\partial w}{\partial y}\right)$
(c) $\frac{1}{2}\left(\frac{\partial v}{\partial x} - \frac{\partial u}{\partial y}\right)$ (d) $\frac{1}{2}\left(\frac{\partial v}{\partial x} - \frac{\partial w}{\partial y}\right)$

3.23 If ω_z = component of rotation of a fluid about z-axis, the *vorticity* along that axis is usually defined as ζ_z =
(a) $\frac{1}{2}\,\omega_z$ (b) $2\omega_z$
(c) $\partial\omega_z/\partial x$ (d) $\int \omega_z dz$

3.24 A two-dimensional flow in x–y plane is rotational if
(a) $\frac{\partial u}{\partial x} = \frac{\partial v}{\partial y}$ (b) $\frac{\partial u}{\partial x} = \frac{\partial u}{\partial y}$
(c) $\frac{\partial v}{\partial x} = \frac{\partial u}{\partial y}$ (d) $\frac{\partial v}{\partial x} = \frac{\partial v}{\partial y}$

3.25 Vorticity in z-direction is given by
(a) $\left[\frac{\partial u}{\partial x} + \frac{\partial v}{\partial y}\right]$ (b) $\left[\frac{\partial u}{\partial x} - \frac{\partial v}{\partial y}\right]$
(c) $\left[\frac{\partial v}{\partial x} + \frac{\partial u}{\partial y}\right]$ (d) $\left[\frac{\partial v}{\partial x} - \frac{\partial u}{\partial y}\right]$

3.26 If ψ_2 and ψ_1 are the values of stream function at points 2 and 1 respectively, the volume rate of flow per unit depth across an element Δs connecting 2 and 1 is given by
(a) $\frac{\Delta\psi}{\Delta s}$ (b) $\Sigma\,\Delta\psi\cdot\Delta s$
(c) $\frac{1}{\Delta\psi}$ (d) $\Delta\psi$

3.27 A two-dimensional flow is described by velocity components $u = 2\,x$ and $v = -\,2y$. The discharge between points (1,1) and (2,2) is equal to
(a) 9 units (b) 8 units
(c) 7 units (d) 6 units

3.28 If ψ is a stream function then the velocity u and v are given by

(a) $u = \partial\psi/\partial y$, $v = -\partial\psi/\partial x$
(b) $u = -\partial\psi/\partial y$, $v = -\partial\psi/\partial x$
(c) $u = \partial\psi/\partial y$, $v = \partial\psi/\partial x$
(d) $u = \partial\psi/\partial x$, $v = -d\psi/\partial y$

3.29 The stream function in a two-dimensional flow field is given by $\psi = x^2 - y^2$. The magnitude of the velocity at point (1, 1) is

(a) 2
(b) $2\sqrt{2}$
(c) 4
(d) 8

3.30 In a two-dimensional, incompressible flow, if the fluid velocity components are $u = x - 4y$ and $v = -y$ then the stream function ψ is given by

(a) $x^2 - xy + 2y^2$
(b) $2x^2 + 2xy + y^2$
(c) $2x^2 + xy - 2y^2$
(d) $2x^2 - xy + 2y^2$

3.31 Stream function is defined for

(a) Flow of a perfect fluid only
(b) All 2–D incompressible flows
(c) All 3–D flows
(d) Irrotational flows only

3.32 Velocity potential exists for

(a) Flow of a perfect fluid only
(b) Stready, irrotational flow only
(c) All irrotational flows
(d) All 3–D flows

3.33 A velocity potential exists

(a) whenever the real fluid flow exists
(b) when the flow is real and rotational
(c) when the flow satisfies the conditions of irrotational motion.
(d) when the flow satisfies the equations of continuity.

3.34 If ϕ is a potential function in two-dimensional flow the velocity components u and v are defined as

(a) $u = \partial\phi/\partial y$, $v = -\partial\phi/\partial x$
(b) $u = \partial\phi/\partial x$, $v = \partial\phi/\partial y$
(c) $u = \partial\phi/\partial x$, $v = -\partial\phi/\partial y$
(d) $u = \partial\phi/\partial y$, $v = \partial\phi/\partial x$

3.35 The velocity potential function for a line source varies with radial distance r as

(a) $1/r$
(b) $1/r^2$
(c) r
(d) $\ln r$

3.36 Lines of constant ϕ

(a) are parallel streamlines
(b) are parallel to the streamlines
(c) are normal to the streamlines
(d) can intersect each other

3.37 In 2-D radial co-ordinates the velocities v_r and v_θ are expressed in terms of ϕ as

(a) $v_r = \dfrac{1}{r}\dfrac{\partial\phi}{\partial\theta}$, $v_\theta = \dfrac{\partial\phi}{\partial r}$
(b) $v_r = -\dfrac{\partial\phi}{\partial r}$, $v_\theta = \dfrac{1}{r}\dfrac{\partial\phi}{\partial\theta}$
(c) $v_r = \dfrac{1}{r}\dfrac{\partial\phi}{\partial\theta}$, $v_\theta = -\dfrac{\partial\phi}{\partial r}$
(d) $v_r = \dfrac{\partial\phi}{\partial r}$, $v_\theta = \dfrac{1}{r}\dfrac{\partial\phi}{\partial\theta}$

3.38 If $\phi = 3xy$, the x and y components of velocity at the point (1, 3) will be

(a) $u = -9$, $v = -3$
(b) $u = -3$, $v = -9$
(c) $u = 9$, $v = -3$
(d) $u = 9$, $v = 9$

3.39 If $u = x^2 + y^2$ and $v = 2xy$, the velocity potential ϕ = constant +

(a) $x\left(\dfrac{x^2}{3} + y^2\right)$
(b) $x\left(\dfrac{x^2}{3} + y^3\right)$
(c) $x\left(\dfrac{x^2}{6} + \dfrac{y^3}{2}\right)$
(d) $x + y$

3.40 If for a flow, a stream function ψ exists and satisfies the Laplace equation, then

(a) the flow is rotational
(b) the flow is irrotational but does not necessarily satisfy continuity equation
(c) the flow satisfies continuity equation but does not necessarily satisfy condition for irrotational flow
(d) the continuity equation is satisfied and the flow is irrotational

3.41 If a stream function ψ exists it implies that

(a) the function ψ represents a possible flow field
(b) the flow is irrotational
(c) the flow is steady, incompressible
(d) the potential function also exists

3.42 The potential function exists for

(a) irrotational motion of incompressible fluids only
(b) irrotatinal motion of fluids whether compressible or incompressible
(c) for two-dimensional irrotational flow only
(d) for steady flows only

3.43 Cauchy-Reimann equations relating ϕ and ψ are

(a) $\dfrac{\partial\phi}{\partial x} = \dfrac{\partial\phi}{\partial y}$; $\dfrac{\partial\psi}{\partial x} = -\dfrac{\partial\psi}{\partial y}$

(b) $\frac{\partial \phi}{\partial x} = -\frac{\partial \phi}{\partial y}$; $\frac{\partial \phi}{\partial y} = -\frac{\partial \psi}{\partial x}$

(c) $\frac{\partial \phi}{\partial x} = \frac{\partial \psi}{\partial x}$; $\frac{\partial \phi}{\partial y} = -\frac{\partial \psi}{\partial y}$

(d) $\frac{\partial \phi}{\partial x} = -\frac{\partial \psi}{\partial y}$; $\frac{\partial \phi}{\partial y} = \frac{\partial \psi}{\partial x}$

3.44 Indicate the *incorrect* statement:
A flow net
(a) is applicable to irrotational fluid flow
(b) for a given boundary is the same whether the flow is in one direction or the other
(c) for a given boundary is applicable to one chosen direction of flow; if the flow is reversed the flow net will change
(d) will be no constructed that the size of the mesh is inversely proportional to the local velocity

3.45 The velocity potential ϕ at any point for a two-dimensional, steady, irrotational flow in polar coordinates is given by

$$\phi = \frac{\lambda \cos \theta}{r}$$

This equation represents a
(a) vortex (b) sink
(c) source (d) doublet

3.46 The velocity potential ϕ at any point for a two-dimensional, steady, irrotational flow is given by $\phi = K\,\theta$.
This equation represents a
(a) vortex (b) sink
(c) source (d) doublet

3.47 Inviscid, incompressible flow about a stationary cylinder in uniform flow can be simulated by superposition of uniform flow and
(a) a sink and a vortex
(b) a doublet
(c) a vortex
(d) a doublet and a vortex.

PROBLEMS

3.1 Water is pumped into the tank shown in Fig. 3.22 at 100 L/s. Both kerosene (relative) density = 0.8) and oil relative density = 0.90) are driven out. If 30 L/s oil is driven out estimate the volume of kerosene coming out of tank per second.
(*Ans.* Q = 91.25 L/s)

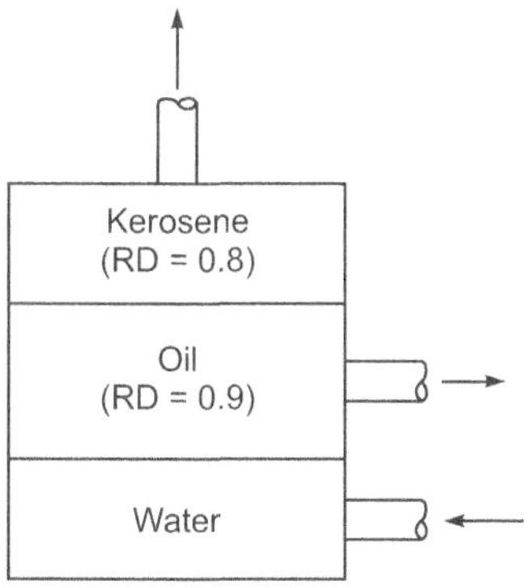

Fig. 3.22

3.2 A water tank has a 3 cm diameter inlet at A, a 4 cm diameter outlet at B and a 3 cm diameter controllable inlet at C, (Fig. 3.23). If the velocity of water at the inlet A is 2.0 m/s and the velocity of flow going out at B = 1.85 m/s, what should be the velocity at the inlet at C to see that the water level in the tank does not change?
(*Ans.* V = 1.285 m/s)

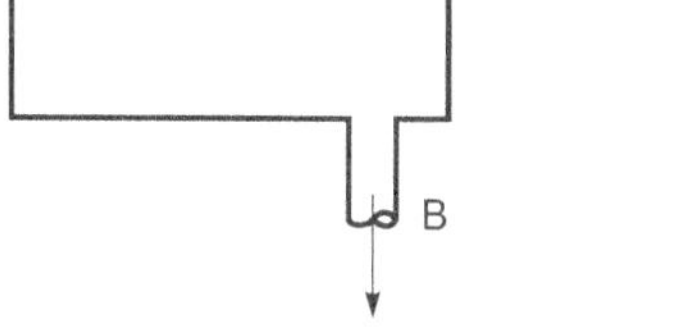

Fig. 3.23

3.3 Figure 3.24 shows a Y-junction of three pipes. The velocity distribution in pipe 1 is uniform while in pipes 2 and 3 it is given by $u = u_m \left(1 - \frac{r}{R}\right)^{1/7}$.
Pipes 1, 2 and 3 have diameters 8 cm, 6 cm and 4

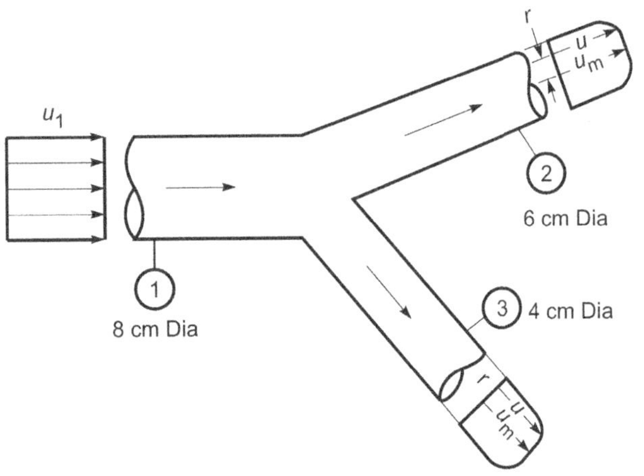

Fig. 3.24 *Problem 3.3*

cm respectively. If the maximum velocity u_m in pipes 2 and 3 are 0.8 m/s and 0.6 m/s respectively, estimate the value of the uniform velocity u_1 in pipe 1.

(*Ans.* $u_1 = 0.49$ m/s)

3.4 A two-dimensional duct 10 cm high carries an incompressible fluid flow. At the entrance section 1 (Fig. 3.25), the velocity can be assumed to be as shown in the figure.

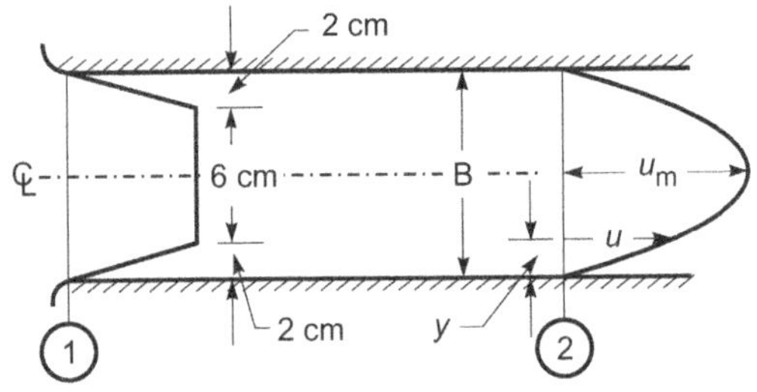

Fig. 3.25

After the flow is fully developed, at a section 2, the velocity u at a distance y from the boundary is given by

$$\frac{u}{u_m} = 4\left(\frac{y}{B} - \frac{y^2}{B^2}\right)$$

where B = height of the duct and u_m = velocity at the centreline of the duct. If the maximum velocity at entrance section 1 = 0.8 m/s, estimate the value of u_m at section 2.

(*Ans.* $u_m = 0.96$ m/s)

3.5 Figure 3.26 shows a flow of incompressible fluid of density ρ flowing past an impervious plate AD. BC is an imaginary plane at a height h above the plate. The flow approaching the plate is of uniform velocity U and the velocity profile at a

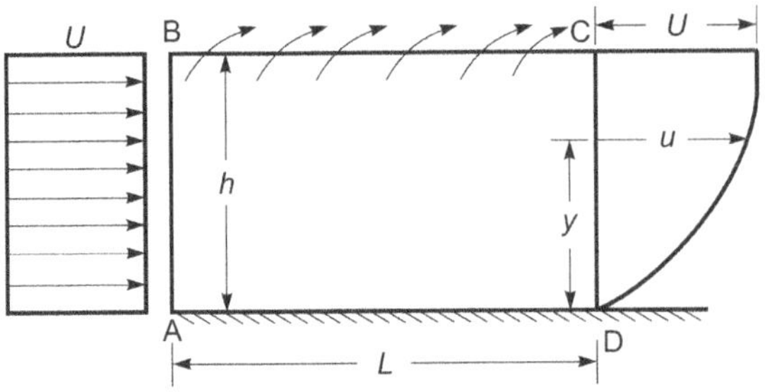

Fig. 3.26 *Problem 3.5*

section DC of plate is $\frac{u}{U} = (y/h)^{1/7}$. Estimate the mass rate of outflow across BC per unit width of the plate.

$\left(Ans.\ q_{BC} = \frac{1}{8}\rho U h\right)$

3.6 Water flows in a pipe network shown in Fig. 3.27. Fill in the missing discharges so that continuity equation is satisfied.

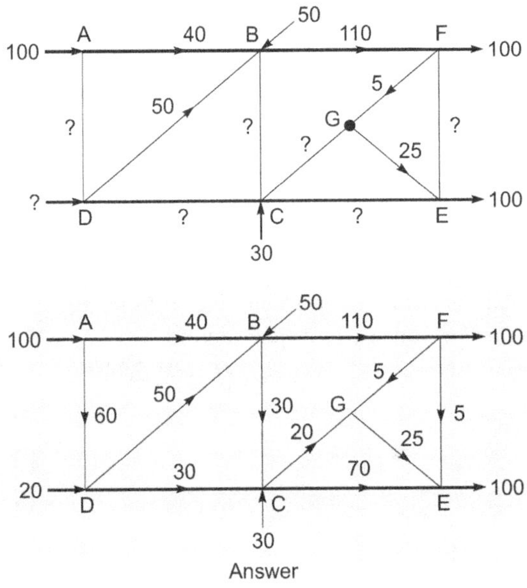

Fig. 3.27

3.7 The velocity in a flow is found to vary as $v_x = V_0/\left(1 - \frac{x}{L}\right)^2$. Determine the acceleration at a point $x = 0.5$ m when $L = 1.5$ m and $V_0 = 3.0$ m/s.

(*Ans.* $a = 71.40$ m/s^2)

3.8 A two-dimensional duct contains a straight sided contraction as in Fig. 3.28. The depth is constant at 30 cm. At a certain time an incompressible fluid flows in the duct at a rate of 0.3 m^3/s and is decreasing at the rate of 0.1 m^3/s per second.

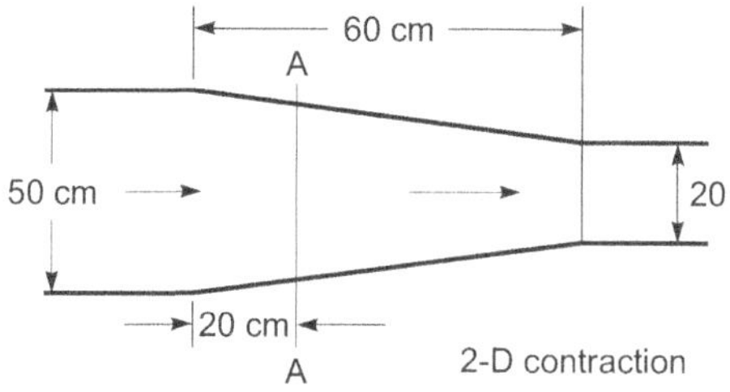

Fig. 3.28

Estimate the acceleration at the section AA 20 cm from the inlet.

(*Ans.* $a = -8.648$ m/s^2)

3.9 A 90 cm diameter pipe is reduced to 30 cm diameter in a length of 1.5 m. Water flows through this pipe at a rate of 280 L/s. If at an instant the discharge is found to decrease at the rate of 60 L/s per second, estimate the total acceleration at a distance of 75 cm from the 90 cm diameter inlet.

(*Ans.* $a = 1.0953$ m/s^2)

3.10 A 40 cm diameter pipe is reduced uniformly to 20 cm diameter in a length of 0.5 m. If the steady discharge of water through this pipe is 100 L/s, determine the acceleration at 10 cm from the 20 cm diameter end (i) when the flow is steady and (ii) when the flow is increasing at a rate of 600 L/s per minute.

(*Ans.* (i) $a = 16.289$ m/s^2, (ii) $a = 16.51$ m/s^2)

3.11 For the following flows find the equation of the streamline passing through the indicated point:

(i) $\mathbf{V} = 2x\mathbf{i} - y\mathbf{j} - z\mathbf{k}$ ..... passing through (1, 1, 1)

(ii) $\mathbf{V} = 4y\mathbf{i} - 3x\mathbf{j}$ ...... passing through (– 1, 2)

(iii) $\mathbf{V} = -x\mathbf{i} + 2y\mathbf{j} + (3 - z)\mathbf{k}$ ...... passing through (1, 1, 2)

(*Ans.* $x^{1/2}y = 1$ and $x^{1/2}z = 1$, (ii) $3x^2 + 4y^2 = 19$, (iii) $xy^{1/2} = 1$ and $x = 3 - z$)

3.12 Calculate the circulation per unit area around a closed contour not including the centre in the following types of vortex motion: (i) $v_s = cr$ and (ii) $v_s = c/r$, where v_s = tangential velocity at a radial distance r and c = a constant.

(*Ans.* (i) $\Gamma/A = 2C$ and (ii) $\Gamma/A = 0$)

313 For the following set of velocity components verify whether the continuity equation is satisfied. If so, determine the (i) vorticity vector and (ii) acceleration vector at point A (1, 1, 1)

$$u = 2x^2 + 3y$$
$$v = -2xy + 3y^3 + 3zy$$
$$w = -\frac{3}{2}z^2 - 2xz - 9y^2 z$$

(*Ans.* (i) $\zeta_z = -(2y + 3)$; $\zeta_x = -(18yz + 3y)$, $\zeta_y = 2z$, (ii) $a_x = 32$, $a_y = -7.5$ and $a_z = 93$)

3.14 Check whether the following sets of velocity components satisfy the continuity equation of steady incompressible flow.

(i) $u = 4x + 2y - 3$, $v = 2x + 4y + 3$ (ii) $u = 4xy + y^2$, $v = 6xy + 3x + 2$

(iii) $u = 2x^2 + y^2$, $v = -4xy$ (iv) $u = x^3 + y^3$, $v = x - 3x^2 y$

(v) $u = \dfrac{C(y^2 - x^2)}{(x^2 + y^2)^2}$, $v = \dfrac{-2Cxy}{(x^2 + y^2)^2}$

(*Ans.* (i) not satisfied, (ii) not satisfied, (iii) satisfied (iv) satisfied, (v) not satisfied)

3.15 Calculate the unknown velocity component in the following so that the equation of continuity is satisfied.

(i) $u = \dfrac{2}{3}xy^3 - x^2 y$, $v = ?$ (ii) $u = Aye^x$, $v = ?$

$\left(Ans.\ \text{(i) } v = -\dfrac{1}{6}y^4 + xy^2 + f(x),\ \text{(ii) } v = -\dfrac{Ay^2}{2}e^x + f(x)\right)$

3.16 For a two-dimensional incompressible flow, the x component of velocity and boundary conditions are as follows:

$u = x^2 y^2 + 2xy$ and at $y = 0$, $u = v = 0$

Determine the y component of the velocity.

$\left(Ans.\ v = -\dfrac{2}{3}xy^3 - y^2\right)$

3.17 Given the components of a velocity vector in a three-dimensinal flow, as below, determine the missing component.

(i) $u = 2x^2 + 2xy$, $v = 4yz$, $w = ?$

(ii) $u = 3x^2$, $v = 4yxz$, $w = ?$

(iii) $u = x^3 + y^2 + 2z^2$, $v = -x^2 y - yz - xy$, $w = ?$

{(*Ans.* (i) $w = -(4xz + 2yz + 2z^2) + f(x, y)$, (ii) $w = -(6xz + 2xz^2) + f(x, y)$, (iii) $w = 2x^2 z - \dfrac{z^2}{2} - xz + f(x, y)$}

3.18 If the velocity u in a two-dimensional turbulent boundary layer over a flat plate is given by $\dfrac{u}{U_0} = (y/\delta)^{1/7}$ where U_0 = constant and δ = boundary layer thickness = $K\,x^{4/5}$, obtain an expression for the velocity component v in the y direction.

$$\left(Ans.\ v = \frac{1}{10}\frac{KU_0(y/\delta)^{8/7}}{(\delta/K)^{1/4}}\right)$$

3.19 For the following sets of velocity components obtain the relevant stream functions.

(i) $u = 6y$ (ii) $u = \frac{y^3}{3} + 2x - x^2y$

$v = 6x$ $v = xy^2 - 2y - \frac{x^2}{3}$

(iii) $u = -A \ln x$

$v = A\frac{y}{x} + x^2$

$$\left(Ans.\ \text{(i)}\ \psi = 3(y^2 - x^2) + C,\right.$$

$$\text{(ii)}\ \psi = \frac{y^4}{12} - \frac{x^2y^2}{2} + 2xy + \frac{x^3}{9},$$

$$\left.\text{(iii)}\ \psi = -Ay\ln x - \frac{x^3}{3} + C\right)$$

3.20 The velocity components of a steady, two-dimensional flow of an ideal fluid, are $u = 2xy$ and $v = a^2 + x^2 - y^2$. Show that the velocity potential exists and determine the same.

(*Ans.* $\phi = a^2y + x^2y - y^3/3$ + constant)

3.21 If the velocity potential $\phi = (x^3/3) - x^2 - xy^2 + y^2$, determine the corresponding stream function.

(*Ans.* $\psi = x^2y - 2xy - (y^3/3)$ + constant)

3.22 Find the velocity potential if the velocity field is

$$u = \frac{x}{x^2 + y^2};\ v = \frac{y}{x^2 + y^2}$$

$$\left(Ans.\ \phi = \frac{1}{2}\log(x^2 + y^2)\right)$$

3.23 Verify whether the following flow fields are rotational. If so, determine the components of rotation about various axes.

(i) $u = xyz$ (ii) $u = xy$

$v = zx$ $v = \frac{1}{2}(x^2 - y^2)$

$w = \frac{1}{2}yz^2 - xy$

$$\left(Ans.\ \text{(i) Rotational, } \omega_z = \frac{1}{2}z(1 - x),\right.$$

$$\omega_x = \frac{1}{2}(z^2/2 - 2x),\ \omega_y = \frac{1}{2}y(x + 1)$$

$$\left.\text{(ii) Irrotational, } \omega_x = \omega_y = 0\right)$$

3.24 Calculate the velocity at the point (3, 3) for the following stream function.

(i) $\psi = -x \ln xy + x$

(ii) $\psi = \frac{1}{2}(y^2 - x^2) + xy + 6$

(*Ans.* (i) $V = 2.42$ units,
(ii) $V = 6$ units)

3.25 Given the following stream functions, determine the corresponding potential functions. Also, estimate the discharge, per unit depth, in the z direction passing between the streamlines through the points (1, 3) and (3, 3).

(i) $\psi = 3xy$

(ii) $\psi = \frac{3}{2}(y^2 - x^2)$

$$\left(Ans.\ \phi = \frac{3}{2}(x^2 - y^2) + c; 18\text{ units},\right.$$

$$\left.\text{(ii)}\ \phi = 3xy; -12\text{ units}\right)$$

3.26 Verify whether the following functions are valid potential functions.

(i) $\phi = y^3 - 3x^2y$ (ii) $\phi = y^4 - 6x^2y^2$

(iii) $\phi = x^2 - 3x^2y$ (iv) $\phi = x^3 - y^3$

(*Ans.* (i) Yes, (ii) No, (iii) No, (iv) No.)

3.27 Determine the corresponding conjugate functions relating to the following functions representing irrotational fluid flows.

(i) $\phi = Ur\cos\theta$

(ii) $\psi = -m\theta + k\ln r$

(*Ans.* (i) $\psi = Ur\sin\theta + C$,
(ii) $\phi = -k\theta - m\ln r$)

3.28 The potential flow around a cylinder with circulation is given by

$$\phi = U\left(r + \frac{a^2}{r}\right)\cos\theta - C\theta$$

where a = radius of the cylinder, U = velocity far away from the cylinder and C = a measure of circulation. Calculate the corresponding stream function and the velocities v_r and v_θ.

$$\left(Ans.\ \psi = U\left(r - \frac{a^2}{r}\right)\sin\theta - C\ln r;\right.$$

$$v_r = U\left(1 - \frac{a^2}{r^2}\right)\cos\theta,$$

$$\left.v_\theta = -U\sin\theta\left(1 + \frac{a^2}{r^2}\right) - \frac{C}{r}\right)$$

4

Energy Equation and Its Applications

CONCEPT REVIEW

4.1 BERNOULLI EQUATION

Euler equation: For the frictionless flow along a streamline of an incompressible fluid the relationship among the pressure, elevation and velocity is given by the Euler equation.

$$\frac{\partial V}{\partial t} + \frac{1}{\rho}\frac{\partial p}{\partial s} + g\frac{\partial z}{\partial s} + V\frac{\partial V}{\partial s} = 0 \qquad (4.1)$$

Bernoulli equation: Integration of the Euler equation for steady, incompressible fluid flow, without friction, yields the Bernoulli equation

$$\frac{p}{\gamma} + \frac{V^2}{2g} + Z = \text{constant} = \text{H} \qquad (4.2)$$

It can be shown that the Bernoulli equation is applicable across the streamlines also if the flow is irrotational.

In Eq. (4.2) the term $V^2/2g$ represents kinetic energy of the flow per unit weight of the fluid. Similarly, Z represents potential energy per unit weight. The term p/γ represents flow work, i.e. the work done by the fluid on the surroundings. All the terms in Eq. 4.2 have unit of [L] = (N.m/N) of fluid. The constant H is called the *total energy*. For any two points in a steady irrotational flow field of an ideal fluid,

$$\frac{(p_1 - p_2)}{\gamma} + \left(\frac{V_1^2}{2g} - \frac{V_2^2}{2g}\right) + (Z_1 - Z_2) = \text{H}_1 - \text{H}_2 = 0 \qquad (4.3)$$

4.2 PRACTICAL APPLICATIONS OF BERNOULLI EQUATION

In practical applications of Bernoulli equation the restriction of frictionless flow is accommodated by introducing a loss of energy term and the restriction of irrotational flow is waived in most of the cases. Equation 4.2 is used as a special case of the general energy equation. The general energy equation dealing with the conservation of energy is written for steady, incompressible fluid flow between two sections 1 and 2 as

$$\text{H}_1 + \text{H}_E - \text{H}_L = \text{H}_2 \qquad (4.3a)$$

where H_1 = total energy at section 1

H_E = energy input to the system between sections 1 and 2

H_L = energy loss due to friction, etc. between sections 1 and 2

H_2 = total energy at section 2.

Energy is transferred to the system as mechanical work done on the fluid by a pump. Similarly, energy is extracted from the system by a turbine. For incompressible fluid flow all non-recoverable energy such as change of internal energy and heat transfer are usually clubbed under a common term *energy loss.*

Thus for a fluid flow system shown in Fig. 4.1 the Bernoulli equation is

$$\text{H}_1 + \text{H}_E - \text{H}_L = \text{H}_2$$

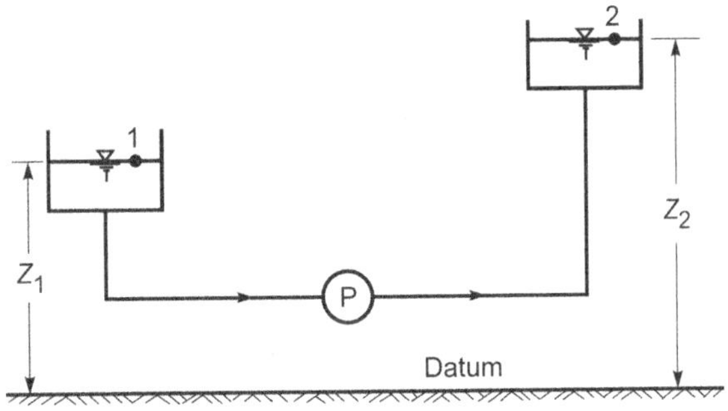

Fig. 4.1

where
$$H_1 = \frac{p_1}{\gamma} + \frac{V_1^2}{2g} + Z_1$$

$$H_2 = \frac{p_2}{\gamma} + \frac{V_2^2}{2g} + Z_2$$

$H_E = H_P$ = energy input per unit weight of fluid per second by the pump

H_L = energy loss between points 1 and 2

4.3 ENERGY EQUATION

The general equation for conservation of energy for an incompressible fluid flow can be written as

$$\left(\frac{p_1}{\gamma} + \frac{V_1^2}{2g} + Z_1\right) + q_w + H_E = \left(\frac{p_2}{\gamma} + \frac{V_2^2}{2g} + Z_2\right) + (e_2 - e_1) \tag{4.4}$$

where q_w = heat added per unit weight of fluid

e_1, e_2 = internal energy per unit weight of fluid at the respective states

H_E = external work done (i.e. shaft work added) on the fluid per unit weight of fluid from a device such as a pump.

If the total head $H = \frac{p}{\gamma} + \frac{V^2}{2g} + Z$, then Eq. 4.4 is written as

$$H_1 + H_E - [(e_2 - e_1) - q_w) = H_2$$

The term:

$$(e_2 - e_1) - q_w = \text{(reversible + irreversible) head}$$

In incompressible fluid flow irreversible head is called *head loss* H_L and represents energy loss per unit weight of fluid due to friction and other causes. Thus for an incompressible fluid

$$\begin{bmatrix}\text{Total head}\\\text{at section 1}\end{bmatrix} + \begin{bmatrix}\text{Head added due}\\\text{to a machine such}\\\text{as a pump}\end{bmatrix} - [\text{Head loss}] = \begin{bmatrix}\text{Total head}\\\text{at section 2}\end{bmatrix}$$

or
$$\boxed{H_1 + H_E - H_L = H_2} \tag{4.3a}$$

When a pump is used $H_E = H_P$ (a positive quantity), and when a turbine is used $H_E = H_t$ (a negative quantity).

4.3.1 Hydraulic Grade Line

A line joining the piezometric heads at various points in a flow is known as the *hydraulic grade line* (HGL). As the piezometric head $h = \frac{p}{\gamma} + Z$, the HGL represents the variation of $h\left(= \frac{p}{\gamma} + Z\right)$ measured above a datum, (Fig. 4.2).

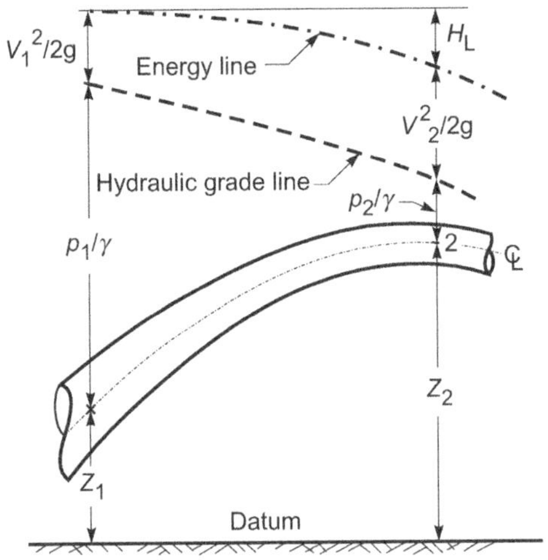

Fig. 4.2

4.3.2 Energy Line

The total energy $$H = \left(\frac{p}{\gamma} + Z\right) + \frac{V^2}{2g} = h + \frac{V^2}{2g}$$

A line joining the elevation of total energy of a flow measured above a datum is known as *energy line*, (Fig. 4.2). The energy line lies above the HGL by an amount of $V^2/2g$.

4.3.3 Kinetic Energy Correction Factor, α

In one-dimensional method of analysis, the average velocity V is used to represent the velocity at a cross section. The actual velocity distribution in the cross section may be non-uniform. Hence, the kinetic energy calculated by using V must be multiplied by a correction factor to obtain proper kinetic energy at the cross section due to non-uniform velocity distribution.

Thus the velocity head in the Bernoulli equation will be $\alpha \frac{V^2}{2g}$ where

$$\boxed{\alpha = \frac{1}{A}\int\left(\frac{v}{V}\right)^3 dA} \qquad (4.5)$$

The term α is called the *kinetic energy correction factor*. For uniform velocity distribution $\alpha = 1.0$ and in all other cases it will be greater than 1.0. Greater the non uniformity in velocity distribution larger will be the value of α. For laminar flow through a pipe, $\alpha = 2.0$ and for turbulent flow through a pipe its value varies from 1.01 to 1.20. In the absence of specific information about the value of α, it is usual practice to assume its value as unity.

4.4 POWER

In the case of work done over a fluid the power input into the flow is

$$\boxed{\mathbf{P} = \gamma Q H_m \text{ in watts}} \qquad (4.6)$$

where γ = unit weight of fluid in N/m^3,
Q = discharge in m^3/s and
H_m = head added to the flow, in m

In a pump $H_m = H_p$ is positive. In a turbine $H_m = H_t$ is negative and power is extracted from the flow.
If η_p = efficiency of the pump, the power input required at the pump is

$$\boxed{\mathbf{P}_{in} = \frac{\gamma Q H_m}{\eta_p}} \qquad (4.7)$$

In the case of a turbine, in η_t is the efficiency of the turbine, power delivered by the turbine is

$$\boxed{\mathbf{P}_{out} = \gamma Q H_m \eta_t} \qquad (4.8)$$

WORKED EXAMPLES

4.1 A pipeline is 15 cm in diameter and is at an elevation of 100.00 m at section A. At section B it is at an elevation of 107.00 m and has a diameter of 30 cm. When a discharge of 50 L/s of water is passed through this pipe the pressure at section A is observed to be 30 kPa. The energy loss in the pipe is 2 m. Calculate the pressure at B when the flow is (i) from A to B and (ii) from B to A.

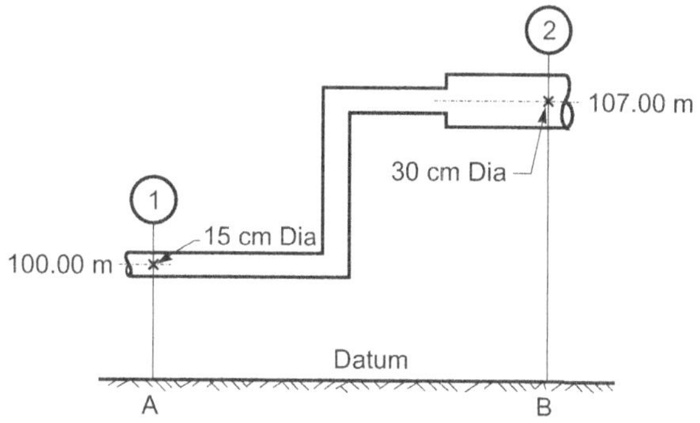

Fig. 4.3

Solution:

$$Q = 0.05 \text{ m}^3/\text{s}$$

Let suffixes 1 and 2 refer to sections A and B respectively.

$$V_1 = \frac{Q}{A_1} = \frac{0.05}{\frac{\pi}{4}\times(0.15)^2} = 2.829 \text{ m/s}$$

$$V_2 = \frac{Q}{A_2} = \frac{0.05}{\frac{\pi}{4}\times(0.30)^2} = 0.7074 \text{ m/s}$$

γ = unit weight of water = 998 × 9.81/1000
= 9.79 kN/m^3

(i) When the flow is from A to B:
Taking the atmospheric pressure as zero

$$\left(\frac{p_1}{\gamma} + \frac{V_1^2}{2g} + Z_1\right) = \frac{p_2}{\gamma} + \frac{V_2^2}{2g} + Z_2 + H_L$$

$$\frac{30}{9.79} + \frac{(2.829)^2}{2\times 9.81} + 100.00$$

$$= \frac{p_2}{\gamma} + \frac{(0.7074)^2}{2\times 9.81} + 107.00 + 2.0$$

$$3.064 + 0.4080 + 100.00$$

$$= \frac{p_2}{\gamma} + 0.0255 + 107.0 + 2.0$$

$$\frac{p_2}{\gamma} = -5.554 \text{ m (gauge)}$$

$$p_2 = -5.554 \times 9.79$$

$$= \mathbf{-54.37 \text{ kPa (gauge)}}$$

(ii) When the flow is from B to A:
Taking atmospheric pressure as zero,

$$\left(\frac{p_2}{\gamma} + \frac{V_2^2}{2g} + Z_2\right) - H_L = \left(\frac{p_1}{\gamma} + \frac{V_1^2}{2g} + Z_2\right)$$

$$\frac{p_2}{\gamma} + \frac{(0.7074)^2}{2 \times 9.81} + 107.00 - 2.00$$

$$= \frac{30}{9.79} + \frac{(2.829)^2}{2 \times 9.81} + 100.00$$

$$\frac{p_2}{\gamma} + 0.0255 + 107.00 - 2.00$$

$$= 3.064 + 0.4080 + 100.00$$

$$p_2/\gamma = -1.554 \text{ m (gauge)}$$

$$p_2 = -1.554 \times 9.79$$

$$= \mathbf{-15.21 \text{ kPa (gauge)}}$$

4.2 In a smooth pipe of uniform diameter 25 cm a pressure of 50 kPa was observed at section 1 which was at elevation 10.00 m. At another section 2 at elevation 12.00 m the pressure was 20 kPa and the velocity was 1.25 m/s. Determine the direction of flow and the head loss between these two sections. The fluid in the pipe is water.

Solution: Refer to Fig. 4.4

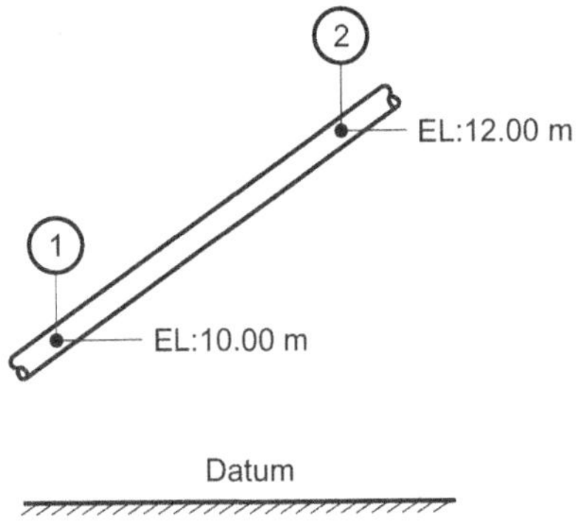

Fig. 4.4

Discharge $Q = A_1V_1 = A_2V_2$

$$Q = \frac{\pi}{4} \times (0.25)^2 \times 1.25 = 0.0614 \text{ m}^3/\text{s}$$

$$= 61.4 \text{ L/s}$$

As $A_1 = A_2$, $V_1 = 1.25$ m/s

$$\gamma = 998 \times 9.81/1000 = 9.79 \text{ kN/m}^3$$

Taking atmospheric pressure as zero,

Total energy at section 1,

$$H_1 = \frac{p_1}{\gamma} + \frac{V_1^2}{2g} + Z_1$$

$$H_1 = \frac{50.0}{9.79} + \frac{(1.25)^2}{2 \times 9.81} + 10.00$$

$$= 5.107 + 0.0796 + 10.00 = 15.187 \text{ m}$$

Total energy at section 2,

$$H_2 = \frac{p_2}{\gamma} + \frac{V_2^2}{2g} + Z_2$$

$$H_2 = \frac{20.0}{9.79} + \frac{(1.25)^2}{2 \times 9.81} + 12.00$$

$$= 2.043 + 0.0796 + 12.00 = 14.123 \text{ m}$$

As $H_1 > H_2$, the flow is from section 1 to 2.

Head loss between the sections 1 and 2, i.e.

$$H_L = H_1 - H_2 = 15.187 - 14.123 = \mathbf{1.064 \text{ m}}$$

4.3 Figure 4.5 shows a nozzle at the end of a pipe discharging oil from a tank to atmosphere. Estimate the discharge from the nozzle when the head H in the tank is 4.0 m. The loss in the pipe can be taken as 20 $V^2/2g$, where V = velocity in the pipe. The loss of energy in the nozzle can be assumed to be zero. Also determine the pressure at the base of the nozzle.

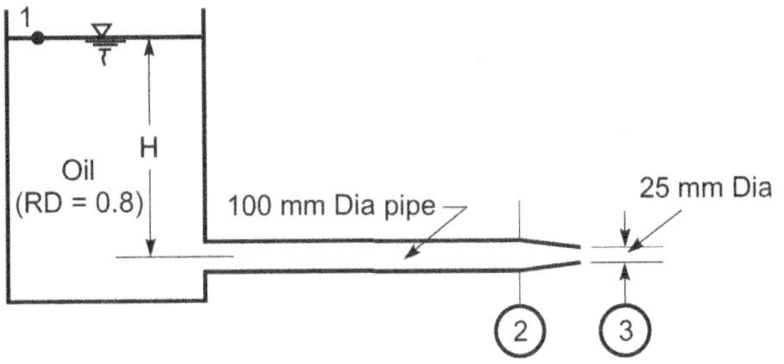

Fig. 4.5

Solution: By continuity criterion,

$$V_2 \frac{\pi}{4} (D)^2 = V_3 \frac{\pi}{4} (d)^2 = Q$$

Velocity in the pipe $V_2 = V_3 \cdot \left(\frac{d}{D}\right)^2 = \left(\frac{25}{100}\right)^2 V_3$

$$= \frac{1}{16} V_3$$

Apply Bernoulli equation to points 1 and 3 with the centre line of the pipe as datum and atmospheric pressure as zero. The velocity at point 1 can be taken as zero.

$$\frac{p_1}{\gamma} + \frac{V_1^2}{2g} + Z_1 = \frac{p_3}{\gamma} + \frac{V_3^2}{2g} + Z_3 + H_L$$

$$0 + 0 + H = 0 + \frac{V_3^2}{2g} + 0 + 20 \frac{V_2^2}{2g}$$

$$H = \frac{V_3^2}{2g} + 20 \times \left(\frac{1}{16}\right)^2 \frac{V_3^2}{2g}$$

$$= 1.07813 \frac{V_3^2}{2g}$$

As H = 4.0 m, $V_3 = \left(\frac{2 \times 9.81 \times 4.0}{1.07813}\right)^{1/2}$

$$= 8.5319 \text{ m/s}$$

Discharge $Q = \frac{\pi}{4} \times (0.025)^2 \times 8.5319$

$= 4.1881 \times 10^{-3}$ m^3/s

= **4.1881 L/s**

V_2 = Velocity in the pipe

$= \frac{1}{16} \times 8.53188 = 0.5332$ m/s

Loss of head in the pipe

$$H_L = 20 \times \frac{V_2^2}{2g}$$

$$= 20 \times \frac{(0.5332)^2}{2 \times 9.81} = 0.2899 \text{ m}$$

Applying Bernoulli equation to points 1 and 2.

$$0 + 0 + H = \frac{p_2}{\gamma} + \frac{V_2^2}{2g} + 0 + H_L$$

$$4.0 = \frac{p_2}{\gamma} + \frac{(0.5332)^2}{2 \times 9.81} + 0 + 0.2899$$

$$\frac{p_2}{\gamma} = 4.0 - 0.3044 = 3.6956 \text{ m}$$

$\gamma = (0.8 \times 998 \times 9.81/1000)$

$= 7.832$ kN/m^3

p_2 = pressure at the base of the nozzle

$= 3.6956 \times 7.832 =$ **28.944 kPa**

4.4 Water flows up a tapered pipe as shown in Fig. 4.6 Find the magnitude and direction of the deflection h of the differential mercury manometer corresponding to a discharge of 120 L/s. The friction in the pipe can be completely neglected.

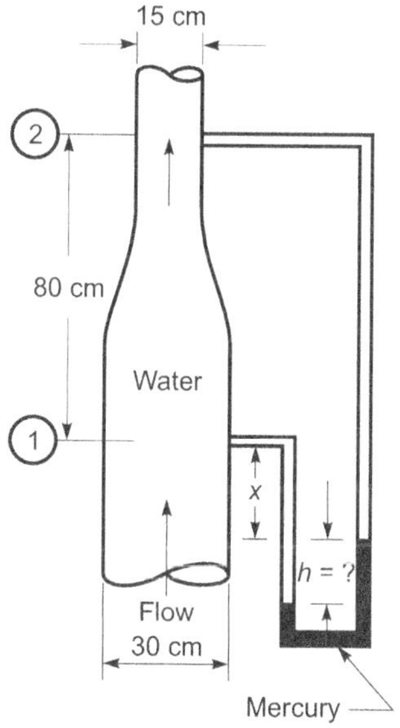

Fig. 4.6

Solution:

Let S = Relative density of mercury.

For the manometer:

Considering the elevation of section 1 as datum

$$\frac{p_1}{\gamma} + x + h = \frac{p_2}{\gamma} + 0.8 + x + Sh$$

$$\left(\frac{p_1}{\gamma} - \frac{p_2}{\gamma}\right) - 0.8 = (S - 1)h$$

$$= (13.6 - 1)h = 12.6\,h \qquad (1)$$

By continuity criterion,

$$Q = \frac{\pi}{4} \times (0.30)^2 \times V_1$$

$$= \frac{\pi}{4} \times (0.15)^2 \times V_2 = 0.120$$

$V_1 = 1.6977$ m/s, $V_2 = 6.79$ m/s

By Bernoulli equation for points 1 and 2,

$$\frac{p_1}{\gamma} + \frac{V_1^2}{2g} + Z_1 = \frac{p_2}{\gamma} + \frac{V_2^2}{2g} + Z_2$$

$$\left(\frac{p_1}{\gamma} - \frac{p_2}{\gamma}\right) + 0 - 0.8 = \frac{V_2^2 - V_1^2}{2g}$$

$$= \frac{(6.79)^2 - (1.6977)^2}{2 \times 9.81} = 2.2034$$

$$\left(\frac{p_1 - p_2}{\gamma}\right) - 0.8 = 12.6h = 2.2034$$

Therefore $h = \frac{2.2034}{12.6} = 0.175$ m

= **17.5 cm**

Deflection: As indicated in Fig. 4.6, the manometer limb connected to section 1 will be having a smaller column of mercury than the other limb.

4.5 Kerosene ($\rho = 800$ kg/m^3) flows through the system shown in Fig. 4.7. The outflow is to the atmosphere.

$Z_0 = 20.0$ m, $D_1 = 10$ cm

and $h_1 = 4$ cm

$D_2 = 7$ cm

$D_3 = D_4 = 10$ cm

$D_5 = 5$ cm

The pipe is horizontal. The fluid in the manometer is mercury. Neglecting friction, calculate (i) discharge Q (ii) velocity V_5 (iii) Z_3 and (iv) h_4

Solution:

By continuity consideration, $V_1 D_1^2 = V_2 D_2^2$

Hence $V_1 = V_2(D_2/D_1)^2 = V_2\left(\frac{7.0}{10.0}\right)^2$

$= 0.49\ V_2$

$\gamma = 0.800 \times 9.81 = 7.848$ kN/m^3

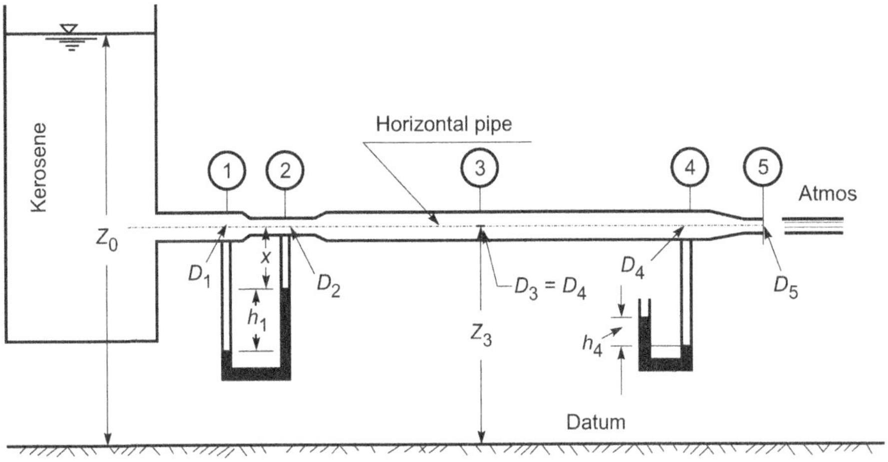

Fig. 4.7

Relative density of kerosene $= \dfrac{800}{998} = 0.8016$

For manometer connected to 1 and 2:

$$\frac{p_1}{\gamma} + h_1 + x = \frac{p_2}{\gamma} + \left(\frac{13.6}{0.8016}\right) h_1 + x$$

$$(p_1 - p_2)/\gamma = 15.966 \times 0.04$$

$$= 0.6386 \text{ m of kerosene.}$$

By applying Bernoulli equation to sections 1 and 2,

$$\frac{p_1}{\gamma} + \frac{V_1^2}{2g} + Z_1 = \frac{p_2}{\gamma} + \frac{V_2^2}{2g} + Z_2$$

As $Z_1 = Z_2$, $\dfrac{(p_1 - p_2)}{\gamma} = \dfrac{V_2^2 - V_1^2}{2g}$

$$= \frac{V_2^2}{2g}[1 - (0.49)^2] = 0.7599\,\frac{V_2^2}{2g}$$

$\therefore \quad 0.7599\,\dfrac{V_2^2}{2g} = 0.6386$

$$V_2 = 4.06 \text{ m/s,}$$

$$V_1 = 0.49 \times 4.06 = 1.9899 \text{ m/s}$$

(i) Discharge $\quad Q = 4.06 \times \dfrac{\pi}{4} \times (0.07)^2$

$$= 0.015625 \text{ m}^3/\text{s}$$

$$= \mathbf{15.625\ L/s}$$

(ii) Velocity $\quad V_5 = \dfrac{0.015625}{\frac{\pi}{4}\times(0.05)^2} = 7.9577 \text{ m/s}$

(iii) $\quad V_4 = V_1 = 1.9899 \text{ m/s}$

By applying Bernoulli equation to sections 4 and 5,

$$\frac{p_4}{\gamma} + \frac{V_1^2}{2g} + Z_4 = \frac{P_{atm}}{\gamma} + \frac{V_5^2}{2g} + Z_5$$

$$P_{atm} = 0, \text{ and } Z_5 = Z_4$$

$$\frac{p_4}{\gamma} = (V_2^5 - V_1^2)/2g$$

$$= \frac{(7.9577)^2 - (1.9899)^2}{2 \times 9.81}$$

$$= 3.0258 \text{ m}$$

Total energy at point 4:

$$H_4 = \frac{p_4}{\gamma} + \frac{V_1^2}{2g} + Z_4$$

$$= 3.0258 + \frac{(1.9899)^2}{2 \times 9.81} + Z_3$$

As there is no loss of energy,

$$H_4 = Z_0 = 20.00$$

Hence $\quad Z_3 = 20.0 - 3.0258 - \dfrac{(1.9899)^2}{2 \times 9.81}$

$$= \mathbf{16.7724\ m}$$

(iv) For the manometer at point 4

$$\frac{p_4}{\gamma} = h_4\left(\frac{13.60}{0.8016} - 1\right) = 3.0258$$

$$h_4 = \frac{3.0258}{15.97} = 0.18946 \text{ m}$$

$$= \mathbf{0.189\ m}$$

4.6 A siphon consisting of a pipe of 15 cm diameter is used to empty kerosene oil (RD = 0.80) from a tank A. The siphon discharges to the atmosphere at an elevation of 1.00 m. The oil surface in the tank is at an elevation of 4.00 m. The centreline of the siphon pipe at its highest point C is at an elevation of 5.50 m. Estimate (i) the discharge in the pipe, and (ii) pressure at point C. The losses in the pipe can be assumed to be 0.5 m up to the summit and 1.2 m from the summit to the outlet.

Solution: Consider points 1 and 2 at the surface of the oil in tank A and at the outlet as in Fig. 4.8. The velocity V_1 can be assumed to be zero. Between points 1 and 2, by Bernoulli equation we have

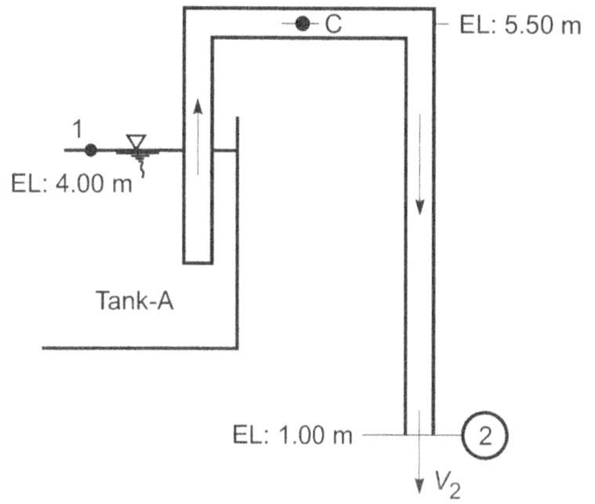

Fig. 4.8

$$\frac{p_1}{\gamma} + \frac{V_1^2}{2g} + Z_1 = \frac{p_2}{\gamma} + \frac{V_2^2}{2g} + Z_2 + H_{L(1-2)}$$

$$0 + 0 + 4.00 = 0 + \frac{V_2^2}{2g} + 1.00 + (0.5 + 120)$$

$$\frac{V_2^2}{2g} = 4.00 - 2.70 = 1.30$$

$$V_2 = (2 \times 9.81 \times 1.30)^{1/2} = 5.05 \text{ m/s}$$

(i) Discharge $= Q = \frac{\pi}{4} \times (0.15)^2 \times 5.05$

$= 0.0892 \text{ m}^3/\text{s}$

$=$ **89.2 L/s**

(ii) Using suffix 3 to denote the conditions at the summit C, by applying Bernoulli equation to points 1 and 3,

$$\frac{p_1}{\gamma} + \frac{V_1^2}{2g} + Z_1 = \frac{p_3}{\gamma} + \frac{V_3^2}{2g} + Z_3 + H_{L(1-3)}$$

$V_3 = V_2 = 5.05$ m/s as the pipe is of uniform cross section.

$$\therefore\ 0 + 0 + 4.0 = \frac{p_3}{\gamma} + \frac{(5.05)^2}{2 \times 9.81} + 5.50 + 0.5$$

$$\frac{p_3}{\gamma} = 4.0 - 1.30 - 5.50 - 0.5 = -3.3 \text{ m}$$

$$p_3 = -3.3 \times (0.8 \times 998 \times 9.81)/1000$$

$=$ **– 25.85 kPa (gauge)**

4.7 The outlet at the bottom of a tank is so formed that the velocity of water at point A (see Fig. 4.9) is 2.0 times the mean velocity within the outlet pipe.

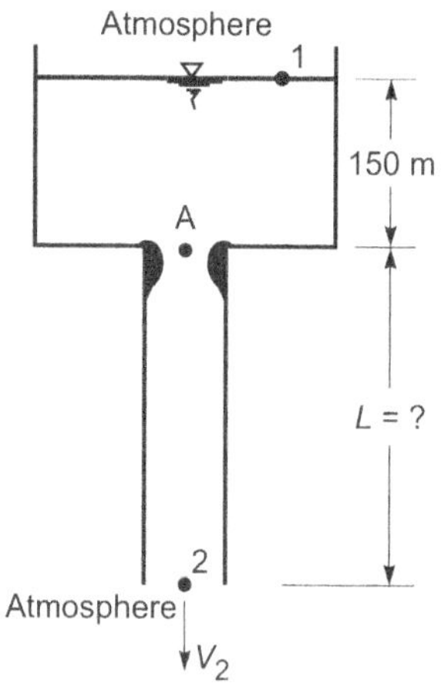

Fig. 4.9

What is the greatest length of pipe L which may be used without producing cavitation? Atmospheric pressure = 95.480 kPa (abs) and vapour pressure = 4.00 kPa (abs). Neglect all other losses.

Solution:

p_A = vapour pressure = 4.00 kPa (abs)

$p_1 = p_2$ = atmospheric pressure = 95.48 kPa (abs)

By applying Bernoulli equation to points 1 and A.

$$\frac{p_1}{\gamma} + \frac{V_1^2}{2g} + Z_1 = \frac{p_A}{\gamma} + \frac{V_A^2}{2g} + Z_A$$

$$\left(\frac{95.48}{9.79}\right) + 0 + 1.5 = \left(\frac{4.00}{9.79}\right) + \frac{V_A^2}{2g} + 0$$

$$\therefore\ \frac{V_A^2}{2g} = \frac{95.48 - 4.00}{9.79} + 1.5 = 10.844 \text{ m}$$

$$V_A = (2 \times 9.81 \times 10.844)^{1/2}$$

$$= 14.59 \text{ m/s}$$

$$V_2 = \frac{1}{2} V_A = 14.59/2 = 7.29 \text{ m/s}$$

By applying Bernoulli equation to points 1 and 2, with datum at point 2,

$$\frac{p_1}{\gamma} + \frac{V_1^2}{2g} + Z_1 = \frac{p_2}{\gamma} + \frac{V_2^2}{2g} + Z_2$$

$$\frac{(95.48)}{9.79} + 0 + (L + 1.50)$$

$$= \left(\frac{95.48}{9.79}\right) + \frac{(7.29)^2}{2 \times 9.81} + 0$$

$L = 2.709 - 1.50 =$ **1.21 m**

4.8 A closed tank containing water is partly filled with water and the air space above it is under pressure. A 5 cm hose connected to the tank discharges water to atmosphere on to the roof of a building 3.0 m above the level of water in the tank (Fig. 4.10). If friction losses are 1.5 m of water, what air pressure must be maintained in the tank to deliver 15 L/s to the roof?

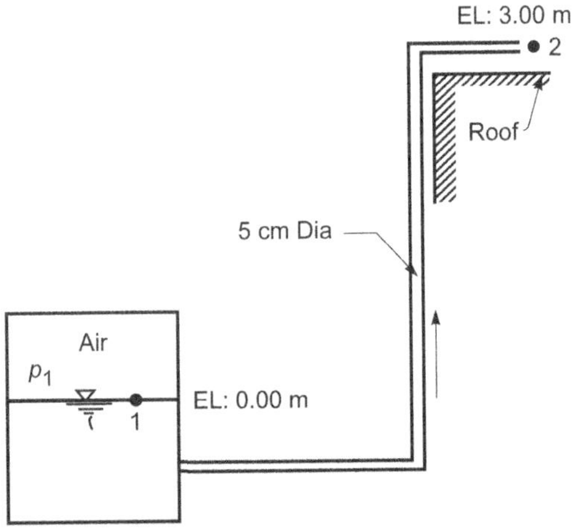

Fig 4.10

Solution:

$$\text{Discharge } Q = 0.015 \text{ m}^3/\text{s} = \frac{\pi}{4} \times (0.05)^2 \times V_2$$

$$V_2 = 7.639 \text{ m/s}$$

By applying Bernoulli equation to points 1 and 2,

$$\frac{p_1}{\gamma} + \frac{V_1^2}{2g} + Z_1 = \frac{p_2}{\gamma} + \frac{V_2^2}{2g} + Z_2 + H_L$$

$$\frac{p_1}{\gamma} + 0 + 0 = 0 + \frac{(7.639)^2}{2 \times 9.81} + 3.00 + 1.5$$

$$\frac{p_1}{\gamma} = 7.475 \text{ m and } \gamma = 9.79 \text{ kN/m}^3$$

Hence, $p_1 = 9.79 \times 7.475 =$ **73.18 kPa**

4.9 The loss of head in sudden expansion of a pipe is given by $H_L = (V_1 - V_2)^2/2g$ where suffixes 1 and 2 refer to sections before and after the expansion. A 10 cm pipe carrying oil (RD = 0.85) undergoes a sudden expansion to 20 cm diameter in a horizontal stretch of the pipeline. If the pressure difference between sections 2 and 1 is recorded as 660 Pa, estimate the discharge through the pipe and sketch the energy line and piezometric head line at the expansion, by considering the pressure at section 1, p_1 = 3.0 kPa.

Solution: Refer to Fig. 4.11.

$$D_1 = 10 \text{ cm}, A_1 = \frac{\pi}{4} \times (0.1)^2 = 7.854 \times 10^{-3} \text{ m}^2$$

$$D_2 = 20 \text{ cm}, A_2 = \frac{\pi}{4} \times (0.2)^2 = 3.142 \times 10^{-2} \text{ m}^2$$

$$Q = V_1 A_1 = V_2 A_2. \text{ Hence } V_2 = \left(\frac{D_1}{D_2}\right)^2 V_1 = \frac{1}{4} V_1$$

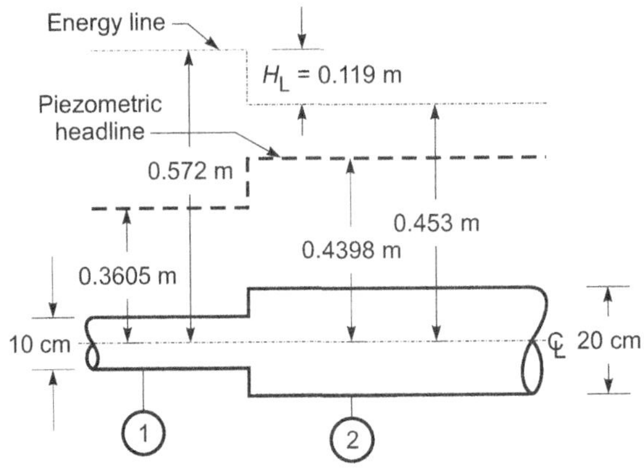

Fig. 4.11

Head loss at the expansion $H_L = (V_1 - V_2)^2/2g$

$$= \frac{(V_1 - V_1/4)^2}{2g} = \frac{9}{16}\frac{V_1^2}{2g}$$

By applying Bernoulli equation to sections 1 and 2

$$\frac{p_1}{\gamma} + \frac{V_1^2}{2g} + Z_1 = \frac{p_2}{\gamma} + \frac{V_2^2}{2g} + Z_2 + H_L$$

$$\left(\frac{p_2}{\gamma} - \frac{p_1}{\gamma}\right) = \frac{V_1^2}{2g} - \frac{V_2^2}{2g} + Z_1 - Z_2 - H_L$$

$$\frac{660}{(0.85 \times 9790)} = \frac{V_1^2}{2g}\left(1 - \frac{1}{16} - \frac{9}{16}\right) + 0 = \frac{3}{8}\frac{V_1^2}{2g}$$

$$V_1^2 = \frac{8}{3} \times 2 \times 9.81 \times 0.0793 = 4.1496$$

$$V_1 = 2.037 \text{ m/s}$$

$$\text{Discharge } Q = A_1 V_1 = 7.854 \times 10^{-3} \times 2.037 \text{ m}^3/\text{s}$$

$$= 0.016 \text{ m}^3/\text{s} = \mathbf{16.0 \text{ L/s}}$$

$$V_1 = 2.037 \text{ m/s}, V_1^2/2g = 0.2115$$

$$V_2 = \frac{2.037}{4} = 0.509 \text{ m/s},$$

$$V_2^2/2g = 0.0132 \text{ m}$$

$$(V_1 - V_2) = 1.528 \text{ m/s},$$

$$H_L = \frac{(V_1 - V_2)^2}{2g} = 0.1190 \text{ m}$$

$$\frac{p_1}{\gamma} = \frac{3000}{0.85 \times 9790} = 0.3605 \text{ m}$$

$$\frac{p_2}{\gamma} = 0.3605 + 660/(0.85 \times 9790)$$

$$= 0.4398 \text{ m}$$

With centre line of pipe as datum, the elevations of energy and piezometric head are as follows:

At Section 1:

$$\text{Energy head} = H_1 = \frac{p_1}{\gamma} + \frac{V_1^2}{2g} + Z_1$$

$$= 0.3605 + 0.2115 + 0 = 0.572 \text{ m}$$

Piezometric head $h_1 = \frac{p_1}{\gamma} + Z_1 = 0.3605$ m

At Section 2:

$$\text{Energy head} = H_2 = \frac{p_2}{\gamma} + \frac{V_2^2}{2g} + Z_2$$
$$= 0.4398 + 0.0132 = 0.453 \text{ m}$$

Piezometric head $h_2 = 0.4398$ m

Energy loss: $H_1 - H_2 = 0.572 - 0.453 = 0.119$ m $= (V_1 - V_2)^2/2g$

The energy line and the piezometric head line (hydraulic grade line) are shown in Fig. 4.11.

4.10 Gasoline which has a vapour pressure of 5.5×10^4 Pa (abs) and density $\rho = 680$ kg/m^3 flows through a constriction in a pipe where the diameter is reduced from 20 cm to 10 cm. The pressure in the 20 cm pipe just upstream of the constriction is 50 kPa. If the atmospheric pressure is 75 cm of mercury, calculate the maximum discharge that can be passed through this constriction without cavitation occurring.

Solution: Refer to Fig. 4.12.

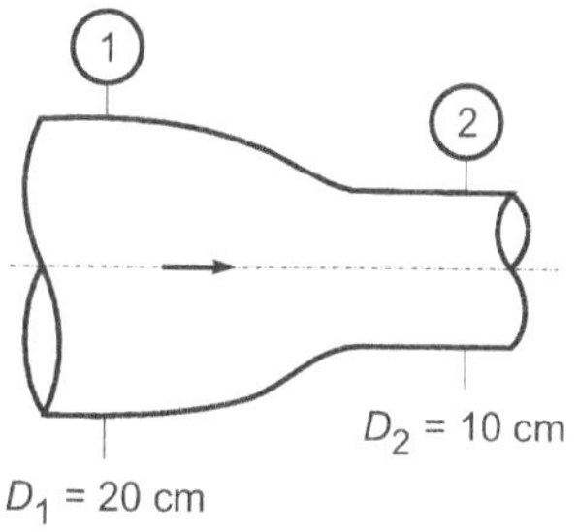

Fig. 4.12

Cavitation occurs when the local pressure reaches the vapour pressure of the liquid. Here atmospheric pressure

$p_a = 75$ cm of mercury $= 0.75 \times 13.6 \times 9.79$
$= 99.86$ kPa

$p_1 = 50.0 + 99.86 = 149.86$ kPa (abs)

p_2 = Vapour pressure = 55 kPa (abs)

By continuity consideration,

$$V_1 \frac{\pi}{4} D_1^2 = V_2 \frac{\pi}{4} D_2^2$$

Hence $V_2 = V_1(D_1^2/D_2^2) = V_1(20/10)^2 = 4V_1$

$$\gamma = \rho \times g = 680 \times 9.81/1000$$
$$= 6.67 \text{ kN/m}^3$$

By applying Bernoulli equation to points 1 and 2

$$\frac{p_1}{\gamma} + \frac{V_1^2}{2g} + Z_1 = \frac{p_2}{\gamma} + \frac{V_2^2}{2g} + Z_2$$

$$\frac{(149.86)}{6.67} + \frac{V_1^2}{2g} + 0 = \frac{55.0}{6.67} + \frac{(4V_1)^2}{2g} + 0$$

$$\frac{V_1^2}{2g}(16 - 1) = 14.22$$

$$V_1 = 4.313 \text{ m/s}$$

and $Q = 4.313 \times \frac{\pi}{4} \times (0.2)^2$
$= 0.1355$ m^3/s
$=$ **135.5 L/s**

4.11 A liquid jet issues out of a nozzle into atmosphere at an angle of 60° above the horizontal and with a velocity of 6.0 m/s. At the nozzle exit the jet has a diameter of 5 cm. Assuming the jet to be unbroken throughout the trajectory, determine (i) the equation of the jet trajectory, (ii) the maximum height attained by the jet and its size at that location, and (iii) the horizontal distance travelled by the jet at the elevation of its exit from the nozzle.

Solution:

(i) In a free jet the pressure is atmospheric throughout the trajectory. Referring to Fig. 4.13.

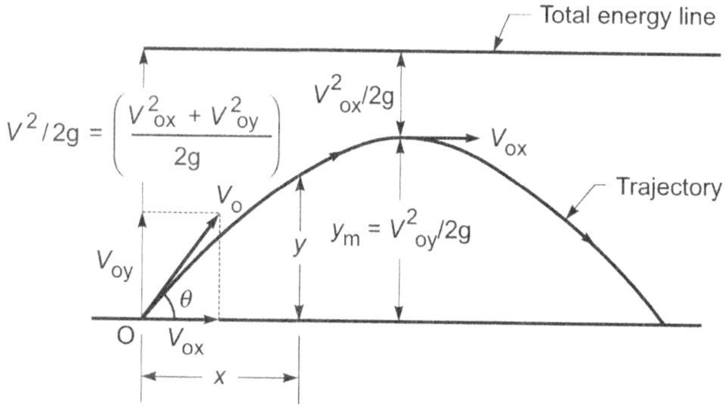

Fig. 4.13

$$V_{ox} = V_o \cos\theta = \text{constant} = V_x \quad (1)$$
$$V_{oy} = V_o \sin\theta \quad (2)$$
$$x = V_{ox}\, t \quad (3)$$
$$y = V_{oy}\, t - \frac{1}{2} g t^2 \quad (4)$$

$t = \frac{x}{V_{ox}}$ and substituting in Eq. 4

$$y = x\frac{V_{oy}}{V_{ox}} - \frac{1}{2} g x^2/(V_{ox})^2$$

$$y = x \tan\theta - \frac{g x^2}{2 V_o^2 \cos^2\theta} \quad (5)$$

Equation of the trajectory:

$$y = x \tan 60° - \frac{(9.81)x^2}{2 \times (6)^2 \times \cos^2 60} \quad (6)$$

$$\mathbf{y = 1.732x - 0.545x^2}$$

Equation (6) is the equation of the trajectory.

(ii) At the point of maximum elevation, $V_y = 0$ and hence

$$\frac{V_{ox}^2}{2g} + y_m = \frac{V_{ox}^2}{2g} + \frac{V_{oy}^2}{2g} = H$$

or $$y_m = \frac{V_{oy}^2}{2g}$$

In the present case

$$V_o = 6.0 \text{ m/s and } \theta = 60°$$
$$V_{ox} = 6.0 \cos 60° = 3.0 \text{ m/s}$$

If d_m is the diameter of the jet at the point of maximum elevation

$$V_{ox}\left(\frac{\pi}{4} d_m^2\right) = V_o\left(\frac{\pi}{4}\right)\left(\frac{5}{100}\right)^2$$

$$d_m = 5\sqrt{\left(\frac{V}{V_{ox}}\right)} = 5\sqrt{\frac{6.0}{3.0}} = 7.07 \text{ cm}$$

$$V_{oy} = 6.0 \sin 60° = 5.196 \text{ m/s}$$

Maximum elevation of the jet $= y_m = \dfrac{V_{oy}^2}{2g}$

$$= \frac{(5.196)^2}{2 \times 9.81} = 1.376 \text{ m above the jet exit level}$$

(iii) For maximum horizontal distance at jet exit level: Putting $y = 0$ in the trajectory equation

$$1.732\, x - 0.545\, x^2 = 0$$
$$x(1.732 - 0.545x) = 0$$
$$x = 0 \text{ or } x = 1.732/0.545 = 3.178 \text{ m}$$

Hence maximum horizontal distance at jet exit level is $\boldsymbol{x_m = 3.178}$ **m.**

4.12 A jet of water from a nozzle is to reach a window on a vertical wall. The window is at 15 m above the ground. The nozzle exit is at 1.5 m above the ground level. If the velocity of the jet at the nozzle exit is 25 m/s, calculate the maximum horizontal distance from the wall at which the nozzle can be kept to achieve the objective. What is the corresponding angle of inclination of the nozzle?

Solution: Refer to Fig. 4.14.

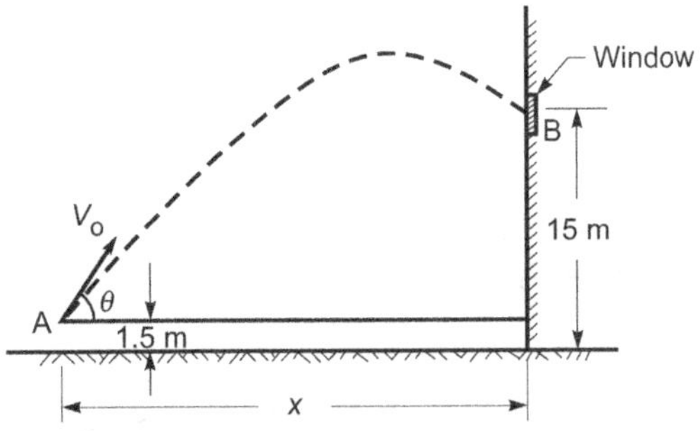

Fig. 4.14

Let θ = angle of inclination of the nozzle.

$$V_0 = 25 \text{ m/s}$$
$$x = V_o \cos \theta \cdot t$$
$$y = V_o \sin \theta \cdot t - \frac{1}{2} g t^2$$

Eliminating t, the equation of the trajectory is

$$y = x \tan \theta - \frac{g x^2}{2V_o^2} \sec^2 \theta \qquad (1)$$

Substituting $y = 15 - 1.5 = 13.5$ m,

$$13.5 = x \tan \theta - \frac{9.81}{2 \times (25)^2} x^2 \sec^2 \theta$$

$$x \tan \theta - 7.848 \times 10^{-3} x^2 \sec^2 \theta - 13.5 = 0 \qquad (2)$$

The maximum value of x is obtained by differentiating with respect to θ and putting $dx/d\theta = 0$.

On differentiation with respect to θ.

$$\left(x \sec^2 \theta + \tan \theta \frac{dx}{d\theta}\right) - 7.848 \times 10^{-3}$$

$$\times \left(x^2 \cdot 2 \sec^2 \theta \cdot \tan \theta + 2x \sec^2 \theta \cdot \frac{dx}{d\theta}\right) = 0$$

Putting $\dfrac{dx}{d\theta} = 0$

$$x \sec^2 \theta - 7.848 \times 10^{-3} (2x^2 \sec^2 \theta \cdot \tan \theta) = 0$$

$$1 - 0.0157\, x \tan \theta = 0 \text{ or } x = \frac{63.71}{\tan \theta}$$

Substituting this in Eq. 2,

$$63.71 - 7.848 \times 10^{-3}\left[\frac{(63.71)^2}{\tan^2 \theta} \sec^2 \theta\right] - 13.5 = 0$$

$$63.71 - \frac{31.855}{\sin^2 \theta} - 13.5 = 0$$

$$\sin \theta = 0.7965 \quad \text{and} \quad \boldsymbol{\theta = 52.8°}$$
$$\tan \theta = 1.3174$$

$$x = \frac{63.71}{\tan \theta} = \frac{63.71}{1.3174} = \mathbf{48.36 \text{ m}}$$

4.13 A tank has two outlets (i) a rounded entry orifice A of diameter D and (ii) a pipe B with well rounded entry and of length L, as shown in Fig. 4.15. For a head of water H in the tank determine the (i) discharge from the outlets A and B, (ii) velocities in the two outlets at levels 1 and 2 indicated in Fig. 4.15.

Solution:

Rounded entry orifice Applying Bernoulli equation to a point on the water surface 3 and point 1.

$$0 + 0 + H = \frac{p_1}{\gamma} + \frac{V_{1a}^2}{2g} + 0$$

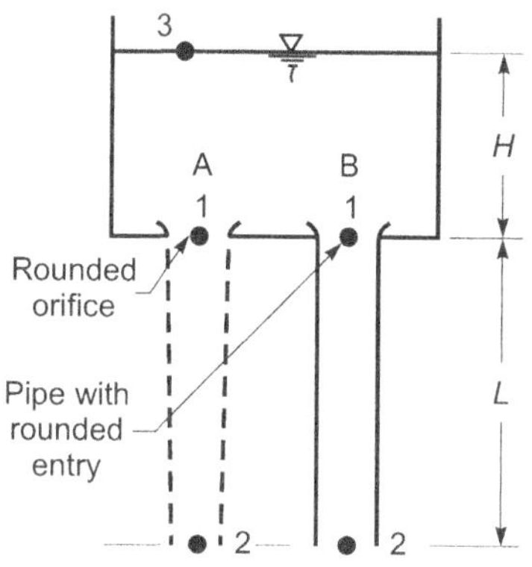

Fig. 4.15

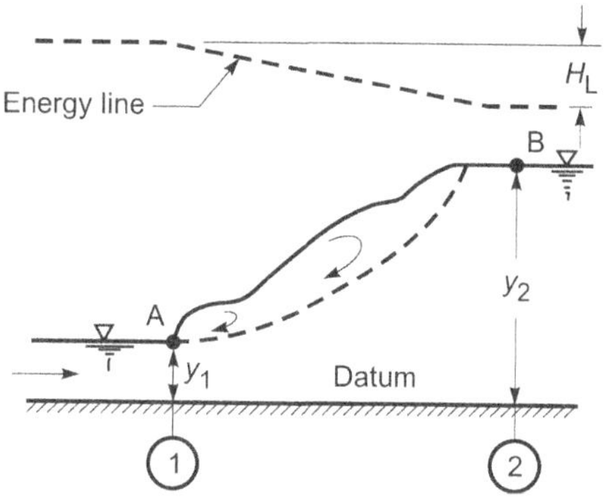

Fig. 4.16 *Hydraulic Jump*

As the orifice discharges to atmosphere, $\frac{p_1}{\gamma} = 0$

and $V_{1a} = \sqrt{2gH}$. The discharge $Q_a = \frac{\pi}{4} D^2 \sqrt{2gH}$.

At point 2, the pressure is atmospheric and hence by applying Bernoulli equation between points 3 and 2.

$$0 + 0 + (H + L) = 0 + \frac{V_{2a}^2}{2g} + 0$$

or $$V_{2a} = \sqrt{2g(H+L)}$$

As the discharge is Q_a, the diameter at 2 will be smaller than D.

Pipe:

By applying Bernoulli equation between points 3 and 2.

$$0 + 0 + (H + L) = 0 + \frac{V_{2b}^2}{2g} + 0$$

or $$V_{2b} = \sqrt{2g(H+L)}$$

As the pipe size is uniform from point 1 to 2, by continuity criterion

$$V_{1b} = V_{2b} = \sqrt{2g(H+L)}$$

Thus the results are

	For Orifice	For Pipe
Velocity at 1 =	$\sqrt{2gH}$	$\sqrt{2g(H+L)}$
Velocity at 2 =	$\sqrt{2g(H+L)}$	$\sqrt{2g(H+L)}$
Discharge Q =	$\frac{\pi}{4} D^2 \sqrt{2gH}$	$\frac{\pi}{4} D^2 \sqrt{2g(H+L)}$

4.14 A horizontal rectangular channel has a discharge of 8.0 m^3/s/metre width of the channel. In a hydraulic jump occurring in this channel the depth before and after the jump is 0.50 m and 4.75 m respectively. Calculate the energy loss in the jump.

Solution: Refer to Fig. 4.16.

Consider unit width of the channel.

$$V_1 = \frac{q}{y_1} = \frac{8.0}{0.5} = 16 \text{ m/s}$$

$$V_2 = \frac{8.0}{4.75} = 1.684 \text{ m/s}$$

Applying Bernoulli equation to points A and B, with the channel bed as the datum

$$0 + \frac{V_1^2}{2g} + y_1 = 0 + \frac{V_2^2}{2g} + y_2 + H_L$$

$$\frac{(16.0)^2}{2\times 9.81} + 0.5 = 0 + \frac{(1.684)^2}{2\times 9.81} + 4.75 + H_L$$

$$13.548 = 4.895 + H_L$$

$$H_L = \text{Energy loss in the jump}$$

$$= \mathbf{8.653\ m}$$

4.15 A rectangular duct of width 25 cm has a two-dimensional irrotational flow. It has an elbow made up of circular areas of radius 40 cm and 65 cm for the inner and outer walls respectively. Assuming the flow in the elbow to be an irrotational vortex (Vr = constant), calculate the discharge per unit width of the duct when the difference in pressure between outer and inner walls in the elbow is 30 kPa.

Solution: For an irrotational vortex $Vr = C$ and Bernoulli equation can be used across the streamlines also. Considering the duct to be in a horizontal plane, applying Bernoulli equation to points 1 and 2 shown in Fig. 4.17,

$$\frac{p_2}{\gamma} + \frac{V_2^2}{2g} + Z_2 = \frac{p_1}{\gamma} + \frac{V_1^2}{2g} + Z_1$$

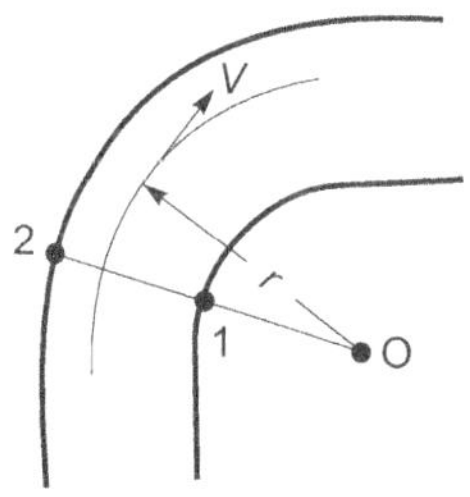

Fig. 4.17

As $Z_2 = Z_1$ and $V = C/r$

$$(p_2 - p_1) = \frac{\rho}{2}(V_1^2 - V_2^2) = \frac{\rho C^2}{2}\left(\frac{1}{r_1^2} - \frac{1}{r_2^2}\right)$$

Thus

$$30 \times 10^3 = \frac{998 \times C^2}{2}\left[\frac{1}{(0.4)^2} - \frac{1}{(0.65)^2}\right]$$

$$= 1937.7\ C^2$$

$$C^2 = 15.48 \text{ and } C = 3.9348$$

Discharge per unit depth

$$q = \int_{r_1}^{r_2} v\,dr$$

$$= \int_{r_1}^{r_2} \frac{C}{r}\,dr = C \ln r_2/r_1$$

$$= 3.9348 \ln \frac{0.65}{0.40}$$

$$= \mathbf{1.910\ m^3/s\ per\ metre\ width}$$

4.16 An irrotational two-dimensional flow leaves a straight duct and enters a bend of the same width, in the horizontal plane as shown in Fig. 4.18. Assuming the velocity distribution in the bend like that in a potential vortex (vr = constant), find the velocities at the inner and outer walls of the bend. Also, estimate the difference in the pressure head between the outer and the inner wall.

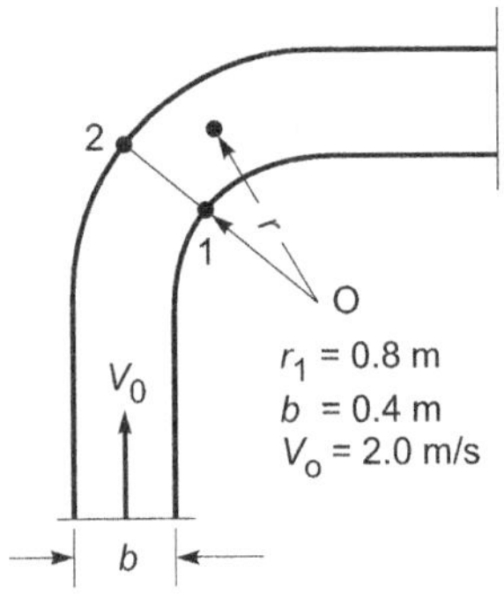

Fig. 4.18

Solution: Consider a unit depth of the duct.

$$v = C/r$$

$$q = bV_0 = \int_{r_1}^{r_2} v\,dr$$

$$= \int_{r_1}^{r_2} C/r\,dr = C \ln \frac{r_2}{r_1}$$

$$C = \frac{bV_0}{\ln(r_2/r_1)} = \frac{0.4 \times 2.0}{\ln(1.2/0.8)} = 1.973$$

V_1 = Velocity at the inner wall = C/r_1

$$= \frac{1.973}{0.80} = \mathbf{2.466\ m/s}$$

V_2 = Velocity at the outer walls = C/r_2

$$= \frac{1.973}{1.20} = \mathbf{1.644\ m/s}$$

Since the flow is irrotational, Bernoulli theorem can be applied across the streamlines also. Applying Bernoulli's theorem to points 1 and 2, located on the inner and outer wall, as in Fig. 4.16,

$$\frac{p_2}{\gamma} + \frac{V_2^2}{2g} + Z_2 = \frac{p_1}{\gamma} + \frac{V_1^2}{2g} + Z_1$$

Since $Z_1 = Z_2$ $\left(\frac{p_2 - p_1}{\gamma}\right) = \frac{V_1^2 - V_2^2}{2g}$

$$= \frac{1}{2 \times 9.81}[(2.446)^2 - (1.644)^2]$$

Therefore the difference in pressure head between outer and inner wall is **0.172 m.**

4.17 Figure 4.19 shows pump P pumping 60 L/s of water from a tank. (i) What will be the pressures at points A and B when the pump delivers 10 kW of powers to the flow. Assume losses in the system to be negligible. (ii) What will be the pressure at B when the loss in the inlet up to the pump is negligible and between the pump and the point B, a loss equal to 2 times the velocity head at B takes place.

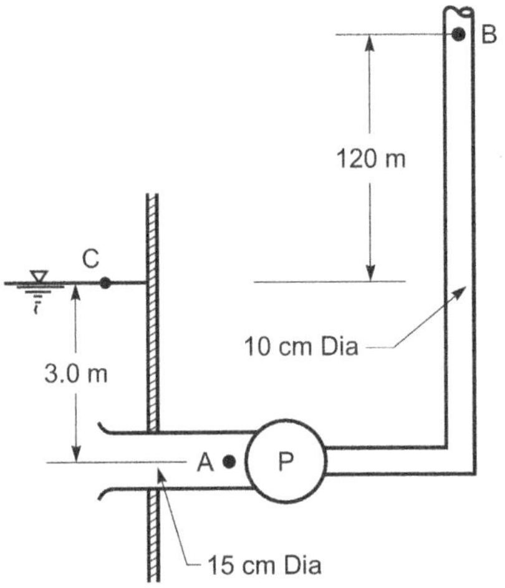

Fig. 4.19

Solution:

Discharge $Q = 0.060\ m^3/s = A_a V_a = A_b V_b$

$$V_a = \frac{0.06}{\frac{\pi}{4} \times (0.15)^2} = 3.395\ m/s, \quad \frac{V_a^2}{2g} = 0.588\ m$$

$$V_b = \frac{0.06}{\frac{\pi}{4}\times(0.10)^2} = 7.639 \text{ m/s}, \quad \frac{V_b^2}{2g} = 2.975 \text{ m}$$

Power delivered by pump $P = \gamma Q H_p = 10$ kW

$$9.79 \times 0.060 \times H_p = 10$$

H_p = head delivered by the pump = 17.02 m

(i) By applying Bernoulli equation to points C and A,

$$0 + 0 + 3 = \frac{p_a}{\gamma} + \frac{V_a^2}{2g} + 0$$

$$\frac{p_a}{\gamma} = 3 - 0.588 = 2.412 \text{ m}$$

$$p_a = 9.79 \times 2.412 = \mathbf{23.61\ kPa}$$

By applying Bernoulli equation between C and B with level at A as datum:

$$\frac{p_c}{\gamma} + \frac{V_c^2}{2g} + Z_c + H_p = \frac{p_b}{\gamma} + \frac{V_b^2}{2g} + Z_b$$

$$0 + 0 + 3.0 + 17.02 = \frac{p_b}{\gamma} + 2.975 + (3.0 + 1.2)$$

$$\frac{p_b}{\gamma} = 12.845 \text{ m}$$

and $p_b = 9.79 \times 12.845 = \mathbf{125.75\ kPa}$

(ii) *When losses are considered:* By applying Bernoulli equation between C and B with level at A as datum,

$$\frac{p_c}{\gamma} + \frac{V_c^2}{2g} + Z_c + H_p = \frac{p_b}{\gamma} + \frac{V_b^2}{2g} + Z_b + H_L$$

$$0 + 0 + 3.0 + 17.02 = \frac{p_b}{\gamma} + 2.975 + (3.0 + 1.2) + (2 \times 2.975)$$

$$\frac{p_b}{\gamma} = 6.895 \text{ m}$$

and $p_b = 9.79 \times 6.895$ $= \mathbf{67.50\ kPa}$

4.18 A pump is 2.5 m above the water level in the sump and has a pressure of –22 cm of mercury at the suction side. The suction pipe is of 25 cm diameter and the delivery pipe is a short 30 cm dia pipe ending in a nozzle of 10 cm diameter. If the nozzle is directed vertically upwards at an elevation of 4.0 m above the sump water level, determine (i) the discharge, (ii) power input into the flow by the pump and the (iii) elevation, above the sump water level, to which the jet would reach. (Neglect all losses).

Solution: Applying Bernoulli equation to points 1 and 2 (Fig. 4.20).

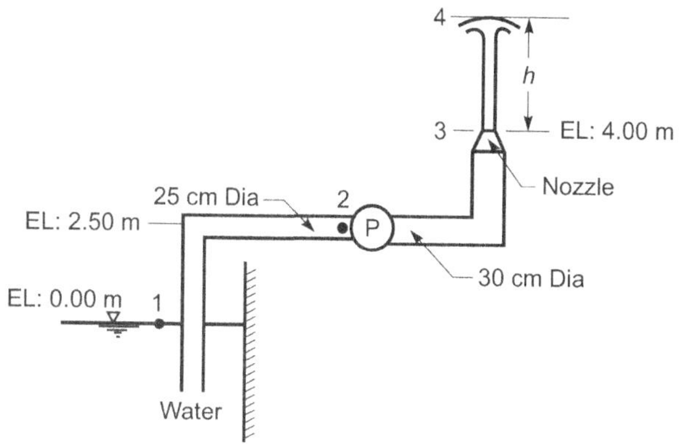

Fig. 4.20

$$0 + 0 + 0 = (-0.22 \times 13.6) + \frac{V_1^2}{2g} + 2.5$$

$$V_1^2/2g = 0.492 \text{ m and } V_1 = 3.107 \text{ m/s}$$

(i) Discharge $Q = A_1V_1 = \frac{\pi}{4} \times (0.25)^2 \times 3.107$

$= 0.1525$ m^3/s

$= \mathbf{152.5\ L/s}$

(ii) To find the power delivered by the pump: Apply Bernoulli equation to points 1 and 3:

$$0 + 0 + 0 + H_p = 0 + V_3^2/2g + 4.00$$

$$H_p = 19.22 + 4.00 = 23.22 \text{ m}$$

Power delivered by pump $\mathbf{P} = \gamma Q H_p$

$$\mathbf{P} = 9.79 \times 0.1525 \times 23.22$$

$$= \mathbf{34.67\ kW}$$

(iii) $V_3 = V_1 \times (D_1/D_3)^2 = 3.107 \times (25/10)^2$

$= 19.419$ m/s

$$V_3^2/2g = \frac{(19.419)^2}{2 \times 9.81} = 19.22 \text{ m}$$

Hence, the height to which the jet will reach h = 19.22 m. The elevation of point 4, the summit of the jet, is

$= 4.00 + 19.22 = \mathbf{23.22\ m}$

4.19 A turbine is set 40 m below the water level of a reservoir and is fed by a 60 cm diameter pipe. A short pipe of 45 cm diameter discharges the water from the turbine to atmosphere (Fig. 4.21).

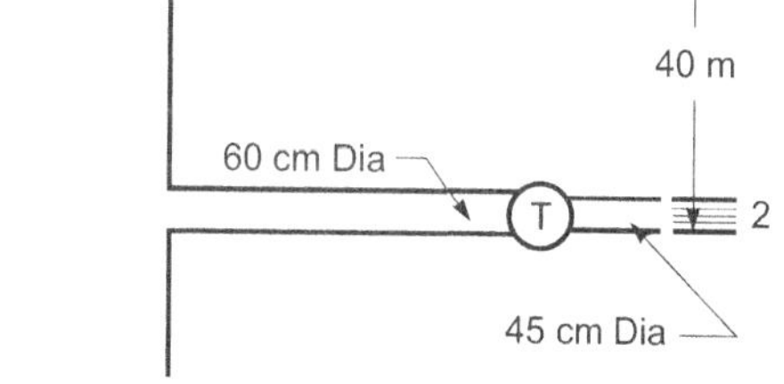

Fig. 4.21

(i) Neglecting friction, estimate the power extracted by the turbine when the discharge is 0.8 m^3/s.

(ii) If a total frictional loss of 10 m is assumed and the turbine efficiency 85%, estimate the power output.

Solution:

(i) Water friction loss is neglected:

Discharge $Q = 0.8\ m^3/s$.

$$V_2 = \text{Velocity at outlet} = \frac{0.8}{\frac{\pi}{4}\times(0.45)^2}$$

$= 5.03$ m/s.

Applying Bernoulli equations to points 1 and 2,

$$\frac{p_1}{\gamma} + \frac{V_1^2}{2g} + Z_1 = H_T + \frac{V_2^2}{2g} + Z_2$$

$$0 + 0 + 40 = H_T + \frac{(5.03)^2}{2\times 9.81} + 0$$

H_T = Head extracted by the turbine

$= 40 - 1.29 = 38.71$ m

Power extracted by the turbine $p = \gamma Q H_T$

$= 9.79 \times 0.8 \times 38.71 =$ **303.2 kW**

(ii) When losses are included:

By applying Bernoulli equation to points 1 and 2

$$0 + 0 + 40 = H_T + \frac{(5.03)^2}{2\times 9.81} + 10.0 + 0$$

$H_T = 28.71$ m

Power extracted by the turbine $= \gamma Q H_T$

Power output of the turbine $\mathbf{P}_n = \gamma Q H_T \eta$

where η = efficiency of the turbine

$\therefore$ $\mathbf{P}_n = 9.79 \times 0.8 \times 28.71 \times 0.85$

$=$ **191.1 kW**

4.20 A turbine has a supply line of diameter 40 cm and a tapering draft tube as shown in Fig. 4.22. When the flow in the pipe is 0.5 m^3/s the pressure head at point A upstream of the turbine is 30 m and at point B in the draft tube, where the diameter is 60 cm, the pressure head is (–4 m). Point B is 2.0 m below point A. Estimate the power output of the turbine by assuming 90% efficiency.

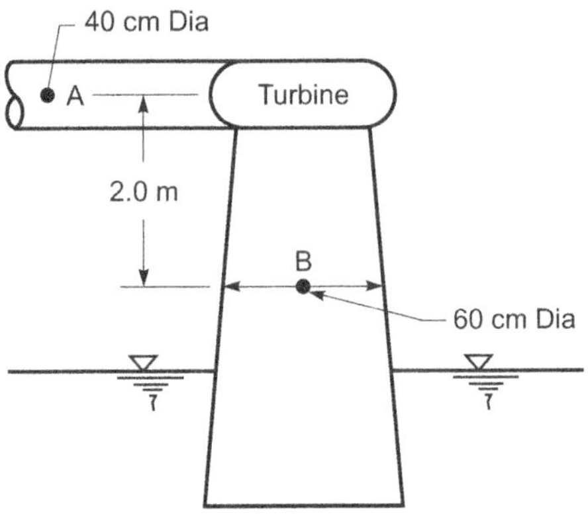

Fig. 4.22

Solution:

$$V_a = \frac{Q}{A_a} = \frac{0.5}{\frac{\pi}{4}\times(0.4)^2} = 3.98 \text{ m/s}$$

$$V_b = \frac{0.5}{\frac{\pi}{4}\times(0.6)^2} = 1.768 \text{ m/s}$$

Applying Bernoulli equation to points A and B:

$$\frac{p_a}{\gamma} + \frac{V_a^2}{2g} + Z_a = \frac{p_b}{\gamma} + \frac{V_b^2}{2g} + Z_b + H_t$$

$$30.0 + \frac{(3.98)^2}{2\times 9.81} + 2.0 = -4 + \frac{(1.768)^2}{2\times 9.81} + 0 + H_t$$

$H_t = 32.807 + 3.841 = 36.65$ m

Power $\mathbf{P} = \gamma Q H_t \times \eta$

$= 9.79 \times 0.5 \times 36.65 \times 0.90$

$=$ **161.45 kW**

4.21 Calculate the kinetic energy correction factor α for the following velocity distributions in a circular pipe of radius r_0.

(i) $\frac{u}{u_m} = \left(1 - \frac{r}{r_0}\right)$ (ii) $\frac{u}{u_m} = [1 - (r/r_0)^2]$

Solution:

(i) $\frac{u}{u_m} = \left(1 - \frac{r}{r_0}\right)$

Refer to Fig. 4.23(a)

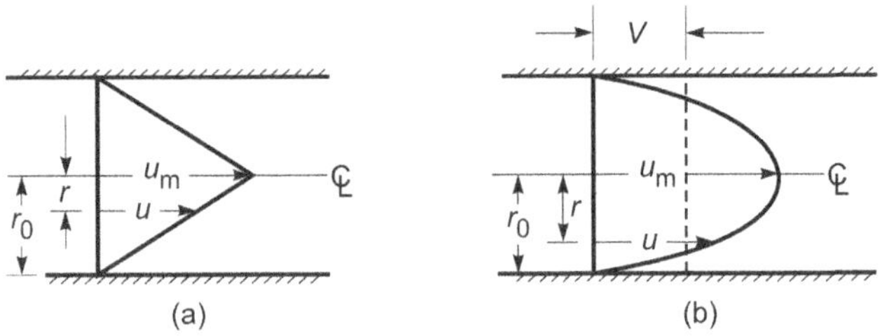

Fig. 4.23

The kinetic energy correction factor

$$\alpha = \frac{1}{V^3 A}\int_A u^3\, dA$$

where V = average velocity.

$$Q = \pi r_0^2 V = \int_0^{r_0} u 2\pi r dr$$

$$= \int_0^{r_0} u_m \left(1 - \frac{r}{r_0}\right) 2\pi r dr$$

$$= 2\pi u_m \left(\frac{r_0^2}{2} - \frac{r_0^2}{3}\right) = \frac{\pi r_0^2}{3} u_m$$

$$V = \frac{1}{3} u_m$$

Now $$\alpha = \frac{1}{V^3 \pi r_0^2} \int_0^{r_0} u_m^3 \left(1 - \frac{r}{r_0}\right)^3 2\pi r dr$$

$$= \frac{2\pi u_m^3}{\left(\frac{1}{3} u_m\right)^3 \pi r_0^2} \int_0^{r_0} \left(1 - \frac{r}{r_0}\right)^3 r dr$$

$$= \frac{54}{r_0^2}\left[r_0^2\left(\frac{1}{2} - \frac{3}{3} + \frac{3}{4} - \frac{1}{5}\right)\right] = \frac{54}{20} = \mathbf{2.7}$$

(ii) $$\frac{u}{u_m} = \left[\left(1 - \frac{r}{r_0}\right)^2\right]$$

Refer to Fig. 4.23(b)

$$Q = \pi r_0^2 V = \int_0^{r_0} u 2\pi r\, dr$$

$$= \int_0^{r_0} u_m \left(1 - \frac{r^2}{r_0^2}\right) 2\pi r dr$$

$$= 2\pi\, u_m \left(\frac{r_0^2}{2} - \frac{r_0^2}{4}\right) = \frac{\pi u_m r_0^2}{2}$$

$$V = \frac{u_m}{2}$$

$$\alpha = \frac{1}{V^3 \pi r_0^2} \int_0^{r_0} u_m^3 \left(1 - \frac{r^2}{r_0^2}\right) 2\pi r dr$$

$$= \frac{2\pi u_m^3}{(u_m/2)^3 \pi r_0^2} \times$$

$$\int_0^{r_0} \left(r - 3\frac{r^3}{r_0^2} + 3\frac{r^5}{r_0^4} - \frac{r^7}{r_0^6}\right) dr$$

$$= \frac{16}{r_0^2}\left[r_0^2\left(\frac{1}{2} - \frac{3}{4} + \frac{3}{6} - \frac{1}{8}\right)\right]$$

$\boldsymbol{\alpha = 2.0}$

4.22 A two-dimensional conduit has the velocity distributions shown in Fig. 4.24 (a) and 4.22 (b). Calculate the kinetic energy correction factor α for these distributions.

Solution: Distribution (a): $u = \frac{u_1}{B} y$. Consider unit width of the conduit. Refer to Fig. 4.24(a).

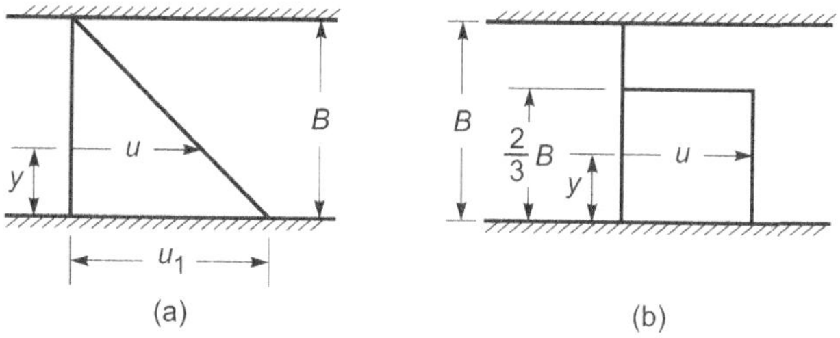

Fig. 4.24

Average velocity $$V = \frac{u_1}{2}$$

$$\alpha = \frac{1}{V^3 B} \int_0^B u^3 dy$$

$$= \frac{1}{\left(\frac{u_1}{2}\right)^3 B} \int_0^B \left(\frac{u_1}{B} \cdot y\right)^3 dy$$

$$= \frac{8}{B}\left(\frac{B^4}{4B^3}\right) = 2.0$$

Distribution (b):

u = constant for $0 \le y \le 2B/3$ and $u = 0$ for $y > 2B/3$. Refer to Fig. 4.24(b). Consider unit width of the conduit.

Discharge $$q = V \cdot B = u \times \frac{2}{3} B + 0$$

Average velocity $$V = \frac{2}{3} u$$

$$\alpha = \frac{1}{V^3 B}\left[\int_0^{2/3B} u^3 dy + \int_{2/3B}^{B} u^3 dy\right]$$

$$= \frac{1}{(2/3\, u)^3 B} \times u^3 \, [y]_0^{2/3\, B}$$

$$= \frac{27}{8B} \frac{2}{3} B = \frac{9}{4}$$

$\boldsymbol{\alpha = 2.25}$

4.23 A tapering pipe has a diameter of 25 cm at point 1 (elevation 25.00 m) and diameter of 35 cm at point 2 (elevation 20.00 m) (Fig. 4.25). If the pressure at point 1 is 120 kPa, calculate the pressure at point 2 for a discharge of 0.20 m^3/s of water. The kinetic energy correction factors for sections 1 and 2 are 1.1 and 1.5 respectively. The loss of head through the pipe can be assumed to be 1.2 $(V_1 - V_2)^2/2g$. The flow is from section 1 to section 2.

Solution:

$$Q = 0.20 \text{ m}^3/\text{s}$$

$$V_1 = \frac{0.20}{\frac{\pi}{4} \times (0.25)^2} = 4.074 \text{ m/s}$$

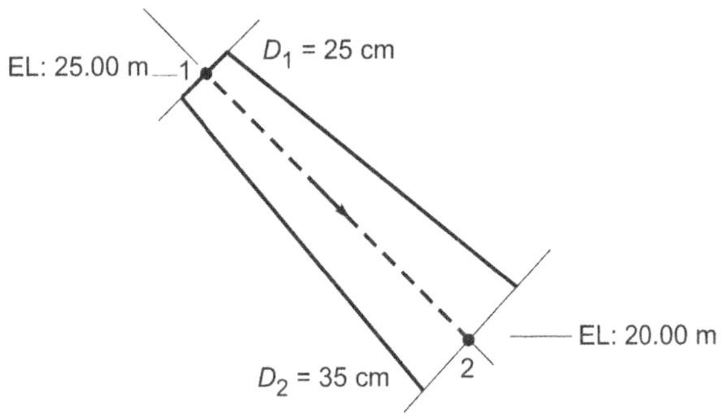

Fig. 4.25

$$V_2 = \frac{0.20}{\frac{\pi}{4} \times (0.35)^2} = 2.079 \text{ m/s}$$

$$(V_1 - V_2) = 1.995 \text{ m/s}$$

$$\alpha_1 = 1.1 \text{ and } \alpha_2 = 1.5$$

By applying Bernoulli equation to points 1 and 2,

$$\frac{p_1}{\gamma} + \alpha_1 \frac{V_1^2}{2g} + Z_1 = \frac{p_2}{\gamma} + \alpha_2 \frac{V_2^2}{2g} + Z_2 + H_L$$

$$\left(\frac{120}{9.79}\right) + 1.1 \times \frac{(4.074)^2}{2 \times 9.81} + 25.00$$

$$= \frac{p_2}{\gamma} + 1.5 \times \frac{(2.079)^2}{2 \times 9.81} + 20.00 + \frac{1.2 \times (1.995)^2}{2 \times 9.81}$$

$$12.257 + 0.931 + 25.00 = \frac{p_2}{\gamma} + 0.330 + 20.00 + 0.243$$

$$\frac{p_2}{\gamma} = 38.188 - 20.573 = 17.615 \text{ m}$$

$$p_2 = 9.79 \times 17.615 = \mathbf{172.45 \text{ kPa}}$$

4.24 The stream function ψ for the uniform flow past a circular cylinder of radius a in polar coordinates is given by

$$\psi = U_\infty \left[1 - \frac{a^2}{r^2}\right] r \sin \theta$$

where U_∞ is the free stream velocity parallel to the horizontal axis. (i) Determine the velocity on the cylinder. (ii) If p_∞ and p are the pressures at the free stream far away from the cylinder and the pressure on the cylinder (Fig. 4.26), obtain the values of the pressure coefficient $C_p = (p - p_\infty)/\frac{1}{2} \rho U_\infty^2$ on the cylinder at $\theta = 0, \pi/6, \pi/2, 5\pi/6$ and π.

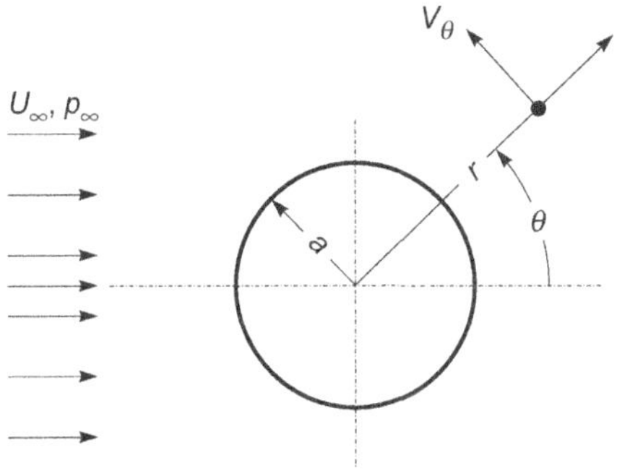

Fig. 4.26

Solution:

$$\psi = U_\infty \left[1 - \frac{a^2}{r^2}\right] r \sin \theta$$

$$v_r = \frac{1}{r}\frac{\partial \psi}{\partial r} \text{ and } v_\theta = -\frac{\partial \psi}{\partial r}$$

On the cylinder $\quad v_r = 0$ and $v_\theta = V = -\left.\frac{\partial \psi}{\partial r}\right|_{r=a}$

Hence $\quad V = -U_\infty \left(1 + \frac{a^2}{r^2}\right) \sin \theta \Big|_{r=a}$

$$V = -2U_\infty \sin \theta$$

By applying Bernoulli equation to a point on the free stream and a point on the cylinder at θ, for a 2-D flow

$$\frac{p_\infty}{\gamma} + \frac{V_\infty^2}{2g} = \frac{p}{\gamma} + \frac{V^2}{2g}$$

$$\frac{p - p_\infty}{\gamma} = \frac{1}{2g}(U_\infty^2 - 4U_\infty^2 \sin^2 \theta)$$

$$C_p = \frac{p - p_\infty}{\frac{1}{2}\rho U_\infty^2} = 1 - 4 \sin^2 \theta$$

Values of C_p for various values of θ are as shown in the following:

θ	0	$\pi/6$	$\pi/2$	$5\pi/6$	π
C_p	1	0	–3	0	1

OBJECTIVE QUESTIONS

4.1 The Bernoulli equation is written with usual notation as $p/\gamma + V^2/2g + Z$ = constant. In this equation each of the terms represents
(a) energy in kg.m/kg mass of fluid
(b) energy in N.m/kg mass of fluid
(c) energy in N.m/N weight of fluid
(d) power in kW/kg mass of fluid

4.2 The velocity in a cross section is non-uniform. In one-dimensional method of analysis the kinetic energy of fluid per unit weight is
(a) $V^2/2g$ (b) $\alpha V^2/2g$
(c) $\beta V^2/2g$ (d) p/γ

4.3 Bernoulli equation is applicable between any two points
(a) in any rotational flow of an incompressible fluid
(b) in any type of irrotational flow of a fluid
(c) in steady rotational flow of an incompressible fluid
(d) in steady, irrotational flow of an incompressible fluid

4.4 Indicate the *incorrect* statement:
For Bernoulli equation as $p/\gamma + V^2/2g + Z$ = constant to be applicable
(a) the flow must be steady
(b) the fluid must be an ideal gas
(c) the flow must be irrotational
(d) the fluid must be incompressible

4.5 The piezometric head of a flow is
(a) the sum of the velocity head and datum head
(b) the sum of the pressure head and datum head
(c) the sum of the pressure head and velocity head.
(d) the sum of the velocity head, pressure head and datum head.

4.6 In a flow of a real fluid with no addition of energy
(a) the energy line will be horizontal or sloping upward in the direction of flow
(b) the energy line can never be horizontal or slopping upward in the direction of the flow.
(c) the piezometric line can never be horizontal or sloping downward in the direction of the flow.
(d) the centre line of the pipe can never be above the energy line

4.7 Indicate the *incorrect* statement:
(a) The hydraulic grade line and the energy line are coincident for a body of fluid at rest.
(b) Whenever the hydraulic grade line falls below the centre line of a conduit carrying a liquid the local pressures are less than the reference atmospheric pressure.
(c) The hydraulic grade line in a flow of real fluid with addition of energy will have an upward slope in the direction of flow.
(d) The energy line in a flow of an ideal fluid with no addition of energy will always be horizontal.

4.8 The total head in a flow is the sum of
(a) piezometric head and datum head
(b) piezometric head and pressure head
(c) piezometric head and velocity head
(d) piezometric head, velocity head and datum head.

4.9 The difference between the total head line and the hydraulic grade line represents
(a) the velocity head
(b) the piezometric head
(c) the pressure head
(d) the elevation head

4.10 In a pipeline the hydraulic grade line is above the pipe centre line in the longitudinal section at point A and below the pipe centre line at another point B. From this it can be inferred that
(a) vacuum pressures prevail at B
(b) vacuum pressures prevail at A
(c) the flow is from A to B
(d) the flow is from B to A

4.11 The kinetic energy correction factor α is defined as α =
(a) $\frac{1}{A^3V^3}\int v^3 dA$ (b) $\frac{1}{A}\int v^3 dA$
(c) $\frac{1}{AV^3}\int v\, dA$ (d) $\frac{1}{AV^3}\int v^3\, dA$
where V = average velocity in the cross section.

4.12 In a two-dimensional duct flow, air flows in the bottom half of the duct with uniform velocity and there is no flow in the upper half. The value of the kinetic energy correction factor α for this flow is
(a) 2.0 (b) 2¼
(c) 4.0 (d) 3.0

4.13 A 15 cm diameter pipe carries a flow of 70 L/s of an oil (RD = 0.75). At a section 12 cm above the datum the pressure is vacuum of 2 cm of mercury. If the kinetic energy correction factor α for this section is 1.1, the total head at the section in metres of oil is

(a) 0.648 (b) 0.728
(c) 0.557 (d) 0.637

4.14 In connection with the flow of a fluid in a pipe the following statement regarding the kinetic energy correction factor α is correct:

(a) $\alpha_{\text{laminar flow}} < \alpha_{\text{turbulent flow}}$
(b) $\alpha_{\text{laminar flow}} = \alpha_{\text{turbulent flow}}$
(c) $\alpha_{\text{laminar flow}} = 2.0$
(d) $\alpha_{\text{laminar flow}} = 1.33$

4.15 A 20 cm diameter horizontal pipe is attached to a tank containing water. The water level in the tank is 7 m above the pipe outlet and the pipe discharges into the atmosphere. Assuming a total loss of 3 m in the pipe and the kinetic energy correction factor α of the jet issuing from the pipe to be 1.20, the discharge in the pipe is L/s is

(a) 254 (b) 278
(c) 368 (d) 305

4.16 Water flows steadily down a vertical pipe of constant cross section. Neglecting friction, according to Bernoulli's equation,

(a) pressure is constant along the length of the pipe
(b) velocity decreases with height
(c) pressure decreases with height
(d) pressure increases with height

4.17 In the siphon shown in Fig. 4.27 assuming ideal flow, pressure p_B

(a) $= p_A$ (b) $< p_A$
(c) $> p_A$ (d) $= p_C$

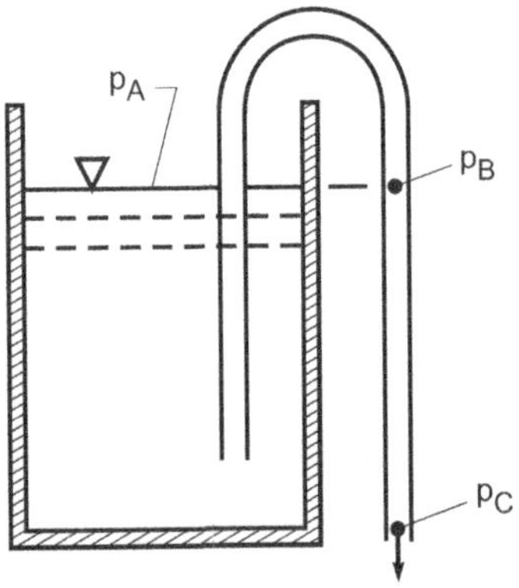

Fig. 4.27

4.18 A jet of air is directed over a circular cylinder as shown in Fig. 4.28. Neglecting viscous effects, the force acting on the cylinder due to air motion is

(a) upwards
(b) downwards
(c) in the direction of jet flow
(d) zero

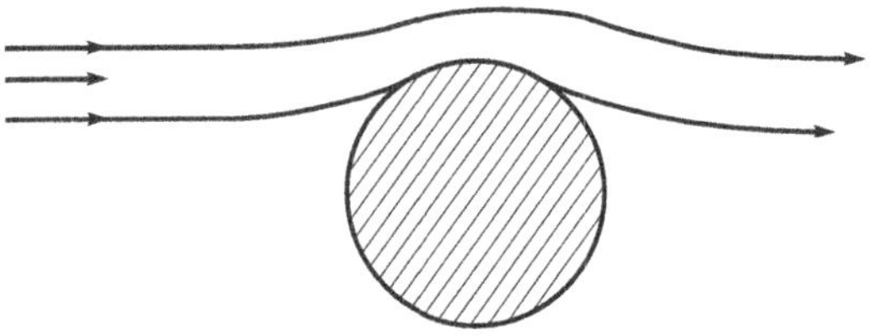

Fig. 4.28

4.19 Figure 4.29 shows a tank being emptied by a pipe of length L. If the friction of the pipe is neglected, the pressure at a point A at the pipe inlet would

(a) increase if the length L is increased
(b) be constant and equal to γL
(c) remain constant at γH for all lengths L
(d) decrease with an increase in L

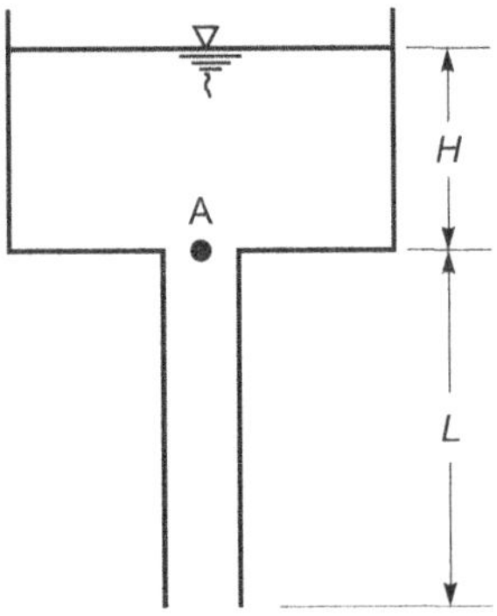

Fig. 4.29

4.20 In a fluid flow, point A is at a higher elevation than point B. The head loss between these points is H_L. The total heads at A and B are H_a and H_b respectively. The flow will take place

(a) from A to B if $H_a + H_L = H_b$
(b) from B to A if $H_a + H_L = H_b$
(c) always from A to B
(d) from B to A if $H_b + H_L = H_a$

4.21 In a siphon the summit is 4 m above the water level in the reservoir from which the flow is being discharged out. If the head loss from the inlet of the siphon to the summit is 2 m and the velocity head at the summit is 0.5 m the pressure at the summit is

(a) – 63.64 kPa
(b) – 9.0 m of water
(c) 6.5 m of water (abs)
(d) – 39.16 kPa

4.22 A liquid jet issues out from a nozzle inclined at an angle of 60° to the horizontal and directed upwards. If the velocity of the jet at the nozzle is 18 m/s the maximum vertical distance attained by the jet, measured above the point of exit from the nozzle is
(a) 14.30 m (b) 16.51 m
(c) 4.12 m (d) 12.39 m

4.23 A nozzle emits a 5 cm diameter liquid jet at 20 m/s in to air at an angle of elevation of 30° to horizontal. At a point of maximum elevation, if the jet is assumed to be unbroken throughout, the diameter of the jet is
(a) 5.373 cm (b) 5.000 cm
(c) 4.653 cm (d) 2.582 cm

4.24 A water jet with a velocity of 20 m/s is directed upwards at an angle of 45° to the horizontal. If air resistance is neglected, it will reach a maximum elevation at a horizontal distance x from the nozzle. The value of x in metres is
(a) 20.38 (b) 40.77
(c) 10.19 (d) 14.41

4.25 A nozzle directs a liquid jet at an angle of elevation of 45°. The hydraulic grade line for the jet
(a) coincides with the centre line of the jet
(b) will be horizontal at the level of the jet
(c) will be horizontal at the level of the energy line
(d) coincides with the energy line

4.26 A pump delivers 50 L/s of water and delivers 7.5 kW of power to the system. The head developed by the pump is
(a) 7.5 m (b) 5.0 m
(c) 1.53 m (d) 15.32 m

4.27 In a pump the suction and delivery pipes are of the same size and are at the same level. At a given discharge the loss of head between a point A on the suction side and a point B on the delivery side is 3.0 m. If the pressure at point B is 120 kPa and the head developed by the pump is 10 m, the pressure at point A is
(a) 51.5 kPa (b) – 7.25 kPa
(c) – 188.6 kPa (d) 29.37 kPa

4.28 In a hydro-project a turbine has a head of 50 m. The discharge in the feeding penstock is 3.0 m^3/s. If a head loss of 5 m takes place due to losses, and a power of 1000 kW is extracted, the residual head downstream of the turbine is
(a) 5.0 m (b) 10.95 m
(c) 15.95 m (d) 20.95 m

4.29 In a turbine having a flow of 1.2 m^3/s the net head is 120 m. If the efficiency of the turbine is 90% the shaft power developed, in kW, is
(a) 1440 (b) 160
(c) 1566 (d) 1269

4.30 In a two-dimensional irrotational vortex flow the velocity V at a radial distance r from the centre is $V = C/r$ where C = constant. The difference in pressure between any two points 1 and 2 is given by $(p_1 - p_2) =$

(a) $\frac{\rho}{2}(r_2^2 - r_1^2)$ (b) $2\rho\left(\frac{1}{r_2^2} - \frac{1}{r_1^2}\right)$

(c) $\frac{\rho}{2}\left(\frac{1}{r_2^2} - \frac{1}{r_1^2}\right)$ (d) $\frac{\rho}{2}\left(\frac{1}{r_1^2} - \frac{1}{r_2^2}\right)$

4.31 At a distance of 10 cm from the axis of a whirlpool in an ideal liquid, the velocity is 5 m/s. At a radius of 30 cm the depression of the free surface below the surface of the liquid at a very large distance is
(a) 7.98 cm (b) 3.33 cm
(c) 14.16 cm (d) 21.37 cm

4.32 In a flow net for the flow in a two-dimensional constriction, the size of the mesh in the uniform flow upstream of the constriction is 4 mm and at a point B in the constriction the mesh surrounding the point has a size of 3 mm. With the usual notations the pressure coefficient C_p =

$(p_b - p_0)/\left(\frac{1}{2}\rho U_0^2\right)$ at point B is

(a) 7/9 (b) 4/9
(c) –1/3 (d) –7/9

4.33 In an irrotational flow past a body the free stream velocity and pressure are V_0 and p_0 respectively. The stagnation pressure p_s is given by p_s =

(a) $p_0 + \frac{\rho}{2}V_0^2$ (b) $p_0 - \frac{\rho}{2}V_0^2$

(c) $p_0/\left(\frac{\rho}{2}V_0^2\right)$ (d) $\frac{\rho}{2}V_0^2$

PROBLEMS

4.1 For the pipeflow system shown in Fig. 4.30 the following data are available:

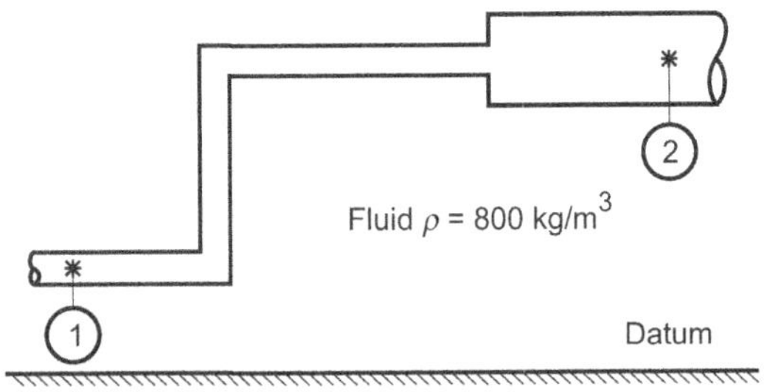

Fig. 4.30

Item	*Point* 1	*Point* 2
Diameter	20 cm	30 cm
Elevation (m)	103.00	106.00
Pressure	55 kPa	75 kPa
Velocity	2.5 m/s	

Determine the direction of flow and the loss of energy between these two points.

(*Ans.* The flow is from 2 to 1. H_L = 5.292 m)

4.2 A 15 cm diameter pipe is reduced to 7.5 cm diameter through gradual contraction. At this contraction the difference between the piezometric heads at the main and the contracted section is 4 cm of mercury. By neglecting losses, calculate the discharge of water.

(*Ans.* Q = 14.9 L/s)

4.3 For the flow system shown in Fig. 4.31 diameter $D_1 = D_3$ = 20 cm, D_2 = 10 cm, D_4 = 7.5 cm. The fluid in the manometer is mercury. The mercury water differential head h_3 = 10 cm. Assuming zero energy loss, find (i) the discharge, (ii) manometer differential head h_1, (iii) velocity of flow at section 4 and (iv) head of water H in the tank.

(*Ans.* Q = 22 L/s; h_1 = 3.03 cm; V_4 = 5.022 m/s; H = 1.285 m)

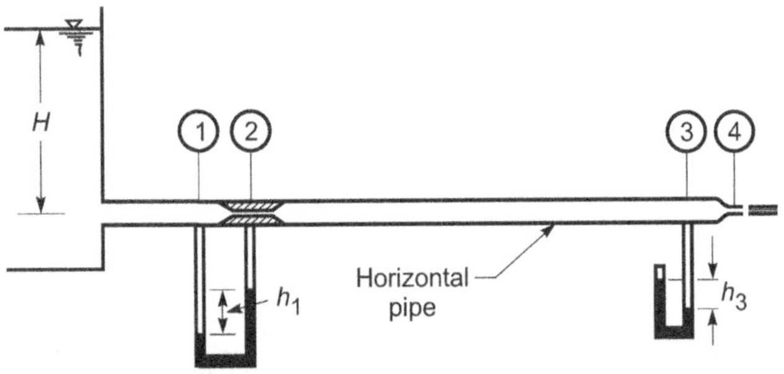

Fig. 4.31

4.4 Two points A and B are located in a long 20 cm diameter pipe. When a downstream valve is completely closed the difference in pressure between B and A, $p_B - p_A$ = 100 kPa. When the valve is open and a discharge of 70 L/s of water is flowing, $p_A - p_B$ = 50 kPa. Calculate the head loss between A and B.

(*Ans.* H_L = 15.32 m)

4.5 A siphon consisting of a 3 cm diameter tube is used to drain water from a tank. The outlet end of the tube is 2.0 m below the water surface in the tank. Neglecting friction, calculate the discharge. If the summit of the siphon is 1.4 m above the water surface in the tank, estimate the pressure at the summit of the siphon.

(*Ans.* Q = 4.428 L/s, p = – 33.29 kPa)

4.6 For the system shown in Fig. 4.32, find the height h to which the jet from the nozzle would rise. If the nozzle and the pipe were to have diameters of 10 cm and 20 cm respectively, calculate the discharge and the velocity in the pipe. Neglect frictional losses.

(*Ans.* h = 2.361 m, Q = 53.5 L/s, V_p = 1.702 m/s)

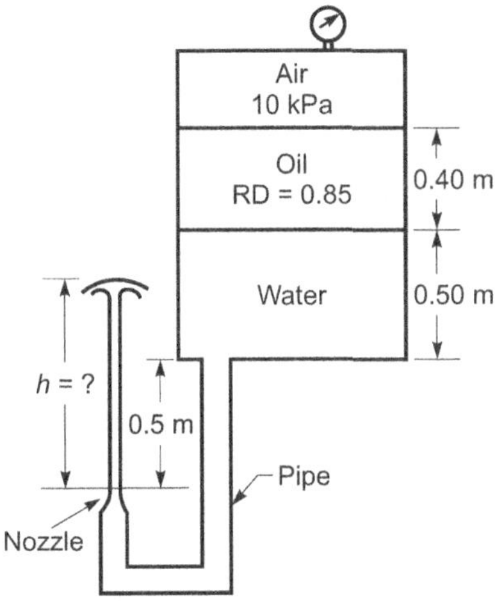

Fig. 4.32

4.7 A 10 cm long nozzle of exit diameter 10 cm is attached to a pipe of 30 cm diameter. The nozzle is vertical. If a water jet issuing out of the nozzle reaches a height of 4.5 m above the nozzle exit, calculate the discharge. Also, by assuming a head loss in the nozzle equal to 10% of the exit velocity head, calculate the pressure at the base of the nozzle.

(*Ans.* Q = 73.8 L/s, p_1 = 48.895 kPa)

4.8 Water is discharged from a 10 cm diameter nozzle held at a an angle of elevation of 30°. The tip of the nozzle is 2.0 m above the ground level. Estimate the discharge if the jet strikes the ground at 5 m away from the nozzle.

(*Ans.* Q = 45.4 L/s)

4.9 A fire hose nozzle is at a height of 1.8 m above the ground level and is issuing a water jet of velocity 32 m/s. It is to be used to throw the jet to reach a window at a height of 22 m above the ground and at a horizontal distance of 30 m from the nozzle. Determine the angles of inclination of the jet which will achieve this result. Neglect all frictional losses.

(*Ans.* θ_1 = 80.56°, θ_2 = 43.40°)

4.10 A nozzle at the end of a hose has a diameter of 5 cm. The inclination of the nozzle is at 45° to the horizontal and is directed upwards. A point on the jet axis is at a distance of 3 m from the nozzle and 2.0 m above it. Estimate the discharge from the nozzle.

(*Ans.* Q = 18.45 L/s)

4.11 A fire extinguishing service is trying to train a fire hose at A on to a window B as shown in Fig. 4.33. The velocity of the jet is 30 m/s. Calculate the maximum distance x of the nozzle at which this could be achieved. What is the corresponding angle θ of the nozzle?

(*Ans.* x_{max} = 71.51 m, θ = 52.06°)

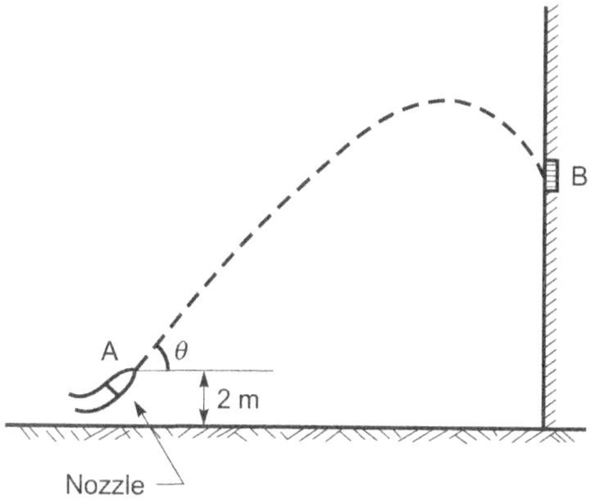

Fig. 4.33

4.12 Water flows radially outwards between two horizontal circular plates of diameter 0.80 m. The plates are 3 cm apart and the flow is supplied through a 20 cm diameter pipe at the centre of the plates (Fig. 4.34). If the discharge, from between the plates to atmosphere is 40 L/s. Calculate the pressure at point A in the pipe 0.8 m above the plates.

(*Ans.* p_a = – 8.5 kPa)

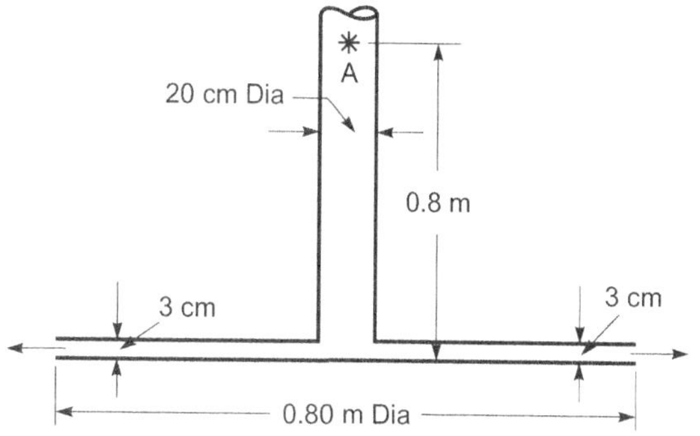

Fig. 4.34 *Problem 4.12*

4.13 A conical tube is fixed vertically with its larger diameter at the top and forms a part of a pipeline carrying kerosene (RD = 0.80). The velocity at the smaller end is 3.0 m/s and at the larger end it is 1.5 m/s. The tube is 2.0 m long. At the bottom of the tube the pressure is 50 kPa. The head loss in the tube an be assumed to be 0.35 times the difference in the velocity heads at the two ends. Estimate the pressure at the top of the tube when the flow is upwards.

(*Ans.* p_2 = 36.09 kPa)

4.14 Two streams of water at the same pressure p_1 but with velocities V_{11} = 25 m/s and V_{12} = 5 m/s enter a mixing chamber, and after complete mixing emerge as a single stream with uniform properties [See Fig. 4.35]. If no loss of any kind occurs in the flow, determine the exit velocity V_3 and the pressure difference $(p_3 - p_1)$. Given are A_{11} = 20 cm^3, A_3 = 40 cm^2. ρ = 1000 kg

(*Ans.* $(p_3 - p_1)$ = 158.72 kPa)

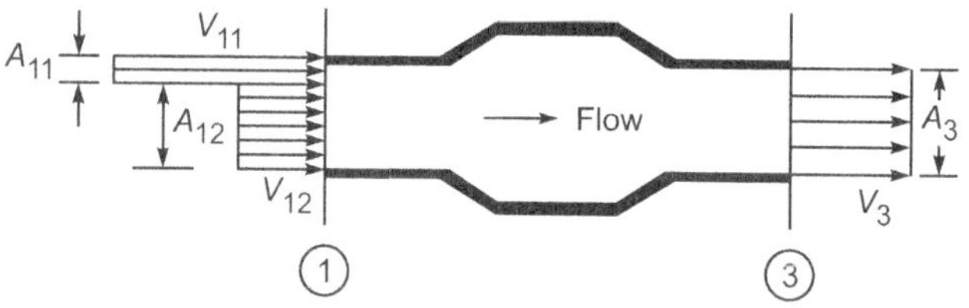

Fig. 4.35

4.15 A 15 cm diameter pipe is expanded to 25 cm diameter suddenly at a section. The head loss at a sudden expansion from section 1 to 2 is given by

$h_L = (V_1 - V_2)^2 /2g$. For a discharge of 45 L/s for a pipeline set-up shown in Fig. 4.36. Calculate the reading h of the mercury-water differential manometer.

(*Ans.* h = 1.21 cm)

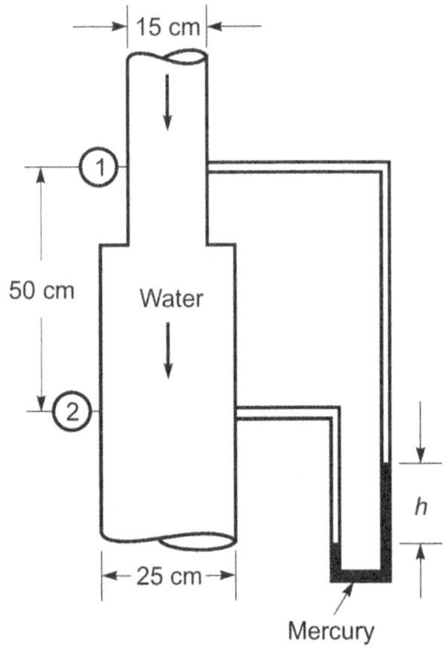

Fig. 4.36 *Problem 4.15*

4.16 For the flow situation shown in Fig. 4.37 calculate the discharge of oil (RD = 0.75) when the oil-mercury differential manometer reading h = 10 cm. Neglect all losses in the flow system.

(*Ans.* Q = 47 L/s)

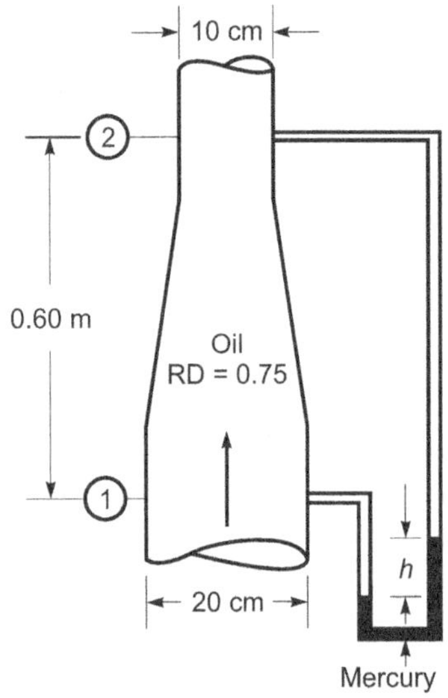

Fig. 4.37

4.17 Show that the average velocity V and the kinetic energy correction factor α for the following velocity distribution in a pipe of radius r_0:

$$\frac{u}{u_m} = \left(1 - \frac{r}{r_0}\right)^m$$

are given by

$$V = \frac{2u_m}{(1+m)(2+m)}$$

$$\alpha = \frac{(1+m)^3(2+m)^3}{4(1+3m)(2+3m)}$$

In the above u_m = maximum velocity and u = velocity at any radius r, and m = a coefficient.

4.18 Find the kinetic energy correction factor α for the velocity distributions in a two-dimensional duct, as shown in Fig. 4.38(a), (b) and (c).

$$\left[\textit{Ans.}\ \text{(a)}\ \alpha = \frac{10}{9}, \text{(b)}\ \alpha = 4\frac{2}{3}, \text{(c)}\ \alpha = 1.543\right]$$

4.19 A conical pipe has diameters 0.40 m and 0.80 m at its two ends. The smaller end is 2 m above the larger end. For a flow of 0.30 m^3/s of water the pressure at the lower end is 10 kPa. Assuming a head loss of 2 m and kinetic energy correction factor α = 1.1 and 1.5 at the smaller and larger ends respectively, estimate the pressure at the smaller end.

(*Ans.* p_1 = 7.144 kPa)

4.20 A pipeline has the following data at its two sections A and B:

Item	*Section A*	*Section B*
Diameter	30 cm	45 cm
Elevation (m)	10.000	16.000
Pressure	40.0 kPa	30 kPa
Kinetic energy correction factor, α	1.08	1.25

Assume a head loss equal to 20 times the velocity head at A and calculate the discharge of water through this pipeline when flowing from B to A.

(*Ans.* Q_a = 147.5 L/s)

4.21 A pump has a 30 cm diameter suction pipe and 25 cm diameter delivery pipe. When 220 L/s of water was being pumped, the pressure on the suction side of the pump was 4 m of vacuum and on the delivery side the pressure was 100 kPa. Assuming an efficiency of 50% for the pump-motor set, estimate the electrical power consumed.

(*Ans.* **P** = 63.51 kW)

4.22 A mercury-water differential manometer connected to the 15 cm diameter suction pipe and 12 cm delivery pipe of a pump shows a deflection of 40 cm. The centerlines of the suction and delivery

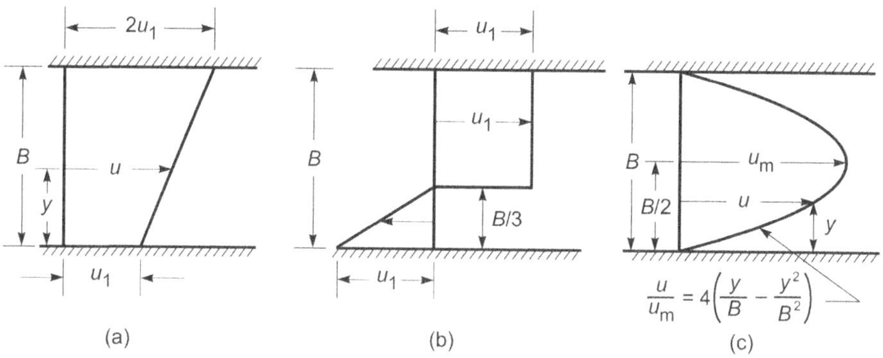

Fig. 4.38 *Problem 4.18*

pipes are at the same level. If the pump is discharging 70 L/s of water, estimate the head developed by the pump.

(*Ans.* H_p = 6.193 m)

4.23 A pump draws from a sump whose water surface is 1.5 m below the centre of pump and discharges it freely to atmosphere at 1.2 m above the pump centreline. The suction and delivery pipes are 20 cm and 25 cm in diameter respectively. If the pressure at the suction end of the pump is (–2.0) cm of mercury (gauge) determine the discharge and power imparted by the pump. Neglect all losses.

(*Ans.* Q = 153.7 L/s, **P** = 4.81 kW)

4.24 Determine the shaft power for a 70% efficient pump to discharge 1.5 m^3/min of oil of RD = 0.90 from a tank with oil surface elevation 100.00 m to another with oil surface elevation at 120.00 m. The pipeline is of 15 cm diameter. The headloss in the pipe can be taken to be 10 times the velocity head in the pipeline.

(*Ans.* $\mathbf{P}_s$ = 6.615 kW)

4.25 For a hydraulic machine shown in Fig. 4.39 the following data are available:

Flow	:	From A to B
Diameters	:	at A: 20 cm; at B : 30 cm
Elevation (m)	:	at A: 105.00; at B: 100.00
Pressures	:	at A: 100 kPa; at B: 200 kPa
Discharge	:	200 L/s of water.

Is this machine a pump or a turbine? Calculate the power input or output depending on whether it is pump or a turbine.

(*Ans.* The machine is a pump. **P** = 6.965 kW)

4.26 A 20 cm diameter pipe leading water from a reservoir ends in a nozzle of exit diameter of 10 cm at elevation 90.00 m. The water surface in the reservoir is at elevation 100.00 m (Fig. 4.40). The energy loss in the system can be assumed as 12 times the velocity head in the pipe. Calculate the discharge. If the discharge is to be increased by 75%, calculate the power input through a pump introduced in the pipeline at the base of the nozzle.

(*Ans.* Q_0 = 83.16 L/s; **P** = 12.236 kW)

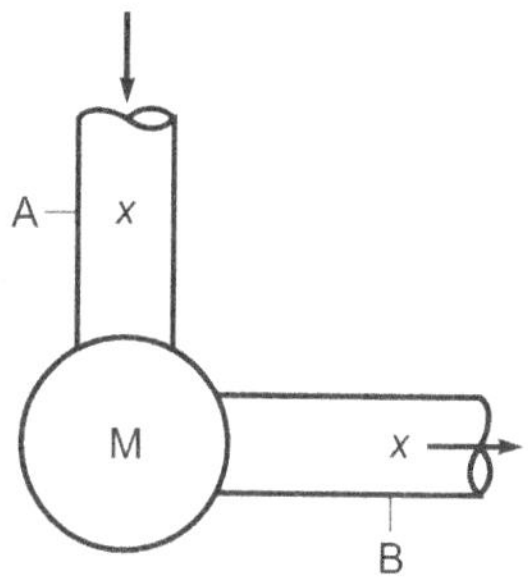

Fig 4.39 *Problem 4.25*

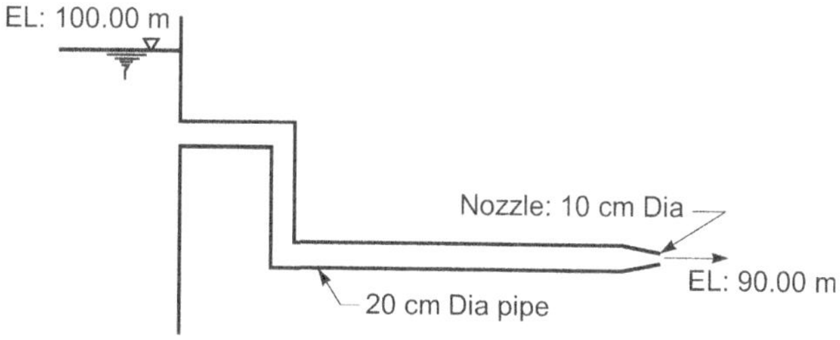

Fig 4.40 *Problem 4.26*

4.27 A turbine discharges 2.0 m^3/s of water into a vertical draft tube as shown in Fig. 4.41. The diameter of the tube is 0.8 m at A. If the head loss in the draft tube can be assumed as 1.5 times the velocity head at A, estimate the pressure at A.

(*Ans.* p_a = – 30.32 kPa)

4.28 A reaction turbine has a supply pipe of 0.80 m diameter and a draft tube with a diameter of 1.2 m at the turbine and expanding gradually downwards. The tailwater surface is 4.0 m below the centreline

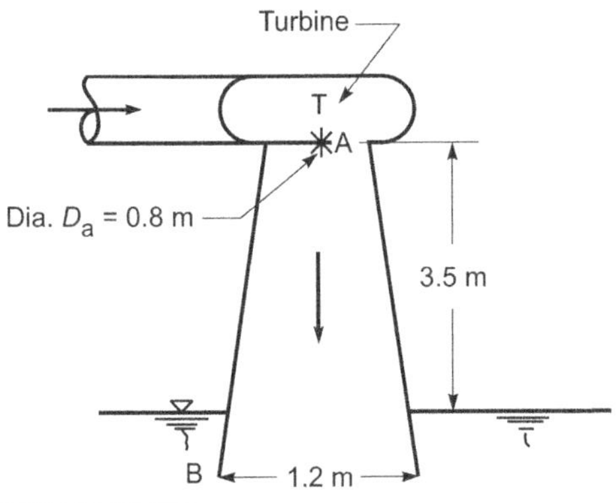

Fig 4.41

of the supply pipe at the turbine. For a discharge of 1.1 m^3/s, the pressure head just upstream of the turbine is 40 m. Estimate the power output of the turbine by assuming 85% efficiency. Also, determine the pressure at the commencement of the draft tube, next to the turbine, which is 3.50 m above the tailwater, (Fig. 4.41)

(*Ans.* **P** = 405 kW; P_b = – 34.79 kPa)

4.29 A pipeline delivering water from a reservoir is shown in Fig. 4.42. A pump M adds energy to the flow and 45 L/s of water is discharged to atmosphere at the outlet. Calculate the power delivered by the pump. Assume the head loss in the pipe as

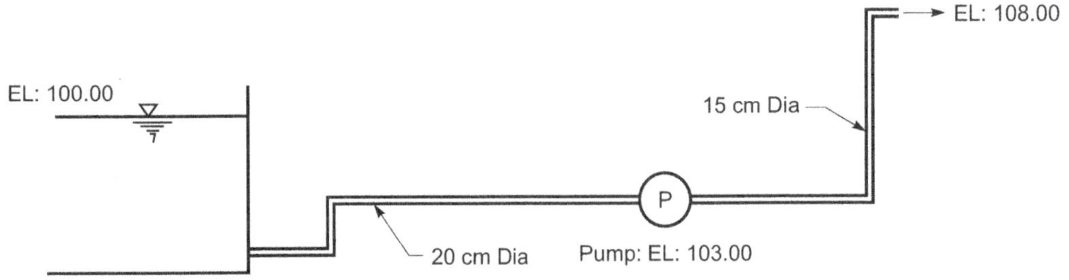

Fig 4.42

two times the velocity head at the suction side and 10 times the velocity head in the delivery pipe. Draw a neat sketch showing energy line and hydraulic grade lines.

(*Ans.* **P** = 5.218 kW)

4.30 Air (ρ = 1.08 kg/m^3) flows past a stationary cylinder. Far away from the cylinder the air velocity V_0 = 40 km/h. Calculate the stagnation pressure on the cylinder. If the velocity v at any point on the cylinder is v = $2V_0 \sin\ \theta$ where θ = angle measured from the front stagnation point, calculate the pressure on the surface at angles θ = 30°, 60° and 90°. (Take atmospheric pressure as reference pressure).

[*Ans.* p_s = 66.67 Pa (gauge)
θ = 30°, p = 0,
θ = 60°, p = – 133.33 Pa (gauge)
θ = 90°, p = – 220 Pa (gauge)]

4.31 For the uniform flow of velocity U_∞ past a circular cylinder of radius a (Fig. 4.43) and circulation Γ the velocity on the cylinder is given by

$$v_\theta = -\,2U_\infty \sin\,\theta - \frac{\Gamma}{2\pi a}$$

Obtain an expression for the variation of the pressure coefficient $C_p = (P - p_\infty)/\left(\frac{1}{2}\rho U_\infty^2\right)$ on the cylinder. Also obtain an expression for the location of the stagnation point on the cylinder.

$$\left(Ans.\ C_p = \left(\frac{p - p_\infty}{\frac{1}{2}\rho U_\infty^2}\right) = \left[1 - \left(2\sin\theta + \frac{\Gamma}{2\pi U_\infty R}\right)^2\right]\right.$$

$$\left.\text{At stagnation point, } \sin\theta = \frac{\Gamma}{4\pi U_\infty a}\right)$$

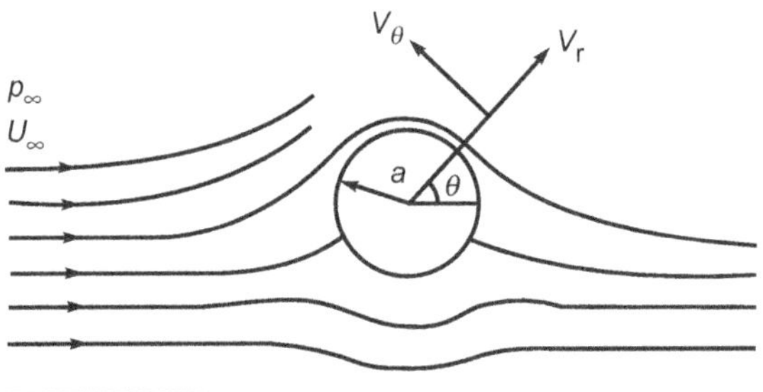

Fig 4.43

5

Momentum Equation and Its Applications

CONCEPT REVIEW

The momentum principle is derived from Newton's second law and is considered under two categories as (i) Linear momentum equation and (ii) Moment of momentum equation. Both the equations are applicable to a control volume and are vector equations.
A *control volume* is defined as a volume fixed in space.

5.1 LINEAR MOMENTUM EQUATION

This equation states that the vector sum of all external forces acting on a control volume in a fluid flow equals the time rate of change of linear momentum vector of the fluid mass in the control volume.
The external forces are of two kinds, viz. boundary (surface) forces and body forces. *Boundary forces* consist of

1. Pressure intensities acting normal to a boundary, F_p, and
2. Shear stresses acting tangential to a boundary, F_s.

Body forces are those that depend upon the mass of the fluid in the control volume, for example weight, F_b.
The linear momentum equation in a general flow can be written for any direction x as

$$\sum F_x = F_{px} + F_{sx} + F_{bx} = \frac{\partial}{\partial t}(M_x)_{cv} + M_{x_{out}} - M_{x_{in}} \tag{5.1}$$

where M_x = momentum flux in x-direction = $\rho Q V_x$.

Suffixes *out* represent the flux going out of the control volume and *in* represent the flux coming into the control volume.
F_{px}, F_{sx} and F_{bx} represent x-component of pressure force, shear force and body force respectively acting on the control volume surface.

$\frac{\partial}{\partial t}(M_x)_{cv}$ = rate of change of x-momentum within the control volume. This component is zero in a steady flow.

Thus for a steady flow, in the x-direction,

$$F_{px} + F_{sx} + F_{bx} = (M_x)_{out} - (M_x)_{in} = (\rho Q V_x)_{out} - (\rho Q V_x)_{in} \tag{5.2}$$

Similar momentum equations are applicable to other coordinate directions, y and z also.

5.1.1 Application to One-dimensional Flow

Momentum correction factor In one-dimensional flow analysis the flow characteristic in one major direction, say longitudinal axis direction, is considered and the variation in other directions neglected. Thus, for example, in the two-dimensional transition shown in Fig. 5.1, the velocity distribution of u with y is accounted for by taking average velocity $V = \frac{1}{B}\int u\,dy$ and V is used in the analysis. The discharge $Q = VA$.

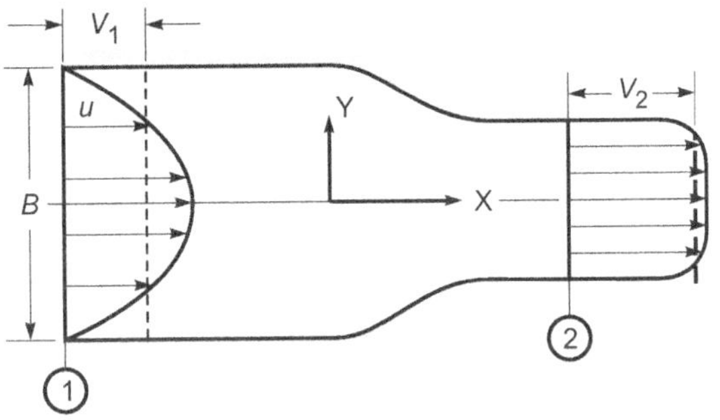

Fig. 5.1

A momentum correction factor

$$\beta = \frac{1}{V^2 A}\int u^2 \mathrm{d}A \qquad (5.3)$$

is used to account for the variation of the velocity across the area in the calculation of the momentum flux. Thus the momentum flux at section 1 is

$$M_1 = \beta_1 \rho Q V_1 \qquad (5.4a)$$

and the momentum flux at section 2 is

$$M_2 = \beta_2 \rho Q V_2 \qquad (5.4b)$$

For uniform velocity distribution $\beta = 1$ and for all other cases $\beta > 1.0$. In laminar flow through a circular tube, $\beta = 1.33$ and for turbulent flow through pipes $\beta \approx 1.05$. By definition β depends upon the nature of the velocity distribution; larger the non-uniformity, greater will be the value of β. If no other information is given, it is usual practice to assume $\beta = 1.0$.

Control volume In the application of the linear momentum equation the control volume can be assumed arbitrarily. It is usual practice to draw a control volume in such a way that (Fig. 5.2).

(i) Its boundaries are normal to the direction of flow at inlets and outlets.
(ii) It is inside the flow boundary and has the same alignment as the flow boundary.
(iii) Wherever the magnitudes of the boundary forces (due to pressure and shear stresses) are not known, their resultant is taken as a reaction force R (with components, R_x, R_y and R_z) on the control volume. This reaction R is the *Force acting on the fluid* in the control volume due to reaction from the boundary. The *Force F of the fluid on the boundary* will be *equal and opposite* to the reaction R.

Reaction of the boundary, R As indicated above, the reaction of the boundary R, with component R_x and R_y is the force exerted by the boundary on the

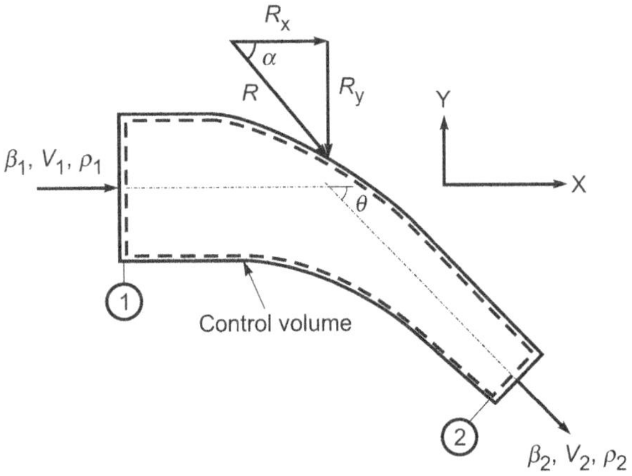

Fig. 5.2

fluid. In most of the applications, R is an unknown to be determined. As such, R_x and R_y are assumed to act in chosen directions and the momentum equation written. Upon solving for R_x and R_y, depending upon the sign of the answer, the assumption is corrected, if need be. Thus, R_x and R_y can be *assumed* to be in positive or negative direction of x and y respectively and upon solving, the final answer will emerge out with the proper direction of the reaction force, R. Also,

$$R = \sqrt{R_x^2 + R_y^2} \qquad (5.5)$$

and its inclination α to x-axis is

$$\tan \alpha = \frac{R_y}{R_x} \qquad (5.6)$$

Momentum correction factor β When β at a section is given, the momentum flux past the section in the chosen x-direction is given by

$$M_x = \beta \rho Q V_x \qquad (5.7)$$

In Fig. 5.2, the momentum flux in various directions are:

at 1, in x-direction:	$M_{x1} = \beta_1 \rho Q V_1$
at 1, in y-direction:	$M_{y1} = 0$
at 2, in x-direction:	$M_{x2} = \beta_2 \rho Q V_2 \cos\theta$
at 2, in y-direction:	$M_{y2} = \beta_2 \rho Q V_2 \sin\theta$
Discharge	$Q = AV = A_1 V_1 = A_2 V_2$

5.1.2 Forces on Moving Blades

A major application of the momentum equation relates to impact of liquid jets on blades. Figure 5.3 shows a liquid jet of velocity V impacting on a curved blade moving at a velocity u.

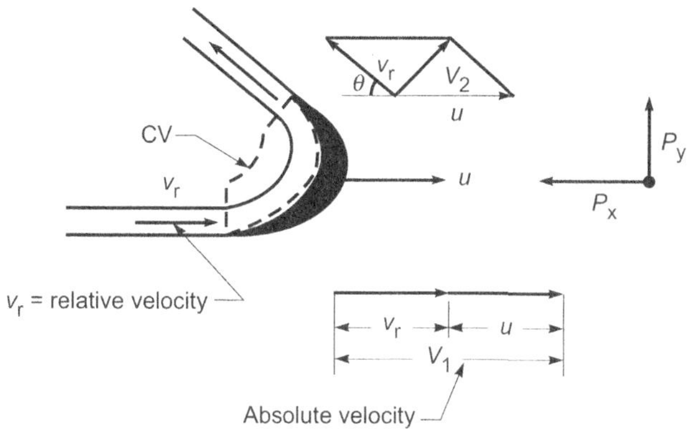

Fig. 5.3

The static pressure is atmospheric everywhere.

Relative velocity of water entering the blade = $v_r = V_1 - u$.

where V_1 = absolute velocity of the jet.

If there is no friction, the relative velocity will remain constant all over the blade. At the exist of the blade, the relative velocity $V_{r2} = V_r = V_1 - u$. The absolute velocity V_2 is obtained as vector sum of V_r and u as in Fig. 5.3.

$$\boxed{\mathbf{V}_2 = \boldsymbol{v}_r + \boldsymbol{u}} \qquad (5.8)$$

The relative velocity is *always* assumed to leave the blade tangentially. Hence, the momentum equation can be applied to the relative velocities.

If P_x is the reaction of the blade on the fluid in the control volume.

$$0 - P_x = \rho Q_r (-v_r \cos\theta - v_r) \qquad (5.8)$$
$$0 - P_x = -\rho A v_r^2 (\cos\theta + 1)$$
$$P_x = \rho A (V_1 - u)^2 (1 + \cos\theta) \qquad (5.9)$$

Force on the blade = + F_x = $|P_x|$ in the positive x-direction

$$\text{Power developed} = F_x u \qquad (5.10)$$

If a series of vanes are so arranged on a wheel that the entire jet is intercepted by one blade or other, the discharge to be used in Eq. (5.8) is the actual discharge of the jet Q instead of Q_r.

This principle is used in pelton turbines. In reaction turbines, the pressure on the blade is not atmospheric and the velocity triangles have to be written for both inlet and outlet of the blades. Details about turbines and pumps are presented in Chapter 16.

5.1.3 Momentum Equation for Steady Flow

For a control volume lying in a horizontal plane, shown in Fig. 5.2, the linear momentum equation for steady flow is written as outlined below.

Let R_x along positive x-direction and R_y in negative y-direction be the reaction of the boundary on the fluid of the control volume (*cv*). Then in x-direction:

$$\begin{bmatrix}\text{The resultant of all forces}\\ \text{on } cv \text{ in x-direction}\end{bmatrix} = \begin{bmatrix}\text{x-Momentum flux}\\ \text{going } out \text{ of } cv\end{bmatrix} - \begin{bmatrix}\text{x-Momentum flux}\\ \text{going } into\ cv\end{bmatrix}$$

Thus

$$p_1 A_1 - p_2 A_2 \cos\theta + R_x = M_{x2} - M_{x1}$$
$$= (\beta_2 \rho Q V_2 \cos\theta - \beta_1 \rho Q V_1) \qquad (5.11)$$

Similarly in y-direction,

$$0 - p_2 A_2 \sin\theta - R_y = M_{y2} - M_{y1}$$
$$= \beta_2 \rho Q V_2 \sin\theta \qquad (5.12)$$

For any direction, that does not lie in a horizontal plane, the component of the body force (weight of fluid in *cv*) should be suitably included among the forces on cv.

In the solution of Eqs 5.11 and 5.12 often, depending upon the data, the continuity equation.

$$A_1 V_1 = A_2 V_2 \qquad (5.13)$$

and the Bernoulli equation

$$\frac{p_1}{\rho g} + \frac{\alpha_1 V_1^2}{2g} + Z_1 = \frac{p_2}{\rho g} + \frac{\alpha_2 V_2^2}{2g} + Z_2 \qquad (5.14)$$

will have to be used.

5.2 THE MOMENT OF MOMENTUM EQUATION

The moment of momentum equation is based on Newton's second law applied to a rotating fluid mass system. Moment of momentum about an axis is known as *angular momentum*. The moment of a force about a point is *torque*. The moment of momentum principle states that in a rotating system the torque exerted by the resultant force on the body with respect to an axis is equal to the time rate of change of angular momentum.

In a steady flow rotating system, i.e. when the rotating speed is constant,

$$\begin{bmatrix}\text{Torque exerted on the fluid}\\ \text{by the rotating element}\end{bmatrix} = \begin{bmatrix}\text{Angular momentum of}\\ \text{fluid leaving out of } cv\end{bmatrix} - \begin{bmatrix}\text{Angular momentum of}\\ \text{fluid entering the } cv\end{bmatrix}$$

$$T = \rho Q [(\Sigma V_r)_{out} - (\Sigma V_r)_{in}] \qquad (5.15)$$

where Q = discharge, V = absolute velocity, *out* and *in* denote items leaving or entering a control volume (*cv*) respectively.

Equation (5.15) finds considerable application in the analysis of roto dynamic machines, viz., turbines, pumps, propellors, etc. details of which can be had in Chapter 16. In the following section, the details of reaction with rotation with a typical application to a lawn sprinkler is given.

5.2.1 Reaction with Rotation

The reaction of fluid discharging in a rotating system would generate force and hence a torque. This is clearly illustrated in a rotating arm of a lawn sprinkler. Figure 5.4 shows a fluid entering the arm of a sprinkler normal to the arm at 1 and discharging at the outlet 2.

$$T = -\rho Q r (u_2 + v_2 \cos\beta) \tag{5.17}$$

where $u_2 = \omega r_2$.

(a) Thus when the retarding torque = T, the angular velocity ω is

$$\omega = -\frac{v_2 \cos\beta}{r} - \frac{T}{\rho Q r^2} \tag{5.18}$$

(b) The maximum speed of a sprinkler (run away speed) is when $T = 0$ and hence

$$\omega_{max} = -\frac{v_2 \cos\beta}{r} \tag{5.19}$$

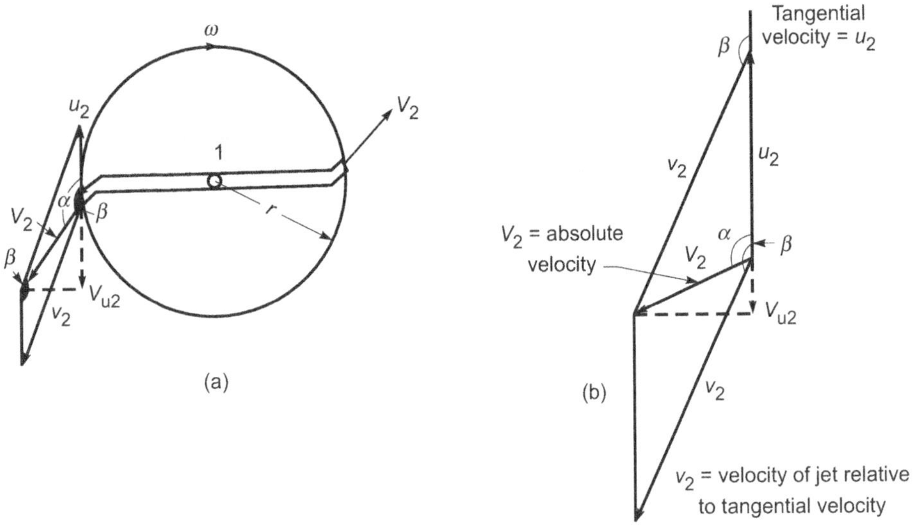

Fig. 5.4

If Q = discharge and a = area of the outlets (two in number here) then $Q/2a = v_2$ = relative velocity of exit. Also β = angle of the jet with the direction of rotation and u_2 = velocity of the arm = $r\omega$ where ω = angular velocity. The absolute velocity is V_2. Its tangential component = $V_{u2} = V_2 \cos\alpha$. It is the change in the absolute velocity that causes change in angular momentum.

From the velocity triangle (in vector notation)

$$\mathbf{V}_2 - \mathbf{u}_2 = \mathbf{v}_2 = \text{relative velocity} \tag{5.16}$$

i.e., $u_2 + v_2 \cos\beta = V_{u2} = V_2 \cos\alpha$

Force exerted by the fluid on the system,

$$F = -\rho Q(V_{u2})$$
$$= -\rho Q(u_2 + v_2 \cos\beta)$$

The retarding torque (due to bearing friction, etc.) is

Also, $V_{u2} = V_2 \cos\alpha = 0$ and α = inclination of the absolute velocity with the tangential direction = 90°.

(c) For a sprinkler with β = 180°:

$$T = -\rho Q r(\omega r - v_2) \tag{5.20a}$$

and when $T = 0$ (frictionless system)

$$\omega = v_2/r \tag{5.20b}$$

(d) The torque required to hold the sprinkler in fixed position (i.e. to prevent it from rotating) is T for $\omega = 0$, i.e.

$$T_0 = \rho Q r v_2 \cos\beta \tag{5.21}$$

WORKED EXAMPLES

A: Linear Momentum

5.1 A 7.5 cm diameter water jet having a velocity of 12 m/s impinges on a plane, smooth plate at an angle of 60° to the normal to the plate. What will be the impact force when (i) the plate is stationary and (ii) when moving in the direction of the jet at 6 m/s. Estimate the work done per unit time on the plate in each case.

Solution: Consider the normal and tangential directions and a control volume as in Fig. 5.5. Let R_n be the normal reaction on the fluid in the control volume. Consider the normal direction n. The pressure in the jet is atmospheric.

(i) When the plate is stationary:

$$0 - R_n = \rho Q(0 - V_1 \cos \theta)$$
$$R_n = \rho Q V_1 \cos \theta$$
$$= 998 \times \left[\frac{\pi}{4} \times (0.075)^2 \times 12\right] \times 12 \cos 60°$$
$$= 317.45 \text{ N}$$

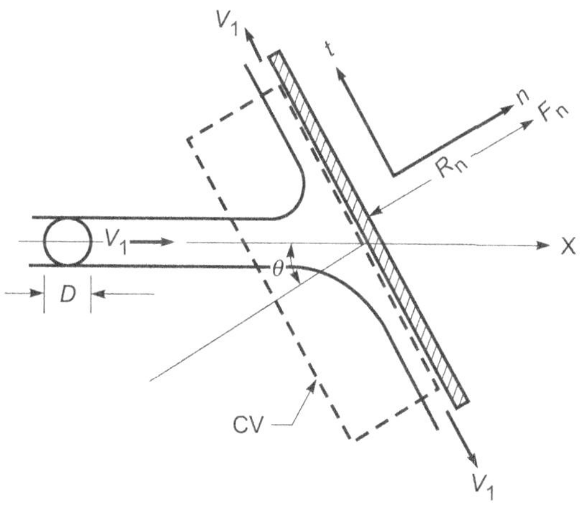

Fig. 5.5 *Jet Impingement on a Plate*

The normal force of the jet on plate is $\boldsymbol{F_n = 317.45}$ **N in the positive n direction** (opposite to R_n).

(ii) When the plate moves in the x-direction with $u = 6$ m/s.

Considering normal direction and relative velocities:

$$-R_n = \rho Q_r (0 - V_{1r} \cos \theta)$$
$$R_n = \rho A V_{1r}^2 \cos \theta$$
$$V_{1r} = 12.0 - 6.0 = 6.0 \text{ m/s}$$
$$R_n = 998 \times \frac{\pi}{4} \times (0.075)^2 \times 6^2 \times \cos 60°$$
$$= 79.36 \text{ N}$$

The normal force of the water jet on the plate F_n will be equal and opposite to R_n. Hence $\boldsymbol{F_n = 79.36}$ ***N*** **and acts in positive n direction.**

Work done $W = F_n \times u_n = F_n \cos 60° \times u$

In case (i): $u = 0$, $\boldsymbol{W = 0}$

(ii) $u = 6$ m/s,

$W = 79.36 \times 0.5 \times 6.0 =$ **238.0 N.m/s**

5.2 A vertical jet is issuing upwards from a nozzle with a velocity of 11 m/s. The fluid is oil ($\rho = 800$ kg/m^3) and the nozzle exist diameter is 7.5 cm. A flat horizontal plate bearing a total load of 400 N is supported only by the impact of the Jet (Fig. 5.6). Determine the equilibrium height of the plate above the nozzle exit. Neglect all losses.

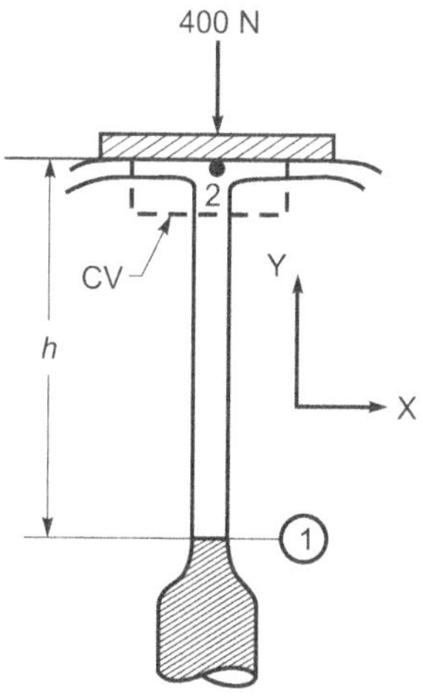

Fig. 5.6

Solution:

$$Q = \text{discharge} = A_1 V_1$$
$$= \frac{\pi}{4} \times (0.075)^2 \times 11.0 = 0.0486 \text{ m}^3/\text{s}$$

Since the jet is issuing into atmosphere $p_1 = p_2 = 0$.

Let h = equilibrium height of the plate above 1

V_2 = velocity of the plate just before impact.

By Bernoulli equation between points 1 and 2,

$$0 + \frac{V_1^2}{2g} + 0 = 0 + \frac{V_2^2}{2g} + h$$

Hence $V_2^2 = V_1^2 - 2gh$

or $V_2 = \sqrt{(V_1^2 - 2gh)}$

By momentum equation in y-direction,
R_y = Reaction force on the *cv* in (– *y*)-direction = weight supported by the jet = 400 N

Hence $\quad 0 - R_y = 0 - \rho Q V_2$

$$R_y = 400 = 800 \times 0.0486 \times [(11)^2 - 2 \times 9.81 \times h]^{1/2}$$

$$121 - 19.62\,h = 105.86$$

$$h = 0.0772 \text{ m}$$

The equilibrium height of the plate is **0.772 m**

5.3 A jet of oil (RD = 0.80) issues from nozzle of 15 cm diameter with a velocity of 12 m/s. A smooth cone with vertex angle of 90° deflects the jet. The jet is horizontal and the vertex of the cone points towards the jet. Calculate the force required to hold the cone in position.

Solution:
Consider a control volume as shown in Fig. 5.7. Let R_x = reaction of the cone on the fluid in the control volume. The pressure is everywhere atmospheric. As the cone is smooth, by neglecting friction the velocity of the sheet of water over the cone is V everywhere. The inclination of this velocity V to x-axis is 90/2 = 45°.

$$A = \frac{\pi}{4} \times (0.15)^2 = 0.01767 \text{ m}^2,$$

$$\rho = 0.8 \times 998 = 798.4 \text{ kg/m}^3$$

$$Q = AV = 0.01767 \times 12 = 0.2121 \text{ m}^3\text{/s}$$

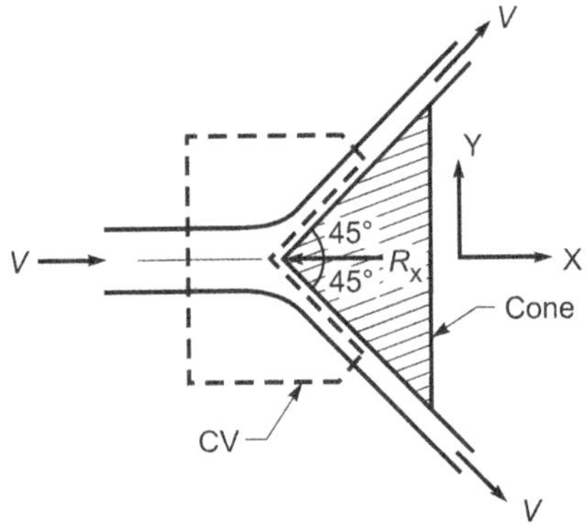

Fig. 5.7

By momentum equation in x-direction:

$$0 - R_x = \rho Q\,(V \cos 45° - V)$$

$$R_x = \rho Q V\,(1 - \cos 45°)$$

$$= 798.4 \times 0.2121 \times 12 \times (1 - \cos 45°)$$

$$= 595 \text{ N}$$

By symmetry $R_y = 0$

Hence the resultant reaction force on the fluid $R = R_x$. Thus the force required to hold the cone in position is $F = R = R_x$ = **595 N along (– x)-direction.**

5.4 A 10 cm diameter jet of water strikes a curved vane with a velocity of 25 m/s. The inlet angle of the vane is zero and the outlet angle is 30°. Determine the resultant force on the vane (a) When the vane is stationary and (b) When the vane is moving in the direction of the jet at 10 m/s velocity.

Solution:
(a) When the vane is stationary:
Let R_x and R_y, as shown in Fig. 5.8(a), be the reaction forces on the fluid in the control volume.

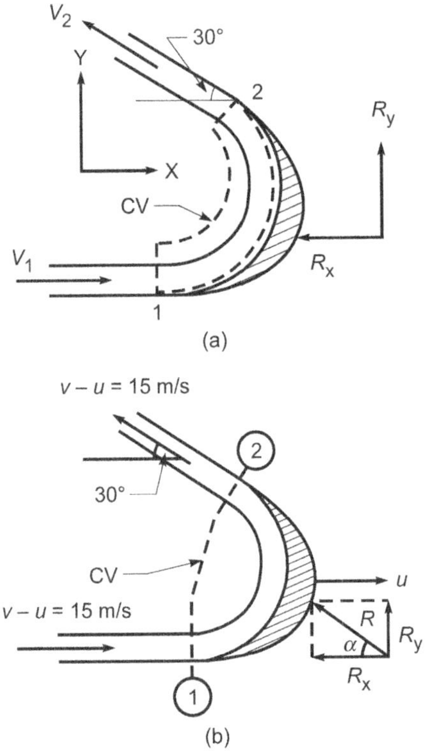

Fig. 5.8

The pressure is atmospheric at inlet and outlet.
By momentum equation in the x-direction,

$$-R_x = \rho Q\,(-V_2 \cos 30° - V_1)$$

$V_2 = V_1$ = 25 m/s as there are no losses.

$$-R_x = 998 \times \frac{\pi}{4} \times (0.10)^2 \times 25\,(-25 \cos 30° - 25)$$

$$= -195.95 \times (46.65)$$

$$R_x = 9141.5 \text{ N}$$

By momentum equation in the y-direction,

$$R_y = \rho Q\,(V_2 \sin 30° - 0)$$

$$= 998 \times \frac{\pi}{4} \times (0.10)^2 \times 25 \times 25 \sin 30°$$

$$= 195.95 \times 25 \times 0.5 = 2449.5 \text{ N}$$

Resultant Reaction

$$R = \sqrt{(9141.5)^2 + (2449.5)^2} = 9464 \text{ N}$$

acting at an angle α to the (– x)-direction given by

$$\alpha = \tan^{-1}\frac{R_y}{R_x} = \tan^{-1}\frac{2449.5}{9141.5} = 15°$$

The force on the blade is equal and opposite to R. Hence **F = 9464 N and acts at (360 – 15) = 345° to the x-direction**.

(b) When the vane moves at 10 m/s in the x-direction. The situation is shown in Fig. 5.8 (b).

The relative velocity entering the vane V_{r1} = the relative velocity leaving the vane

$$V_{r1} = V_{r2} = V_1 - u = 25 - 10 = 15 \text{ m/s}$$

The relative discharge acting on the vane = Q_r

$$= A \times (V - u) = \frac{\pi}{4} \times (0.10)^2 \times 15$$

$$= 0.1178 \text{ m}^3/\text{s}$$

By momentum equation in x-direction,

$$-R_x = \rho Q_r(-V_{r2}\cos 30° - V_{r1})$$
$$= 998 \times 0.1178\,(-15\cos 30° - 15)$$
$$R_x = 3290.7 \text{ N}$$

By momentum equation in y-direction,

$$R_y = \rho Q_r\,(V_{r2}\sin 30° - 0)$$
$$= 998 \times 0.1178 \times (15 \sin 30°)$$
$$R_y = 881.7 \text{ N}$$

Resultant reaction R on the water in the control volume:

$$R = \sqrt{(3290.7)^2 + (881.7)^2} = 3406.7 \text{ N}$$

Inclination of R with (– x)-direction $\alpha = \tan^{-1}\frac{R_y}{R_x}$

$$\alpha = \tan^{-1}\frac{881.7}{3290.7} = 15°$$

The resultant force on the vane will be equal and opposite to R. **Hence R = 3406.7 N and inclined at (360° – 15°) = 345° to the x-direction.**

5.5 A jet of water with a velocity of 30 m/s impinges on a moving vane of velocity 12 m/s at 30° to the direction of motion. The vane angle at the outlet is 18°. Find

(i) the blade angle at inlet so that the water enters without shock.

(ii) the work done on the vane per unit weight of water per second entering the vane, and

(iii) the efficiency

Solution: Here the jet velocity = absolute velocity

$$= V_1 = 30 \text{ m/s}$$

Velocity of blade = u = 12 m/s

The inlet and outlet velocity triangles are shown in Fig. 5.9.

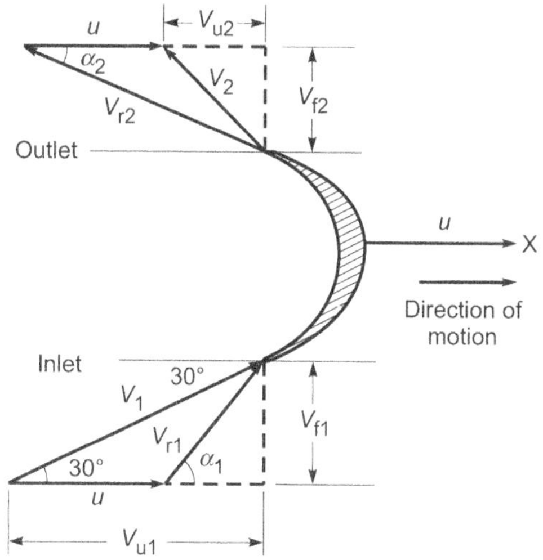

Fig. 5.9 *Jet Impingement on a Moving Vane*

(i) Consider the inlet velocity triangle.

$$V_1 = 30 \text{ m/s}, \; V_{f1} = V_1 \sin 30°$$
$$= 30 \times 0.5 = 15 \text{ m/s}$$
$$V_{u1} = V_1 \cos 30° = 30 \times 0.866 = 25.98 \text{ m/s}$$

If α_1 = angle of the relative velocity with the direction of the vane = vane angle at inlet.

$$\tan\alpha_1 = \frac{V_{f1}}{(V_{u1} - u)} = \frac{15}{(25.98 - 12.0)} = 1.073$$
$$\alpha_1 = 47°$$

The relative velocity $V_{r1} = V_{r2} = \dfrac{V_{f1}}{\sin 47°}$

$$= 20.505 \text{ m/s}$$

From the outlet velocity triangle,

$$\alpha_2 = 18°$$
$$V_{u2} = V_{r2}\cos\alpha_2 - u_2$$

As $u_2 = u_1 = 12$ m/s

$$V_{u2} = 20.505\cos 18° - 12.0$$

= 7.50 m/s (in the – x-direction)

(ii) Considering the direction of motion, x, force exerted by the jet in the x-direction

$$F_x = \rho Q_r[V_{u2} + V_{u1}]$$

Work done/second

$$\mathbf{P} = F_x u$$
$$= \rho Q_r(V_{u2} + V_{u1})\,u$$

Work done/second/unit weight of liquid = $\mathbf{P}/\gamma Q$

$$= \text{Head extracted} = \frac{1}{g}(V_{u2} + V_{u1})\,u$$

$$= \frac{1}{9.81}(7.50 + 25.98) \times 12 = \mathbf{40.95 \text{ m}}$$

(iii) Initial energy head of the jet

$$= \text{Velocity head} = V_1^2/2g = (30)^2/2 \times 9.81 = 45.87 \text{ m}$$

$$\text{Efficiency} = \frac{\text{Head extracted}}{\text{Initial energy head}} = 40.95/45.87 = \mathbf{89.27\%}$$

5.6 A discharge of 0.06 m^3/s flows through a horizontal bend as shown in Fig. 5.10 Calculate the force on the bolts in section 1.

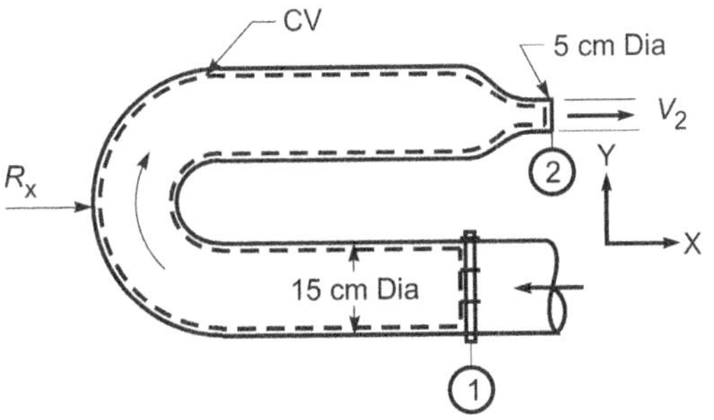

Fig. 5.10

Solution: The control volume is shown in dotted lines. The Reaction on the control volume fluid is shown as R_x in positive x-direction.

$$\text{Discharge} \quad Q = \frac{\pi}{4}(D_2)^2 V_2 = 0.06 \text{ m}^3/\text{s}$$

$$V_2 = \frac{0.06}{\frac{\pi}{4} \times (0.05)^2} = 30.56 \text{ m/s}$$

$$V_1 = V_2 (D_2/D_1)^2 = 30.56 \times \left(\frac{5}{15}\right)^2 = 3.395 \text{ m/s}$$

By applying Bernoulli equation to sections 2 and 1, by assuming the bend to be in the horizontal plane,

$$0 + \frac{(30.56)^2}{2 \times 9.81} = \frac{p_1}{\gamma} + \frac{(3.395)^2}{2 \times 9.81} \quad \ldots (p_2 = \text{atmospheric})$$

$$\frac{p_1}{\gamma} = 47.59 - 0.59 = 47.00 \text{ m}$$

$$p_1 = 47.00 \times 9.79 = 460.2 \text{ kPa}$$

By momentum equation in the x-direction,

$$-p_1A_1 + R_x - 0 = \rho Q\,[V_2 - (-V_1)]$$

$$-(460.2 \times 10^3) \times \frac{\pi}{4} \times (0.15)^2 + R_x = 998 \times 0.06 \times (30.56 + 3.395)$$

$$R_x = 8132 + 2033 = 10165 \text{ N}$$

The force F exerted by the fluid on the pipe, and hence on the bolts in Section 1, is equal and opposite to R_x.

Thus F = **10165 N and acts to the left, i.e. in the negative x-direction**, as a pull (tension) on the joint.

5.7 A 30 cm diameter pipe is bifurcated into two nozzles at a Y-junction as shown in Fig. 5.11. The nozzle discharge to atmosphere and have a velocity of 10 m/s each.

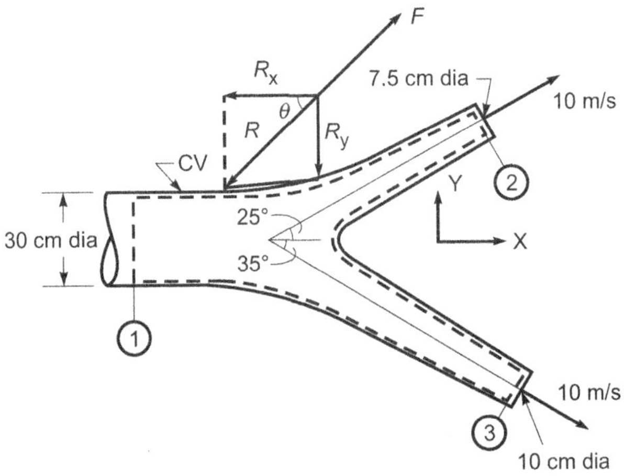

Fig. 5.11

The junction is in a horizontal plane and the friction can be neglected. Determine the magnitude and direction of the resultant force on the Y-junction.

Solution: Consider the control volume and reaction forces on the fluid in the control volume as in Fig. 5.11.

$$A_1 = \frac{\pi}{4} \times (0.30)^2 = 0.07069 \text{ m}^2$$

$$A_2 = \frac{\pi}{4} \times (0.075)^2 = 0.004418 \text{ m}^2,\ V_2 = 10 \text{ m/s}$$

$$A_3 = \frac{\pi}{4} \times (0.10)^2 = 0.007854 \text{ m}^2,\ V_3 = 10 \text{ m/s}$$

$$Q = A_1V_1 = Q_1 + Q_2$$

$$Q_2 = A_2V_2 = (0.004418 \times 10) = 0.04418 \text{ m}^3/\text{s}$$

$$Q_3 = A_3V_3 = (0.007854 \times 10) = 0.07854 \text{ m}^3/\text{s}$$

$$Q = Q_1 + Q_2 = 0.1227 \text{ m}^3/\text{s}$$

$$V_1 = 0.1227/0.07069 = 1.736 \text{ m/s}$$

By applying Bernoulli theorem to sections 2 and 1

$$\frac{p_2}{\gamma} + \frac{V_2^2}{2g} = \frac{p_1}{\gamma} + \frac{V_1^2}{2g}$$

$$0 + \frac{(10)^2}{2 \times 9.81} = \frac{p_1}{\gamma} + \frac{(1.736)^2}{2 \times 9.81}$$

$$p_1/\gamma = 5.0968 - 0.1536 = 4.943 \text{ m}$$

$$p_1 = 9790 \times 4.943 = 48394 \text{ Pa}$$

Applying momentum equation in the x-direction:

$$p_1A_1 - R_x = \rho[Q_2V_2 \cos 25° + Q_3V_3 \cos 35° - Q_1V_1]$$

$(48394 \times 0.07069) - R_x$
$= 998\,[0.04418 \times 10 \times \cos 25° + 0.07854 \times 10 \times \cos 35° - 0.1227 \times 1.736]$

$3421 - R_x = 998\,(0.4004 + 0.6434 - 0.2130)$

$R_x = 3421 - 829 = 2592$ N

By applying momentum equation in the y-direction:

$0 - R_y = \rho(Q_2V_2 \sin 25° - Q_3V_3 \sin 35°)$
$= 998\,(0.04418 \times 10 \times \sin 25° - 0.07854 \times 10 \times \sin 35°)$
$= 998\,(0.1867 - 0.4505) = -263$

$R_y = 263$ N

The reaction R on the fluid in the control volume is

$$R = \sqrt{R_x^2 + R_y^2} = \sqrt{(2592)^2 + (263)^2}$$
$$= 2605 \text{ N}$$

Inclination θ of R with the (–x) direction

$$\theta = \tan^{-1}\frac{R_y}{R_x} = \tan^{-1}\frac{263}{2592} = 5.79°$$

Force F on the body of the junction exerted by the fluid will be equal and opposite to R. Hence **F = 2605 N and will be directed at 5.79° to the positive x-direction**.

5.8 A reducer bend having an outlet diameter of 15 cm discharges freely. The bend, connected to a pipe of 20 cm diameter, has a deflection of 60° and lies in a horizontal plane. Determine the magnitude and direction of force on the anchor block supporting the pipe when a discharge of 0.3 m^3/s passes through the pipe.

Solution: Consider the control volume as shown by the dotted line in Fig. 5.12 (a). At section 2: $p_2 = 0$ = atmospheric pressure.

$$V_2 = \frac{0.3}{\frac{\pi}{4} \times (0.15)^2} = 16.98 \text{ m/s}$$

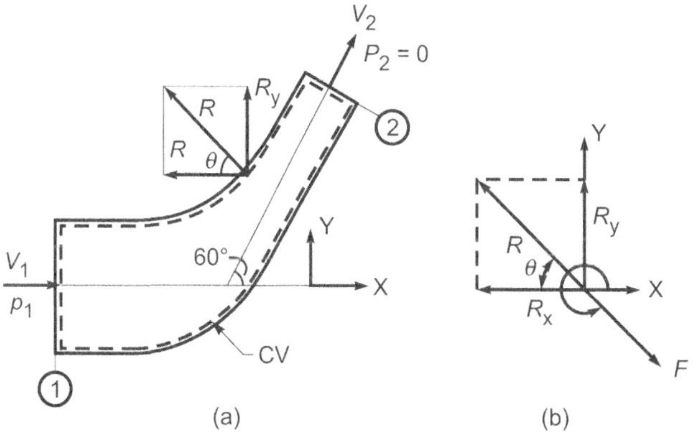

Fig. 5.12

At section 1: $V_1 = V_2\left(\frac{D_2}{D_1}\right)^2 = 16.98 \times \left(\frac{15}{20}\right)^2$
$= 9.55$ m/s

By applying Bernoulli equation to sections 2 and 1,

$$0 + (16.98)^2/(2 \times 9.81) = \frac{p_1}{\gamma} + (9.55)^2/(2 \times 9.81)$$

$$\frac{p_1}{\gamma} = 14.689 - 4.648 = 10.041 \text{ m}$$

$$p_1 = 9790 \times 10.041 = 98301 \text{ Pa}$$

Let R_x and R_y be the reaction of the pipe on the fluid in the control volume in (–x) and y-directions, respectively [Fig. E-5.12 (b)]. By applying momentum equation in x-direction),

$$p_1A_1 - R_x = \rho Q(V_2 \cos 60° - V_1)$$

$$98301 \times \frac{\pi}{4} \times (0.2)^2 - R_x$$
$$= 998 \times 0.3 \times (16.98 \cos 60° - 9.55)$$
$$R_x = 3088.2 + 317.4 = 3405.6 \text{ N}$$

By momentum equation in y-direction,

$$0 + R_y = \rho Q(V_2 \sin 60° - 0)$$
$$R_y = 998 \times 0.3 \times 16.98 \times 0.866 = 4402.7 \text{ N}$$

Resultant $R = \sqrt{R_x^2 + R_y^2}$

$$= \sqrt{(3405.6)^2 + (44027)^2} = 5564.3 \text{ N}$$

Inclined at an angle θ such that $\tan\theta = \frac{R_y}{R_x}$

$$\theta = \tan^{-1}\frac{4402.7}{3405.6} = 52.28°$$

The force F on the pipe is equal and opposite to R and hence **F = 5564 N inclined at (360 – θ) = 307.72° to positive x-axis** (see Fig. 5.12 (b)].

5.9 Water flows upwards through a vertical nozzle of inlet and outlet diameters 0.5 m and 0.25 m respectively as shown in Fig. 5.13. The pressure at section 1 is 15 kPa and the volume rate of flow is 0.5 m^3/s. Find (i) the velocities at section 1 and 2, (ii) pressure at section 2 and (iii) total force acting on the walls of the nozzle [Neglect frictional resistance]

Solution: Refer to Fig. 5.13.

$$A_1 = (\pi/4)\,(0.5)^2 = 0.1964 \text{ m}^2$$
$$A_2 = (\pi/4)\,(0.25)^2 = 0.0491 \text{ m}^2$$
$$V_1 = 0.5/0.1964 = 2.546 \text{ m/s}$$
$$V_2 = 0.5/0.04951 = 10.185 \text{ m/s}$$

By applying Bernoulli's theorem to sections 1 and 2

$$\frac{p_1}{\gamma} + Z_1 + \frac{V_1^2}{2g} = \frac{p_2}{\gamma} + Z_2 + \frac{V_2^2}{2g}$$

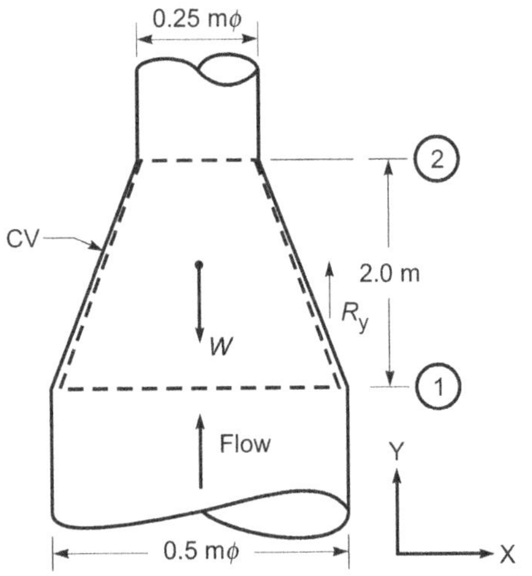

Fig. 5.13

$$\frac{15.0}{9.79} + 0 + \frac{(2.546)^2}{2 \times 9.81} = \frac{p_2}{\gamma} + 2.0 + \frac{(10.185)^2}{2 \times 9.81}$$

$$1.532 + 0.3304 = \frac{p_2}{\gamma} + 2.0 + 5.287$$

$$\frac{p_2}{\gamma} = -5.425 \text{ m}$$

and $\quad p_2 = -53.11$ kPa (gauge)

Now, consider a control volume encompassing section 1 and 2 shown in Fig. 5.13.

W = weight of water in the control volume which is in the shape of frustum of cone

$= \gamma(\pi/3)h\,[r_1^2 + r_1 r_2 + r_2^2]$

$= 9.79\,(\pi/3) \times 2.0\,[(0.25)^2 + (0.25)\,(0.125) + (0.125)^2]$

$= 2.243$ kN

Applying linear momentum equation in the vertical direction to *cv* with R_y = reaction force on the water in the control volume.

$$\Sigma\,(\text{Forces})_{Y\text{-direction}} = -W + R_y + p_1A_1 - p_2A_2 = \rho Q(V_2 - V_1)$$

$-2.243 + R_y + (15.0 \times 0.1964) - (-53.11 \times 0.0491)$

$\quad = 0.998 \times 0.5 \times (10.185 - 2.546)$

$R_y - 2.243 + 2.945 + 2.607 = 3.812$

$R_y = 0.503$ kN

F = Net force on the nozzle walls is equal and opposite to R_y.

Hence F = 0.503 kN acting vertically downwards

[***Note:*** *Remember to include the component of body force (i.e. weight of fluid in the control volume) in the application of linear momentum equation in a direction that does not lie in a horizontal plane.*]

5.10 The diameter of a pipe bend is 30 cm at inlet and 15 cm at outlet and the flow is turned through 120° in a vertical plane. The axis at inlet is horizontal and the centre of the outlet section is 1.5 m below the centre of the inlet section. Total volume of water in the bend is 0.9 m³. Neglecting friction, calculate the magnitude and direction of the force exerted on the bend by water flowing through it at 250 L/s and when the inlet pressure is 0.15 N/mm².

Solution: Refer to Fig. 5.14

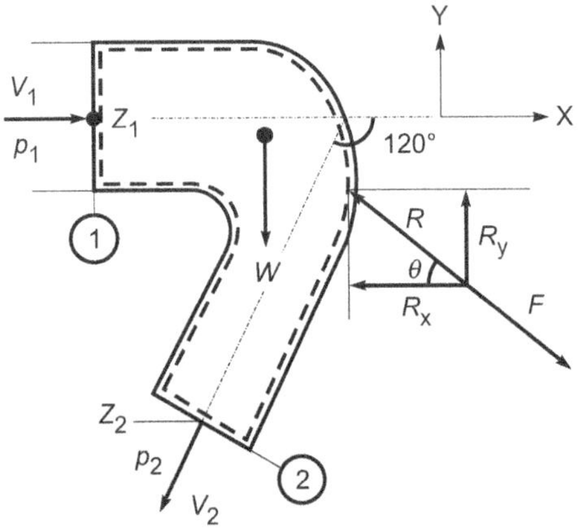

Fig. 5.14

$$A_1 = \frac{\pi}{4} \times (0.3)^2 = 0.07069 \text{ m}^2$$

$$A_2 = \frac{\pi}{4} \times (0.15)^2 = 0.01767 \text{ m}^2$$

$$V_1 = \frac{0.250}{0.07069} = 3.537 \text{ m/s},$$

$$V_2 = \frac{0.25}{0.01767} = 14.148 \text{ m/s}$$

$Z_1 = 1.5$ m, $Z_2 = 0$,

$p_1 = 0.15 \times 10^6$ Pa $= 150{,}000$ Pa

By applying energy equation to sections 1 and 2,

$$\frac{p_1}{\gamma} + \frac{V_1^2}{2g} + Z_1 = \frac{p_2}{\gamma} + \frac{V_2^2}{2g} + Z_2$$

$$\frac{0.15 \times 10^6}{9790} + \frac{(3.537)^2}{2 \times 9.81} + 1.5 = \frac{p_2}{\gamma} + \frac{(14.148)^2}{2 \times 9.81} + 0$$

$$\frac{p_2}{\gamma} = 15.3218 + 0.6376 + 1.5 - 10.2021$$

$$= 7.257 \text{ m}$$

$$p_2 = 7.257 \times 9790 = 71046 \text{ Pa}$$

Apply momentum equation in the x-direction:

Consider the reaction components R_x and R_y as shown in Fig. 5.14.

$p_1A_1 - R_x + p_2A_2 \cos 60° = \rho Q\,[-V_2 \cos 60° - V_1]$

$(0.15 \times 10^6 \times 0.07069) - R_x + (71046 \times 0.01767 \times 0.5)$

$\quad = 998 \times 0.250\,[(-14.148 \times 0.5) - 3.537]$

$R_x = 10603.5 + 627.7 + 2647.4 = 13879$ N

By momentum equation in the vertical direction:

$$0 - W + R_y + p_2A_2 \sin 60° = \rho Q[-V_2 \sin 60° - 0]$$

$$-(9790 \times 0.9) + R_y + (71046 \times 0.01767 \times 0.866) = 998 \times 0.25\,(-14.148 \times 0.866)$$

$$-8811 + R_y + 1087 = -3057$$

$$R_y = 4667 \text{ N}$$

Resultant reaction force on the fluid

$$R = \sqrt{R_x^2 + R_y^2} = [(13879)^2 + (4667)^2]^{1/2}$$

$$= 14642.7 \text{ N}$$

Inclination of R to the (−x)-direction, $\theta = \tan^{-1}\dfrac{R_y}{R_x}$

$$\theta = \tan^{-1}\frac{4667}{13879} = 18.59°$$

The force F exerted by the fluid on the bend is equal and opposite to R. Hence, **F = 14642.7 N and is inclined at (360 − θ) = 341.41° to the (+ x) direction.**

5.11 Given that a laminar boundary layer at zero pressure gradient is described by the velocity profile

$$\frac{u}{U_0} = \sin\left(\frac{\pi}{2}\cdot\frac{y}{\delta}\right)$$

determine the momentum correction coefficient and energy correction coefficient.

Solution: V = average velocity = $\dfrac{1}{A}\displaystyle\int_A u\,dA$

Considering unit width

$$V = \frac{1}{\delta}\int_0^{\delta} U_0 \sin\frac{\pi y}{2\delta}\,dy$$

$$= U_0\left[\left(-\cos\frac{\pi}{2}\cdot\frac{y}{\delta}\right)\times\frac{2\delta}{\pi}\right]_0^{\delta} = \frac{2\delta}{\pi}U_0$$

Momentum correction coefficient

$$\beta = \frac{1}{V^2A}\int_A u^2\,dA$$

$$\beta = \frac{\pi^2}{4U_0^2\delta}\int_0^{\delta} U_0^2 \sin^2\left(\frac{\pi}{2}\cdot\frac{y}{\delta}\right)dy$$

$$= \frac{\pi^2 U_0^2}{4U_0^2\delta}\left[\left(\frac{1}{2}\cdot\frac{\pi}{2}\frac{y}{\delta} - \frac{1}{4}\cdot\sin\frac{2\pi y}{2\delta}\right)\times\frac{2\delta}{\pi}\right]_0^{\delta}$$

$$= \frac{\pi^2}{4\delta}\left(\frac{\pi}{4}\right)\frac{2\delta}{\pi} = \frac{\pi^2}{8} = \mathbf{1.2337}$$

Kinetic energy correction coefficient

$$\alpha = \frac{1}{V^3A}\int_A u^3\,dA$$

$$\alpha = \frac{\pi^3}{8U_0^3\delta}\int_0^{\delta} U_0^3 \sin^3\left(\frac{\pi y}{2\delta}\right)dy$$

$$= \frac{\pi^3}{8\delta}\left[-\frac{1}{3}\cos\frac{\pi y}{2\delta}\left(\sin^2\frac{\pi y}{2\delta} + 2\right)\times\frac{2\delta}{\pi}\right]_0^{\delta}$$

$$= \frac{\pi^3}{8\delta}\left(\frac{1}{3}\times 2\times\frac{2\delta}{\pi}\right) = \frac{\pi^2}{6} = \mathbf{1.6449}$$

5.12 The velocity distribution in a pipe of radius r_0 is given by

$$\frac{u}{u_m} = \left(1 - \frac{r}{r_0}\right)^m$$

with $m < 1$. Determine the momentum correction factor β and calculate the value of β for $m = 1/7$.

Solution: Average velocity

$$V = \frac{1}{\pi r_0^2}\int_0^{r_0} u\cdot 2\pi r\,dr$$

$$V = \frac{2u_m}{r_0^2}\int_0^{r_0} r\left(1 - \frac{r}{r_0}\right)^m dr$$

$$= \frac{2u_m}{r_0^2}\left[\frac{r_0^2}{(m+2)}\left(1 - \frac{r}{r_0}\right)^{m+2} - \frac{r_0^2}{(m+1)}\left(1 - \frac{r}{r_0}\right)^{m+1}\right]_0^{r_0}$$

$$= \frac{2u_m}{r_0^2}\left[\frac{r_0^2}{m+1} - \frac{r_0^2}{m+2}\right]$$

$$V = \frac{2u_m}{(m+1)(m+2)}$$

Momentum correction factor

$$\beta = \frac{1}{V^2\pi r_0^2}\int_0^{r_0} u^2\,2\pi r\,dr$$

$$\beta = \frac{2}{V^2 r_0^2}\int_0^{r_0} u_m^2\cdot\left(1 - \frac{r}{r_0}\right)^{2m}\cdot r\,dr$$

$$= \frac{2u_m^2}{V^2 r_0^2}\left[\frac{r_0^2}{(2m+2)}\left(1 - \frac{r}{r_0}\right)^{2m+2} - \frac{r_0^2}{(2m+1)}\left(1 - \frac{r}{r_0}\right)^{2m+1}\right]_0^{r_0}$$

$$= \frac{2u_m^2}{V^2 r_0^2}\left[\frac{r_0^2}{2m+1} - \frac{r_0^2}{2m+2}\right]$$

$$= \frac{2u_m^2}{V^2}\cdot\frac{1}{(2m+1)(2m+2)}$$

Substituting for V,

$$\beta = \frac{2u_m^2 (m+1)^2 (m+2)^2}{4u_m^2} \cdot \frac{1}{2(m+1)(2m+1)}$$

$$= \frac{1}{4} \frac{(m+2)^2 (m+1)}{(2m+1)}$$

For $m = 1/7$,

$$\beta = \frac{1}{4} \frac{(2+1/7)^2 (1+1/7)}{(1+2/7)}$$

$$= \frac{225 \times 8}{4 \times 9 \times 49} = \mathbf{1.02}$$

5.13 A transition in a 30 cm diameter pipe reduces the diameter to 15 cm without any deflection of the direction of flow. The transition is horizontal. Velocity measurements indicated the momentum correction factor β and kinetic energy correction factor α in the 30 cm pipe to be 1.30 and 1.90 respectively. Corresponding values of β and α for the 15 cm pipe are 1.05 and 1.15 respectively. If the pressure and the mean velocity in the 15 cm pipe at the end of the transition are to be 15 kPa and 6.0 m/s respectively, calculate the resultant force on the transition due to the flow of water. Neglect any losses in the transition.

Solution: Consider the control volume as shown by the dotted line in Fig. 5.15.

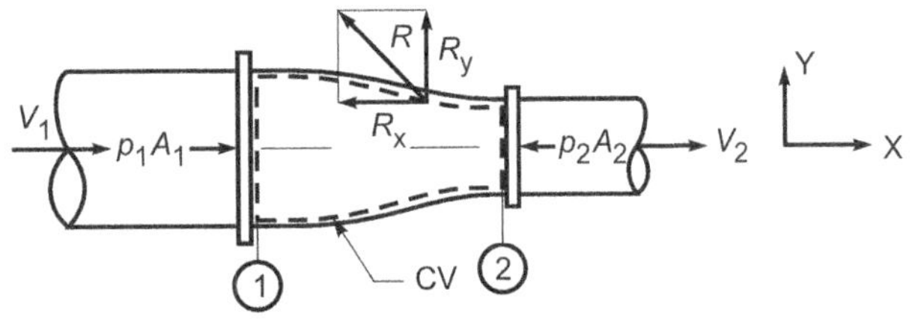

Fig. 5.15

R is the reaction on the fluid in the control volume.

$$A_1 = \frac{\pi}{4} \times (0.30)^2 = 0.07069 \text{ m}^2$$

$$A_2 = \frac{\pi}{4} \times (0.15)^2 = 0.01767 \text{ m}^2$$

$$V_2 = 6.0 \text{ m/s}, \ V_1 = V_2 (D_2/D_1)^2 = \frac{1}{4} V_2 = 1.5 \text{ m/s}$$

$$Q = A_2 V_2 = 0.01767 \times 6.0 = 0.106 \text{ m}^3\text{/s}$$

$$p_2 = 15 \text{ kPa}, \quad Z_2 = Z_1$$

By applying Bernoulli equation to sections 1 and 2,

$$\frac{p_1}{\gamma} + \alpha_1 \frac{V_1^2}{2g} + Z_1 = \frac{p_2}{\gamma} + \alpha_2 \frac{V_2^2}{2g} + Z_2$$

$$\frac{p_1}{\gamma} + \frac{1.90 \times (1.5)^2}{2 \times 9.81} = \frac{15.0}{9.79} + \frac{1.15 \times (6.0)^2}{2 \times 9.81}$$

$$\frac{p_1}{\gamma} = 1.532 + 2.110 - 0.218 = 3.424$$

$$p_1 = 9.79 \times 3.424 = 33.521 \text{ kPa}$$

There is no change in momentum flux in y-direction. Hence $R_y = 0$.

By momentum equation in x-direction:

$$p_1A_1 - R_x - p_2A_2 = \rho Q(\beta_2 V_2 - \beta_1 V_1)$$

$$(33521 \times 0.07069) - R_x - (15000 \times 0.01767)$$
$$= 998 \times 0.106 \times (1.05 \times 6.00 - 1.30 \times 1.5)$$
$$2370 - R_x - 265 = 460$$

$R_x = 1645$ N = Reaction force on the fluid in the control volume. Hence **the resultant force on the transition which is equal and opposite to R_x is $F = 1645$ N in the positive x-direction.**

5.14 A sluice gate in an open channel is shown in Fig. 5.16(a). Estimate the force on the unit width of the gate. Neglect frictional force on the channel bottom.

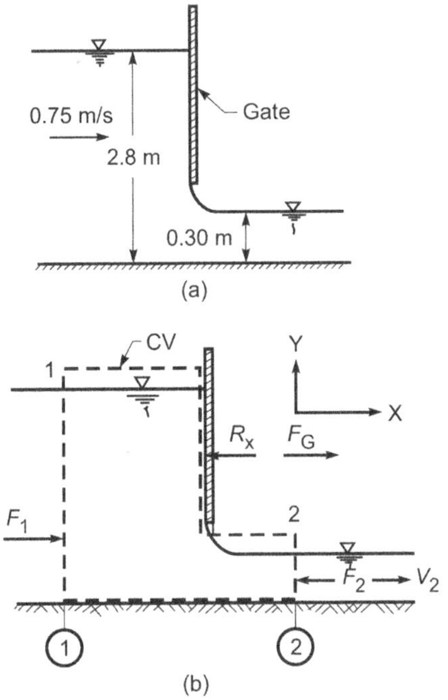

Fig. 5.16 *Sluice Gate*

Solution: Consider a unit width of the gate and the control volume (cv) as shown in Fig. 5.16(b). The frictional force on the channel bottom is neglected.

The forces on the control volume surfaces are:

F_1 = pressure force on the section 1 = $\gamma y_1^2/2$
(by assuming hydrostatic pressure distribution)

F_2 = pressure force on the section 2 = $\gamma y_2^2/2$

R_x = reaction of the gate on the water in the control volume acting in the (– x)-direction.

Here

$y_1 = 2.8$ m, $y_2 = 0.3$ m, $V_1 = 0.75$ m/s

q = discharge per unit width of channel = $y_1V_1 = y_2V_2$ = 0.75 × 2.8 = 2.1 m^3/s/m

$V_2 = q/y_2 = 2.1/0.3 = 7.0$ m/s

From momentum equation to the control volume in the x-direction.

$$F_1 - F_2 - R_x = \rho Q(V_2 - V_1)$$

$$\frac{1}{2} \times 9.79 \times (2.8)^2 - \frac{1}{2} \times 9.79 \times (0.3)^2 - R_x$$

$$= \frac{998}{1000}\,(2.1)\,(7.0 - 0.75)$$

$$38.38 - 0.44 - R_x = 13.10$$

$$R_x = 24.84 \text{ kN}$$

The force F_G on the gate due to water is equal and opposite to R_x. **Thus F_G = 24.84 kN and acts in the positive x-direction**.

5.15 Figure 5.17(a) shows a submerged flow over a rectangular weir in a rectangular channel. If the discharge per unit width of the channel is 1.0 m^3/s/m, estimate the force on unit width of the wier plate.

Solution: Consider a control volume as shown in Fig. 5.17(b). Hydrostatic pressure is assumed at sections 1 and 2. The reaction force on the fluid in the control volume is R_x in (– x)-direction as shown in Fig. 5.17(b).

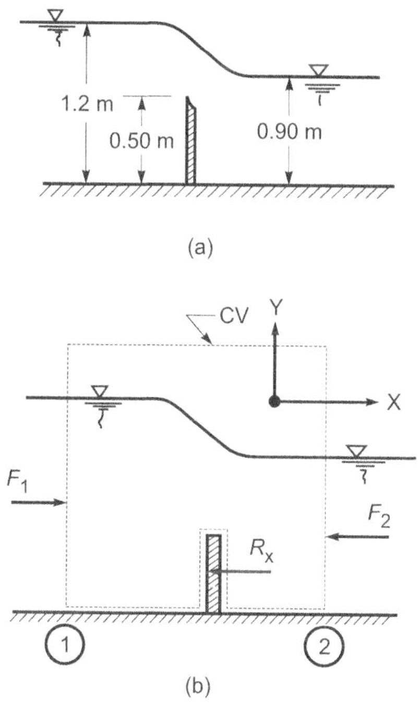

Fig. 5.17

The friction in the bed of the channel is ignored.

$$V_1 = \frac{1.0}{1.2} = 0.833 \text{ m/s}$$

$$V_2 = \frac{1.0}{0.9} = 1.111 \text{ m/s}$$

$$F_1 = \gamma y_1^2/2 = 9.79 \times (1.2)^2/2 = 7.049 \text{ kN}$$

$$F_2 = \gamma y_2^2/2 = 9.79 \times (0.9)^2/2 = 3.965 \text{ kN}$$

By momentum equation in the x-direction:

$$F_1 - F_2 - R_x = \rho Q(V_2 - V_1)$$

$$7.049 - 3.965 - R_x = \frac{998}{1000} \times 1.0 \times (1.111 - 0.833)$$

$$= 0.277$$

$$R_x = 2.807 \text{ kN}$$

As there is no change of momentum in y-direction,

$$R_y = 0.$$

Hence $R_x = R$.

The force F on the weir plate is equal and opposite to R_x. **Thus F = 2.807 kN acting in the (+ x)-direction.**

5.16 A jet of water with velocity V_1 and area of cross section A_1 enters a stream of slow moving water in a pipe of area A_2 and velocity V_2. The two streams enter with the same pressure p_1. After thoroughly mixing in the pipe the stream emerges as a single stream with velocity V_3 and pressure p_2 (see Fig. 5.18) Assuming no losses in the pipe, determine $(p_2 - p_1)$ for $V_1 = 20$ m/s and $V_2 = 10$ m/s, $A_1 = 0.01$ m^2, $A_2 = 0.02$ m^2 and density of water $\rho = 1000$ kg/m^3.

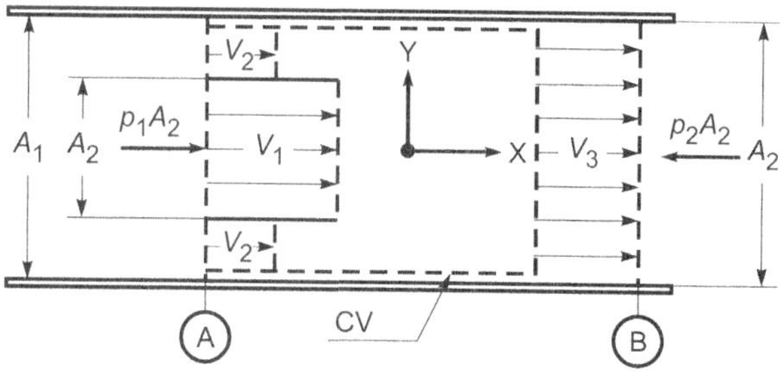

Fig. 5.18

Solution: Consider the control volume as shown in Fig. 5.18. The pressure at section A is p_1 and at B it is p_2. There are no frictional losses and hence no additional force on *cv*. Pressures p_1 and p_2 act over the cross sectional area A_2.

$$Q_1 = V_1A_1 = 20 \times 0.01 = 02 \text{ m}^3/\text{s}$$

$$Q_2 = V_2(A_2 - A_1) = 10 \times (0.02 - 0.01) = 0.1 \text{ m}^3/\text{s}$$

$$Q = Q_1 + Q_2 = \text{total flow} = 0.2 + 0.1 = 0.3 \text{ m}^3/\text{s}$$

$$V_3 = Q/A_2 = 0.3/0.02 = 15 \text{ m/s}$$

By momentum equation in the x-direction:

$$\begin{aligned} p_1A_2 - p_2A_2 &= M_{out} - M_{in} \\ &= \rho(Q \cdot V_3 - Q_1V_1 - Q_2V_2) \\ (p_1 - p_2) \times 0.02 &= 1000\,[(0.3 \times 15) - (0.2 \times 20) \\ &\quad - (0.1 \times 10)] \\ &= -500 \end{aligned}$$

$$(p_2 - p_1) = \frac{500}{0.02} = 25000 \text{ Pa} = \mathbf{25\ kPa}$$

5.17 In a steady entrance flow in a pipe of diameter D (Fig. 5.19) the flow develops from uniform flow at section 1 to a parabolic profile at section 2. If the momentum correction factor at section 2 is 4/3, show that the wall drag F is given by

$$F = \frac{\pi D^2}{4}\left(p_1 - p_2 - \frac{1}{3}\rho U_0^2\right)$$

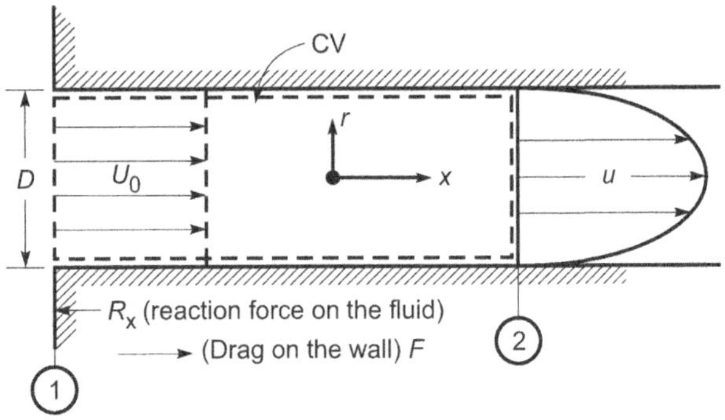

Fig. 5.19

Solution: Consider a control volume encompassing section 1 and 2 as shown in Fig. 5.19. Let R_x = reaction force on the fluid in the control volume (*cv*). Average velocity at section 1 = $V_1 = U_0 = V_2$ = average velocity at section 2. At section 1, the velocity distribution is uniform. Hence, $\beta_1 = 1.0$. It is given $\beta_2 = 4/3$.

By momentum equation in x-direction.

$$\Sigma\,(\text{Forces})_x = -R_x + (p_1 - p_2)A = \rho Q\,[\beta_2 U_0 - \beta_1 U_0]$$

$$R_x = (p_1 - p_2)A + \rho A U_0^2\left(1 - \frac{4}{3}\right)$$

$$= \frac{\pi D^2}{4}\left[(p_1 - p_2) - \frac{1}{3}\rho U_0^2\right]$$

The wall drag F on the pipe is equal and opposite to R_x.

Thus $F = \dfrac{\pi D^2}{4}\left[(p_1 - p_2) - \dfrac{1}{3}\rho U_0^2\right]$ **and acts in the (+ x) direction.**

B: Moment of Momentum

5.18 Figure 5.20 shows a lawn sprinkler with two jets, each located at 30 cm from the centre. The jets are of 1 cm diameter. Assuming no friction, find the speed of rotation for a discharge of 2.5 L/s.

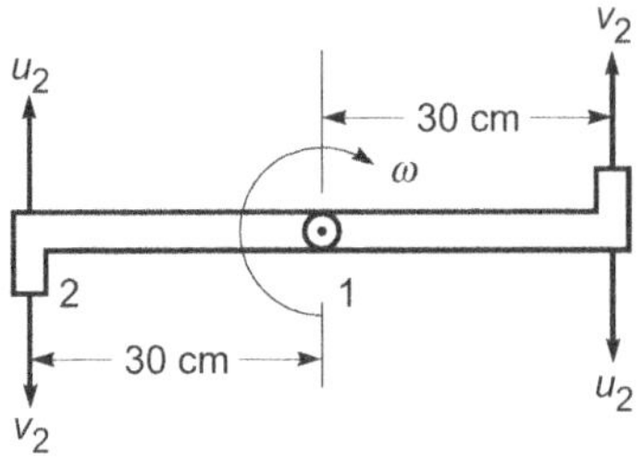

Fig. 5.20

Solution: Let

$$\begin{aligned} \omega &= \text{angular velocity} \\ u_2 &= \omega r \\ a &= \frac{\pi}{4} \times (1)^2 = 0.7854 \text{ cm}^2 \\ Q &= 2500 \text{ cm}^3/\text{s} \\ v_2 &= \frac{2500}{2 \times 0.7854} = 1519.5 \text{ cm/s} \\ &= 15.20 \text{ m/s} = \text{relative velocity of jet.} \end{aligned}$$

Torque $\quad T = -\rho Q r (u_2 - v_2) = 0$

or $\quad v_2 = u_2 = \omega r = 15.20$ m/s

$$\omega = \frac{15.20}{0.30} = 50.67 \text{ rad./s}$$

Speed of rotation per minute,

$$N = \frac{\omega}{2\pi} \times 60$$

$$\therefore \quad N = \frac{50.67}{2\pi} \times 60 = \mathbf{483.8\ rpm}$$

5.19 Find the torque required to hold the sprinkler in Example 5.18 stationary.

Solution: When the sprinkler is stationary,

$\omega = 0$ and torque $T_0 = \rho Q r v_2$

$$= 998 \times \frac{2.5}{1000} \times 0.30 \times 15.20$$

$$= \mathbf{11.38\ N.m}$$

5.20 For the lawn sprinkler of Example 5.18 what will be its steady rotation rate if it has a retarding friction torque of 1.5 N.m?

Solution:

$$T = -\rho Q r (u_2 - v_2)$$

$$\begin{aligned} 1.5 &= -998 \times \frac{2.5}{1000} \times 0.30 \times (\omega r - 15.20) \\ &= -0.7485\,(\omega \times 0.3 - 15.20) \end{aligned}$$

$$\omega = \frac{1}{0.3}\left(15.20 - \frac{1.5}{0.7485}\right) = 43.99$$

$$N = \frac{43.99}{2\pi} \times 60 = \mathbf{420\ rpm}$$

5.21 Figure 5.21 shows a lawn sprinkler with unequal arms. The jets issuing out of the sprinkler are of 0.8 cm in diameter and the total discharge is 1.2 L/s. (a) Assuming zero friction, determine the rotational speed of the sprinkler. (b) What torque would be required to hold the sprinkler stationary?

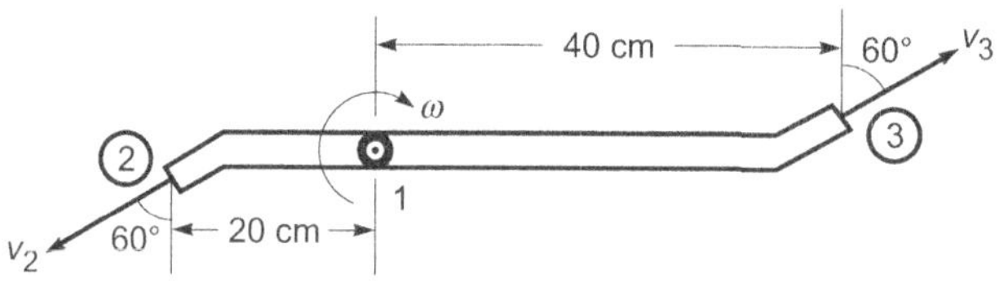

Fig. 5.21

Solution:

$$a = \text{area of the jet} = \frac{\pi}{4} \times (0.8)^2 = 0.5027 \text{ cm}^2$$

$$Q_n = \text{discharge from each nozzle} = \frac{1200}{2} = 600 \text{ cm}^3/\text{s}$$

$$v_2 = v_3 \text{ relative velocity of jets} = \frac{1200}{2 \times 0.5027}$$

$= 1194$ cm/s $= 11.94$ m/s

$\beta =$ Inclination of v_2 with the positive direction sprinkler arm

$= 180 - 60° = 120°$

$r_2 = 0.20$ m and $r_3 = 0.40$ m

Torque $T = -\rho Q_n (r_2(u_2 + v_2 \cos\beta) + r_3(u_3 + v_3 \cos\beta)]$

(a) For zero friction $T = 0$. Hence

$$r_2(\omega r_2 + v_2 \cos\beta) + r_3(\omega r_3 + v_3 \cos\beta) = 0$$

$$0.2\,(0.2 + 11.94 \cos 120°)$$

$$+ 0.4\,(0.4\omega + 11.94 \cos 120°) = 0$$

$$0.04\,\omega - 1.194 + 0.16\omega - 2.388 = 0$$

$$\omega = 17.91 \text{ rad./s}$$

or $$N = \frac{17.91 \times 60}{2\pi} = \mathbf{171 \text{ rpm}}$$

(b) When the sprinkler is stationary, $\omega = 0$

Torque required to hold the sprinkler stationary

$$= T_0 = -\rho Q_n (r_2 v_2 \cos\beta + r_3 v_3 \cos\beta)$$

$$= -\rho Q_n\, v_2 \cos\beta\,(r_2 + r_3)$$

$$T_0 = 998 \times \frac{0.6}{1000} \times 11.94 \times (0.5)\,(0.2 + 0.4)$$

$= \mathbf{2.145 \text{ N.m}}$

5.22 A sprinkler with unequal arms and jets of area 0.8 cm² facing in the same direction is shown in

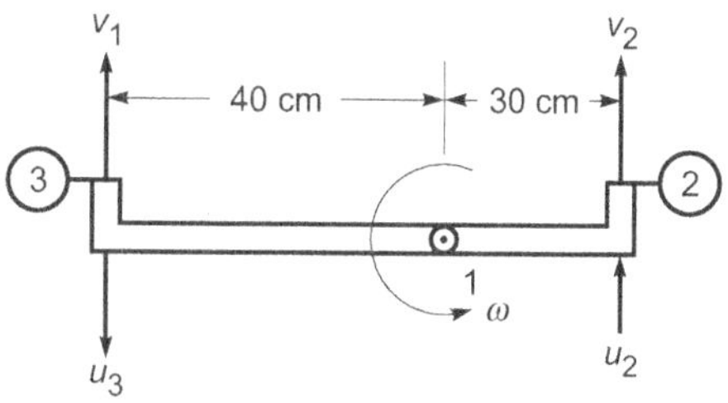

Fig. 5.22

Fig. 5.22. A flow of 1.5 L/s enters the assembly normal to the rotating arm. (i) Assuming the frictional resistance to be zero, calculate its speed of rotation, (ii) What torque is required to hold it from rotating?

Solution:

$$Q_n = \text{discharge from each nozzle} = \frac{1.5}{2} = 0.75 \ l/\text{s}$$

$$v = \frac{1.5 \times 1000}{0.8 \times 2} = 937.5 \text{ cm/s} = 9.375 \text{ m/s}$$

Let ω = angular velocity of the arm. Designating the jet at shorter arm by 2 and at the longer arm by 3, relative velocities

$$v_2 = v_3 = 9.375 \text{ m/s}$$

Tangential velocity

$$u_2 = \omega r_2 = 0.3\omega$$

$$u_3 = \omega r_3 = 0.4\omega$$

(Note that u_2 and v_2 are in the same direction)

Absolute velocity:

$$V_2 = v_2 + \omega r_2 = 9.375 + 0.3\,\omega$$

$$V_3 = v_3 - \omega r_3 = 9.375 - 0.4\,\omega$$

(i) Torque on the arm

$$= T_0 = -\rho Q_n[0 - (r_3 V_3 - r_2 V_2)]$$

For zero frictional resistance, $T = 0$ and hence

$$r_3 V_3 = r_2 V_2$$

$$0.4\,(9.375 - 0.4\,\omega) = 0.3\,(9.375 + 0.3\,\omega)$$

$$\omega = \frac{0.9375}{0.25} = 3.75 \text{ rad./s}$$

or $$N = \frac{3.75 \times 60}{2\pi} = \mathbf{35.81 \text{ rpm}}$$

(ii) When the arm is stationary, $\omega = 0$

Torque $T_0 = -\rho Q_n(-r_3 V_3 + r_2 V_2)$

$= -\rho Q_n(-v_3 r_3 + v_2 r_2)$

$$= 998 \times \frac{0.75}{1000} \times 9.375 \times (0.4 - 0.3)$$

$= \mathbf{0.702 \text{ N.m}}$

OBJECTIVE QUESTIONS

Linear Momentum

5.1 The linear momentum equation is based on
(a) Newton's law of viscosity
(b) Newton's first law
(c) Newton's second law
(d) Newton's third law

5.2 A control volume is
(a) the volume of fluid flowing per unit of time.
(b) a volume fixed in space.
(c) the volume in which a control device is situated.
(d) the volume of the fluid controlling device.

5.3 The linear momentum equation is
(a) a scaler relation.
(b) an approximate relation for engineering analysis.
(c) a relation applicable to incompressible fluids only.
(d) a vector relation.

5.4 In steady, incompressible, fluid flow with uniform velocity distribution, the momentum flux in a given x-direction past a given section is expressed as M_x =
(a) ρQV (b) $\rho V^2/2$
(c) ρQV_x (d) Q^2/A

5.5 The linear momentum equation applied to a control volume in a flow through a nozzle yielded the resultant reaction force R on the fluid in the control volume. The force required to keep the nozzle in position is
(a) the same as in magnitude and direction.
(b) equal to R but opposite in direction
(c) equal to the x-component of R
(d) equal to R minus the friction force

5.6 Identify the *incorrect* statement:
The linear momentum equation in x-direction.

$$\Sigma F_x = \rho Q(V_{x2} - V_{x1})$$

has the following assumptions:
(a) The flow is steady
(b) The fluid is incompressible
(c) The flow is uniform
(d) The velocity distribution at the end sections 1 and 2 is uniform.

5.7 A jet of oil (RD = 0.8) has an area of 0.02 m^2 and a velocity of 10 m/s. If it strikes a plate normally, the force exerted on the plate is
(a) 1597 N (b) 1996 N
(c) 15665 N (d) 19581 N

5.8 A water jet has an area of 0.03 m^3 and impinges normally on a plate. If a force of 1 kN is produced as a result of this impact, the velocity of the jet, in m/s, is
(a) 15 (b) 33.4
(c) 3.4 (d) 5.78

5.9 A water Jet 0.015 m^2 in area has a velocity of 15 m/s. If this jet impinges normally on a plate which is moving at a velocity of 5 m/s in the direction of the jet, the force on the plate due to this impact is
(a) 3368 N (b) 2246 N
(c) 1497 N (d) 14686 N

5.10 Uniform flow of a real fluid takes place in a horizontal pipe of diameter D. If p_1 and p_2 are the pressures at the upstream and downstream sections of a stretch of length L off the pipe, the boundary shear stress τ_0 could be expressed by the momentum equation as τ_0 =
(a) $(p_1 - p_2)\pi D^2/4L$ (b) $(p_1 - p_2)\, D/4L$
(c) $(p_1 - p_2)\, 4L/D$ (d) $(p_1 - p_2)\, D/\rho gL$

5.11 A fire house has a nozzle attached to it and the nozzle discharges a jet of water into the atmosphere at 20 m/s. This places the joint of the nozzle.
(a) in compression
(b) in tension
(c) in a state of zero stress
(d) in bending stress

5.12 A two-dimensional jet strikes a fixed two-dimensional plane at 45° to the normal to the plane. This causes the jet to split into two streams whose discharges are in the ratio
(a) 1.0 (b) 2.41
(c) 5.83 (d) 1.414

5.13 A jet of water with a velocity of 20 m/s impinges on a single vane at 5.0 m/s in the direction of the jet and transmits a power P_1. If the same jet drives a series of similar vanes mounted on a wheel under similar velocity conditions, the power transmitted is P_2. The ratio of P_1 and P_2 is
(a) 0.25 (b) 0.33
(c) 0.50 (d) 0.75

5.14 The momentum correction factor is given by $\beta =$

(a) $\dfrac{1}{V^3 A}\displaystyle\int_A u^3 \, dA$ (b) $\dfrac{1}{VA}\displaystyle\int_A u \, dA$

(c) $\dfrac{1}{A^2}\displaystyle\int_A A^2 \, du$ (d) $\dfrac{1}{V^2 A}\displaystyle\int_A u^2 \, dA$

5.15 The momentum correction factor β for triangular velocity distribution for flow between two parallel plates as shown in Fig. 5.23 is

(a) 2.0 (b) 1.50
(c) 1.33 (d) 2.50

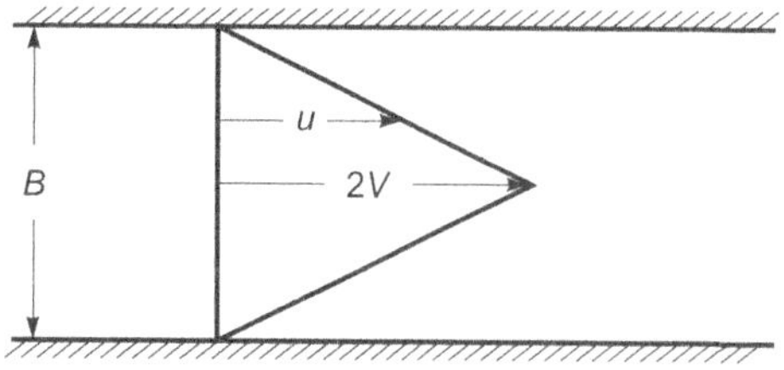

Fig. 5.23

5.16 A nozzle discharging under a head H has an area a and a discharge coefficient $C_d = 1.0$. A vertical plate is acted upon by the fluid force F_j when held across the free jet and by the fluid force F_n when held against the nozzle to stop the flow. The ratio F_j/F_n is

(a) 1/2 (b) 1
(c) $\sqrt{2}$ (d) 2

5.17 Consider the following velocity profiles in a pipeline (Fig. 5.24)

Among these profiles the momentum correction factor would be

(a) least in 4
(b) highest in 1
(c) more in 3 than that for 2
(d) the same in 1, 2, 3 and 4

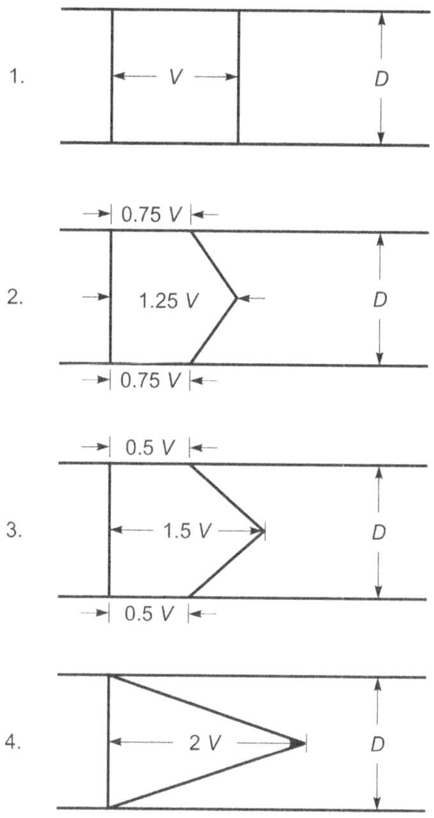

Fig. 5.24 *Question 5.17*

5.18 The velocity distribution over one half of a cross section is uniform and is zero over the remaining half. The momentum correction factor for this cross section is

(a) 2.0 (b) 4.0
(c) 1.0 (d) 3.0

5.19 A jet strikes a stationary plate normally with a velocity of 8 m/s and the plate suffers a force of 120 N. The power obtained, in kW is

(a) 0.96 (b) 9.4
(c) zero (d) 958

5.20 If all the energy in a jet of velocity 20 m/s issuing as a jet of 15 cm diameter could be extracted, the power available, in kW, is

(a) 70.54 (b) 7.21
(c) 20.39 (d) 705

5.21 A horizontal venturimeter has a head loss of 1/10 the upstream pressure head. If the flow is in x-direction, this causes on the venturimeter.

(a) a net force in (– x)-direction
(b) zero force
(c) a net force in x-direction
(d) an oscillating force

Moment of Momentum

5.22 The moment of momentum principle states that in a rotating system

(a) the resultant force exerted by the fluid on the body is equal to the rate of change of angular momentum
(b) the torque exerted by the resultant force is equal to the time rate of change of angular momentum
(c) the torque exerted by the resultant force is equal to the time rate change of linear momentum
(d) the angular moment is conserved

5.23 In a rotating fluid flow system, **V** = absolute velocity of a jet, **v** = relative velocity of the same jet with respect of its nozzle, and **u** = absolute velocity of the nozzle. The relationship between these vectors is,

(a) **V** = **v** + **u** (b) **V** = **v** – **u**
(c) **v** = **V** + **u** (d) **u** = **V** + **v**

5.24 A lawn sprinkler has two nozzles of area 0.75 cm^2 each, on either side of an arm capable of rotating about its midpoint. A discharge of 1.5 L/s is introduced at its axis to be discharged out at its nozzles, and the sprinkler arm rotates at a constant speed. The jet issuing from the nozzle will have

(a) an absolute velocity of 10.0 m/s
(b) an absolute velocity of 20.0 m/s
(c) a relative velocity of 10.0 m/s
(d) a relative velocity of 20.0 m/s

5.25 Figure 5.25 shows a rotating water sprinkler. The area of the nozzle at each end of the arm is 1.2 cm^2. If the discharge flowing out of the sprinkler is 2.4 L/s and the angular velocity is 7.69 rad/s, the absolute velocity of water emanating from A (at the end of the shorter arm), in m/s, is

(a) 7.69 (b) 8.462
(c) 10.00 (d) 11.538

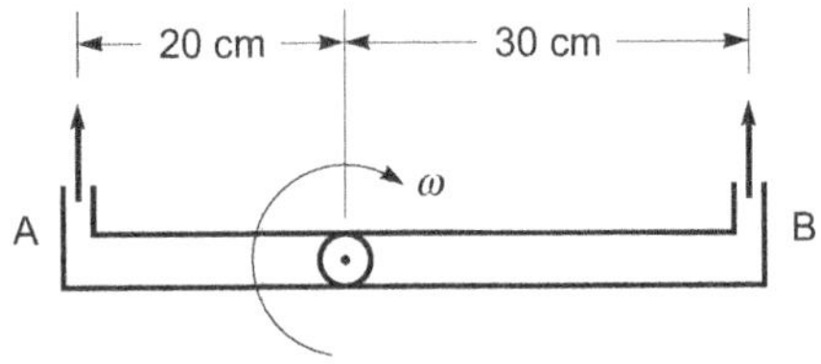

Fig. 5.25

5.26 For the sprinkler of Question 5.25 the absolute velocity of water issuing out from nozzle B (at the end of the longer arm), in m/s, is

(a) 10.0 (b) 7.693
(c) 12.307 (d) zero

5.27 Consider the lawn sprinkler shown in Fig. 5.26 with v_2 = relative velocity and u_2 = tangential velocity of the sprinkler jet. For a frictionless system, the angular velocity is given by

(a) $\omega = v_2/r$
(b) $\omega = (v_2 \cos \beta + u_2)/r$
(c) $\omega = v_2 \sin \beta/r$
(d) $\omega = v_2 \cos \beta/r$

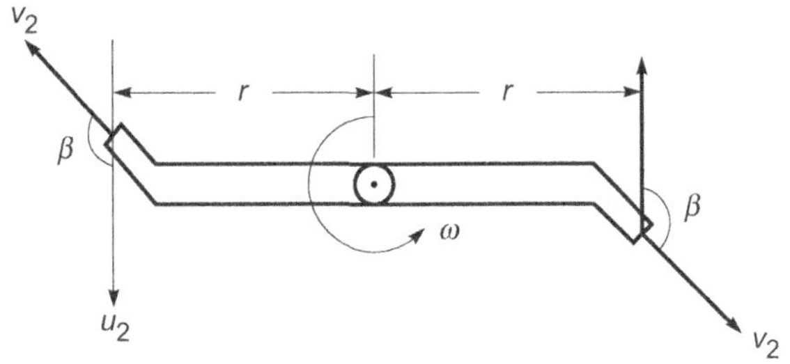

Fig. 5.26 *Problem 5.27*

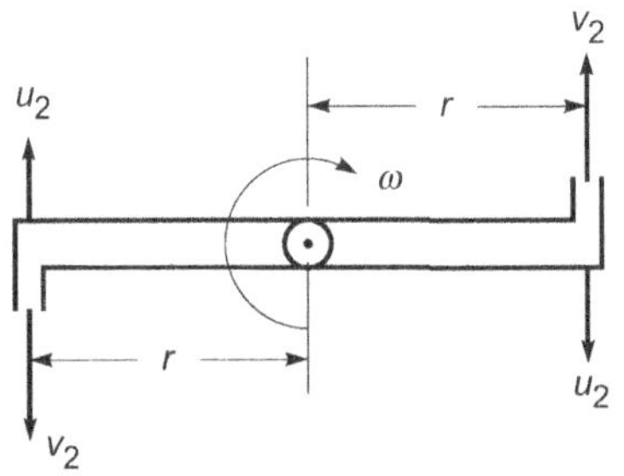

Fig. 5.27

5.28 For a lawn sprinkler shown in Fig. 5.27, if T = torque due to friction at bearings, etc., and Q = discharge, then the angular velocity ω is related to relative velocity v_2 as

(a) $\omega = \dfrac{v_2}{r}$ (b) $\omega = \dfrac{v_2}{r} - \dfrac{T}{\rho Q r^2}$

(c) $\omega = \dfrac{T}{\rho Q r^2}$ (d) $\omega = \dfrac{v_2}{r} + \dfrac{T}{\rho Q r^2}$

5.29 A sprinkler, such as shown in Fig. 5.27 is frictionless and requires a torque of 10 N.m to keep it stationary when a discharge of 1.5 L/s is passing through the system. If the discharge is doubled, the torque required to keep the sprinkler stationary, in N.m, is

(a) 40 (b) 20
(c) 10 (d) 5

PROBLEMS

Linear Momentum

5.1 A two-dimensional jet of water impinges on a plane at an angle θ to the normal to the plane. If the jet splits into two streams of discharges in the ratio 1 : 2, calculate the angle θ.

(*Ans.* $\theta = 19.47°$)

5.2 A two-dimensional jet of liquid issuing from a long slot strikes a plate at an angle of 60° with the plate. This causes the flow to divide into two parts q_1 and q_2 on either side of the impact zone. Calculate the ratio of q_1/q_2.

(*Ans.* $q_1/q_2 = 3.0$)

5.3 A 15 cm diameter jet of water with a velocity of 15 m/s strikes a plane normally. If the plate is moving with a velocity of 6 m/s in the direction of the jet calculate the work done per second on the plate and the efficiency (η) of energy transfer.

(*Ans.* **P** = 8571 N.m/s, $\eta = 28.8\%$)

5.4 A 15 cm diameter jet of oil (RD = 0.8) strikes a stationary flat plate at an angle of 35° to the normal. Calculate the force exerted on the plate when the velocity of the jet is 16 m/s.

(*Ans.* $F = 2958.7$ N)

5.5 A 20 cm diameter jet of oil (RD = 0.9) strikes a flat plate at an angle of 25° to the normal. The plate is moving at a velocity of 3 m/s in the direction of the jet. Calculate the absolute velocity of the jet if the resultant force exerted on the plate is 2500 N.

(*Ans.* $V = 12.89$ m/s)

5.6 A rectangular plate 30 cm long weighs 200 N. It is hung at the mid point of the top edge and is capable of swinging freely. A horizontal jet of water 2.5 cm in diameter and having a velocity of 12 m/s strikes the plate normally at the centre of the plate. Find the force to be applied at the middle of the lower edge to keep the plate vertical. If the plate is allowed to swing freely find the inclination (θ) to the vertical which the plate will assume due to impact of the jet.

(*Ans.* $F_b = 35.27$ N, $\theta = 20.65°$)

5.7 A vertical jet of oil ($\rho = 900$ kg/m^3) issues out of a 10 cm diameter nozzle at a velocity of 15 m/s. The jet is directed upwards and is deflected by a horizontal fixed plate kept at a height of 3.0 m above the nozzle exit. Estimate the force of impact of the jet on the plate.

(*Ans.* $F = 1367$ N, upwards)

5.8 A 20 cm long nozzle having an outlet of 5 cm diameter is attached to a 10 cm diameter pipe and is directed vertically downwards. The pressure in the pipe at the base of the nozzle is 10 kPa. The jet is discharged to atmosphere and strikes a horizontal plate at a depth of 1.5 m below the nozzle exit. Calculate the force on the plate due to impact of the jet.

(*Ans.* $F_y = 69.2$ N)

5.9 A tank shown in Fig. 5.28 has a nozzle of exit diameter D_1 at a depth H_1 below the free surface. At the side opposite to that of nozzle 1, another nozzle is proposed at a depth $H_1/2$. What should be the diameter D_2 in terms of D_1 so that the net horizontal force on the tank is zero?

(*Ans.* $D_2 = \sqrt{2}\,.D_1$)

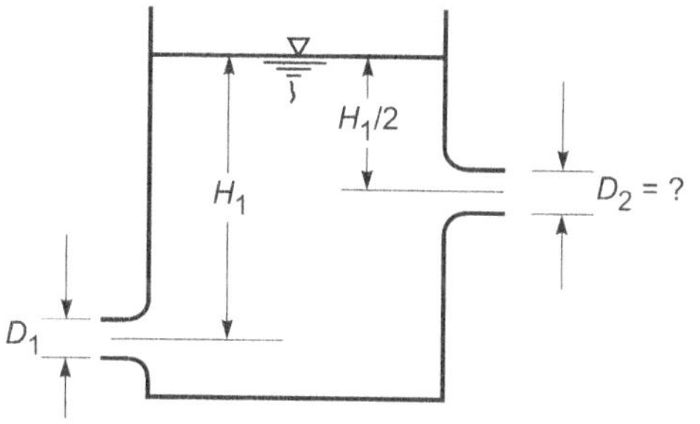

Fig. 5.28

5.10 A stationary curved plate deflects a 10 cm diameter water jet through an angle of 120° in the horizontal plane. Calculate the force required to hold the plate in position of the velocity of the jet is 15 m/s.

(*Ans.* $F = 3054$ N at 30° to – ve x-direction)

5.11 A jet of water 15 cm in diameter strikes a curved blade at 20 m/s velocity. The inlet angle and the outlet angles of the vane are 0° and 45° respectively. Determine the resultant force exerted on the blade when (i) the jet is stationary and (ii) the blade moves against the direction of the water at 5 m/s. Neglect friction along the blade.

(*Ans.* (i) $F = 13035$ N inclined at 337.5° to x-direction. (ii) $F = 20367$ N inclined at 337.5° to x-direction)

5.12 A 10 cm diameter jet having a velocity of 18 m/s impinges on a moving vane of velocity 8 m/s at 25° to the direction of motion. The vane angle at the outlet is 30°. Find the (i) blade angle at inlet so that the water enters without shock, (ii) the component of force exerted by the jet on the vane in the direction of motion of the vane.

(*Ans.* (i) $\alpha_1 = 42.46°$, (ii) $F_X = 1597$ N in x-direction)

5.13 A jet of oil (RD = 0.90) issues out horizontally from a nozzle and has a size of 12 cm diameter. A 60° vertex angle cone with its vertex pointing to the jet and its axis aligned in the direction of flow deflects the jet. If a force of 600 N is required to hold the cone in position, calculate the velocity of the jet.

(*Ans.* $V = 21$ m/s)

5.14 For Problem 4.14 in Chapter 4, calculate the reaction force on the mixing chamber.

(*Ans.* $F_R = 533.3$ N (+ x) direction)

5.15 A reducer bend has an inlet of 30 cm diameter and an outlet of 15 cm diameter. The outlet is turned at a deflection angle of 45° and the bend lies in a horizontal plane. Water flows through the bend at a rate of 90 L/s with a pressure of 20 kPa at the outlet end. Determine the magnitude and direction of the force required to keep the bend in position.

(*Ans.* $F = 1900.5$ N inclined at 197.53° to x-direction.)

5.16 Two pipes 0.5 m and 0.25 m in diameter are connected by a smooth transition bend making an angle of 75° between the two pipes. When a discharge of 0.4 m^3/s of water is flowing through this bend the pressure on the upstream larger diameter pipe is 200 kPa. Calculate the force on the bend. Assume the bend to be horizontal.

(*Ans.* $F = 38.72$ kPa inclined at 196.74° to x-direction.)

5.17 A 20 cm diameter pipe has a 90° bend (to the right) in the horizontal plane. When a discharge of 150 L/s of oil (RD = 0.8) is sent in this pipe, the pressure at the beginning of the bend is found to be 0.5 m of oil. Estimate the resultant force exerted by the oil on the bend.

(*Ans.* $F = 1612.5$ N inclined at 41.9° to x-direction.)

5.18 A reducer bend at the end of a pipe having inlet diameter of 30 cm and inlet axis horizontal is turned through 60° and discharges freely with an exit diameter of 20 cm. The bend lie in a vertical plane and the centre of the exit section is 60 cm higher than the axis at the inlet. The water inside the bend weighs 700 N. If the exit velocity is 10 m/s, determine the force required to hold the bend in position.

(*Ans.* $R = 4593$ N at $\theta = 131.95°$ to x-direction.)

5.19 Water flows through a 180° vertical reducing bend shown in Fig. 5.29. The pressure

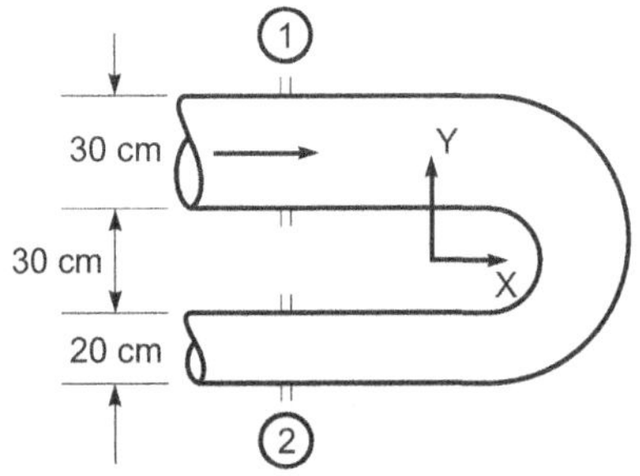

Fig. 5.29

at the inlet pipe is 20 kPa, and the discharge is 0.4 m^3/s. If the bend volume is 0.8 m^3, calculate the force required to hold the bend in place. (The bend is in a vertical plane).

(*Ans.* $F = 10854$ N at 136.19° to x-direction)

5.20 Water flows through the Y-joint as shown in Fig. 5.30. Find the horizontal and vertical components of the force acting on the joint because of the flow of water. Neglect energy loss and body forces. [Take $\rho = 1000$ kg/m^3]

(*Ans.* $F_x = 27.5$ kN in (– x) direction, $F_y = 16.65$ kN in (– y) direction)

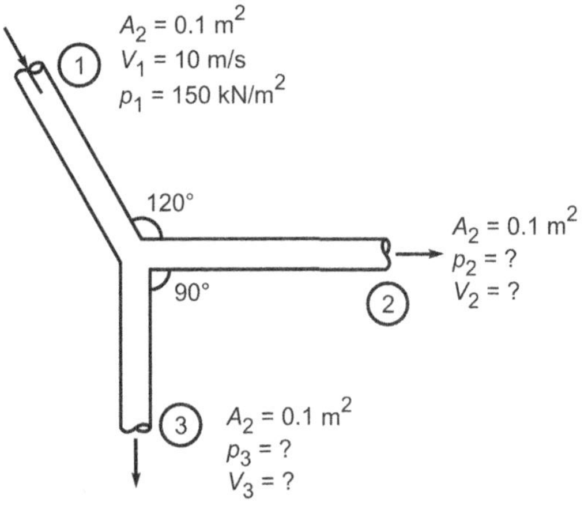

Fig. 5.30

5.21 0.28 m^3/s of water flows up a vertical enlargement, one meter high, 30 cm in diameter at the bottom and 60 cm diameter at the top. Calculate the magnitude and direction of force exerted by the flow on the enlargement.

(*Ans.* F = 21.30 kN acting vertically downwards)

5.22 A sluice gate in a rectangular channel carrying water is so opened to create depths of flow of 1.70 m and 0.25 m on the upstream and downstream of the gate respectively. The discharge intensity is 1.30 m^3/s per metre width and the flow from the sluice gate is in free flow mode. Estimate the force per metre width on the gate.

(*Ans.* F = 8.09 kN)

5.23 In a hydraulic jump occurring in a rectangular open channel the discharge is 5.4 m^3/s per metre width. The depth of flow before the jump is 0.3 m. Estimate the depth of flow after the jump by neglecting friction on the channel bottom.

[**Hint:** Use momentum equation and continuity equation].

(*Ans.* y_2 = 4.304 m)

5.24 For the velocity distribution in a two dimensional duct, shown in Fig. 5.31 (a) and (b) calculate the value of the momentum correction factor.

(*Ans.* (a) 1.333, (b) 1.5)

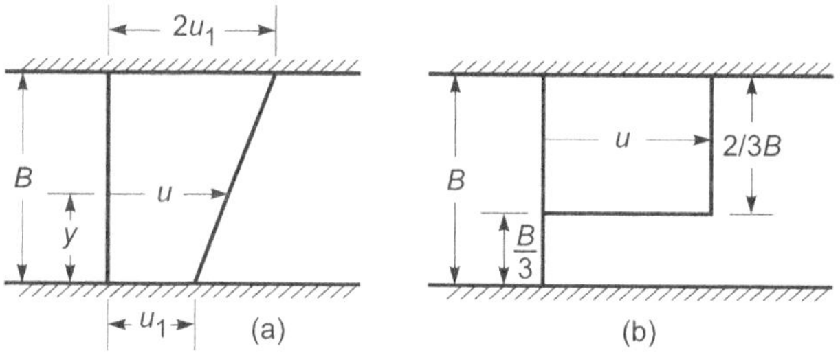

Fig. 5.31

5.25 Determine the momentum correction factor β for the following velocity distributions in a rectangular open channel of depth of flow y_0.

(a) $u/u_m = y/y_0$ for $y = 0$ to $y = y_0$.

(b) $u/u_m = (y/y_0)^n$ for $y = 0$ to $y = y_0$.

(*Ans.* (a) 1.333, (b) $(n + 1)^2/(2n + 1)$)

5.26 A 20 cm diameter pipe has a nozzle with 10 cm diameter exit attached to it. A jet of water with 15 m/s velocity is discharged, horizontally, from this nozzle into the atmosphere. The momentum correction factor β for the pipe flow is 1.25 and 1.0 for the jet. The kinetic energy correction factor α for the pipe flow is 1.75 and it is 1.0 for the jet. Calculate the force on the nozzle at its junction with the pipe.

(*Ans.* F = 1929 N in x-direction)

5.27 A sluice gate in a rectangular channel is discharging a flow of water in free flow mode. The depths of flow are 2.20 m and 0.40 m on the upstream and downstream of the gate respectively. Assuming zero frictional losses estimate the discharge per unit width of the channel and the force per unit width of the gate.

(*Ans.* q = 2.417 m^3/s/m and F_x = 11.98 kN in x-direction)

Moment of Momentum

5.28 A lawn sprinkler has 1.5 cm diameter nozzles at the two ends of a rotating arm. The nozzles make an angle of 150° with the tangential direction of rotation. The flow enters the sprinkler normally at the centre. The nozzle ends lie on a diameter of 0.8 m and the discharge of water entering the nozzle is measured as 2.0 L/s. Calculate the speed of rotation by

(i) assuming zero frictional resistance

(ii) assuming the torque due to friction as 2.0 N.m

(*Ans.* (i) N = 117 rpm, (ii) N = 57.2 rpm)

5.29 A four-arm lawn sprinkler, shown in Fig. 5.32 has provision for admitting water at the axis of rotation. If the discharge is 1.5 L/s, calculate the steady rotational speed by neglecting friction.

(*Ans.* N = 237.5 rpm)

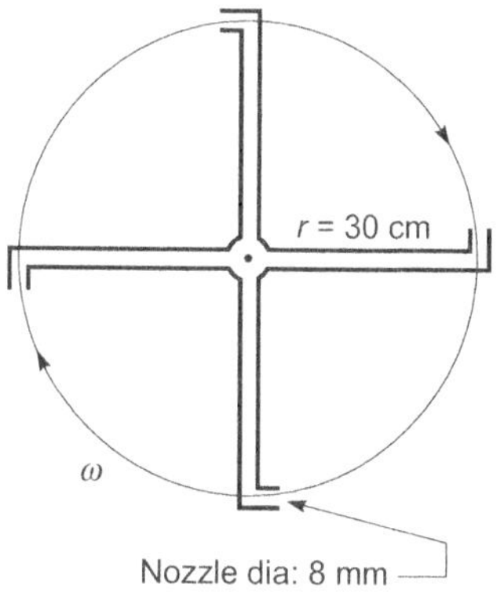

Fig. 5.32

5.30 For the water sprinkler, shown in Fig. 5.33 β = 158°, radius r = 0.50 m, jet diameter = 1.1 cm. If the speed of rotation is 240 rpm calculate the discharge supplied to the sprinkler. Assume zero frictional losses.

(*Ans.* Q = 2.758 L/s)

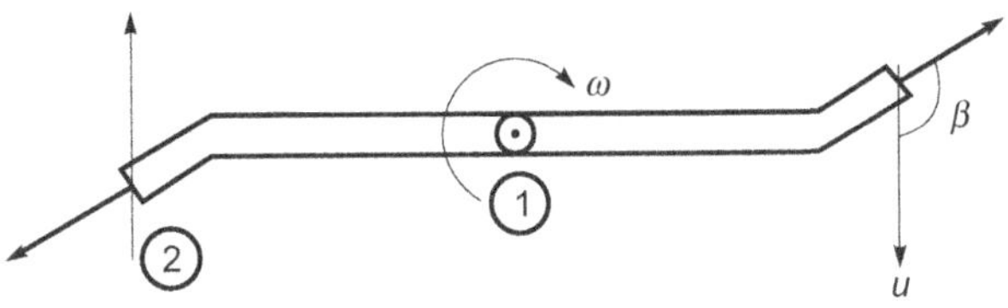

Fig. 5.33 *Problem 5.30*

5.31 In Problem 5.30, what external torque is required to make the sprinkler stationary?

(*Ans.* T_0 = 17.3 N.m)

5.32 A lawn sprinkler has two jets of 1.5 cm diameter on a rotating arm which can describe a circle of 60 cm diameter (Fig. 5.34). If the rotating speed is 210 rpm and the discharge is 3.0 L/s. Calculate the torque due to friction at the axis of rotation.

(*Ans.* T = 1.698 N.m)

5.33 A lawn sprinkler has unequal arms and has 1.25 cm nozzles at the ends as shown in Fig. 5.35. If the discharge through the sprinkler is 2.8 L/s calculate the constant speed of rotation (i) by neglecting friction and (ii) by assuming frictional torque as 6 Nm.

(*Ans.* (i) N = 512.7 rpm, (ii) N = 30.22 rpm)

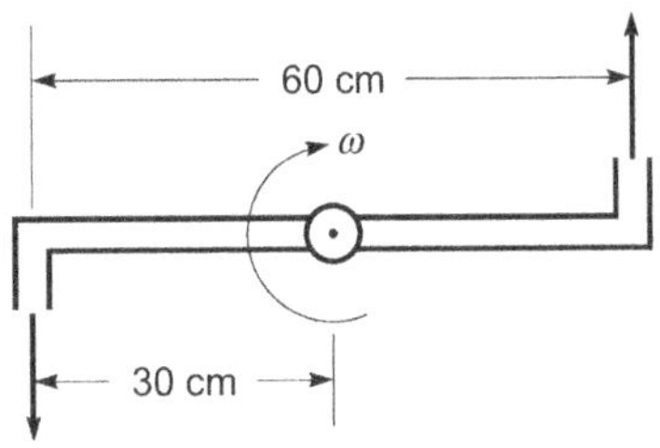

Fig. 5.34 *Problem 5.32*

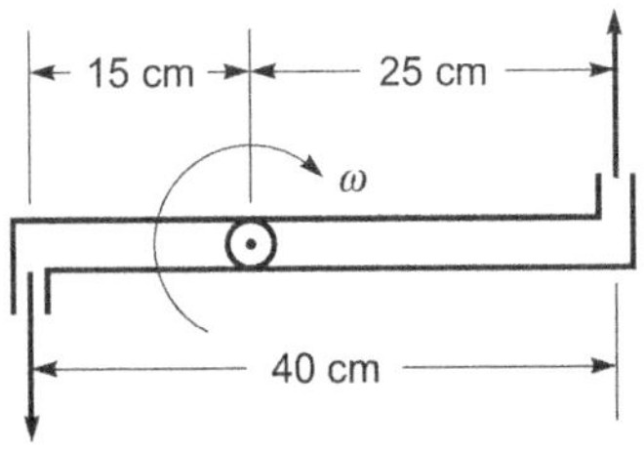

Fig. 5.35 *Problem 5.33*

5.34 For the sprinkler of Problem 5.33 what torque is required to keep the sprinkler stationary when a discharge of 1.5 L/s of water is passing through the sprinkler? (Neglect all frictional forces).

(*Ans.* T_0 = 1.83 N.m)

6

Dimensional Analysis and Similitude

CONCEPT REVIEW

6.1 INTRODUCTION

Dimensional analysis is a type of compacting technique to reduce the number of variables to be studied in an experimental investigation of a physical phenomenon. All physical phenomena are expressible in terms of a set of basic or fundamental dimensions. In fluid mechanics these are usually mass **M**, length **L**, time **T** and temperature θ in what is known as the **MLT** θ system. Sometimes the force **F** is used in place of **M** to get **F**, **L**, **T** and θ as the basic dimensions in what is known as the engineering system.

6.2 COMMON VARIABLES IN FLUID FLOW

The commonly occurring variables in fluid mechanics together with their common notations, units of measurement in SI units and their dimensions are given in Table 6.1. This table will be very helpful in performing dimensional analysis and as such warrants careful study.

6.3 DIMENSIONAL HOMOGENEITY

An equation which expresses the proper relationship between the variables in a physical phenomenon will be dimensionally homogeneous. This means that each of the additive terms in an equation should have the same dimension.

The principle is useful in checking the proper form of the equation, and in converting the equations having

Table 6.1 *Common Variables in Fluid Flow*

Quantity	*Notation*	*Dimensions*	
		(MLT θ) System	*(FLT θ) System*
Length	L	L	L
Area	A	L^2	L^2
Volume	$\forall$	L^3	L^3
Angle	α	$M^0 L^0 T^0$	$F^0 L^0 T^0$
Angular velocity	ω	T^{-1}	T^{-1}
Frequency	f	T^{-1}	T^{-1}
Discharge	Q	$L^3 T^{-1}$	$L^3 T^{-1}$
Velocity	$U.V.$	LT^{-1}	LT^{-1}
Mass density	ρ	ML^{-3}	$FL^{-4} T^2$
Dynamic viscosity	μ	$ML^{-1} T^{-1}$	$FL^{-2} T$
Kinematic viscosity	ν	$L^2 T^{-1}$	$L^2 T^{-1}$
Surface tension	σ	MT^{-2}	FL^{-1}
Volume modulus of elasticity	K	$ML^{-1} T^{-2}$	FL^{-2}
Specific weight	γ	$ML^{-2} T^{-2}$	FL^{-3}
Relative density	RD	$M^0 L^0 T^0$	$F^0 L^0 T^0$
Force	F	$ML T^{-2}$	F
Moment, Torque	M, T	$ML^2 T^{-2}$	FL
Momentum	M	MLT^{-1}	FT
Work, Energy	W, E	$ML^2 T^{-2}$	FL
Power	**P**	$ML^2 T^{-3}$	FLT^{-1}
Rotation	rpm	T^{-1}	T^{-1}
Strain	–	$M^0 L^0 T^0$	$F^0 L^0 T^0$
Strain rate	–	T^{-1}	T^{-1}
Stress, Pressure	p	$ML^{-1} T^{-2}$	FL^{-2}
Temperature	T	θ	θ
Specific heat	C_p, C_v	$L^2 T^{-2} \theta^{-1}$	$L^2 T^{-2} \theta^{-1}$
Thermal conductivity	k	$MLT^{-3} \theta^{-1}$	$FT^{-1} \theta^{-1}$

dimensional constants from one system to another (Examples 6.1 and 6.2).

6.4 DIMENSIONAL ANALYSIS

In the dimensional analysis of a physical phenomenon the relationship between the dependent and independent variables is studied in terms of their basic dimensions to obtain the information about the functional relationship between the dimensionless parameters that control the phenomenon. There are several methods of reducing the number of dimensional variables into a smaller number of dimensionless parameters. Two of the commonly used methods are the (i) Raleigh's method and (ii) Buckingham Pi theorem method.

6.4.1 Raleigh's Method

If A_1 is a dependent variable and A_2, A_3, ... A_n are independent variables in a phenomenon, A_1 is expressed as

$$\boxed{A_1 = k A_2^a A_3^b A_4^c \ldots A_n} \qquad (6.1)$$

where k is a dimensionless constant. The dimensions of each of the quantities $A_1, A_2, A_3, \ldots A_n$ are written and the sum exponents of each of **M**, **L**, and **T** on both sides are equated. Solution of the equations on simplification yields dimensionless groups controlling the phenomenon. Example 6.3, 6.4 and 6.5 illustrate this method. While this method is simple for a small number of parameters, it becomes rather cumbersome when a large number of parameters are involved.

6.4.2 Buckingham Pi Theorem

The Buckingham Pi theorem states that if there are m primary dimensions involved in the n variables controlling a physical phenomenon, then the phenomenon can be described by $(n - m)$ independent dimensionless groups (known as π s). The word Pi here refers to a product of variables and the Greek letter π is used to indicate these products, for example π_1, π_2 ....

In the application of this method, m number of repeating variables are selected and dimensionless groups obtained by each one of the remaining variables one at a time. Raleigh's method is used in this part of the operation. Examples 6.6 to 6.10 illustrate the use of Buckingham Pi theorem. This method is also known as the *method of repeating variables.*

Care is needed in selecting the repeating variables.

(i) They must have amongst themselves all the basic dimensions involved in the problem.

(ii) The dependent variable must *not* be chosen as a repeating variable.

(iii) Usually a length parameter (such as a diameter D or head over a weir, H); a typical velocity V and the fluid density ρ are convenient set of repeating variables.

6.5 SIMILITUDE

In hydraulic and aeronautical engineering valuable results are obtained at a relatively small cost by performing tests on small scale models of full size systems (prototypes). Similarity laws help us interpret the results of model studies. Similitude, the relation between model and a prototype, is classified into three kinds as follows:

6.5.1 Geometric Similarity

If the ratios of corresponding lengths in a model and the prototype are the same, the model is said to be geometrically similar model. In such models if

$$\frac{L_{\text{model}}}{L_{\text{prototype}}} = \frac{L_m}{L_p} = L_r$$

then
$$\frac{\text{Area}_{\text{model}}}{\text{Area}_{\text{prototype}}} = \frac{A_m}{A_p} = \frac{(L_m)^2}{(L_p)^2} = L_r^2 \qquad (6.2)$$

and
$$\frac{(\text{Volume})_{\text{model}}}{(\text{Volume})_{\text{prototype}}} = \frac{\forall_m}{\forall_p} = \frac{(L_m)^3}{(L_p)^3} = L_r^3 \qquad (6.3)$$

6.5.2 Kinematic Similarity

Kinematic similarity means geometric similarity and in addition the ratio of velocities at all corresponding points in the flow is the same.

If $L_r = \dfrac{L_m}{L_p}$ = length ratio, and if

$$\frac{(\text{Velocity})_{\text{model}}}{(\text{Velocity})_{\text{prototype}}} = \left(\frac{V_m}{V_p}\right) = V_r$$

then (i) time ratio $= \dfrac{t_m}{t_m} = t_r = \dfrac{L_r}{V_r}$ (6.4)

(ii) acceleration ratio $= \dfrac{a_m}{a_p} = a_r = \dfrac{V_r^2}{L_r} = \dfrac{L_r}{T_r^2}$ (6.5)

(iii) discharge ratio $= \dfrac{Q_m}{Q_p} = Q_r = \dfrac{L_r^3}{T_r}$ (6.6)

6.5.3 Dynamic Similarity

Two systems are dynamically similar, if geometric and kinematic similarities exist and further the ratios of all corresponding forces in the two systems are the same. If forces due to

Gravity = F_G
Viscosity = F_v

Elasticity = F_E
Surface tension = F_T
Inertia = F_I

and suffixes m and p stand for model and prototype respectively, strict dynamic similarity means

$$\boxed{\frac{F_{Gm}}{F_{Gp}} = \frac{F_{vm}}{F_{vp}} = \frac{F_{Em}}{F_{Ep}} = \frac{F_{Tm}}{F_{Tp}} = \frac{F_{Im}}{F_{Ip}} = \text{constant}}$$

From the above the following relationships can be derived.

$$\frac{(\text{Inertia force})_m}{(\text{Viscous force})_m} = \frac{(\text{Inertia force})_p}{(\text{Viscous force})_p} = \text{Constant 1}$$

$$\frac{(\text{Inertia force})_m}{(\text{Gravity force})_m} = \frac{(\text{Inertia force})_p}{(\text{Gravity force})_p} = \text{Constant 2}$$

and so on for all the forces.

6.6 IMPORTANT DIMENSIONLESS FLOW PARAMETERS

The following dimensionless parameters representing ratios of forces per unit volume are of great significance in the analysis of fluid flow:

(1) *Reynolds number* = $Re = \dfrac{\text{Inertial force}}{\text{Viscous force}}$

$$= \frac{\rho VL}{\mu} = \frac{VL}{\nu}$$

For dynamic similarity where viscous forces are predominant.

$$\left(\frac{VL}{\nu}\right)_m = Re_m = Re_p = \left(\frac{VL}{\nu}\right)_p$$

(2) *Froude number* = $Fr = \dfrac{(\text{Inertia force})^{1/2}}{(\text{Gravity force})^{1/2}}$

$$= \frac{V}{\sqrt{gL}}$$

For dynamic similarity where gravity forces are predominant.

$$\left(\frac{V}{\sqrt{gL}}\right)_m = (Fr)_m = (Fr)_p = \left(\frac{V}{\sqrt{gL}}\right)_p$$

(3) *Mach number* = $M = \dfrac{(\text{Inertial force})^{1/2}}{(\text{Compressibility force})^{1/2}}$

$$= \frac{V}{\sqrt{E/\rho}} = \frac{V}{C}$$

where C = velocity of sound in the medium.
Where compressibility effects predominate, for dynamic similarity, we have the relations:

$$\left(\frac{V}{C}\right)_m = M_m = M_p = \left(\frac{V}{C}\right)_p$$

(4) *Weber number* = $W = \dfrac{(\text{Inertial force})}{(\text{Surface tension})} = \dfrac{\rho V^2 L}{\sigma}$

(5) *Euler number* = $E = \dfrac{(\text{Inertial force})^{1/2}}{(\text{Pressure force})^{1/2}} = \dfrac{V}{\sqrt{\Delta p/\rho}}$

6.7 MODEL SCALES

Fluid flow models are usually designed to account for one most dominant force, and occasionally for two dominant forces. Thus, if the dominant force is the gravity force, then the Froude number must be the same in the model and prototype. Thus,

$$\frac{V_m}{\sqrt{gL_m}} = \frac{V_p}{\sqrt{gL_p}}$$

As g is the same for both model and prototype,

$$\boxed{\frac{V_m}{V_p} = V_r = \sqrt{L_r}}$$

From this other scales ratios, for such Froude law modelling are developed as shown in Table 6.2.

Table 6.2 *Similitude Scale Ratios*

		Scale Ratios for Laws of	
Parameter	*Dimension*	*Froude*	*Reynolds*
Geometric			
Length	L	L_r	L_r
Area	L^2	L_r^2	L^2
Volume	L^3	L_r^3	L_r^3
Kinematic			
Velocity	LT^{-1}	$(L_r)^{1/2}$	$(\mu_r/L_r\rho_r)$
Time	T	$(L_r)^{1/2}$	$(L_r^2\rho_r/\mu_r)$
Acceleration	LT^{-2}	1	$(\mu_r^2/\rho_r^2 L_r^3)$
Discharge	L^3T	$(L_r)^{5/3}$	$(L_r\mu_r/\rho_r)$
Dynamic			
Mass	M	$(L_r^3\rho_r)$	$(L_r^3\rho_r)$
Force	MLT^{-2}	$(L_r^3\rho_r)$	(μ_r^2/ρ_r)
Pressure	$ML^{-1}T^{-2}$	$(L_r\rho_r)$	$(\mu_r^2/L_r^2\rho_r)$
Momentum	MLT^{-1}	$(L_r^{7/2}\rho_r)$	$(L_r^2\mu_r)$
Energy of work	ML^2T^{-2}	$(L_r^4\rho_r)$	$(L_r\mu_r^2/\rho_r)$
Power	ML^2T^{-3}	$(L_r^{7/2}\rho_r)$	$(\mu_r^3/L_r\rho_r^2)$

Similarly, if viscous forces are predominant the Reynolds number (VL/ν) must be the same in the model and prototype. The scale ratios for Reynolds number modeling are also shown in Table 6.2.
Similarly appropriate scales can be developed for Mach law and Weber number law.

- Froude number similarity is used when there is
  (i) Wave action as in breakwaters and ships.

(ii) Free surface flow (open channel flow).
(iii) Hydraulic structures such as spillways, stilling basins, weirs and notches.
(iv) Forces on bridge piers and offshore structures.

- Reynolds number similarity is used when there is dominant action of viscous forces. Typical examples includes
  (i) Completely submerged flow. (air planes, torpedo, etc.)
  (ii) Completely enclosed flow. (flow through pipes and over plates)
  (iii) Viscous flow. (Settling of particles in fluids)
  (iv) Flow in flowmeters in pipes. (Venturi meter, Orifice meter, etc.)

6.8 DISTORTED MODELS

Hydraulic modeling of rivers, harbors, estuaries, etc. which have longitudinal slope and large arial spread, pose many practical problems if strict similitude is attempted. To overcome most of these problem, they are usually modeled by using distorted scales. The vertical flow dimension (viz. depth) is used to simulate Froude's law while the other two dimensions (viz. length and width) are scaled to suit available space. Thus, vertical scales of 1/100 and horizontal scales of 1/200 to 1/1000 are common. In such *distorted models*,

Horizontal scale is L_r = length ratio = (L_m/L_p)
= width ratio = (B_m/B_p)
Vertical scale is h_r = depth ratio = (y_m/y_p)

In the above prototype and model quantities are denoted by suffixes p and m respectively. Based on these two scales, various scaling ratios for physical parameters are obtained as follows:

Cross sectional area ratio = (area in model)/(area in prototype) = $(By)_m/(By)_p = L_r h_r$

Froude number ratio = $(F_m/F_p) = 1$

Hence $\dfrac{(V_m/V_p)^2}{(y_m/y_p)} = \dfrac{V_r^2}{h_r}$. Thus the velocity ratio

$$V_r = (V_m/V_p) = \sqrt{h_r}$$

Discharge ratio Q_r = (area ratio) (velocity ratio) = $L_r h_r^{1.5}$

Slope ratio $S_r = S_m/S_p = h_r/L_r$.

Time ratio $T_r = L_r/V_r = L_r/\sqrt{h_r}$

Manning's roughness ratio: By Manning's formula, considering the channel to be wide,

$$V_r = \frac{1}{n_r} h_r^{2/3} S_r^{1/2}$$

Substituting the various ratios derived above

$$\sqrt{h_r} = \frac{1}{n_r} h_r^{2/3} (\sqrt{h_r}/\sqrt{L_r})$$

Hence,
$$\boxed{n_r = \frac{h_r^{2/3}}{L_r^{1/2}}}$$

Similarly, scale ratio for any other physical quantity or flow phenomenon can be derived. Similitude scale ratios of some commonly used parameters in distorted scale models are listed in Table 6.3.

Table 6.3 *Similitude Scale Ratios in Distorted Models*

Parameter	*Symbol*	*Scale ratio*
Length	L	L_r
Width	B	L_r
Depth	y	h_r
Area of flow	A	$L_r h_r$
Volume	¥	$L_r^2 h_r$
Velocity	V	$\sqrt{h_r}$
Time	T	$L_r/\sqrt{h_r}$
Acceleration	a	h_r/L_r
Discharge	Q	$L_r h_r^{1.5}$
Slope	S	h_r/L_r
Manning's roughness	n	$\dfrac{h_r^{2/3}}{L_r^{1/2}}$
Reynolds number	R_e	$h_r^{1.5}$

WORKED EXAMPLES

Dimensional Homogeneity

6.1 Manning's formula for velocity of flow V in uniform flow in a open channel is written as

$$V = \frac{1}{n} R^{2/3} S_0^{1/2}$$

where R = hydraulic radius, S_0 = bed slope and n = Manning's coefficient. Determine the dimensions of n.

Solution:

Rewriting the equation explicitly for n,

$$n = \frac{R^{2/3} S_0^{1/2}}{V}$$

Writing the dimensions of the terms on the right hand side and noting that S_0 does not have any dimensions,

$$[n] = \frac{(L)^{2/3}}{(LT^{-1})} = \mathbf{[L^{-1/3}\ T]}$$

6.2 Lacey's formula for wetted perimeter P in an alluvial channel is given in FPS units as

$$P = 2.67\sqrt{Q}$$

where Q = discharge in ft^3/s. Find the dimensions of the numerical constant and convert this formula into SI units.

Solution:

Rewriting the given equation as

$$2.67 = \frac{P}{\sqrt{Q}}$$

In terms of dimensions, the right hand side has

$$\frac{L}{[L^3\ T^{-1}]^{1/2}} = [L^{-1/2}\ T^{+1/2}]$$

1 ft = 0.305 m.

The constant in SI units is

$$= 2.67\ (0.305)^{-1/2} = 4.83$$

The formula in SI units would now read as

$$\mathbf{P = 4.83\sqrt{Q}}$$

Dimensional Analysis: Raleigh Method

6.3 The drag force F_D on a sphere in laminar flow is known to depend on its diameter D, velocity of flow V, density of fluid ρ and coefficient of viscosity μ. Obtain an expression for F_D using Raleigh's method.

Solution:

Since $F_D = \text{fn}\ (D, V, \rho, \mu)$

$F_D = KD^a V^b \rho^c \mu^d$

where K is a dimensionless constant. Using F, L, T as primary units,

$$F = [L]^a\ (LT^{-1})^b\ [FL^{-4}\ T^2]^c\ [FL^{-2}\ T]^d$$

Equating the powers of F, L and T on both sides:

For F: $1 = c + d$ (1)

for L: $0 = a + b - 4c - 2d$ (2)

for T: $0 = -b + 2c + d$ (3)

Since there are 3 equations and 4 unknowns, three variables can be expressed in terms of the fourth.

from (1) $c = 1 - d$

from (3) $b = 2c + d = 2 - d$

from (2) $a = -b + 4c + 2d = -(2 - d) + 4 - 4d + 2d$
$= 2 - d$

$\therefore$ $F = K{\cdot}D^{2-d}{\cdot}V^{2-d}{\cdot}\rho^{1-d}{\cdot}\mu^d$

$$= K\rho D^2V^2\left(\frac{\mu}{VD\rho}\right)^d$$

$$\mathbf{F_D = \rho D^2V^2\ fn\left(\frac{\mu}{VD\rho}\right)}$$

6.4 For laminar flow in a pipe, the drop in pressure Δp is a function of the pipe length L, its diameter D, mean velocity of flow V and the dynamic viscosity μ. Using Raleigh's method, develop an expression for Δp.

Solution:

$$\Delta p = \text{fn}\ (L, D, V, \mu)$$

This can be written in terms of a dimensionless constant K as

$$\Delta p = KL^a D^b V^c \mu^d$$

Using the MLT system of basic units

$$[ML^{-1}\ T^{-2}] = [L]^a\ [L]^b\ [LT^{-1}]^c\ [ML^{-1}\ T^{-1}]^d$$

Equating the

Powers of M: $d = 1$

Powers of T: $-2 = -c - d = -c - 1;\ c = 1$

Powers of L: $-1 = a + b + c - d$
$= a + b + 1 - 1;\ a = -1 - b$

$\therefore$ $$\Delta p = KL^{-1-b}\ D^b V\mu = K\left(\frac{V\mu}{L}\right)\left(\frac{D}{L}\right)^b$$

$\therefore$ $$\mathbf{\Delta_p = \frac{\mu V}{L}\cdot fn\left(\frac{D}{L}\right)}$$

6.5 The drag force F on a body in supersonic flow is considered to depend on the length L, velocity V, diameter D, and fluid properties like density ρ, dynamic viscosity μ and bulk modulus of elasticity E. Derive an expression for F.

Solution:

$$F = \text{fn}\ (L, V, D, \rho, \mu, E)$$

Hence $F = KL^aV^bD^c\rho^d\mu^eE^f$

Using the MLT system of basic units,

$$[MLT^{-2}] = [L]^a\ [LT^{-1}]^b\ [L]^c\ [ML^{-3}]^d\ [ML^{-1}\ T^{-1}]^e\ [ML^{-1}\ T^{-2}]^f$$

Equating powers of

M: $1 = d + e + f$ (1)

L: $1 = a + b + c - 3d - e - f$ (2)

T: $-2 = -b - e - 2f$ (3)

Note in this case there are 6 variables and 3 equations. Values of three unknowns are expressed in term of other three (viz. c, e, and f).

From 1: $d = -e - f + 1$

From 3: $b = 2 - e - 2f$

From 2: $a = 1 - b - c + 3d + e + f$

$= 1 - 2 + e + 2f - c - 3e - 3f + 3 + e + f$

$= 2 - c - e$

$\therefore \quad F = KL^{2-c-e} \cdot V^{2-e-2f} \cdot D^c \cdot \rho^{-e-f+1} \cdot \mu^e \cdot E^f$

$$F = K\rho V^2 L^2 \cdot \left[\frac{D}{L}\right]^c \cdot \left[\frac{\mu}{\rho VL}\right]^e \left[\frac{E}{\rho V^2}\right]^f$$

$$\boldsymbol{F = \rho V^2 L^2 \, \varphi_1\left[\left(\frac{D}{L}\right), \left(\frac{\mu}{\rho VL}\right), \left(\frac{E}{\rho V^2}\right)\right]}$$

[Note: It is very important *not* to make a mistake in writing the basic dimensions of the variables at the beginning of the problem. As such it is essential to know the dimensions of parameters given in Table 6.1 thoroughly].

Dimensional Analysis: Buckingham Pi theorem

6.6 The resistance force F of a ship is a function of its length L, velocity V, acceleration due to gravity g and fluid properties like density ρ and viscosity μ. Write this relationship in a dimensionless form.

Solution:

$$F = \text{fn}\,(L, V, g, \rho, \mu)$$

For using Buckingham Pi theorem, list the dimensions of each variable.

F	L	V	g	ρ	μ
[MLT^{-2}]	[L]	[LT^{-1}]	[LT^{-2}]	[ML^{-3}]	[ML^{-1} T^{-1}]

There are a total of six variables ($n = 6$) and 3 primary dimensions. Hence $m = 3$.

As such there are $n - m = 3$ dimensionless Pi terms. Using L, V and ρ as repeating variables:

$$\pi_1 = [\text{M}^0\, \text{L}^0\, \text{T}^0] = FL^a V^b \rho^c$$

or $\quad [\text{M}^0\text{L}^0\text{T}^0] = [\text{MLT}^{-2}]\,[\text{L}]^a\,[\text{LT}^{-1}]^b\,[\text{ML}^{-3}]^c$

Hence by equating powers of M, L and T

$$1 + c = 0$$
$$1 + a + b - 3c = 0$$
$$-2 - b = 0$$
$$c = -1,\ b = -2,\ a = -2$$
$$\pi_1 = \frac{F}{\rho V^2 L^2}$$

π_2 term: $\quad \pi_2 = gL^a\, V^b\, \rho^c$

$$[\text{M}^0\text{L}^0\text{T}^0] = [\text{LT}^{-2}]\,[\text{L}]^a\,[\text{LT}^{-1}]^b\,[\text{ML}^{-3}]^c$$
$$c = 0$$
$$1 + a + b = 0$$
$$-2 - b = 0$$

or $\quad a = 1,\ b = -2,\ c = 0$

$$\therefore \quad \pi_2 = \frac{gL}{V^2}$$

π_3 term: $\quad \pi_3 = \mu L^a\, V^b\, \rho^c$

$$[\text{M}^0\text{L}^0\text{T}^0] = [\text{ML}^{-1}\,\text{T}^{-1}]\,[\text{L}]^a\,[\text{LT}^{-1}]^b\,[\text{ML}^{-3}]^c$$
$$1 + c = 0$$
$$-1 + a + b - 3c = 0$$
$$-1 - b = 0$$
$$\therefore \quad c = -1,\ b = -1,\ a = -1$$
$$\pi_3 = \frac{\mu}{LV\rho}$$

Hence $\quad \dfrac{F}{\rho V^2 L^2} = \text{fn}\left[\dfrac{gL}{V^2}, \dfrac{\mu}{\rho VL}\right]$

$$\boldsymbol{\frac{F}{\rho V^2 L^2} = \varphi[Fr, Re]}$$

where $\quad Fr$ = Froude number = $V/\sqrt{gL}$

and $\quad Re$ = Reynold's number = $\rho VL/\mu$.

6.7 The discharge Q over a small rectangular weir is known to depend upon the head H over the weir, the weir height P, gravity g, width of the weir L and fluid properties density ρ, dynamic viscosity μ and surface tension σ. Express the relationship between the variables in dimensionless form.

Solution:

$$Q = \text{fn}\,(H, P, g, L, \rho, \mu, \sigma)$$

For using Buckingham Pi theorem list the dimensions of each variable.

Q	H	P	g	L	ρ	μ	σ
[L^3 T^{-1}]	[L]	[L]	[LT^{-2}]	[L]	[ML^{-3}]	[ML^{-1}T^{-1}]	[MT^{-2}]

In this case, the number of variables $n = 8$

number of primary variables $m = 3$

Hence number of dimensionless terms = 8 – 3 = 5

Selecting ρ, H and g as the repeating variables

$$\pi_1 = QH^a g^b \rho^c$$
$$[\text{M}^0\text{L}^0\text{T}^0] = [\text{L}^3\,\text{T}^{-1}]\,[\text{L}]^a\,[\text{LT}^{-2}]^b\,[\text{ML}^{-3}]^c$$

Equating exponents of M, L, T on both sides,

$$c = 0$$
$$3 + a + b - 3c = 0$$
$$-1 - 2b = 0$$
$$\therefore \quad c = 0,\ b = -1/2 \text{ and } a = -5/2$$

Hence $\quad \pi_1 = \dfrac{Q}{H^{5/2} g^{1/2}}$

π_2 term: $\quad \pi_2 = PH^a\, g^b\, \rho^c$

$$[\text{M}^0\text{L}^0\text{T}^0] = [\text{L}]\,[\text{L}]^a\,[\text{LT}^{-2}]^b\,[\text{ML}^{-3}]^c$$

By inspection $b = 0$, $c = 0$ and $a = -1$

$\therefore \quad \pi_2 = P/H$

π_3 term: $\quad \pi_3 = LH^a g^b \rho^c$

$$[M^0L^0T^0] = [L]\,[L]^a\,[LT^{-2}]^b\,[ML^{-3}]^c$$

By inspection, here also

$$b = 0,\ c = 0 \text{ and } a = -1$$

$$\pi_3 = L/H$$

π_4 term: $\quad \pi_4 = \mu H^a g^b \rho^c$

$$[M^0L^0T^0] = [ML^{-1}\,T^{-1}]\,[L]^a\,[LT^{-2}]^b\,[ML^{-3}]^c$$

By inspection, we have

$$1 + c = 0$$

$$-1 + a + b - 3c = 0$$

$$-1 - 2b = 0$$

$$\therefore \quad c = -1,\ b = -\frac{1}{2} \text{ and } a = 1 - b + 3c = -\frac{3}{2}$$

$$\pi_4 = \frac{\mu}{H^{3/2}\cdot g^{1/2}\rho}$$

π_5 term: $\quad \pi_5 = \sigma H^a g^b \rho^c$

$$[M^0L^0T^0] = [MT^{-2}]\,[L]^a\,[LT^{-2}]^b\,[ML^{-3}]^c$$

$$1 + c = 0$$

$$a + b - 3c = 0$$

$$-2 - 2b = 0$$

$$c = -1,\ b = -1 \text{ and } a = 3c - b = -2$$

Thus $\quad \pi_5 = \dfrac{\sigma}{H^2\cdot g\cdot \rho}$

$$\therefore \quad \frac{Q}{gH^{5/2}} = \text{fn}\left[\frac{P}{H}, \frac{L}{H}, \frac{\mu}{H^{3/2}g^{1/2}\rho}, \frac{\sigma}{\rho g H^2}\right]$$

6.8 A small sphere of density ρ_s and diameter D settles at a terminal velocity V in a liquid of density ρ_f and dynamic viscosity μ. Gravity g is known to be a parameter. Express the functional relationships between these variables in a dimensionless form.

Solution:

$$V = \text{fn}\,[\rho_s, D, \rho_f, \mu, g]$$

List the dimensions of each variable as follows:

V	ρ_s	D	ρ_f	μ	g
$[LT^{-1}]$	$[ML^{-3}]$	$[L]$	$[ML^{-3}]$	$[ML^{-1}\,T^{-1}]$	$[LT^{-2}]$

There are six variables, $n = 6$

and 3 basic dimensions, $m = 3$

Hence there are $(6 - 3) = 3$ dimensionless terms.

Select D, ρ_f and g as the repeating variables

I term: $\quad \pi_1 = VD^a \rho_f^b g^c$

Following the usual procedure

$$[M^0L^0T^0] = [LT^{-1}]\,[L]^a\,[ML^{-3}]^b\,[LT^{-2}]^c$$

$$b = 0$$

$$1 + a - 3b + c = 0$$

$$-1 - 2c = 0$$

Hence $\quad b = 0,\ c = -1/2,\ a = -1/2$

$$\therefore \quad \pi_1 = \frac{V}{\sqrt{gD}}$$

II term: $\quad \pi_2 = \rho_s D^a \rho_f^b g^c$

By inspection, it is easy to see $\pi_2 = \dfrac{\rho_s}{\rho_f}$

III term: $\quad \pi_3 = \mu D^a \rho_f^b g^c$

In this case, we have

$$[M^0L^0T^0] = [ML^{-1}\,T^{-1}]\,[L]^a\,[ML^{-3}]^b\,[LT^{-2}]^c$$

$$1 + b = c$$

$$-1 + a - 3b + c = 0$$

$$-1 - 2c = 0$$

$$b = -1,\ c = -\frac{1}{2} \text{ and } a = 3b - c + 1$$

$$= -\frac{3}{2}$$

$$\therefore \quad \pi_3 = \frac{\mu}{D^{3/2}\rho_f g^{1/2}}$$

Hence $\quad \dfrac{V}{\sqrt{gD}} = \varphi\left[\dfrac{\rho_s}{\rho_f}, \dfrac{\mu}{\rho_f D\sqrt{gD}}\right]$

6.9 The pressure drop Δp generated by a pump of a given geometry is known to depend upon the impeller diameter D, the rotational speed N, the fluid discharge Q, the fluid density ρ and viscosity μ. Obtain the dimensionless form of the functional relationship.

Solution: $\quad \Delta p = \text{fn}\,(D, N, Q, \rho, \mu)$

List the dimensions of each variable as:

Δp	D	N	Q	ρ	μ
$[ML^{-1}\,T^{-2}]$	$[L]$	$[T^{-1}]$	$[L^3\,T^{-1}]$	$[ML^{-3}]$	$[ML^{-1}T^{-1}]$

There are a total of six variables: $n = 6$

Number of primary dimensions $m = 3$

Hence there are $(6 - 3) = 3$ dimensionless Pi terms.

Select D, N and ρ as repeating variables.

I term: $\quad \pi_1 = \Delta p D^a N^b \rho^c$

The dimensional equation will be

$$[M^0L^0T^0] = [ML^{-1}\,T^{-2}]\,[L]^a\,[T^{-1}]^b\,[ML^{-3}]^c$$

from which we get

$$1 + c = 0$$

$$-1 + a - 3c = 0$$
$$-2 - b = 0$$
or $$c = -1, b = -2 \text{ and } a = -2$$
$$\therefore \quad \pi_1 = \frac{\Delta p}{D^2 N^2 \rho}$$

II term: $\pi_2 = QD^a N^b \rho^c$

Equating dimensions on either side

$$[M^0L^0T^0] = [L^3T^{-1}]\,[L]^a\,[T^{-1}]^b\,[ML^{-3}]^c$$

we have

$$c = 0$$
$$3 + a - 3c = 0$$
$$-1 - b = 0$$
or $$a = -3, b = -1, c = 0$$
$$\therefore \quad \pi_2 = \frac{Q}{D^3 N}$$

III term:

$$\pi_3 = \mu D^a N^b \rho^c$$

As usual,

$$[M^0L^0T^0] = [ML^{-1}T^{-1}]\,[L]^a\,[T^{-1}]^b\,[ML^{-3}]^c$$
$$1 + c = 0$$
$$-1 + a - 3c = 0$$
$$-1 - b = 0$$
or $$c = -1, b = -1, a = -2$$
$$\therefore \quad \pi_3 = \frac{\mu}{D^2 N\rho}$$

Thus the functional relationship can be expressed as

$$\frac{\Delta p}{\rho N^2 D^2} = \text{fn}\left(\frac{Q}{ND^3}, \frac{\mu}{\rho N D^2}\right)$$

6.10 The time period T of water surface waves is known to depend on the wave length λ, depth of flow D, density of the fluid ρ, acceleration due to gravity g and surface tension σ. Obtain the dimensionless form of the functional relationship.

Solution:

$$T = \text{fn}\,(\lambda, D, \rho, g, \sigma)$$

List the dimensions of each variable as follows:

T	λ	D	ρ	g	σ
[T]	[L]	[L]	$[ML^{-3}]$	$[LT^{-2}]$	$[MT^{-2}]$

There are a total of six variables; $n = 6$
Number of primary dimensions; $m = 3$
Hence there are $(6 - 3) = 3$ dimensionless Pi terms.
Select λ, g, and ρ as repeating variables.

I term: $\pi_1 = T\lambda^a g^b \rho^c$

$$[M^0L^0T^0] = [T]\,[L]^a\,[LT^{-2}]^b\,[ML^{-3}]^c$$
$$1 - 2b = 0 \quad \therefore \quad b = 1/2,$$
$$c = 0$$
$$a + b - 3c = 0; \quad a = -1/2$$
$$\therefore \quad \pi_1 = \frac{T\sqrt{g}}{\sqrt{\lambda}}$$

II term:

$$\pi_2 = D\lambda^a g^b \rho^c$$

By inspection it is easy to see that $\pi_2 = \dfrac{D}{\lambda}$

III term:

$$\pi_3 = \sigma\lambda^a g^b \rho^c$$

Again

$$[M^0L^0T^0] = [MT^{-2}]\,[L]^a\,[LT^{-2}]^b\,[ML^{-3}]^c$$
$$1 + c = 0 \quad \therefore \quad c = -1$$
$$-2 - 2b = 0; \quad b = -1$$
$$a + b - 3c = 0; \quad a = -2$$
$$\pi_3 = \frac{\sigma}{\lambda^2 g\rho}$$

Hence $$T\sqrt{\frac{g}{\lambda}} = \text{fn}\left[\frac{D}{\lambda}, \frac{\sigma}{\lambda^2 g\rho}\right]$$

Similitude

6.11 A fluid flow phenomenon is to be studied in a model which is to be constructed by using Reynolds model law. Find the expressions for model to prototype ratios of velocity, discharge, pressure, work and power.

Solution: Using the subscripts m for model, p for prototype and r for the ratio of model to prototype:

In Reynolds law $$Re = \left(\frac{\rho VL}{\mu}\right)_m = \left(\frac{\rho VL}{\mu}\right)_p$$

Let $L_r = \dfrac{L_m}{L_p}$, then

Velocity ratio $$\frac{V_m}{V_p} = V_r = \frac{\mu_r}{\rho_r L_r} = \left(\frac{v_r}{L_r}\right)$$

where v_r = ratio of the kinematic viscosities = $\dfrac{\mu_r}{\rho_r}$

Discharge ratio $$\frac{Q_m}{Q_p} = Q_r = V_r L_r^2$$
$$= \frac{\mu_r L_r}{\rho_r} = (v_r L_r)$$

Pressure ratio $$\frac{p_m}{p_p} = p_r$$

Nothing that Force $\propto \rho L^2 V^2$

$$p_r = \left(\frac{\text{Force}}{\text{Area}}\right)_r = \left(\frac{\rho L^2 V^2}{L^2}\right)_r = (\rho_r V_r^2)$$

$$= \frac{\rho_r \mu_r^2}{\rho_r^2 L_r^2} = \left(\frac{\mu_r^2}{\rho_r L_r^2}\right) = \frac{\mathbf{v_r^2 \rho_r}}{\mathbf{L_r^2}}$$

$$\text{Work ratio} = \text{Energy ratio} = \frac{E_m}{E_p} = E_r$$

$$= (\text{Force} \times \text{distance})_r = (\rho L^3 V^2)_r$$

$$= \frac{\rho_r L_r^3 \mu_r^2}{\rho_r^2 L_r^2} = \left(\frac{\mu_r^2 L_r}{\rho_r}\right) = (\mathbf{v^2 \rho_r L_r})$$

$$\text{Power ratio} = \frac{\mathbf{P}_m}{\mathbf{P}_p} = \mathbf{P}_r = (\text{Force} \times \text{velocity})_r$$

$$= (\rho L^2 V^3)_r$$

$$\mathbf{P}_r = \frac{\rho_r L_r^3 \mu_r^3}{\rho_r^3 L_r^3} = \left(\frac{\mu_r^3}{\rho_r^2 L_r}\right) = \frac{\mathbf{v_r^2 \rho_r}}{\mathbf{L_r}}$$

[***Note:*** A full list of the various ratios by Reynolds model law is given in Table 6.2.]

6.12 For Froude model law, find the ratios of velocity, discharge, force, work and power in terms of the length scale.

Solution: In Froude model law the model and prototype Froude numbers are the same. Hence

$$\text{Froude number } (Fr)_m = \frac{V_m}{\sqrt{gL_m}} = (Fr)_p = \frac{V_p}{\sqrt{gL_p}}$$

The gravity g is same for both model and prototype. Hence if the length ratio $L_m/L_p = L_r$ and $\rho_m/\rho_p = \rho_r$

Velocity ratio $V_r = \sqrt{\mathbf{L_r}}$

Discharge ratio $Q_r = (\text{Velocity} \times \text{area})_r = V_r L_r^2 = (\mathbf{L_r^{5/2}})$

Force ratio $F_r = (\rho L^2 V^2)_r = (\boldsymbol{\rho_r L_r^3})$

Work ratio = Energy ratio = E_r

$= (\text{Force} \times \text{distance})_r = (\boldsymbol{\rho_r L_r^4})$

Power ratio = $(\text{Force} \times \text{velocity})_r = \mathbf{P}_r$

$= \rho_r L_r^3 L_r^{1/2} = (\boldsymbol{\rho_r L_r^{7/2}})$

[***Note:*** A full list of the various ratios by Froude model law is given in Table 6.2.]

6.13 A 1:50 spillway model has a discharge of 1.25 m³/s. What is the corresponding prototype discharge? If a flood phenomenon takes 12 hours to occur in the prototype, how long should it take in the model?

Solution: Since this is the case of a free surface phenomenon affected by gravity, Froude model law is appropriate.

$$Fr_m = \frac{V_m}{\sqrt{gL_m}} = \frac{V_p}{\sqrt{gL_p}} = Fr_p$$

$\therefore$ If $L_r = \dfrac{L_m}{L_p}$,

Discharge ratio = $Q_r = V_r L_r^2 = L_r^{5/2}$

As $L_r = \dfrac{1}{50}$, $\dfrac{Q_m}{Q_p} = \dfrac{125}{Q_p} = \left(\dfrac{1}{50}\right)^{2.5}$

Prototype discharge

$$Q_p = (50)^{2.5} \times 1.25 = \mathbf{22097\ m^3/s}$$

$$\text{Time ratio} \quad T_r = \frac{L_r}{V_r} = \frac{L_r}{\sqrt{L_r}} = \sqrt{L_r}$$

$$\frac{t_m}{t_p} = \frac{t_m}{12} = \sqrt{1/50}$$

$$t_m = \frac{12}{\sqrt{50}} = \mathbf{1.697\ hours}$$

6.14 A model boat, 1/100 size of its prototype has 0.12 N of resistance when simulating a speed of 5 m/s of the prototype. Water is the fluid in both cases. What is the corresponding resistance in the prototype? [The frictional forces can be neglected].

Solution: The resistance offered at the free surface is the significant force and as such Froude model law is appropriate.

$$Fr_m = \frac{V_m}{\sqrt{gL_m}} = \frac{V_p}{\sqrt{gL_p}} = Fr_p$$

If $L_r = \dfrac{L_m}{L_p}$; $V_r = \sqrt{L_r}$ and

$$\frac{(\text{Force})_m}{(\text{Force})_p} = \rho_r L_r^2 V_r^2 = \rho_r L_r^3$$

Since the same fluid is used in the model and prototype, $\rho_m = \rho_p$ and $\rho_r = 1$. Hence,

$$\frac{F_m}{F_p} = L_r^3$$

$$F_p = \text{Prototype force} = (0.12)/\left(\frac{1}{100}\right)^3$$

$$= 120000\ \text{N} = \mathbf{120\ kN}$$

6.15 A model of an open channel is built to a scale of 1/100. If the model has a Manning's $n = 0.013$, to what value of prototype roughness coefficient would this correspond?

Solution: Froude model law is applicable here

$$\frac{V_m}{\sqrt{gL_m}} = \frac{V_p}{\sqrt{gL_p}}$$

If $\frac{L_m}{L_p} = L_r$, $\quad V_r = \frac{V_m}{V_p} = \sqrt{L_r}$

By Manning formula

$$V = \frac{1}{n} \cdot R^{2/3} S_0^{1/2}$$

The dimension of $R = [L]$

$$S_0 = [M^0\ L^0\ T^0]$$

As $\quad n_r = \frac{R_r^{2/3}}{V_r}$

$$\frac{n_m}{n_p} = n_r = \frac{L_r^{2/3}}{\sqrt{L_r}} = L_r^{1/6}$$

∴ $\quad n_p = n_m/L_r^{1/6} = \frac{0.013}{(1/100)^{1/6}} = \mathbf{0.028}$

6.16 Estimate for a 1/20 model of a spillway
(i) the prototype velocity corresponding to a model velocity of 1.5 m/s
(ii) the prototype discharge per unit width corresponding to a model discharge per unit width of 0.2 m^3/s per metre
(iii) the pressure head in the prototype corresponding to a model pressure head of 5 cm of mercury at a point
(iv) The energy dissipated per second in the model corresponding to a prototype value of 1 kW.

Solution: For dynamic similarity Froude number must be the same in the model and prototype. If L_r is the length ratio, then

(i) $\quad V_r = \frac{V_m}{V_p} = \sqrt{L_r}$

$$V_p = V_m/\sqrt{L_r} = 1.5\sqrt{20} = \mathbf{6.71\ m/s}$$

(ii) ratio of discharge per unit width $= q_r = \frac{(Q/L)_m}{(Q/L)_p}$

$$q_r = \frac{Q_r}{L_r} = V_r L_r = L_r^{3/2}$$

$$q_p = q_m/L_r^{3/2} = 0.20 \times (20)^{3/2}$$

$$= \mathbf{17.89\ m^3/s/m}$$

(iii) pressure ratio $p_r = (L_r \rho_r)$

Assume $\quad \rho_m = \rho_p$, i.e. $\rho_r = 1.0$

Hence $\quad p_r = L_r$

∴ $\quad p_p = p_m/L_r = 5 \times 20$

$= \mathbf{100\ cm\ of\ mercury}$

(iv) Power ratio = (Energy loss/second)$_r$

$= [L_r^{7/2}\ \rho_r]$

As $\quad \rho_r = 1.0$ (assumed), $\mathbf{P}_r = L_r^{7/2}$

$$\mathbf{P}_m = \mathbf{P}_p \cdot L_r^{7/2}$$

$$= 1000 \times \left(\frac{1}{20}\right)^{7/2} = \mathbf{0.028\ W}$$

6.17 Oil of density 917 kg/m^3 and dynamic viscosity 0.29 Pa.s flows in a pipe of diameter 15 cm at a velocity of 2.0 m/s. What would be the velocity of flow of water in a 1.0 cm diameter pipe, to make the two flows dynamically similar? The density and viscosity of water can be taken as 998 kg/m^3 and 1.31×10^{-3} Pa.s respectively.

Solution:
Reynolds similarity law is applicable.

$$(Re)_m = \frac{V_m d_m}{\nu_m} = (Re)_p = \frac{V_p d_p}{\nu_p}$$

∴ $\quad V_r = \frac{V_m}{V_p} = \frac{\nu_r}{L_r} = \frac{\mu_r}{L_r \rho_r}$

$$\frac{V_m}{V_p} = \frac{\mu_m}{\mu_p} \frac{1}{\left(\frac{L_m}{L_p}\right)\left(\frac{\rho_m}{\rho_p}\right)}$$

Referring to oil with a subscript p and water with a suffix m

$$\frac{V_m}{V_p} = \frac{1.31 \times 10^{-3}}{0.29} \times \frac{1}{\left(\frac{1.0}{15.0}\right)\left(\frac{998}{917}\right)}$$

$= 0.0623$

V_m = Velocity of water flow $= V_p \times 0.0623$

$= 2 \times 0.0623 = \mathbf{0.1246\ m/s}$

6.18 A 1 : 6 scale model of a passenger car is tested in a wind tunnel. The prototype velocity is 60 km/h. If the model drag is 250 N what is the drag and the power required to overcome the drag in the prototype? The air in the model and prototype can be assumed to have the same properties.

Solution: Reynolds similarity law is applicable.

$$(Re)_m = \frac{V_m L_m}{\nu_m} = (Re)_p = \frac{V_p L_p}{\nu_p}$$

∴ $\quad V_r = \frac{\nu_r}{L_r}$

If $\quad \nu_r = 1$ (i.e. $\rho_m = \rho_p$, $\mu_m = \mu_p$)

then $\quad V_r = 1/L_r$,

$V_m = V_p/L_r = 60 \times 6 = 360$ km/h

$= 100$ m/s

Force ratio $\quad \frac{F_m}{F_p} = \frac{\mu_r^2}{\rho_r}$

If $\rho_r = 1$, and $\mu_r = 1$, $\frac{F_m}{F_p} = 1.0$

∴ F_p = **250 N** (same as in the model)

Power to overcome drag in the prototype:

$$\mathbf{P}_p = F_p \cdot V_p = 250 \times \left(\frac{60 \times 10^3}{3600} \right) = 4167 \text{ W}$$

$$= \mathbf{4.167\ kW}$$

6.19 The pressure drop in a flow meter in which an oil flows at an upstream velocity of 0.9 m/s is to be estimated by model studies. A 1 : 6 scale model using water is used. If the pressure drop in the model is 450 Pa, what will be the prototype pressure drop? If the prototype discharge is 200 L/s what is the model discharge? The following data are relevant:

Item	*Prototype*	*Model*
Density	900 kg/m^3	998 kg/m^3
Viscosity	0.104 Pa.s	1 × 10^{-3} Pa.s

Solution: The Reynolds model law is applicable.

$$(Re)_m = \frac{\rho_m V_m L_m}{\mu_m} = \frac{\rho_p V_p L_p}{\mu_p}$$

$$L_m/L_p = L_r = 1/6$$

$$V_r = \frac{\mu_r}{L_r \rho_r}$$

Pressure ratio $\quad p_r = \rho_r V_r^2 = \dfrac{\mu_r^2}{L_r^2 \rho_r}$

and discharge ratio $\quad Q_r = V_r \cdot L_r^2 = \dfrac{\mu_r L_r}{\rho_r}$

In the present problem:

$$\mu_r = \frac{1 \times 10^{-3}}{0.104} = 9.615 \times 10^{-3}$$

$$\rho_r = \frac{998}{900} = 1.109$$

$$L_r = 1/6$$

$$\therefore \quad p_r = \frac{(9.615 \times 10^{-3})^2}{\left(\frac{1}{6}\right)^2 \times 1.109} = 3 \times 10^{-3}$$

$$\therefore \quad p_p = \frac{p_m}{p_r} = \frac{450}{3 \times 10^{-3}} = 150{,}000 \text{ Pa}$$

$$= \mathbf{150\ kPa}$$

$$Q_r = \frac{(9.615 \times 10^{-3}) \times 1/6}{1.109} = 1.445 \times 10^{-3}$$

$$Q_m = Q_p \times 1.445 \times 10^{-3}$$

$$= 200 \times 1.445 \times 10^{-3} = \mathbf{0.289\ L/s}$$

6.20 An underwater device is 1.5 m long, and is to move at 3.5 m/s. A geometrically similar model 30 cm long is tested in a variable pressure wind tunnel at a speed of 35 m/s. Calculate the pressure of air in the model. If the model exhibits a drag force 40 N, calculate the prototype drag force.
[Assume ρ_{water} = 998 kg/m^3, ρ_{air} at standard atmospheric pressure = 1.17 kg/m^3, $\mu_{air} = 1.90 \times 10^{-5}$ Pa.s, at local atmospheric pressure and $\mu_{water} = 1.0 \times 10^{-3}$ Pa.s].

Solution: Reynolds model law is applicable. Hence

$$(Re)_m = \frac{\rho_m V_m L_m}{\mu_m} = (Re)_p = \frac{\rho_p V_p L_p}{\mu_p}$$

$$\therefore \text{ If } \quad \frac{L_m}{L_p} = L_r,\ \frac{\rho_m}{\rho_p} = \rho_r \text{ and } \frac{\mu_m}{\mu_p} = \mu_r$$

$$\frac{V_m}{V_p} = V_r = \frac{\mu_r}{\rho_r L_r}$$

In the present case

$$L_r = \frac{0.30}{1.50} = \frac{1}{5}$$

$$V_r = \frac{35}{3.5} = 10$$

$$\mu_r = \frac{1.90 \times 10^5}{1 \times 10^{-3}} = 1.9 \times 10^{-2}$$

Hence $\quad \rho_r = \dfrac{\mu_r}{V_r L_r} = \dfrac{1.9 \times 10^{-2}}{10 \times 1/5} = 0.0095$

Hence $\quad \rho_{air} = \rho_m = 998 \times 0.0095 = \mathbf{9.481}$

This is about 8 times larger than the density at atmospheric pressure. Since at constant temperature, by the equation of state p/ρ = constant.

Hence $\quad \dfrac{(\text{pressure})_{model}}{(\text{Atmospheric pressure})} = \dfrac{\rho_{model}}{\rho_{atmospheric}}$

$$p_{model} = \frac{9.481}{1.17} \times p_a = 8.103\ p_a$$

= 8.103 times local atmospheric pressure

$$\text{Force ratio} = F_r = \rho_r V_r^2 L_r^2$$

$$= \frac{\mu_r^2}{\rho_r} = \frac{(1.9 \times 10^{-2})^2}{0.0095}$$

$$= 0.038$$

$$F_p = \frac{F_m}{0.038} = \frac{40}{0.038} = 1053 \text{ N}$$

$$= \mathbf{1.053\ kN}$$

6.21 A pipe of diameter 1.5 m is required to transport an oil of relative density 0.9 and kinematic viscosity = 3 × 10^{-2} stoke at a rate of 3.0 m^3/s. If a 15 cm diameter pipe with water at 20°C (ν = 0.01 stoke) is used to model the above flow, find the velocity and discharge in the model.

Solution: The Reynolds number must be the same in the model and prototype for similar pipe flows.

$$\frac{V_p D_p}{\nu_p} = \frac{V_m D_m}{\nu_m}$$

$$V_m = V_p \frac{D_p}{D_m} \frac{\nu_m}{\nu_p}$$

$$V_p = \frac{Q}{\pi D^2/4} = \frac{3.0}{\frac{\pi}{4} \times (1.5)^2} = 1.6977 \text{ m/s}$$

$$V_m = 1.6977 \times \frac{1.5}{0.15} \times \frac{0.01}{0.03} = \mathbf{5.659 \text{ m/s}}$$

$$Q_m = \text{discharge in the model} = \frac{\pi}{4} D_m^2 \times V_m$$

$$= \frac{\pi}{4} \times (0.15)^2 \times (5.659) = \mathbf{0.1 \text{ m}^3/\text{s}}$$

6.22 A 1/10 model of an airplane is tested in a variable density wind tunnel. The prototype plane is to fly at 400 km/h speed under atmospheric conditions The pressure used in the wind tunnel is 10 times the atmospheric pressure. Calculate the velocity of air in the model. To what prototype value would a measured drag of 500 N in the model correspond? If some vortices are shed at a frequency of 25 Hz in the model, what would be the corresponding prototype frequency?

Solution: The Reynolds number in the model and the prototype must be the same.

$$(Re)_m = \frac{\rho_m V_m L_m}{\mu_m} = (Re)_p = \frac{\rho_p V_p L_p}{\mu_r}$$

Here $\frac{L_m}{L_p} = L_r = 1/10.$

Since pressure does not affect the viscosity appreciably $\mu_m = \mu_p$ and hence $\mu_r = 1.0$. Further, at constant temperature p/ρ = constant.

Pressure ratio $\frac{p_m}{p_a} = 10 = \frac{\rho_m}{\rho_p} = \rho_r$

$$\therefore \quad V_m = V_p \times \frac{\mu_r}{\rho_r L_r} = \frac{400}{10 \times \frac{1}{10}}$$

$$= \mathbf{400 \text{ km/h}}$$

Model velocity is the same as prototype velocity,

i.e. $V_r = 1.0$

$$\text{Force ratio} = F_r = L_r^2 \rho_r V_r^2$$

$$= \frac{1}{(10)^2} \times 10 \times (1) = \frac{1}{10}$$

$$F_p = \frac{F_m}{F_r} = 500 \times 10 = \mathbf{5000 \text{ N}}$$

Frequency ratio $f_r = \frac{T_p}{T_m} = \frac{1}{T_r} = \frac{V_r}{L_r}$

$$f_r = \frac{f_m}{f_p} = \frac{1}{1/10} = 10$$

Hence, $f_p = \frac{f_m}{10} = \frac{25}{10} = \mathbf{2.5 \text{ Hz}}$

6.23 Obtain an expression for the scale of a model, which has to satisfy both Froude's model law and Reynolds model law.

Solution: In this case,

$$\frac{V_m}{\sqrt{gL_m}} = \frac{V_p}{\sqrt{gL_p}}$$

i.e. $V_r = L_r^{1/2}$

Also by Reynolds model law,

$$\frac{\rho_m V_m L_m}{\mu_m} = \frac{\rho_p V_p L_p}{\mu_p}$$

$$\therefore \quad L_r = \frac{\mu_r}{V_r \rho_r} = \frac{\mu_r}{(L_r)^{1/2} \cdot \rho_r}$$

$$\therefore \quad (L_r)^{3/2} = \frac{\mu_r}{\rho_r} = \nu_r$$

$$\text{Hence scale} = L_r = \left(\frac{\mu_r}{\rho_r}\right)^{2/3} = \mathbf{(\nu_r)^{2/3}}$$

6.24 A 1 : 25 scale model of a ship has a submerged surface area of 6 m², a length of 5 m and experiences a total drag of 25 N when towed through water with a velocity of 1.2 m/s. Estimate the total drag on the prototype when cruising at the corresponding speed. The skin friction force can be estimated by $F_s = C_f A \rho V^2/2$ where the coefficient $C_f = 0.0735/(Re)^{1/5}$.
Assume $\mu_{water} = 1 \times 10^{-3}$ Pa.s and $\rho_{water} = 1030$ kg/m³ for both model and the prototype.

Solution: The total drag is the sum of the wave resistance and the skin friction. The Froude criterion of similarity is used for the gravity-affected component of the drag, namely the wave resistance. The skin friction is estimated by the formula separately. For the model:

Reynolds number $Re = \frac{\rho V L}{\mu}$

$$= \frac{1030 \times 1.2 \times 5}{1 \times 10^{-3}} = 6.18 \times 10^6$$

$$C_f = \frac{0.0735}{(6.18 \times 10^6)^{1/5}} = 3.222 \times 10^{-3}$$

Skin friction resistance

$$F_s = C_f A\rho V^2/2$$
$$= (3.222 \times 10^{-3}) \times 6 \times 1030 \times (1.2)^2/2$$
$$= 14.34 \text{ N}$$

Total model drag = 25.00 N

Hence, the model wave resistance = 25.00 – 14.34

$$(F_w)_m = 10.66 \text{ N}$$

The corresponding wave resistance in the prototype is calculated by Froude's law of similarity.

For the prototype, by Froude's law, the wave resistance $(F_w)_p$ is given by

$$\frac{(F_w)_p}{(F_w)_m} = (F_w)_r = \rho_r V_r^2 L_r^2$$

$$\text{Froude number} = \frac{V_m}{\sqrt{gL_m}} = (Fr)_m = \frac{V_p}{\sqrt{gL_p}} = (Fr)_p$$

$$V_r = \sqrt{L_r} \text{ and } (F_w)_r = \rho_r L_r^3$$

In the present case,

$$\rho_r = 1.0$$
$$(F_w)_r = L_r^3$$

$$\therefore \quad (F_w)_p = \frac{(F_w)_m}{L_r^3} = \frac{10.66}{(1/25)^3} = 166.6 \times 10^3 \text{ N}$$
$$= 166.6 \text{ kN}$$

Skin friction for the prototype:

Prototype Reynolds number

$$= \frac{\rho_p V_p L_p}{\mu_p} = (Re)_p$$

Since $\quad V_p = V_m/\sqrt{L_r}$ and $L_p = \dfrac{L_m}{L_r}$

Also $\quad \rho_p = \rho_m$ and $\mu_p = \mu_m$

$$(Re)_p = (Re)_m \frac{1}{L_r^{3/2}} = (25)^{3/2} \times 6.18 \times 10^6$$
$$= 7.73 \times 10^8$$

$$C_f = \frac{0.0735}{(7.73 \times 10^8)^{1/5}} = 1.2266 \times 10^{-3}$$

Skin friction resistance

$$(F_s)_p = (C_f A\rho V^2/2)_p$$
$$= (1.2266 \times 10^{-3}) \times \{6 \times (25^2)\} \times 1030 \times [(1.2) \times \sqrt{25}]^2/2$$
$$= 85279 \text{ N} = 85.3 \text{ kN}$$

Total prototype resistance

$$= (F_w)_p + (F_s)_p$$
$$= 166.6 + 85.3 = \mathbf{251.9 \text{ kN}}$$

Distorted Models

6.25 A proposed model of a river stretch of 15 km is to have a horizontal scale of 1/200 and vertical scale of 1/40. If the normal discharge, width and depth of the river are 152 m^3/s, 90 m and 2 m respectively, estimate the corresponding model quantities. What value of Manning's roughness n is to be provided in the model to represent a prototype roughness value of 0.025?

Solution: Horizontal scale = L_r = 1/200

Vertical scale = h_r = 1/40

1. Discharge $Q_m = Q_p (L_r)(h_r)^{1.5}$
   $= 152 \times (1/200)(1/40)^{1.5}$
   $= \mathbf{0.03 \text{ m}^3/\text{s}}$
2. Depth $y_m = y_p (h_r) = 2.0/40 = \mathbf{0.05 \text{ m}}$
3. Width $B_m = B_p (L_r) = 90/200 = \mathbf{0.045 \text{ m}}$
4. Manning's $n_m = n_p (h_r)^{2/3}/(L_r)^{1/2}$
   $= (0.025)(1/40)^{2/3}/(1/200)^{1/2} = \mathbf{0.03}$

[**Note** that the model has to be rougher than the prototype]

6.26 In a tidal model, the horizontal scale ratio is 1/500. The vertical scale is 1/50. What model period would correspond to a prototype period of 12 hours?

Solution: L_r = 1/500 and h_r = 1/50

Time ratio $\quad T_r = L_r/\sqrt{h_r} = \dfrac{1/500}{\sqrt{1/50}} = 0.01414$

Model period $\quad T_m = T_p T_r = (12 \times 60 \times 60) \times 0.01414$
$= 610$ s
$=$ **10 minutes 10 seconds**

6.27 For a river model of horizontal scale 1/250 and vertical scale 1/25, estimate the model value corresponding to a prototype slope of 0.0002. If the model velocity and discharges are 0.50 m/s and 0.02 m^3/s respectively, estimate the corresponding prototype velocity and discharge values.

Solution: L_r = 1/250 and h_r = 1/25

(1) Slope ratio $\quad S_r = S_m/S_p = h_r/L_r$
$= (1/25)/(1/250) = 10$
$S_m = S_p S_r = 0.0002 \times 10 = \mathbf{0.002}$

(2) Velocity ratio $\quad = V_r = (V_m/V_p) = \sqrt{h_r}$
$= (1/25)^{0.5} = 1/5$
$V_p = V_m/V_r = 0.50/(1/5) = \mathbf{2.5 \text{ m/s}}$

(3) Discharge ratio $Q_r = Q_m/Q_p = L_r\, h_r^{1.5}$
$= (1/250)(1/25)^{1.5} = 1/31250$
$Q_p = Q_m/Q_p = 0.02/(1/31250)$
$= \mathbf{625 \text{ m}^3/\text{s}}$

[***Note*** that the model has to be steeper than the prototype.]

OBJECTIVE QUESTIONS

Dimensional Analysis

6.1 Which of the following is *not* a dimensionless number:
(a) Darcy-Weisbach friction factor f
(b) Coefficients of drag C_D
(c) Manning's coefficient n
(d) coefficient of velocity C_v

6.2 Which of the following is a dimensionless number:
(a) Manning's coefficient n
(b) Pipe friction factor f
(c) Chezy coefficient C
(d) Hazen-William coefficient C_H

6.3 The dimensions of surface tension σ are
(a) ML^{-1} (b) $L^2 T^{-1}$
(c) $ML^{-1} T^{-2}$ (d) MT^{-2}

6.4 The dimensions of dynamic viscosity μ are
(a) $FL^{-2} T$ (b) L^2T^{-1}
(c) FL^{-1} (d) $FL^{-2} T^{-2}$

6.5 The dimensions of volume modulus of elasticity K are
(a) $FL^{-1} T^{-2}$ (b) $FL^{-2} T$
(c) FL^{-2} (d) $FL^{-4} T^{-2}$

6.6 The dimensions of specific heat (C_v or C_p) are
(a) $L^2 \theta^{-2}$ (b) $L^2 T^{-2} \theta^{-1}$
(c) $FT^{-1} \theta^{-1}$ (d) $L^2\theta^{-1}$

6.7 The Euler number E_n is written as E_n =
(a) $V/\sqrt{K/\rho}$ (b) $\rho V^2 L/\sigma$
(c) $V\rho/\sqrt{\Delta p}$ (d) $V/\sqrt{\Delta p/p}$

6.8 Assuming the thrust T of a propeller depends upon the diameter D, speed of advance V, angular velocity ω, dynamic viscosity μ, and density ρ, which of the following dimensionless parameters can be derived by dimensional analysis?

(1) $\dfrac{T}{\rho D^2 V^2}$ (2) $\dfrac{VD\mu}{\rho}$

(3) $\dfrac{D\omega}{V}$ (4) $\dfrac{VD\rho}{\mu}$

Select the correct answer using the codes given below:
(a) 1, 2 and 3 (b) 2, 3 and 4
(c) 1, 3 and 4 (d) 1, 2 and 4

6.9 There are four variables in a fluid flow. viz., K (volume modulus of elasticity), p (pressure), μ (dynamic viscosity) and g (acceleration due to gravity. They are associated with Mach, Euler, Froude and Reynolds number, respectively, in the order
(a) K, p, μ, g (b) K, p, g, μ
(c) p, K, g, μ (d) p, μ, g, K

Similitude

6.10 Dynamic similarity exists when the model and prototype have the same
(a) length scale ratio and time scale ratio
(b) length scale ratio and velocity scale ratio
(c) length scale ratio, time scale ratio and velocity scale ratio
(d) length scale ratio, velocity scale ratio and force scale ratio

6.11 Both Reynolds number and Froude number assume significance in one of the following example:
(a) Motion of submarine at large depths
(b) Motion of ship in deep seas
(c) Cruising of a missile in air
(d) Flow over a spillway

6.12 In hydraulic modelling of flow pattern around a body submerged in a fluid the non-dimensional number which has to be kept the same in the model and prototype is
(a) Weber number
(b) Froude number
(c) Reynolds number
(d) Strouhal number

6.13 The time scale ratio for a model based on Froude law criterion in terms of length scale ratio L_r is
(a) L_r (b) $\sqrt{L_r}$
(c) $1/\sqrt{L_r}$ (d) $L_r^{1.5}$

6.14 If Froude law of similitude exists between a model and a prototype, then the force ratio F_r =
(a) L_r^3 (b) $L_r\rho_r$
(c) $L_r^3\rho_r$ (d) $L_r^3\rho_r^{-1}$

6.15 In a model experiment with a weir, if the dimensions of the model weir are reduced by a factor k, the flow rate through the model weir is the following fraction of the flow rate through the prototype:
(a) $k^{5/2}$ (b) k^2
(c) 1 (d) k^{-2}

6.16 In the Froude law of similitude the acceleration ratio a_r =

(a) L_r^2 (b) 1.0

(c) $1/L_r$ (d) $L_r^{-3/2}$

6.17 If Reynolds law of similitude exists between a model and a prototype, then the force ratio F_r =

(a) $L_r^3 \rho_r$ (b) $\mu_r^2 \rho_r^{-1}$

(c) $L_r \mu_r \rho_r^{-1}$ (d) $L_r^{-2} \mu_r \rho_r^{-1}$

6.18 In Reynolds law of similitude the discharge ratio Q_r =

(a) $L_r \mu_r / \rho_r$ (b) $L_r^3 \rho_r$

(c) $L_r^{5/2}$ (d) $L_r \rho_r / \mu_r$

6.19 A hydraulic model of a spillway is constructed with a scale 1 : 16. If the prototype discharge is 2048 m^3/s, then the corresponding discharge in m^3/s for which the model should be tested is

(a) 1 (b) 2

(c) 4 (d) 8

6.20 In the model of a highway bridge constructed to a scale of 1:25, the force of water on the pier was measured as 5 N. The force on the prototype pier will be

(a) 75015 N (b) 76225 N

(c) 78125 N (d) 79165 N

6.21 The fall velocity of a sand grain in water is to be modeled by using particles of the same relative density as sand and a liquid whose kinematic viscosity is 100 times larger than that of water. The diameters of the particles in the model that will have the same fall velocity as the prototype will be

(a) 10 times smaller

(b) 10 times larger

(b) 100 times smaller

(d) 100 times larger

6.22 In a model built on Froude law of similarity a phenomenon lasts for 20 min. If the model scale is 1/25, the duration of the phenomenon in the prototype, in minutes, is

(a) 50 (b) 100

(c) 2500 (d) 4

6.23 If the same fluid is used both in the model and prototype, and if it is desired to have equal Reynolds number and Froude number in the model and prototype, the scale of the model is

(a) V_r (b) $V_r^{1/2}$

(c) 1.0 (d) $V_r^{-1/2}$

where V_r = velocity ratio.

6.24 A harbor model has a horizontal scale of 1/150 and a vertical scale of 1/75. The interval between successive daily high tides in the model will be nearly

(a) 90 min (b) 40 min

(c) 15 min (d) 5 hours

6.25 In the distorted model of a river, the horizontal and vertical scales are L_r and h_r respectively. The discharge ratio will be

(a) $L_r^{1/2} h_r^2$ (b) $L_r h_r^{3/2}$

(c) $L_r^2 h_r^{1/2}$ (d) $L_r^3 h_r^{1/2}$

6.26 A river model is constructed to a horizontal scale of 1 : 1000 and a vertical scale of 1 : 100. If the model discharge were 0.1 m^3/s, then the discharge in the prototype would be

(a) 10^3 m^3/s (b) 10^4 m^3/s

(c) 10^5 m^3/s (d) 10^2 m^3/s

6.27 A model of a weir made to a horizontal scale of 1/40 and vertical scale of 1/9 passes a discharge of 1 Litre/s. The corresponding discharge in the prototype would be

(a) 10.8 Lps (b) 108 Lps

(c) 1080 Lps (d) 10800 Lps

PROBLEMS

Dimensional Analysis

6.1 The Chezy formula for velocity V in an open channel is given by

$$V = C\sqrt{RS_0}$$

where R = hydraulic radius, S_0 = longitudinal slope of the channel and C = Chezy coefficient. If for a channel the value of C in FPS units is 108, what is its value in SI units? (*Ans.* C = 59.6)

6.2 The following is a formula for calculating the maximum flood discharge Q from a catchment of area A:

$$Q = \frac{7000A}{\sqrt{A+4}}$$

In this Q is in cubic feet per second and A is in square miles. Convert this formula to have Q in m^3/s and A in km^2.

$$\left(Ans.\ Q = \frac{123.2A}{\sqrt{A+10.4}}\right)$$

6.3 The capillary rise h of a fluid of density ρ and surface tension σ in a tube of diameter D depends upon the contact angle θ and gravity g. Obtain an expression for h by Raleigh's method.

$$\left(Ans.\ \frac{h}{D} = \text{fn}\left(\frac{\sigma}{\rho g D^2}, \theta\right)\right)$$

6.4 The critical depth y_c in a trapezoidal channel depends upon the discharge Q, the side slope of the channel m, the bottom width B and the gravity g. Obtain an expression for y_c by Raleigh's method.

$$\left(Ans.\ \frac{y_c}{B} = \text{fn}\left(m, \frac{Q}{B^2\sqrt{gB}}\right)\right)$$

6.5 The stagnation pressure p_s in an air flow depends upon the static pressure p_0, the velocity V of the free stream and density ρ of the air. Derive a dimensionless expression for p_s.

$$\left(Ans.\ \frac{p_s}{p_0} = \text{fn}\left(\frac{p_0}{\rho V^2}\right)\right)$$

6.6 The discharge Q over a V-shaped notch is known to depend on the angle θ of the notch, the head H of the water surface, the velocity of approach V_0 and the acceleration due to gravity g. Determine the dimensionless form of the discharge equation.

$$\left(Ans.\ \frac{Q}{H^2\sqrt{gH}} = \text{fn}\left(\theta, \frac{V_0}{\sqrt{gH}}\right)\right)$$

6.7 The discharge Q through a sluice gate depends upon the gate opening height d, head of water H, acceleration due to gravity g, density ρ and viscosity μ of water. Obtain an expression for Q in dimensionless form.

$$\left(Ans.\ \frac{Q}{g^{1/2}H^{5/2}} = \text{fn}\left(\frac{d}{H}, \frac{\mu}{\rho g^{1/2}H^{3/2}}\right)\right)$$

6.8 The power **P** required to drive a propeller is known to depend on the diameter D and angular velocity ω of the propeller; the density ρ, viscosity μ and bulk modulus of elasticity K of the fluid; and the free stream velocity V. Drive the functional relationship for **P** in a dimensionless form.

$$\left(Ans.\ \frac{\mathbf{P}}{\rho V^3 D^2} = \text{fn}\left(\frac{\omega D}{V}, \frac{\mu}{\rho V D}, \frac{K}{\rho V^2}\right)\right)$$

6.9 The head loss h_L due to fluid friction in a pipe is known to depend on the diameter D, length L and roughness magnitude ε of the pipe; the velocity of flow V, the gravity g; and fluid density ρ and viscosity μ. Derive an expression for h_L in dimensionless form.

$$\left(Ans.\ \frac{h_L}{D} = \text{fn}\left(\frac{L}{D}, \frac{\varepsilon}{D}, \frac{gD}{V^2}, \frac{\mu}{\rho V D}\right)\right)$$

6.10 In laminar flow through a tube the discharge Q is a function of diameter D, the fluid viscosity μ and the pressure gradient dp/dx. Obtain an expression for Q in a dimensionless form.

$$\left(Ans.\ \left(\frac{Q\mu}{D^4 \frac{dp}{dx}}\right) = \text{Constant}\right)$$

6.11 The flow velocity u very near a rotating disk depends on the angular velocity ω of the disk, the radial distance r, vertical distance z and kinematic viscosity of the fluid ν. Derive a relation for u in a dimensionless form.

$$\left(Ans.\ \left(\frac{\mu}{\omega r}\right) = \text{fn}\left(\frac{z}{r}, \frac{\omega r^2}{\nu}\right)\right)$$

6.12 The shear stress τ_0 at the bed of a rough channel depends upon the depth of flow y, velocity of the fluid V, roughness height ε of the bed and fluid density ρ and viscosity μ. Derive an expression for τ_0 in dimensionless form.

$$\left(Ans.\ \frac{\tau_0}{\rho V^2} = \text{fn}\left(\frac{\mu}{\rho V y}, \frac{\varepsilon}{y}\right)\right)$$

6.13 The shear stress τ_0 on the walls of a triangular channel depends upon the vertex angle θ, depth of flow y, longitudinal slope S, density ρ and acceleration due to gravity g. Obtain an expression for τ_0 in dimensionless form.

$$\left(Ans.\ \frac{\tau_0}{\rho g y} = \text{fn}\,[\theta, S]\right)$$

6.14 The terminal velocity of descent V of a hemispherical parachute is found to depend on its diameter D, weight W, acceleration due to gravity g, density of air ρ_a and viscosity of air μ. Obtain an expression for V in dimensionless form.

$$\left(Ans.\ \frac{V}{\sqrt{gD}} = \text{fn}\left(\frac{W}{\rho D^3 g}, \frac{\mu}{\rho D\sqrt{gD}}\right)\right)$$

6.15 The lift force F on an airfoil is a function of the angle of attack, α, velocity of flow V, chord length C, span L, density ρ, viscosity μ, and bulk modulus of elasticity E. Obtain the dimensionless form of the functional relationship.

$$\left(Ans. \frac{F}{\rho V^2 C^2} = \text{fn}\left(\frac{\rho VC}{\mu}, \frac{V\sqrt{\rho}}{\sqrt{E}}, \frac{L}{C}, \alpha\right)\right)$$

6.16 The variables controlling the motion of a floating vessel through water are the drag force F, speed V, length L, acceleration due to gravity g, fluid density ρ and viscosity μ. Derive an expression for F by dimensional analysis.

$$\left(Ans. \frac{F}{\rho V^2 L^2} = \text{fn}\left(\frac{\sqrt{gL}}{V}, \frac{\mu}{\rho VL}\right)\right)$$

6.17 In the study of the vortex shedding phenomenon due to a bluff body in an open channel flow the following parameters are found to be important: velocity of flow = V, depth of flow = D, density of fluid = ρ, acceleration due to gravity = g, dynamic viscosity = μ, width of body = B, thickness of body = d and frequency of vortex shedding = n. Obtain the dimensionless parameters governing the phenomenon.

$$\left(Ans. \left(\frac{nD}{V}\right) = \text{fn}\left(\frac{\sqrt{gD}}{V}, \frac{\mu}{\rho VD}, \frac{B}{D}, \frac{d}{D}\right)\right)$$

6.18 The discharge Q from a centrifugal pump is dependent upon the pump speed N (rpm), diameter of the impeller D, head H, acceleration due to gravity g, density of he fluid ρ and viscosity μ. Derive an expression for Q in dimensionless form.

$$\left(Ans. \frac{Q}{D^2\sqrt{gD}} = \text{fn}\left(\frac{N\sqrt{D}}{\sqrt{g}}, \frac{H}{D}, \frac{\mu D^2}{\rho Q}\right)\right)$$

Similitude

6.19 A spillway model is constructed on a scale of 1 : 25. Calculate:

(a) the prototype discharge corresponding to a model discharge of 0.12 m^3/s;

(b) the model velocity corresponding to a prototype velocity of 3.5 m/s;

(c) the discharge per metre width in the prototype when the model discharge is 0.15 m^3/s and the length of the spillway model is 40 cm.

(*Ans.* (a) $Q_p = 375$ m^3/s, (b) $V_m = 0.7$ m/s, (c) $q_p = 46.88$ m^3/s per metre length)

6.20 A 1 : 36 model of a spillway crest records an acceleration of 1.5 m/s^2, a velocity of 0.5 m/s and a force of 0.30 N at a certain area of the model. What would be the values of the corresponding parameters in the prototype?

(*Ans.* $a_p = 1.5$ m/s^2, $V_p = 3.0$ m/s, $F_p = 14.0$ kN)

6.21 A spillway model is to be built to a scale of 1/50 across a flume of 60 cm width. The prototype is 15 m high and the maximum head on it is expected to be 1.5 m.

(i) What height of model and what head on the model should be used?

(ii) If the flow over the model at a particular head is 12 L/s, what flow per metre length of the prototype is expected?

(iii) If the maximum negative pressure head in the model is 20 cm, what is the corresponding negative pressure head predicted for the prototype? Is it practicable?

(*Ans.* Height of model = 0.30 m,
Head on the model = 0.03 m,
$q_p = 7071$ m^3/s per metre length,
$\left(\frac{p}{\gamma}\right)_p = -10.0$ m; not practicable)

6.22 A concrete open channel has Manning's $n = 0.014$. A 1/64 model of this channel is needed. Find the value of n for the model. Comment on the result.

[*Ans.* $n_m = 0.007$. It is not possible to get such a low value of n. It is better to go for a bigger model.]

6.23 A geometrically similar model of a spillway built to 1/50 scale is tested. The discharge and velocity of flow over the model were measured as 2.5 m^3/s and 1.5 m/s respectively. Find the corresponding discharge and velocity of flow in the prototype.

(*Ans.* $Q_p = 44194$ m^3/s, $V_p = 10.61$ m/s)

6.24 Model tests are to be conducted on a seawall constructed on a scale of 1 : 25. If the wave period in the prototype is 12 seconds what should be the corresponding wave period in the model? To what prototype force per metre length of wall would a model value of 200 N/m correspond? Assume $\rho_{model} = \rho_{prototype}$.

(*Ans.* $T_m = 2.4$ s, $F_p = 125$ kN/m)

6.25 A gravity fed lock in a navigational channel is to be studied with a 1/75 scale model.

(a) If the model lock fills up in 1.15 minutes estimate the corresponding time for the prototype.

(b) If the pressure at a point in the model is 0.5 kPa, what is the corresponding pressure in the prototype?

(*Ans.* $T_p = 9.96$ min, $p_p = 37.5$ kPa)

6.26 An offshore platform is known to encounter waves 4.5 m high at a frequency of 0.15 Hz and a steady current of 1.5 m/s. If a 1/30 model of the platform is to be built, determine the values of the above parameters in that model.

(*Ans.* $h_m = 0.15$ m, $f_m = 0.82$ Hz, $V_m = 0.274$ m/s)

6.27 A hydraulic jump occurring in a stilling basin is to be studied in a 1 : 36 scale model. The prototype jump has an initial velocity of 8 m/s, an entrance Froude number of 6.5 and a power loss of 1.7 kW per metre width of the basin. Calculate the corresponding model velocity, model Froude number and power loss per metre width of the model.

(*Ans.* $V_m = 1.33$ m/s, $(Fr)_m = 6.5$, $\mathbf{P}_m = 0.219$ W/metre width)

6.28 The resistance offered to the movement of a 2.0 m long ship model is a towing tank full of fresh water while moving with a speed of 1.5 m/s was 450 N.

(i) If the prototype is 60 m in length what will be the corresponding speed?

(ii) What would be the force required to drive at a corresponding speed, a prototype of 80 m length in sea water of relative density 1.025?

(*Ans.* $V_p = 8.22$ m/s, $F_p = 29520$ kN)

6.29 The drag of a submerged mine due to ocean currents is studied in a wind tunnel on a scale of 1 : 5. What wind velocity is needed to simulate a 2.0 m/s current? What prototype value would correspond to a model resistance of 10 N? [$\rho_{air} = 1.23$ kg/m^3, $\rho_{water} = 998$ kg/m^3, $\mu_{air} = 1.71 \times 10^{-5}$ Pa.s, $\mu_{water} = 1.0 \times 10^{-3}$ Pa.s].

(*Ans.* $V_m = 139$ m/s, $F_p = 42.1$ N)

6.30 Characteristics of a small underwater craft are studied under dynamic similarity conditions in a variable density wind tunnel on a model scale of 1 : 12. What prototype speed and power are indicated by model values of 100 m/s of velocity and 30 N of drag force? The model is operated at 8 atm pressure. [Take ρ_{air} at atmospheric pressure = 0.95 kg/m^3, $\mu_{air} = 2.17 \times 10^{-5}$ Pa.s, $\rho_{water} = 998$ kg/m^3 and $\mu_{water} = 1 \times 10^{-3}$ Pa.s].

[*Ans.* $V_p = 2.93$ m/s, (Power)$_p$ = 1422 W,

[***Hint:*** $\rho_m = 8 \times \rho_{atmos} = 8 \times 0.95$ kg/m^3]

6.31 If a 1.0 m diameter pipe carrying air at a velocity of 3.8 m/s is to be modelled for dynamic similarity by a water pipe of diameter 10 cm, what would be the velocity of water?

(*Ans.* $V_m = 1.335$ m/s)

6.32 A component of a airplane of length 3.0 m is tested in a variable density wind tunnel. The model has a length of 60 cm. If the model has the same speed as the prototype, what pressure in the wind tunnel, relative to local atmospheric pressure (p_a), is needed?

$$\left(Ans.\ \frac{p_m}{p_a} = 5\right)$$

6.33 A flowmeter to be installed in a 0.80 m pipe is to be tested in the laboratory by using a 1/4 model. If the same fluid as in the prototype is used in the dynamically similar model what would be the prototype pressure drop corresponding to a model value of 6 kPa? If the model discharge is 0.2 m^3/s, what would be the prototype discharge?

(*Ans.* $\Delta p_p = 375$ Pa, $Q_p = 0.8$ m^3/s)

6.34 A venturimeter fixed in a 60 cm pipe is being modelled to scale of 1 : 5 in a model using air as the working fluid. If $\rho_{air} = 1.1$ kg/m^3, $\rho_{water} = 998$ kg/m^3, $\mu_{air} = 1.95 \times 10^{-5}$ Pa.s $\mu_{water} = 1 \times 10^{-3}$ Pa.s, estimate the prototype discharge corresponding to a model discharge of 5 m^3/s.

(*Ans.* $Q_p = 1.413$ m^3/s)

6.35 Obtain expressions for the velocity ratio and force ratio similitude for a model which obeys Mach's law of similarity.

(*Ans.* $V_r = C_r = \sqrt{K_r / \rho_r}$, $F_r = L_r^2 K_r$)

6.36 If a model is so constructed as to have the same Weber number in the model and prototype, obtain expressions for the velocity ratio and force ratio similitude.

$$\left(Ans.\ V_r = \left(\frac{\sigma_r}{L_r \rho_r}\right), F_r = \sigma_r L_r\right)$$

6.37 If acceleration due to gravity, viscosity and surface tension are equally important in a model, show that for dynamic similarity the relationship between viscosity ratio μ_r, surface tension ratio σ_r and model scale ratio L_r is given by

$$\frac{\mu_r L_r^{1/2}}{\sigma_r} = 1$$

6.38 The pressure drop in an air duct depends on the length and diameter of the duct, the mass density, viscosity of the fluid and the velocity of the flow. Obtain an expression for the pressure drop in dimensionless form. Estimate the pressure drop in a 20 m long air duct if a model of the duct operating with water produces a pressure drop of 10 kN/m^2 over 10 m length. The scale ratio is 1 : 50.

$$\rho_{\text{water}} = 1000 \text{ kg/m}^3$$
$$\rho_{\text{air}} = 1.2 \text{ kg/m}^3$$
$$\mu_{\text{water}} = 0.001 \text{ N.s/m}^2$$
$$\mu_{\text{air}} = 0.0002 \text{ N.s/m}^2$$

$$\left(Ans.\ \frac{\Delta p}{\rho V^2} = \text{fn}\left(\frac{\mu}{\rho V D}, \frac{L}{D} \right); \Delta p_{\text{p}} = 133.3 \text{ N/m}^2 \right)$$

6.39 A 1/20 model of a ship having a submerged surface area of 5 m^2 and length of 8 m has a total drag of 20 N when towed through sea water at a velocity' of 1.5 m/s. Calculate the total drag on the prototype when moving at the corresponding speed. The skin friction can be estimated by $F_s = C_f A\rho V^2/2$ where the coefficient $C_f = \dfrac{0.0735}{(Re)^{1/5}}$.
[Take $\mu = 1.07 \times 10^{-3}$ Pa.s and $\rho = 1025$ kg/m^3].

(*Ans.* Total drag = 82.17 kN)

6.40 An ocean-going vessel has a length of 80 m and a submerged surface area of 1200 m^2. Its cruising speed is 30 km/h. A 1/20 model of this vessel is tested in a towing tank using fresh water and Froude law of similarity. A total drag force of 18.5 N was measured in the model at a velocity corresponding to the cruising speed. Calculate the total drag on the prototype at the cruising speed. The skin friction is estimated by $F_s = C_f A\rho V^2/2$ where

$$C_f = \frac{0.074}{(Re)^{1/5}}.\ \text{Use}$$

for fresh water:

$$\rho = 998 \text{ kg/m}^3$$
$$\mu = 0.001 \text{ Pa.s}$$

and for sea water $\rho = 1025$ kg/m^3

$$\mu = 1.07 \times 10^{-3} \text{ Pa.s}$$

(*Ans.* Total drag = 73.36 kN)

6.41 A model propeller with diameter $d = 0.5$ m is tested in water tunnel at a speed of $n = 360$ rpm when the flow velocity V = 2.5 m/s. The model produces a thrust of 250 N at a torque of 22 Nm. The prototype has a diameter of 4 m and operates at 100 rpm in water. If the significant non-dimensional group is (V/*nd*) only, determine (a) the flow velocity, (b) thrust produced and (c) the applied torque for the prototype.

(*Ans:* (a) $V_p = 0.556$ m/s, (b) $F_p = 460.8$ kN, (c) $T_p = 324.4$ kN.m)

Distorted Models

6.42 A distorted model of a river has a horizontal scale of 1/750. The slope in the model is 0.0025. If the prototype slope is 0.0001, estimate the prototype discharge corresponding to a model discharge of 0.01 m^3/s.

(*Ans:* $Q_p = 1232$ m^3/s)

6.43 A model of a river has a horizontal scale of 1/500 and vertical scale of 1/50. Fill up the following table pertaining to the above model—prototype pair.

SI. No.	Parameter	Prototype Value	Model Value
1	Cross-section area	100 m^2	?
2	Longitudinal slope	0.0001	?
3	Manning's roughness coefficient	?	0.05
4	Wave period	1 hour	?
5	Volume of water	?	0.5 m^3
6	Width	?	0. 2 m
7	Velocity	1.5 m/s	?
8	Depth	2.5 m	?

(*Ans:* (1) 0.04 m^2, (2) 0.001, (3) 0.030, (4) 50.9 s, (5) 6.25 Mm^3, (6) 100 m, (7) 0.353 m/s, (8) 0.05 m)

6.44 A river carries a discharge of 16,000 m^3/s of water at a depth of 8.0 and slope of 0.0025 when its width is 400 m. Above 15 km reach of this river is to be reproduced in the laboratory where 30 m long space is available. Determine the appropriate horizontal and vertical scales for this model. Also, determine the roughness scale and the model slope.

(*Ans:* $L_r = 1/500$, $h_r = 1/30$, $n_r = 2.316$ and $S_m = 0.04167$)

7

Laminar Flow

CONCEPT REVIEW

7.1 INTRODUCTION

The laminar flow of a fluid is one in which the fluid moves in layers (laminas), each layer sliding over the adjacent one. The flow is orderly and the shear stress is due to the action of fluid viscosity as given by Newton's law relating the shear stress, velocity gradient and viscosity. The viscous forces play a relatively prominent role on the fluid motion and any small disturbance that is produced is quickly damped out by viscous shear forces.

Example of laminar fluid motion are found in lubricating mechanisms; in flow of fluids in minute capillaries, blood veins and porous media; in flow of highly viscous fluids such as oils in tubes and conduits.

Critical Reynolds Number The Reynolds number value below which the flow can be definitely considered to the laminar is known as *Critical Reynolds Number*. The value depends upon the flow conditions and geometry of the flow. For circular conduits a value of $Re_{\text{crit}} = \left(\frac{VD}{\nu}\right)_{\text{crit}} = 2000$ is generally taken as the Critical Reynolds Number.

7.2 BASIC EQUATIONS

The basic equations which govern the motion of incompressible viscous fluid in laminar motion are called as *Navier–Stokes equations*. In Cartesian coordinates, for two-dimensional flow, these are:

$$\boxed{\rho\left(\frac{\partial u}{\partial t}+u\frac{\partial u}{\partial x}+v\frac{\partial u}{\partial y}\right)=X-\frac{\partial p}{\partial x}+\mu\left(\frac{\partial^2 u}{\partial x^2}+\frac{\partial^2 u}{\partial y^2}\right)} \tag{7.1}$$

$$\boxed{\rho\left(\frac{\partial v}{\partial t}+u\frac{\partial v}{\partial x}+v\frac{\partial v}{\partial y}\right)=Y-\frac{\partial p}{\partial y}+\mu\left(\frac{\partial^2 v}{\partial x^2}+\frac{\partial^2 v}{\partial y^2}\right)} \tag{7.2}$$

The continuity equation is:

$$\boxed{\frac{\partial u}{\partial x}+\frac{\partial v}{\partial y}=0} \tag{7.3}$$

These equations can be solved exactly for only a few simple flow situations.

An important result that can be obtained from the above for the two-dimensional, steady, uniform flows in the X-direction is

$$\boxed{\frac{\partial p}{\partial x}=\frac{\partial \tau}{\partial y}} \tag{7.4}$$

which states that in steady uniform flow the pressure gradient depends upon the existence of viscous shear stress and its variation across the flow.

7.2.1 Flow in Circular Conduits

Consider a horizontal circular pipe carrying an incompressible fluid in laminar motion, as illustrated in Fig. 7.1. The following relationships for the velocity distribution, shear stress and its distribution and for the head loss have been established analytically.

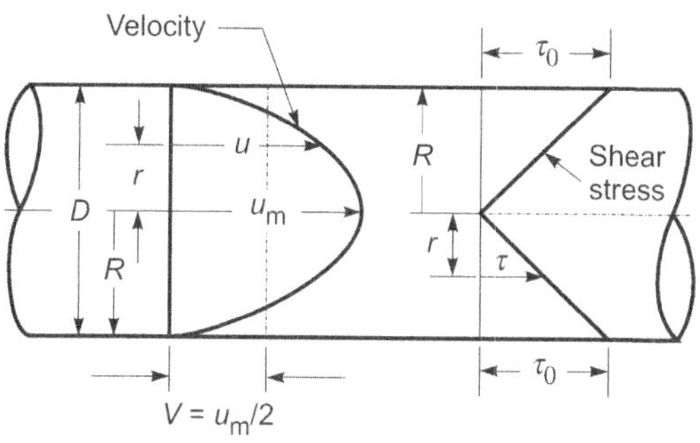

Fig. 7.1 *Laminar Flow in a Circular Conduit*

Velocity distribution: $$u = \left(-\frac{1}{4\mu}\frac{dp}{dx}\right)(R^2 - r^2) \quad (7.5)$$

Maximum velocity: $$u_m = \left(-\frac{1}{4\mu}\frac{dp}{dx}\right)R^2$$

Hence $$u = u_m\left[1 - \left(\frac{r}{R}\right)^2\right] \quad (7.6)$$

Mean velocity: $$V = \frac{u_m}{2} = \left(-\frac{1}{8\mu}\frac{dp}{dx}\right)R^2 \quad (7.7)$$

Shear stress at the boundary: $$\tau_0 = \frac{R}{2}\left(-\frac{dp}{dx}\right)$$

or $$\tau_0 = \frac{8\mu V}{D} \quad (7.8)$$

Variation of the shear stress: $$\tau = \tau_0 \frac{r}{R} \quad (7.9)$$

Pressure gradient $\left(-\frac{dp}{dx}\right)$:

For a horizontal pipe, for two sections 1 and 2 distance L apart,

$$\left(-\frac{dp}{dx}\right) = \left(\frac{p_1 - p_2}{L}\right) = -\frac{\Delta p}{L}$$

For inclined pipes, replace

$$\left(-\frac{dp}{dx}\right) \text{ by } \left(-\frac{d}{ds}(p + \gamma Z)\right)$$

i.e. by $\left(-\frac{\gamma dh}{ds}\right)$ where $h = p/\gamma + Z$ = piezometric head.

Here $$\left(-\frac{dh}{ds}\right) = \frac{h_1 - h_2}{L} = \frac{\left(\frac{p_1}{\gamma} + Z_1\right) - \left(\frac{p_2}{\gamma} + Z_2\right)}{L}$$

$$= -\frac{\Delta h}{L}$$

Head loss, h_f

Designating $h_f = -\Delta h$ = head loss in a length L

$$-\frac{dh}{ds} = \frac{h_f}{L}$$

Note that for a uniform flow the velocity is same everywhere and hence the energy loss = head loss = drop in piezometric head.

In general, the variation of the head loss h_f due to uniform laminar flow in a length L of a pipe of diameter D is given by,

$$h_f = \frac{32\mu VL}{\gamma D^2} \quad (7.10)$$

This equation is known as *Hagen–Poiseuille equation*. Since the mean velocity

$$V = \frac{Q}{\frac{\pi}{4}D^2} \quad \text{where } Q = \text{discharge}$$

$$h_f = \frac{128\mu QL}{\gamma\pi D^4} \quad (7.11)$$

Power, P

Power required to overcome a head H is

$$\mathbf{P} = \gamma QH \text{ watts.} \quad (7.12)$$

Hence, in laminar flow the power required to overcome frictional resistance in a pipe of length L and diameter D, carrying a discharge Q of a fluid of specific weight γ and viscosity μ is

$$\mathbf{P} = \gamma Q h_f = \frac{128\mu Q^2 L}{\pi D^4} \quad (7.13)$$

Friction factor, f

It is usual to designate the frictional resistance to flow in a pipe by *Darcy–Weisbach equation* as

$$h_f = \frac{fLV^2}{2gD} \quad (7.14)$$

where f = friction factor.

For laminar flow

$$h_f = \frac{32\mu VL}{\gamma D^2} = \frac{fLV^2}{2gD}$$

Hence $$f = \frac{32\mu VL}{\gamma D^2}\cdot\frac{2gD}{LV^2}$$

$$= \frac{64\mu}{\rho V D} = \frac{64}{Re} \quad (7.15)$$

or

$$\boxed{f = \frac{64}{Re}}$$

where $Re = \dfrac{VD}{\nu}$ = Reynolds number

7.2.2 Flow Between Two Stationary Parallel Plates

For uniform laminar flow between two stationary parallel plates separated by a distance B, (Fig. 7.2) an exact solution of the Navier–Stokes equations yields:

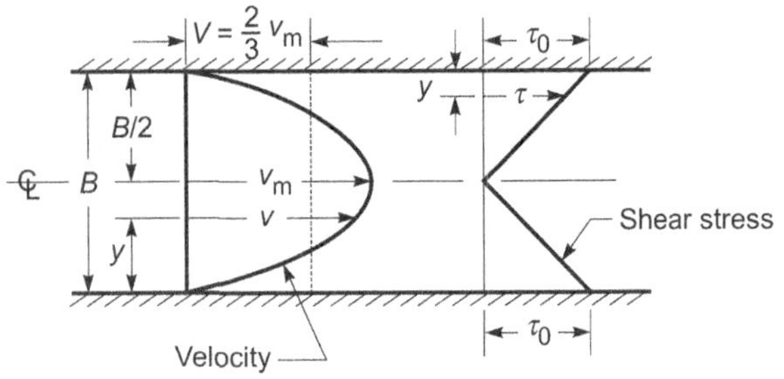

Fig. 7.2 *Laminar Flow Between Stationary Parallel Plates*

Velocity distribution

$$v = \left(-\frac{1}{2\mu}\frac{dp}{dx}\right)(By - y^2) \quad (7.16)$$

$$= v_m\left[2\left(\frac{y}{B/2}\right) - \left(\frac{y}{B/2}\right)^2\right] \quad (7.16(a))$$

Maximum velocity

$$v_m = \left(-\frac{1}{8\mu}\frac{dp}{dx}\right)B^2 \quad (7.17)$$

Average velocity

$$V = \frac{2}{3}v_m = \left(-\frac{dp}{dx}\right)\frac{B^2}{12\mu} \quad (7.18)$$

Shear stress at the boundary

$$\tau_0 = \left(-\frac{dp}{dx}\right)\frac{B}{2} = \frac{6\mu V}{B} \quad (7.19)$$

Variation of the shear stress

$$\tau = \left(-\frac{dp}{dx}\right)\left(\frac{B}{2} - y\right)$$

$$= \tau_0\left(1 - \frac{y}{B/2}\right) \text{ for } y < B/2 \quad (7.20(a))$$

and

$$\tau = \tau_0\left(\frac{y}{B/2} - 1\right) \text{ for } y > B/2 \quad (7.20(b))$$

The head loss h_f in a length L is

$$\boxed{h_f = \frac{12\mu VL}{\gamma B^2}} \quad (7.21)$$

[*Note:* As in laminar pipe flow case, for inclined flow use $\left(-\dfrac{d(p+\gamma Z)}{ds}\right)$ in place of $\left(-\dfrac{dp}{dx}\right)$ in the various formulae.]

7.2.3 Viscous Flow with a Free Surface

When a viscous uniform flow takes place in laminar regime down an inclined plane with a free surface (Fig. 7.3), the flow is similar to flow between two parallel plates. Here the depth of flow $d = B/2$ = half the spacing between the plates.

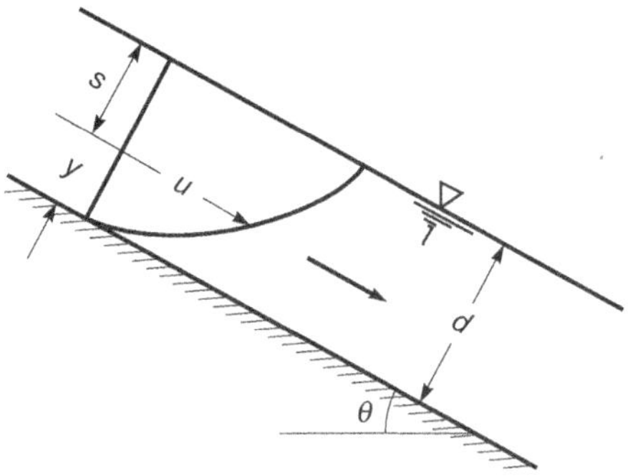

Fig. 7.3

On this basis the various parameters of the flow, viz. the velocity distribution and shear stress distribution can be estimated (see Example 7.20). For the head loss equation,

$$\boxed{\frac{h_f}{L} = S_0 = \sin\theta = \frac{3\mu V}{\gamma d^2}} \quad (7.22)$$

where S_0 = slope of the inclined plane
= slope of the liquid surface
= slope of the hydraulic grade line

7.2.4 Creeping Motion

Very slow motion of an object in an infinite expanse of a viscous fluid is known as creeping motion. For the case of a sphere of diameter D moving with a velocity V_0 in a viscous fluid, the creeping motion occurs at the Reynolds number

$$\boxed{Re = \frac{V_0 D}{\nu} \le 1.0} \quad (7.23)$$

Through an analytical procedure Stokes has shown that the net longitudinal force F exerted upon the sphere is

$$\boxed{F = 3\pi D\mu V_0} \qquad (7.24)$$

This equation, known as *Stokes Equation*, finds application in the determination of the fall velocity of small particles [For details see Chapter 9].

7.3 VISCOMETERS

A viscometer is a device for determining the viscosity of a liquid. Many of these instruments use laminar flow situations to estimate the viscosity of the liquid. The capillary tube viscometer utilizes the Hagen–Poiseuille equation to estimate the coefficient of viscosity μ of the liquid.

WORKED EXAMPLES

7.1 For a steady laminar flow in a horizontal circular pipe derive expressions for (i) the shear stress at any radial distance from the axis, (ii) the pressure drop along the pipe in terms of the boundary shear stress.

Solution:

(i) Consider a cylindrical element of fluid as shown in Fig. 7.4. Since the flow is

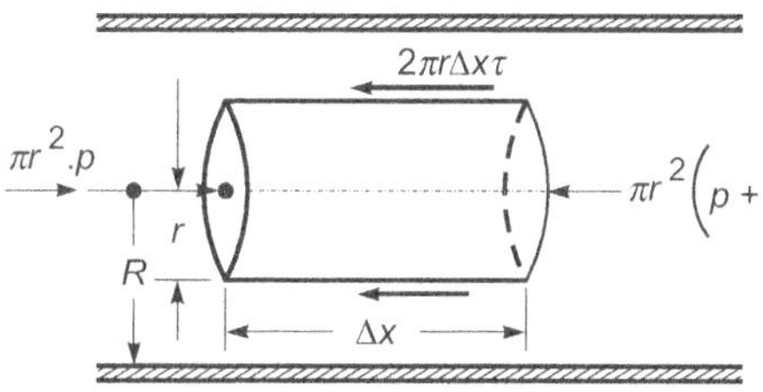

Fig. 7.4

steady and uniform, there is no acceleration and, as such, the sum of horizontal forces on the element must be zero. Hence

$$\pi r^2 p - \pi r^2 \left(p + \frac{dp}{dx}\Delta x\right) - 2\pi r \Delta x \tau = 0$$

Note that the shear stress τ at a radial distance r acts over the surface of the cylindrical element, the surface area of which is $(2\pi r \cdot \Delta x)$.

Simplifying $\qquad \tau = -\dfrac{dp}{dx}\dfrac{r}{2}$

At the boundary $r = R$, $\qquad \tau = \tau_0 = -\dfrac{dp}{dx}\dfrac{R}{2}$

At the centre line $r = 0$ $\qquad \tau_c = 0$

The variation of the shear stress τ with r is shown in Fig. 7.1.

(ii) $\dfrac{dp}{dx}$ is constant. Hence $\dfrac{dp}{dx} = \dfrac{\Delta p}{\Delta x} = \dfrac{\Delta p}{\Delta L}$ over a length L.

The pressure drop $-\Delta P = \dfrac{2\tau_0 L}{R} = \dfrac{4\tau_0 L}{D}$

7.2 Derive the Darcy–Weisbach equation for head loss due to a flow in a pipe in terms of the friction factor f. For laminar flow develop relations for the estimation of f and boundary shear stress in terms of flow parameters.

Solution: Considering two sections 1 and 2, distance L apart, as in Fig. 7.5,

$$\frac{p_1}{\gamma} + Z_1 + \frac{V_1^2}{2g} = \frac{p_2}{\gamma} + Z_2 + \frac{V_2^2}{2g} + h_f$$

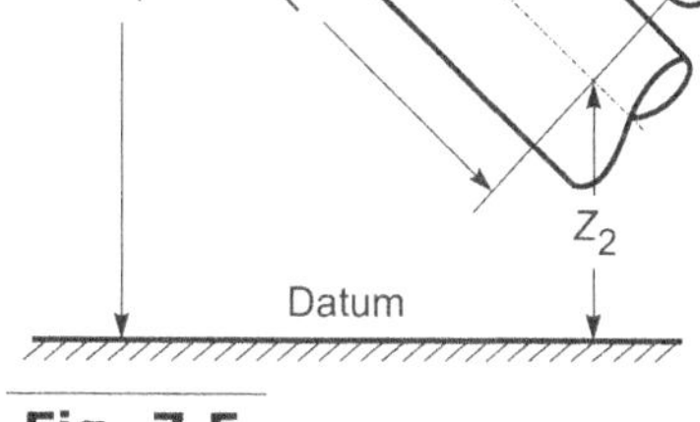

Fig. 7.5

As the flow is uniform $\qquad V_1 = V_2 = V$

$$\left(\frac{p_1}{\gamma} + Z_1\right) - \left(\frac{p_2}{\gamma} + Z_2\right) = h_f = -\Delta\left(\frac{p}{\gamma} + Z\right)$$

Considering the force balance between sections 1 and 2

$$p_1\frac{\pi D^2}{4} - p_2\frac{\pi D^2}{4} + \gamma(\pi R^2)L\sin\theta - \tau_0 2\pi R L = 0$$

But $L \sin \theta = (Z_1 - Z_2)$. Hence on simplifying

$$\left(\frac{p_1}{\gamma} + Z_1\right) - \left(\frac{p_2}{\gamma} + Z_2\right) = h_f = \frac{2\tau_0 L}{\gamma R} = \frac{4\tau_0 L}{\gamma D}$$

The shear stress τ_0 is expressed as

$$\tau_0 = \frac{f}{8}\rho V^2$$

where f = Darcy–Weisbach friction factor

$$\therefore \quad h_f = 2 \times \frac{f}{8}\rho V^2 \times L \times \frac{1}{\gamma D/2}$$

$$\mathbf{h_f = \frac{fLV^2}{2gD}}$$

This is known as *Darcy–Weisbach equation* and is valid for both laminar and turbulent flows.
In laminar flow by Hagen–Poiseuille equation

$$h_f = \frac{32\mu VL}{\gamma D^2}$$

Thus,

$$\frac{fLV^2}{2gD} = \frac{32\mu VL}{\gamma D^2}$$

By solving for f:

(i) $$f = \frac{64}{\rho VD/\mu} = \mathbf{\frac{64}{Re}}$$

(ii) Shear stress $$\tau_0 = \frac{f}{8}\rho V^2$$

$$= \frac{64}{\rho VD/\mu}\cdot\frac{1}{8}\cdot\rho V^2$$

$$\boldsymbol{\tau_0 = \frac{8\mu V}{D}}$$

7.3 Laminar flow takes place in a circular tube. At what distance from the boundary does the local velocity equal the average velocity?

Solution: The velocity distribution in laminar flow in a circular tube is given by

$$v = \left(-\frac{\gamma}{4\mu}\frac{dh}{ds}\right)(R^2 - r^2)$$

Average velocity $$V = \frac{1}{\pi R^2}\int_0^R v(2\pi r)dr$$

$$= \frac{2}{R^2}\left(-\frac{\gamma}{4\mu}\frac{dh}{ds}\right)\left(\frac{R^4}{2} - \frac{R^4}{4}\right)$$

$$= \frac{R^2}{8\mu}\left(-\gamma\frac{dh}{ds}\right)$$

Equating $V = v$

$$\frac{R^2}{8\mu}\left(-\gamma\frac{dh}{ds}\right) = (R^2 - r^2)\cdot\frac{1}{4\mu}\left(-\gamma\frac{dh}{ds}\right)$$

i.e. $$r^2 = \frac{R^2}{2}$$

or $$r = \frac{R}{\sqrt{2}} = 0.707\,R$$

Distance from the boundary $y = R - r$
$= \mathbf{0.293\,R}$

7.4 A viscous liquid was flowing in laminar regime in a 6 cm diameter circular pipe. A pilot tube at a radial distance of 2 cm from the axis indicated a velocity of 0.6 m/s. Calculate (a) the maximum velocity (b) the mean velocity and (c) the discharge in the pipe.

Solution: For a laminar flow $u = \left(-\frac{\gamma}{4\mu}\frac{dh}{ds}\right)(R^2 - r^2)$
$= K(R^2 - r^2)$

At $R = 0.06$ m and $r = 0.02$ m, $u = 0.6$ m/s
$0.6 = K(0.06^2 - 0.02^2)$
$K = 187.5$

(a) Maximum velocity $u_m = \left(-\frac{\gamma}{4\mu}\frac{dh}{ds}\right)R^2$
$= KR^2$
$= 187.5 \times (0.06)^2$
$= \mathbf{0.675\ m/s}$

(b) Mean velocity $= \frac{u_m}{2} = \mathbf{0.3375\ m/s}$

(c) Discharge $Q = \frac{\pi}{4} \times (0.06)^2 \times (0.3375)$
$= 9.5426 \times 10^{-4}\ m^3/s$
$= \mathbf{0.954\ L/s}$

7.5 Determine the kinetic energy correction factor for laminar flow in a round pipe.

Solution: $u = K(R^2 - r^2)$

where $K = \left(-\frac{1}{4\mu}\frac{dp}{dx}\right)$ for horizontal pipes

and $K = \left(-\frac{\gamma}{4\mu}\frac{d(p/\gamma + Z)}{ds}\right)$ for inclined pipes

Average velocity $$V = \frac{1}{\pi R^2}\int_0^R u2\pi r dr$$

$$= \frac{2\pi K}{\pi R^2}\int_0^R (R^2 r - r^3)dr$$

$$= \frac{2K}{R^2}\left[\frac{R^4}{2} - \frac{R^4}{4}\right] = \frac{K}{2}R^2$$

Kinetic energy correction factor

$$\alpha = \frac{1}{V^3 A}\int u^3 \mathrm{d}A$$

$$= \frac{1}{\left(\frac{K}{2}R^2\right)^3 \pi R^2}\int_0^R K^3 (R^2 - r^2)^3\, 2\pi r \mathrm{d}r$$

$$= \frac{16}{R^8}\int_0^R (R^2 - r^2)^3\, r\,\mathrm{d}r$$

$$= \frac{16}{R^8}\int_0^R [R^6 r - r^7 - 3R^4 r^3 + 3R^2 r^5]\mathrm{d}r$$

$$= 16\left[\frac{1}{2} - \frac{1}{8} - \frac{3}{4} + \frac{3}{6}\right]$$

$$= 16\left[\frac{12 - 3 - 18 + 12}{24}\right]$$

$$\alpha = \frac{16 \times 3}{24} = \mathbf{2.0}$$

7.6 Oil having a viscosity of 0.08 kg/ms flows through 100 mm diameter pipe. The laminar flow velocity distribution $u(r)$ is given by

$$u = 1.4\,[1 - (r/R)^2]$$

Where R = radius of the pipe and r = radial co-ordinate. (a) Determine the shear stress at the wall (b) Calculate the pressure drop for 1 m length of pipe.

Solution: $u = 1.4\,[1 - (r/R)^2]$

(a) $\tau = \mu \frac{\mathrm{d}u}{\mathrm{d}y} = -\mu \frac{\mathrm{d}u}{\mathrm{d}r}$ (Distance y is from the boundary and $= R - r$; $dy = -dr$)

$= 2\mu\,(1.4)r/R^2$

At the boundary, $r = R$ and

$$\tau_0 = \frac{2 \times 0.08 \times 1.4}{0.05} = \mathbf{4.48\ N/m^2}$$

(b) Maximum velocity u_m occurs at $r = 0$

Hence $u_m = 1.4$ m/s

Mean velocity $V = u_m/2 = 0.7$ m/s

Pressure drop $(-\Delta p) = (32\ \mu VL)/D^2$

Pressure drop per unit length of pipe.

$(-\Delta p)_1 = (32 \times 0.08 \times 0.7 \times 1.0)/(0.1)2$

$= \mathbf{179.2\ N/m^2}$

Alternate method for Part (b)

Pressure drop $(-\Delta p) = \frac{4\tau_0 L}{D}$ [See Examples 7.1 and 7.2]

Pressure drop per unit length of pipe,

$(-\Delta p)_1 = (4 \times 4.48 \times 1)/0.01 = \mathbf{179.2\ N/m^2}$

7.7 Oil of absolute viscosity 1.5 poise and relative density 0.85 flows through a 30 cm diameter pipe. If the head loss in 300 m length of pipe is 20 m, estimate

(a) the shear stress at the pipe wall.

(b) shear stress at a radial distance of 10 cm from the pipe axis.

(c) the friction factor f by assuming the flow to be laminar.

Solution: $\mu = 1.5$ poise $= 0.15$ Pa.s

$\rho = 0.85 \times 998 = 848.3$ kg/m^3

(a) Wall shear stress $\tau_0 = \frac{R}{2}\left(\frac{\gamma h_f}{L}\right)$

$$= \frac{0.30}{2}\left(\frac{848.3 \times 9.81 \times 20}{3000}\right)$$

$= 8.32$ Pa

(b) Shear stress τ at $r = 10$ cm

$$\tau = \frac{\tau_0}{R}r = \frac{8.32 \times 0.10}{0.15}$$

$= \mathbf{5.548\ Pa}$

(c) If the flow is laminar

$$h_f = \frac{32\mu VL}{\gamma D^2}$$

$$V = \frac{h_f}{L}\frac{\gamma D^2}{32\mu}$$

$$= \frac{20}{3000} \times \frac{848.3 \times 9.81 \times (0.30)^2}{32 \times 0.15}$$

$= 1.04$ m/s

Reynolds number $Re = \frac{VD\rho}{\mu} = \frac{1.04 \times 0.30 \times 848.3}{0.15}$

$= 1764.5 < 2000$

The assumption of laminar flow is therefore correct.

In laminar flow $f = \frac{64}{Re}$

$= 64/1764.5$

$= \mathbf{0.03627}$

7.8 Glycerine ($\mu = 1.50$ Pa.s and $\rho = 1260$ kg/m^3) flows at a velocity of 5.0 m/s in a 10 cm diameter pipe. Estimate: (a) the boundary shear stress in the pipe due to the flow (b) Head loss in a length of 12 m of pipe and (c) Power expended by the flow in a distance of 12 m.

Solution: Reynolds number

$$Re = \frac{\rho V D}{\mu} = \frac{1260 \times 5.0 \times 0.10}{1.50}$$

$$= 420$$

(a) As this value is less than 2000, the flow is laminar. In laminar flow in a conduit

$$\tau_0 = \frac{8\mu V}{D}$$

$$= \frac{8 \times 1.50 \times 5.0}{0.10} = \mathbf{600\ Pa}$$

(b) In laminar flow the head loss

$$h_f = \frac{32\mu V L}{\gamma D^2}$$

$$= \frac{32 \times 1.50 \times 5.0 \times 12}{(1260 \times 9.81)(0.1)^2} = \mathbf{23.3\ m}$$

(c) Power expended

$$\mathbf{P} = \gamma Q h_f$$

Discharge $Q = AV = \frac{\pi \times (0.1)^2}{4} \times 5.0$

$$= 0.03927\ m^3/s$$

Power $\mathbf{P} = (1260 \times 9.81) \times 0.03927 \times 23.3$

$$= 11309.8\ W = \mathbf{11.31\ kW}$$

7.9 Two tanks containing an oil of $\rho = 950$ kg/m^3 and $\mu = 8 \times 10^{-2}$ Pa.s are connected by a 15 cm diameter pipeline of length 200 m. Calculate the maximum difference in the elevation in the oil surfaces that can exist while maintaining laminar flow in the pipe.

Solution:

The maximum Reynolds number = Re_{crit}

$$= 2000 = \frac{VD}{\nu}$$

$$V = 2000 \times \frac{8 \times 10^{-2}}{950} \times \frac{1}{0.15}$$

$$= 1.123\ m/s$$

Head loss $h_f = \frac{32\mu V L}{\gamma D^2}$

$$= \frac{32 \times 8 \times 10^{-2} \times 1.123 \times 200}{(950 \times 9.81)(0.15)^2}$$

= **2.742 m**

= maximum difference in oil surface elevations

7.10 Oil is pumped through a 7.5 cm diameter pipe 1 km long. The dynamic viscosity of the oil is 0.1 N.s/m^2. A power input of 5.4 kW with an overall efficiency of 60% at the pump is noted. Determine the quantity of oil pumped per minute. Also find the Reynolds number of the flow if the relative density of the oil is 0.90.

Solution:

P = Power spent in fluid friction = 5.4 × 0.6 = 3.24 kW

$$= \frac{128\mu Q^2 L}{\pi D^4}$$

Therefore $$3240 = \frac{128 \times 0.1 \times Q^2 \times 1000}{\pi \times (0.075)^4}$$

$$Q^2 = 2.51611 \times 10^{-5}$$

$$Q = 0.00502\ m^3/s$$

$$= 0.00502 \times 1000 \times 60$$

$$= \mathbf{301.2\ L/min}$$

Velocity $$V = \frac{Q}{\frac{\pi}{4} \times D^2} = \frac{0.00502}{\frac{\pi}{4} \times (0.075)^2}$$

$$= 1.136\ m/s$$

Reynolds number $$Re = \frac{VD\rho}{\mu}$$

$$= \frac{1.136 \times 0.075 \times (0.90 \times 998)}{0.10}$$

= **765**

7.11 An oil of dynamic viscosity 1.5 poise and relative density 0.9 flows through a 3 cm vertical pipe. Two pressure gauges are fixed 20 m apart. The gauge A fixed at the top records 200 kPa and the gauge B fixed at the bottom records 500 kPa. Find the direction of flow and the rate of flow.

Solution: μ = 1.5 poise = 0.15 Pa.s

$$\rho = 0.9 \times 998 = 898.2\ kg/m^3$$

$$\gamma = 898.2 \times 9.81 = 8811\ N/m^3 = 8.811\ kN/m^3$$

Piezometric head at *A*

$$= h_a = \frac{P_a}{\gamma} + Z_a$$

$$h_a = \frac{200}{8.811} + 20$$ (with level at gauge B as the datum)

$$= 42.699$$

h_b = piezometric head at $B = \frac{p_b}{\gamma} + Z_b$

$$= \frac{500}{8.811} + 0 = 56.747\ m$$

Since $h_b > h_a$, **the flow is from B towards A, i.e. upwards.**

Head loss $= h_f = h_b - h_a = 56.747 - 42.699$

$= 14.048$ m

Assuming the flow as laminar

$$h_f = \frac{32\mu VL}{\gamma D^2}$$

$$14.048 = \frac{32 \times 0.15 \times V \times 30}{8811 \times (0.03)^2} = 18.159\ V$$

$$V = 0.7736 \text{ m/s}$$

Reynolds number $Re = \frac{VD\rho}{\mu} = \frac{0.7736 \times 0.03 \times 898.2}{0.15}$

$= 139$

As $Re < 2000$, the flow is laminar and our initial assumption is correct.

Discharge $Q = AV$

$$= \frac{\pi}{4} \times (0.03)^2 \times 0.7736$$

$$= 0.5468 \times 10^{-3} \text{ m}^3\text{/s}$$

$= \mathbf{0.5468\ L/s}$

7.12 Oil of relative density 0.9 and dynamic viscosity 2.5 poises is pumped through a 100 mm diameter pipe 500 m long at a rate of 2 L/s (a) Find the Reynolds number of the flow. (b) Calculate the pressure required at the pump if the outlet end, which is free, is at 20 m above the pump level. (c) What should be the power input if the overall efficiency of the pump set is 65%?

Solution:

$\mu = 2.5$ poise $= 0.25$ Pa.s

$\rho = 0.9 \times 998 = 898.2$ kg/m^3

$D = 100$ mm $= 0.1$ m

$Q = 2$L/s $= 0.002$ m^3/s

$$V = \frac{0.002}{\frac{\pi}{4} \times (0.1)^2} = 0.2546 \text{ m/s}$$

(a) Reynolds number

$$Re = \frac{\rho VD}{\mu}$$

$$= \frac{898.2 \times 0.2546 \times 0.1}{0.25}$$

$= \mathbf{91.49}$

As $Re < 2000$, the flow is laminar.

(b) Head loss due to friction:

$$h_f = \frac{32\mu VL}{\gamma D^2}$$

$$= \frac{32 \times 0.25 \times 0.2546 \times 500}{(898.2 \times 9.81)(0.1)^2}$$

$= 11.55$ m

If A is the pump end and B is the outlet, then

$$\frac{p_A}{\gamma} + 0 + \frac{V^2}{2g} = \frac{p_B}{\gamma} + 20 + \frac{V^2}{2g} + h_f$$

But $p_B = 0$ = atmospheric, as the outlet is free.

Also $h_f = 11.55$ m

$$\frac{p_A}{\gamma} = 20.0 + 11.55 = 31.55 \text{ m}$$

$p_A = 31.55 \times (898.2 \times 9.81)$

$= 278000$ Pa

$= \mathbf{278\ kPa}$

(c) Power required for pumping the fluid

$\mathbf{P} = \gamma QH$

where H = overall head

= static head + friction head

$= 31.55$ m

$\mathbf{P} = (898.2 \times 9.81) \times 0.002 \times 31.55$

$= 556$ W

As the overall efficiency of the pump set is 65%

Power input required

$$= \mathbf{P}_i = \frac{556}{0.65} = \mathbf{855.4\ W}$$

7.13 An 8 cm diameter pipe, 300 m long conveys oil of kinematic viscosity 1.5 stokes and mass density 900 kg/m^3. Assuming laminar flow, (a) find the rate of flow if it takes 5.0 kW of power input to a pump set of overall efficiency 60% to drive the flow, (b) verify whether the assumption of laminar flow is correct.

Solution:

(a) Power delivered to the fluid

$= 60\% \times 5000 = 3000$ W

Also, power $\mathbf{P} = \gamma Q h_f$

$$= \frac{\gamma Q 128\mu QL}{\gamma \pi D^4}$$

$$= \frac{128\mu Q^2 L}{\pi D^4}$$

Kinematic viscosity

$\nu = 1.5$ stokes $= 1.5 \times 10^{-4}$ m^2/s

Dynamic viscosity

$\mu = \rho\nu = 900 \times 1.5 \times 10^{-4}$

$= 0.135$ Pa.s

$$\mathbf{P} = \frac{128 \times 0.135 \times Q^2 \times 300}{\pi \times (0.08)^4} = 3000$$

$Q^2 = 7.4467 \times 10^{-5}$

$Q = 8.63 \times 10^{-3}$ m^3/s $= \mathbf{8.63\ L/s}$

(b) $$V = \frac{8.63 \times 10^{-3}}{\frac{\pi}{4} \times (0.08)^2} = 1.717 \text{ m/s}$$

$$Re = \frac{VD}{\nu} = \frac{1.717 \times 0.08}{1.5 \times 10^{-4}} = \mathbf{915.7}$$

As $Re < 2000$, the flow is laminar.

7.14 An oil of density 917 kg/m^3 is being pumped in a pipe of diameter 15 cm. The discharge is measured as 850 L/min. The drop in pressure in a stretch of 800 m of pipeline, both ends of which are at the same elevation, is measured as 95 kPa. Estimate the absolute viscosity of the oil.

Solution:

$$Q = 850 \text{ L/min} = \frac{850}{1000 \times 60}$$
$$= 0.01417 \text{ m}^3\text{/s}$$

$$V = \frac{0.01417}{\frac{\pi}{4} \times (0.15)^2} = 0.8017 \text{ m/s}$$

$$\text{Head loss} = \frac{\Delta p}{\gamma} = \frac{95{,}000}{917 \times 9.81} = 10.56 \text{ m}$$

Assuming the flow as laminar

$$h_f = \frac{32 \mu V L}{\gamma D^2}$$

$$10.56 = \frac{32 \times \mu \times 0.8017 \times 800}{(917 \times 9.81) \times (0.15)^2}$$
$$= 101.4\ \mu$$

$$\mu = \frac{10.56}{101.4} = \mathbf{0.1041 \text{ Pa.s}}$$

Reynolds number

$$Re = \frac{VD\rho}{\mu} = \frac{0.8017 \times 0.15 \times 917}{0.104}$$
$$= \mathbf{1059}$$

As this value of Re is less than 2000, the flow is laminar as assumed initially.

7.15 A large tank shown in Fig. 7.6 has a vertical pipe 70 cm long and 2 cm in diameter. The tank contains an oil of density 920 kg/m^3 and viscosity 1.5 poise. Find the discharge through the tube when the height of oil level of the tank is 0.80 m above the pipe inlet.

Solution:

$$\mu = 1.5 \text{ poise} = 0.15 \text{ Pa.s}$$

By energy equation between sections A and B,

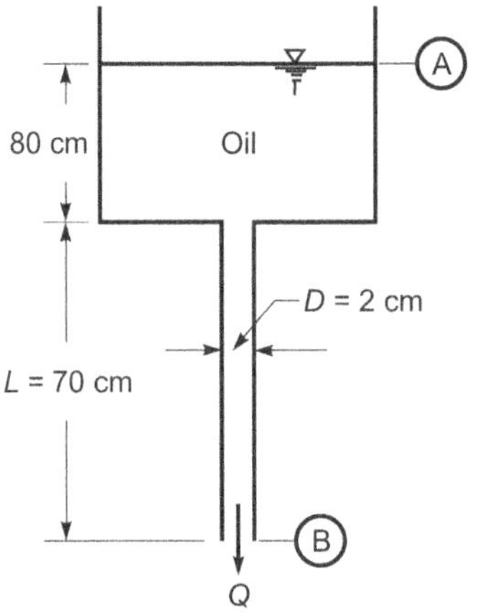

Fig. 7.6

$$\frac{p_a}{\gamma} + Z_a + \frac{V_A^2}{2g} = \frac{p_b}{\gamma} + Z_b + \frac{V_b^2}{2g} + h_f$$

where h_f = head lost in friction.

As the tank is large $V_A = 0$,

$$p_a = p_b = 0 = \text{atmospheric pressure}$$
$$Z_a - Z_b = 1.50 \text{ m}$$
$$h_f = 1.5 - \frac{V_b^2}{2g}$$

Assuming laminar flow in the pipe,

$$h_f = \frac{32 \mu V_b L}{\gamma D^2}$$

$$\therefore \quad 1.5 - \frac{V_b^2}{2 \times 9.81} = \frac{32 \times 0.15 \times 0.70\, V_b}{(920 \times 9.81) \times (0.02)^2}$$

(Note the length of the tube is 70 cm).

$$1.5 - \frac{V_b^2}{19.62} = 0.9307\ V_b$$

Solving for V_b, $V_b = 1.49$ m/s

Reynolds number $$Re = \frac{VD\rho}{\mu} = \frac{1.49 \times 0.02 \times 920}{0.15}$$
$$= 182.8$$

As this value of Re is less than 2000, the flow is laminar as assumed. Discharge

$$Q = AV$$
$$= \frac{\pi}{4} \times (0.02)^2 \times 1.49$$
$$= 4.681 \times 10^{-4} \text{ m}^3\text{/s}$$
$$= \mathbf{28.086 \text{ L/min}}$$

7.16 A lubricating oil of relative density 0.92 and dynamic viscosity 0.085 Pa.s is to be pumped at a rate of 40 L/s with energy gradient in laminar flow not exceeding 0.04. Determine the least diameter of the pipe to satisfy the above requirement.

Solution:

$$\rho = 0.92 \times 998 = 918.2 \text{ kg/m}^3$$

Assuming laminar flow with

$$\frac{h_f}{L} = 0.04$$

Then

$$\frac{h_f}{L} = \frac{128\mu Q}{\pi\gamma D^4}$$

$$0.04 = \frac{128 \times 0.085 \times 0.040}{\pi \times 9.81 \times 918.2 \times D^4}$$

$$= 1.5379 \times 10^{-5}/D^4$$

$$D^4 = 3.845 \times 10^{-4} \text{ or } D = 0.14 \text{ m}$$

$$= 14 \text{ cm}$$

$$V = \frac{0.040}{\frac{\pi}{4} \times (0.14)^2} = 2.5984 \text{ m/s}$$

$$Re = \frac{2.5984 \times 0.14 \times 918.2}{0.085} = 3929.7$$

As this value is > 2000, the flow is not laminar. Hence the required diameter is controlled by the critical Reynolds number, $Re_{(crit)} = 2000$.

$$Re = 2000 = \frac{Q}{\frac{\pi}{4}D^2} \times \frac{D \times \rho}{\mu}$$

$$= \frac{0.04}{\left(\frac{\pi}{4}\right)D} \times \frac{918.2}{0.085} = \frac{550.16}{D}$$

$$D = 0.275 \text{ m}$$

$$\frac{h_f}{L} = \frac{128\mu Q}{\pi\gamma D^4}$$

$$= \frac{128 \times 0.085 \times 0.04}{\pi \times 9.81 \times 918.2 \times (0.275)^4}$$

$$= 0.00269$$

As this is less than permissible gradient of 0.04 the diameter is acceptable.

Hence the least diameter = **27.5 cm**

7.17 A tank of surface area A containing a liquid is being emptied through a pipe of diameter D and length L. Obtain an expression for the time taken to lower the liquid surface from an elevation H_1 to H_2 measured above the outlet end (Fig. 7.7). The flow can be assumed to be laminar. Indicate any assumptions made in the derivation.

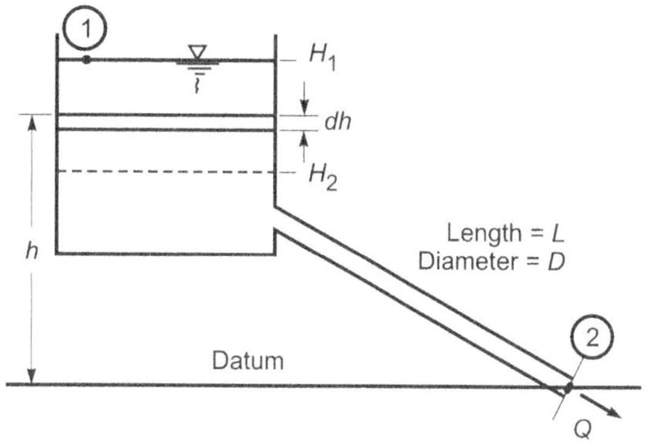

Fig. 7.7

Solution: Let at any instant t the liquid surface be at an elevation h above the datum drawn at the level of the outlet. In a time dt the liquid surface will drop by a height dh.

By neglecting (1) the velocity of the free surface in the tank, (2) velocity head at the outlet, and (3) all other minor losses, the total head loss due to friction $h_f = h$

Hence $$h_f = h = \frac{32\mu VL}{\gamma D^2} = \frac{128\mu QL}{\pi D^4\gamma}$$

By continuity

$$-A\,dh = Q\,dt$$

$$= \frac{\pi D^4\gamma}{128\mu L} h\,dt = Kh\,dt$$

where $K = \dfrac{\pi D^4\gamma}{128\mu L}$

$$-\frac{A}{K}\frac{dh}{h} = dt$$

Integrating

$$\int_0^T dt = T = \frac{A}{K}\int_{H_1}^{H_2} -\frac{dh}{h} = \frac{A}{K}\ln\frac{H_1}{H_2}$$

Hence $$\boldsymbol{T = \frac{128\mu AL}{\pi D^4\gamma} \cdot \ln\frac{H_1}{H_2}}$$

7.18 Calculate the diameter of a vertical pipe needed for flow of a liquid at a Reynolds number of 1200 when the pressure remains constant throughout the pipe. Kinematic viscosity of the fluid $\nu = 1.92 \times 10^{-3}$ m²/s.

Solution: Reynolds number

$$Re = \frac{VD}{\nu} = 1200$$

$$VD = 1200 \times 1.92 \times 10^{-3}$$

$$= 2.304 \qquad (1)$$

$$S_e = \text{Energy gradient}$$

$$= \left\{\left(\frac{p_1}{\gamma} + Z_1\right) - \left(\frac{p_2}{\gamma} + Z_2\right)\right\} \Big/ L$$

When pressure is constant $p_1 = p_2$

Also for a vertical pipe $(Z_1 - Z_2) = L$

$$S_e = (Z_1 - Z_2)/(Z_1 - Z_2) = 1.0$$

For laminar flow

$$S_f = \frac{h_f}{L} = \frac{32\mu V}{\gamma D^2}$$

$$= \frac{32\nu V}{gD^2}$$

In the present case

$$\frac{32\nu V}{gD^2} = 1$$

or

$$\frac{V}{D^2} = \frac{g}{32\nu} = \frac{9.81}{32 \times 1.92 \times 10^{-3}}$$

$$= 159.67 \qquad (2)$$

From (1) and (2)

$$\frac{V}{D^2} \Big/ (VD) = \frac{1}{D^3} = \frac{159.67}{2.304} = 69.3$$

$$\mathbf{D = 0.243\ m}$$

7.19 A water tank containing 5 m depth of water has a horizontal crack of thickness 0.12 mm extending all that way through the 50 cm thickness of the wall at that location. The water seeps through this crack to atmosphere. Estimate the leakage rate per minute per metre width of the crack.
[Kinematic viscosity of water = 0.01 stoke]

Solution: The flow can be considered as laminar flow between parallel plates of spacing B = 0.12 mm.

$$L = 0.50 \text{ m}$$

$$h_f = \text{head loss} = 5.0 \text{ m}$$

$$= \frac{12\mu VL}{\gamma B^2}$$

$$5.0 = \frac{12 \times (998 \times 0.01 \times 10^{-4}) \times V \times 0.50}{9790 \times (0.12 \times 10^{-3})^2}$$

$$= 42.475\ V$$

$$V = 0.1177 \text{ m/s}$$

Discharge per metre width

$$= q = 1 \times 0.1177 \times 0.12 \times 10^{-3}$$

$$= 1.4126 \times 10^{-5} \text{ m}^3\text{/s}$$

q = 0.848 L/min per metre width of crack

7.20 Considering the laminar flow of a liquid down an inclined plane with a free surface as half of the flow between two parallel plates obtain expressions for the velocity distribution, the mean velocity and the boundary shear stress.

Solution: Consider the laminar flow between two inclined plates spaced B apart and inclined at θ to the horizontal, as shown in Fig. 7.8. The velocity u at any y from the boundary is.

$$u = \frac{1}{2\mu}\left[-\frac{d(p+\gamma Z)}{dl}\right](By - y^2)$$

$$= \frac{\gamma}{2\mu}\left(-\frac{d\left(\frac{p}{\gamma} + Z\right)}{dl}\right)(By - y^2)$$

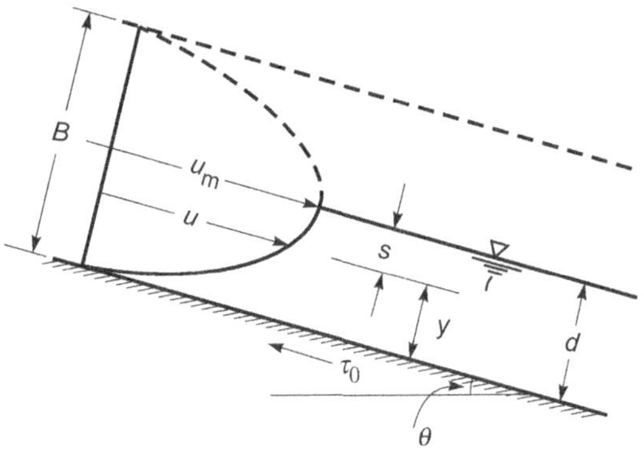

Fig. 7.8

For a free surface flow of depth d, the maximum velocity will be at $y = d$. Hence the depth d can be considered as $d = B/2$. Also on the free surface the pressure is atmospheric and hydraulic grade line coincides with the free surface.

Thus
$$\frac{-d\left(\frac{p}{\gamma} + Z\right)}{dl} = \frac{h_f}{L} = \sin\theta = S_0$$

Thus
$$u = \frac{\gamma}{2\mu}\sin\theta\,(2dy - y^2)$$

or if s = depth below the surface is used,

$$s = d - y \text{ and } (2dy - y^2) = (d^2 - s^2)$$

Thus
$$u = \frac{\gamma}{2\mu}(d^2 - s^2)\sin\theta$$

Maximum velocity

$$u_m = \frac{\gamma}{2\mu}d^2\sin\theta = \text{Surface velocity}$$

Mean velocity
$$= \frac{\gamma}{2\mu}\sin\theta\left[\frac{1}{d}\int_0^d (d^2 - s^2)\,ds\right]$$

$$V = \frac{\gamma}{3\mu} d^2 \sin \theta$$

Discharge per unit width

$$= Vd = \frac{\gamma}{3\mu} d^3 \sin \theta \qquad (1)$$

Shear stress on the bed

$$\tau_0 = \gamma d \sin \theta$$

Using result in Eq. 1,

$$\tau_0 = \frac{3\mu V}{d}$$

7.21 Oil of relative density 0.92 and dynamic viscosity 1.05 poise flows between two fixed parallel plates kept 1.2 cm apart. If the mean velocity is 1.40 m/s calculate (a) the maximum velocity (b) the boundary shear stress (c) velocity and shear stress at a distance of 0.2 cm from of the plates and (d) head loss in a distance of 25 m.

Solution:

(a) For two-dimensional laminar flow between parallel plates

$$u_m = \text{maximum velocity} = \frac{3}{2} V$$

$$= \frac{3}{2} \times 1.40 = \mathbf{2.10\ m/s}$$

(b) Since $V = \left(-\frac{dp}{dx}\right) \frac{B^2}{12\mu}$

$$\left(-\frac{dp}{dx}\right) = \frac{12\mu V}{B^2} = \frac{12 \times 0.105 \times 1.40}{(0.012)^2}$$

$$= 12250$$

Boundary shear stress

$$\tau_0 = \left(-\frac{dp}{dx}\right) \frac{B}{2}$$

$$= 12250 \times \frac{0.012}{2}$$

$$= \mathbf{73.5\ Pa}$$

(c) Shear stress τ at any y from the boundary

$$\tau = \left(-\frac{dp}{dx}\right)\left(\frac{B}{2} - y\right)$$

At $y = 0.002$ m:

$$\tau = (12250)\left(\frac{0.012}{2} - 0.002\right) = \mathbf{49\ Pa}$$

Velocity $v = \frac{1}{2\mu}\left(-\frac{dp}{dx}\right)(By - y^2)$

$$= \frac{1}{2 \times 0.105} \times 12250\,[0.012 \times 0.002 - (0.002)^2]$$

$$v = \mathbf{1.167\ m/s}$$

(d) Head loss $h_f = \frac{12\mu VL}{\gamma B^2}$

$$= \frac{12 \times 0.105 \times 1.4 \times 25}{(0.92 \times 998 \times 9.81)(0.012)^2}$$

$$= \mathbf{34.0\ m}$$

OBJECTIVE QUESTIONS

7.1 The equations of motion for laminar flow of a real fluid are known as
(a) Euler's equations
(b) Bernoulli equation
(c) Navier–Stokes equation
(d) Hagen–Poiseuille equation

7.2 Identify the *incorrect* statement
(a) Laminar flow is rotational
(b) Loss of head in laminar flow in a pipe is proportional to square of the velocity
(c) In laminar flow the friction factor f is directly proportional to the Reynolds number.
(d) In laminar flow in a conduit the shear stress varies linearly across the pipe.

7.3 In a two-dimensional, steady, horizontal, uniform laminar flow the shear gradient in the normal direction is equal to
(a) the velocity gradient in the normal direction.
(b) the velocity gradient in the longitudinal direction
(c) the pressure gradient in the normal direction.
(d) the pressure gradient in the direction of flow.

7.4 An oil of kinematic viscosity 0.25 stokes flows through a pipe of diameter 10 cm.
The flow is critical at a velocity of
(a) 7.2 m/s (b) 5.0 m/s
(c) 0.5 m/s (d) 0.72 m/s

7.5 Air ($\rho = 1.2$ kg/m^3 and $\mu = 1.80 \times 10^{-5}$ Pa.s) flows with a velocity of 20 m/s in a 10 cm diameter pipe. If the friction factor $f = 0.02$, the shear stress at the wall is
(a) Zero (b) 153.6 Pa
(c) 2.4 Pa (d) 1.2 Pa

7.6 In a steady flow of an oil in a pipe in laminar regime the shear stress is
(a) constant across the pipe
(b) maximum at the centre and decreases parabolically towards the sides
(c) zero at the boundary and increases linearly towards the centre
(d) zero at the centre and increases linearly towards the boundary.

7.7 Oil of viscosity 1.5 Pa.s and relative density 0.9 flows through a circular pipe a diameter 5 cm with a mean velocity of 1.2 m/s. The shear stress at the wall in Pa is
(a) 360 (b) 288
(c) 180 (d) 144

7.8 In a circular pipe of certain length carrying oil at a Reynolds number 100, it is proposed to triple the discharge. If the viscosity remains unchanged, the power input will have to be
(a) decreased to 1/3 its original value
(b) increased by 100%
(c) increased to 3 times the original value
(d) increased to 9 times its original value

7.9 In a steady laminar flow of a given discharge of a liquid through a circular tube of diameter D, the hydraulic gradient varies directly as
(a) D^{-4} (b) D^4
(c) D^{-2} (d) $D^{1/2}$

7.10 For the same discharge the power consumed in unit length in laminar friction varies directly as
(a) D^4 (b) D^{-4}
(c) D^{-2} (d) D^2

7.11 A liquid flowing in a pipe has a head loss of 2 m in a pipe length of 10 m. The Reynolds number of the flow is 100. If the flow rate is doubled and all other fluid properties remain the same, the head loss in m is
(a) 0.5 (b) 8.0
(c) 4.0 (d) 2.0

7.12 The pressure drop in an 8 cm horizontal pipe is 75 kPa in a distance of 15 m. The shear stress at the pipe wall, in kPa is
(a) 0.2 (b) 2.0
(c) 5.0 (d) 0.4

7.13 The Reynolds number for flow of an oil in a certain pipe is 640. The Darcy–Weisbach friction factor f for this flow is
(a) 0.02 (b) 0.01
(c) 0.1 (d) 0.064

7.14 The minimum value of friction factor f that can occur in laminar flow through a circular pipe is
(a) 0.025 (b) Zero
(c) 0.064 (d) 0.032

7.15 A 20 cm diameter pipe carries a fluid of relative density 0.9. If the boundary shear stress in the pipe is 0.50 Pa, the head loss in a length of 100 m of the pipe line is
(a) 11.35 m (b) 4.54 m
(c) 0.36 m (d) 9.08 m

7.16 The friction factor f in a laminar pipe flow was found to be 0.04. The Reynolds number of the flow was
(a) 2000 (b) 1000
(c) 800 (d) 1600

7.17 The shear stress in a fully developed laminar flow in a circular pipe is
(a) constant over the cross section
(b) varies parabolically across the section
(c) maximum at the pipe wall
(d) maximum at the pipe center

7.18 In a laminar flow through a circular pipe of diameter 20 cm, the maximum velocity is found to be 1 m/s. The velocity at a radial distance of 5 cm from the axis of the pipe will be
(a) 0.25 m/s (b) 0.50 m/s
(c) 0.75 m/s (d) 0.10 m/s

7.19 In a laminar flow through a circular tube of diameter D, carrying a laminar flow, the average velocity is indicated as the local velocity at a radial distance, measured from the centre, of
(a) $0.253D$ (b) $0.051D$
(c) $0.203D$ (d) $0.354D$

7.20 In a circular tube of radius R carrying a laminar flow, the ratio of average velocity to the maximum velocity in the conduit is
(a) 0.50 (b) 1.00
(c) 0.67 (d) 0.33

7.21 In uniform laminar flow through a conduit the hydraulic gradient varies
(a) as the square of the diameter
(b) inversely as the dynamic viscosity of the fluid
(c) inversely as the kinematic viscosity of the fluid
(d) as the velocity

7.22 The momentum correction factor β for laminar flow through a circular pipe is
(a) 1.5 (b) 2.0
(c) 1.67 (d) 1.33

7.23 The kinetic energy correction factor α for laminar flow through a circular pipe is
(a) 1.54 (b) 2.0
(c) 1.67 (d) 2.33

7.24 In a uniform laminar flow through a two-dimensional passage, the ratio of maximum velocity to the average velocity is
(a) 2.0 (b) 1.67
(c) 1.5 (d) 1.33

7.25 A laminar motion between two vertical parallel plates occurs in such a manner that the hydraulic grade line is vertical. This indicates that
(a) the viscosity is negligibly small
(b) the flow is with a free surface
(c) the pressure is same at all the sections
(d) the flow has stopped

7.26 A wall shear stress of 28 Pa exists in a laminar flow in an 8 cm diameter pipe. At a radial distance of 3 cm from the axis, the shear stress, in Pa, is
(a) 21.0 (b) 28.0
(c) 7.8 (d) 12.25

7.27 If the maximum velocity in a laminar flow through two parallel static plates is 9 m/s, then the average velocity of the flow will be
(a) 3.0 m/s (b) 4.5 m/s
(c) 6.0 m/s (d) 7.5 m/s

7.28 In a laminar flow between two parallel plates separated by a distance B, the head loss varies.
(a) directly as B
(b) inversely as B^2
(c) directly as B^2
(d) inversely as B^3

7.29 In a laminar flow between two parallel plates with a separation distance of 6 mm, the centre line velocity is 1.8 m/s. The velocity at a distance of 1 mm from the boundary is
(a) 0.15 m/s (b) 1.0 m/s
(c) 0.55 m/s (d) 0.75 m/s

7.30 In laminar flow between two fixed parallel plates, the shear stress is
(a) constant across the passage
(b) maximum at centre and zero at the boundary
(c) zero all through the passage
(d) maximum at the boundary and zero at the centre.

7.31 A fluid (RD = 0.9 and μ = 1.2 Pa.s) flows in laminar regime between two parallel plates fixed 3 cm apart. If the discharge is 600 cm^3/s/cm width of plate, the shear stress on the boundary, in Pa, is
(a) 800 (b) 640
(c) 480 (d) 240

7.32 In the laminar flow of a liquid down an inclined plane, the surface velocity is found to be 30 cm/s. The average velocity of the flow, in cm/s, is
(a) 20 (b) 30
(c) 15 (d) 10

7.33 In laminar flow between two parallel plates, the hydraulic grade line was found to be 0.05. If the discharge remains the same but the viscosity increases by 50%, the value of the new hydraulic grade line will be
(a) larger by 25% (b) smaller by 50%
(c) larger by 100% (d) larger by 50%

7.34 The velocity distribution for laminar flow between two parallel plate
(a) is constant over the whole cross section
(b) zero at the boundary and increases linearly towards the centre line
(c) varies linearly across the section with a maximum at the centre line
(d) varies parabolically across the section with a maximum at the centre line

7.35 The velocity profile for laminar flow of water between two parallel plates shown in Fig. 7.9 is given as

$$u = 0.01\,[1 - 1000\, y^2]$$

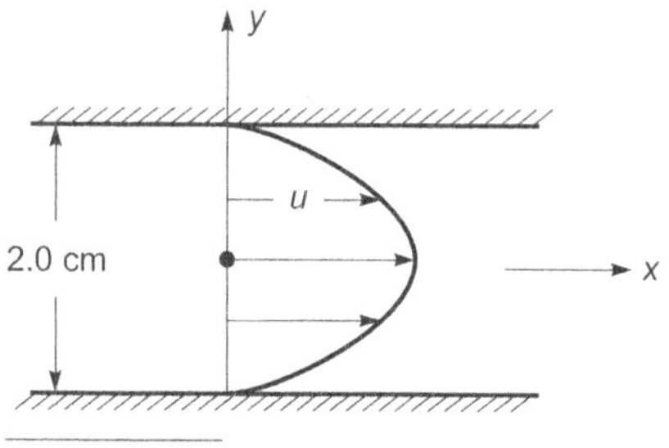

Fig. 7.9

where u is in m/s and y is in m. The viscosity of water can be assumed to be 10^{-3} Ns/m^2. The shear stress on each plate will be
(a) 2.0 N/m^2 (b) 0.002 N/m^2
(c) 0.04 N/m^2 (d) 0.004 N/m^2

7.36 The creeping motion obeys Stokes law up to a critical Reynolds number of value
(a) 0.001 (b) 1.0
(c) 100 (d) 2000

7.37 The Darcy law for flow through porous media is applicable to
(a) ideal fluids only
(b) turbulent flows only
(c) laminar flows only
(d) any type of liquid flow

7.38 The Darcy law relating means velocity and hydraulic gradient in a porous medium is applicable to
(a) water only
(b) ideal fluid only
(c) flow of all liquids in all regimes of flow
(d) laminar regime flow only.

PROBLEMS

7.1 An oil of dynamic viscosity 0.008 Pa.s and relative density 0.86 is to flow in a 6 cm diameter pipe. What is the maximum discharge that can be achieved while maintaining laminar flow? If crude oil (RD = 0.925 and dynamic viscosity = 0.09 Pa.s) is used instead, at the same velocity, would the flow be laminar?

(*Ans.* Q = 0.878 L/s; flow would still be laminar)

7.2 An oil of relative density 0.92 and dynamic viscosity 0.082 Pa.s flows in an 80 mm diameter pipe. In a distance of 20 m the flow has a head loss of 2 m. Calculate (i) the mean velocity, (ii) discharge, (iii) velocity and shear stress at a radial distance of 38 mm from the pipe axis and (iv) boundary shear stress.

(*Ans.* (i) V = 2.197 m/s. (ii) Q = 11.04 L/s; (iii) u = 0.4284 m/s; τ = 17.114 Pa (iv) τ_0 = 18.02 Pa)

7.3 It is required to maintain a shear stress of 3 Pa at the wall when water ($\nu = 1 \times 10^{-6}$ m²/s) flows with a head loss of 10 cm in 1000 m. What diameter pipe would achieve this? Is this applicable only to laminar flow?

(*Ans.* D = 12.26 cm; this is true for all regimes of flow.)

7.4 With laminar flow in a circular pipe, at what radial distance from the centre line does the local velocity equal one third the maximum velocity?

(*Ans.* r = 0.8165 R)

7.5 With laminar flow between two flat plates, at what distance from the centre line does the local velocity equal the mean velocity?

$$\left(Ans.\ \frac{y'}{B} = 0.2887 \text{ on either side of centre line}\right)$$

7.6 Show that the momentum correction factor for laminar flow in a circular tube is 1.33.

7.7 Determine the kinetic energy correction factor and momentum correction factor for laminar flow between two fixed parallel plates.

(*Ans.* α = 1.543, β = 1.20)

7.8 Estimate the power required in kW per kilometre length of a pipeline to overcome the viscous resistance to flow of crude oil (μ = 0.1 N.s/m²) through a horizontal 8.0 cm diameter pipe at the rate of 5 L/s.

(*Ans.* **P** = 2.487 kW)

7.9 What power will be required per kilometre length of a pipeline to overcome viscous resistance to the flow of an oil of viscosity 2.0 poises through a horizontal 10 cm diameter pipe at the rate of 200 L/min? Find the Reynolds number of the flow if the relative density of the oil is 0.92.

(*Ans.* Re = 194.8; **P** = 0.905 kW)

7.10 An oil of relative density 0.90 flows at a rate of 10.0 L/s through a horizontal pipe of 7.5 cm diameter. The pressure drop over a length of 300 m of pipe is found to be 40 N/cm². Estimate the viscosity of the oil. What is the Reynolds number of the flow?

(*Ans.* μ = 0.1035 Pa.s, Re = 1473)

7.11 A capillary tube of 1.5 mm diameter and 15 cm long is connected horizontally to a tank 6 cm in diameter. The tank contains oil up to a height of 9.0 cm above the axis of the capillary tube. When the oil is allowed to discharge through the capillary to atmosphere, it takes 10 min to discharge 85 cm³ of oil. Estimate the kinematic viscosity of the oil.

(*Ans.* $\nu = 1.77 \times 10^{-6}$ m²/s)

7.12 A capillary viscometer has a capillary of diameter 2.0 mm and length 50 cm. A liquid of density 850 kg/m³ and dynamic viscosity 0.5 poise is sent through the capillary under a constant pressure difference of 6 kPa. Find the time taken to collect 50 cm^3 of liquid at the capillary outlet.

(*Ans.* T = 8 min 50.5 s)

7.13 A liquid of dynamic viscosity 0.07 Pa.s and relative density 0.86 flows through an inclined pipe of 2 cm diameter. A discharge of 13 L/min is to be sent through the pipe in such a manner that the pressure along the length is constant. Find the required inclination of the pipe.

$\left(\textit{Hint}\text{: For constant pressure the slope} = \sin\theta = \frac{h_f}{L} \right)$

(*Ans.* $\theta = 27° \, 18' \, 17''$)

7.14 Determine a suitable diameter for a horizontal pipeline which is to convey 50 L/s of an oil through a distance of 250 m, if the pressure loss is not to exceed 150 kPa. At the operating temperature the relative density of the oil is 0.9 and the dynamic viscosity 1.3 Pa.s.

(*Ans.* D = 25.7 cm; in practice the next larger standard size is used.)

7.15 Calculate the least diameter of a pipe to carry 10 L/s of an oil of density 900 kg/m^3 and dynamic viscosity 1.5 poise with a permissible energy gradient in laminar flow regime of 0.03.

(*Ans.* D = 12.3 cm)

7.16 An oil of relative density 0.92 and dynamic viscosity 0.9 poise flows though a 10 cm diameter pipe 30 m long. Determine the largest flow that can be passed through this pipe while maintaining laminar regime. What is the head loss between the two ends of the pipe under this flow?

(*Ans.* Q = 15.4 L/s and h_L = 1.88 m)

7.17 A liquid of relative density 0.85 flows in a pipe 8 cm is diameter with a velocity of 0.8 m/s in laminar regime. If two pressure gauges located at the ends of a 12 m long horizontal stretch of pipe record a pressure difference of 60 kPa, estimate the (i) viscosity of the fluid and (ii) Reynolds number of the flow.

(*Ans.* μ = 1.25 Pa.s; Re = 43.4)

7.18 For a steady fully developed laminar flow of an oil of density ρ_f through two pipes in series as shown in Fig. 7.10 find the ratio h_1/h_2 of the manometer fluid deflections Consider only friction losses in the pipes.

$$\left(\textit{Ans.} \frac{h_1}{h_2} = \left(\frac{D_2}{D_1}\right)^4 \left(\frac{L_1}{L_2}\right) \right)$$

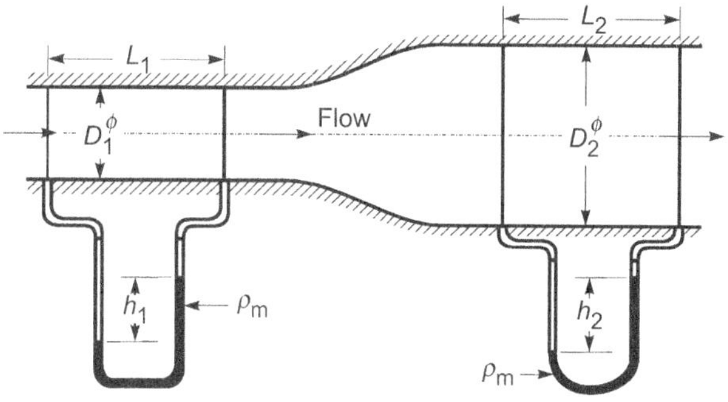

Fig. 7.10

7.19 Calculate the force F required on the piston to discharge 500 mm^3/s of water through a syringe (Fig. 7.11) taking into account the frictional loss in the syringe needle only. Assume fully developed laminar flow in the needle. (Take $\mu_{water} = 10^{-3}$ Ns/m^2)

(*Ans.* F = 1.024 N)

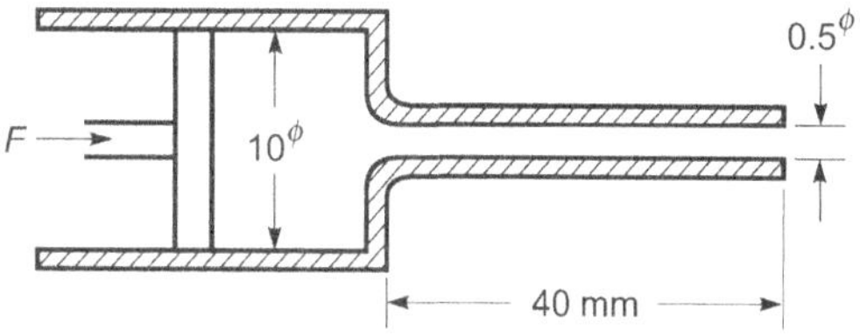

Fig. 7.11

7.20 A piston of area A is forced with a force F into a cylinder full of oil of density ρ and viscosity μ. If there are n tubular openings of diameter d in the piston, show that the rate of descent of the piston is

$$V = \frac{n\pi F d^4}{128 \mu A^2 L}$$

where L = length of the piston. Assume that no oil leaks between the piston and the cylinder. Further, laminar flow can be assumed through the tubular openings.

7.21 A pump delivers a lubricating oil of relative density 0.90 and dynamic viscosity 0.85 poise through 25 m of 30 mm pipe to a tank whose oil surface is 10 m higher than the oil surface of the supply tank. The pump efficiency is 70%. (a) Estimate the power input required to pump oil at a rate of 250 L/min (b) What power input is required to pump the liquid at a Reynolds number of 2000 in this system?

(*Ans.* (a) **P** = 3.18 kW; (b) **P** = 3.6 kW)

7.22 A 5 cm diameter pipe carries lubricating oil of relative density 0.92 and dynamic viscosity 2.0 poise in a vertical pipe. Two pressure gauges are connected 25 m apart. The upper gauge records 220 kPa and the lower gauge records 350 kPa. Find the direction and rate of flow.

(*Ans.* Flow is upwards: Q = 175 L/min)

7.23 A flow of 60 L/s per metre width of glycerine of relative density 1.25 and dynamic viscosity 1.5 Pa.s takes place between two parallel plates having a gap of 25 mm between them. Calculate the (i) maximum velocity, (ii) boundary shear stress and (iii) energy gradient.

(*Ans.* u_m = 3.6 m/s; τ_0 = 864 Pa and h_f/L = 5.648)

7.24 A masonry wall of a water tank is 0.90 m thick. At the bottom a crack of thickness 0.3 mm and 60 cm wide has developed and the crack extends to the entire thickness of the wall. If the tank contains 4 m of water above the crack and the other end of the crack is at atmospheric pressure, estimate the leakage volume per day from the crack (ν = 1 centistoke.)

(*Ans.* Q = 5.09 m^3/day)

7.25 Solve Example 7.15 by neglecting the exit velocity. Estimate the error in the discharge due to this assumption.

(*Ans.* E_Q = 8.12%)

7.26 An oil having a viscosity of 0.098 N.s/m^2 and a relative density of 1.59 flows through a horizontal pipe of 5 cm diameter with a pressure drop of 0.3 N/cm^2 per metre length of pipe. Determine the (i) rate of flow, (i) shear stress at the pipe wall and (iii) power required for 100 m of pipe to maintain the flow.

(*Ans.* (i) Q = 4.696 L/s; (ii) τ_0 = 37.51 Pa; (iii) **P** = 1.41 kW)

7.27 A fluid film of RD = 0.90 and thickness 2.0 mm flows down a vertical surface at a surface velocity of 0.45 m/s. Estimate the (i) discharge in cm^3/s/cm width, (ii) viscosity of the fluid and (iii) the boundary shear stress.

(*Ans.* (i) 6 cm^3/s/cm; (ii) μ = 0.0392 Pa.s; (iii) τ_0 = 17.62 Pa)

7.28 A film of liquid moves down a plane inclined at 60° to the horizontal. The surface velocity of the film, of thickness 3 mm, is found to be 3.2 cm/s. If the dynamic viscosity of the liquid is 1.5 Pa.s, calculate the (i) boundary shear stress and (ii) specific weight of the liquid.

(*Ans.* (i) τ_0 = 31.95 Pa; (ii) γ = 12317 N/m^3)

8

Boundary Layer Concepts

CONCEPT REVIEW

8.1 INTRODUCTION

A *boundary layer* is a very narrow region of flow near a boundary in which, for a fluid flow with relatively large Reynolds number, all the viscous effects can be considered to be confined. The flow outside the boundary layer can be considered to be ideal fluid flow. Figure 8.1 shows the flow past a sharp edged smooth plate. The layer in which the flow is retarded, and consequently the local velocity u at any y above the plate is less than U in the boundary layer. A typical velocity distribution in a boundary layer is shown in Fig. 8.2.

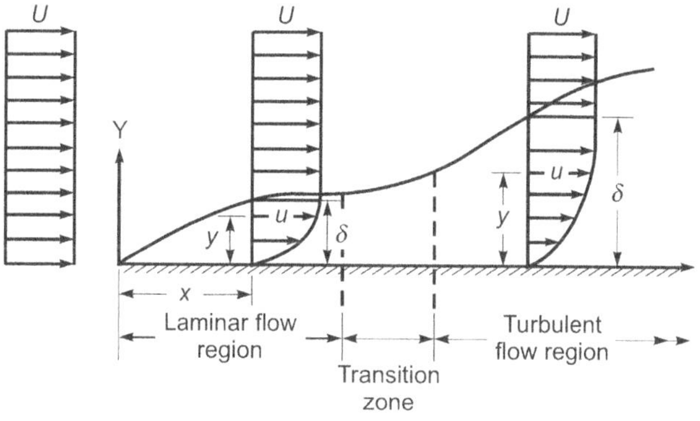

Fig. 8.1 *Boundary Layer Growth*

As u reaches the free stream velocity U asymptotically, the boundary layer thickness δ is defined as the value of y at which $u = 0.99\ U$.

$Re_x = \dfrac{Ux}{\nu}$ is called the local Reynolds number, where

ν = kinematic viscosity of the fluid.

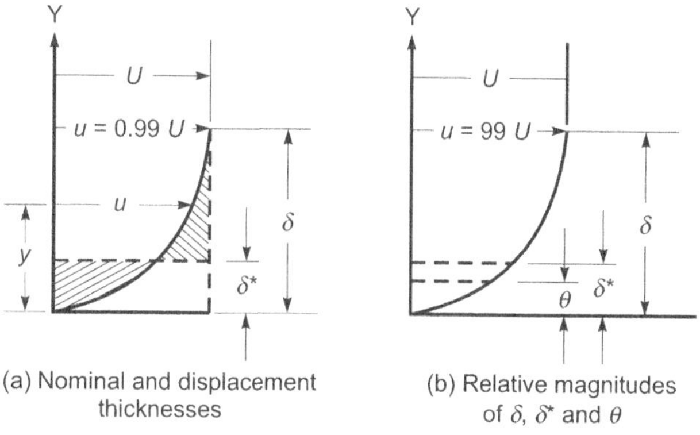

(a) Nominal and displacement thicknesses

(b) Relative magnitudes of δ, δ^* and θ

Fig. 8.2 *Boundary Layer Thickness*

Initially the boundary layer will be laminar. But around a value of $Re_x = 5 \times 10^5$ the flow in the boundary layer will undergo a transition phase and soon becomes turbulent. A boundary layer in which the flow is turbulent is called *turbulent boundary layer*. In a laminar boundary layer the velocity distribution (i.e. variation of u with y) is parabolic, while it is logarithmic in a turbulent boundary layer.

In boundary layer theory the following thicknesses of the boundary layer are defined and used:

Nominal thickness, δ: It is the thickness, measured from the boundary at which the x-component velocity attains 99% of the free stream velocity U, i.e. the value of y at which $u = 0.99U$.

Displacement thickness, δ^*: It is defined as

$$\boxed{\delta^* = \int_0^{\delta}\left(1 - \frac{u}{U}\right)dy}$$

Momentum thickness, θ: It is defined as

$$\theta = \int_0^{\delta} \frac{u}{U}\left(1 - \frac{u}{U}\right) dy$$

Shape factor, *H*: The ratio $H = \delta^* / \theta$ is said to be the shape factor of the boundary layer.

8.2 BOUNDARY CONDITIONS

For a laminar boundary layer, the boundary conditions are:

1. At the wall $y = 0$, $u = 0$ and $v = 0$.
2. At the outer edge $y = \delta$, $u = U$.
3. Shear stress at the wall, $\tau_0 = \mu \left.\frac{\partial u}{\partial y}\right|_{y=0}$

The flow over a flat plate which is described in this section is a particular case in which U = constant or the pressure gradient $\frac{dp}{dx}$ is zero. This case is also known as *zero pressure gradient flow*.

There are situations in which the pressure gradient can be favourable $\left(\frac{dp}{dx} \text{ is } -\text{ve}\right)$ or adverse $\left(\frac{dp}{dx} \text{ is } +\text{ve}\right)$ in which $U = f(x)$. These are beyond the scope of this book. The boundary layer thickness δ and the local shear stress τ_0 are functions of x.

8.3 LAMINAR BOUNDARY LAYER OVER A FLAT PLATE

For laminar flow over a flat plate, Blasius solved the basic boundary layer equations and obtained analytical solution which have been verified experimentally to be remarkably accurate. The classic *Blasius solution* for laminar bound-ary layer are:

$$\frac{\delta}{x} = \frac{5.0}{\sqrt{Re_x}} \qquad (8.1)$$

where $Re_x = \frac{Ux}{\nu}$

By defining $\tau_0 = C_f \frac{\rho U^2}{2}$ (8.2)

where C_f = local shear stress coefficient.

we have

$$C_f = \frac{0.664}{\sqrt{Re_x}} \qquad (8.3)$$

If the total drag force on one side of a plate of length L and width B is defined as,

$$F_D = B \int_0^L \tau_0 \, dx = C_{Df}(L \cdot B) \frac{\rho U^2}{2}$$

then

$$C_{Df} = \frac{1.328}{\sqrt{Re_L}} \qquad (8.4)$$

where $Re_L = \frac{UL}{\nu}$ and C_{Df} = total frictional drag coefficient.

8.4 KARMAN MOMENTUM INTEGRAL FORMULATION

This is an approximate but simple method of solving boundary layer equations. By the application of momentum principle to a steady boundary layer over a flat plate it can be shown that,

$$\frac{\tau_0}{\rho U^2} = \frac{1}{2} C_f = \frac{\partial}{\partial x}\left(\int_0^{\delta} \frac{u}{U}\left(1 - \frac{u}{U}\right) dy\right) = \frac{\partial \theta}{\partial x} \qquad (8.5)$$

Putting $\frac{u}{U} = f(\eta)$ where $\eta = y/\delta$,

$$\tau_0 = \rho U^2 \frac{\partial \delta}{\partial x} \int_0^1 f(\eta)(1 - f(\eta))\, d\eta$$

Let

$$\int_0^1 f(\eta)\,(1 - f(\eta))\, d\eta = \alpha$$

then

$$\tau_0 = \rho U^2 \alpha \frac{\partial \delta}{\partial x} \qquad (8.6)$$

From the boundary conditions for a laminar boundary layer,

$$\tau_0 = \mu \left(\frac{\partial u}{\partial y}\right)_{y=0} = \frac{\mu U}{\delta}\left[\frac{df(\eta)}{d\eta}\right]_{\eta=0}$$

Putting

$$\left.\frac{df(\eta)}{d\eta}\right|_{\eta=0} = \beta$$

$$\tau_0 = \frac{\mu U \beta}{\delta} \qquad (8.7)$$

Equating the two expressions for τ_0,

$$\rho U^2 \alpha \frac{d\delta}{dx} = \frac{\mu U \beta}{\delta}$$

or

$$\delta\, d\delta = \frac{\mu}{\rho} \frac{1}{U}\left(\frac{\beta}{\alpha}\right) dx$$

Integrating with the boundary condition ($x = 0$, $\delta = 0$)

$$\delta = \sqrt{2(\beta/\alpha)}\,\frac{x}{\sqrt{Re_x}}$$

or

$$\frac{\delta}{x} = \frac{\sqrt{2(\beta/\alpha)}}{\sqrt{Re_x}} \tag{8.8}$$

Substituting the value of δ in Eq. (8.7) for τ_0, and simplifying,

$$C_f = \frac{\tau_0}{\rho U^2/2} = \sqrt{2\alpha\beta}\,\frac{1}{\sqrt{Re_x}} \tag{8.9}$$

Knowing that the drag force on side of the plate, for a plate of unit width

$$F_D = \int_0^L \tau_0 \, dx$$

$$F_D/(0.5\,\rho U^2 L) = C_{Df} = \sqrt{2\alpha\beta}\,/\sqrt{Re_L} \tag{8.10}$$

Example 8.5 illustrates in detail the use of this method for a specific $f(\eta)$.

In Table 8.1, some of the commonly adopted forms of $u/U = f(\eta)$ and the corresponding boundary layer parameters δ, C_f, C_{Df}, obtained by using the Karman momentum integral equations, are given. After studying the Example 8.5, the reader is advised to derive all the elements listed in Table 8.1 as a good exercise.

Table 8.1 *Laminar Boundary Layer Solutions for Various Velocity Profiles*

($y/\delta = \eta$)

$u/U = f(\eta)$	$(\delta/x)\sqrt{Re_x}$	$(\delta^*/x)\sqrt{Re_x}$	$C_f\sqrt{Re_x}$	$C_{Df}\sqrt{Re_L}$
Exact (Blasius)	5.00	1.729	0.664	1.328
$2\eta - 2\eta^3 + \eta^4$	5.84	1.752	0.686	1.372
$\frac{3}{2}\eta - \frac{1}{2}\eta^3$	4.64	1.740	0.626	1.292
$2\eta - \eta^2$	5.48	1.826	0.730	1.460
$\sin\left(\frac{\pi}{2}\eta\right)$	4.80	1.741	0.654	1.308
η	3.46	1.730	0.577	1.154

8.5 BOUNDARY CONDITIONS FOR A PROPER $f(\eta)$

A proper function $u/U = f(\eta)$ must satisfy the following essential and desirable boundary conditions:

		Essential	Desirable
At the wall,	$y = 0$;	$u = 0$	$\partial^2 u/\partial y^2 = 0$
At	$y = \delta$;	$u = U$	$\partial u/\partial y = \partial^2 u/\partial y^2 = 0$

8.6 TRANSITION FROM LAMINAR BOUNDARY LAYER

As the flow passes down the plate, i.e. as Re_x increases the boundary layer thickness increases and soon it becomes unstable. Turbulence persists and grows in the boundary layer at higher values of Re_x. It is generally believed that the transition from laminar to turbulent boundary layer takes place between $Re_x = 1.3 \times 10^5$ and 4×10^6, with the mean value of $Re_x = (Re_x)_{crit} = 5 \times 10^5$ taken as the commonly accepted *critical Reynolds number.*

In a flow past a long plate, the initial part in the boundary layer up to x_{crit} will be laminar and then onwards the flow of the boundary layer will be turbulent.

8.7 TURBULENT BOUNDARY LAYER

The turbulent boundary layer will have much more steeper velocity gradients at the boundary than the laminar boundary layer. The velocity distribution is logarithmic and could be conveniently expressed in the form of a power law, $u/U = (y/\delta)^{1/n}$ over a range of Reynolds number. The power n can be 5 to 10, depending on the Reynolds number range.

Next to the boundary, in a turbulent boundary layer over a smooth bed, there exists a thin layer called as *laminar sublayer.*

For Re_x between 5×10^6 and 2×10^7 the velocity distribution can be expressed by the 1/7 power law, $u/U = (y/\delta)^{1/7}$. The turbulent boundary layer characteristics found by experiments and analytical calculations, to be valid for $5 \times 10^5 < Re_x < 2 \times 10^7$, are

$$\boxed{\delta/x = 0.377/Re_x^{1/5}} \tag{8.11}$$

$$\boxed{C_f = 0.059/Re_x^{1/5}} \tag{8.12}$$

$$\boxed{C_{Df} = 0.074/Re_L^{1/5}} \tag{8.13}$$

The above formulae assume the boundary layer to be turbulent from $x = 0$. To account for initial laminar boundary layer, C_{Df} can be calculated by

$$\boxed{C_{Df} = (0.074/Re_L^{1/5}) - (1700/Re_L)} \tag{8.14}$$

For higher Reynolds numbers ($10^7 < Re_x < 10^9$), the logarithmic form of velocity distribution in the turbulent boundary layer is more appropriate. The boundary layer shear coefficients are expressed by the following formulae given by *Schlichting*:

$$\boxed{C_f = 0.370/(\log Re_x)^{2.58}} \tag{8.15}$$

$$\boxed{C_{Df} = 0.455/(\log Re_L)^{2.58}} \tag{8.16}$$

The boundary layer thickness δ is estimated by

$$\delta / x = 0.22 / Re_x^{1/6} \tag{8.17}$$

If correction for initial laminar boundary layer is applied then,

$$C_{Df} = 0.455/(\log Re_L)^{2.58} - 1700/Re_L \tag{8.18}$$

The term $1700/Re_L$ is so small that omitting it does not cause any appreciable error.

In Fig. 8.3, the values of C_{Df} are plotted against the Reynolds number Re_L as obtained from various equations, e.g. 8.4, 8.13, 8.14 and 8.16 for various regimes of flow.

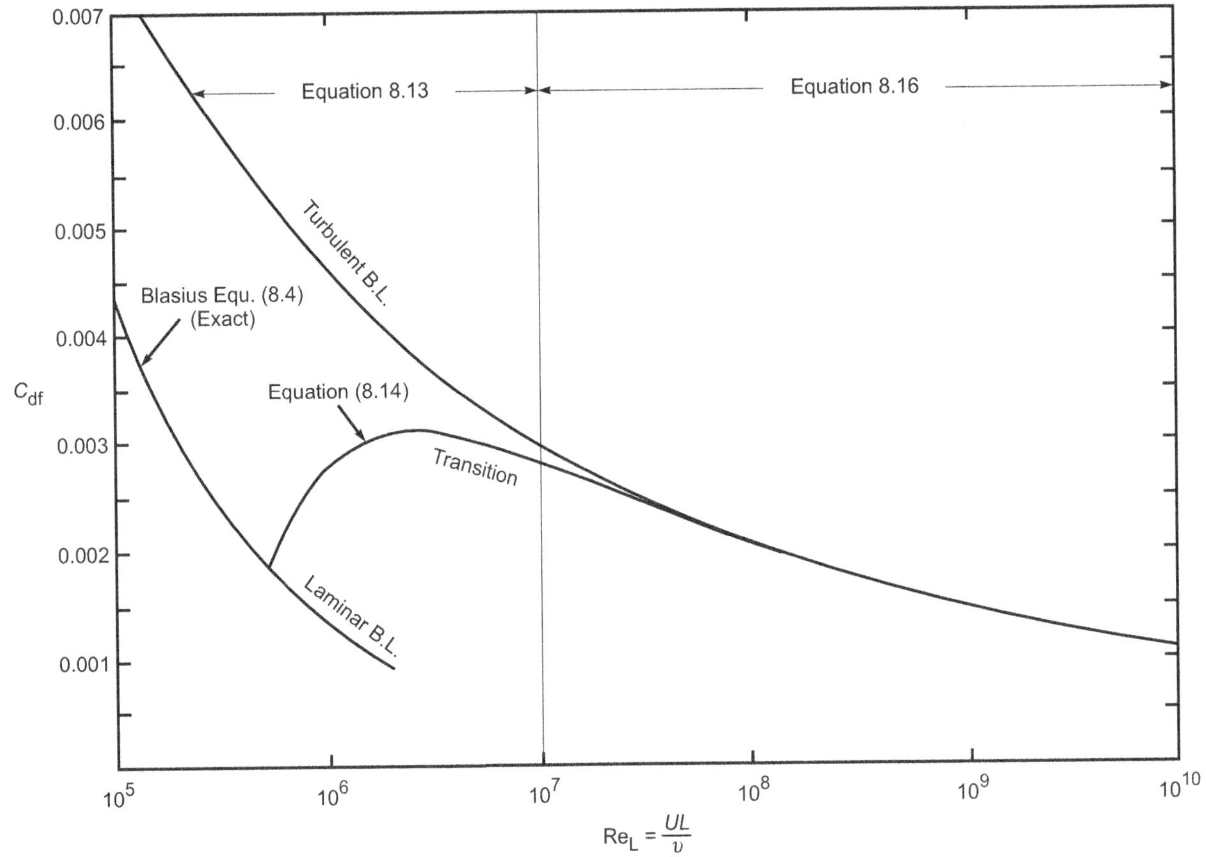

Fig. 8.3 *Drag Coefficients for Plates Kept Parallel to Flow*

8.8 LAMINAR SUBLAYER

The laminar sublayer is usually very thin and its thickness δ' is found by experiments to be

$$\delta' = 11.6\, \nu / u_* \tag{8.19}$$

where $u_* = \sqrt{\tau_0/\rho}$ = shear velocity.

If the roughness magnitude of a surface ε is very small compared to δ', i.e. $\varepsilon << \delta'$, then such a surface is said to be *hydrodynamically smooth*. Roughness does not have any influence in such flows while the viscous effects predominate. Usually $\varepsilon/\delta' < 0.25$ is taken as the criterion for hydrodynamically smooth surface (Fig. 8.4). If the laminar sublayer thickness δ' is very small compared to roughness height ε, (i.e. $\varepsilon >> \delta'$), in such flows viscous effects are not important and the boundary is said to be *hydrodynamically rough*. Usually $\varepsilon/\delta' > 6$ is taken as the criterion for hydrodynamically rough boundaries.

In the region $0.25 < \varepsilon/\delta' < 6$, the boundary is in the transition regime and both viscosity and roughness control the flow.

Rough flat plate: For flow on a completely rough flat plate the local friction coefficient C_f and total drag coefficient C_{Df} are given by

$$C_f = 1/(2.87 + 1.58 \log (x/\varepsilon))^{2.5} \tag{8.20}$$

$$C_{Df} = 1/(1.89 + 1.62 \log (L/\varepsilon))^{2.5} \tag{8.21}$$

8.9 ESTABLISHMENT OF FLOW IN A PIPE

When a flow enters a pipe from a reservoir a boundary layer forms in the pipe at the entrance. The thickness of

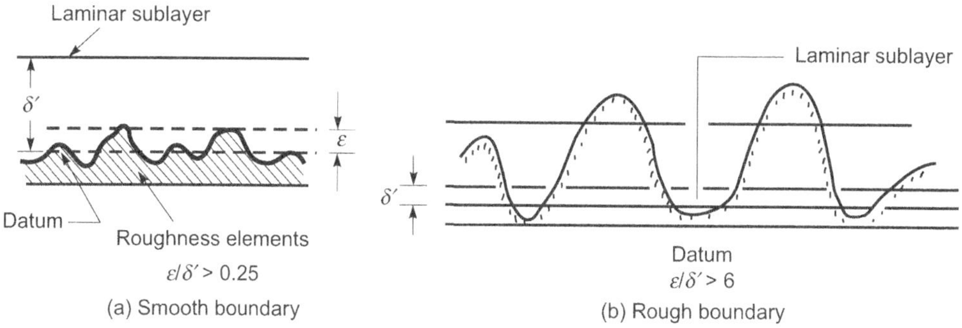

Fig. 8.4 *Smooth and Rough Surfaces*

the boundary layer in the radial direction grows along the length of the pipe till it merges at the centre line at a distance L_e known as entrance length. The flow is uniform beyond L_e. The establishment length in the laminar and turbulent flow is given by the following formulae:

In laminar flow: $\boxed{L_e/D = 0.07\ R_e}$ (8.22)

In turbulent flow: $\boxed{L_e/D = 50}$ (8.23)

8.10 BOUNDARY LAYER SEPARATION

8.10.1 Separation Phenomenon

The flow past a flat plate held parallel to the flow is a case of boundary layer with zero pressure gradient. Flows in converging boundaries are examples of favourable pressure gradient and flows in diverging conduits or diverging boundaries are examples of adverse pressure gradient flows.

In adverse pressure gradient boundary layer flow the boundary layer may at some section leave the boundary. This is called as *separation* and downstream of the separation section turbulent eddies exist and this disturbed region is called as a *wake* (Fig. 8.5). Separation can take place in both laminar and turbulent boundary layers. The location of the separation section on the surface of a body and the size of the wake have important bearing on the total drag force experienced by the body.

At the separation point, the shear stress is zero and the velocity gradient $\partial u/\partial y = 0$.

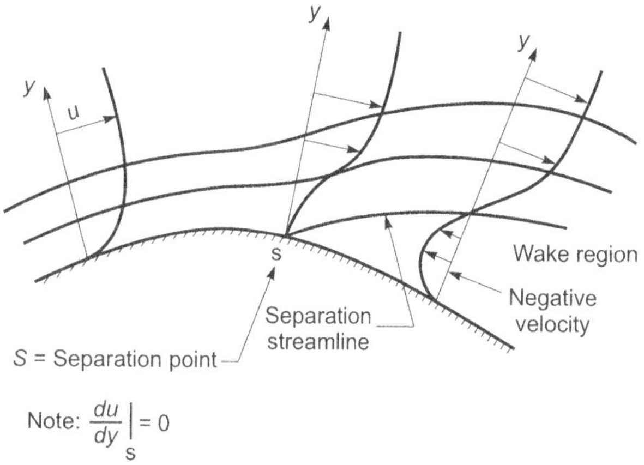

Fig. 8.5 *Separation of Boundary Layer*

8.10.2 Control of Separation

Separation of flow from the boundary leads to inefficiency of the flow unit. In the lifting surfaces such as aerofoils, it may cause reduction of lift and even stalling. Diffusers, conduit transitions, pump and turbine blades and aerofoils are some common flow units where separation may impair the performance.

Common procedures to control separation are based on the following methodologies:

- Streamlining of blunt body shapes
- Fluid injection into the boundary layer
- Suction of fluid from the boundary layer
- Creating a motion of the boundary wall.

WORKED EXAMPLES

8.1 For the following velocity profiles in the boundary layer on a flat plate, calculate the displacement and momentum thickness in terms of the nominal boundary layer thickness δ.

(a) $u/U = \eta$

(b) $u/U = 2\eta - \eta^2$

where $\eta = y/\delta$.

Solution: The displacement thickness is given by

$$\delta^* = \int_0^\delta \left(1 - \frac{u}{U}\right) dy$$

since $\eta = y/\delta$, $dy = \delta\, d\eta$

when $y = 0$, $\eta = 0$

$y = \delta$, $\eta = 1$

$$\delta^* = \delta \int_0^1 (1 - f(\eta))\, d\eta$$

where $f(\eta) = u/U$

The momentum thickness

$$\theta = \frac{u}{U}\left(1 - \frac{u}{U}\right) dy$$

$$= \delta \int_0^1 f(\eta)(1 - f(\eta))\, d\eta$$

(a) when $u/U = f(\eta) = \eta$

$$\delta^* = \delta \int_0^1 (1-\eta)\, d\eta = \delta\left[\eta - \frac{\eta^2}{2}\right]_0^1$$

$$= \boldsymbol{\delta/2}$$

$$\theta = \delta \int_0^1 \eta(1-\eta)\, d\eta = \delta\left[\frac{\eta^2}{2} - \frac{\eta^2}{3}\right]_0^1$$

$$= \boldsymbol{\delta/6}$$

(b) when $u/U = f(\eta) = 2\eta - \eta^2$

$$\delta^* = \delta \int_0^1 (1 - 2\eta + \eta^2)\, d\eta$$

$$= \delta\left[\eta - \eta^2 + \frac{\eta^3}{3}\right]_0^1 = \boldsymbol{\delta/3}$$

$$\theta = \delta \int_0^1 (2\eta - \eta^2)(1 - 2\eta + \eta^2) d\eta$$

$$= \delta \int_0^1 (2\eta - 5\eta^2 + 4\eta^3 - \eta^4) d\eta$$

$$= \delta\left[\eta^2 - \frac{5}{3}\eta^3 + \eta^4 - \frac{\eta^5}{5}\right]_0^1 = \boldsymbol{\frac{2}{15}\delta}$$

8.2 For the velocity profile $u/U = (y/\delta)^{1/7}$ calculate the shape factor H.

Solution: The shape factor $H = \delta^*/\theta$

Putting $u/U = f(\eta) = \eta^{1/7}$ where $\eta = y/\delta$

$$\delta^* = \delta \int_0^1 (1 - \eta^{1/7}) d\eta$$

$$= \delta\left[\eta - \frac{7}{8}\eta^{8/7}\right]_0^1 = \delta/8$$

$$\theta = \delta \int_0^1 \eta^{1/7}(1 - \eta^{1/7}) d\eta$$

$$= \delta\left[\frac{7}{8}\eta^{8/7} - \frac{7}{9}\eta^{9/7}\right]_0^1 = \frac{7}{72}\delta$$

$$H = \delta^*/\theta = \frac{1}{8}\left(\frac{72}{7}\right) = \mathbf{1.286}$$

8.3 For the following velocity distributions, verify whether the essential and desirable boundary conditions to be the velocity distributions in a laminar boundary layer are satisfied:

(a) $u/U = \sin\left(\frac{\pi y}{2\delta}\right)$

(b) $u/U = 1 + \eta - 2\eta^2$ where $\eta = y/\delta$

Solution: The requisite boundary conditions are:

		Essential	Desirable
At	$y = 0$,	$u = 0$,	$\frac{\partial^2 u}{\partial y^2} = 0$
At	$y = \delta$,	$u = U$,	$\frac{\partial u}{\partial y} = \frac{\partial^2 u}{\partial y^2} = 0$

(a) $u/U = \sin\left(\frac{\pi y}{2\delta}\right)$

$$\frac{\partial u}{\partial y} = \frac{U\pi}{2\delta}\cos\left(\frac{\pi y}{2\delta}\right)$$

$$\frac{\partial^2 u}{\partial y^2} = -\frac{U\pi^2}{4\delta^2}\sin\left(\frac{\pi y}{2\delta}\right)$$

At $y = 0$, $u = U \sin(0) = 0$

$$\frac{\partial^2 u}{\partial y^2} = -\frac{U\pi^2}{4\delta^2}(0) = 0$$

At $y = \delta$, $u = U \sin(\pi/2) = U$

$$\frac{\partial u}{\partial y} = \left(\frac{U\pi}{2\delta}\right)\cos(\pi/2) = 0$$

$$\frac{\partial^2 u}{\partial y^2} = -\frac{U\pi^2}{4\delta^2}\sin(\pi/2) \neq 0$$

Hence all essential and one desirable boundary condition are satisfied.

(b) $u = U[1 + (y/\delta) - 2(y/\delta)^2]$

$$\frac{\partial u}{\partial y} = U\left[\frac{1}{\delta} - \frac{4y}{\delta^2}\right]$$

$$\frac{\partial^2 u}{\partial y^2} = U[-4/\delta^2]$$

At $y = 0$, $u = U \neq 0$

$$\frac{\partial^2 u}{\partial y^2} = -4U/\delta^2 \neq 0$$

At $y = \delta$, $u = 0 \neq U$

$$\frac{\partial u}{\partial y} = -3U/\delta \neq 0$$

$$\frac{\partial^2 u}{\partial y^2} = -4U/\delta^2 \neq 0$$

None of the essential and desirable boundary conditions is satisfied. **Hence this is not a proper velocity distribution in a laminar boundary layer.**

8.4 If the velocity distribution in a laminar boundary layer on a flat plate is

$$u/U = a + b\eta + c\eta^2 + d\eta^3$$

where $\eta = y/\delta$, determine the coefficients a, b, c and d.

Solution:

Let $u/U = f(\eta)$

The requisite boundary conditions are:

At $y = 0$, i.e. $\eta = 0$ $\quad u = 0$ i.e. $f(\eta) = 0$

$$\frac{\partial^2 u}{\partial y^2} = 0 \quad \text{i.e. } f''(\eta) = 0$$

At $y = \delta$, i.e. $\eta = 1$, $\quad u = U$ i.e. $f(\eta) = 1$

$$\frac{\partial u}{\partial y} = 0 \quad \text{i.e. } f'(\eta) = 0$$

$$\frac{\partial^2 u}{\partial y^2} = 0 \quad \text{i.e. } f''(\eta) = 0$$

Here
$$f(\eta) = a + b\eta + c\eta^2 + d\eta^3$$
$$f'(\eta) = b + 2c\eta + 3d\eta^2$$
$$f''(\eta) = 2c + 6d\eta$$

Therefore

at $\eta = 0$, $\quad f(\eta) = 0$ i.e. $\mathbf{a = 0}$

$\quad f''(\eta) = 0$ i.e. $\mathbf{c = 0}$

at $\eta = 1$, $\quad f(\eta) = 1$ i.e. $a + b + c + d = 1$

i.e. $b + d = 1$

$\quad f'(\eta) = 0$ i.e. $b + 2c + 3d = 0$

i.e. $b + 3d = 0$

Solving $\quad \mathbf{b = \frac{3}{2}}$ and $\mathbf{d = -\frac{1}{2}}$

Thus $\quad \frac{u}{U} = \frac{3}{2}\eta - \frac{1}{2}\eta^3$ where $\eta = y/\delta$

8.5 Given the velocity distribution in a laminar boundary layer on a flat plate as

$$\frac{u}{U} = 2(y/\delta) - 2(y/\delta)^3 + (y/\delta)^4$$

obtain expressions for the boundary layer thickness, shear intensity and force on one side of the plate.

Solution: By Karman momentum integral equation

$$\tau_0 = \rho U^2 \frac{\partial}{\partial x}\left(\int_0^\delta \frac{u}{U}\left(1 - \frac{u}{U}\right)dy\right)$$

put $y/\delta = \eta$, $dy = \delta\, d\eta$ and the limits of η are 0 and 1.

$$\frac{u}{U} = f(\eta) = 2\eta - 2\eta^3 + \eta^4$$

Substituting,

$$\tau_0 = \rho U^2 \frac{d\delta}{dx}\int_0^1 f(\eta)(1 - f(\eta))d\eta$$

But,

$$\int_0^1 f(\eta)(1 - f(\eta))d\eta$$

$$= \int_0^1 (2\eta - 2\eta^3 + \eta^4)(1 - 2\eta - 2\eta^3 + \eta^4)d\eta$$

$$= \int_0^1 (2\eta - 4\eta^2 - 2\eta^3 + 9\eta^4 - 4\eta^5 + 4\eta^6 - 4\eta^7 - \eta^8)d\eta$$

$= 37/315$

Therefore, $\quad \tau_0 = \frac{37}{315}\rho U^2 \frac{d\delta}{dx}$

From the boundary conditions for a laminar boundary layer

$$\tau_0 = \mu\left(\frac{\partial u}{\partial y}\right)_{y=0} = \frac{uU}{\delta}\left[\frac{df(\eta)}{d\eta}\right]_{\eta=0}$$

$$f(\eta) = 2\eta - 2\eta^2 + \eta^4$$

$$\left[\frac{df(\eta)}{d\eta}\right]_{\eta=0} = [2 - 4\eta + 4\eta^3]_{\eta=0} = 2$$

$\therefore \quad \tau_0 = 2\mu\, U/\delta$

Equating the two expressions for τ_0,

$$\frac{2\mu U}{\delta} = \frac{37}{315}\rho U^2 \frac{d\delta}{dx}$$

$$\delta\, d\delta = \frac{630}{37}\frac{\mu}{\rho U} dx$$

On integration,

$$\delta^2/2 = \frac{630}{37}\frac{\mu}{\rho U} x + \text{constant}$$

Since at $x = 0,\ \delta = 0,$ constant = 0

$$\therefore \quad \delta^2 = \frac{1260}{37}\frac{\mu}{\rho U} x = 34.05 \frac{x^2}{(\rho U x/\mu)}$$

or $$\frac{\boldsymbol{\delta}}{\boldsymbol{x}} = \frac{\mathbf{5.835}}{\sqrt{\boldsymbol{Re_x}}} \quad \text{where } Re_x = \rho U\, x/\mu$$

Shear stress: Substituting the value of δ in the second expression for τ_0,

$$\tau_0 = 2\mu U/\delta = \frac{2\mu U}{5.835x}\sqrt{(\rho U x/\mu)}$$

$$= 0.3428 \frac{\mu^{1/2} U^{3/2} \rho^{1/2}}{x^{1/2}}$$

$$\therefore \quad \tau_0 = 0.6855 \frac{\rho U^2}{2}\left(\frac{\mu}{\rho U x}\right)^{1/2}$$

i.e. $$\frac{\tau_0}{(\rho U^2/2)} = C_f = \frac{\mathbf{0.6855}}{\sqrt{\boldsymbol{Re_x}}}$$

Force on one side, F_D: Consider a plate of unit width and length L.

$$F_D = \int_0^L \tau_0 dx = \frac{\rho U^2}{2}\int_0^L \frac{0.6855}{(\rho U x/\mu)^{1/2}} dx$$

$$= \frac{\rho U^2}{2}\frac{0.6855}{(\rho U/\mu)^{1/2}}(2L^{1/2})$$

$$\frac{F_D}{\frac{1}{2}\rho U^2 L} = C_{Df} = \frac{1.3710}{(\rho U L/\mu)^{1/2}} = \frac{\mathbf{1.371}}{\sqrt{\boldsymbol{Re_L}}}$$

8.6 Calculate the friction drag on a flat plate 15 cm wide and 45 cm long placed longitudinally in a stream of oil of relative density 0.925 and kinematic viscosity 0.9 stoke, flowing with a free stream velocity of 6.0 m/s. Also, find the thickness of the boundary layer and shear stress at the trailing edge.

Solution: Reynolds number at the trailing edge is

$$Re_L = \frac{UL}{\nu} = \frac{6.0 \times 0.45}{0.9 \times 10^{-4}}$$

$$= 3.0 \times 10^4$$

(Note: 1 stoke = 10^{-4} m²/s)

Since Re_L is less than $Re_{(crit)} = 5.0 \times 10^5$, the boundary layer is laminar.

Using the Blasius' results,

(i) Boundary layer thickness:

$$\frac{\delta}{x} = \frac{5.0}{\sqrt{Re_x}}$$

At the trailing edge $x = L = 0.45$ m

$$\delta_L = \frac{5.0 \times 0.45}{\sqrt{Re_L}} = \frac{5.0 \times 0.45}{\sqrt{3 \times 10^4}}$$

$$= 0.01299 \text{ m}$$

$$= \mathbf{1.3\ cm}$$

(ii) Shear stress at the trailing edge, τ_L

$$\tau_L = \frac{\rho U^2}{2} C_f(L) = \frac{\rho U^2}{2}\left(\frac{0.664}{\sqrt{Re_L}}\right)$$

$$= \frac{0.925 \times 1000}{2} \times (6)^2 \times \frac{0.664}{\sqrt{3 \times 10^4}}$$

$$= \mathbf{63.8\ N/m^2}$$

(iii) Drag on one side of the plate

$$F_D = C_{Df}(L \cdot B)\, \rho U^2/2$$

where $$C_{Df} = \frac{1.328}{\sqrt{Re_L}}$$

$$C_{Df} = \frac{1.328}{\sqrt{3 \times 10^4}} = 7.667 \times 10^{-3}$$

$$F_D = (7.667 \times 10^{-3}) \times (0.45 \times 0.15) \times \frac{0.925 \times 1000}{2} \times 6^2$$

$$= \mathbf{8.617\ N}$$

On both sides of the plate drag force

$$F_{2D} = 2 \times F_D = 2 \times 8.617$$

$$= \mathbf{17.23\ N}$$

8.7 A smooth flat plate 2.0 m wide and 2.5 m long is towed in oil (RD = 0.8) at a velocity of 1.5 m/s along its length. Find the thickness of the boundary layer and shear stress at (a) the centre and (b) the trailing edge of the plate. Also, find the power required for towing the plate. [$\nu_{oil} = 10^{-4}$ m²/s]

Solution:

(a) At the centre of the plate

$x = 1.25$ m $\quad U = 1.5$ m/s

$$Re_x = \frac{1.5 \times 1.25}{10^{-4}} = 1.875 \times 10^4$$

This is less than $Re_{(crit)} = 5 \times 10^5$ and hence the boundary layer is laminar.

$$\frac{\delta}{x} = \frac{5.0}{\sqrt{Re_x}}$$

$$\delta_m = \delta \text{ at midpoint} = \frac{5.0 \times 1.25}{\sqrt{1.875 \times 10^4}}$$

$$= 0.0456 \text{ m}$$

$$= \mathbf{4.56 \text{ cm}}$$

$$C_f = \text{local friction coefficient} = \frac{0.664}{\sqrt{Re_x}}$$

$$= \frac{0.664}{\sqrt{1.875 \times 10^4}} = 4.849 \times 10^{-3}$$

Shear stress

$$\tau_{0m} = C_f \frac{\rho U^2}{2}$$

$$= 4.849 \times 10^{-3} \times \frac{0.8 \times 1000}{2} \times (1.5)^2$$

$$= \mathbf{4.364 \text{ N/m}^2}$$

(b) At the trailing edge,

$$x = L = 2.5 \text{ m}$$

$$Re_L = \frac{1.5 \times 2.5}{10^{-4}} = 3.75 \times 10^4$$

$$Re_L < Re_{L(crit)} = 5 \times 10^5$$

The boundary layer is laminar

$$\delta_L = \frac{5.0L}{\sqrt{Re_L}} = \frac{5.0 \times 2.5}{\sqrt{3.75 \times 10^4}} = 0.0645 \text{ m}$$

$$= \mathbf{6.45 \text{ cm}}$$

$$C_f = \frac{0.664}{\sqrt{Re_L}} = \frac{0.664}{\sqrt{3.75 \times 10^4}}$$

$$= 3.429 \times 10^{-3}$$

and shear stress at trailing edge τ_{0L} is

$$\tau_{0L} = C_f \frac{\rho U^2}{2}$$

$$= (3.429 \times 10^{-3}) \times \frac{(0.8 \times 1000)(1.5)^2}{2}$$

$$= \mathbf{3.086 \text{ N/m}^2}$$

(c) Total force (on both sides of the plate)

$$F = C_{Df} \times (\text{area}) \times \rho U^2/2$$

$$C_{Df} = \frac{1.328}{\sqrt{Re_L}} = \frac{1.328}{\sqrt{3.75 \times 10^4}}$$

$$= 6.858 \times 10^{-3}$$

$$F = (6.858 \times 10^{-3})(2 \times 2.5 \times 2.0) \times \frac{(0.8 \times 1000) \times (1.5)^2}{2}$$

$$= \mathbf{61.72 \text{ N}}$$

Power required to tow the plate

$$\mathbf{P} = F \times U$$

$$= 61.72 \times 1.5 = \mathbf{92.58 \text{ W}}$$

8.8 A flat plate 1.5 m wide and length L is kept parallel to a uniform stream of air of velocity 3.0 m/s in a wind tunnel. If it is desired to have a laminar boundary layer only on the plate, what is the maximum length L of the plate? For the maximum length plate, determine the drag force on one side of the plate. [$\rho_{air} = 1.2$ kg/m^3 and $\nu_{air} = 1.45 \times 10^{-5}$ m^2/s]

Solution: The maximum length of plate corresponds to the critical Reynolds number which can be taken as,

$$Re_{L(crit)} = \frac{UL}{\nu} = 5 \times 10^5$$

Thus,

$$\frac{3.0 \times L}{1.45 \times 10^{-5}} = 5 \times 10^5$$

$$\mathbf{L = 2.417 \text{ m}}$$

For this length of plate in a laminar sublayer

$$C_{Df} = \frac{1.328}{\sqrt{Re_L}} = \frac{1.328}{\sqrt{5 \times 10^5}}$$

$$= 1.878 \times 10^{-3}$$

Drag force on one side of the plate

$$F_D = C_{Df} \times \text{area} \times \frac{\rho U^2}{2}$$

$$= 1.878 \times 10^{-3} \times (1.5 \times 2.417) \times \frac{2 \times (3.0)^2}{2}$$

$$= \mathbf{0.03677 \text{ N}}$$

8.9 A smooth flat rectangular plate is placed edgewise in a stream of fluid. At what fraction of the length from the leading edge would the drag force on the front portion be equal to half of the total drag force? Assume the boundary layer to be laminar.

Solution: Let x be the distance from the leading edge such that the drag force in distance x is half of the total drag force.

$$F_{Dx} = \frac{1}{2} F_{DL}$$

$$= C_{Dfx} (Bx) \frac{\rho U^2}{2}$$

$$F_{DL} = C_{DfL} (B L) \frac{\rho U^2}{2}$$

$$\therefore \quad \frac{C_{Dfx}}{C_{DfL}} \cdot \frac{x}{L} = \frac{F_{Dx}}{F_{DL}} = \frac{1}{2}$$

But from $\quad C_{Dfx} = \dfrac{1.328}{\sqrt{(Ux/\nu)}}$

$$C_{DfL} = \frac{1.328}{\sqrt{(UL/\nu)}}$$

$$\therefore \quad \left(\frac{L}{x}\right)^{1/2} \cdot \frac{x}{L} = \frac{1}{2}$$

or $\quad \left(\dfrac{x}{L}\right)^{1/2} = \dfrac{1}{2}$ i.e. $\boldsymbol{x = \dfrac{1}{4}L.}$

8.10 In a turbulent boundary layer over a flat plate it is found that $\dfrac{u}{U} = \left(\dfrac{y}{\delta}\right)^{1/7}$ and the shear stress τ_0 is given by

$$\tau_0 = 0.0228\, \rho U^2 \left(\frac{\nu}{U\delta}\right)^{1/4}$$

Using Karman momentum integral equation obtain expressions for (i) δ and C_f in terms of Re_x and (ii) C_{Df} in terms of Re_L.

Solution: Putting $\eta = y/\delta$,

$u/U = f(\eta) = \eta^{1/7}$

(i) By Karman momentum integral equation (see Example 8.3)

$$\tau_0 = \rho U^2 \frac{d\delta}{dx} \int_0^1 f(\eta)[1 - f(\eta)]d\eta$$

$$= \rho U^2 \frac{d\delta}{dx} \int_0^1 \eta^{1/7}(1 - \eta^{1/7})d\eta$$

$$= \frac{7}{72} \rho U^2 \frac{d\delta}{dx} \qquad (1)$$

It is given that

$$\tau_0 = 0.0228 \rho U^2 \left(\frac{\nu}{U\delta}\right)^{1/4} \qquad (2)$$

Equating (1) and (2),

$$\delta^{1/4} d\delta = 0.2345 \left(\frac{\nu}{U}\right)^{1/4} dx$$

On integrating

$$\delta^{5/4} = 0.2931 \left(\frac{\nu}{U}\right)^{1/4} x + C$$

Using the boundary condition $\delta = 0$ when $x = 0$, $C = 0$

Thus $\quad \boldsymbol{\dfrac{\delta}{x} = \dfrac{0.375}{\left(\dfrac{Ux}{\nu}\right)^{1/5}}} \qquad (3)$

Substituting Eq. (3) in Eq. (2),

$$\tau_0 = 0.0228 \rho U^2 \left(\frac{\nu}{U}\right)^{1/4} \frac{1}{(0.375x)^{1/4}} \left(\frac{Ux}{\nu}\right)^{1/20}$$

$$= 0.02775 \rho U^2 \left(\frac{\nu}{Ux}\right)^{1/5} \qquad (4)$$

Local shear stress coefficient

$$C_f = \frac{\tau_0}{(\rho U^2/2)} = \frac{(0.002775) \times 2}{Re_x^{1/5}}$$

$$= \mathbf{0.0555\, Re_x^{-1/5}}$$

(ii) Force on one side of a plate of unit width and length L

$$F_D = \int_0^L \tau_0 dx = \frac{\rho U^2}{2} \int_0^L \frac{0.0555}{\left(\frac{Ux}{\nu}\right)^{\frac{1}{5}}} dx$$

$$\frac{F_D}{\rho U^2/2} = C_{Df} = \frac{0.0555}{(U/\nu)^{\frac{1}{5}}} \left(\frac{5}{4} L^{\frac{4}{5}}\right)$$

$$\boldsymbol{C_{Df} = \frac{0.0694}{Re_L^{1/5}}} \quad \text{where } Re_L = \frac{UL}{\nu}$$

8.11 A smooth flat plate 1.5 m wide and 20 m long is subjected to flow of water along its length with a velocity of 2 m/s. Find the

(i) extent of the laminar boundary layer on the plate

(ii) the thickness of the boundary layer at the edge of the laminar boundary layer and at the trailing edge.

(iii) the shear stress at the trailing edge.

[$\rho = 998$ kg/m^3 and $\nu = 1 \times 10^{-6}$ m^2/s]

Solution: Reynolds number

$$Re_L = \frac{UL}{\nu} = \frac{2 \times 20}{1 \times 10^{-6}} = 4 \times 10^7$$

The boundary layer is turbulent at the trailing edge.

(i) Taking the critical Reynolds number

$Re_{(crit)} = 5 \times 10^5$,

$$\frac{Ux_{crit}}{\nu} = \frac{2 \times x_{crit}}{1 \times 10^{-6}} = 5 \times 10^5$$

$x_{crit} = 0.25$ m = **25 cm**

Laminar boundary layer exists in the first 25 cm of the plate.

(ii) At x_{crit}, $\frac{\delta}{x_{crit}} = \frac{5.0}{\sqrt{Re_{(crit)}}}$

The boundary layer thickness at the edge of the laminar boundary layer

$$\delta_c = \frac{5.0 \times 0.25}{\sqrt{5 \times 10^5}} = 1.768 \times 10^{-3} \text{m}$$

$= \mathbf{1.768\ mm}$

At the trailing edge, as the Reynolds number is $> 10^7$, the thickness of the turbulent boundary layer is obtained by putting $x = L = 20$ m in the equation

$$\frac{\delta_t}{x} = \frac{0.22}{(Re_x)^{1/6}}$$

$$\delta_t = \frac{0.22 \times 20}{(4 \times 10^7)^{1/6}} = 0.238 \text{ m}$$

$= \mathbf{23.8\ cm}$

(iii) The shear stress coefficient in the turbulent boundary with $Re > 10^7$ is

$$C_f = \frac{0.370}{(\log Re_x)^{2.58}}$$

Here $x = L = 20$ m and $Re_x = Re_L = 4 \times 10^7$

$$C_f = \frac{0.370}{(\log 4 \times 10^7)^{2.50}} = 1.9742 \times 10^{-3}$$

$$\tau_o = C_f \frac{\rho U^2}{2}$$

$$= 1.9742 \times 10^{-3} \times \frac{998 \times 2^2}{2} = \mathbf{3.94\ Pa}$$

8.12 A 1.80 m wide and 5.0 m long plate moves through a stationary air of density 1.22 kg/m^3 and viscosity 1.8×10^{-4} poise at a velocity of 1.75 m/s parallel to its length. Determine the drag force on one side of the plate by assuming (i) laminar flow condition and (ii) turbulent flow condition.

Solution:

$\mu = 1.8 \times 10$ poise $= 1.8 \times 10$ Pa.s

$\nu = \mu/\rho = 1.8 \times 10^{-5}/1.22$

$= 1.475 \times 10^{-5}$ m^2/s

Reynolds number,

$$Re_L = \frac{UL}{\nu} = \frac{1.75 \times 5}{1.475 \times 10^{-5}} = 5.932 \times 10^5$$

(i) *For laminar boundary layer*

Drag coefficient

$$C_{Df} = \frac{1.328}{\sqrt{Re_L}} = \frac{1.328}{\sqrt{5.932 \times 10^5}}$$

$= 1.724 \times 10^{-3}$

$$F_D = C_{Df}(\text{area}) \frac{\rho U^2}{2}$$

$$= 1.724 \times 10^{-3} \times (1.8 \times 5.0) \times \frac{1.22 \times (1.75)^2}{2}$$

$= \mathbf{0.029\ N}$

(ii) *For a completely turbulent boundary layer:* Since Re_L is between 5×10^5 and 10^7, the 1/7th power law-is applicable. Thus the drag coefficient

$$C_{Df} = \frac{0.074}{Re_L^{1/5}} = \frac{0.074}{(5.932 \times 10^5)^{1/5}} = 5.183 \times 10^{-3}$$

Drag force on one side of the plate

$$F_D = C_{Df} \times \text{area} \times \frac{\rho U^2}{2}$$

$$= 5.183 \times 10^{-3} \times (1.8 \times 5.0) \times \frac{1.22 \times (1.75)^2}{2}$$

$= \mathbf{0.0871\ N}$

8.13 A flat barge 35 m long and 10 m wide moves at 1.5 m/s in sea water ($\rho = 1025$ kg/m^3 and $\mu = 0.001$ Pa.s). Estimate the frictional drag on the bottom of the barge for (a) a smooth boundary and (b) a rough boundary with roughness height of 2.0 mm.

Solution: The relative flow past the bottom of the barge is similar to the flow past a flat surface.

$$Re_L = \frac{UL}{\nu} = \frac{1.5 \times 35.0}{(0.001/1025)} = 5.38 \times 10^7$$

The flow is turbulent in the boundary layer.

(a) Smooth surface

The drag coefficient, appropriate to this Reynolds number, is

$$C_{Df} = \frac{0.455}{(\log Re_L)^{2.58}} - \frac{1700}{Re_L}$$

$$= \frac{0.455}{(\log 5.38 \times 10^7)^{2.58}} - \frac{1700}{5.38 \times 10^7}$$

$= 2.3248 \times 10^{-3} - (3.16 \times 10^{-5}) = 0.00229$

F_D = drag force on the bottom of the barge

$$= C_{Df} \times \text{area} \times \frac{\rho U^2}{2}$$

$$= 0.00229 \times (35 \times 10) \times \frac{1025 \times (1.5)^2}{2}$$

$= \mathbf{924\ N}$

(b) For a rough surface

$$\varepsilon = 0.002 \text{ m}, \quad \frac{L}{\varepsilon} = \frac{35}{0.002} = 17500$$

C_{Df} for a rough flat boundary is

$$C_{Df} = \frac{1}{\left(1.89 + 1.62 \log \frac{L}{\varepsilon}\right)^{2.5}}$$

$$= \frac{1}{(1.89 + 1.62 \log 17500)^{2.5}} = 4.4 \times 10^{-3}$$

Drag force

$$F_D = C_{Df} \times \text{area} \times \frac{\rho U^2}{2}$$

$$= 4.4 \times 10^{-3} \times (35 \times 10) \times \frac{1025 \times (1.5)^2}{2}$$

$$= \mathbf{1775\ N}$$

8.14 A 20 km/h wind blows over a 6 m flat plate. If the density and kinematic viscosity of air are 1.2 kg/m^3 and 1.5×10^{-5} m^2/s respectively, calculate the force per metre width of the plate. Also estimate the thickness of the boundary layer at the trailing edge.

Solution:
Wind velocity

$$U = \frac{20 \times 1000}{3600} = 5.56 \text{ m/s}$$

Reynolds number

$$Re_L = \frac{UL}{\nu} = \frac{5.56 \times 6}{1.5 \times 10^{-5}} = 2.22 \times 10^6$$

The boundary layer is turbulent and the appropriate drag coefficient is the one corresponding to the power law, viz.

$$C_{Df} = \frac{0.074}{Re_L^{1/5}} - \frac{1700}{Re_L}$$

$$= \frac{0.074}{(2.22 \times 10^6)^{1/5}} - \frac{1700}{2.22 \times 10^6} = 0.003215$$

Drag force on one side of the plate per unit metre width:

$$F_D = C_{Df} \times \text{area} \times \frac{\rho U^2}{2}$$

$$= 0.003215 \times (6.0 \times 1.0) \times \frac{1.2 \times (5.56)^2}{2}$$

$$= \mathbf{0.3578\ N}$$

The turbulent boundary layer thickness at the trailing edge is given by putting $x = L = 6.0$ m in

$$\frac{\delta}{x} = \frac{0.377}{Re_L^{1/5}}$$

$$\delta = L \times \frac{0.377}{Re_L^{1/5}} = 6.0 \times \frac{0.377}{(2.22 \times 10^6)^{1/5}}$$

$$= 0.1217 \text{ m} = \mathbf{12.17\ cm}$$

8.15 A 3.0 m long smooth model of a ship is towed in fresh water at 3.0 m/s with a total measured drag of 70 N. The wetted surface of the model is 3.50 m^2. Estimate (a) the skin friction drag and (b) the wave drag. [$\rho = 998$ kg/m^3 and $\nu = 1.0 \times 10^{-6}$ m^2/s]

Solution: The wetted surface of the model is considered as an equivalent flat plate 3.0 m long and having a surface area of 3.50 m^2

Reynolds number

$$Re_L = \frac{UL}{\nu} = \frac{3.0 \times 3.0}{1.0 \times 10^6} = 9 \times 10^6$$

The boundary layer is turbulent and the drag coefficient appropriate to this $Re_L = 9 \times 10^6$ is

$$C_{Df} = \frac{0.074}{Re_L^{1/5}} - \frac{1700}{Re_L}$$

$$= \frac{0.074}{(9.00 \times 10^6)^{1/5}} - \frac{1700}{9.00 \times 10^6} = 2.82 \times 10^{-3}$$

Drag force due to skin friction

$$F_{Df} = C_{Df} \times \text{area} \times \frac{\rho U^2}{2}$$

$$= 2.82 \times 10^{-3} \times 3.5 \times \frac{998 \times (3)^2}{2} = \mathbf{44.3\ N}$$

Total measured drag = Skin friction drag + wave drag

$$\therefore \quad 70.0 = 44.3 + F_{DW}$$

Wave drag

$$F_{DW} = 70.0 - 44.3 = \mathbf{25.7\ N}$$

8.16 Find the ratio of skin friction drag on the front two thirds and rear one third of a smooth flat plate in a uniform stream at zero incidence. Assume the boundary layer to be turbulent and the Reynolds number to be of the order of 10^6.

Solution: Consider a distance x from the leading edge. The total drag force on the plate of length x on one side of it is

$$F_{Dx} = C_{Dfx}\,(Bx)\,\frac{\rho U^2}{2}$$

Similarly, the drag force on one side of a plate of length L is

$$F_{DL} = C_{DfL}\,(BL)\,\frac{\rho U^2}{2}$$

$$\frac{F_{Dx}}{F_{DL}} = \frac{C_{Dfx}}{C_{DfL}}\frac{x}{L}$$

However, as the boundary layer is turbulent, the plate is smooth and $Re_x \approx 10^6$,

$$C_{Dfx} = 0.074/Re_x^{1/5}$$

and $$C_{DfL} = 0.074/Re_L^{1/5}$$

Thus $$C_{Dfx}/C_{DfL} = (L/x)^{1/5}$$

and $$F_{Dx}/F_{DL} = (L/x)^{1/5}\,(x/L)$$

If $x = \frac{2}{3}L$, as in the present case, F_{Dx} = drag on the first 2/3 of the plate = F_1

and $$F_1/F_{DL} = (3/2)^{1/5}\,(2/3) = 0.723$$

F_2 = drag on the rear 1/3 of the plate = $F_{DL} - F_1$

$$F_2/F_{DL} = 1 - (F_1/F_{DL}) = 1 - 0.723 = 0.277$$

Thus $$F_1/F_2 = \frac{F_1/F_{DL}}{F_2/F_{DL}} = 0.723/0.277 = \mathbf{2.61}$$

8.17 A thin rectangular plate 10 m long and 2 m wide is to be towed in sea water at a steady speed of 5 m/s. Both the surfaces of the plate are rough, whose roughness magnitude is assumed as 5 mm. Estimate the power required to tow the plate. [$\rho = 1020$ kg/m^3 and $\mu = 0.0018$ Pa.s].

Solution: Reynolds number

$$Re_L = \frac{\rho UL}{\mu} = \frac{1020 \times 5.0 \times 10.0}{0.0018} = 2.83 \times 10^7$$

The boundary layer is turbulent.

Relative roughness $L/\varepsilon = 10.0/0.005 = 2000$

The drag coefficient for fully rough turbulent boundary layer flow is

$$C_{Df} = \frac{1}{\left(1.89 + 1.62 \log \frac{L}{\varepsilon}\right)^{2.5}}$$

$$= \frac{1}{(1.89 + 1.62 \log 2000)^{2.5}} = 0.007096$$

Drag force on both sides of the plate

$$F_{Df} = C_{Df} \times \text{area} \times \frac{\rho U^2}{2}$$

$$= 0.007096 \times (2 \times 2 \times 10) \times \frac{1020 \times (5)^2}{2}$$

$$= 3619 \text{ N}$$

Power $$\mathbf{P} = F_D \times U = 3619 \times 5 = 18095 \text{ W}$$

$$= \mathbf{18.1\ kW}$$

8.18 A 1:30 scale model of a ship was tested in a towing tank. The model had the same Froude number as the prototype and seawater was used in the model. If the prototype is to have a speed of 10 m/s, a length of 100 m and waterline wetted area of 1600 m^2, what would be the prototype drag corresponding to a measured total drag of 20 N in the model? [$\rho = 1025$ kg/m^3 and $\mu = 1.07 \times 10^{-3}$ Pa.s].

Solution: The total resistance of the model and the prototype consists of two parts: Surface drag and wave drag. The wave drag follows the Froude law of similarity. The surface drag being dependent on Reynolds number is not modelled. It has to be estimated separately for the model as well the prototype.

Let subscripts *m* and *p* stand for the model and prototype respectively.

Froude number

$$Fr_m = \frac{U_m}{\sqrt{gL_m}} = Fr_p = \frac{U_p}{\sqrt{gL_p}}$$

$$U_m = U_p\sqrt{(L_m/L_p)} = 10.0 \times \sqrt{1/30}$$

$$= 1.826 \text{ m/s}$$

$$L_m = 100 \times \frac{1}{30} = 3.3333 \text{ m}$$

$$A_m = 1600 \times \left[\frac{1}{30}\right]^2 = 1.778 \text{ m}^2$$

Reynolds number:

$$Re_{Lp} = \frac{\rho U_p L_p}{\mu} = \frac{1025 \times 10 \times 100}{1.07 \times 10^{-3}}$$

$$= 9.58 \times 10^8$$

$$Re_{Lm} = \frac{1025 \times 1.826 \times 3.333}{1.07 \times 10^{-3}} = 5.83 \times 10^6$$

The surface resistance in the model F_{sm} is first calculated. Since $Re_{Lm} = 5.83 \times 10^6$ the boundary layer is turbulent and

$$C_{Dfm} = \frac{0.074}{Re_{Lm}^{1/5}} - \frac{1700}{Re_{Lm}}$$

$$= \frac{0.074}{(5.83 \times 10^6)^{1/5}} - \frac{1700}{5.83 \times 10^6} = 2.99 \times 10^{-3}$$

$$F_{sm} = C_{Dfm} \times (\text{area})_m \times \frac{\rho U_m^2}{2}$$

$$= 2.99 \times 10^{-3} \times 1.778 \times \frac{1025 \times (1.826)^2}{2}$$

$$= 9.085 \text{ N}$$

(Wave drag)$_m$ = (Total measured drag)$_m$ – (surface drag)$_m$

$$F_{wm} = 20.0 - 9.085 = 10.915 \text{ N}$$

This F_{wm} is modelled by Froude's law. Hence

$$(\text{Wave drag})_p = F_{wp} = (F_{wm})/(\rho_r L_r^3)$$

Since $\rho_m = \rho_p$, $\rho_r = 1.0$

$$\therefore \quad F_{wp} = \frac{10.915}{(1/30)^3} = 294705 \text{ N}$$

$$= 294.7 \text{ kN} = \text{Prototype wave drag}$$

For the prototype, the surface drag F_{sp} is

$$F_{sp} = C_{Dfp} \times (\text{area})_p \times \frac{\rho U_p^2}{2}$$

Since $R_{eLp} = 9.58 \times 10^9$

$$C_{Dfp} = \frac{0.455}{(\log Re_{Lp})^{2.58}} - \frac{1700}{Re_{Lp}}$$

$$= \frac{0.455}{(\log 9.58 \times 10^8)^{2.58}} - \frac{1700}{9.58 \times 10^8}$$

$$= 1.577 \times 10^{-3}$$

$$\therefore \quad F_{sp} = 1.577 \times 10^{-3} \times 1600 \times \frac{1025 \times (10)^2}{2}$$

$$= 129{,}314 \text{ N} = 129.3 \text{ kN}$$

Total prototype drag

$$F_p = F_{wp} + F_{sp}$$

$$= 294.7 + 129.3$$

$$= \mathbf{424.0 \text{ kN}}$$

OBJECTIVE QUESTIONS

8.1 The nominal distance of a boundary layer is defined as the distance from the wall to a point
(a) where the velocity is 99% less than the asymptotic limit
(b) where the velocity ceases to be laminar
(c) where the velocity is within 90% of the asymptotic limit
(d) where the velocity is 99% of its asymptotic limit

8.2 The displacement thickness of a boundary layer is
(a) The distance to the point where $u/U_0 = 0.99$
(b) The distance where $u = u_*$ where u_* = shear velocity
(c) The distance by which the main flow is to be shifted from the boundary to maintain the continuity equation
(d) One half of the actual thickness of the boundary layer

8.3 In a two-dimensional boundary layer over a flat surface
(a) The longitudinal pressure gradient is important and transverse pressure gradient can be neglected
(b) The transverse pressure gradient is important and longitudinal pressure gradient can be neglected
(c) Both the longitudinal and transverse pressure gradients can be neglected
(d) Both the longitudinal and transverse pressure gradients are important

8.4 The displacement thickness δ^* of a boundary layer is defined as δ^* =
(a) $\int_0^\delta \frac{u}{U}\left(1-\frac{u}{U}\right)dy$ (b) $\int_0^\delta u/U\, dy$
(c) $\int_0^\delta \left(1-\frac{u}{U}\right)dy$ (d) $\int_0^\delta \left(1-\frac{u}{U}\right)^2 dy$

8.5 A laminar boundary layer has a velocity distribution given by $u/U = y/\delta$. The displacement thickness δ^* for this boundary layer is
(a) δ (b) $\delta/2$
(c) $\delta/4$ (d) $\delta/6$

8.6 If the velocity distribution in a laminar boundary layer can be assumed as $u/U = y/\delta$ the ratio of momentum thickness θ to nominal thickness δ is given by θ/δ =
(a) 1/2 (b) 1/3
(c) 1/6 (d) 1.25

8.7 The following boundary conditions exist at the wall ($y = 0$) in a boundary layer.
(a) $u = U$ (b) dp/dx = – ve
(c) $\tau_0 = 0$ (d) $u = 0, v = 0$

8.8 Identify the *incorrect statement*:
At the outer edge of a laminar boundary layer ($y = \delta$) over a flat plate
(a) $u = 0$ (b) $u = U$
(c) $\partial^2 u/\partial y^2 = 0$ (d) $\tau_0 = 0$

8.9 If the velocity distribution in a laminar boundary layer over a flat plate is to be expressed as $u/U = \sin(A\pi y/\delta)$ where δ = thickness of the boundary layer, the appropriate value of A is
(a) 1.0 (b) –1/2
(c) 1/2 (d) 2

8.10 If the velocity distribution in a turbulent boundary layer is $u/U = (y/\delta)^{1/10}$ the displacement thickness δ^* is given by $\delta^*/\delta =$

(a) 0.3 (b) 0.125
(c) 0.091 (d) 0.0758

8.11 In a boundary layer at a certain location the boundary layer thickness $\delta = 0.5$ cm, displacement thickness $\delta^* = 0.15$ cm and the momentum thickness $\theta = 0.0585$ cm. The shape factor H at this location is

(a) 2.564 (b) 0.39
(c) 3.333 (d) 8.547

8.12 In a laminar boundary layer the shear stress τ_0 at a location x is given by $\tau_0 =$

(a) $\mu \left.\dfrac{\partial u}{\partial x}\right|_{y=0}$ (b) $\mu \left.\dfrac{\partial u}{\partial y}\right|_{y=0}$

(c) $\mu \left.\dfrac{\partial u}{\partial y}\right|_{y=\delta}$ (d) $\mu \left.\dfrac{\partial u}{\partial x}\right|_{y=x}$

8.13 In a laminar boundary layer the nominal thickness varies with the longitudinal distance x as

(a) $x^{-1/2}$ (b) $x^{-1/5}$
(c) x (d) $x^{1/2}$

8.14 In a laminar boundary layer over a flat plate the ratio of shear stresses τ_1 and τ_2 at two sections 1 and 2 at distances from the leading edge such that $x_2 = 5\,x_1$, is given by $\tau_1/\tau_2 =$

(a) $\sqrt{5}$ (b) $\dfrac{1}{\sqrt{5}}$
(c) 1.0 (d) 5.0

8.15 If δ_1 and δ_2 denote boundary layer thicknesses at a point distance x from the leading edge when the Reynolds numbers are 100 and 256 respectively, then the ratio of δ_1/δ_2 will be

(a) 0.625 (b) 1.6
(c) 2.56 (d) 4.90

8.16 The mean drag coefficient C_{Df} for a laminar boundary layer over a flat plate was found to be 0.015. If all other flow factors remain the same and the length of the plate is decreased to 1/4 of its original value, the drag coefficient C_{Df} would be equal to

(a) 0.015 (b) 0.060
(c) 0.030 (d) 0.0075

8.17 In the case of flow over a flat plate the growth of the boundary layer δ/x

(a) decreases with an increase in the kinematic viscosity
(b) increases with an increase in the free stream velocity
(c) decreases with an increase in the free stream velocity only if the boundary layer is laminar
(d) increases with an increase in the kinematic viscosity in both laminar and turbulent boundary layers.

8.18 If the velocity u in a turbulent boundary layer varies as $y^{1/7}$, the growth of the boundary layer thickness δ/x varies as

(a) $Re_x^{-1/5}$ (b) $Re_x^{-1/2}$
(c) $Re_x^{-4/5}$ (d) Re_x^{-1}

8.19 The rate of growth of the boundary layer thickness with the longitudinal distance on a flat plate

(a) is faster when the boundary layer is laminar than when it is turbulent
(b) is the same whether the boundary layer is laminar or turbulent
(c) is faster in a turbulent boundary layer when compared to that in a laminar boundary layer
(d) depends only on the aspect ratio of the plate

8.20 In a turbulent boundary layer on a flat plate the velocity distribution is given by the 1/7th power law. If the boundary layer over the whole plate is turbulent, the ratio of the shear stresses τ_1 and τ_2 at two sections 1 and 2 at distances from the leading edge $x_2 = 5\,x_1$ is given by $\tau_1/\tau_2 =$

(a) 1.379 (b) 1.258
(c) 2.236 (d) 1.308

8.21 The ratio of the coefficient of friction drag in laminar boundary layer compared to that in turbulent boundary layer is proportional to

(a) $R_L^{1/2}$ (b) $R_L^{1/5}$
(c) $R_L^{3/10}$ (d) $R_L^{3/10}$

8.22 In a turbulent boundary layer over a flat plate with Reynolds number $10^7 < Re_L < 10^9$ the growth of the boundary layer thickness is given by d/x is proportional to

(a) $Re_x^{1/5}$ (b) $Re_x^{-1/5}$
(c) $Re_x^{-1/6}$ (d) $Re_x^{-3/10}$

8.23 The laminar sublayer exists

(a) only in laminar boundary layers
(b) in all turbulent boundary layers
(c) only in smooth turbulent boundary layers
(d) only in rough fully developed turbulent boundary layers

8.24 On a flat plate, point A is at the mid-section and point B is at the trailing edge. The shear stresses τ_A at A and τ_B at B are such that

(a) $\tau_A > \tau_B$
(b) $\tau_B > \tau_A$

(c) $\tau_A = \tau_B$
(d) $\tau_A = \tau_B$ if the boundary layer is laminar

8.25 The laminar sublayer is
(a) a boundary layer that occurs before the formation of a regular laminar boundary layer
(b) is the first 1/8 of a laminar boundary layer, next to a boundary
(c) is a region where the wall roughness predominates
(d) is a region next to the wall where the laminar motion persists while the rest of the flow is turbulent

8.26 In a hydrodynamically smooth surface the roughness magnitude ε and laminar sublayer thickness δ' are related as
(a) $\varepsilon/\delta' > 1.0$ (b) $\varepsilon/\delta' < 0.25$
(c) $\varepsilon/\delta' \geq 6.0$ (d) $\varepsilon/\delta' = 1/30$

8.27 The thickness of laminar sublayer δ' is given by
(a) $11.6\ u_*/\nu$ (b) $u_*/(11.6\nu)$
(c) $11.6\ \nu/u_*$ (d) ν/u_*

8.28 In a boundary layer flow the parameter $u_*\varepsilon/\nu$ was equal to 1.5. The boundary can be classified hydrodynamically as
(a) rough
(b) in transition from rough to smooth
(c) in transition from smooth to rough
(d) smooth
[Here u_* = shear velocity, ε = equivalent roughness]

8.29 In a boundary layer flow the parameter $u_*\varepsilon/\nu$ was equal to 12. The boundary can be classified hydrodynamically as
(a) smooth (b) rough
(c) in transition (d) unstable

8.30 A turbulent boundary layer occurs over a flat plate practically right from the leading edge.
(a) it is possible for the surface to be hydrodynamically rough upstream and yet hydrodynamically smooth downstream.
(b) it is possible for the surface to be hydrodynamically smooth upstream and yet hydrodynamically rough downstream.
(c) if the flow is rough on the upstream it will have to rough on the downstream also.
(d) if the plate has roughness projections it will behave as rough plate everywhere.

8.31 The separation of boundary layer takes place when the pressure gradient is
(a) negative (b) positive
(c) zero (d) constant

8.32 At the point of separation
(a) Velocity is negative
(b) Shear stress is zero
(c) Shear stress is maximum
(d) Pressure gradient is zero

8.32 Separation of boundary layer takes place when
(a) $(\partial u/\partial y)_{y=0} > 0$ (b) $(\partial^2 u/\partial y^2)_{y=0} > 0$
(c) $(\partial u/\partial y)_{y=0} = 0$ (d) $(\partial u/\partial y)_{y=\delta} > 0$

8.34 The separation of a boundary layer occurs when
(a) the flow is accelerated past a boundary
(b) the boundary layer comes to rest
(c) any adverse pressure is encountered
(d) the fluid is ideal

8.35 To reduce separation a commonly used boundary layer control method is
(a) suction of accelerating fluid within the boundary layer
(b) using large divergence angles in the boundary
(c) suction of retarded fluid within the boundary layer
(d) use of smooth boundaries

8.36 It is required to have laminar flow in a 6 cm diameter pipe at a Reynolds number of 1500. The entrance length required for fully developed laminar flow to exist in the pipe is
(a) 105 cm (b) 630 cm
(c) 257 cm (d) 900 cm

8.37 An 8 cm diameter pipe is to carry water at a Reynolds number of 10^5. The entrance length required for the establishment of turbulent flow is about
(a) 4.0 m (b) 560 cm
(c) 10.0 m (d) 8.7 m

8.38 In the flow of a fluid past a circular cylinder the angle θ from the front stagnation point to the location at which separation takes place is about
(a) 82° if the boundary layer is laminar
(b) 82° if the boundary layer is turbulent
(c) 110° if the boundary layer is laminar
(d) 90° whether the boundary layer is laminar or turbulent.

PROBLEMS

8.1 Calculate the displacement thickness and momentum thickness in terms of δ, the nominal boundary layer thickness, for the following velocity distributions.

(a) $u/U = 2\eta - 2\eta^3 + \eta^4$

(b) $u/U = \dfrac{3}{2}\eta - \eta^{3/2}$

(c) $u/U = \eta^{1/7}$

where $\eta = y/\delta$.

(*Ans.* (a) $\delta^* = 0.3\ \delta$, $\theta = 0.117\ \delta$; (b) $\delta^* = 0.375\ \delta$, $\theta = 0.135\ \delta$; (c) $\delta^* = 0.125\ \delta$, $\theta = 0.097$)

8.2 For the velocity profile $u/U = \sin(\pi y/2\delta)$, calculate the shape factor H. How does it compare with the shape factor of the laminar boundary layer obtained by the exact solution of Blasius?

$$\left(Ans.\ H = \frac{\delta^*}{\theta} = 2.662;\quad H_{\text{Blasius}} = 2.604\right)$$

8.3 If the velocity distribution in a turbulent boundary layer is expressed as $u/U = (y/\delta)^{1/m}$, show that $\delta^*/\delta = 1/(m + 1)$ and $\theta/\delta = m/(m + 1)(m + 2)$.

8.4 For the laminar boundary layer over a flat plate the velocity distribution is given by $u/U = a + b\eta + c\eta^2$, where $\eta = y/\delta$. Determine the coefficients a, b and c.

(*Ans.* $a = 0$, $b = 2$, $c = -1$; $u/U = 2\eta - \eta^2$)

8.5 For a laminar boundary layer on a flat plate, derive the expressions for δ, C_f and C_{Df} when $u/U = f(\eta)$ given by

(a) $f(\eta) = 3\eta/2 - \eta^3/2$

(b) $f(\eta) = 2\eta - \eta^2$

(c) $f(\eta) = \sin(\pi\eta/2)$

(d) $f(\eta) = \eta$

(*Ans.* See Table 8.1)

8.6 A laminar boundary layer on a flat plate has a velocity distribution given by $u/U = 2\eta - 2\eta^3 + \eta^4$ where $\eta = y/\delta$. At a location the boundary layer thickness is 1.6 cm and the free stream velocity is 1.25 m/s. If $\mu = 1.136 \times 10^{-5}$ Pa.s, calculate the shear stress at that location.

(*Ans.* $\tau_0 = 1.775 \times 10^{-3}$ Pa)

8.7 A thin plate 2 m × 2 m is placed edgewise in a flow of oil. Calculate the boundary layer thickness and the shear stress at the trailing edge when the free stream velocity is 2 m/s. (RD of oil = 0.85 and $\nu = 10^{-5}$ m^2/s).

(*Ans.* $\tau_{0L} = 1.785$ Pa)

8.8 A sharp edged flat plate 1.5 m along the direction of flow and 3 m across is placed parallel to the flow of air. Find the drag on one side of the plate. Also find δ, δ^* and θ at the trailing edge for flow of air at 3 m/s past the plate. ($\rho = 1.23$ kg/m^3 and $\nu = 1.45 \times 10^{-5}$ m^2/s)

(*Ans.* $\delta = 0.01346$ m, $\delta^* = 0.023$ m, $\theta = 1.788 \times 10^{-3}$ m, $F_D = 0.0594$ N)

8.9 If the boundary layer over a flat plate, kept parallel to the flow, is laminar find the ratio of the skin friction drags on the front half of the plate to the rear half.

$$\left(Ans.\ \frac{F_1}{F_2} = 2.414\right)$$

8.10 A smooth flat plate 1.5 m wide and 2 m long is subjected to flow of water along its length at a uniform velocity of 2 m/s. Find (a) the extent of the laminar boundary layer on the plate, (b) the thickness of the boundary layer at the edge of the laminar boundary layer and at the trailing edge and (c) the shear stress at the trailing edge. (ρ = 998 kg/m^3, $\nu = 1 \times 10^{-6}$ m^2/s).

(*Ans.* $x_{crit} = 25$ cm, $\delta_c = 1.768$ mm, $\delta_L = 3.6$ cm, $\tau_0 = 5.631$ Pa)

8.11 A thin plate is moving in a direction parallel to its length in still air at a velocity of 4.0 m/s. The length of the plate is 0.5 m and width is 0.6 m. Taking $\nu_{air} = 1.5 \times 10^{-5}$ m^2/s and $\rho_{air} = 1.25$ kg/m^3, calculate (a) the boundary layer thickness at the end of the plate, (b) shear stress at 20 cm from the leading edge and (c) drag force on one side of the plate.

(*Ans.* $\delta = 6.847$ mm, $\tau_0 = 0.02875$ Pa, $F_D = 0.0109$ N)

8.12 Find the ratio of friction drags on the front half and rear half of a plate kept in a stream at zero angle of incidence. Assume the boundary layer to be turbulent over whole plate and the Reynolds number to be of the order of 10^6.

$$\left(Ans.\ \frac{F_1}{F_2} = 1.349\right)$$

8.13 A smooth flat plate is kept at zero angle of incidence in a stream and the Reynolds number in terms of the length of the plate is of the order of 10^6. At what fraction of the total length, measured from the leading edge, would the drag force on the front portion would be equal to half of the total drag force on the plate? Assume the boundary layer to be turbulent over the whole plate.

$$\left(Ans.\ \frac{x}{L} = 0.420\right)$$

8.14 Wind at 100 km/h blows along a long flat surface 50 m long. Estimate (i) the thickness of the boundary layer at distances of 5 m and 50 m from the leading edge, and (ii) shear stress at 5 m and 50 m from the leading edge. [$\rho = 1.22$ kg/m^3 and $\nu = 1.5 \times 10^{-5}$ m^2/s]

(*Ans.* $\delta_5 = 7.62$ cm, $\delta_{50} = 51.7$ cm; $\tau_{05} = 1.123$ Pa, $\tau_{050} = 0.8236$ Pa)

8.15 A barge has a rectangular bottom 30 m long and 10 m wide. Calculate (i) the frictional force on the bottom when the barge moves at a velocity of 1.5 m/s, and (ii) the thickness of the boundary layer and the shear stress at the trailing edge. ($\rho = 998$ kg/m^3 and $\nu = 1 \times 10^{-6}$ m^2/s).

(*Ans.* (i) $F_D = 722$ N; (ii) $\delta = 33.3$ cm, $\tau_0 = 1.952$ Pa)

8.16 A train is 250 m long and its surface area of top and bottom add up to 15 m^2 per metre length of the train. If the train moves at a speed of 120 km/h, calculate the power required to overcome surface resistance. The surfaces can be assumed to be smooth. Take $\rho_{air} = 1.2$ kg/m^3 and $\mu_{air} = 1.80 \times 10^{-5}$ Pa.s

(*Ans.* **P** = 140.7 kW)

8.17 Determine the skin friction drag on a plate 3 m long and 4.5 m wide subjected to a wind of velocity 40 m/s blowing along its length, under the following conditions:

(a) the boundary layer on the plate being laminar in the initial part of the plate.
(b) the boundary layer is turbulent over the entire length of the plate.
(c) the boundary layer is turbulent over the entire surface which is hydrodynamically rough with a roughness size $\varepsilon = 3$ mm.
($\rho = 1.18$ kg/m^3 and $\nu = 1.50 \times 10^{-5}$ m^2/s).

(*Ans.* (a) $F_D = 36.55$ N, (b) $F_D = 39.25$ N, (c) $F_D = 107.65$ N)

8.18 The length of a submarine is 80 m and its surface area is 3000 m^2. If the submarine is moving with a velocity of 4 m/s, determine the frictional drag, considering

(a) the boundary layer to be turbulent over the entire surface
(b) the turbulent boundary layer is preceded by a laminar sublayer
(c) the surface of the submarine is rough with roughness height $\varepsilon = 3$ mm.
[$\mu = 1.0 \times 10^{-3}$ Pa.s and $\rho = 1040$ kg/m^3]

(*Ans.* (a) $F_D = 45.13$ kN, (b) $F_D = 45.003$ kN, (c) $F_D = 101.02$ kN)

8.19 In a flow over a smooth flat plate the Reynolds number at the trailing edge is 10^6. If the critical Reynolds number is 5×10^5, what fraction of the total frictional force occurs in the laminar boundary layer?

(*Ans.* $F_{lam} = 31.6\%$ of the total drag)

8.20 An air stream flows over a smooth flat plate with a terminal Reynolds number of 10^6. The boundary layer can be assumed to be turbulent over the entire plate. If the length of the plate is increased by 10%, keeping all other factors same, what is the percentage change in the (a) total drag coefficient and (b) total drag force?

(Ans. (a) 1.9% decrease in C_{Df}; (b) 7.92% increase in F_D)

8.21 In a turbulent boundary layer over a flat plate it is found that $u/U = (y/\delta)^{1/7}$ and $C_f = 0.02\, Re_\delta^{-1/6}$. By using Karman momentum integral equation obtain expressions (i) for δ and C_f in terms of Re_x and (ii) for C_{Df} in terms of Re_L
[$Re_\delta = U\,\delta/\nu$, $Re_x = U\,x/\nu$, and $Re_L = UL/\nu$]

$$\left(Ans.\ C_f = \frac{0.027}{Re_x^{1/6}},\quad C_{Df} = \frac{0.031}{Re_L^{1/6}},\quad \frac{\delta}{x} = \frac{0.16}{Re_x^{1/7}}\right)$$

8.22 Assuming laminar boundary layer flow and flat plate boundary layer results to be valid, develop expressions for the friction drag of the following objects immersed parallel to a stream of uniform velocity U:

(a) Thin small disk of diameter D
(b) The equilateral triangular plate of sides a

$$\left(Ans.\ (a)\ C_{Df} = \frac{2.90}{\sqrt{UD/\nu}}\ (b)\ C_{Df} = \frac{4.09}{\sqrt{Ua/\nu}}\right)$$

8.23 A 1/25 model of an ocean-going ship was towed in a towing tank containing fresh water. The model had the same Froude number as the prototype. The model had a length of 3 m, wetted surface area of 2 m^2 and was towed at a speed to reproduce the prototype speed of 9 m/s. What is the prototype drag corresponding to a measured total drag of 16 N in the model?
[For sea water: ρ_w = 1025 kg/m^3 and μ_s = 1.07 × 10^{-3} Pa.s
For fresh water ρ_w = 998 kg/m^3 and μ_w = 1.00 × 10^{-3} Pa.s].

(*Ans.* F_{Dp} = 186.1 kN)

9

Drag and Lift on Immersed Bodies

CONCEPT REVIEW

9.1 INTRODUCTION

A fluid moving relative to a rigid boundary exerts a force on the boundary. Shear stresses acting on the surfaces and pressure acting normal to the boundary give rise to forces which would add up vectorially to give a resultant force on the body (Fig. 9.1). The component of the resultant force in the direction of the relative velocity V_0 of the flow past the body is called the *drag*. The component of the resultant normal to relative velocity is known as *lift*.

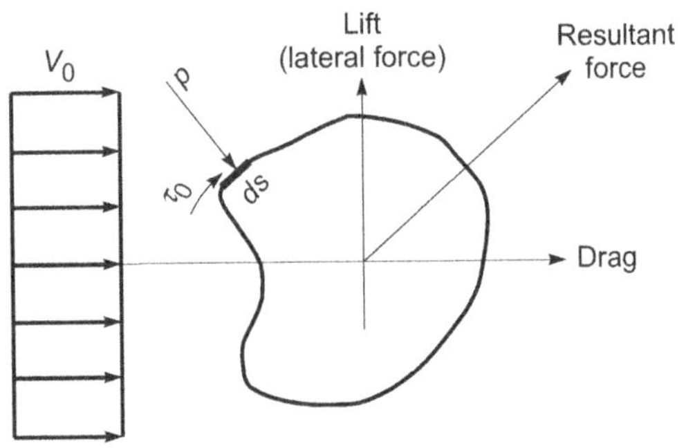

Fig. 9.1 *Definition Sketch for Flow Induced Forces*

9.2 DRAG

The total drag (sometimes called as *profile drag*) of a body is made up of two parts: frictional drag and pressure drag (or *form drag*), depending on whether the shear stresses or pressure differences cause it. The relative proportion of these two parts depends upon the shape of the body and flow condition. In practical applications, it is the total drag rather than its constituent components that is important. Hence the total drag F_D on an immersed body in a relative free stream velocity V_o is expressed as

$$\boxed{F_D = C_D A \frac{\rho V_0^2}{2}} \qquad (9.1)$$

where C_D = total drag coefficient

A = a characteristic area of the body

In general C_D = fn (geometry, Reynolds number, Froude number, Mach number).

For incompressible flow and in the absence of free surface effects

$$\boxed{C_D = \text{fn (geometry, } Re)}$$

The characteristic area A is one of the following, as per convention:

(1) A = frontal area = projected area on a plane normal to V_0.
This is used for blunt (bluff) shaped objects such as spheres, cylinders, cars, trains, projectiles and missiles.

(2) A = planform area = area of the body as seen from above. This is used for thin, flat surfaces where

frictional forces are predominant as for example: flat plate flow, airplane wings and hydrofoils.

(3) A = wetted area. Customarily this is used for water crafts such as boats, barges and ships.

The variation of the drag coefficient for certain interesting and commonly used objects is discussed in the following sub-sections.

9.2.1 Sphere

The variation of the drag coefficient C_D with Reynolds number $Re = \frac{V_0 D}{\nu}$ where D = diameter of the sphere, is illustrated in Fig. 9.2. The variation is studied in three regimes:

(i) *Very small Reynolds numbers*

For $Re < 1.0$, the fluid motion is known as *creeping motion*. The total drag is given by Stokes' law, as

$$\boxed{F_D = 3\pi D \mu V_0} \qquad (9.2)$$

Writing
$$F_D = C_D \left(\frac{\pi D^2}{4}\right)\frac{\rho V_0^2}{2} \qquad (9.3)$$

$$\boxed{C_D = \frac{24\rho}{V_0 D/\mu} = \frac{24}{Re}}$$

Stokes' law is valid for $Re \le 1.0$ and its application to determine fall velocities of small particles is given in Examples 9.1 to 9.4.

(ii) *Reynolds number:* $1 \le Re \lesssim 2 \times 10^5$

In this regime, increasing Reynolds number reflects decreasing role of viscosity and C_D drops from a value of 24 at $Re = 1$ to about 0.4 at $Re \approx 10^4$. Then onwards, C_D is essentially constant, and at around $Re = 10^5$, $C_D = 0.5$. The region $10^4 \lesssim Re \lesssim 2 \times 10^5$ marks laminar boundary layer flow. In the range $1 < Re < 100$, C_D can be approximated by

$$\boxed{C_D = \frac{24}{Re}\left(1 + \frac{3}{16} Re\right)^{1/2}} \qquad (9.4)$$

(iii) *Transition Region:* ($Re \approx 2 \times 10^5$)

Around $Re = 2 \times 10^5$ the laminar boundary layer undergoes transition to turbulent boundary layer. Consequently, the boundary layer separation line which occurred at around 85° angle shifts downstream to an angular position of about 120° causing a reduction in wake area. This is reflected in a rapid reduction in the value of C_D from 0.5 to 0.2. The critical Reynolds number depends upon boundary roughness and free stream turbulence, and an average value of 2×10^5 is generally adopted.

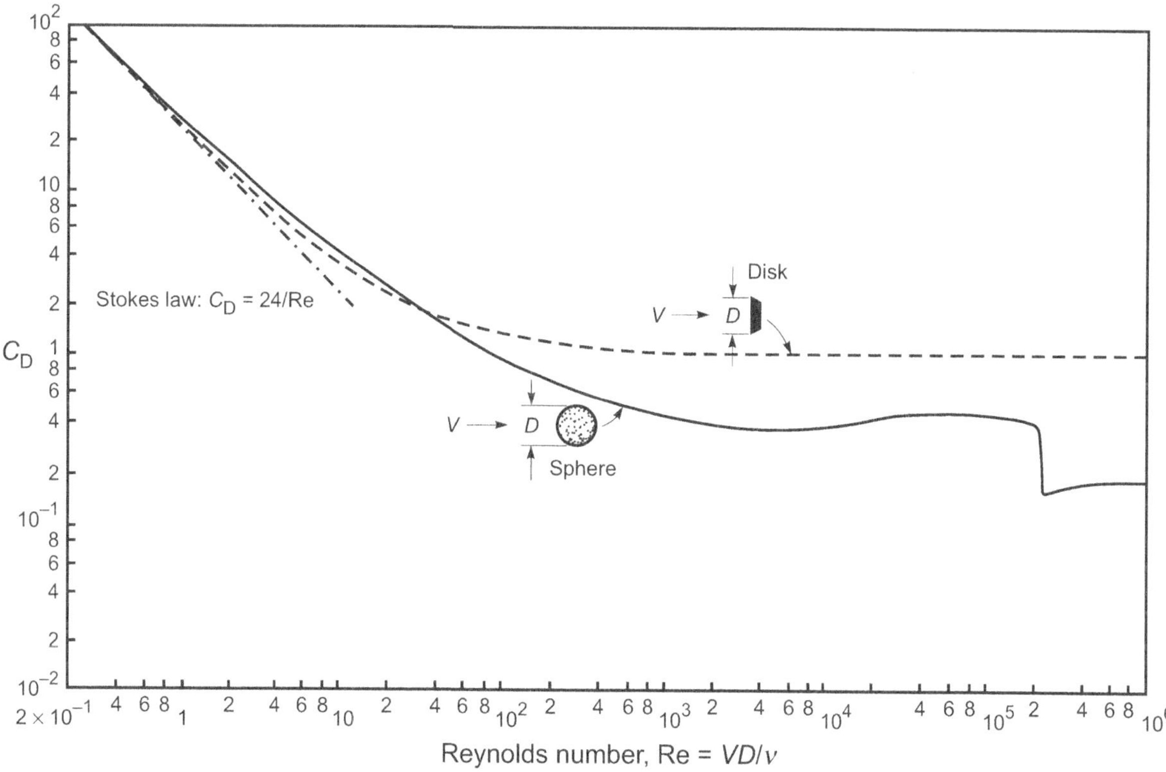

Fig. 9.2 *Variation of C_D of a Sphere*

(iv) *Reynolds number:* $Re > 2 \times 10^5$
The boundary layer is turbulent and the drag coefficient assumes an essentially constant value of 0.2.

9.2.2 Cylinder

A cylinder is a typical example of a two-dimensional bluff body. The nature of variation of C_D with Reynolds number Re is essentially similar to that of the sphere. The values of C_D are, however, different and depend upon L/D also in addition to Re. For $L/D \approx \infty$, $C_D = 1.20$ in laminar boundary layer region ($Re \approx 10^5$) and it drops to 0.33 at $Re > 5 \times 10^5$.
An interesting feature of flow past two-dimensional bodies is the formation and alternate release of vortices behind the cylinder for $Re > 30$. This vortex shedding leads to lateral vibration of two-dimensional bodies. The frequency of vortex shedding n is expressed as *Strouhal number*

$$S = \frac{nD}{V_0} \tag{9.6}$$

For cylinders with $Re > 10^3$ the Strouhal number is essentially constant at $S = 0.21$. For $Re < 10^3$, the value of S can be estimated by the formula due to Taylor, given by

$$S = 0.20\left(1 - \frac{20}{Re}\right) \tag{9.7}$$

9.2.3 Miscellaneous Bodies

The drag coefficients of a variety of commonly used three-dimensional and two-dimensional body shapes are given in Table 9.1.

Table 9.1 *Drag Coefficient C_D for Various Bodies*

Form of body	*L/D*	*Re*	C_D
1. Sphere		10^5	0.50
		$> 3 \times 10^5$	0.20
2. Hemisphere:			
Hollow upstream		$> 10^3$	1.33
Hollow downstream		$> 10^3$	0.34
3. Ellipsoid (1 : 2, major axis along flow)		$> 2 \times 10^5$	0.07
4. Circular cylinder (axis along flow)	0	$> 10^3$	1.12
	1		0.91
	2		0.85
	4		0.87
	7		0.99
5. Circular cylinder (axis normal to flow)	1	10^5	0.63
	5		0.74
	20		0.90
	∞		1.20
	5	$> 5 \times 10^5$	0.35
	∞		0.33
6. Rectangular plate (L = length, D = width)	1	$> 10^3$	1.16
	5		1.20
	20		1.50
	∞		1.90
7. Circular disk		$> 10^3$	1.12

9.3 LIFT

The lift force F_L, which occurs normal to the direction of relative motion V_0 is expressed as

$$F_L = C_L A \frac{\rho V_0^2}{2} \tag{9.8}$$

where C_L = Lift coefficient and
A = a characteristic area.

For lifting bodies such as airfoils, hydrofoil vanes, the practice is to have large surface areas with small thickness and large chord lengths (Fig. 9.3). For such lifting bodies the characteristic area A in the definition of the lift coefficient is the platform area $A = A_p = CL$. Note that the drag coefficients for lifting bodies area also defined in terms of A_p.

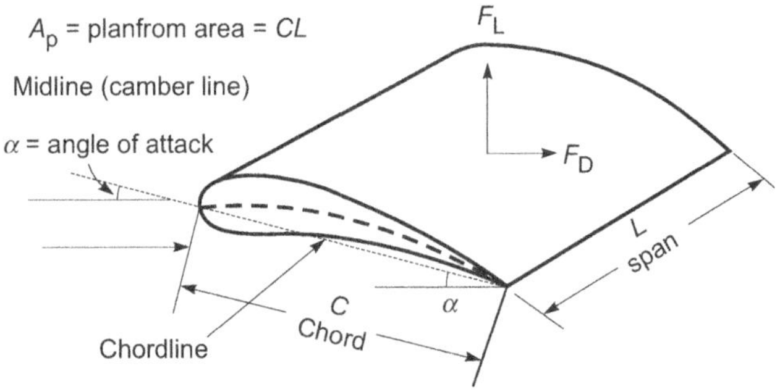

Fig. 9.3 *Definition Sketch for a Lifting Vane*

Thus

$$\text{Lift coefficient (for a lifting body)} \quad C_L = \frac{F_L}{A_p \cdot (\rho V_0^2/2)} \tag{9.9}$$

$$\text{Drag coefficient (for a lifting body)} \quad C_D = \frac{F_D}{A_p \cdot (\rho V_0^2/2)} \tag{9.10}$$

For incompressible flow

$$C_L \text{ (or } C_D) = \text{fn}(\alpha, Re_c) \tag{9.11}$$

where α = angle of attack and

Re_c = Reynolds number = $\dfrac{V_0 C}{\nu}$ where C = Chord length.

9.3.1 Rotating Cylinders

For a cylinder of radius r rotating with an angular velocity ω in an ideal fluid flowing with a velocity V_0:

V_c = tangential velocity on the circumference of the cylinder = ωr

The circulation $\Gamma = 2\pi r V_c$

The lift force $F_L = L\rho V_0 \Gamma$

Expressing $F_L = C_L = \rho(\text{Area})\,\dfrac{V_0^2}{2}$,

where Area = LD,

$$L\rho V_0 \Gamma = C_L \rho L D\,\frac{V_0^2}{2}$$

or
$$\boxed{C_L = 2\Gamma / V_0 D = \frac{2\pi D V_c}{D V_0} = \frac{2\pi V_c}{V_0}} \qquad (9.12)$$

The production of lift due to a rotating body is known as *Magnus effect*. The resulting streamline pattern round such a rotating cylinder is shown in Fig. 9.4 (b, c and d). It will be noticed from Fig. 9.4(b) that the stagnation points S_1 and S_2 are not at 0 and 180° to flow direction, but are at angle $\theta > 180°$. The value of θ is given by

$$2\sin\theta = -\frac{V_c}{V_0}$$

or
$$\boxed{\theta = \sin^{-1}\left(-\frac{1}{2}\frac{V_c}{V_0}\right)}$$

Thus the two stagnation points merge and occur at θ = 270° when $V_c = 2V_0$, (Fig. 9.4(c)). For $V_c > 2V_0$, the stagnation point is removed from the cylinder surface and a ring of fluid is dragged around the cylinder, (Fig. 9.4(d)). The above are theoretical results. In practice, due to viscosity, there will be considerable variation in C_L and θ values.

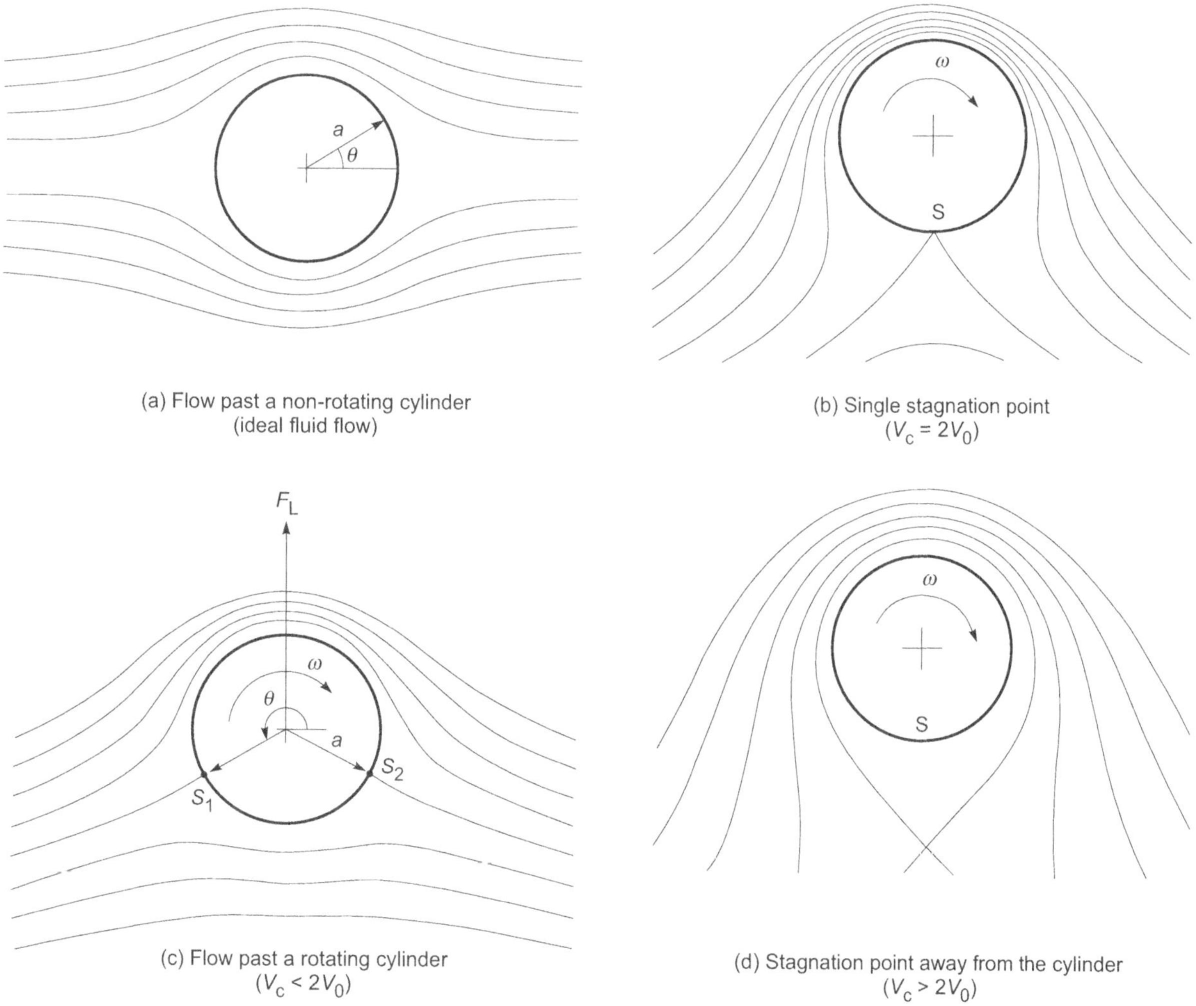

(a) Flow past a non-rotating cylinder (ideal fluid flow)

(b) Single stagnation point ($V_c = 2V_0$)

(c) Flow past a rotating cylinder ($V_c < 2V_0$)

(d) Stagnation point away from the cylinder ($V_c > 2V_0$)

Fig. 9.4

9.3.2 Lift and Stall in an aerofoil

In a lifting surface such as an aerofoil, at small angles of attack the rounded leading edge prevents flow separation. However, the sharp trailing edge causes flow separation. This separation at the tail generates a vortex which in turn causes realignment of stream lines resulting in lift. The lift increases with the angle of attack up to a limit of about 15°–20°. At this limit, the flow separates completely from the leading edge and the aerofoil is said to be *stalled*. At the stalling point, the lift drops off markedly and the drag increases very significantly making the aerofoil incapable of flying (Fig. 9.5).

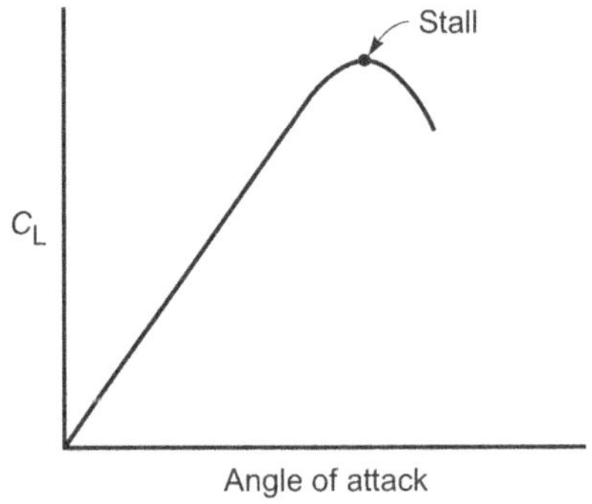

Fig. 9.5 *Variation of C_L and Location of Stall in an aerofoil*

WORKED EXAMPLES

9.1 A small sphere of diameter D and density ρ_s is dropped into a large body of fluid of density ρ_f. Calculate the terminal velocity of the sphere.

(a) by assuming that the Stokes' law is applicable;
(b) in terms of the drag coefficient C_D.

Solution: At terminal velocity, since there is no change in the velocity, the net force on the body is zero. Hence, (weight of cylinder – buoyant force) = drag (see Fig. 9.6), i.e.

$$\frac{\pi}{6}\cdot D^3\,\rho_s g - \frac{\pi}{6}\,D^3\rho_f g = F_D$$

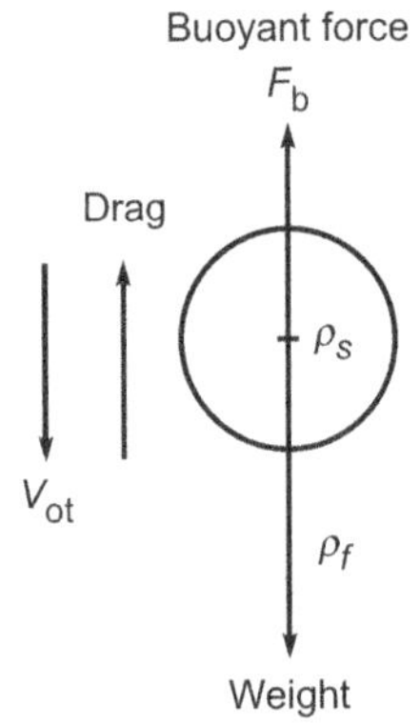

Fig. 9.6

(a) If the Reynolds number is very small ($Re < 1$) Stokes' law is applicable and $F_D = 3\pi D\mu V_{0t}$. where V_{0t} = terminal velocity.

$$\frac{\pi}{6}D^3\,(\rho_s - \rho_f)\,g = 3\pi D\mu V_{0t}$$

$$V_{0t} = \frac{D^2}{18}\,\frac{g(\rho_s - \rho_f)}{\mu} = \frac{D^2\,(\gamma_s - \gamma_f)}{18\mu}$$

where $\gamma_f = \rho_f g$ = specific weight of the liquid
and $\gamma_s = \rho_s g$ = specific weight of the sphere.

Hence terminal velocity $\quad V_{0t} = \dfrac{D^2\,(\gamma_s - \gamma_f)}{18\,\mu}$

and this is valid for $\quad Re = \dfrac{V_{0t}\,D}{\nu} \le 1.0$

(b) In general $\quad F_D = C_D A \rho_f \dfrac{V_{0t}^2}{2}$

where C_D = drag coefficient and A = frontal area = $\dfrac{\pi D^2}{4}$ for a sphere.

Hence $\quad \dfrac{\pi}{6}D^3\,(\rho_s - \rho_f)\,g = C_D\left(\dfrac{\pi D^2}{4}\right)\rho_f\,\dfrac{V_{0t}^2}{2}$

and on solving $\quad V_{0t} = \left[\dfrac{4}{3}\dfrac{gD}{C_D}\left(\dfrac{\rho_s}{\rho_f} - 1\right)\right]^{1/2}$

9.2 A sphere 20 cm in diameter and of relative density 2.8 is dropped into a large body of water. Calculate the terminal velocity of the sphere. [ρ_{water} = 998 kg/m³, $\nu = 1 \times 10^{-6}$ m²/s] [C_D of a sphere: for $10^4 < Re < 3 \times 10^5$, $C_D = 0.5$ and for $Re > 3 \times 10^5$, $C_D = 0.20$].

Solution: Assume $Re > 3 \times 10^5$. Then $C_D = 0.20$.
At terminal velocity V_{0t}, submerged weight = Drag

$$\frac{\pi}{6}D^3\,(\gamma_s - \gamma_f) = F_D = C_D A\,\frac{\rho_f V_{0t}^2}{2}$$

$$V_{0t}^2 = \frac{\pi D^3(\gamma_s - \gamma_f)}{3(\pi D^2/4)\,\rho_f C_D} = \frac{4}{3}\frac{D}{C_D}g\left(\frac{\rho_s}{\rho_f} - 1\right)$$

$$= \frac{4}{3} \times \frac{0.2}{0.2} \times 9.81 \left(\frac{998 \times 2.8}{998} - 1 \right)$$

$$V_{0t}^2 = 23.544$$

$$\mathbf{V_{0t} = 4.852\ m/s}$$

$$Re = \frac{V_0 D}{\nu} = \frac{4.852 \times 0.2}{1 \times 10^{-6}}$$

$$= 9.70 \times 10^5 > 3 \times 10^5$$

Hence the initial assumption is correct and $V_{0t} = 4.852$ m/s.

9.3 What is the terminal velocity of the maximum size of sand particles (RD = 2.65) that will settle in water according to Stokes' law? What is its size? ($\rho_{water} = 998$ kg/m^3, $\mu = 1 \times 10^{-3}$ Pa.s).

Solution: Stokes' law is valid up to $Re = 1.0$. For maximum size particles that obey Stokes' law,

$$(Re)_{max} = \frac{\rho_f V_{0t} \cdot D}{\mu} = 1.0$$

$$D = \frac{\mu}{\rho_f V_{0t}}$$

Also the fall velocity

$$V_{0t} = \frac{1}{18} \frac{D^2}{\mu} (\gamma_s - \gamma_f)$$

Substituting for D,

$$V_{0t} = \frac{1}{18} \frac{\mu^2}{\rho_f^2 V_{0t}^2 \mu} (\gamma_s - \gamma_f)$$

$$V_{0t}^3 = \frac{1}{18} g \left(\frac{\mu}{\rho_f} \right) \left(\frac{\gamma_s}{\gamma_f} - 1 \right)$$

$$= \frac{1}{18} \times 9.81 \times \left(\frac{1 \times 10^{-3}}{998} \right) \left(\frac{2.65 \times 998}{998} - 1 \right)$$

$$= 9.01 \times 10^{-7}$$

$$V_{0t} = 9.66 \times 10^{-3} \text{ m/s} = \mathbf{9.66\ mm/s}$$

$$D = \frac{\mu}{\rho_f V_{0t}} = \frac{1 \times 10^{-3}}{998 \times 9.66 \times 10^{-3}}$$

$$= 1.0373 \times 10^{-4} \text{ m} = \mathbf{0.1037\ mm}$$

9.4 A glass sphere (RD = 2.60) 2.0 mm in diameter is observed to have a fall velocity of 1.25 cm/s in a oil of density 917 kg/m^3. Estimate the coefficient of dynamic viscosity of the oil.

Solution: Assuming the validity of Stokes' law,

$$\text{Fall velocity } V_{0t} = \frac{1}{18} \frac{D^2 (\gamma_s - \gamma_f)}{\mu} \quad (1)$$

Using Eq. 1

$$\mu = \frac{D^2 (\gamma_s - \gamma_f)}{18 V_{0t}}$$

$$= \frac{(2 \times 10^{-3})^2 (2.60 \times 998 \times 9.81 - 917 \times 9.81)}{18 \times 0.0125}$$

$$= \mathbf{0.293\ Pa.s}$$

Reynolds number

$$Re = \frac{\rho V_{0t} D}{\mu}$$

$$= \frac{917 \times 0.0125 \times 2 \times 10^{-3}}{0.293}$$

$$= 0.078 < 1.0$$

Hence the assumed Stokes' law is valid.

9.5 At standard atmospheric pressure (101.3 kPa) and temperature ($\rho_a = 1.20$ kg/m^3) calculate the terminal velocities of 0.2 mm and 2 mm rain drop given their drag coefficients (C_D) as 4.20 and 0.517 respectively.

Solution: The terminal velocity

$$V_{0t} = \left[\frac{4}{3} \frac{gD}{C_D} \left(\frac{\rho_s}{\rho_f} - 1 \right) \right]^{1/2}$$

For 0.2 mm rain drop:

ρ_s = density of water = 998 kg/m^3

ρ_f = density of air = 1.20 kg/m^3

$$V_{0t} = \left[\frac{4 \times 9.81 \times (0.2 \times 10^{-3})}{3 \times 4.20} \left(\frac{998}{1.20} - 1 \right) \right]^{1/2}$$

$$= \mathbf{0.72\ m/s}$$

For 2.0 mm rain drop:

$$V_{0t} = \left[\frac{4 \times 9.81 \times (2 \times 10^{-3})}{3 \times 0.517} \left(\frac{998}{1.20} - 1 \right) \right]^{1/2}$$

$$= \mathbf{6.48\ m/s}$$

9.6 A car has a frontal projected area of 1.6 m^2 and travels at 60 km/h. It has a drag coefficient of 0.35 based on frontal area.

(a) Calculate the power required to overcome wind resistance by the car.

(b) If the drag coefficient is reduced to 0.30 by streamlining, for the same power expended in overcoming air resistance, what speed of the car is possible? [$\rho_{air} = 1.2$ kg/m^3]

Solution:

$V_0 = 60$ km/h = 16.67 m/s

(a) F_D = Drag force = $C_D A \dfrac{\rho V_0^2}{2}$

$= 0.35 \times 1.6 \times 1.2 \times \dfrac{(16.67)^2}{2} = 93.33$ N

Power = $F_D \times V_0 = 93.33 \times 16.67$

= 1555.6 W = **1.556 kW**

(b) Power = $C_D A \dfrac{\rho V_0^3}{2} = 1555.6$

$V_0^3 = \dfrac{1555.6 \times 2}{0.3 \times 1.6 \times 1.2} = 5401.4$

$V_0 = 17.545$ m/s = **63.16 km/h**

9.7 A cup anemometer used to measure wind velocities consists of two hemispherical cups of diameter 10 cm mounted in opposite directions on a horizontal connecting rod which can turn freely on a vertical axis (Fig. 9.7) midway. What torque would be required to hold the rod from rotating in a wind of velocity (a) 45 km/h (b) 60 km/h?
[C_D for hemispherical bodies: C_D (Hollow upstream) = 1.33, C_D (Hollow downstream) = 0.34; ρ_{air} = 1.2 kg/m^3]

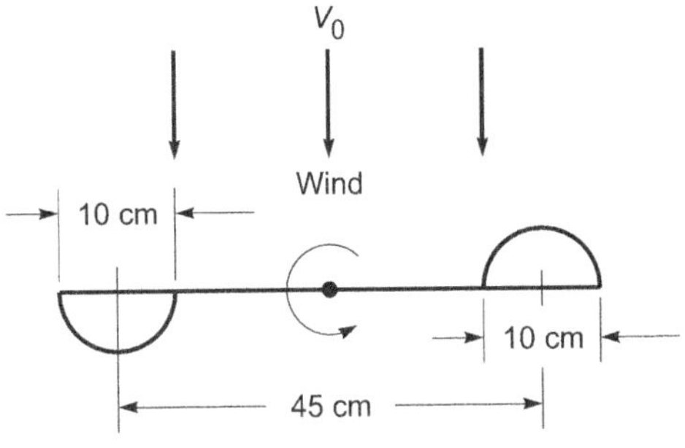

Fig. 9.7

Solution: Refer Fig. 9.7.

F_D = Drag force on a cup = $C_D A \dfrac{\rho V_0^2}{2}$

Since both the cups are subjected to same V_0, torque about the vertical axis,

$T = (F_{D1} - F_{D2}) \times r$ where r = lever arm

$= (C_{D1} - C_{D2})\, A \dfrac{\rho V_0^2}{2} \times r$

$= (1.33 - 0.34) \times \dfrac{\pi \times (0.10)^2}{4} \times \dfrac{1.2}{2} \times V_0^2 \times \dfrac{0.45}{2}$

$= 1.04968 \times 10^{-3}\, V_0^2$

(a) When V_0 = 45 km/h = 12.5 m/s

Torque = $T = 1.04968 \times 10^{-3} \times (12.5)^2$

= **0.164 N.m**

(b) When V_0 = 60 km/s = 16.67 m/s

$T = 1.04968 \times 10^{-3} \times (16.67)^2$

= **0.2916 N.m**

9.8 A cup anemometer consists of two hemispherical cups of 12 cm diameter mounted in opposite directions on a horizontal connecting rod as shown in Fig. 9.8. The attachment rotates about a vertical axis at the middle of the horizontal rod. Assuming zero friction at the bearings, calculate the rotational speed of the anemometer in a wind velocity of 60 km/h. [C_D = 1.33 and 0.34 depending on the frontal shape of the hemisphere, ρ_{air} = 1.2 kg/m^3].

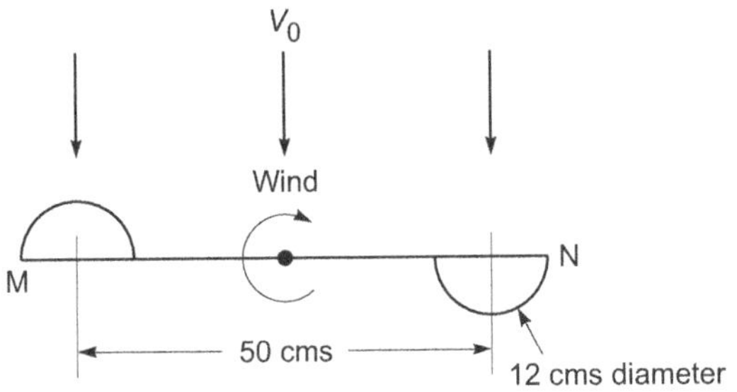

Fig. 9.8

Solution: Let ω = steady angular velocity in a wind velocity of V_0. For constant ω, the net torque about the axis of rotation is zero.

$V_0 = 60$ km/h = 16.67 m/s

v = velocity of the cup = $\omega r = 0.25\,\omega$

Relative velocity of the concave cup

$N = V_0 - 0.25\omega$

Relative velocity of the convex cup

$M = V_0 + 0.25\omega$

C_D for N = 1.33

C_D for M = 0.34

Drag force

$F_{DN} = 1.33 \times \dfrac{\pi D^2}{4} \dfrac{\rho}{2} (V_0 - 0.25\,\omega)^2$

$F_{DM} = 0.34 \times \dfrac{\pi D^2}{4} \dfrac{\rho}{2} (V_0 + 0.25\,\omega)^2$

Torque = $(F_{DN} - F_{DM}) \times r = 0$

i.e. $F_{DN} = F_{DM}$ (as $r \neq 0$)

$\therefore$ $1.33\,(V_0 - 0.25\,\omega)^2 = 0.34\,(V_0 + 0.25\,\omega)^2$

$\left(\dfrac{1.33}{0.34}\right)^{1/2} (V_0 - 0.25\,\omega) = V_0 + 0.25\,\omega$

$0.9778\, V_0 = 0.7445\,\omega$

$\omega = 1.31346\, V_0 = 1.31346 \times 16.667 = 21.89$ rad/s

Rotations per minute rpm $= \dfrac{60\omega}{\pi(2r)}$

$$= \frac{60 \times 21.89}{\pi \times 0.50} = \mathbf{836\ rpm}$$

[*Note* the difference in calculating the drag in this and the previous problem. Here the cups are rotating and hence the relative velocity is used. In Example 9.7 the cups are held stationary and as such the relative velocity is equal to the velocity of wind.]

9.9 It is desired to provide a para–trooper with a hemispherical parachute which will provide a terminal fall velocity no greater than that caused by a jump from a 2 m high wall. The parachute is to be designed for a total load of 1.0 kN. Estimate the minimum size of the parachute. [$\rho_a = 1.2$ kg/m^3; C_D for a hemisphere = 0.34 and 1.33 depending on a convex or concave frontal surface to the flow].

Solution:

Terminal velocity $= V_{0t}$

$= $ Velocity due to 2 m jump $= \sqrt{2gh}$

$= \sqrt{2 \times 9.81 \times 2} = 6.26$ m/s

For a parachute in terminal velocity the C_D of a hemisphere with concave frontal surface is appropriate. Hence $C_D = 1.33$. The total weight is balanced by the drag.

$$F_D = 1000 = C_D \times A \times \frac{\rho V_0^2}{2}$$

$$1000 = 1.33 \times \left(\frac{\pi \times D^2}{4}\right) \times \frac{1.2 \times (6.26)^2}{2}$$

$$D^2 = 40.71$$

$$D = \mathbf{6.38\ m,\ say\ 6.5\ m}$$

9.10 A high tension cable 5 cm in diameter is strung out between two towers. At a wind velocity of 80 km/h, calculate the,

(i) drag per unit length of the cable and

(ii) frequency of vortex shedding.

[$\rho_{air} = 1.2$ kg/m^3, $\nu = 1.5 \times 10^{-5}$ m^2/s]

[C_D of a cylinder: for $10^4 < Re \le 5 \times 10^5$, $C_D = 1.20$ for $Re > 5 \times 10^5$, $C_D = 0.35$]

Solution:

$$V_0 = 80 \text{ km/h} = 22.22 \text{ m/s}$$

$$Re = \frac{V_0 D}{\nu} = \frac{22.22 \times 0.05}{1.5 \times 10^{-5}} = 7.4 \times 10^4$$

Hence $C_D = 1.20$

(i) Drag force for unit length of cable: ($A = LD$)

$$F_D = C_D A \frac{\rho V_0^2}{2}$$

$$= 1.2 \times (1.0 \times 0.05) \times \frac{1.2}{2} \times (22.22)^2$$

$$= \mathbf{17.78\ N/metre\ length\ of\ cable}$$

(ii) Strouhal number $S = 0.21$ for $Re > 10^3$

$$\therefore \quad S = \frac{nD}{V_0} = 0.21$$

$$n = 0.21 \frac{V_0}{D} = \frac{0.21 \times 22.22}{0.05}$$

$= \mathbf{93.3\ Hz}$ = frequency of vortex shedding

9.11 A chimney has a diameter of 2.5 m and is 50 m tall. Calculate the bending moment at the base of the chimney when a gale at 80 km/h blows across it. [$C_D = 1.20$ and 0.33 with critical Reynolds number at 5×10^5, $\rho_{air} = 1.2$ kg/m^3, $\nu_{air} = 1.5 \times 10^{-5}$ m^2/s]

Solution:

$$V_0 = 80 \text{ km/h} = 22.22 \text{ m/s}$$

Reynolds number

$$Re = \frac{V_0 D}{\nu} = \frac{22.22 \times 2.5}{1.5 \times 10^{-5}}$$

$$= 3.7 \times 10^6$$

This is larger than the critical Reynolds number and, as such, $C_D = 0.33$.

Force on the chimney

$$F_D = C_D \cdot (LD) \frac{\rho V_0^2}{2}$$

$$= 0.33 \times (50 \times 2.5) \times \frac{1.2 \times (22.22)^2}{2}$$

$$= 12{,}222 \text{ N} = 12.222 \text{ kN}$$

Bending moment at the base

$$M_0 = F\frac{L}{2} = 12.222 \times \frac{50}{2} = \mathbf{305.6\ kN.m}$$

9.12 A spherical balloon weighs 135 N and contains helium with a density of 0.22 kg/m^3. The surrounding air has a density of 1.22 kg/m^3.

(a) What diameter of the balloon permits a terminal rising velocity of 2.0 m/s?

(b) If the balloon is tied to ground by a cable, what would be the inclination of the cable when a wind of 10 km/h blows past the balloon?

[$\nu_{air} = 1.5 \times 10^{-5}$ m^2/s; For $10^4 \le Re$ 3×10^5, $C_D = 0.5$ and for $Re > 3 \times 10^5$, $C_D = 0.2$]

Solution: When the balloon is rising:

(a) Assume $Re > 3 \times 10^5$ so that $C_D = 0.2$

Buoyant force = weight + drag

$$(\rho_{air} - \rho_{He})\, g \frac{\pi}{6} D^3 = 135 + C_D A \frac{\rho V_0^2}{2}$$

$$(1.22 - 0.22) \times 9.8 \times \frac{\pi}{6} \times D^3 = 135 + 0.2 \times \left(\frac{\pi}{4} D^2\right) \times \frac{1.22 \times 2^2}{2}$$

$$5.1365\, D^3 = 135 + 0.3833\, D^2$$

By trial and error

D = 3.0 m

$$Re = \frac{2.0 \times 3.0}{1.5 \times 10^{-5}} = 4 \times 10^5 > Re \text{ (critical)}$$

Hence assumed $C_D = 0.2$ is correct.

(b) *When the Balloon is tethered* Referring to Fig. 9.9,

V_0 = 10 km/h = 2.78 m/s

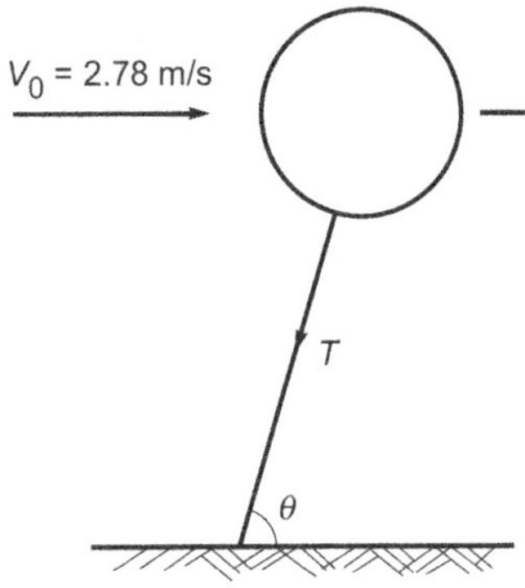

Fig. 9.9

Let θ = Inclination of the cable to the horizontal.

$$T \cos\theta = \text{Drag force} = C_D A \frac{\rho V_0^2}{2}$$

$$T \sin\theta = (\rho_{air} - \rho_{He})\, g \frac{\pi}{6} D^3 - W$$

where T = tension in the cable.

D = 3.0 m. Assume $C_D = 0.2$

$$T \sin\theta = 5.136\, D^3 - 135 = 3.672$$

$$T \cos\theta = 0.2 \times \left(\frac{\pi}{4} \times 3^2\right) \times 1.22 \times \frac{(2.78)^2}{2} = 6.654$$

$$\tan\theta = \frac{3.672}{6.654} = 0.5518$$

$\theta = 28.89° =$ **28° 53′ 30″**

$$\text{Tension in the cable} = \frac{3.672}{\sin\theta} = \frac{3.672}{0.48313} = \mathbf{7.6\ N}$$

9.13 A sphere 3 cm in diameter and of relative density 2.5 is attached to a string and is suspended from the roof of a wind tunnel. If an air stream of 25 m/s flows past the sphere, find the inclination of the string to horizontal and the tension in the string. [ρ_{air} = 1.25 kg/m^3; $\nu_{air} = 1.40 \times 10^{-5}$ m^2/s. For $10^4 < Re \le 3 \times 10^5$; $C_D = 0.5$. For $Re \ge 3 \times 10^5$, $C_D = 0.2$]

Solution: Neglecting the weight and drag of the string, and referring to Fig. 9.10,

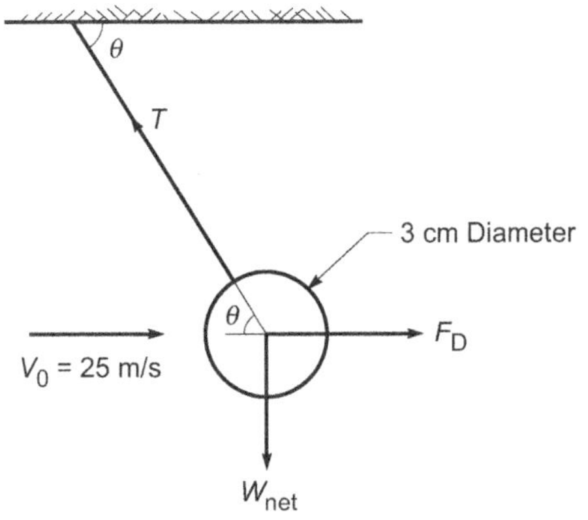

Fig. 9.10

$$T \cos\theta = F_D = C_D A \frac{\rho V_0^2}{2}$$

$$T \sin\theta = W_{net}$$

$$\therefore \quad \tan\theta = \frac{W_{net}}{C_D A \rho V_0^2/2}$$

$$Re = \frac{V_0 D}{\nu} = \frac{25 \times 0.03}{1.4 \times 10^{-5}} = 53571$$

As $10^4 < Re < 3 \times 10^5$; $C_D = 0.5$

$$W_{net} = W = (2.5 \times 9790) \times \frac{\pi}{6} \times (0.03)^3 = 0.346 \text{ N}$$

$$F_D = 0.5 \times \frac{\pi}{4} \times (0.03)^2 \times 1.25 \times \frac{(25)^2}{2} = 0.138 \text{ N}$$

$$\tan\theta = \frac{0.346}{0.138} = 2.506$$

$\theta = \tan^{-1}(2.506) =$ **68° 14′ 51″**

$$\text{Tension in the string} = T = \frac{W_{net}}{\sin\theta} = \frac{0.346}{\sin(68.246)^0} = \mathbf{0.3725\ N}$$

9.14 A flat plate 2 m long and 1.5 m wide is towed at 30 km/h in water. The drag and lift coefficient are found to be 0.20 and 0.60 respectively. Calculate (a) the resultant force on the plate and (b) the power required to keep the plate in motion.

Solution:

$$V_0 = 30 \text{ km/h} = \frac{30 \times 10^3}{3600} = 8.33 \text{ m/s}$$

$$F_D = \text{Drag force} = C_D A \frac{\rho V_0^2}{2}$$

$$= 0.20 \times (2.0 \times 1.5) \times 998 \times \frac{(8.33)^2}{2}$$

$$= 20790 \text{ N} = 20.79 \text{ kN}$$

$$F_L = \text{Lift force} = C_L A \frac{\rho V_0^2}{2}$$

$$= 0.60 \times (2.0 \times 1.5) \times 998 \times \frac{(8.33)^2}{2}$$

$$= 62370 \text{ N} = 62.37 \text{ kN}$$

Resultant force

$$F = \sqrt{F_D^2 + F_L^2}$$

$$= \sqrt{(20.79)^2 \times (62.37)^2}$$

$$= \mathbf{65.74\ kN}$$

Power required to tow the plate

$$\mathbf{P} = F_D \times V_0 = 20.79 \times 8.33 \text{ kW}$$

$$= \mathbf{173.2\ kW}$$

9.15 A kite is in the form of a rectangular airfoil with a chord length of 60 cm and a width of 45 cm and weighs 0.8 N. It is maintained at an angle of 10° to horizontal and the string makes an angle of 30° to the vertical. If the wind speed is 15 km/h and C_D is 0.25 (Fig. 9.11) estimate the tension in the string and the lift coefficient. [ρ_{air} = 1.2 kg/m^3]

Solution:

$$V_0 = 15 \text{ km/h} = 4.167 \text{ m/s}$$

$$\text{Drag force} = F_D = C_D A \frac{\rho V_0^2}{2}$$

$$\text{Lift force} = F_L = C_L A \frac{\rho V_0^2}{2}$$

Weight of kite = W

Resolving the tension T into components in horizontal and vertical directions:

$$T \cos 30° + W = F_L$$

$$T \sin 30° = F_D$$

$$\therefore \quad T = \frac{F_D}{\sin 30°} = \frac{0.25 \times (0.60 \times 0.45) \times 1.2 \times (4.167)^2}{\sin 30° \times 2}$$

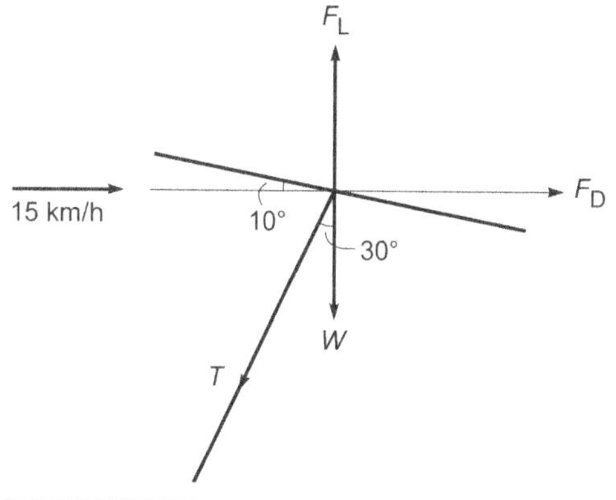

Fig. 9.11

Tension in the string T = **1.406 N**

Considering the vertical component of T

$$(1.406) \cos 30° + 0.8 = C_L \times (0.6 \times 0.45) \times \frac{1.2 \times (4.167)^2}{2}$$

$$2.0176 = 2.81295\ C_L$$

$$\mathbf{C_L = 0.717}$$

[***Note*** that the planform area A = 0.6 × 0.45 has been used in the calculation of drag and lift forces, as per the convention relating to lifting surfaces.]

9.16 A kite has a planform area of 0.25 m^2 and is flying in a wind of velocity 25 km/h. The kite has a net weight of 1.2 N. When the string is inclined at an angle of 15° to the vertical, the tension in the string was found to be 3.0 N. Evaluate the coefficients of lift and drag. [ρ_{air} = 1.15 kg/m^3].

Solution: Referring to Fig. 9.12

$$V_0 = 25 \text{ km/h} = 6.94 \text{ m/s}$$

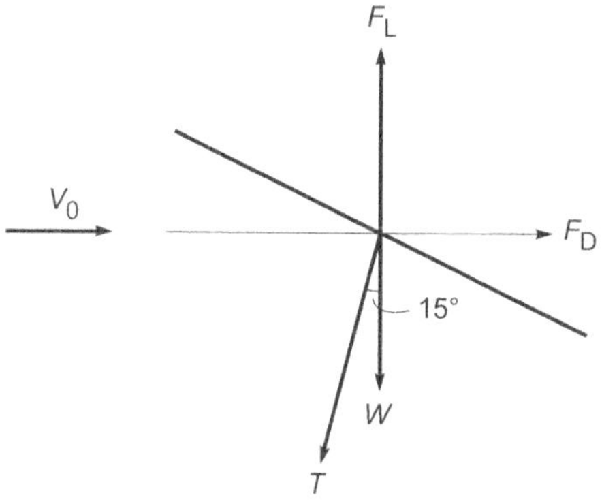

Fig. 9.12

Resolve the tension T into horizontal and vertical components. For the vertical component we have:

$$T \cos 15° + W = F_L = C_L A \frac{\rho V_0^2}{2}$$

$$3.0 \cos 15° + 1.2 = C_L \times 0.25 \times \frac{1.15 \times (6.94)^2}{2}$$

$$C_L = \frac{4.098}{6.932} = \mathbf{0.591}$$

For the horizontal component of T,

$$T \sin 15° = F_D = C_D A \frac{\rho V_0^2}{2}$$

$$3.0 \sin 15° = C_D \times (0.25) \times 1.15 \times \frac{(6.94)^2}{2}$$

$$C_D = \frac{0.77646}{6.932} = \mathbf{0.112}$$

9.17 Experiments were conducted in a wind tunnel with a wind speed of 50 km/h on a flat plate of size 2 m long and 1.2 m wide. The density of air is 1.20 kg/m^3. The plate is kept at an angle and the coefficients of lift and drag are 0.75 and 0.15 respectively. Determine (i) the lift force, (ii) drag force, (iii) resultant force, (iv) direction of resultant force, and (v) power expended in overcoming resistance of the plate.

Solution:

V_0 = 50 km/h = 13.89 m/s

A = planform area = 2 × 1.2 = 2.4 m^2

(i) Lift force

$$F_L = C_L A \frac{\rho V_0^2}{2}$$

$$= 0.75 \times (2.0 \times 1.2) \times \frac{1.2 \times (13.89)^2}{2}$$

$$= \mathbf{208.3\ N}$$

(ii) Drag force

$$F_D = C_D A \frac{\rho V_0^2}{2}$$

$$= 0.15 \times (2.0 \times 1.2) \times \frac{1.2 \times (13.89)^2}{2}$$

$$= 41.67\ \text{N}$$

(iii) Resultant force

$$F = \sqrt{F_L^2 + F_D^2}$$

$$= \sqrt{(208.3)^2 + (41.67)^2}$$

$$= \mathbf{212.43\ N}$$

(iv) Inclination of F with free stream = θ

$$\tan \theta = F_L/F_D = 5$$

$$\theta = \tan^{-1} 5 = \mathbf{78° \ 42'}$$

(v) Power expended

$$\mathbf{P} = F_D \times V_0$$

$$= 41.67 \times 13.89 = 578.8\ W$$

9.18 A cylinder 1.2 m in diameter and 9 m long rotates at 210 rpm with its axis normal to an air stream of velocity 10 m/s. Calculate the theoretical (i) circulation around the cylinder, (ii) theoretical lift coefficient and lift force and (iii) location of stagnation points. [ρ_{air} = 1.2 kg/m^3]

Solution:

V_c = tangential velocity due to rotation

$$= \frac{\pi DN}{60} = \frac{\pi \times 1.2 \times 210}{60} = 13.19\ \text{m/s}$$

V_0 = 10 m/s

$$\frac{V_c}{V_0} = \frac{13.19}{10.0} = 1.319$$

(i) Circulation

$$\Gamma = 2\pi r V_c = 2 \times \pi \times 0.6 \times 13.19$$

$$= \mathbf{49.73\ m^2/s}$$

(ii) Lift force

$$F_L = L\rho V_0 \Gamma$$

$$= 9.0 \times 1.2 \times 10.0 \times 49.73$$

$$= 5371\ \text{N} = \mathbf{5.371\ kN}$$

Lift coefficient

$$C_L = 2\pi \frac{V_c}{V_0}$$

$$= 2\pi \times 1.319 = \mathbf{8.29}$$

(iii) Stagnation point location

$$\theta = \sin^{-1}\left(-\frac{1}{2}\frac{V_c}{V_0}\right)$$

$$= \sin^{-1}\left(-\frac{1.319}{2}\right)$$

$$= -(41.26)° = \mathbf{318.74°}$$

(Stagnation point S_2)

and $\theta = 180 + 41.26° = \mathbf{221.26°}$

(Stagnation point S_1)

(See Fig. 9.4 (b) for direction of measurement of θ).

OBJECTIVE QUESTIONS

9.1 The drag force on a body
(a) is the net frictional force on the body
(b) is the net pressure force on the body in the direction of the relative velocity.
(c) is the component of the resultant force in the direction of the relative velocity.
(d) is the component of the resultant force in a direction perpendicular to the direction of gravity

9.2 The lift force on a body
(a) is due to buoyant force
(b) is always in the direction of the gravity
(c) is the component of the resultant force in a vertical direction
(d) is the component of the resultant force in a direction normal to relative velocity

9.3 In calculating the drag force using C_D the area used is
(a) always the frontal area
(b) the planform area when the body is flat like an airfoil
(c) the planform area when the body is bluff like a sphere
(d) always the planform area

9.4 Pressure drag results due to
(a) formation of wake
(b) turbulence in the wake
(c) existence of stagnation point in the front of a body
(d) high Reynolds numbers

9.5 In calculating the lift force
(a) always the frontal area is used
(b) always the planform area is used
(c) planform area is used if the body is a lifting surface
(d) actual surface area of the body is used

9.6 Indicate the *incorrect* statement
When there is a flow past a submerged body such as
(a) an airfoil the viscous forces are predominant
(b) a plate held normal to the flow the skin friction drag predominates
(c) a streamlined body the skin friction drag is predominant
(d) a bluff body the pressure drag is more than the skin friction drag.

9.7 If a streamlined body A and a sphere B both having the same maximum cross sectional area are subjected to laminar flow past them, then
(a) the body drag on A will be smaller than on B
(b) the total drag on A will be larger than on B
(c) the total drag on both A and B will be identical
(d) initially A will have larger drag but later on, both the objects will experience the same drag

9.8 A streamlined body with a round nose and a tapering back is generally best suited for
(a) creeping motion
(b) turbulent sub-sonic flow
(c) supersonic flow
(d) laminar flow with low Reynolds number

9.9 Indicate the *incorrect* statement:
In the case of flow past a sphere, the critical Reynolds number at which the transition from laminar to turbulent boundary layer takes place, depends on
(a) free stream turbulence
(b) relative roughness of the sphere
(c) localised large roughness in the forward region
(d) buoyant force on the sphere

9.10 For a sphere falling at terminal velocity in the Stokes' law range, the drag coefficient C_D is given by
(a) 24 Re (b) 64/Re
(c) 24/Re (d) 24 (1 + 3/16 Re)/Re
where Re is Reynolds number of the flow.

9.11 Stokes' law is valid up to a maximum Reynolds number of
(a) 0.1 (b) 1.0
(c) 2000 (d) 5×10^5

9.12 A very tiny sphere is settling down in a viscous liquid at a Reynolds number of 0.2. Its drag coefficient is
(a) 320 (b) 120
(c) 80 (d) 32

9.13 Body M has twice the weight, twice the projected area and twice the drag coefficient of body N. The terminal velocity of body M in air would be x times that of N, where x is
(a) 8 (b) 2
(c) $\sqrt{2}$ (d) $1/\sqrt{2}$

9.14 For a spherical sand particle in Stokes' law range the fall velocity V_0 is related to the diameter D as
(a) V_0 increases as D
(b) V_0 varies inversely as D
(c) V_0 varies as D^2
(d) V_0 varies inversely as D^2

9.15 For a solid sphere falling under gravity at terminal velocity in a fluid
(a) buoyant force = drag
(b) weight of the body = buoyant force
(c) weight of the sphere = buoyant force + drag
(d) drag = weight

9.16 In very low Reynolds number flow the deformation drag
(a) consists of frictional drag only
(b) consists of pressure drag only
(c) is essentially zero
(d) consists of both pressure drag and friction drag

9.17 The drag coefficient of a cylinder at small Reynolds number ($Re < 10$)
(a) increases with increase in the Reynolds number
(b) decreases with increase in the Reynolds number
(c) is essentially constant
(d) increases with the Reynolds number in the range $0 < Re < 1$ and then decreases for higher Reynolds numbers

9.18 When compared to a streamlined body, a bluff body will have
(a) more pressure drag but less friction drag
(b) more pressure drag and more friction drag
(c) less pressure drag and less friction drag
(d) less pressure drag but more friction drag

9.19 An approximate value of the drag coefficient of a hemispherical parachute is
(a) 2.35 (b) 1.33
(c) 0.30 (d) 0.07

9.20 A very long circular cylinder at a Reynolds number of $Re > 5 \times 10^5$ will have a drag coefficient C_D
(a) 1.20 (b) 0.50
(c) 0.33 (d) 0.20

9.21 At a Reynolds number $Re > 10^3$, one can expect a long rectangular plate held normal to the flow to have a drag coefficient C_D
(a) 0.50 (b) 2.40
(c) 1.00 (d) 1.90

9.22 In the case of flow past a circular cylinder at the critical Reynolds number, the drag coefficient drops from about
(a) 0.03 to 0.015 (b) 0.0 to 0.3
(c) 1.20 to 0.33 (d) 2.0 to 1.0

9.23 In the case of flow past a sphere at the critical Reynolds number, the drag coefficient drops from about
(a) 1.2 to 0.5 (b) 0.5 to 0.2
(c) 0.2 to 0.07 (d) 2.1 to 1.2

9.24 The drag coefficient C_D of a sphere undergoes a sudden drop in its value at the critical Reynolds number of about.
(a) 1.0 (b) 2000
(c) 2×10^5 (d) 5×10^6

9.25 The critical Reynolds number at which there is a sudden drop in the C_D value for a cylinder is about
(a) 2000 (b) 5×10^4
(c) 5×10^5 (d) 5×10^6

9.26 The Karman vortex trail occurs
(a) in all shapes and at all Reynolds numbers
(b) in 2-dimensional body shapes and in a range of Reynolds numbers
(c) in 2-dimensional bodies and at all Reynolds numbers > 30
(d) in circular cylinder only

9.27 The Strouhal number S is defined as S =
(a) V_0/nd (b) $V_0 d/g$
(c) nd^2/g (d) nd/V_0
where V_0 = free stream velocity and d = diameter of the cylinder and n = vortex shedding frequency

9.28 For circular cylinders with Reynolds numbers $Re > 10^3$, the Strouhal number S
(a) has a constant value of 0.21
(b) varies as $Re^{1/4}$
(c) increases linearly with Re
(d) is essentially constant at a value of 0.12

9.29 When a cylinder rotates in a fluid
(a) only one stagnation point is possible
(b) always two stagnation points occur
(c) no stagnation point is formed
(d) either two or one stagnation point is formed depending upon the ratio of free stream and rotational velocity

9.30 When an airfoil reaches the stall angle
(a) the drag coefficient is zero
(b) the lift decreases rapidly for any further increase in the angle of attack
(c) the drag decreases rapidly for any further increase in the angle of attack
(d) the lift coefficient is zero.

9.31 The overall drag of an aircraft of weight W and wing area S is given by

$$C_D = a + bC_L^2$$

where a and b are constants. The maximum drag in horizontal flight will be

(a) $6W\sqrt{ab}$ (b) $4W\sqrt{ab}$
(c) $2W\sqrt{ab}$ (d) $W\sqrt{ab}$

9.32 The optimum efficiency of a lifting vane is limited by the

(a) onset of stall
(b) separation from the leading edge
(c) separation from the leading edge
(d) more rapid increase in C_D than in C_L

9.33 A streamlined body is one for which

(a) the skin friction is zero
(b) the skin friction is minimum
(c) the thickness of the body is minimum
(d) the separation point occurs on the far downstream part of the body

PROBLEMS

9.1 Find the terminal settling velocity of a 0.8 mm diameter sediment particle (RD = 2.65) in water ($\nu = 1.03 \times 10^{-6}$ m²/s). The drag coefficient for $1.0 < Re < 200$ can be taken as

$$C_D = 24\left(1 + \frac{3}{16}Re\right)^{1/2} / Re$$

(*Ans.* V_{0t} = 12.7 cm/s)

9.2 If a raindrop of 1 mm diameter has a terminal velocity of 4.0 m/s, (1) what is the Reynolds number of the flow? (2) What is the magnitude of the drag coefficient C_D? [Air: $\rho_a = 1.20$ kg/m³, $\mu_a = 1.80 \times 10^{-5}$ Ns/m²
Water: $\rho_w = 998$ kg/m³, $\mu_w = 1.0 \times 10^{-3}$ Ns/m²]

(*Ans*: Re_D = 267, C_D = 0.679)

9.3 Determine the largest diameter and the corresponding terminal velocity of particle that will obey Stokes' law in the following cases:

(a) Polysterene (RD = 1.05) spheres settling in air ($\rho_a = 1.2$ kg/m³, $\nu_a = 1.5 \times 10^{-5}$ m²/s)
(b) Polysterene (RD = 1.05) spheres settling in water ($\nu_w = 1 \times 10^{-6}$ m²/s)
(c) Aluminum (RD = 2.8) spheres settling in glycerene ($\rho_g = 1260$ kg/m³, $\mu_g = 1.49$ Pa.s)

(*Ans.* (a) D_m = 0.078 mm; V_{0t} = 0.1925 m/s,
(b) D_m = 0.332 mm; V_{0t} = 3 mm/s
(c) D_M = 1.282 cm; V_{0t} = 9.22 cm/s)

9.4 A 2 mm sphere made of stainless steel (RD = 7.8) is observed to have a fall velocity of 9.5 mm/s in a liquid of density 1260 kg/m³. Estimate the kinematic viscosity of the liquid.

(*Ans.* $\nu = 1.188 \times 10^{-3}$ m²/s)

9.5 Determine the terminal velocity at which an air bubble of 3 mm diameter will rise in SAE 30 oil. What is the corresponding Reynolds number? ($\rho_f = 917$ kg/m³ and $\mu_f = 0.29$ Pa.s]

(*Ans.* V_{0t} = 1.55 cm/s;
Re = 0.147]

9.6 A helium filled spherical balloon rises in air at a terminal velocity of 3 m/s. Estimate its diameter if the total weight of the balloon is 210 N. [$\rho_{air} = 1.2$ kg/m³, $\rho_{helium} = 0.20$ kg/m³, $\nu_a = 1.5 \times 10^{-5}$ m²/s. For $10^4 < Re < 3 \times 10^5$, $C_D = 0.5$ and for $Re > 3 \times 10^5$, $C_D = 0.2$]

(*Ans.* D = 3.5 m)

9.7 A hydrogen balloon is 2.0 m in diameter and contains hydrogen of density $\rho_h = 0.066$ kg/m³. The balloon is found to rise at a terminal speed in air of density 0.95 kg/m³ when the combined weight of empty balloon and payload is 24.5 N. Estimate the terminal speed. [$\nu_a = 2.3 \times 10^{-5}$ m²/s. $C_D = 0.5$ for $10^4 \le Re \le 3 \times 10^5$ and $C_D = 0.2$ for $Re > 3 \times 10^5$]

(*Ans.* V_{0t} = 6.3 m/s)

9.8 Calculate the weight of ball of diameter 15 cm which is just supported by a vertical air stream of velocity 10 m/s, $\rho_a = 1.25$ kg/m³ and $\nu_a = 1.5$ stoke. The variation of C_D with Reynolds number Re is as follows:

Re	10^4	10^5	$> 3 \times 10^5$
C_D	0.4	0.5	0.2

(*Ans.* W = 0.4418 N)

9.9 What is the maximum load that can be carried by a 7.0 m diameter hemispherical parachute without exceeding the terminal velocity of 5 m/s? [ρ_a = 1.2 kg/m³. For a hemisphere C_D = 0.34 and 1.33 depending upon the aspect of the hemisphere].

(*Ans.* F_D = 768 N)

9.10 A sky diver and his pack weigh 1 kN and has a drag area $C_D A = 0.9$ m^2 when spread-eagled and 0.12 m^2 when falling feet first. Calculate the minimum and maximum terminal speeds of this sky diver. [Average value of $\rho_a = 1.15$ kg/m^3].

(*Ans.* $V_{0t1} = 43.96$ m/s; $V_{0t2} = 120.39$ m/s)

9.11 A 1940 model car has a drag coefficient C_D = 0.95. (a) What is the aerodynamic drag on this car when it travels at 30 km/h. Assume frontal area of 1.8 m^2. (b) What is the power spent in overcoming this drag? [$\rho_a = 1.22$ kg/m^3].

(*Ans.* $F_D = 72.44$ N; **P** = 603.6 W)

9.12 A radio antenna having a diameter of 6 mm extends 1.5 m upwards into air in the front part of a car. If the car moves at 100 km/h, estimate the bending moment at the base of the antenna. [$\rho_a = 1.2$ kg/m^3. $C_D = 1.2$].

(*Ans.* $M = 0.375$ Nm)

9.13 A parachute is to lower a box of machinery weighing 1450 N. The vertical component of landing velocity is not to exceed 4.0 m/s. How many parachutes of 6 m diameter are to be used for the purpose? Assume $C_D = 1.33$ for the parachutes. [$\rho_a = 1.22$ kg/m^3].

(*Ans.* 4 parachutes)

9.14 A sphere of 4 cm diameter made of aluminium (RD = 2.8) is attached to a string and suspended from the roof of a wind tunnel test section. If an air stream of 30 m/s flows past the sphere, find the inclination of the string and the tension in the string. [$\rho_a = 1.2$ kg/m^3, $\nu_a = 1.5 \times 10^{-5}$ m^2/s. $C_D = 0.5$ for $10^4 < Re \le 3 \times 10^5$ and $C_D = 0.2$ for $Re > 3 \times 10^5$]. Neglect the drag of the string.

(*Ans.* $T = 0.979$ N; $\theta = 69° 43' 39''$)

9.15 A wind anemometer has two 6 cm diameter hemispherical cups on a horizontal rod, the centre to centre distance between the cups being 30 cm. The arrangement rotates about a vertical axis at the middle of the rod. If the frictional resistance at the middle of the horizontal bearing can be neglected, compute the speed of rotation of the anemometer in a wind of 10 km/h [C_D of hemispherical cups = 1.33 and 0.34. $\rho_a = 1.2$ kg/m^3].

(*Ans.* $N = 387$ rpm)

9.16 A television antenna is supported on a vertical cylindrical support of 7.5 cm diameter extending 10 m above a flat roof. If a wind of 60 km/h blows past it, estimate the drag force on the support. [ρ_a = 1.2 kg/m^3 and $\nu_a = 1.5 \times 10^{-5}$ m^3/s. $C_D = 0.5$ for $10^4 < Re \le 3 \times 10^5$ and $C_D = 0.2$ for $Re > 3 \times 10^5$].

(*Ans.* $F_D = 62.5$ N)

9.17 A horizontal pipeline 15 cm diameter crosses a deep canal at mid-depth. If the velocity of flow in the canal is 3.0 m/s, estimate the frequency of vortex shedding from the pipe. [$\nu = 1.0 \times 10^{-6}$ m^2/s].

(*Ans.* n – 4.2 Hz)

9.18 At what wind velocity would an overhead transmission wire of 1.5 cm diameter attain a vibration of frequency equal to 100 Hz? [$\nu_a = 1.5 \times 10^{-5}$ m^2/s].

(*Ans.* $V_0 = 25.7$ km/h)

9.19 An advertisement hoarding has a height of 3.0 m and a length of 10 m. If a gale with wind velocity 90 km/h is expected, what would be the force on the hoarding if the wind blows normal to it? [$\nu_a = 1.2 \times 10^{-5}$ m^2/s]. The variation of C_D with L/D is as follows:

L/D	1.0	5	20	∞
C_D	1.16	1.20	1.5	1.9

(*Ans.* $F_D = 13.309$ kN)

9.20 If the power to overcome aerodynamic drag of an aircraft remains the same, what percentage increase in velocity is reflected by 15% reduction in the drag coefficient?

(*Ans.* 5.57% increase in the velocity)

9.21 A kite 0.8 m × 0.8 m weighing 4 N assumes an angle of 12° to the horizontal. The string attached to the kite makes an angle of 45° to the horizontal. The pull on the string is 25 N when the wind is blowing at a speed of 30 km/h. Find the coefficients of lift and drag? [$\rho_a = 1.2$ kg/m^3].

(*Ans.* $C_L = 0.813$; $C_D = 0.663$)

9.22 A kite having a weight of 0.5 N soars at an angle to the horizontal. The string holding the kite makes an angle of 35° to the horizontal and has a tension of 5 N. Calculate the ratio of lift to drag force on the kite?

(*Ans.* $F_L/F_D = 0.822$)

9.23 An airplane weighs 25 kN and has a wing area of 25 m^2. If the plane travels at a steady speed of 200 km/h in the horizontal plane, calculate its lift coefficient? [$\rho_a = 1.2$ kg/m^3].

(*Ans.* $C_L = 0.54$)

9.24 A small airplane has a weight of 10 kN, a wing area of 25 m^2 and a take off speed of 100 km/h. The lift and drag coefficients of the wing for small angles of attack can be approximated as $C_L = 0.11$

α and $C_D = 0.002 + 0.0025\ \alpha$ where α = angle of attack in degrees. Find the angle of attack needed and the power required at the take off? [$\rho_a = 1.2$ kg/m^3].

(*Ans.* $\alpha = 7.85°$; **P** = 6.96 kW)

9.25 An airfoil has a planform area of 10 m^2 and travels at 200 km/h in air [$\rho_a = 1.2$ kg/m^3]. If the lift and the drag coefficients at the particular angle of attack are 0.8 and 0.005 respectively, calculate (a) the lift force (b) drag force and (c) the resultant force.

(*Ans.* (a) $F_L = 14.815$ kN; (b) $F_D = 0.0926$ kN; (c) $F_R = 14.82$ kN)

9.26 A 2.0 m diameter cylinder is 10 m long and rotates at 300 rpm about its axis which is normal to an air stream of velocity 20 m/s. Calculate the (a) theoretical lift force per unit length (b) the position of stagnation points and (c) actual lift, drag and resultant force on the cylinder. [For actual lift and drag take $C_L = 3.40$ and $C_D = 0.65$] $\rho_{air} = 1.25$ kg/m^3).

(*Ans.* (a) $F_L = 4.935$ kN/m; (b) $\theta_1 = -51.77°$, $\theta_2 = 231.77°$; (c) $F_{aL} = 17$ kN, $F_{aD} = 3.25$ kN, $\alpha = 79.18°$)

9.27 A cylinder of 1.5 m diameter and 6 m long rotates inside a stream of water. The velocity of water is 2 m/s and the rotational speed is such that a double stagnation point is formed. Find the theoretical lift force and the corresponding lift coefficient.

(*Ans.* $F_L = 225.7$ kN; $C_L = 12.57$)

10

Turbulent Pipe Flow

CONCEPT REVIEW

10.1 TURBULENT FLOW

10.1.1 Critical Reynolds Number

When the Reynolds number $Re = \dfrac{VD}{\nu}$ of a pipe flow exceeds a critical value, the flow becomes turbulent. Even though the critical Reynolds number can assume a value within a range depending upon many flow parameters, for practical purposes $Re_{crit} = 2000$ is usually adopted.

10.1.2 Shear Stress

In turbulent flow the shear stress τ_t is expressed as

$$\boxed{\tau_t = \tau_{lam} + \tau_{turb} = \mu \frac{du}{dy} + \eta \frac{du}{dy} = (\mu + \eta)\frac{du}{dy}} \quad (10.1)$$

where μ = dynamic viscosity and η = eddy viscosity which is not a fluid property but depends upon turbulence conditions of the flow. Different models are proposed for the estimation of the turbulent shear stress $\tau_t = \eta \dfrac{du}{dy}$

Prandtl's model assumes

$$\boxed{\tau_t = \rho l^2 \left(\frac{du}{dy}\right)^2} \quad (10.2)$$

or

$$\boxed{\eta = \rho l^2 \left(\frac{du}{dy}\right)} \quad (10.3)$$

where mixing length

$$l = ky \quad (10.4)$$

in which k is the Karman's coefficient = 0.4

Karman's model assumes the mixing length to be

$$\boxed{l = k \frac{du/dy}{(d^2u/dy^2)}} \quad (10.5)$$

and

$$\boxed{\tau_t = \rho l^2 \left(\frac{du}{dy}\right)^2}$$

10.1.3 Turbulent Flow Near a Wall

For turbulent flow near a wall typical shear stress and the velocity distributions are shown in Fig. 10.1. Three important regions are to be noted:

1. Laminar sublayer (viscous wall layer) — where τ_{lam} predominates
2. Overlap region — where both τ_{lam} and τ_{turb} are important
3. Outer layer — where flow is turbulent and τ_{turb} predominates

The shear stress at the wall τ_0 is an important flow parameter. $\sqrt{\tau_0/\rho} = u_*$ is called *shear velocity*. In the laminar sublayer

$$\frac{u}{u_*} = \frac{yu_*}{\nu} \quad (10.6)$$

and the thickness of laminar sublayer δ' is taken as

$$\delta' = \frac{11.6\,\nu}{u_*} \quad (10.7)$$

10.2 PIPE FLOW

Fully developed pipe flow is called as *equilibrium boundary layer flow*. Here boundary layer thickness

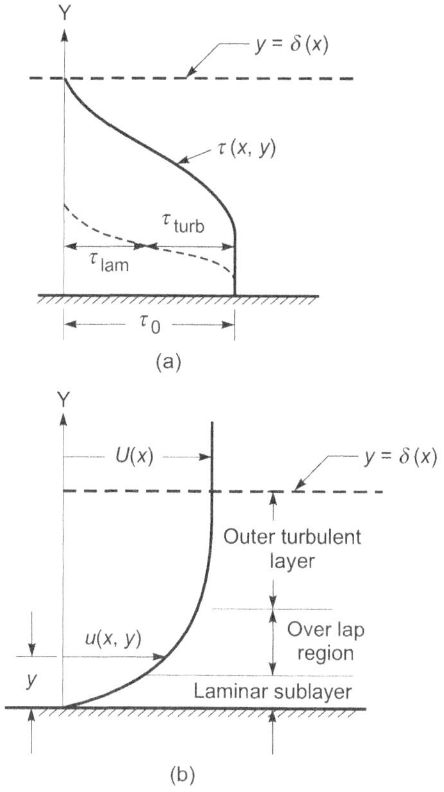

Fig. 10.1 *Turbulent Flow Near a Wall*

$\delta = r_0 = D/2$ = constant. The shear stress varies linearly with the distance from the boundary to become zero at the centre

$$\boxed{\tau = \tau_0 \,(1 - y_0/r_0)} \qquad (10.8)$$

where r_0 = radius of the pipe.

In a horizontal pipe of diameter D carrying a steady flow there will be pressure drop $(p_2 - p_1)$ in a length L of the pipe. Equating the frictional resistance to the difference in pressure forces.

$$(p_1 - p_2)\,\frac{\pi D^2}{4} = \tau_0\, \pi D L \qquad (10.9)$$

or $$\tau_0 = (p_1 - p_2)\,\frac{D}{4L} = \left(-\frac{\Delta p \cdot D}{4L}\right) \qquad (10.10)$$

Designating $$\tau_0 = \frac{f}{4}\,\frac{\rho V^2}{2} \qquad (10.11)$$

where f = Darcy–Weisbach friction factor,

$$\sqrt{\tau_0/\rho} = u_* = V\sqrt{f/8} \qquad (10.12)$$

and $$(p_1 - p_2) = (-\Delta p) = f\,\frac{L}{D}\,\frac{\rho V^2}{2} \qquad (10.13)$$

For sloping pipes, $(-\Delta p)$ will be replaced by change in piezometric head and Eq. (10.13) can be written as

$$\boxed{\left(\frac{p_1}{\gamma} + Z_1\right) - \left(\frac{p_2}{\gamma} + Z_2\right) = h_f = \frac{f\,LV^2}{2gD}} \qquad (10.14)$$

Here h_f = drop in piezometric head in a distance L of the pipe due to friction. Equation (10.14) is known as *Darcy–Weisbach formula*.

10.2.1 Hydrodynamically Smooth and Rough Pipes

On the basis of relative magnitudes of surface protrusions ε and thickness of laminar sublayer δ', a pipe surface is classified as follows:

(a) Hydrodynamically smooth:

$$\varepsilon/\delta' \le 0.25 \qquad (10.15)$$

i.e $$\boxed{\frac{u_* \varepsilon}{\nu} \le 3.0} \qquad (10.15a)$$

(b) Surface in transition:

$$0.25 < \frac{\varepsilon}{\delta'} < 6.0 \qquad (10.16)$$

i.e. $$\boxed{3 < \frac{u_* \varepsilon}{\nu} < 70} \qquad (10.16a)$$

(c) Hydrodynamically rough:

$$\varepsilon/\delta' > 6.0 \qquad (10.17)$$

i.e. $$\boxed{\frac{u_* \varepsilon}{\nu} \ge 70} \qquad (10.17a)$$

It may be noted that the laminar sublayer thickness δ' could be expressed in terms of pipe radius, as

$$\frac{\delta'}{r_0} = \frac{11.6\nu}{r_0 u_*} = \frac{11.6\nu}{\frac{D}{2} V\sqrt{f/8}} = \frac{65.6}{Re\sqrt{f}} \qquad (10.18)$$

Hence $$\boxed{\frac{\varepsilon}{\delta'} = \frac{\varepsilon}{r_0}\,\frac{r_0}{\delta'} = \frac{Re\sqrt{f}}{(r_0/\varepsilon)\,(65.6)}} \qquad (10.19)$$

Thus $\dfrac{\varepsilon}{\delta'} < 0.25$ corresponds to $\dfrac{Re\sqrt{f}}{(r_0/\varepsilon)} < 17$

$\dfrac{\varepsilon}{\delta'} \ge 6.0$ corresponds to $\dfrac{Re\sqrt{f}}{(r_0/\varepsilon)} \ge 400$

10.2.2 Velocity Distribution

(a) *Local velocity*: For turbulent flow in a pipe of diameter D (radius $r_0 = D/2$), the velocity distribution at any distance y from the boundary, the local velocity u is given by the following expressions:

(i) for a smooth boundary

$$\frac{u}{u_*} = 5.75 \log \frac{y u_*}{\nu} + 5.5 \tag{10.20}$$

(ii) for a rough boundary

$$\frac{u}{u_*} = 5.75 \log \frac{y}{\varepsilon} + 8.5 \tag{10.21}$$

(b) *Mean velocity*: The mean velocity V in a pipe of radius r_0 is obtained by integrating the velocity u over the entire area of the pipe and dividing by πr_0^2. This gives For *smooth pipes*:

$$\frac{V}{u_*} = 5.75 \log \frac{r_0 u_*}{\nu} + 1.75 \tag{10.20-Av}$$

For *rough pipes*

$$\frac{V}{u_*} = 5.75 \log \frac{r_0}{\varepsilon} + 4.75 \tag{10.21-Av}$$

(c) *Maximum velocity* u_m: For *both smooth and rough boundaries*, the maximum velocity u_m is at $y = r_0$ and is given by

$$\frac{u - u_m}{u_*} = 5.75 \log \frac{y}{r_0} \tag{10.22}$$

(d) *Relations among*: V, u, u_m, u_* and f:

(i) In terms of mean velocity $V = [Q/(\pi D^2/4)]$, the velocity u and the maximum velocity u_m are given for *both smooth and rough pipes* as:

$$\frac{u - V}{u_*} = 5.75 \log \frac{y}{r_0} + 3.75 \tag{10.23a}$$

Replacing $u_* = V\sqrt{f/8}$

$$\frac{u - V}{V\sqrt{f}} = 2 \log \frac{y}{r_0} + 1.32 \tag{10.23b}$$

(ii) In terms of centreline velocity u_{max}, the mean velocity V and f are related by the expression.

$$\frac{u_{max}}{V} = 1.43\sqrt{f} + 1 \tag{10.24}$$

This equation is obtained from Eq. (10.23(b)) after making a small change in the coefficients to fit the experimental data better.

(iii) In terms of u_{max}, V and u_* are related as

$$\frac{u_{max} - V}{u_*} = 3.75 \tag{10.25}$$

10.2.3 Frictional Resistance

The energy loss due to friction in pipe flow is expressed by Darcy–Weisbach formula as

$$h_f = \frac{f L V^2}{2gD} \quad \text{[From Eq. 10.14]}$$

and the variation of the friction factor f is as follows:

(a) *Laminar flow*: Up to Re ≤ 2000

$$f = \frac{64}{Re} \tag{10.26}$$

(b) *Smooth turbulent flow*: $\left(\text{When } \frac{Re\sqrt{f}}{(r_0/\varepsilon)} \le 17\right)$

(i) $$f = \frac{0.316}{Re^{1/4}} \quad \text{...Blasius equation} \tag{10.27}$$

valid for $10^4 < Re < 10^5$

(ii) $$\frac{1}{\sqrt{f}} = 2 \log Re\sqrt{f} - 0.80 \tag{10.28}$$

for all Reynolds numbers in the turbulent range in smooth pipes.
For explicit relationship of f one can use either

$$f = 0.0032 + \frac{0.221}{Re^{0.237}} \tag{10.29}$$

or

$$\frac{1}{\sqrt{f}} = 1.8 \log Re - 1.5186 \tag{10.30}$$

to get approximate but adequate value of f.
(c) *Rough turbulent flow*:

$$\frac{1}{\sqrt{f}} = 2 \log \frac{r_0}{\varepsilon} + 1.74 \tag{10.31}$$

for $\frac{\varepsilon}{\delta'} \ge 6.0$ i.e. $\frac{Re\sqrt{f}}{(r_0/\varepsilon)} \ge 400$

(c) *Transitional regime in turbulent flow*:

$$\frac{1}{\sqrt{f}} - 2 \log \frac{r_0}{\varepsilon} = 2 \log \frac{Re\sqrt{f}}{(r_0/\varepsilon)} - 0.8 \tag{10.32}$$

valid for uniform sand grain roughness only.

10.3 COMMERCIAL PIPES

The rough turbulent flow Eqs. 10.31 and 10.32 mentioned in Sub-section 10.2.3 were based on Nikuradse's classical experiments on uniform sand grain roughness. However, in commercial pipes the roughness magnitude, shape and distribution vary widely. To overcome the difficulty of describing all these parameters the concept of *equivalent sand grain roughness* is used. The roughness of a uniform sand grain coated pipe of the sand size as the given commercial pipe which gives the same value of f in the completely rough turbulent regime is termed as the equivalent sand grain roughness, and also as *effective roughness* (ε_s). Some of the usual ranges of equivalent sand grain roughness of commercial pipes are given in Table 10.1.

Table 10.1 *Values of ε_s for Some Common Commercial Pipes*

Pipe material	ε_s *(mm)*
1. Glass, drawn brass piping plastic, fibre glass	Smooth (0.0025)
2. Wrought iron, Steel	0.045
3. Asphalted cast iron	0.120
4. Galvanised iron	0.150
5. Cast iron	0.260
6. Concrete	0.30 to 3.0
7. Precast concrete	0.90 to 9.0

10.3.1 Colebrook Equation

For commercial pipes Colebrook found a semi-empirical relationship amongst f, Re and r_0/ε_s covering the smooth pipe, transition and rough pipe turbulent flow regimes, as

$$\frac{1}{\sqrt{f}} - 2\log\frac{r_0}{\varepsilon_s} = 1.74 - 2\log\left(1 + 18.7\frac{r_0/\varepsilon_s}{Re\sqrt{f}}\right) \tag{10.33}$$

which simplifies to

$$\frac{1}{\sqrt{f}} = 1.14 - 2\log\left[\frac{\varepsilon_s}{D} + \frac{9.35}{Re\sqrt{f}}\right] \tag{10.34}$$

This equation gives a smooth transition which is asymptotic to smooth pipe and rough pipe relationships of $f = \text{fn}\,(Re, r_0/\varepsilon_s)$. The Moody diagram is a graphical plot of this relationship.

As the Colebrook equation is too complex for practical application, many empirical equations are available for easy and explicit solution of f.

Three of these are given below. In the following three equations D = diameter of the pipe. Re = Reynolds number = VD/ν.

(1) Moody equation:

$$f = 0.0055\left[1 + \left(20000\frac{\varepsilon_s}{D} + \frac{10^6}{\text{Re}}\right)^{1/3}\right] \tag{10.35}$$

valid for Reynolds numbers between 4×10^3 and 1×10^7 and ε_s/D up to 0.01. This is claimed to give values within ±5% of the Colebrook equation.

(2) Swamee and Jain equation:

$$\frac{1}{\sqrt{f}} = 1.14 - 2\log\left(\frac{\varepsilon_s}{D} + \frac{21.25}{\text{Re}^{0.9}}\right) \tag{10.36}$$

valid for $5000 \le Re \le 10^8$ and $10^{-6} < \varepsilon_s/D < 10^{-2}$. This equation is claimed to yield values of f within 1% of the Colebrook equation.

(3) Haaland equation:

$$\frac{1}{\sqrt{f}} = -1.8\log\left[\frac{6.9}{\text{Re}} + \left(\frac{\varepsilon_s/D}{3.7}\right)^{1.11}\right] \tag{10.37}$$

This equation yields f values within 2% of Colebrook equation.

10.3.2 Moody Diagram

It is a chart showing the variation of $f = \text{fn}\,(Re, \varepsilon_s/D)$ for the full range of Reynolds numbers. It is a graphical representation of the Colebrook formula. Moody diagram is accurate to about 15% for design calculations and finds a large number of applications. It can be used for non-circular conduits and also for open channels, by replacing D with $4R_h$ where R_h is the hydraulic radius of the conduit. Figure 10.2 is a reproduction of the commonly used form of Moody diagram.

10.4 AGING OF PIPES

All commercial pipes change their roughness characteristics due to continuous usage. The increase in the equivalent sand grain roughness ε_s with age is usually taken to be linear as

$$\varepsilon_s = \varepsilon_{s0} + \alpha t \tag{10.38}$$

where t is the number of years of use after ε_{s0} was recorded.

10.5 SIMPLE PIPELINE DESIGN PROBLEMS

When the pipe friction is the only loss, the variables in a pipeline flow are:

$$Q,\ L,\ D,\ h_f,\ \varepsilon_s \text{ and } \nu$$

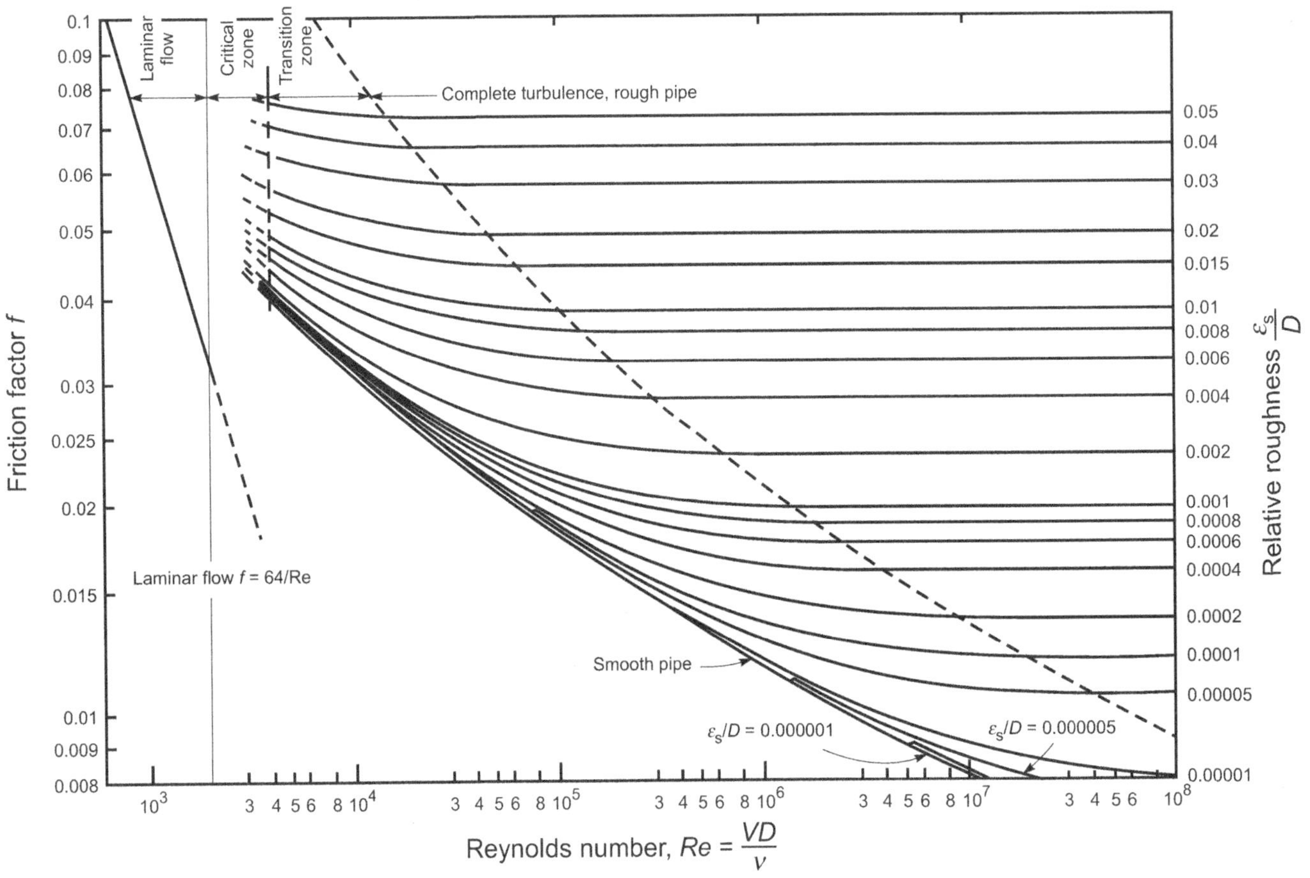

Fig. 10.2 *Moody Diagram*

Out of these ε_s, ν and L are known or can be determined. Thus one has the following three types of problems:

Type	Given	Required
I	$Q, L, D, \nu, \varepsilon_s$	h_f
II	$h_f, L, D, \nu, \varepsilon_s$	Q
III	$Q, h_f, L, \nu, \varepsilon_s$	D

In each of these types of problems the head loss h_f is found by the Darcy–Weisbach formula

$$\boxed{h_f = \frac{f\,LV^2}{2gD}}$$

The friction factor f is given by the Colebrook formula or its empirical equivalent or by the Moody diagram.

Type I problem: To find h_f

This is a straight forward problem as is shown in Example 10.17.

Type II problem: To find Q

Here ε_s/D is known. Also $Re\sqrt{f}$ is given by

$$Re\sqrt{f} = \frac{VD}{\nu}\left(\frac{2gh_f D}{V^2 L}\right)^{1/2} = \frac{D^{3/2}}{\nu}\left(\frac{2gh_f}{L}\right)^{1/2} \qquad (10.39)$$

Using these two parameters, f is found from the relevant resistance equation (smooth pipe, rough pipe or Colebrook equation) or from Moody diagram. Knowing f and h_f, V and (hence) Q are determined. [See Example 10.19).

Type III problem: To find D

This problem requires a trial and error solution.

As $$h_f = \frac{f\,LV^2}{2gD} = \frac{f\,LQ^2}{2g(\pi/4)^2 D^5} \qquad (10.40)$$

$$D^5 = C_1 f \qquad (10.41)$$

where $$C_1 = \frac{8LQ^2}{h_f g\pi^2}$$

Also $$Re = \frac{VD}{\nu} = \frac{QD}{\left(\frac{\pi}{4}D^2\right)\nu} = \frac{C_2}{D} \qquad (10.42)$$

where $$C_2 = \left(\frac{4Q}{\pi\nu}\right)$$

Thus by using $D^5 = C_1 f$ (10.43)

and $Re = C_2/D$ (10.44)

D is solved by trial and error. The procedure involves the following steps:

(i) Assume f.
(ii) Find D from Eq. 10.40
(iii) Find Re from Eq. 10.41. Also calculate ε_s/D.
(iv) By Colebrook (or its empirical equivalent) formula or by Moody chart find f for Re and ε_s/D of step (iii).
(v) Use new f of step (iv) and repeat the procedure till satisfactory value of f is obtained.
(vi) Use the next larger standard pipe size.

10.6 VELOCITY DISTRIBUTION IN THE NEIGHBOURHOOD OF FLAT SURFACES

The turbulent flow velocity distribution in pipe flow is essentially applicable to describe the velocity distribution near boundaries other than pipes also. Thus,

(a) Near hydrodynamically smooth boundaries:

(i)
$$\boxed{\frac{u}{u_*} = 5.75 \log \frac{u_* y}{\nu} + 5.5} \qquad (10.45)$$

valid in a region $\frac{u_* y}{\nu} > 70$, i.e. a region well beyond the laminar sublayer.

(ii) In the laminar sublayer, i.e. in the region $\frac{u_* y}{\nu} < 5.0$ the velocity distribution is given by

$$\boxed{\frac{u}{u_*} = \frac{u_* y}{\nu}} \qquad (10.46)$$

(iii) In the transition zone $\left(70 > \frac{u_* y}{\nu} > 5.0\right)$ an approximate relation is

$$\boxed{\frac{u}{u_*} = 11.5 \log \frac{u_* y}{\nu} - 3.0} \qquad (10.47)$$

(b) Near hydrodynamically rough boundaries:

$$\boxed{\frac{u}{u_*} = 5.75 \log \frac{y}{\varepsilon_s} + 8.5} \qquad (10.48)$$

where ε_s = equivalent sand grain roughness (also known as effective roughness height).

(c) General equation:
The general form of equation for both smooth and rough surfaces in the turbulent flow zone (i.e. a region beyond the laminar sublayer) is

$$\boxed{\frac{u}{u_*} = 5.75 \log \frac{y}{y'}} \qquad (10.49)$$

where y' = a parameter that depends upon laminar sublayer thickness δ' and roughness magnitude ε_s (see Fig 10.3).

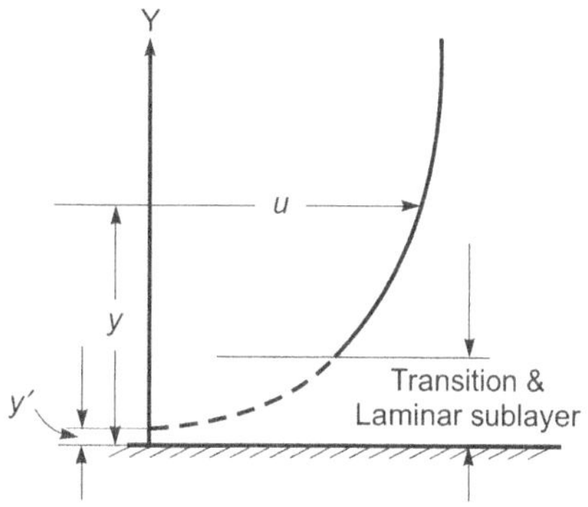

Fig. 10.3

For $\varepsilon_s > 10\,\delta'$,
$$\boxed{y' = \frac{\varepsilon_s}{30}} \qquad (10.50)$$

and for $\varepsilon_s < 0.25\,\delta'$,
$$\boxed{y' = \frac{\delta'}{107} = \frac{0.108\,\nu}{u_*}} \qquad (10.51)$$

These equations are useful in evaluating velocity profiles over flat surfaces, e.g. wind blowing over a ground, flow over a flat plate and other similar turbulent boundary layer flows.

REFERENCE

[10.1] White, F.M., *Fluid Mechanics*, 3rd Ed., McGraw-Hill Inc., New York, USA, 1994.

WORKED EXAMPLES

10.1 Water flows in a 2 cm diameter pipe. Taking the kinematic viscosity of water as 0.0098 stoke, calculate the largest discharge for which the flow will be definitely laminar. What is the boundary shear stress at this discharge?

Solution: The largest discharge corresponds to the critical Reynolds number.

$$(Re)_{crit} = 2000 = \frac{VD}{\nu}$$

$$V = \frac{2000 \times 0.0098 \times 10^{-4}}{0.02}$$
$$= 0.098 \text{ m/s}$$
$$= \frac{\pi}{4} \times (0.02)^2 \times 0.098$$
$$= 3.079 \times 10^{-5} \text{ m}^3\text{/s}$$
$$= 0.03079 \text{ L/s}$$

The required largest discharge = **1.847 L/min**

$$f = \frac{64}{Re} = \frac{64}{2000} = 0.032$$

$$u_* = V\sqrt{f/8} = 0.098\sqrt{\frac{0.032}{8}}$$
$$= 6.198 \times 10^{-3}$$
$$\tau_0 = \rho u_*^2 = 998 \times (6.198 \times 10^{-3})^2$$
$$= \mathbf{0.0383 \text{ Pa}}$$

10.2 In a turbulent flow through a pipe of radius r_0, at what distance from the boundary would the local velocity
(a) be equal to the mean velocity,
(b) be equal to half the mean velocity if the shear velocity is 1/20 of the mean velocity.

Solution: For both smooth and rough turbulent flows

$$\frac{u - V}{u_*} = 5.75 \log \frac{y}{r_0} + 3.75$$

(a) When $u = V$

$$5.75 \log \frac{y}{r_0} = -3.75$$

$$\log \frac{y}{r_0} = -\frac{3.75}{5.75} - 0.65217$$

$$\frac{y}{r_0} = 0.22275 \quad \text{or} \quad \mathbf{y = 0.22275\ r_0}$$

(b) When $u = \frac{V}{2}$ and $u_* = \frac{1}{20} V$

$$\left(\frac{0.5\,V - V}{1/20\,V}\right) = 5.75 \log \frac{y}{r_0} + 3.75$$

$$5.75 \log \frac{y}{r_0} = -10.0 - 3.75 = -13.75$$

$$\log \frac{y}{r_0} = -2.391$$

$$\frac{y}{r_0} = 0.00406 \text{ or } \mathbf{y = 0.00406\ r_0}$$

10.3 In a turbulent flow in a pipe of radius r_0 the friction factor f was found to be 0.02. Estimate
(a) the shear velocity u_* and the maximum velocity in terms of the mean velocity V.
(b) the ratio of the local velocity to mean velocity at a radial distance of 0.3 r_0 from the axis of the pipe.

Solution:

(a) In turbulent flow the shear velocity

$$u_* = V\sqrt{f/8}$$
$$= V\sqrt{\frac{0.02}{8}} = \mathbf{0.05\ V}$$

The maximum velocity u_m is given by

$$\frac{u_m}{V} = 1.43\sqrt{f} + 1$$
$$= 1.43\sqrt{0.02} + 1 = 1.20223$$
$$\mathbf{u_m = 1.20223\ V}$$

(b) Here y = distance from the wall = $r_0 - r$

$$y = r_0 - 0.3\ r_0 = 0.7\ r_0$$

$$\frac{y}{r_0} = 0.7$$

$$\frac{u - V}{u_*} = 5.75 \log \frac{y}{r_0} + 3.75$$
$$= 5.75 \log (0.7) + 3.75 = 2.8593$$

But $u_* = 0.05\ V$

$$\therefore \quad \frac{u}{0.05V} - \frac{1}{0.05} = 2.8593$$

or

$$\frac{u}{V} = \mathbf{1.14297}$$

10.4 Show that in a turbulent flow in a pipe of radius r_0 the variation of the difference between the maximum velocity u_m and the local velocity u at any distance y from the boundary follows the same variation with respect to the relative distance y/r_0 in both the smooth and rough pipes.

Solution: For smooth pipes,

$$\frac{u}{u_*} = 5.75 \log \frac{yu_*}{\nu} + 5.5$$

At the centreline where $y = r_0$, $u = u_m$, so that

$$\frac{u_m}{u_*} = 5.75 \log \frac{r_0 u_*}{\nu} + 5.5$$

$$\therefore \quad \frac{u_m - u}{u_*} = 5.75 \log \frac{r_0}{y} \qquad (1)$$

In rough pipes:

$$\frac{u}{u_*} = 5.75 \log \frac{y}{\varepsilon} + 8.5$$

At $\quad y = r_0,\ u = u_m$

$$\therefore \quad \frac{u_m}{u_*} = 5.75 \log \frac{r_0}{\varepsilon} + 8.5$$

$$\frac{u_m - u}{u_*} = 5.75 \log \frac{r_0}{y} \qquad (2)$$

It is seen that Eqs (1) and (2) are identical. Hence, the same equation is valid for both smooth and rough pipe flow.

10.5 A turbulent flow in a 10 cm diameter pipe is found to have a friction factor $f = 0.035$ at a Reynolds number of 10^5. (a) Estimate the thickness of the laminar sublayer. (b) If the equivalent sand grain roughness of the pipe is 0.80 mm, is the pipe rough, smooth or in transition?

Solution:

(a) The thickness of the laminar sublayer δ' is

$$\frac{\delta'}{r_0} = \frac{65.6}{Re\sqrt{f}}$$

$$\frac{\delta'}{0.05} = \frac{65.6}{10^5\sqrt{0.035}}$$

$$\delta' = 1.7532 \times 10^{-4} \text{ m} = \mathbf{0.175\ mm}$$

(b) $\varepsilon_s = 0.80$ mm

$$\frac{\varepsilon_s}{\delta'} = \frac{0.80}{0.175} = 4.57$$

This value being less than 6.0 and greater than 0.25, the pipe flow is in **transition regime**.

10.6 A 200 mm diameter pipeline has equivalent sand grain roughness magnitude or 0.2 mm. At what velocity of flow will the roughness (a) cause the flow to depart from the smooth pipe regime and (b) cause the flow to become fully rough turbulent. [$\nu = 1 \times 10^{-6}$ m²/s]

Solution:

(a) The transition from the smooth pipe regime begins at

$$\frac{Re\sqrt{f}}{(r_0/\varepsilon)} = 17$$

$$Re\sqrt{f} = 17 \times \left(\frac{0.10}{0.0002}\right) = 8500$$

The friction factor in smooth-turbulent flow is

$$\frac{1}{\sqrt{f}} = 2.0 \log Re\sqrt{f} - 0.8$$

$$= 2.0\ (\log 8500) - 0.8$$

$$= 7.0588$$

$$\therefore \quad Re = 8500 \times 7.0588 = 6 \times 10^4$$

$$\frac{VD}{\nu} = \frac{V \times 0.2}{1 \times 10^{-6}} = 6 \times 10^4$$

$$V = \frac{6 \times 10^4 \times 10^{-6}}{0.2} = \mathbf{0.3\ m/s}$$

Hence the velocity at the upper limit of smooth pipe regime is 0.3 m/s.

(b) Fully rough turbulent flow begins at

$$\frac{Re\sqrt{f}}{(r_0/\varepsilon)} = 400$$

$$Re\sqrt{f} = 400 \times \frac{0.1}{0.0002} = 200{,}000$$

At rough turbulent flow

$$\frac{1}{\sqrt{f}} = 2.0 \log \frac{r_0}{\varepsilon} + 1.74$$

$$\frac{1}{\sqrt{f}} = 2.0\left(\log \frac{0.1}{0.0002}\right) + 1.74$$

$$= 7.138$$

$$Re\sqrt{f} \times \frac{1}{\sqrt{f}} = \text{Re} = 200{,}000 \times 7.138$$

$$= 1.428 \times 10^6$$

Therefore, $$V = \frac{Re\,\nu}{D} = \frac{1.428 \times 10^6 \times 10^{-6}}{0.2}$$

$$= \mathbf{7.14\ m/s}$$

The flow will be at fully-rough turbulent regime at $V = 7.14$ m/s.

10.7 If the velocities in a 60 cm pipe carrying oil are 4.50 m/s and 4.20 m/s on the centreline and at a radial distance of 10 cm from the axis, calculate the discharge in the pipe.

Solution: For smooth as well as rough pipe

$$\frac{u_m - u}{u_*} = 5.75 \log \frac{r_0}{y}$$

$$\left(\frac{4.50 - 4.20}{u_*}\right) = 5.75 \log \frac{0.30}{(0.30 - 0.10)}$$

$$= 1.01252$$

$$u_* = 0.2963 \text{ m/s}$$

The mean velocity V is related to u_m as

$$\frac{u_m - V}{u_*} = 3.75$$

$$\frac{4.50 - V}{0.2963} = 3.75$$

$$V = 3.389 \text{ m/s}$$

The discharge $Q = \dfrac{\pi D^2}{4} V$

$$= \frac{\pi \times (0.6)^2}{4} \times 3.389$$

$$= \mathbf{0.958\ m^3/s}$$

10.8 In a fully rough turbulent flow in a 15 cm diameter pipe the centre line velocity is 2.50 m/s and the local velocity at mid-radius is 2.28 m/s. Find the discharge and the height of the roughness projections.

Solution: For rough turbulent flow

$$\frac{u_m - u}{u_*} = 5.75 \log \frac{r_0}{y}$$

$$\frac{2.50 - 2.28}{u_*} = 5.75 \log \left(\frac{r_0}{r_0/2}\right) = 1.731$$

$$u_* = 0.1271 \text{ m/s}$$

The mean velocity V is related to u_m as

$$\frac{u_m - V}{u_*} = 3.75$$

$$2.50 - V = 3.75 \times (0.1271) = 0.4766$$

$$V = 2.0234 \text{ m/s}$$

The discharge $Q = \left(\dfrac{\pi}{4} D^2\right) \times V$

$$Q = \frac{\pi}{4} \times (0.15)^2 \times 2.0234$$

$$= 0.0358 \text{ m}^3/\text{s}$$

$$= \mathbf{35.8\ L/s}$$

For rough turbulent flow

$$\frac{u_m}{u_*} = 5.75 \log \frac{r_0}{\varepsilon} + 8.5$$

$$\frac{2.50}{0.1271} = 5.75 \log \frac{r_0}{\varepsilon} + 8.5$$

$$\log \frac{r_0}{\varepsilon} = 1.9425$$

$$\frac{r_0}{\varepsilon} = 87.61,$$

$$\varepsilon = \frac{(0.15/2)}{87.61} = \mathbf{8.56 \times 10^{-4}\ m}$$

The height of roughness projections $\varepsilon = 0.856$ mm

10.9 A pipe of 25 cm diameter carries air ($\rho = 1.22$ kg/m^3, $\nu = 1.5 \times 10^{-5}$ m^2/s) at an average velocity of 8 m/s. The equivalent sand grain roughness of the pipe is 0.50 mm. Calculate the friction factor f by assuming the flow to be fully rough-turbulent. What is the shear stress at the boundary?

Solution: Reynolds number

$$Re = \frac{VD}{\nu} = \frac{8.0 \times 0.25}{1.5 \times 10^{-5}} = 1.33 \times 10^5$$

The friction factor f in fully rough-turbulent flow is

$$\frac{1}{\sqrt{f}} = 2.0 \log \frac{r_0}{\varepsilon} + 1.74$$

$$= 2.0 \log \left(\frac{0.125}{0.50 \times 10^{-3}}\right) + 1.74$$

$$= 6.536$$

$$f = 0.0234$$

Shear stress at the boundary

$$u_* = \sqrt{\tau_0/\rho} = V_0\sqrt{f/8}$$

$$\tau_0 = \frac{\rho V_0^2 f}{8} = \frac{1.22 \times (8.0)^2 \times 0.0234}{8}$$

$$= \mathbf{0.2285\ Pa}$$

10.10 A 30 cm diameter pipeline carries water at a velocity of 3.5 m/s. If the friction factor $f = 0.028$, (a) determine the roughness height ε by assuming fully-rough turbulent flow to exist, (b) Find the shear stress at the wall, and (c) Find at a radial distance of 5 cm from the axis, the shear stress and the value of the velocity gradient. [$\nu = 0.8 \times 10^{-6}$ m^2/s].

Solution:

(a) For rough-turbulent flow:

$$\frac{1}{\sqrt{f}} = 2 \log \frac{r_0}{\varepsilon} + 1.74$$

$$\frac{1}{\sqrt{0.028}} = 2 \log \frac{r_0}{\varepsilon} + 1.74$$

$$2 \log \frac{r_0}{\varepsilon} = 4.236,$$

$$\frac{r_0}{\varepsilon} = 131.24$$

$$\varepsilon = \frac{0.150}{131.24} = 1.1429 \times 10^{-3} \text{ m}$$

$$\boldsymbol{\varepsilon = 1.14 \text{ mm}}$$

(b) Shear velocity

$$u_* = \sqrt{\tau_0/\rho} = V\sqrt{f/8}$$

$$\sqrt{\tau_0/\rho} = 3.5 \times \sqrt{0.028/8} = 0.20706$$

$$\tau_0 = (998)\,(0.20706)^2$$

$$= \mathbf{42.79 \text{ Pa}}$$

(c) At $r = 5$ cm, $y = r_0 - r = 15 - 5 = 10$ cm $= 0.10$ m
The shear stress varies linearly with y with zero at the centre and as such

$$\tau = \tau_0\,(1 - y/r_0) = 42.79\left(1 - \frac{0.10}{0.15}\right)$$

$$= \mathbf{14.26 \text{ Pa}}$$

For the velocity gradient, the Prandtl formula is used.

$$\frac{u}{u_*} = 5.75 \log \frac{y}{\varepsilon} + 8.5$$

$$= 2.5 \ln \frac{y}{\varepsilon} + 8.5$$

$$\frac{du}{dy} = u_*\,(2.5)\,\frac{\varepsilon}{y} \times \frac{1}{\varepsilon} = 2.5\, u_*/y$$

$$= 2.5 \times 0.20706/0.10$$

$$= \mathbf{5.177 \text{ s}^{-1}}$$

10.11 Fully rough turbulent flow occurs in a 15 cm diameter pipe carrying oil of density $\rho = 917$ kg/m^3. If at a distance of 2.5 cm from the wall the velocity is 1.5 m/s and the velocity gradient is 7 s^{-1}, determine the (i) boundary shear stress, (ii) equivalent sand grain roughness, (iii) discharge and (iv) friction factor.

Solution: For fully rough-turbulent flow

$$\frac{u}{u_*} = 5.75 \log \frac{y}{\varepsilon_s} + 8.5 \qquad (1)$$

and

$$\frac{du}{dy} = 2.5\, u_*\, \frac{\varepsilon_s}{y} \cdot \frac{1}{\varepsilon_s} = 2.5\, u_*/y \qquad (2)$$

(i) From Eq. (2) $7 = 2.5\, u_*/0.025$

$$u_* = 0.07 \text{ m/s},$$

$$\tau_0 = u_*^2\, \rho = (0.07)^2 \times 917$$

$$\boldsymbol{\tau_0 = 4.493 \text{ Pa}}$$

(ii) Substituting the known values in Eq. (1)

$$\frac{1.5}{0.07} = 5.75 \log \frac{0.025}{\varepsilon_s} + 8.5$$

$$0.025/\varepsilon_s = 177.19$$

$$\varepsilon_s = 1.411 \times 10^{-4} \text{ m}$$

$$= \mathbf{0.141 \text{ mm}}$$

(iii)

$$\frac{u - V}{u_*} = 5.75 \log y/r_0 + 3.75 \qquad (3)$$

and

$$\frac{u}{u_*} = 5.75 \log y/\varepsilon_s + 8.5 \qquad (1)$$

Hence

$$V/u_* = 5.75 \log \frac{r_0}{\varepsilon_s} + 4.75$$

$$\frac{V}{0.07} = 5.75 \log\left(\frac{0.075}{1.411 \times 10^{-4}}\right) + 4.75$$

$$= 20.422$$

$$V = 1.4295 \text{ m/s}$$

Discharge

$$Q = V \times \frac{\pi}{4} D^2 = 1.4295 \times \frac{\pi}{4} \times (0.15)^2$$

$$= 0.0253 \text{ m}^3/\text{s} = \mathbf{25.3 \text{ L/s}}$$

(iv) Since $u_* = V\sqrt{f/8}$

$$f = 8\left(\frac{u_*}{V}\right)^2 = 8\left(\frac{0.07}{1.4295}\right)^2$$

$$= \mathbf{0.0192}$$

[***Note:*** f could also have been evaluated by the friction factor formula for rough turbulent flow: $\frac{1}{\sqrt{f}} = 2.0 \log \frac{r_0}{\varepsilon_s} + 1.74$.]

10.12 A pipeline 30 cm in diameter carries 300 L/s of petrol ($\rho = 680$ kg/m^3, $\mu = 2.9 \times 10^{-4}$ Pa.s). Calculate the (i) friction factor (ii) shear stress at the boundary, (iii) shear stress and velocity at a radial distance of 5 cm from the pipe axis (iv) maximum velocity and (v) thickness of the laminar sublayer. Assume the pipe to be hydrodynamically smooth.

Solution:

Mean velocity

$$V = \frac{Q}{A} = \frac{0.300}{\frac{\pi}{4} \times (0.30)^2} = 4.244 \text{ m/s}$$

Reynolds number

$$Re = \frac{VD\rho}{\mu}$$

$$= \frac{4.244 \times 0.30 \times 680}{2.9 \times 10^{-4}}$$

$$= 2.985 \times 10^6$$

(i) By using the explicit formula for f in smooth pipe flow, Eq. 10.29

$$f = 0.0032 + \frac{0.221}{(2.985 \times 10^6)^{0.237}}$$

$$= \mathbf{0.009654}$$

(ii)

$$u_* = \sqrt{\tau_0/\rho} = V\sqrt{f/8}$$

$$\tau_0 = \frac{\rho V^2 f}{8}$$

$$= \frac{680 \times (4.244)^2 \times 0.009654}{8}$$

$$= \mathbf{14.78\ Pa}$$

$$u_* = \sqrt{\frac{14.78}{680}} = \mathbf{0.14743\ m/s}$$

(iii) At $r = 5$ cm, $y = r_0 - r = 15 - 5 = 10$ cm

$$\tau = \tau_0 \frac{r}{r_0} = 14.78 \times \frac{5}{15} = \mathbf{4.927\ Pa}$$

The velocity at $y = 0.1$ m is given by the smooth pipe formula:

$$\frac{u}{u_*} = 5.75 \log \frac{u_* y}{\nu} + 5.5$$

Noting that $\nu = \mu/\rho$

$$\frac{u}{u_*} = 5.75 \log \left[\frac{0.14743 \times 0.1 \times 680}{2.9 \times 10^{-4}}\right] + 5.5$$

$$= 31.597$$

$$u = u_* (31.597) = 0.14743 \times 31.597$$

$$= \mathbf{4.658\ m/s}$$

(iv) The maximum velocity u_m is related to V as

$$\frac{u_m - V}{u_*} = 3.75$$

$$u_m = V + 3.75\, u_*$$

$$= 4.244 + 3.75 \times 0.14743$$

$$= \mathbf{4.797\ m/s}$$

(v) Laminar sublayer thickness

$$\delta' = \frac{11.6 \nu}{u_*}$$

$$= 11.6 \times \left(\frac{2.9 \times 10^{-4}}{680}\right) \times \frac{1}{0.14743}$$

$$= 3.356 \times 10^{-5} \text{ m} = \mathbf{0.0336\ mm}$$

10.13 The velocity of flow in a rough 10 cm diameter pipe is found to increase 10% as a pitot tube is moved from a point 1.5 cm from the wall to a point 3.0 cm from the wall. Estimate the relative roughness r_0/ε and friction factor f for the pipe.

Solution: The flow is rough-turbulent, and as such

$$\frac{u}{u_*} = 5.75 \log \frac{y}{\varepsilon} + 8.5$$

Hence $$\frac{u_2 - u_1}{u_*} = 5.75 \log \frac{y_2}{y_1}$$

$$u_2 = 1.10\, u_1 \text{ and } y_2 = 3 \text{ cm}$$

and $$y_1 = 1.5 \text{ cm}$$

$$0.10 \frac{u_1}{u_*} = 5.75 \log \frac{0.03}{0.015} = 1.731$$

$$\frac{u_1}{u_*} = 17.31$$

Thus $$\frac{u_1}{u_*} = 17.31 = 5.75 \log \frac{0.015}{\varepsilon} + 8.5$$

$$\frac{0.015}{\varepsilon} = 34.04$$

or $$\varepsilon = 4.41 \times 10^{-4} \text{ m} = 0.441 \text{ mm}$$

Relative roughness

$$\frac{r_0}{\varepsilon} = \frac{0.05}{4.41 \times 10^{-4}} = \mathbf{113.5}$$

For rough-turbulent flow, the friction factor f is given by

$$\frac{1}{\sqrt{f}} = 2 \log \frac{r_0}{\varepsilon} + 1.74$$

$$= 2 \log (113.5) + 1.74$$

$$f = \mathbf{0.0292}$$

10.14 Kerosene ($\rho = 804$ kg/m^3 and $\mu = 1.92 \times 10^{-3}$ Pa.s) is to flow in a 20 cm diameter pipe at a velocity of 3 m/s with a minimum loss of energy. Determine the permissible height of surface roughness of the pipe wall.

Solution:

Reynolds number

$$Re = \frac{\rho V D}{\mu} = \frac{804 \times 3.0 \times 0.20}{1.92 \times 10^{-3}}$$

$$= 2.513 \times 10^5$$

At the permissible height of roughness in the pipe the flow will be in smooth-turbulent regime. Hence the friction factor f for $Re > 10^5$ is given by Eq. 10.29 as

$$f = 0.0032 + \frac{0.221}{Re^{0.237}}$$

$$= 0.0032 + \frac{0.221}{(2.513 \times 10^5)^{0.237}}$$

$$= 0.0148$$

Since the pipe will behave as smooth, at the limiting value

$$\frac{Re\sqrt{f}}{(r_0/\varepsilon)} = 17$$

$$\frac{r_0}{\varepsilon} = \frac{2.513 \times 10^5 \times \sqrt{0.0148}}{17}$$

$$= 1798$$

$$\varepsilon = \frac{0.10}{1798} = 5.56 \times 10^{-5} \text{ m}$$

$$= \mathbf{0.0556\ mm}$$

[***Note:*** A pipe carries a flow with minimum loss of energy when it is in the smooth pipe regime. In other words, for a given Reynolds number the lowest value of f occurs in the smooth pipe regime. (See Moody diagram, Fig. 10.2).]

10.15 Compare the cost of pumping a fluid A between two points at the same volumetric rate in 10 cm and 15 cm diameter pipes having the same absolute roughness magnitude of 0.2 mm. The flow can be assumed to be in fully developed rough-turbulent regime. [Assume the pipes to be horizontal]

Solution: For 10 cm diameter pipe:

$$\frac{1}{\sqrt{f}} = 2 \log \frac{r_0}{\varepsilon} + 1.74$$

$$= 2 \log \frac{0.05}{0.0002} + 1.74$$

$$f_1 = 0.0234$$

For 15 cm diameter pipe:

$$\frac{1}{\sqrt{f_2}} = 2 \log \frac{0.075}{0.0002} + 1.74$$

$$f_2 = 0.0211$$

Power $\quad \mathbf{P} = \gamma Q h_f$

But $\quad h_f = \frac{f L V^2}{2gD} = \frac{f L Q^2}{2g\left(\frac{\pi}{4}\right)^2 D^5}$

$\therefore$ Ratio of powers $\frac{\mathbf{P}_1}{\mathbf{P}_2}$ for same Q, γ and L

$$\frac{\mathbf{P}_1}{\mathbf{P}_2} = \left(\frac{f_1}{f_2}\right)\left(\frac{D_2}{D_1}\right)^5$$

$$= \frac{0.0234}{0.0211} \times \left(\frac{15}{10}\right)^5 = 8.42$$

Cost of pumping is proportional to power consumed.

Hence $\quad \dfrac{\text{Cost of pumping in 10 cm pipe}}{\text{Cost of pumping in 15 cm pipe}} = \mathbf{8.42}$

10.16 A 300 mm diameter water supply pipe had a friction factor of 0.02 when freshly laid. After 10 years of service the friction factor was found to be 0.025. What friction factor can be expected after another 15 years? The pipe can be assumed to be in rough-turbulent flow regime.

Solution: Initially $t = 0$ years: roughness $= \varepsilon_0$ and

$$\frac{1}{\sqrt{f_0}} = 2.0 \log \frac{r_0}{\varepsilon_0} + 1.74$$

$$\frac{1}{\sqrt{0.02}} = 2.0 \log \frac{0.150}{\varepsilon_0} + 1.74$$

$$\log \frac{0.150}{\varepsilon_0} = 2.6655$$

$$\frac{0.150}{\varepsilon_0} = 463$$

and $\quad \varepsilon_0 = 3.24 \times 10^{-4}$ m $= \mathbf{0.324\ mm}$

After 10 years, $t = 10$ years: the roughness $= \varepsilon_1$ and

$$\frac{1}{\sqrt{f_1}} = 2.0 \log \frac{r_0}{\varepsilon_1} + 1.74$$

$$\frac{1}{\sqrt{0.025}} = 2.0 \log \frac{0.15}{\varepsilon_1} + 1.74$$

$$\log \frac{0.15}{\varepsilon_1} = 2.2923,$$

$$\varepsilon_1 = 7.653 \times 10^{-4} \text{ m} = \mathbf{0.765\ mm}$$

The roughness magnitude can be taken to increase linearly with age as

$$\varepsilon_1 = \varepsilon_0 + \alpha t$$

$$(\varepsilon_1 - \varepsilon_0) = \alpha t$$

$$\alpha = \frac{(0.765 - 0.324)}{10} = 0.0441 \text{ mm/year}$$

In another 15 years, $t = 25$ years and

$$\varepsilon_2 = \varepsilon_0 + \alpha t = 0.324 + (0.0441 \times 25)$$

$$= 1.4265 \text{ mm}$$

The friction factor f_2 will be

$$\frac{1}{\sqrt{f_2}} = 2 \log \frac{0.15}{0.0014265} + 1.74 = 5.7836$$

$$f_2 = \mathbf{0.0299}$$

10.17 Determine the head (energy) loss for flow of 200 L/s of Gasoline ($\rho = 680$ kg/m^3, $\mu = 2.92 \times 10^{-4}$ N.s/m^2) through a 30 cm diameter pipe. The pipe is 500 m long and has an equivalent roughness magnitude of 0.20 mm. Express the head loss as equivalent pressure differential.

Solution:

Velocity $\quad V = \frac{Q}{\pi D^2/4} = \frac{0.200}{\frac{\pi}{4} \times (0.3)^2} = 2.83$ m/s

Reynolds number

$$Re = \frac{\rho V D}{\mu} = \frac{680 \times 2.83 \times 0.30}{2.92 \times 10^{-4}}$$

$$= 1.977 \times 10^{-6}$$

Relative roughness

$$\varepsilon_s/D = \frac{0.2 \times 10^{-3}}{0.30} = 6.67 \times 10^{-4}$$

From the empirical equivalent form of Colebrook equation (Swamee and Jain formula, Equation (10.36))

$$\frac{1}{\sqrt{f}} = 1.14 - 2 \log \left(\frac{\varepsilon_s}{D} + \frac{21.25}{Re^{0.9}} \right)$$

$$= 1.14 - 2\log\left(6.67\times10^{-4} + \frac{21.25}{(1.977\times10^{6})^{0.9}}\right)$$

$$= 7.43445$$

$$f = 0.0181$$

[Instead of the above equation one can use Moody chart to get the value of f as above].

Head loss $h_f = \dfrac{fLV^2}{2gD} = \dfrac{0.0181\times500\times(2.83)^2}{2\times9.81\times0.3}$

$$= 12.31 \text{ m of oil}$$

Putting $h_f = \dfrac{\Delta p}{\gamma}$

$$\Delta p = 12.31 \times 680 \times 9.81$$

$$= 82118 \text{ N/m}^2$$

$$= \mathbf{82.12\ kPa}$$

10.18 A badly corroded concrete pipe of diameter 1.50 m has an equivalent sand roughness of ε_s = 15 mm. A 10 mm thick lining is proposed to reduce the roughness value to ε_s = 0.2 mm. For a discharge of 4.0 m^3/s in the pipe calculate the power saved per kilometre of pipe.
[Take $\nu = 1 \times 10^{-6}$ m^2/s]

Solution: Before the lining:

$$V_1 = \frac{4.0}{\frac{\pi}{4}\times(1.5)^2} = 2.264 \text{ m/s}$$

Reynolds number

$$Re_1 = \frac{V_1D_1}{\nu} = \frac{2.264\times1.5}{1.0\times10^{-6}}$$

$$= 3.395 \times 10^6$$

$$\frac{\varepsilon_{s1}}{D_1} = \frac{15\times10^{-3}}{1.5} = 10^{-2}$$

From the empirical equivalent of Colebrook equation (Swamee and Jain equation, (10.36))

$$\frac{1}{\sqrt{f_1}} = 1.14 - 2\log\left(\frac{\varepsilon_{s1}}{D_1} + \frac{21.25}{Re_1^{0.9}}\right)$$

$$= 1.14 - 2\log\left(10^{-2} + \frac{21.25}{(3.395\times10^{6})^{0.9}}\right)$$

$$= 5.1376$$

or $f_1 = 0.0379$

After the lining:

$$D_2 = [1.50 - 2 \times 0.01] = 1.48 \text{ m}$$

$$V_2 = \frac{4.0}{\frac{\pi}{4}\times(1.48)^2} = 2.325 \text{ m/s}$$

Reynolds number

$$Re_2 = \frac{V_2D_2}{\nu} = \frac{2.325\times1.48}{1\times10^{-6}} = 3.44 \times 10^6$$

$$\varepsilon_s/D_2 = \frac{0.2\times10^{-3}}{1.48} = 1.351 \times 10^{-4}$$

Friction factor f_2 is given by

$$\frac{1}{\sqrt{f_2}} = 1.14 - 2\log\left(\frac{\varepsilon_{s2}}{D_2} + \frac{21.25}{Re_2^{0.9}}\right)$$

$$= 1.14 - 2\log\left(1.351\times10^{-4} + \frac{21.25}{(3.44\times10^{6})^{0.9}}\right)$$

$$= 8.7158$$

$$f_2 = 0.0132$$

Head loss $h_f = \dfrac{fLV^2}{2gD}$

$$h_{f1} = \frac{0.0379\times1000\times(2.264)^2}{2\times9.81\times1.50} = 6.600 \text{ m}$$

$$h_{f2} = \frac{0.0132\times1000\times(2.325)^2}{2\times9.81\times1.48} = 2.450 \text{ m}$$

Saving in head:

$$h_s = h_{f1} - h_{f2} = 4.150 \text{ m}$$

Saving in power

$$\mathbf{P}_s = \gamma Q h_s$$

$$= 9.79 \times 4.0 \times 4.150 \text{ kW}$$

$$= \mathbf{162.5\ kW}$$

10.19 A 25 cm diameter C.I. pipe (ε_s = 015 mm) conveys water ($\nu = 1 \times 10^{-6}$ m^2/s) with a hydraulic gradient of 0.025. Estimate the discharge in the pipe and the power required to pump this rate over 100 m of pipeline.

Solution: Relative roughness

$$\frac{\varepsilon_s}{D} = \frac{0.15\times10^{-3}}{0.25} = 6 \times 10^{-4}$$

$$Re\sqrt{f} = \left(\frac{VD}{\nu}\right)\left(\frac{2gh_fD}{V^2L}\right)^{1/2}$$

$$= \frac{D^{3/2}}{\nu}\left(\frac{2gh_f}{L}\right)^{1/2}$$

h_f/L = hydraulic gradient = 0.025

$$Re\sqrt{f} = \frac{(0.25)^{3/2}}{1\times10^{-6}}(2 \times 9.81 \times 0.025)^{1/2}$$

$$= 87545$$

By Colebrook formula

$$\frac{1}{\sqrt{f}} = 1.14 - 2\log\left(\frac{\varepsilon_s}{D} + \frac{9.35}{Re\sqrt{f}}\right)$$

$$\frac{1}{\sqrt{f}} = 1.14 - 2\log\left(6\times10^{-4} + \frac{9.35}{87545}\right)$$

or $\quad f = 0.01806$

$$Re = 87545/\sqrt{0.01806} = 6.515 \times 10^5$$

$$= \frac{V\times(0.25)}{1\times10^{-6}}$$

$\therefore \quad V = 2.606$ m/s

Discharge $\quad Q = \frac{\pi}{4} \times (0.25)^2 \times 2.606 = 0.1279\ \text{m}^3/\text{s}$

$= \mathbf{127.9\ L/s}$

Head lost in 100 m length of pipe = 0.025 × 100 = 2.5 m

Power required

$$\mathbf{P} = \gamma Q h_f$$

$$= 9790 \times 0.1279 \times 2.5$$

$= 3130$ W = **3.13 kW**

10.20 Water is to be conveyed at a rate of 250 L/s over a distance of 1.5 km at a head loss of 15 m. Asphalted C.I. pipe ($\varepsilon_s = 0.12$ mm) is suggested to be used. Taking the kinematic viscosity of water as 1×10^{-6} m²/s, estimate the diameter of the pipe for this purpose.

Solution:

$$h_f = \frac{fLV^2}{2gD} = \frac{8LQ^2}{\pi^2 g}\frac{f}{D^5}$$

$$15.0 = \left[\frac{8\times1500\times(0.250)^2}{\pi^2\times9.81}\right]\frac{f}{D^5}$$

$$= 7.746\,\frac{f}{D^5}$$

$\therefore \quad D^5 = 0.516 f \qquad (1)$

Reynolds number

$$Re = \frac{VD}{\nu} = \left(\frac{4Q}{\pi\nu}\right)\frac{1}{D}$$

$$Re = \frac{4\times0.25}{\pi\times1\times10^{-6}}\frac{1}{D} = 3.183\times10^5\frac{1}{D}$$

A trial and error procedure is adopted:

1st trial: Assume $f = 0.025$

$$D = (0.516 \times 0.025)^{1/5} = 0.419 \text{ m}$$

$$Re = 759665$$

$$\varepsilon_s/D = \frac{0.12\times10^{-3}}{0.419} = 2.864\times10^{-4}$$

By empirical equivalent form of Colebrook equation, Ea. 10.36

$$\frac{1}{\sqrt{f}} = 1.14 - 2\log\left(\frac{\varepsilon_s}{D} + \frac{21.25}{Re^{0.9}}\right)$$

$$= 1.14 - 2\log\left(2.864\times10^{-4}\times\frac{21.25}{(759665)^{0.9}}\right)$$

$$\frac{1}{\sqrt{f}} = 7.9474$$

$$f = 0.0159$$

2nd trial: Using $f = 0.0159$

$$D = (0.516 \times 0.0159)^{1/5} = 0.383 \text{ m}$$

$$Re = 831070$$

$$\varepsilon_s/D = 3.133 \times 10^{-4}$$

$$\frac{1}{\sqrt{f}} = 1.14 - 2\log\left(3.133\times10^{-4}\times\frac{21.25}{(831070)^{0.9}}\right)$$

$$= 7.908$$

$$f = 0.016$$

This is practically the same as the assumed value and as such no further trials are necessary. The required diameter D = **38.3 cm**. In practice, the next larger standard size would be used.

10.21 Design the diameter of a steel pipe to carry water ($\nu = 1 \times 10^{-6}$ m²/s) with a mean velocity of 1.0 m/s. The head loss is to be limited to 10 cm per 100 m length. The effective roughness height can be taken as 0.45 mm. Use the following empirical formula for the friction factor:

$$f = 0.0055\left[1+\left(2000\frac{\varepsilon_s}{D}+\frac{10^6}{Re}\right)^{1/3}\right]$$

Solution: The Reynolds number

$$Re = \frac{VD}{\nu} = \frac{1\times D}{1\times10^{-6}}$$

$$= 10^6\,\text{D}$$

Again

$\varepsilon_s = 0.45 \times 10^{-3}$ m, so

$$f = 0.0055\left[1+\left(\frac{2000\times0.45\times10^{-3}}{D}+\frac{10^6}{10^6 D}\right)^{1/3}\right]$$

$$= 0.0055\left[1+\left(\frac{0.9}{D}+\frac{1}{D}\right)^{1/3}\right] \qquad (1)$$

$$= 0.0055 + 6.812 \times 10^{-3}/D^{1/3}$$

Head loss $h_f = \dfrac{fLV^2}{2gD}$

$$\frac{10}{100} = \frac{f}{D} \times \frac{100 \times 1}{2 \times 9.81}$$

$$f = 0.01962\ D \qquad (2)$$

The value of D is obtained by trial and error from Equation 1 and 2, f is found to be 0.0133 and the corresponding $\boldsymbol{D = 0.678}$ **m**. In practice, the next larger standard size would be used.

10.22 Oil (RD = 0.85) is flowing in a 150 m long smooth pipeline at the rate of 80 L/s. The loss of head in the line is 67 m. If the kinematic viscosity of the oil is 0.1 cm²/s. Calculate the diameter of the pipe. Use Blasius formula for the friction factor f.

Solution:

$$h_f = \frac{fLV^2}{2gD} = \frac{8LQ^2}{\pi^2 g} \cdot \frac{f}{D^5}$$

$$67 = \left(\frac{8 \times 150 \times (0.08)^2}{\pi^2 \times 9.81}\right)\frac{f}{D^5}$$

$$\frac{f}{D^5} = 844.66 \quad \text{or} \quad D = \left(\frac{f}{844.66}\right)^{1/5}$$

Reynolds number

$$= \frac{VD}{\nu} = \left(\frac{4Q}{\pi\nu}\right)\frac{1}{D}$$

$$= \left(\frac{4 \times 0.08}{\pi \times 0.1 \times 10^{-4}}\right)\frac{1}{D} = 10186 \times \frac{1}{D}$$

A trial and error procedure is used.

1st trial: Assume $f = 0.02$

$$D = \left(\frac{0.02}{844.66}\right)^{1/5} = 0.1188 \text{ m}$$

$$Re = \frac{10186}{0.1188} = 85730$$

As $Re < 10^5$, Blasius formula can be used.

By Blasius formula:

2nd trial: $$f = \frac{0.316}{(Re^{1/4})} = \frac{0.316}{(85730)^{1/4}} = 0.0185$$

$$D = \left(\frac{0.0185}{844.66}\right)^{1/5} = 0.11698 \text{ m}$$

$$Re = \frac{10186}{0.11698} = 87077$$

3rd trial: f by Blasius formula $= \dfrac{0.316}{(87077)^{1/4}}$

$= 0.0184$

As this value of f is very nearly same as the assumed value at the beginning of this trial, the required diameter $D = 0.11698$ m = 11.70 cm. In practice the next larger standard size would be used.

10.23 A 40 cm × 25 cm rectangular duct is 200 m long. Air ($\rho = 1.20$ kg/m³ and $\nu = 1.5 \times 10^{-5}$ m²/s) is blown through this duct at a velocity of 20 m/s. Compute the pressure drop, head loss and power required if the blower efficiency is 60%. [Assume ε_s for the boundary as 0.05 mm].

Solution: For non-circular conduits the diameter D_h to be used in pipe flow equations is given by

$$D_h = 4 \times \frac{\text{Area}}{\text{Perimeter}}$$

$= 4 \times$ hydraulic radius

$= 4R_h$

Here Area $= 0.4 \times 0.25 = 0.10$ m²

Perimeter $= 2 \times (0.4 + 0.25) = 1.3$ m

$$D_h = 4 \times \frac{0.1}{1.3} = 0.3077 \text{ m}$$

$$Re_h = \frac{VD_h}{\nu} = \frac{20 \times 0.3077}{1.5 \times 10^{-5}} = 4.10 \times 10^5$$

$$\varepsilon_s/D_h = \frac{0.05 \times 10^{-3}}{0.3077} = 1.625 \times 10^{-4}$$

By empirical equivalent of Colebrook formula, Eq. (10.36),

$$\frac{1}{\sqrt{f}} = 1.14 - 2.0 \log\left\{\frac{\varepsilon_s}{D_h} + \frac{21.25}{(Re_h)^{0.9}}\right\}$$

$$= 1.14 - 2.0 \log\left\{1.625 \times 10^{-4} \times \frac{21.25}{(4.10 \times 10^5)^{0.9}}\right\}$$

or $f = 0.0154$

$$h_f = \text{head loss} = \frac{fLV^2}{2gD_h}$$

$$= \frac{0.0154 \times 200 \times (20)^2}{2 \times 9.81 \times 0.3077}$$

$=$ **204 m (of air column)**

Δp = pressure loss $= \rho g h_f$

$= 1.20 \times 9.81 \times 204$

$= 2402$ Pa $=$ **2.40 kPa**

Power required at 60% efficiency

$$= \frac{\gamma Q h_f}{\eta} = \frac{1.2 \times 9.81 \times (20 \times 0.4 \times 0.25) \times 204}{0.60}$$

$= 8004$ W $=$ **8.0 kW**

10.24 Wind velocity measurements at a height of 5 m and 10 m above the ground in a level area yielded values of 5 m/s and 6.0 m/s respectively. What velocity could be expected at a height of 30 m under the same conditions?

Solution: Using the general equation

$$\frac{u}{u_*} = 5.75 \log y/y'$$

At $y_1 = 5.0$ m $\quad u_1 = 5.0$ m/s

$y_2 = 10.0$ m $\quad u_2 = 6.0$ m/s

$$\frac{u_2 - u_1}{u_*} = 5.75 \log y_2/y_1$$

$$\frac{6.0 - 5.0}{u_*} = 5.75 \log \frac{10.0}{5.0} = 1.730$$

$$u_* = 0.5777 \text{ m/s}$$

At $y_3 = 30.0$ m the velocity y_3 is given by

$$\frac{u_3 - u_1}{u_*} = 5.75 \log \frac{y_3}{y_1}$$

$$\frac{u_3 - 5.0}{0.5777} = 5.75 \log\left(\frac{30}{5}\right)$$

$$u_3 = 2.585 + 5.0 = \mathbf{7.585 \text{ m/s}}$$

10.25 Two horizontal smooth pipes of equal lengths carry air and water respectively at velocities such that the Reynolds numbers and pressure drops for each are the same. Obtain an expression for the ratio of the mean air velocity to that of the water.

Solution: Let suffix *a* stand the air and *w* standard for water.

Reynolds number $Re = \dfrac{V_a D_a}{\nu_a} = \dfrac{V_w D_w}{\nu_w}$

$$\therefore \quad \frac{V_a}{V_w}\left(\frac{D_a}{D_w}\right) = \left(\frac{\nu_a}{\nu_w}\right)$$

Head loss $h_f = \dfrac{fLV^2}{2gD}$

Since the pressure drops are same and the pipes are horizontal,

$$h_f = \frac{f_a L_a V_a^2}{2gD_a} = \frac{f_w L_w V_w^2}{2gD_w}$$

$L_a = L_w$ and since Reynolds numbers are same and the pipes are smooth $f_a = f_w$.

$$\therefore \quad \left(\frac{V_a}{V_w}\right)^2 = \left(\frac{D_a}{D_w}\right)$$

Substituting in Eq. (1) above

$$\left(\frac{V_a}{V_w}\right)^3 = \left(\frac{\nu_a}{\nu_w}\right)$$

or

$$\left(\frac{\mathbf{V_a}}{\mathbf{V_w}}\right) = \left(\frac{\boldsymbol{\nu_a}}{\boldsymbol{\nu_w}}\right)^{\mathbf{1/3}}$$

OBJECTIVE QUESTIONS

10.1 Identify the *incorrect* statement:
(a) In laminar flow the eddy viscosity is zero
(b) In turbulent flow the molecular viscosity is insignificantly small when compared to eddy viscosity
(c) In any flow the eddy viscosity is small across the flow section
(d) The eddy viscosity depends upon the state of turbulent flow

10.2 The intensity of turbulence refers to
(a) Correlation of u' and v'
(b) average kinetic energy of turbulence per unit mass.
(c) root mean square value of turbulent velocity fluctuations
(d) the Reynolds stresses

10.3 The turbulent shear stress in *xy* plane is given by
(a) $\rho\overline{u'^2}$ (b) $\dfrac{1}{\rho}\overline{u'v'}$
(c) $\rho\, u'\, v'$ (d) $-\overline{\rho\, u'\, v'}$

10.4 According to Prandtl mixing length theory the turbulent shear stress $\tau_1 =$
(a) $\rho l^2(du/dy)^2$ (b) $\rho l(du/dy)^2$
(c) $\rho^2\, l^2(du/dy)^2$ (d) $\rho l(du/dy)$

10.5 The turbulent intensity $\sqrt{\overline{u'^2}}$ in a turbulent pipe flow in maximum at
(a) the centreline
(b) at the wall
(c) at a small distance from the wall
(d) at mid-radius

10.6 In pipe flow the Prandtl mixing length l is assumed as equal to
(a) $k\, y^2$ (b) $k\, y$
(c) $k \log y$ (d) $-k/y$

10.7 In a turbulent flow in a pipe the shear stress is
(a) maximum at the centre and decreases linearly towards the wall.
(b) Maximum at the centre and decreases logarithmically towards the wall.

(c) maximum midway between the centreline and the wall.
(d) maximum at the wall and decreases linearly to a zero value at the centre.

10.8 In a horizontal pipe flow if Δp is the difference in pressure between two sections distance L apart in a pipe of diameter D then τ_0 =
(a) $(\Delta p\, D)/4L$
(b) $(\Delta p\, L)/D$
(c) $\frac{1}{8}\Delta\, pLV^2$
(d) $\sqrt{\Delta p/8}$

10.9 Prandtl's mixing length in a pipe flow is
(a) a constant
(b) zero at the pipe wall
(c) a function of the shear stress at the wall.
(d) a function of the Reynolds number.

10.10 In a turbulent pipe flow, the term $\text{Re}\sqrt{f}/(r_0/\varepsilon_s)$ is directly proportional to
(a) δ'/r_0 (b) r_0/δ'
(c) ε_s/δ' (d) δ'/ε_s
where δ' = thickness of laminar sublayer and ε_s = equivalent roughness height.

10.11 In a turbulent pipe flow, inside the laminar boundary layer the velocity distribution is
(a) parabolic
(b) linear
(c) logarithmic
(d) exponential decay type

10.12 In a turbulent flow through smooth pipe of radius r_0 the velocity distribution plotted against y/r_0
(a) will be logarithmic and independent of the Reynolds number
(b) shows increasing fullness with decrease in Reynolds number
(c) will be parabolic and independent of the Reynolds number
(d) shows increasing fullness with increase in Reynolds number

10.13 In a turbulent flow through a fully rough pipe of radius r_0 and relative roughness ε_s/r_0, a plot of u/u_m vs y/r_0
(a) shows that an increase in ε_s/r_0 causes an increase in the peakedness of the profile
(b) shows that a decrease in ε_s/r_0 causes a peaking of the profile
(c) is independent of ε_s/r_0
(d) is a paraboloid of revolution

10.14 The gradient in the normal direction at the wall of velocity in a fully developed turbulent pipe flow is
(a) nearly zero
(b) less than the gradient of velocity in laminar fully developed pipe flow having the same flow rate
(c) more than the gradient of velocity in laminar fully developed pipe flow having he same flow rate
(d) 1/2

10.15 In a turbulent flow through a pipe the centreline velocity is 3.61 m/s and the friction factor f = 0.02. The mean velocity of the flow in m/s is
(a) 4.80 (b) 3.00
(c) 2.21 (d) 0.96

10.16 In a turbulent flow through pipes the mean velocity V and the centreline velocity u_m are related as

$\frac{u_m}{V}$ =

(a) $1.3\sqrt{f} + 1$ (b) 3.75
(c) $1.43\sqrt{f} + 1$ (d) $1.33 \log \frac{1}{\sqrt{f}}$

10.17 Shear velocity is
(a) a non-dimensional quantity
(b) a convenient fictitious quantity
(c) the velocity of fluid at the edge of laminar sublayer
(d) the velocity of fluid at the edge of roughness

10.18 Friction factor f in laminar and turbulent flow in a pipe varies as Re^{-1} and $Re^{-0.25}$ respectively. If V is the average velocity, the pressure drop in a horizontal pipe for laminar and turbulent flow respectively will be proportional to
(a) V and $V^{0.75}$ (b) $V^{0.5}$ and V^2
(c) $V^{0.5}$ and $V^{1.75}$ (d) V and $V^{1.75}$

10.19 A liquid flows in a 30 cm diameter pipe at a Reynolds number of 10^6. If the friction factor is 0.025, the thickness of laminar sublayer, in mm, is
(a) 0.025 (b) 1.00
(c) 0.062 (d) 0.0031

10.20 In a pipe flow the ratio of equivalent sand grain roughness to the thickness of laminar sublayer (ε/δ') is 7.5. The flow can be classified as
(a) in transition from laminar to turbulent flow
(b) fully rough-turbulent flow
(c) smooth boundary turbulent flow
(d) transition from hydrodynamically smooth to rough regime

10.21 In turbulent flow through pipes, the maximum velocity u_m and the mean velocity V are related as $(u_m - V)/u_* =$
(a) 3.75 for smooth boundaries only
(b) 3.75 for both smooth and rough boundaries
(c) 5.75 for both smooth and rough boundaries
(d) 3.75 for rough boundaries only

10.22 In a turbulent flow through a pipe of radius r_0, the radial distance at which the local velocity is equal to the mean velocity is
(a) $r = 0.1\ r_0$ (b) $0.223\ r_0$
(c) $0.777\ r_0$ (d) $0.5\ r_0$

10.23 In a pipe flow the shear velocity u_* is related to friction factor f and mean velocity V as $u_*/V =$
(a) $\sqrt{f/8}$ (b) $\sqrt{8/f}$
(c) $\sqrt{8g/f}$ (d) $\dfrac{1}{\sqrt{f}}$

10.24 The Darcy–Weisbach friction factor f is related to boundary shear τ_0 as $f =$
(a) $\dfrac{8\rho V^2}{\tau_0}$ (b) $\dfrac{\tau_0}{8\rho V^2}$
(c) $\dfrac{8\tau_0}{\rho V^2}$ (d) $\dfrac{\rho V^2}{8\tau_0}$

10.25 The head loss in 100 m length of a 0.1 m diameter pipe (f = 0.02) carrying water is 10 m. The boundary shear stress, in kPa, is
(a) 9.79 (b) 0.298
(c) 0.1958 (d) 0.0245

10.26 In laminar flow through a pipe the Darcy–Weisbach friction factor f is given by $f =$
(a) $\dfrac{64}{Re}$ (b) $\dfrac{24}{Re}$
(c) $\dfrac{16}{Re}$ (d) $\dfrac{3}{16} Re$

10.27 In a turbulent pipe flow the boundary is hydrodynamically smooth if
(a) $\varepsilon_s/\delta' \leq 0.25$ (b) $\varepsilon_s/\delta' < 6.0$
(c) $\delta'/\varepsilon_s \leq 0.25$ (d) $\varepsilon_s/\delta' \geq 6.0$

10.28 The Blasius equation for friction factor f in turbulent flow through pipes relates f to the Reynolds number Re as $f =$
(a) $64/Re$ (b) $0.316/Re^{1/4}$
(c) $1.328/Re^{1/2}$ (d) $0.316/Re^{1/5}$

10.29 In a fully rough-turbulent pipe flow, the friction factor f is
(a) a function of Re and ε_s/D
(b) a function of Re only
(c) a function of ε_s/D only
(d) independent of Re and ε_s/D
where Re = Reynolds number and ε_s/D = relative roughness.

10.30 For hydrodynamically smooth pipes the friction factor f
(a) is a constant
(b) is a function of ε_s/r_0 only
(c) is a function of Re and ε_s/r_0 only
(d) is a function of Re only
where Re = Reynolds number and ε_s/r_0 = relative roughness.

10.31 In a fully developed rough-turbulent regime in pipe flow,
(a) rough and smooth pipes have the same friction factor
(b) the laminar sublayer is thicker than the roughness projections
(c) the friction factor is independent of the Reynolds number
(d) the friction factor is independent of the relative roughness ε_s/r_0

10.32 A turbulent pipe flow is said to be in the transitional regime, if the friction factor f is
(a) independent of Reynolds number
(b) dependent on both Reynolds number and relative roughness
(c) independent of both Reynolds number and relative roughness
(d) independent of the relative roughness

10.33 In a Moody diagram the third parameter is ε_s/D. Here ε_s refers to
(a) the equivalent uniform sand grain roughness
(b) an arbitrarily chosen roughness
(c) median size of a non-uniform sand grain roughness
(d) mean height of actual roughness of commercial pipes

10.34 Two pipes are identical in diameter and carry the same fluid. One of the pipes is rough and the other has a very smooth inside surface. If both the pipes have the same friction factor, when carrying the same discharge, then
(a) the flow is rough-turbulent regime
(b) the roughness magnitudes are much smaller than the laminar sublayer
(c) the flow must be laminar
(d) the flow is in transitional regime from smooth to rough flow

10.35 For the same cross sectional area, discharge, fluid properties and roughness of the boundary between a circular cross sectional conduit and square cross sectional conduit,
(a) the square section offers more resistance
(b) the circular section offers more resistance
(c) both circular and square have identical resistance
(d) the square section offers more resistance at low Reynolds number and offers less resistance at high Reynolds number

10.36 In the Moody diagram the values of f for turbulent flow are based on
(a) Nikuradse's uniform sand grain data
(b) data on non-uniform sand grains coated to a pipe
(c) Colebrook-White data on commercial pipes
(d) Hazen-William pipe flow formula

10.37 The Colebrook–White formula for pipe friction is for
(a) artificially roughened pipes
(b) commercial pipes
(c) pipes coated with non-uniform sand
(d) pipes coated with uniform sand

10.38 The Colebrook–White formula can be used for
(a) turbulent smooth pipe flow only
(b) turbulent rough pipe flow only
(c) laminar flow in commercial pipes only
(d) any kind of turbulent flow regime in commercial pipes

10.39 With increasing aging of pipes, the proportion between maximum velocity and the mean velocity in turbulent flow
(a) initially decreases and then increases
(b) initially decreases and then increases
(c) decreases
(d) increases

PROBLEMS

10.1 Which of the following two pipes carrying water has larger shear stress at the wall? What are the magnitudes of these stresses?
(i) 2.5 cm diameter pipe with a Reynolds number of 1000.
(ii) 25 cm diameter pipe with Reynolds number of 5×10^4 (Take $\rho = 998$ kg/m^3 and $\nu = 1 \times 10^{-6}$ m^2/s).

(*Ans.* 25 cm diameter pipe has larger wall shear stress; (i) $\tau_0 = 0.01277$ Pa, and (ii) $\tau_0 = 0.1054$ Pa)

10.2 In a turbulent flow through a pipe of radius r_0 at what radial distance would the local velocity be equal to half the maximum velocity, if the shear velocity is 1/20 of the maximum velocity?

$$\left(Ans.\ \frac{r}{r_0} = 0.9914\right)$$

10.3 In a turbulent flow in a pipe the friction factor $f = 0.016$. Estimate the ratio of (a) the maximum velocity to mean velocity and (b) the shear velocity to mean velocity.

$$\left(Ans.\ (a)\ \frac{u_m}{V} = 1.1809;\quad (b)\ \frac{u_*}{V} = 0.04472\right)$$

10.4 If the velocity distribution in a pipe of radius R is given by

$$\frac{u - V}{u_*} = 5.75 \log \frac{y}{R} + 3.75$$

show that $$\frac{u_m - V}{V} = -2.03\sqrt{f}\log\left(\frac{y}{R}\right)$$

10.5 In a turbulent flow the friction factor $f = 0.025$. Estimate the ratio of the centreline velocity to shear velocity. If the mean velocity is 2.0 m/s, estimate the value of the centreline velocity.

$$\left(Ans.\ \frac{u_m}{u_*} = 21.93;\quad u_m = 2.452\ \text{m/s}\right)$$

10.6 The centreline velocity in a pipe of 20 cm diameter is found to be 5.0 m/s. Assuming the flow to be in the fully developed rough-turbulent regime, find the discharge in the pipe. The effective roughness of the pipe wall is 0.5 mm.

(*Ans.* Q = 130 L/s)

10.7 The velocities in a 30 cm diameter pipe are found to be 3.0 m/s on the centreline and 2.8 m/s at a radial distance of 10 cm from the pipe axis. Calculate the discharge of the fluid in the pipe.

(*Ans.* Q = 192.7 L/s)

10.8 In a fully developed rough-turbulent flow in a 45 cm diameter pipe the mean velocity was found to be 60% of the maximum velocity. Estimate the equivalent sand grain roughness of the pipe.

(*Ans.* $\varepsilon_s = 0.0343$ mm)

10.9 Crude oil is being pumped through a 40 cm diameter pipeline. The flow is known to be turbulent. If the velocities at radial distances of 5 cm and 10 cm at a section are 3.80 m/s and 3.60 m/s respectively, estimate the discharge of the oil in the pipe.

(*Ans.* $Q = 402$ L/s)

10.10 In a turbulent flow in a pipe the velocity at a distance of 25% of the pipe radius from the wall is 0.80 times the centreline velocity. Determine the ratio of average velocity to the centreline velocity and the value of the friction factor f.

$$\left(Ans.\ \frac{V}{u_m} = 0.783;\quad f = 0.0376 \right)$$

10.11 A 30 cm pipeline carrying water has a centreline velocity of 2.0 m/s. If the velocity at mid-radius is 1.6 m/s, estimate the discharge.

(*Ans.* $Q = 80$ L/s)

10.12 The centreline velocity in a smooth 10 cm pipe carrying water ($\nu = 1 \times 10^{-6}$ m²s) is found to be 3.5 m/s. Determine the discharge and the friction factor f. The flow can be assumed to be hydrodynamically smooth.

(Hint: Find u_* by trial and error method).

(*Ans.* $Q = 23.7$ L/s; $f = 0.0143$)

10.13 A 9.0 cm diameter pipeline carries 10 L/s of oil ($\nu = 2.5 \times 10^{-6}$ m²/s). Calculate (i) the friction factor f, (ii) shear stress at the boundary, (iii) shear stress and velocity at a radial distance of 3.0 cm from the pipe axis and (iv) the thickness of laminar sublayer. Assume the pipe to be hydrodynamically smooth. ($\rho_{oil} = 800$ kg/m³).

(*Ans.* (i) $f = 0.0205$; (ii) $\tau_0 = 5.063$ Pa; (iii) $\tau = 3.375$ Pa, $u = 1.663$ m/s; (iv) $\delta' = 0.365$ mm)

10.14 An oil flows in a 40 cm diameter pipe at a Reynolds number of 10^6. The friction factor f is found to be 0.03. (a) Estimate the thickness of laminar sublayer, (b) If the equivalent sand grain roughness ε_s of the pipe is 0.5 mm determine whether the pipe is rough, smooth or in transition regime.

(*Ans.* (a) $\delta' = 0.0757$ mm; (b) Rough-turbulent regime)

10.15 A pipe 30 cm in diameter carries water ($\nu = 0.9 \times 10^{-6}$ m²/s). If the velocities at the centreline and at a radial distance of 8 cm from the pipe axis are 5.0 m/s and 4.7 m/s respectively, calculate the discharge in the pipe. Identify the pipe flow regime if the equivalent sand grain roughness is 0.3 mm.

(*Ans.* $Q = 311.6$ L/s; the flow is in transition between hydrodynamically rough and smooth boundary turbulent flow.)

10.16 A 60 cm diameter pipe is to carry 1.2 m³/s of oil ($\nu = 2.4 \times 10^{-6}$ m²/s). Estimate the maximum height of surface roughness that will have no effect on the resistance.

(*Ans.* $\varepsilon_{max} = 0.045$ mm)

10.17 Kerosene ($\rho = 804$ kg/m³) flows in a 20 cm diameter pipe with a mean velocity of 6 m/s. The pipe has an equivalent sand grain roughness $\varepsilon_s = 0.55$ mm. If the friction factor $f = 0.022$ find the thickness of the viscous sublayer. Is the pipe behaving as rough, smooth or in transition? ($\mu = 1.92 \times 10^{-3}$ Pa.s).

(*Ans.* $\delta' = 0.088$ mm; The boundary behaves as hydrodynamically rough.)

10.18 A 30 cm diameter pipe is known to have roughness elements of equivalent height 0.25 mm. What is the maximum velocity of flow of oil ($\nu = 2.4 \times 10^{-6}$ m²/s) at which the pipe will behave as a smooth pipe?

(*Ans.* $V = 0.589$ m/s)

10.19 Galvanized iron pipes can be assumed to have equivalent roughness magnitude of 0.15 mm. What minimum size of galvanized iron pipe will be hydrodynamically smooth at a Reynolds number of 2×10^5?

(*Ans.* $D = 0.439$ m)

10.20 A 10 cm diameter pipe is to carry water ($\nu = 1 \times 10^{-6}$ m²/s). The equivalent sand grain roughness of the pipe of 0.5 mm. What is the minimum discharge at which the pipe will have fully developed rough-turbulent flow?

(*Ans.* $Q = 18.0$ L/s)

10.21 Water ($\nu = 1.0 \times 10^{-6}$ m²/s) flows in a 35 cm diameter pipe in fully rough-turbulent regime. If at a radial distance of 15 cm from the pipe surface the velocity is 2.5 m/s and the velocity gradient is 7×10^{4} s^{-1} determine the discharge in the pipe.

(*Ans.* $Q = 175$ L/s)

10.22 Two pipes A and B have the same diameter and carry the same discharge of different fluids. The Reynolds number in pipe A is 1000 while that in B is 50,000. Determine (i) the ratio of maximum velocities in pipes A and B, (ii) the ratio of energy gradients in pipes A and B.

$$\left(\textit{Ans.}\ \text{(i)}\ \frac{u_{mA}}{u_{mB}} = 1.677;\right.$$

$$\left.\text{(ii)}\ (h_f/L)_A/(h_f/L)_B = 3.033\right)$$

10.23 A smooth pipe of 10 cm diameter carries 2.27 m^3/min of water at 20°C. Calculate the friction factor f, maximum velocity and shear stress at the boundary. (Take $\nu = 0.01$ stoke).

(*Ans.* (i) $f = 0.013$ (ii) $u_m = 5.52$ m/s (iii) $u_* = 37.64$ Pa)

10.24 Water is to be pumped from reservoir A to reservoir B. Two alternative pipes are available. Pipe 1 has 15 cm diameter and has an absolute roughness magnitude of 0.5 mm. Pipe 2 has 12 cm diameter and has an absolute roughness magnitude of 0.1 mm. Compare the costs of pumping the same discharge of water through these two pipes. The flow can be assumed to be fully developed rough-turbulent.

$$\left(\textit{Ans.}\ \frac{C_1}{C_2} = 0.47\right)$$

10.25 In a 20 cm diameter pipe carrying water the power required to pump a given discharge over a given length has increased by 20% over last 15 year due to increase in the roughness of the pipe. The absolute roughness magnitude at this stage is 0.5 mm. What additional power would be required to pump the same discharge over the same length at the end of next 15 years. The pipe can be assumed to be in rough-turbulent flow regime.

(*Ans.* Additional power = 12.1% of present power)

10.26 A 25 cm diameter pipe carrying water ($\nu = 1 \times 10^{-6}$ m^2/s) has a velocity of 2.0 m/s. The surface irregularities are estimated as $\varepsilon_0 = 0.01$ mm. It is also estimated that the rate of growth of surface imperfections is 0.007 mm/year. In how many years would this flow cease to be in smooth pipe regime?

(*Ans.* $T = 3.9$ years)

10.27 A pipeline carrying an oil in fully developed rough-turbulent flow has a head loss of 5 m in a certain length. The friction factor f is 0.02. Estimate the head loss if the roughness magnitude is doubled due to corrosion.

(*Ans.* $h_f = 5.975$ m)

10.28 A smooth pipe 20 cm in diameter carries crude oil at a velocity of 2.5 m/s. What is the loss of head in 100 m? What power is required to maintain the flow? (Assume $\nu = 0.4$ cm^2/s and RD = 0.90).

(*Ans.* $h_f = 4.762$ m (of crude oil), **P** = 3.30 kW)

10.29 An aeration device has a 10 cm diameter smooth pipe through which air is supplied. If the Reynolds number of the flow is 2×10^4, calculate the head loss in 100 m of the pipe. Express your answer in cms of water column.

[$\rho_{air} = 1.25$ kg/m^3, $\nu = 1.5 \times 10^{-5}$ m^2/s.]

(*Ans.* $(h_f)_{water} = 1.53$ cm of water column)

10.30 What is the head loss per metre of pipe when oil (RD = 0.85) having a dynamic viscosity of 1×10^{-4} Pa.s flows in a 10 cm diameter pipe at 20 L/s? [Assume effective roughness magnitude as 0.25 mm].

(*Ans.* $f = 0.025$; $h_f = 0.0825$ m of oil per metre length)

10.31 Methanol ($\rho = 971$ kg/m^3 and $\mu = 5.98 \times 10^{-4}$ Pa.s) flows at the rate of 150 L/min through 20 m of 2 cm diameter piping ($\varepsilon_s = 0.001$ mm). Calculate the head loss and power required for the flow to be possible.

(*Ans.* $h_f = 51.10$ m of methanol; **P** = 991.4 W)

10.32 Gasolene ($\rho = 680$ kg/m^3 and $\mu = 2.92 \times 10^{-4}$ Pa.s) flows in a 6 cm diameter pipe ($\varepsilon_s = 0.15$ mm) with a head loss of 5 m per 100 m. Determine the flow rate and power required to maintain this flow over a distance of 200 m.

(*Ans.* $Q = 257$ L/s; **P** = 286 W)

10.33 A horizontal 20 cm C.I. pipe ($\varepsilon_s = 0.26$ mm) is to carry water ($\nu = 1 \times 10^{-6}$ m^2/s) with a maximum allowable pressure drop of 50 kPa per km length of pipe. What maximum discharge is possible under these conditions?

(*Ans.* $Q = 29.87$ L/s)

10.34 Using the empirical formula for the friction factor f,

$$f = 0.0055\left[1 + \left(2000\frac{\varepsilon_s}{D} + \frac{10^6}{Re}\right)^{1/3}\right]$$

design the diameter of a C.I. pipe ($\varepsilon_s = 0.26$ mm) to carry sea water ($\mu = 1.07 \times 10^{-3}$ Pa.s and $\rho = 1025$ kg/m^3) with a velocity of 3.0 m/s. The head loss is limited to 6 m per 1000 m.

(*Ans.* $D = 0.794$ m; Use $D = 0.80$ m)

10.35 Water flows in an 80 mm pipe at Reynolds number of 80,000. The pipe is estimated to have an equivalent sand grain roughness of size 0.16 mm. Determine the head loss expected in 500 m length of the pipe. How much head loss would be expected if this pipe were smooth? Take kinematic viscosity of water as 10^{-6} m^2/s. The following explicit empirical equation for friction factor may be used.

$$\frac{1}{\sqrt{f}} = 1.14 - 2\log_{10}\left(\frac{\varepsilon_s}{D} + \frac{21.25}{R_e^{0.9}}\right)$$

(*Ans.* Rough h_f = 8.18 m, Smooth: h_f = 5.96 m)

10.36 A C.I. pipe (ε_s = 0.3 mm) is to carry 200 L/s of oil of kinematic viscosity 2.5×10^{-6} m^2/s. If the head loss is not to exceed 6.5 m per 1000 m of pipe, determine the required size of the pipe.

(*Ans.* D = 0.40 m)

10.37 Wind velocities over a flat land are observed to be 2.6 m/s and 3.0 m/s at 3.0 m and 9.0 m above ground respectively. Estimate the wind velocity at a height of 30 m above the ground. Assuming a fully developed rough-turbulent flow, determine the effective roughness height of the surface.

(*Ans.* u_3 = 3.438 m/s; ε_s = 71.5 mm)

11

Pipe Flow Systems

CONCEPT REVIEW

11.1 INTRODUCTION

Pipeline systems used in water distribution; industrial application and in many engineering systems may range from simple arrangement to extremely complex ones. This chapter deals with the basic elements of a pipeline and methods of evaluating or including their effects in computations concerning pipe systems. In general, the friction factor f is a function of Reynolds number and relative roughness. The method of evaluating f is described in Chapter 10. However, for purposes of simplifying computations the value of f is assumed to be constant and known in the examples that follow in this chapter.

The basic equation used in the calculation of head loss h_f in a pipe is the *Darcy–Weisbach* equation

$$\boxed{h_f = \frac{fLV^2}{2gD}} \qquad (11.1)$$

where L = length of a pipe of diameter D;
V = mean velocity in the pipe;
f = friction factor = fn (Reynolds number, relative roughness) in general.

The ratio $\frac{h_f}{L} = S_f$ represents the *energy slope* which is equal to the hydraulic gradient in uniform flow. In long pipelines h_f forms a very large part of the total loss.

11.2 MINOR LOSSES

In pipeline systems there will be a large number of pipe fittings such as bends, elbows, joints, valves and transitions. These fittings cause localised energy losses due to their shape and these losses are classified as minor losses. Table 11.1 lists some important minor losses.

Minor losses are usually neglected as insignificant if they are less than 5% of the frictional losses.

11.2.1 Equivalent Length

Equivalent length L_e of a minor loss is that length of the pipe which will have the same head loss for the same discharge.

Thus if the head loss in pipe fittings is expressed as $h_L = k\frac{V^2}{2g}$ then the length L_e of a pipe of diameter D and friction factor f which has the same velocity V is called its equivalent length if

$$k\frac{V^2}{2g} = \frac{fL_eV^2}{2gD} \qquad (11.2)$$

The equivalent length L_e will then be

$$\boxed{L_e = \frac{kD}{f}} \qquad (11.3)$$

11.2.2 Equivalent Pipes

(i) A pipe with length L_1, diameter, D_1 and friction factor f_1 will be equivalent to another pipe of corresponding parameters L_2, D_2 and f_2, if

$$\boxed{\frac{f_1L_1}{D_1^5} = \frac{f_2L_2}{D_2^5}} \qquad (11.4)$$

Table 11.1 *Minor Losses in Pipeline Systems*

Situation	*Head loss = h_L*	*Remarks/Explanation*
1. Sudden expansion	$\frac{(V_1 - V_2)^2}{2g}$	Expansion from section 1 to 2
2. Sudden contraction	$\left(\frac{1}{C_c} - 1\right)^2 \frac{V_2^2}{2g}$	C_c = coefficient of contraction V_2 = velocity in contracted section
3. Square edged entrance to a pipe	$0.5\ \frac{V^2}{2g}$	Square edged entrance from a reservoir
4. Exit from a pipe	$\frac{V^2}{2g}$	
5. Conical expansion	$k\ \frac{(V_1 - V_2)^2}{2g}$	k = fn (θ, D_2/D_1)
6. Bends	$k\ \frac{V^2}{2g}$	k = fn (θ, R/D)
7. Pipe fittings	$k\ V_1^2/2g$	Values of k given in Table 11.2.
8. Nozzle	$\left(\frac{1}{C_v^2} - 1\right) V^2/2g$	C_v = coeff. of velocity of the nozzle

Table 11.2 *Loss at Pipe Fittings ($h_L = kV^2/2g$)—Average Values of k*

Fitting	*k*
1. Globe valve (fully open)	10.0
2. Angle valve (fully open)	5.0
3. Gate valve (fully open)	0.19
4. Gate valve (half open)	5.6
5. Standard tee	1.8
6. Standard elbow	0.90
7. Sharp 90° bend	1.20
8. Sharp 90° bend with vanes	0.20

(ii) If a set of pipes described by (L_1, D_1, f_1), (L_2, D_2, f_2)... are connected in *series* then an equivalent pipe (L_e, D_e, f_e) is related as

$$\boxed{\frac{f_e L_e}{D_e^5} = \frac{f_1 L_1}{D_1^5} + \frac{f_2 L_2}{D_2^5} + \frac{f_3 L_3}{D_3^5} + \ldots} \quad (11.5)$$

(iii) If a set of pipes (L_1, D_1, f_1), (L_2, D_2, f_2), (L_3, D_3, f_3)... are connected in *parallel* between two points, then an equivalent pipe (L_e, D_e, f_e) is related as

$$\boxed{\left(\frac{D_e^5}{f_e L_e}\right)^{1/2} = \left(\frac{D_1^5}{f_1 L_1}\right)^{1/2} + \left(\frac{D_2^5}{f_2 L_2}\right)^{1/2} + \left(\frac{D_3^5}{f_3 L_3}\right)^{1/2} + \ldots} \quad (11.6)$$

11.3 SIMPLE PIPE PROBLEMS

If a pipeline has a large number of minor head losses, the total head loss

$$H_L = h_f + k_1 \frac{V^2}{2g} + k_2 \frac{V^2}{2g} + \ldots$$

or

$$\boxed{H_L = \frac{fLV^2}{2gD} + \left(\sum_1^n k_i\right)\frac{V^2}{2g}} \quad (11.7)$$

Power required to pump the fluid over a length L,

$$\boxed{\mathbf{P} = \gamma Q H_L} \quad (11.8)$$

By energy consideration, if

H_1 = total energy at beginning of the pipe

$$= \frac{p_1}{\gamma} + Z_1 + \frac{V^2}{2g}$$

H_2 = total energy at the end of the pipe (after a length L from section 1) $= \frac{p_2}{\gamma} + Z_2 + \frac{V^2}{2g}$

Then $\boxed{H_1 = H_2 + H_L}$

Also, since the velocity V is same at sections 1 and 2.

$$\left(\frac{p_1}{\gamma}+Z_1\right)-\left(\frac{p_2}{\gamma}+Z_2\right)=H_L \tag{11.9}$$

11.3.1 Siphon

A siphon is a situation where the centreline of the pipe is above the hydraulic grade line (Fig. 11.1 and 11.2).

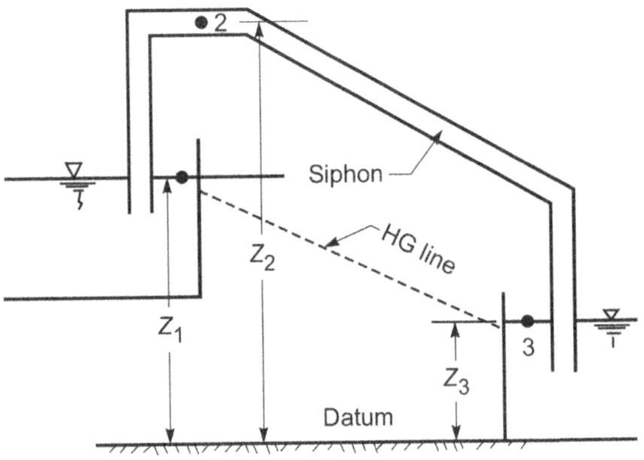

Fig. 11.1 *Siphon*

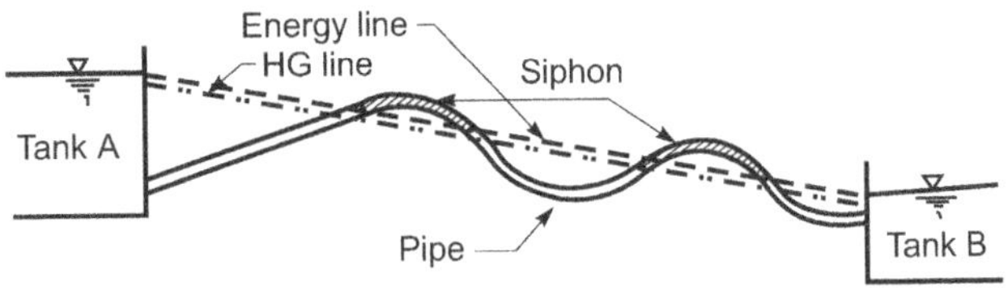

Fig. 11.2 *Siphon*

In a siphon that portion of the pipeline which is above the hydraulic grade line experiences negative pressures. In Fig. 11.1, by taking atmospheric pressure as datum (zero).

$$Z_1+0+0=\frac{p_2}{\gamma}+Z_2+\frac{V_2}{2g}+H_{L(1-2)}=Z_3+H_{L(1-3)}$$

where

$H_{L(1-2)}$ = total head loss between sections 1 and 2 = (friction loss + minor losses) between sections 1 and 2.

It is seen that $H_{L(1\text{-}3)}=Z_1-Z_3$

and $\frac{p_2}{\gamma}=(Z_1-Z_2)-H_{L(1\text{-}2)}$ (gauge pressure) (11.10)

11.3.2 Pipes in Series

When two or more pipes of different diameters or roughnesses are so connected that the full discharge of the fluid from one flows into the others serially, the system represents a series pipeline. Figure 11.3 shows a typical set of pipes in series. The head losses are cumulative.

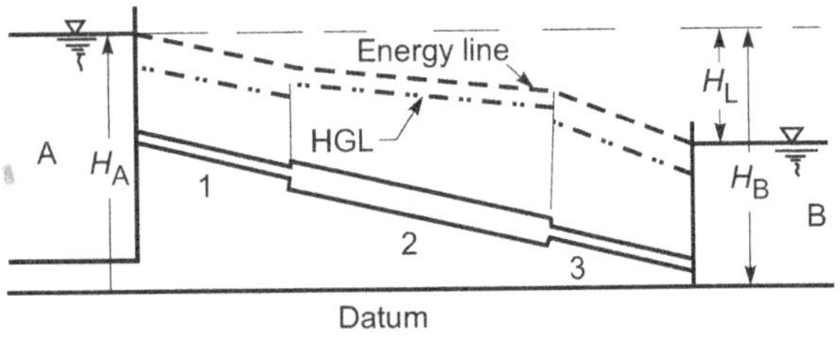

Fig. 11.3 *Pipes in Series*

The total head H at A an B are related as:

$$H_A-H_B=H_L$$

where H_L = total energy loss in pipes 1, 2 and 3

$$=\left((\Sigma k)\frac{V_1^2}{2g}+\frac{f_1L_1V_1^2}{2gD_1}\right)+\left((\Sigma k)\frac{V_2^2}{2g}+\frac{f_2L_2V_2^2}{2gD_2}\right)$$

$$+\left((\Sigma k)\frac{V_3^3}{2g}+\frac{f_3L_3V_3^2}{2gD_3}\right) \tag{11.11}$$

$\left(\Sigma k\frac{V^2}{2g}\right)$ represents the total minor losses in each pipe.

Also by continuity, since the same discharge passes through all the pipes,

$$Q_1=Q_2=Q_3$$

i.e. $$V_1D_1^2=V_2D_2^2=V_3D_3^2 \tag{11.12}$$

The solution of pipes in series is generally relatively straight forward and simple.

11.3.3 Pipes in Parallel

A combination of two or more pipes connected between two points so that the discharge divides at the first junction and rejoins at the next is known as pipes in parallel (Fig. 11.4). Here the head loss between the two junctions (M and N in Fig. 11.4) is same for all the pipes. Thus as total discharge

$$Q=Q_1+Q_2+Q_3$$

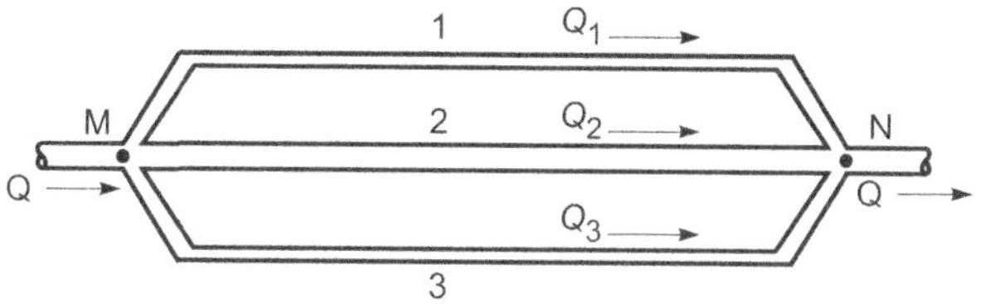

Fig. 11.4 *Pipes in Parallel*

Head loss

$$H_{L1} = H_{L2} = H_{L3} = \left(\frac{p_M}{\gamma} + Z_M\right) - \left(\frac{p_N}{\gamma} + Z_N\right) = h_M - h_N \quad (11.13)$$

It is usual to consider minor losses as equivalent lengths in parallel pipe flow problems and as such

$$H_L = h_f = \frac{fLV^2}{2gD}$$

Two types of problems can be recognised:

(1) Given piezometric heads at M and N (h_M and h_N) to find Q_1, Q_2, ... etc. This is a straight forward problem, especially if f of each pipe is known.

(2) Given total discharge Q, to determine the discharge division (i.e. to find Q_1, Q_2, ... etc.).

Three solution procedures are available to solve this problem

1. Exact method
2. Equivalent Pipe method
3. Trial and Error method

(1) Exact Method

Let there be N pipes in parallel.

Put $h_f = r_1 Q_1^2 = r_2 Q_2^2 = \ldots = r_i Q_i^2 = \ldots = r_n Q_n^2$

In this

$$r_i = \frac{8 f_i L_i}{\pi^2 g D_i^5} \quad \left(\text{obtained from } h_f = \frac{f_i L_i V_i^2}{2g D_i}\right) \quad (11.14)$$

Thus

$$Q_2 = \sqrt{\frac{r_1}{r_2}}\, Q_1$$

$$Q_3 = \sqrt{\frac{r_1}{r_3}}\, Q_1$$

....

$$Q_i = \sqrt{\frac{r_1}{r_i}}\, Q_1 \quad (11.15)$$

By continuity, total discharge

$$Q = Q_1 + Q_2 + Q_3 + \ldots + Q_i + \ldots + Q_n = \sum_1^n Q_i$$

Substituting Q_i from Eq. (11.15)

$$Q = Q_1\left[1 + \sqrt{\frac{r_1}{r_2}} + \sqrt{\frac{r_1}{r_3}} + \ldots + \sqrt{\frac{r_1}{r_i}} + \ldots + \sqrt{\frac{r_1}{r_N}}\right]$$

$$= Q_1 \sqrt{r_1}\left[\sum_1^N \frac{1}{\sqrt{r_i}}\right]$$

Thus
$$Q_1 = \frac{Q}{\sqrt{r_1}}\left[\frac{1}{\sum_1^N \frac{1}{\sqrt{r_i}}}\right] \quad (11.16)$$

Knowing Q_1, other discharges Q_i can be found from Eq. 11.15. In addition, from Eq. 11.14 h_f can be found.

(2) Equivalent Pipe Method

This is an alternative exact method. In this an equivalent pipe is used as a replacement of the set of parallel pipes. Examples 11.12 and 11.16 illustrate the use of this method.

(3) Trial and Error Method

(i) Assume a trial discharge Q_1' in pipe 1, (Fig. 11.4).

(ii) Using Q_1' find $h_f' = \dfrac{f L_1 V_1'^2}{2g D_1}$

(iii) Using h_f' as common head find Q_2' and Q_3'.

(iv) Find $Q' = (Q_1' + Q_2' + Q_3')$

(v) Assume the correct discharge is split among the pipes in the same ratio as $Q_1' : Q_2' : Q_3'$, thus $Q_1 = K\, Q_1'$,

$Q_2 = K\, Q_2'$ and $Q_3 = K\, Q_3'$ where $K = \left(\dfrac{Q}{Q'}\right)$

(vi) Check the accuracy of this assumption by calculating h_{f1}, h_{f2} and h_{f3}. If the difference is greater than 5% repeat step (iv) onwards. The Example 11.12 illustrates this method.

11.3.4 Branching Pipes

In this at a junction three or more pipes lead off to different terminals. A typical branched pipe problem is the three-reservoir problem (Fig. 11.5). Here at the junction J the continuity equation must be satisfied, i.e. the flow into the junction = flow out of the junction.

Three-Reservoir Problem

Referring to Fig. 11.5, three reservoirs A, B and C are connected by three pipes 1, 2 and 3 at a common junction J. The three-reservoir problem is about finding the direction and magnitude of the discharges in the three pipes when the geometric characteristics of the pipes and the water surface elevations of the three reservoirs are known.

Two solution procedures viz. (i) Exact method and (ii) Trial and error method are in use. In this type of problem the calculations are greatly simplified by putting $h = rQ^2$ where $r = \dfrac{8fl}{\pi^2 g D^5}$.

$$H_b = H_j - r_2 Q_2^2$$

$$H_c = H_j - r_3 Q_3^2$$

Put $Q_3 = mQ_2$ resulting in $Q_1 = (1 + m)Q_2$ (11.17)

$$H\,(H_a - H_b) = [r_1(1 + m)^2 + r_2]Q_2^2$$

$$(H_b - H_c) = [r_3 m^2 - r_2]Q_2^2 \qquad (11.18)$$

Dividing Eq. 11.17 by 11.18

$$\frac{r_1(1+m^2)+r_2}{r_3 m^2 - r_2} = \frac{H_a - H_b}{H_b - H_c} = h' \qquad (11.19)$$

Equation 11.19 is a quadratic equation in m. Considering the positive root as the relevant value, Q_2 is found by

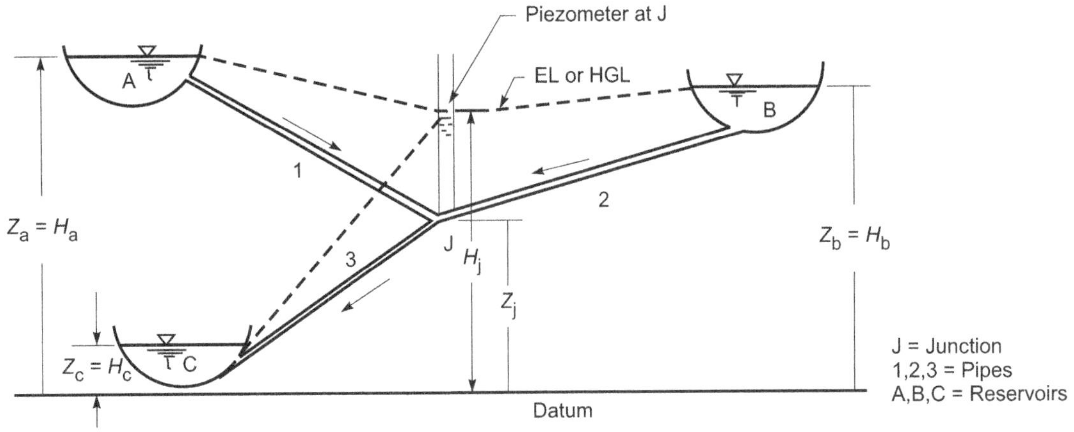

Fig. 11.5 *Three Reservoir Problem*

(1) Exact Method

Let H_j = piezometric head at the junction and H_a, H_b and H_c are the piezometric heads at the highest reservoir (A), intermediate reservoir (B) and lowest reservoir (C) respectively.

(1) To determine the direction of flow:

$$\text{Let } \frac{H_a - H_b}{H_b - H_c} = h' \quad \text{and} \quad \frac{r_1}{r_2} = R_1$$

- If $h' > R_1$, the flow is of **Type-1** with $H_j > H_b$ and $Q_1 = Q_2 + Q_3$.
- If $h' < R_1$, the flow is of **Type-2** with $H_j < H_b$ and $Q_1 + Q_2 = Q_3$.
- If $h' = R_1$, the flow is of **Type-3** with $H_j = H_b$ and $Q_1 = Q_3$. This is a rare situation and the problem degenerates to a case of two pipes in series.

(2) Solution Procedure for Type-1 flow:

By the application of Energy equation between the junction j and the reservoirs A, B and C successively.

$$H_a = H_j + r_1 Q_1^2$$

applying this value of m in Eq. 11.17 or Eq. 11.18 and hence values of Q_3 and Q_1. H_j is found by energy relationship between a reservoir and the junction, e.g., $H_a = H_j + r_1 Q_1^2$.

Example 11.18 illustrates the use of the exact method for Type-1 flow.

(3) Solution Procedure for Type-2 flow:

The energy relationships for this type of flow are:

$$H_a = H_j + r_1 Q_1^2$$

$$H_b = H_j + r_2 Q_2^2$$

$$H_c = H_j - r_3 Q_3^2$$

Put $Q_3 = mQ_2$ resulting in $Q_1 = (\text{m} - 1)Q_2$

$$(H_a - H_b) = [r_1(m - 1)^2 - r_2]Q_2^2 \qquad (11.20)$$

$$(H_b - H_c) = [r_3 m^2 + r_2]Q_2^2 \qquad (11.21)$$

Dividing Eq. 11.20 by 11.21

$$\frac{r_1(m-1^2)-r_2}{r_3 m^2 + r_2} = \frac{H_a - H_b}{H_b - H_c} = h' \qquad (11.22)$$

This quadratic equation in m has two positive roots. Selecting the root where $m > 1$ as relevant (as $m < 1$ gives negative Q_1 values) the discharge Q_2 is found from Eq. 11.20 or Eq. 11.21 and thence Q_3 and Q_1. The head at the junction, H_j, is found from energy relationship between a reservoir and the junction J.

Example 11.19 illustrates the use of the exact method for Type-2 flow.

(2) Trial and Error Method

(i) Assume a trial value of H_j. The first trial H_j may be taken around the average value of the lowest and highest reservoir levels.

(ii) For each H_j calculate Q_i in each pipeline with positive sign if it is towards the junction and negative sign if away from the junction. Find $\Delta Q = \Sigma Q_i$ and also find $\Sigma\, |Q/h_f|$.

(iii) The additive correction, to be added to the assumed value of H_j for purposes of next trial, is

$$\Delta H_j = \frac{2\Delta Q}{\Sigma|Q/h_f|} \qquad (11.23)$$

(iv) New H_j for next trial is H_j = previous $H_j + \Delta H_j$.

(v) Continue till ΔQ is very small.

[Usually two iterations would suffice.]

11.4 PIPE NETWORK

An interconnected system of pipes is called a pipe network. The flow to an outlet may come from different pipes. Figure 11.6 shows a typical network. In a network:

(1) Flow into each junction must be equal to flow out of each junction.

(2) Algebraic sum of head losses round each loop must be zero.

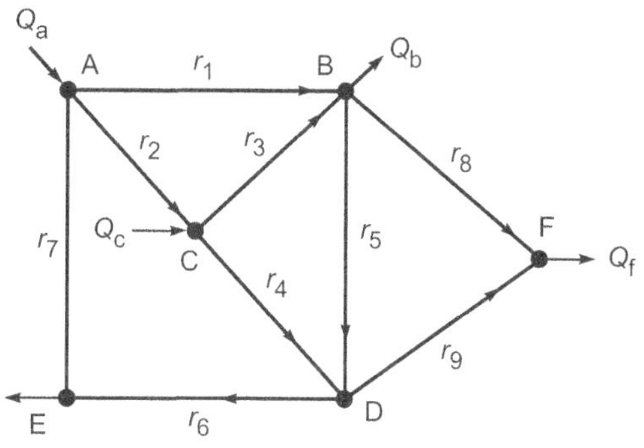

Fig. 11.6 *Typical Pipe Network*

Generally, the solution of real pipe networks, as used in engineering practice, is difficult and needs the help of digital computers. Here, for illustration purposes simplified situation is used. The head loss in each pipe is expressed as $h_f = rQ^n$. The coefficient r depends upon pipe length, diameter and friction factor. For turbulent flow n is of the order of 2.

11.4.1 Hardy–Cross Method of Analysis

(i) In this method a trial distribution of discharges is made, arbitrarily but in such a way as to satisfy continuity at each node.

(ii) Head loss in each pipe is calculated as $h_f = rQ^n = r\,Q\,|Q^{n-1}|$ and also the quantity $(rn\,|\,Q^{n-1}|)$ is calculated for each pipe.

(iii) For each loop, the quantity

$$\Delta Q = \frac{-\Sigma r Q^n}{\Sigma rn|Q^{n-1}|} \qquad (11.24)$$

is calculated. This represents the correction to the assumed discharge.

(iv) Value of ΔQ is calculated for all loops.

(v) Corrections are now applied to each pipe in a loop and to all loops. Clockwise direction is considered positive.

(vi) The procedure is repeated till ΔQ is very small.

[*Note:* The directional sense is important. Flows in a loop which are clockwise are taken positive. Positive ΔQ represents a correction to be added to clockwise (positive) flows. Some pipes which are common to two loops may get two corrections.]

Examples 11.24 and 25 illustrate the Hardy–Cross procedure of network analysis.

11.4.2 Contaminant Propagation

If a contaminant enters a steady state pipe network shown in Fig. 11.6 at any node (say node A) with a concentration C_a, its propagation through the network can be determined by simple mass balance, i.e. continuity relationship extended to include the contaminant.

It is assumed that complete mixing takes place at a node and the contaminant is conservative, i.e. does not die off. Then at any node

$$C_{out} = \frac{\Sigma\, Q_i C_i}{\Sigma\, Q_i} \qquad (11.25)$$

where C_{out} = average (mixed) contaminant concentration going out

Q_i = inflow rates and

C_i = contaminant concentrations of inflows.

Example 11.26 illustrates the method of estimating the propagation of contaminants to various nodes in a steady state network.

11.5 MISCELLANEOUS PROBLEMS

A large number of variations in the basic situation described above are possible. Some of the common types of pipe flow problems of interest are:

(i) Pipelines with pumps/turbines.

(ii) Nozzle at the end of a pipe.

(iii) Nonuniform flow in a pipe, e.g. tapering pipe or pipe with gradually decreasing discharge.

(iv) Combination of series and parallel pipes.

These are illustrated through carefully chosen examples.

WORKED EXAMPLES

11.1 A 6 cm diameter pipe has a discharge of 450 L/min. At a section the pipe has a sudden expansion to a size of 9 cm diameter. If the pressure just upstream of the expansion is 20 kN/m², calculate the pressure just after the expansion. Assume the pipe to be horizontal at the expansion region.

Solution:

$$\text{Discharge } Q = \frac{450}{60} = 7.50 \text{ L/s} = 0.0075 \text{ m}^3\text{/s}$$

$$V_1 = \text{Velocity before expansion}$$

$$= \frac{0.0075}{\frac{\pi}{4}\times(0.06)^2} = 2.653 \text{ m/s}$$

$$V_2 = \text{Velocity downstream of the expansion}$$

$$= V_1\left(\frac{D_1}{D_2}\right)^2 = 2.653\left(\frac{0.06}{0.09}\right)^2 = 1.179 \text{ m/s}$$

Loss of head at sudden expansion

$$H_L = \frac{(V_1 - V_2)^2}{2g}$$

$$= \frac{(2.653 - 1.179)^2}{2\times 9.81} = 0.111 \text{ m}$$

By energy equation:

$$\frac{p_1}{\gamma} + \frac{V_1^2}{2g} + Z_1 = \frac{p_2}{\gamma} + \frac{V_2^2}{2g} + Z_2 + H_L$$

As $Z_1 = Z_2$

$$\frac{p_2}{\gamma} = \frac{p_1}{\gamma} + \frac{V_1^2}{2g} - \frac{V_2^2}{2g} - H_L$$

$$= \frac{20.0}{9.79} + \frac{(2.653)^2}{2\times 9.81} - \frac{(1.179)^2}{2\times 9.81} - 0.111$$

$$= 2.043 + 0.359 - 0.071 - 0.111$$

$$\frac{p_2}{\gamma} = 2.22 \text{ m}$$

$$p_2 = 2.22 \times 9.79 = \mathbf{21.73 \text{ kPa}}$$

11.2 A horizontal pipe of diameter D_1 has a sudden expansion to a diameter D_2. At what ratio D_1/D_2 would the differential pressure on either side of the expansion be maximum? What is the corresponding loss of head and differential pressure head?

Solution: (a) In a sudden expansion, the loss of head

$$h_L = \frac{(V_1 - V_2)^2}{2g}$$

For an expansion in a horizontal pipe

$$\frac{p_1}{\gamma} + \frac{V_1^2}{2g} = \frac{p_2}{\gamma} + \frac{V_2^2}{2g} + \frac{(V_1 - V_2)^2}{2g}$$

$$\left(\frac{p_2}{\gamma} - \frac{p_1}{\gamma}\right) = \frac{\Delta p}{\gamma} = \frac{V_1^2}{2g} - \frac{V_2^2}{2g} - \frac{(V_1 - V_2)^2}{2g}$$

By continuity

$$V_1 D_1^2 = V_2 D_2^2$$

$$V_2 = V_1\left(\frac{D_1}{D_2}\right)^2 = V_1 x^2$$

$$\text{where} \quad x = \left(\frac{D_1}{D_2}\right)$$

$$\therefore \quad \frac{\Delta p}{\gamma} = \frac{V_1^2}{2g}(1 - x^4) - \frac{V_1^2}{2g}(1 - x^2)^2$$

$$= \frac{V_1^2}{2g}[1 - x^4 - (1 - x^2)^2]$$

$$\text{For maximum pressure differential} \quad \frac{d(\Delta p/\gamma)}{dx} = 0$$

$$\therefore \quad -4x^3 - 2(1 - x^2)(-2x) = 0$$

$$-2x^3 + 2x - 2x^3 = 0$$

$$\text{or} \quad (2x^2 - 1) = 0 \quad \text{or} \quad x = \frac{1}{\sqrt{2}}$$

i.e. $D_2 = \sqrt{2}\, D_1$

(b) Head loss $h_L = \dfrac{(V_1 - V_2)^2}{2g} = \dfrac{V_1^2}{2g}(1 - x^2)^2$

$$= \left(1 - \frac{1}{2}\right)^2 \frac{V_1^2}{2g} = \frac{1}{4}\frac{V_1^2}{2g}$$

Differential pressure head

$$= \frac{\Delta p}{\gamma} = \frac{V_1^2}{2g}(1 - x^4) - h_L$$

$$= \frac{V_1^2}{2g}\left(1 - \frac{1}{4} - \frac{1}{4}\right) = \frac{1}{2}\cdot\frac{V_1^2}{2g}$$

11.3 Three pipes with diameter, length and friction factor values of (D_1, L_1, f_1); (D_2, L_2, f_2) and (D_3, L_3, f_3) are connected in parallel between two points A and B in a pipeline. If an equivalent pipe (D_e, L_e, f_e) is to replace the set of parallel pipes, obtain an expression for estimating the equivalent pipe parameters.

Solution: Since there is a common head drop for the parallel pipes

$$h_f = \frac{f_1 L_1 V_1^2}{2g D_1} = \frac{f_2 L_2 V_2^2}{2g D_2} = \frac{f_3 L_3 V_3^2}{2g D_3} \quad (1)$$

The sum of the flows = total discharge

or $\quad Q = Q_1 + Q_2 + Q_3 \quad (2)$

From Eq. (1) $V_1 = \left(\dfrac{D_1}{f_1 L_1}\right)^{1/2} \sqrt{2g \cdot h_f}$

$$Q_1 = \frac{\pi}{4}\left(\frac{D_1^5}{f_1 L_1}\right)\sqrt{2g \cdot h_f}$$

Similarly $\quad Q_2 = \dfrac{\pi}{4}\left(\dfrac{D_2^5}{f_2 L_2}\right)\sqrt{2g \cdot h_f}$

$$Q_3 = \frac{\pi}{4}\left(\frac{D_3^5}{f_3 L_3}\right)\sqrt{2g \cdot h_f}$$

Also for the equivalent pipe

$$Q = \frac{\pi}{4} D_e^2 V_e$$

and the head loss

$$h_f = \frac{f_e L_e}{D_e}\frac{V_e^2}{2g}$$

or $\quad Q = \dfrac{\pi}{4}\left(\dfrac{D_e^5}{f_e L_e}\right)^{1/2}\sqrt{2g \cdot h_f}$

Thus from Eq. (2)

$$\left(\frac{D_e^5}{f_e L_e}\right)^{1/2} = \left(\frac{D_1^5}{f_1 L_1}\right)^{1/2} + \left(\frac{D_2^5}{f_2 L_2}\right)^{1/2} + \left(\frac{D_3^5}{f_3 L_3}\right)^{1/2} \quad (3)$$

[***Note:*** For the equivalent pipe, out of the 3 variables D_e, L_e and f_e, given any two, the third one can be solved by Eq. (3).]

11.4 A compound piping system consists of 1800 m of 50 cm, 1200 m of 40 cm and 600 m of 30 cm diameter pipes of the same material connected in series.

(i) What is the equivalent length of a 40 cm pipe of the same material?

(ii) What is the equivalent size of a pipe 3600 m long?

(iii) If the three pipes are in parallel, what is the equivalent length of a 50 cm pipe?

Solution: It is assumed that f is same for all the pipes.

(i) $$\frac{f_e L_e}{D_e^5} = \frac{f_1 L_1}{D_1^5} + \frac{f_2 L_2}{D_2^5} + \frac{f_3 L_3}{D_3^5}$$

$$f_e = f_1 = f_2 = f_3$$

$$\frac{L_e}{D_e^5} = \frac{L_1}{D_1^5} + \frac{L_2}{D_2^5} + \frac{L_3}{D_3^5}$$

$$\frac{L_e}{(0.4)^5} = \frac{1800}{(0.5)^5} + \frac{1200}{(0.4)^5} + \frac{600}{(0.3)^5}$$

$$\mathbf{L_e = 4318.2\ m}$$

(ii) $$\frac{3600}{D_e^5} = \frac{1800}{(0.5)^5} + \frac{1200}{(0.4)^5} + \frac{600}{(0.3)^5}$$

$$\mathbf{D_e = 38.57\ cm}$$

(iii) Three pipes in parallel:

$$Q_0 = Q_1 + Q_2 + Q_3$$

and $h_{f1} = h_{f2} = h_{f3} = h_{fe}$

$$\therefore \left(\frac{f_1 L_1}{D_1}\right)\frac{V_1^2}{2g} = \left(\frac{f_2 L_2}{D_2}\right)\frac{V_2^2}{2g}$$

$$= \left(\frac{f_3 L_3}{D_3}\right)\frac{V_3^2}{2g} = \left(\frac{f_e L_e}{D_e}\right)\frac{V_e^2}{2g}$$

But $f_1 = f_2 = f_3 = f_e$

$$\therefore \frac{V_2}{V_1} = \sqrt{\frac{L_1}{L_2}\cdot\frac{D_2}{D_1}} = \sqrt{\frac{1800 \times 0.40}{1200 \times 0.50}} = 1.0954$$

Similarly

$$\frac{V_3}{V_1} = \sqrt{\frac{1800 \times 0.3}{600 \times 0.5}} = 1.3416$$

The total discharges Q_0 is given by

$$Q_0 = \left(\frac{\pi}{4}\right)[(0.5)^2 V_1 + (0.4)^2 V_2 + (0.3)^2 V_3]$$

$$= \frac{\pi}{4} V_1[\,0.25 + (0.16)\,(1.095) + (0.09)\,(1.3416)]$$

$$= (0.546)\left(\frac{\pi}{4} V_1\right)$$

For an equivalent pipe $Q_e = Q_0$,
the diameter $D_e = 0.50$ m and length $= L_e$
Since $h_{f1} = h_{fe}$

$$\frac{V_e}{V_1} = \sqrt{\frac{L_1}{L_e}\frac{D_e}{D_1}} = \sqrt{\frac{1800 \times 0.5}{L_e \times 0.5}} = \sqrt{1800/L_e}$$

Also $Q_0 = \left(0.546 \times \frac{\pi}{4} V_1\right) = Q_e = \frac{\pi}{4} \times (0.5)^2\, V_e$

Since $Q_e = Q_0$

$$\frac{V_e}{V_1} = \frac{0.546}{0.250} = 2.184 = \sqrt{\frac{1800}{L_e}}$$

$\therefore$ $\quad$ **L_e = 377.37 m**

Alternatively
By the equivalent pipe formula for parallel pipes.

$$\left(\frac{D_e^5}{f_e L_e}\right)^{1/2} = \left(\frac{D_1^5}{f_1 L_1}\right)^{1/2} + \left(\frac{D_2^5}{f_2 L_2}\right)^{1/2} + \left(\frac{D_3^5}{f_3 L_3}\right)^{1/2} + \ldots$$

Here $D_e = 0.50$ m, $f_1 = f_2 = f_3 = f_e$

$$\left[\frac{(0.5)^5}{L_e}\right]^{1/2} = \left[\left(\frac{(0.5)^5}{1800}\right)^{1/2} + \left(\frac{(0.40)^5}{1200}\right)^{1/2} + \left(\frac{(0.30)^5}{600}\right)^{1/2}\right]$$

$$= [(4.1667 \times 10^{-3}) + (2.9212 \times 10^{-3}) + (2.0125 \times 10^{-3})]$$

$$\frac{0.17678}{(L_e^{1/2})} = 9.1004 \times 10^{-3}$$

L_e = 377.37 m

11.5 Two reservoirs with a difference in water surface elevation of 10 m are connected by a pipeline ABC which consists of two pipes of AB and BC joined in series. Pipe AB is 10 cm in diameter, 20 m long and has a value of $f = 0.02$. Pipe BC is of 16 cm diameter, 25 m long and has an $f = 0.018$. The junctions with the reservoirs and between the pipes are abrupt.

(a) Calculate the discharge.
(b) What difference in reservoir elevations is necessary to have a discharge of 15 L/s? [Include all minor losses].

Solution: Use suffix 1 for pipe AB and suffix 2 for pipe BC.

(i) Entrance loss $h_{L1} = 0.5\; V_1^2/2g$

(ii) Loss at sudden expansion $= h_{eL} = \dfrac{(V_1 - V_2)^2}{2g}$

$$= \frac{V_1^2}{2g}\left(1 - \frac{V_2}{V_1}\right)^2$$

Since

$$\frac{\pi}{4} D_1^2 V_1 = \frac{\pi}{4} D_2^2 V_2 = Q$$

$$\frac{V_2}{V_1} = \left(\frac{D_1}{D_2}\right)^2$$

$$\therefore \quad h_{eL} = \left(1 - \left(\frac{D_1}{D_2}\right)^2\right)^2 \frac{V_1^2}{2g}$$

$$= \left(1 - \left(\frac{0.10}{0.16}\right)^2\right)^2 \frac{V_1^2}{2g} = 0.3714\,\frac{V_1^2}{2g}$$

(iii) Loss at exit

$$h_e = \frac{V_2^2}{2g}$$

(iv) Friction loss

$$h_{f1} = \frac{f_1 L_1}{D_1}\frac{V_1^2}{2g}$$

$$= \frac{0.02 \times 20}{0.10}\frac{V_1^2}{2g} = 4.0\,\frac{V_1^2}{2g}$$

$$h_{f2} = \frac{f_2 L_2}{D_2}\frac{V_2^2}{2g}$$

$$= \frac{0.018 \times 25}{0.16}\frac{V_2^2}{2g}$$

$$= 2.8125\,\frac{V_2^2}{2g}$$

Total loss H_L

$$H_L = 0.5\,\frac{V_1^2}{2g} + 0.3714\,\frac{V_1^2}{2g} + \frac{V_2^2}{2g} + 4.0\,\frac{V_1^2}{2g} + 2.8125\,\frac{V_2^2}{2g}$$

$$H_L = 5.8714\,\frac{V_1^2}{2g} + 2.8125\,\frac{V_2^2}{2g}$$

But $\quad V_2 = \left(\dfrac{D_1}{D_2}\right)^2 V_1 = \left(\dfrac{0.10}{0.16}\right)^2 V_1 = 0.39063\; V_1$

$$H_L = (5.8714 + 0.4292)\frac{V_1^2}{2g} = 6.300\frac{V_1^2}{2g}$$

(a) When $H_L = 10$ m,

$$V_1^2 = \left(\frac{10 \times 2 \times 9.81}{6.300}\right) = 31.14;$$

$V_1 = 5.582$ m/s
$V_2 = 2.18$ m/s

and $Q = \frac{\pi}{4} \times (0.10)^2 \times 5.582 = 0.0438$ m³/s

= 43.8 L/s

(b) When $Q = 15$ L/s $= 0.015$ m³/s

$$V_1 = \frac{0.015}{\frac{\pi}{4} \times (0.1)^2} = \frac{0.015}{0.007854} = 1.91 \text{ m/s}$$

$V_1^2 = 3.648$, $H_L = 6.300\ V_1^2/2g$
$\boldsymbol{H_L = 1.171}$ **m**

11.6 A reservoir discharges its liquid content through a horizontal pipeline into the atmosphere. The pipeline consists of two pipes: One of 10 cm diameter and 25 m long and another of 12 cm diameter and 35 m long, connected in series. The friction factor $f = 0.02$ for both the pipes. The water level in the tank is 10 m above the centre-line of the pipe at the entrance. Considering all the minor losses, (a) Calculate the discharge when the 10 cm dia pipe is joined to the tank (b) Calculate the discharge when the 12 cm dia pipe is joined to the tank (Assume $C_c = 0.70$).

Solution: (a) When 10 cm pipe is connected to the tank:

(i) H_{L1} = loss at entrance $= 0.5\frac{V_1^2}{2g}$

(ii) h_{L2} = Loss at sudden expansion $= \frac{(V_1 - V_2)^2}{2g}$

But $V_1 D_1^2 = V_2 D_2^2$

$$V_2 = V_1\left(\frac{D_1}{D_2}\right)^2 = V_1\left(\frac{10}{12}\right)^2 = 0.694\ V_1$$

$$\therefore \quad h_{L2} = \frac{V_1^2}{2g}(1 - 0.694)^2 = 0.0934\frac{V_1^2}{2g}$$

(iii) Velocity head at exit $= h_{v2} = \frac{V_2^2}{2g} = (0.694)^2\frac{V_1^2}{2g}$

$$= 0.4816\frac{V_1^2}{2g}$$

(iv) h_{f1} = head loss in 10 cm pipe due to friction

$$= \frac{f_1 L_1}{D_1}\frac{V_1^2}{2g} = \frac{0.02 \times 25}{0.1}\frac{V_1^2}{2g}$$

$$= 5\frac{V_1^2}{2g}$$

(v) h_{f2} = head loss in 12 cm pipe due to friction

$$h_{f2} = \frac{f_2 L_2}{D_2}\frac{V_2^2}{2g}$$

$$= \frac{0.02 \times 35}{0.12} \times (0.694)^2\frac{V_1^2}{2g}$$

$$= 2.81\frac{V_1^2}{2g}$$

Total head loss

$$= 0.5\frac{V_1^2}{2g} + 0.0934\frac{V_1^2}{2g} + 5\frac{V_1^2}{2g} + 2.81\frac{V_1^2}{2g} + 0.4816\frac{V_1^2}{2g} = 10.0$$

or $8.885\frac{V_1^2}{2g} = 10.0$

$$\frac{V_1^2}{2g} = \frac{10.0}{8.885} = 1.1255$$

$V_1 = 4.70$ m/s

$$Q = \frac{\pi}{4} \times (0.1)^2 \times 4.70 = 0.0369 \text{ m}^3/\text{s}$$

= 36.9 L/s

(b) When the 12 cm diameter pipe is connected to the tank, suffix 2 is used to represent the same

(i) h_{L1} = loss of head at entrance $= 0.5\frac{V_2^2}{2g}$

(ii) h_{L2} = loss of head at contraction $= \left(\frac{1}{C_c} - 1\right)^2\frac{V_1^2}{2g}$

$$= \left(\frac{1}{0.7} - 1\right)^2\frac{V_1^2}{2g} = 0.1837\frac{V_1^2}{2g}$$

As $V_1 D_1^2 = V_2 D_2^2 \quad \therefore V_1 = V_2\left(\frac{D_2}{D_1}\right)^2$

$$= V_2\left(\frac{12}{10}\right)^2 = 1.44\ V_2$$

$$h_{L2} = 0.1837\,(1.44)^2\frac{V_2^2}{2g} = 0.3809\frac{V_2^2}{2g}$$

(iii) Velocity head at exit $h_{v1} = \frac{V_1^2}{2g} = (1.44)^2 \frac{V_2^2}{2g}$

$$= 2.0736 \frac{V_2^2}{2g}$$

(iv) Loss of head in 12 cm pipe due to friction

$$= \frac{f_2 L_2}{D_2} \frac{V_2^2}{2g} = \frac{0.02 \times 35}{0.12} \frac{V_2^2}{2g} = 5.8333 \frac{V_2^2}{2g}$$

(v) Loss of head in 10 cm pipe due to friction

$$= \frac{f_1 L_1}{D_1} \frac{V_1^2}{2g} = \frac{0.02 \times 25}{0.1} (1.44)^2 \frac{V_2^2}{2g}$$

$$= 10.368 \frac{V_2^2}{2g}$$

By energy balance

$$(0.5 + 0.3809 + 2.0736 + 5.8333 + 10.368) \frac{V_2^2}{2g} = 10.0$$

$$\frac{V_2^2}{2g} = \frac{10.00}{19.1558} = 0.522$$

$$V_2 = 3.20 \text{ m/s}$$

Discharge $Q = \frac{\pi}{4} \times (0.12)^2 \times 3.20$

$= 0.0362 \text{ m}^3\text{/s} =$ **36.2 L/s**

11.7 A concrete pipe 90 cm in diameter and 20 m long is used as a culvert. The pipe is laid on a slope of 1 in 60 and has square entrance. During a heavy flow of 1.50 m^3/s, if the culvert is submerged and the depth of flow over the invert at the downstream end is 1.25 m, estimate the depth of water above the pipe invert at the upstream end. [Assume $f = 0.025$].

Solution: Referring to Fig. 11.7,

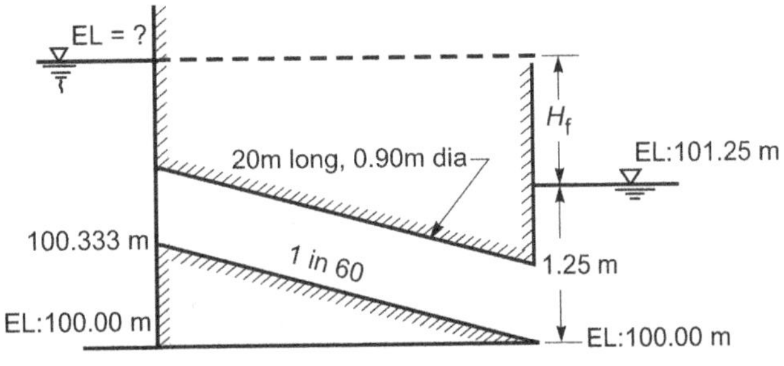

Fig. 11.7

$$h_f = \left[0.5 + \frac{fL}{D} + 1.0\right] \frac{V^2}{2g}$$

$$= \left[0.5 + \frac{0.025 \times 20}{0.9} + 1.0\right] \frac{V^2}{2g}$$

$$= 2.056 \frac{V^2}{2g}$$

$$V = \frac{1.50}{\frac{\pi}{4} \times (0.9)^2} = 2.36 \text{ m/s}$$

$$h_f = 2.056 \times \frac{(2.36)^2}{2 \times 9.81} = 0.583 \text{ m}$$

Elevation of upstream water surface

= 100.0 + 1.250 + 0.583

= 101.833 m

Elevation of upstream pipe invert 100.333 m

∴ Upstream depth = **1.500 m**

11.8 A pipe 6 cm in diameter, 1000 m long and with $f = 0.018$ is connected in parallel between two points M and N with another pipe 8 cm diameter, 800 m long and having $f = 0.020$. A total discharge of 20 L/s enters the parallel pipes through division at M to rejoin at N (Fig. 11.8). Estimate the division of discharge in the two pipes.

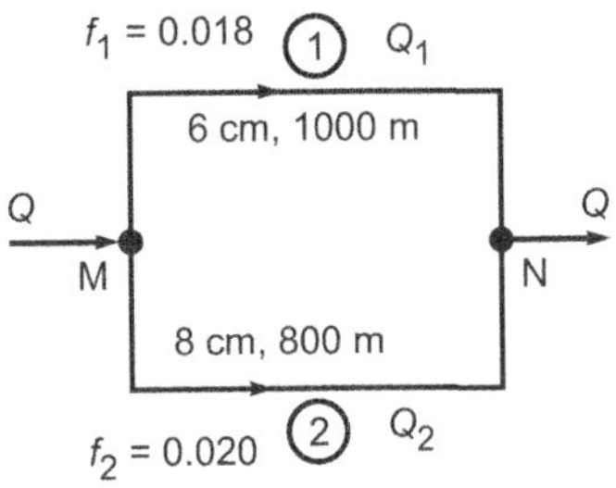

Fig. 11.8

Solution: By continuity consideration,

$A_1V_1 = A_2V_2 = Q$ = Total discharge

$$\frac{\pi}{4} (0.06)^2 V_1 + \frac{\pi}{4} (0.08)^2 V_2 = 0.020$$

$$V_1 + 1.7778 V_2 = 7.074 \quad (1)$$

By considering the head loss between M and N

$$\frac{f_1 L_1 V_1^2}{2g D_1} = \frac{f_2 L_2 V_2^2}{2g D_2}$$

or

$$\frac{0.018 \times 1000}{0.06} V_1^2 = \frac{0.02 \times 800}{0.08} V_2^2$$

$$V_1^2 = \frac{2}{3} V_2^2$$

$$V_1 = 0.8165 V_2 \quad (2)$$

Substituting this in Eq. (1)

$$0.8165\ V_2 = 1.7778\ V_2 = 7.074$$
$$V_2 = 2.727\ \text{m/s}$$

and $$Q_2 = \frac{\pi}{4} \times (0.08)^2 \times 2.727$$
$$= 0.0137\ \text{m}^3/\text{s} = 13.7\ \text{L/s}$$
$$V_1 = 0.8165 \times 2.727$$
$$= 2.227\ \text{m/s}$$

and $$Q_1 = 0.0063\ \text{m}^3/\text{s} = \mathbf{6.3\ L/s}$$

Alternate Solution:

Put $h_f = r_i Q_i^2$ where $r_i = \dfrac{8 f_i L_i}{\pi^2 g D_i^5} = 0.0826\ fL/D^5$.

Pipe	fL/D^5	r
1	23,148,000	1912000
2	4,882,800	403320

Since $$h_f = r_1 Q_1^2 = r_2 Q_2^2$$
$$Q_2 = (r_1/r_2)^{0.5}\ Q_1 = (403320/1912000)^{0.5}\ Q_1$$
$$= 0.4593\ Q_1$$
$$Q = Q_1 + Q_2 = 1.4593\ Q_1 = 0.02$$
$$Q_1 = 0.0137\ \text{m}^3/\text{s and}$$
$$Q_2 = 0.02 - 0.0137 = \mathbf{0.0063\ m^3/s}.$$

11.9 Two pipes A and B are connected in parallel between two points. Pipe A is 150 m long and has a diameter of 15 cm. Pipe B is 100 m long and has a diameter of 12 cm. Both the pipes have the same friction factor of 0.018. A partially closed valve in pipe A causes the discharge in the two pipes to be the same (Fig. 11.9). Estimate the value of the valve coefficient. All other minor losses can be neglected.

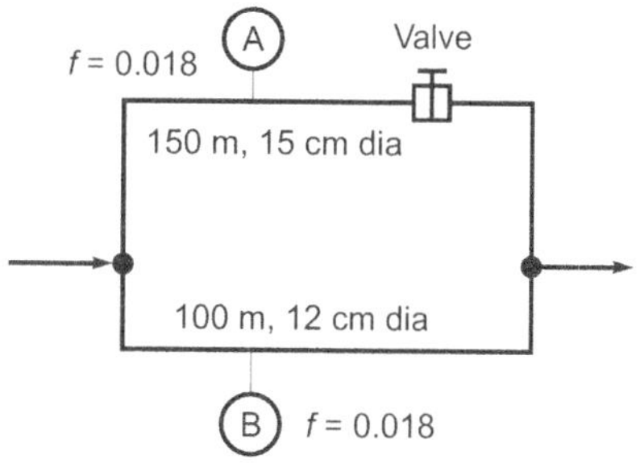

Fig. 11.9

Solution: Since the discharges are same in both the pipes:

$$A_a V_a = A_b V_b$$
$$\frac{\pi}{4} \times (0.15)^2\ V_a = \frac{\pi}{4} \times (0.12)^2\ V_b$$
$$V_b = \left(\frac{0.15}{0.12}\right)^2 V_a = 1.5625\ V_a$$

Let the loss in the valve be $K_L\ V_a^2/2g$.
Head losses in both pipes are the same.

Hence $$\frac{f_a L_a V_a^2}{2gD_a} + K_L \frac{V_a^2}{2g} = \frac{f_b L_b V_b^2}{2gD_b}$$

$$\frac{0.018 \times 150}{0.15} \times \frac{V_a^2}{2g} + K_L \frac{V_a^2}{2g}$$
$$= \frac{0.018 \times 100 \times (1.5625)^2}{0.12} \times \frac{V_a^2}{2g}$$
$$18 + K_L = 36.62$$
$$\mathbf{K_L = 18.62}$$

11.10 Two pipes each of length L and diameters D_1 and D_2 are arranged in parallel; the loss of head when a total quantity of water Q flows through them being h_1. If the pipes are arranged in series and the same quantity of water, Q, flows through them, the loss of head is h_2. If $D_1 = 2D_2$, find the ratio of h_1 to h_2. Neglect minor losses and assume the friction factor f to be constant and to have the same value for both the pipes.

Solution:

Head loss $$h_f = \frac{fLV^2}{2gD} = \frac{fLQ^2}{2gD\left(\frac{\pi}{4}D^2\right)^2} = \frac{8fLQ^2}{\pi^2 g D^5}$$

Case I:

$$h_1 = \frac{8fLQ_1^2}{\pi^2 g D_1^5} = \frac{8fLQ_2^2}{\pi^2 g D_2^5}$$

or $$\left(\frac{Q_1}{Q_2}\right)^2 = \left(\frac{D_1}{D_2}\right)^5$$

or $$\left(\frac{Q_1}{Q_2}\right) = \left(\frac{D_1}{D_2}\right)^{5/2} \quad \text{or} \quad Q_1 = Q_2\left(\frac{D_1}{D_2}\right)^{5/2}$$

Also $Q_1 + Q_2 = Q$

or $$Q_2\left[\left(\frac{D_1}{D_2}\right)^{5/2} + 1\right] = Q$$

Putting $D_1 = 2D_2$, $\quad Q = (2^{5/2} + 1)\ Q_2$
$$= 6.65685\ Q_2$$

$$h_1 = \frac{8fLQ^2}{\pi^2 g D_2^5 (6.65685)^2} = \frac{1}{(44.313)}\frac{8fLQ^2}{\pi^2 g D_2^5} \qquad (1)$$

Case II:

$$h_2 = \frac{8fLQ^2}{\pi^2 g D_1^5} + \frac{8fLQ^2}{\pi^2 g D_2^5}$$

$$= \frac{8fLQ^2}{\pi^2 g}\left(\frac{1}{D_1^5} + \frac{1}{D_2^5}\right) = \frac{8fLQ^2}{\pi^2 g D_2^5}\left(\left(\frac{D_2}{D_1}\right)^5 + 1\right)$$

for $D_1 = 2D_2$; $h_2 = \dfrac{8fLQ^2}{\pi^2 g D_2^5}\left\{\left(\dfrac{1}{2}\right)^5 + 1\right\}$

$$= (1.03125)\frac{8fLQ^2}{\pi^2 g D_2^5} \qquad (2)$$

From Eqs (1) and (2),

$$\frac{h_1}{h_2} = \frac{1}{(44.313)} \times \frac{1}{(1.03125)}$$

$= \mathbf{0.02188}$

11.11 Three pipes are connected in parallel between two reservoirs A and B. The details of the pipes are:

Pipe	*Diameter*	*Length*	*f*
1	10 cm	1000 m	0.022
2	15 cm	800 m	0.018
3	12 cm	950 m	0.020

If the difference in the water level elevations of the two reservoirs is 12 m, estimate the discharge in each pipe.

Solution: The head loss $h_f = 12.0$ m is common to all the three pipes.

Hence $h_f = 12.0 = \dfrac{f_1 L_1}{D_1}\dfrac{V_1^2}{2g} = \dfrac{f_2 L_2}{D_2}\dfrac{V_2^2}{2g} = \dfrac{f_3 L_3}{D_3}\dfrac{V_3^2}{2g}$

Pipe No.	Dia (m)	Area (m²) $= \left(\frac{\pi}{4}\times D^2\right)$	*fL/D*	$V^2/2g = h_f\Big/\left(\frac{fL}{D}\right)$	V (m/s)	Q (m³/s)
1	0.10	7.85×10^{-3}	220	0.05455	1.034	**0.0081**
2	0.15	0.01767	96	0.1250	1.566	**0.0277**
3	0.12	0.01131	158.3	0.07579	1.219	**0.0138**

11.12 Three pipes with details as the following, are connected in parallel between two points.

Pipe	*Diameter*	*Length*	*f*
1	1000 m	20 cm	0.02
2	1200 m	30 cm	0.015
3	800 m	15 cm	0.02

When a total discharge of 0.30 m³/s flows through the system, calculate the distribution of the discharge and the head loss between the junctions.

Solution: A trial and error procedure is adopted.

1st trial:

Assume a trial discharge $Q_1 = 0.05$ m³/s in pipe 1.

$$V_1 = 0.05\Big/\left[\frac{\pi}{4}\times(0.2)^2\right] = 1.592 \text{ m/s}$$

$$h_f = \frac{f_2 L_2 V_1^2}{2gD_2} = \frac{0.02\times1000\times(1.592)^2}{2\times9.81\times0.20}$$

$= 12.91$ m

Using this h_f:

$$\frac{f_2 L_2 V_2^2}{2gD_2} = 12.91$$

$$\frac{0.015\times1200\times V_2^2}{2\times9.81\times0.3} = 12.91$$

$V_2^2 = 4.22157$

$V_2 = 2.055$ m/s

and discharge $Q_2 = \dfrac{\pi}{4}\times(0.3)^2\times2.055 = 0.1453$ m³/s

Also $\dfrac{f_3 L_3 V_3^2}{2gD_3} = 12.91$

$$\frac{0.02\times800\times V_3^2}{2\times9.81\times0.15} = 12.91$$

$V_3^2 = 2.375$ and $V_3 = 1.541$ m/s

Discharge $Q_3 = \dfrac{\pi}{4}\times(0.15)^2\times1.541 = 0.0272$ m³/s

Total discharge $\Sigma Q_i = 0.0500 + 0.1453 + 0.0272$

$= 0.2225$ m³/s

But the actual total discharge is 0.30 m³/s

The discharges in each pipe is now corrected by multiplying the trial discharges by the ratio $\dfrac{0.30}{0.2225} = 1.3483$.

Hence corrected discharges in each pipe are:

Pipe 1 = $Q_1 = 0.05 \times 1.3483$ = **0.0674 m³/s**

Pipe 2 = $Q_2 = 0.1453 \times 1.3483$ = **0.1959 m³/s**

Pipe 3 = $Q_3 = 0.0272 \times 1.3483$ = **0.0367 m³/s**

Total 0.3000 m³/s

2nd trial:

Head loss between the junctions by considering pipe 1:

$$h_f = \frac{f_1 L_1 V_1^2}{2gD} = \frac{0.02\times1000}{2\times9.81\times0.20}\times\frac{(0.0674)^2}{\left(\frac{\pi}{4}\right)^2(0.2)^4} = \mathbf{23.46\ m}$$

Q_2 and Q_3 are found by using this h_f.

By considering pipe 2:

$$h_f = \frac{f_2 L_2 V_2^2}{2gD_2} \times \frac{0.015 \times 1200 \times V_2^2}{2 \times 9.81 \times 0.3}$$
$$= 23.46 \text{ m}$$
$$V_2^2 = 7.6713 \text{ or } V_2 = 2.7697 \text{ m/s}$$

Discharge $Q_2 = \frac{\pi}{4} \times (0.3)^2 \times 2.7697 = \mathbf{0.1958\ m^3/s}$

In pipe 3 $h_f = \frac{f_3 L_3 V_3^2}{2gD_3} = \frac{0.02 \times 800}{2 \times 9.81 \times 0.15} \times V_3^2$

$$= 23.46 \text{ m}$$
$$V_3^2 = 4.3151 \text{ or } V_3 = \mathbf{2.0773\ m/s}$$

Discharge $Q_3 = \frac{\pi}{4} \times (0.15)^2 \times 2.0773 = 0.0367 \text{ m}^3\text{/s}$

Total discharge $= \Sigma Q_i = 0.0674 + 0.1958 + 0.0367$
$= 0.2999 \text{ m}^3\text{/s}$

As this discharge is essentially the same as the given discharge, no more trials are needed. Hence $Q_1 = 0.0674$ m³/s, $Q_2 = 0.1958$ m³/s, $Q_3 = 0.0367$ m³/s and $h_f = 23.46$ m.

[***Note:*** If the total discharge at the end of second trial differed substantially the discharges would have to be once again readjusted by the ratio $Q_{actual}/(\Sigma Q_i)_{trial}$ and the distribution checked by a 3rd trial and so on. Usually 2 trials would be sufficient.]

Alternative method: Using the concept of equivalent pipes the calculations could be greatly simplified and the trial and error procedure eliminated.

If (D_e, L_e, f_e) refer to an equivalent pipe then

$$\left(\frac{D_e^5}{f_e L_e}\right)^{1/2} = \left(\frac{D_1^5}{f_1 L_1}\right)^{1/2} + \left(\frac{D_2^5}{f_2 L_2}\right)^{1/2} + \left(\frac{D_3^5}{f_3 L_3}\right)^{1/2}$$

Here let the equivalent pipe be of $D_e = 0.30$ m and $f_e = 0.02$. Then L_e is found from:

$$\left[\frac{(0.3)^5}{0.02\, L_e}\right]^{1/2} = \left(\frac{(0.2)^5}{0.02 \times 1000}\right)^{1/2} + \left(\frac{(0.3)^5}{0.015 \times 1200}\right)^{1/2} + \left(\frac{(0.15)^5}{0.02 \times 800}\right)^{1/2}$$
$$= (4 \times 10^{-3}) + (11.619 \times 10^{-3}) + (2.1785 \times 10^{-3})$$
$$\frac{0.3486}{L_e^{1/2}} = 0.017798$$

L_e = Length of the equivalent pipe of chosen size = 383.6 m

For a discharge of 0.3 m³/s, the head loss is

$$h_f = \frac{f_e L_e V_e^2}{2gD_e}$$
$$V_e = \frac{0.3}{\frac{\pi}{4} \times (0.3)^2} = 4.2441 \text{ m/s}$$
$$\frac{V_e^2}{2g} = \frac{(4.2441)^2}{2 \times 9.81} = 0.9181$$
$$h_f = \frac{0.02 \times 383.6}{0.3} \times (0.9181) = \mathbf{23.48\ m}$$

The discharge in each branch is found by using this $h_f = 23.48$ m.

Thus $h_f = 23.48 = \frac{f_1 L_1}{D_1} \frac{V_1^2}{2g} = \frac{0.02 \times 1000 \times V_1^2}{0.2 \times 2 \times 9.81}$

$$V_1 = 2.146 \text{ m/s},$$
$$Q_1 = \frac{\pi}{4} \times (0.2)^2 \times 2.146 = \mathbf{0.0674\ m^3/s}.$$
$$h_f = 23.48 = \frac{f_2 L_2}{D_2} \frac{V_2^2}{2g} = \frac{0.015 \times 1200 \times V_2^2}{0.3 \times 2 \times 9.81}$$
$$V_2 = 2.771 \text{ m/s}$$
$$Q_2 = \frac{\pi}{4} \times (0.3)^2 \times 2.771 = \mathbf{0.1959\ m^3/s}$$
$$Q_3 = Q - (Q_1 + Q_2) = 0.3 - (0.0674 + 0.1959) = \mathbf{0.0367\ m^3/s}$$

Second Alternative Solution:

Put $h_f = r_i Q_i^2$ where $r_i = \frac{8 f_i L_i}{\pi^2 g D_i^5} = 0.0826\, fL/D^5$.

Pipe	fL/D^5	r
1	62500	5164
2	7407	612
3	210700	1741

Since $h_f = r_1 Q_1^2 = r_2 Q_2^2 = r_3 Q_3^2$

$$Q_2 = (r_1/r_2)^{0.5}\, Q_1 = (5164/612)^{0.5}\, Q_1 = 2.9048\, Q_1$$
$$Q_3 = (r_1/r_3)^{0.5}\, Q_1 = (5164/1741)^{0.5}$$
$$Q_1 = 0.5446\, Q_1$$
$$Q = Q_1 + Q_2 + Q_3 = [1 + 2.9048 + 0.5446]\, Q_1$$
$$0.30 = 4.4494\, Q_1$$
$$Q_1 = 0.0674 \text{ m}^3\text{/s}$$
$$Q_2 = 2.9048 \times 0.0674 = \mathbf{0.1959\ m^3/s} \text{ and}$$
$$Q_3 = 0.30 - 0.0674 - 0.1959 = \mathbf{0.0366\ m^3/s}$$

11.13 Two pipes 1 and 2, each of 10 cm diameter branch off from a point A in a pipeline and rejoin at B.

Pipe 1 is 400 m long and pipe 2 is 600 m long. The total head at A is 30 m. A short pipe 8 cm diameter is fitted at B and the flow is discharged into atmosphere through it (Fig. 11.10). Assuming f = 0.02 for both the pipes, find the total discharge and division of discharge in pipes 1 and 2.

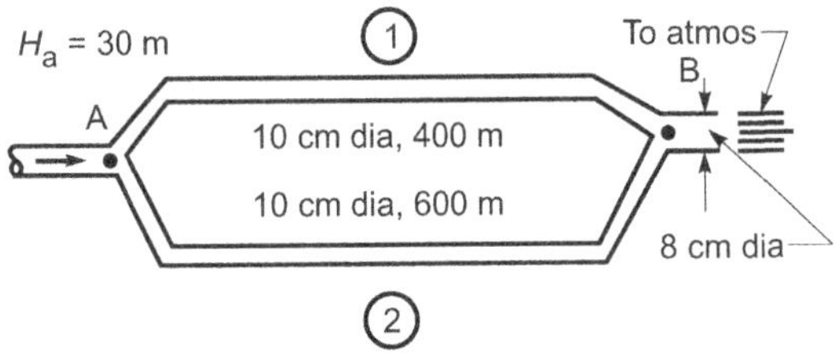

Fig. 11.10 *Example 11.13*

Solution: As the 8 cm pipe is short, the friction loss in it can be neglected.

$$H_B = \text{Head at B} = \frac{V_3^2}{2g}$$

$$H_A - H_B = 30.0 - \frac{V_3^2}{2g}$$

Consider an equivalent pipe $D_e = 0.08$ m and $f_e = 0.02$ to replace the parallel pipes 1 and 2. Then

$$\left(\frac{D_e^5}{f_e L_e}\right)^{1/2} = \left(\frac{D_1^5}{f_1 L_1}\right)^{1/2} + \left(\frac{D_2^5}{f_2 L_2}\right)^{1/2}$$

Since $f_e = f_1 = f_2$ and $D_1 = D_2 = 0.10$ m

$$\frac{(0.08)^{5/2}}{L_e^{1/2}} = (0.1)^{5/2}\left[\frac{1}{\sqrt{400}} + \frac{1}{\sqrt{600}}\right]$$

$$L_e^{1/2} = 6.3026 \text{ and } L_e = 39.72 \text{ m}$$

As $D_e = 0.08$ m, velocity in this pipe = $V_3 = V_e$

∴ Head loss

$$H_A - H_B = 30 - \frac{V_3^2}{2g} = \frac{f_e L_e}{D_e}\frac{V_e^2}{2g}$$

$$= \frac{0.02 \times 39.72}{0.08}\frac{V_3^2}{2g} = 9.93\frac{V_3^2}{2g}$$

$$\frac{V_3^2}{2g} = \frac{30}{10.93} = 2.745$$

$$V_3 = 7.338 \text{ m/s}$$

$$Q = \frac{\pi}{4} \times (0.08)^2 \times 7.338 = 0.0369 \text{ m}^3/\text{s}$$

$$H_A - H_B = 30 - 2.745 = 27.255 \text{ m} = h_{f1} = h_{f2}$$

$$\therefore \quad \frac{f_1 L_1}{D_1}\frac{V_1^2}{2g} = \frac{0.02 \times 400}{0.1} \times \frac{V_1^2}{2g} = 27.255$$

$$V_1 = 2.585 \text{ m/s}$$

$$Q_1 = \frac{\pi}{4} \times (0.1)^2 \times 2.585 = \mathbf{0.0203\ m^3/s}$$

$$\frac{f_2 L_2}{D_2}\frac{V_2^2}{2g} = \frac{0.02 \times 600}{0.1} \times \frac{V_2^2}{2g} = 27.255$$

$$V_2 = 2.111 \text{ m/s}$$

$$Q_2 = \frac{\pi}{4} \times (0.1)^2 \times 2.111 = \mathbf{0.0166\ m^3/s}$$

Check:

$$Q_1 + Q_2 = 0.0203 + 0.0166 = 0.0369 = Q$$

11.14 A city water supply main is 1000 m long and delivers a flow of 100 L/s between two reservoirs with a head difference of 15 m. It is proposed to increase the flow by 30% by adding another pipe of the same diameter as the main from the upstream reservoir in parallel and joining it to the main at a suitable junction. Estimate the length of the additional pipe and the head at the junction relative to the downstream reservoir water surface. Assume the lengths of the two parallel pipes to be same. Also, all pipes can be assumed to have the same friction factor f. [Refer Fig. 11.11].

Solution:

(a) Before introduction of the pipe:

$$h_f = \frac{f L_0 V_0^2}{2gD} = \frac{f L_0 Q_0^2}{2gD^5(\pi/4)^2} = K L_0 Q_0^2$$

where $K = f/[2gD^5 (\pi/4)^2]$

Here $L_0 = 1000$ m, $Q_0 = 0.1$ m³/s, $h_f = 15$ m

$$K = \frac{15}{1000 \times (0.1)^2} = 1.5$$

(b) After the introduction of the parallel pipe: Fig. 11.11 shows the schematic layout. For pipes AJ and BJ

$$h_{f1} = \frac{f_1 L_1 V_1^2}{2gD_1} = \frac{f_2 L_2 V_2^2}{2gD_2}$$

Since $f_1 = f_2$, $L_1 = L_2$, $D_1 = D_2 = D$

$$V_1 = V_2$$

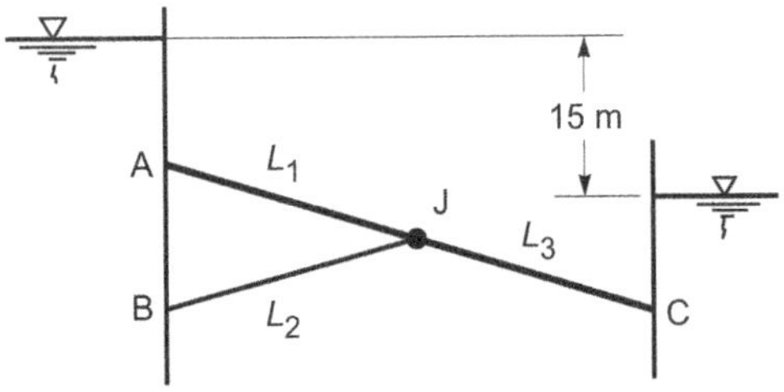

Fig. 11.11

$\therefore \quad Q_1 = Q_2$

Hence $\quad Q_3 = Q_1 + Q_2 = 2Q_1$

Head loss between the two reservoirs

$$H_L = 15 \text{ m} = \frac{f L_1 V_1^2}{2gD} + \frac{f L_3 V_3^2}{2gD}$$

Since $L_1 + L_3 = 1000$ m

$$\frac{f L_1 V_1^2}{2gD} + \frac{f(1000 - L_1)}{2gD} V_3^2 = 15$$

In terms of the new total discharge Q_3

$$\frac{f L_1}{2g} \times \frac{\frac{1}{4} Q_3^2}{\left(\frac{\pi}{4}\right)^2 D^5} + \frac{f(1000 - L_1) Q_3^2}{2g\left(\frac{\pi}{4}\right)^2 D^5} = 15$$

Putting, as before, $\dfrac{f}{2g\left(\frac{\pi}{4}\right)^2 D^5} = K$

$$K\left(\frac{L_1}{4} + 1000 - L_1\right) Q_3^2 = 15$$

$$K\left(1000 - \frac{3}{4} L_1\right) Q_3^2 = 15$$

$$= K \times 1000 \times Q_0^2$$

Hence $\left(1000 - \dfrac{3}{4} L_1\right) = \left(\dfrac{Q_0}{Q_3}\right)^2 \times 1000$

But $Q_3 = 1.30\ Q_0$, and as such

$$1000 - \frac{3}{4} L_1 = 591.7$$

$$L_1 = L_2 = \mathbf{544.4\ m}$$

Head at the junction, H_j (above the downstream reservoir water surface elevation):

$$H_j = h_{f3} = \text{head loss in pipe 3}$$

$$= \frac{f L_3 V_3^2}{2gD} = \frac{f(1000 - 544.4)}{2g\left(\frac{\pi}{4}\right)^2 D^5} Q_3^2$$

$$= K\ (455.6)\ (0.13)^2$$

$$= 1.5 \times (455.6)\ (0.13)^2 = \mathbf{11.55\ m}$$

11.15 A straight 25 cm pipeline 5 km long is laid between two reservoirs having a difference in level of 40 m. To increase the capacity of the system an additional 2.5 km long 25 cm pipe is laid parallel from the first reservoir to the midpoint of the original pipe. Assuming $f = 0.025$ for both the pipes find the increase in discharge due to installation of the new pipe.

Solution: Referring to Fig. 11.12 which shows the schematic layout of the pipes,

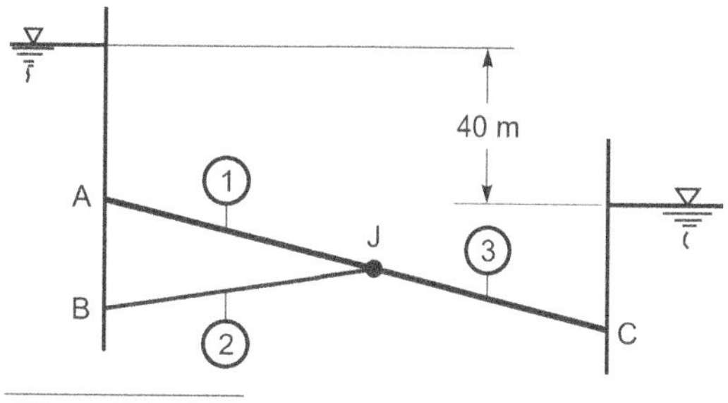

Fig. 11.12

(1) For the single pipe AC:

$$h_f = \frac{f_0 L_0 V_0^2}{2g D_0}$$

$$= \frac{0.025 \times 5000}{2 \times 9.81 \times 0.25} V_0^2 = 40 \text{ m}$$

$V_0 = 1.253$ m/s. Hence original discharge Q_0 is

$$Q_0 = \frac{\pi}{4} D_1^2 V_1 = \frac{\pi}{4} \times (0.25)^2\ (1.253)$$

$$= 0.0615 \text{ m}^3\text{/s}$$

$$Q_0 = 61.5 \text{ L/s}$$

(2) When the parallel pipe BJ is introduced; for pipes AJ and BJ:

$$\frac{f_1 L_1}{D_1}\frac{V_1^2}{2g} = \frac{f_2 L_2}{D_2}\frac{V_2^2}{2g}$$

But $f_1 = f_2$, $L_1 = L_2$, $D_1 = D_2$. Hence $V_1 = V_2$ and $Q_1 = Q_2$

$\therefore \quad Q_3 = Q_1 + Q_2 = 2Q_2 = 2Q_1$

$$\text{Head loss} = 40 = \frac{f_1 L_1}{D_1} \cdot \frac{V_1^2}{2g} + \frac{f_3 L_3}{D_3}\frac{V_3^2}{2g}$$

$$= \frac{0.025 \times 2.5 \times 1000}{0.25} \times \frac{1}{2 \times 9.81} [V_1^2 + V_3^2]$$

$$40 = 12.742\ (V_1^2 + V_3^2)$$

$$V_1 D_1^2 = \frac{1}{2} \times D_3^2 V_3 \qquad \left(\text{Since } Q_1 = \frac{1}{2} Q_3\right)$$

$$V_3 = 2\left(\frac{D_1}{D_3}\right)^2 V_1 = 2V_1$$

$$V_3^2 = 4\ V_1^2$$

$\therefore \quad 12.742\ (5V_1^2) = 40$

$$V_1 = \left(\frac{40}{50 \times 12.742}\right)^{1/2} = 0.7924 \text{ m/s}$$

$$Q_1 = \frac{\pi}{4} \times (0.25)^2 \times 0.7924 = 0.0389 \text{ m}^3\text{/s}$$

$Q_3 = 2 \times 0.0389 = 0.07779 \text{ m}^3/\text{s}$

Hence increase in discharge

$= Q_3 - Q_1 = 0.0779 - 0.0615$

$= 0.0163 \text{ m}^3/\text{s} =$ **16.3 L/s**

11.16 A pipeline carrying water has a diameter of 0.5 m and is 2.0 km long. To increase the delivery another pipeline of the same diameter is introduced parallel to the first pipe in the second half of its length. Find the increase in discharge if the total head loss in both the cases is 15 m. Assume $f = 0.02$ for all the pipes.

Solution: Refer to Fig. 11.13

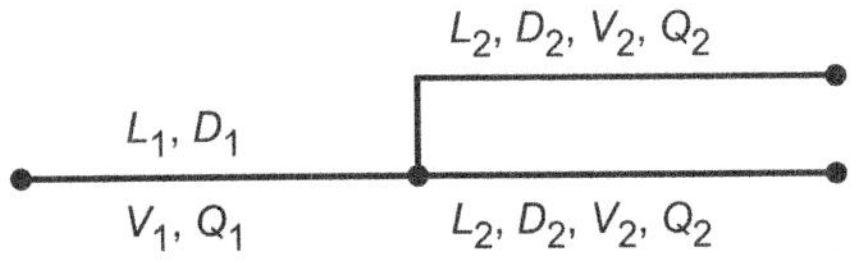

Fig. 11.13

Case I:

$$h_f = \frac{f L V_0^2}{2g D_0}$$

$$15 = \frac{0.02 \times 2000 \times V_0^2}{2 \times 9.81 \times 0.5} = 4.0775\ V_0^2$$

$$V_0^2 = 3.679 \text{ or } V_0 = 1.918 \text{ m/s}$$

Discharge $Q_0 = \frac{\pi}{4} \times (0.5)^2 \times 1.918 = 0.3766 \text{ m}^3/\text{s}$

Case II:

$L_1 = 1000$ m $\quad L_2 = 1000$ m

$D_1 = 0.50$ m $\quad D_2 = 0.50$ m

Total head loss $h_f = 15 \text{ m} = h_{f1} + h_{f2}$

$$h_f = 15 = \frac{f_1 L_1 V_1^2}{2g D_1} + \frac{f_2 L_2 V_2^2}{2g D_2}$$

$$= \frac{0.02 \times 1000}{2 \times 9.81 \times 0.5} V_1^2 + \frac{0.02 \times 1000}{2 \times 9.81 \times 0.5} V_2^2$$

$$h_f = 2.039\,(V_1^2 + V_2^2) = 15 \text{ m}$$

Since the discharge $Q_1 = 2Q_2$

$$D_1^2 V_1 = 2 D_2^2 V_2$$

Since $D_1 = D_2$, $V_1 = 2V_2$

$\therefore \quad h_f = 2.039\,(4V_2^2 + V_2^2) = 15$

$$V_2^2 = 1.4715 \text{ or } V_2 = 1.213 \text{ m/s}$$

$$Q_1 = 2Q_2 = 2 \times \frac{\pi}{4} \times (0.5)^2 \times 1.213 = 0.4763 \text{ m}^3/\text{s}$$

Increase in discharge $= Q_1 - Q_0$

$= 0.4763 - 0.3766 =$ **0.0997 m³/s**

or $\quad \frac{0.0997}{0.3766} \times 100 =$ **26.48%**

Alternative method: For the set of two parallel pipes in the second half, consider an equivalent pipe of 0.5 m diameter and $f = 0.02$.

The equivalent pipe (D_e, f_e, L_e) is related to the two parallel pipes it replaces as

$$\left(\frac{D_e^5}{f_e L_e}\right)^{1/2} = \left[\left(\frac{D_1^5}{f_1 L_1}\right)^{1/2} + \left(\frac{D_2^5}{f_2 L_2}\right)^{1/2}\right]$$

Here $D_1 = 0.50$ m $\quad D_2 = 0.50$ m $\quad D_e = 0.50$ m

$L_1 = 1000$ m $\quad L_2 = 1000$ m $\quad L_e = ?$

$f_1 = 0.02 \quad f_2 = 0.02 \quad f_e = 0.02$

$$\left[\frac{(0.5)^5}{0.02 \times L_e}\right]^{1/2} = \left[2 \times \left\{\frac{(0.5)^5}{0.02 \times 1000}\right\}^{1/2}\right]$$

$$\frac{1}{L_e^{1/2}} = \frac{2}{31.623}$$

and $\quad L_e = 250$ m

Now the total length of pipe $= L = 1000 + 250$

$= 1250$ m

$$h_f = \left(\frac{0.02 \times 1250}{2 \times 9.81 \times 0.5}\right) V^2 = 15.0 \text{ m}$$

$$V^2 = 5.886 \text{ and } V = 2.426 \text{ m/s}$$

Discharge $\quad Q = \frac{\pi}{4} \times (0.5)^2 \times 2.426 = 0.4763 \text{ m}^3/\text{s}$

Increase in discharge $= 0.4763 - 0.3766 = 0.0997 \text{ m}^3/\text{s}$

$=$ **26.48%**

11.17 A 15 cm diameter pipe is 1000 m long and connects two reservoirs having a difference in water surface elevation of 16 m. The elevation of the upstream reservoir water surface is 100.00 m. At a distance of 600 m along the pipe from the upstream reservoir, the pipe is laid at an elevation of 98.00 m.

(a) Calculate the pressure at this summit point, and the discharge in the pipe.

(b) If cavitation conditions are expected at a pressure of 20 kPa (abs) what maximum discharge can pass through the pipe with the upstream reservoir elevation remaining unaltered. [Assume $f = 0.02$ and atmospheric pressure = 10.3 m (abs)].

Solution:

(a) $$h_f = \frac{f L V^2}{2gD} \quad \frac{0.02 \times 1000 \times V^2}{2 \times 9.81 \times 0.15} = 16 \text{ m}$$

$$V_2 = 2.354, \; V = 1.534 \text{ m/s}$$

$$Q = \frac{\pi}{4} \times (0.15)^2 \times 1.534 = 0.0271 \text{ m}^3/\text{s}$$

$= 27.1$ L/s

Head loss up to the summit = h_{f1}

$$h_{f1} = \frac{0.02 \times 600}{2 \times 9.81 \times 0.15} \times (1.534)^2 = 9.6 \text{ m}$$

By energy equation between upstream reservoir water surface and the summit.

Atmospheric pressure = 10.3 m (abs)

$$10.3 + 0 + 100.0 = \frac{p_1}{\gamma} + \frac{V^2}{2g} + 98.00 + h_{f1}$$

$$= \frac{p_1}{\gamma} + \frac{(1.534)^2}{2 \times 9.81} + 98.00 + 9.6$$

$$\frac{p_1}{\gamma} = 110.3 - 107.72 = 2.58 \text{ m (abs)}$$

$$p_1 = 2.58 \times 9.79 = \mathbf{25.26\ kPa\ (abs)}$$

(b) $p_1 = 20.00$ kPa, $\frac{p_1}{\gamma} = \frac{20.00}{9.79} = 2.043$ m

By energy equation

$$10.3 + 100.00 = 2.043 + \frac{V_1^2}{2g} + 98.00 + \frac{fLV_1^2}{2gD_1}$$

$$\left[1 + \frac{fL}{2gD}\right] V_1^2 = 10.257$$

$$\left[1 + \frac{0.02 \times 600}{2 \times 9.81 \times 0.15}\right] V_1^2 = 10.257$$

$$V_1^2 = 2.02 \text{ and } V_1 = 1.421 \text{ m/s}$$

$$Q_1 = \frac{\pi}{4} \times (0.15)^2 \times 1.421 = 0.0251 \text{ m}^3\text{/s}$$

$$= \mathbf{25.1\ L/s}$$

11.18 For the branching system shown in Fig. 11.14, calculate the discharge each pipe. Take $f = 0.02$ for all pipes.

Pipe	*Dia*	*Length (m)*	*Connectivity*
1	15 cm	350	AJ
2	10 cm	200	BJ
3	10 cm	250	JC

(Neglect minor losses.)

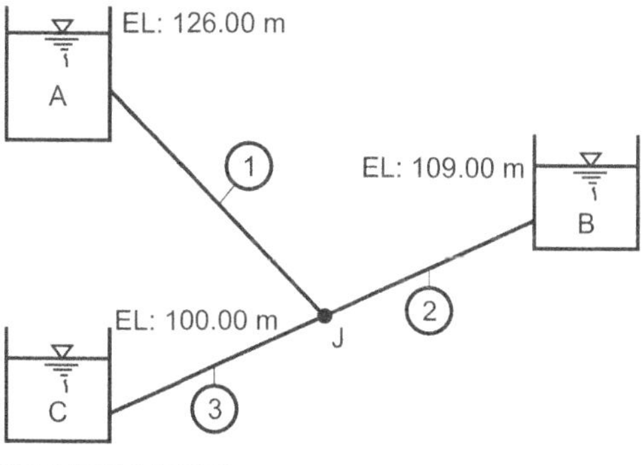

Fig. 11.14

Solution: Putting $h_f = rQ^2$,

$$r = \frac{8fL}{\pi^2 g D^5} = \frac{8 \times 0.02}{\pi^2 \times 9.81} \frac{L}{D^5}$$

$$= 1.6525 \times 10^{-3} \frac{L}{D^5}$$

A trial and error solution method is adopted. Flow away from the junction is taken as negative.

For first trial assume H_j = elevation of hydraulic grade line at the junction $J = 114.00$ m.

First trial

$$H_j = 114.0 \text{ m}$$

Pipe	*r*	*Estimated* h_f *(m)*	$Q = \sqrt{h_f / r}$ *(m^3/s)*	*Q (L/s)*	$\lvert Q/h_f \rvert$ (*Q* in L/s)
1	7617	126.0 – 114.0 = 12.0	+0.0397	+39.7	3.31
2	33051	114.0 – 109.0 = 5.0	–0.0123	–12.3	2.46
3	41313	114.0 – 100.0 = 14.0	–0.0184	–18.4	1.31
				$\Delta Q = 9.0$	$\Sigma\lvert Q/h_f\rvert = 7.08$

$$\Delta H_j = \frac{2 \times \Delta Q}{\Sigma |Q/h_f|} = \frac{2 \times 9.0}{7.08} = 2.54 \text{ m}$$

H_j for next trial = 116.54 m

Second trial

$$H_j = 116.54 \text{ m}$$

Pipe	*r*	h_f *(m)*	*Q (m^3/s)*	*Q (L/s)*	$\lvert Q/hr \rvert$ *(Q in L/s)*
1	7617	126.00 – 116.54 = 9.46	+0.0352	35.2	3.72
2	33051	116.54 – 109.00 = 7.54	–0.0151	–15.1	2.00
3	41313	116.54 – 100.00 = 16.54	–0.020	– 20.0	1.21
				$\Delta Q = 0.1$	$\Sigma\lvert Q/h_f\rvert = 6.93$

As the error in the discharge is 0.1 L/s (0.0001 m^3/s) no further iteration is necessary and the finalised discharges are

$\mathbf{Q_1 = 35.20\ L/s}$

$\mathbf{Q_2 = -15.15\ L/s}$ (towards reservoir B)

$\mathbf{Q_3 = -20.05\ L/s}$ (towards reservoir C)

The elevation of the piezometric head at junction $H_j = 116.54$ m.

Alternate solution procedure (Exact method)

Putting $h_f = rQ^2$

$$r = \frac{8fl}{\pi^2 g D^5} = \frac{8 \times 0.02}{\pi^2 \times 9.81} \times \frac{L}{D^5}$$

$$= 1.6525 \times 10^{-3} \frac{L}{D^5}$$

Pipe	L(m)	D(m)	r
1	350	0.15	7617
2	200	0.10	33051
3	250	0.10	41313

Let H_j = piezometric head at the junction J. Similarly H_a, H_b and H_c are piezometric heads at reservoirs A, B and C respectively.

$$h' = \frac{H_a - H_b}{H_b - H_c} = \frac{126.00 - 109.00}{109.00 - 100.00} = 1.89$$

$$R_1 = \frac{r_1}{r_3} = \frac{7617}{41313} = 0.184 < h'$$

Since $h' > R_1$, the flow is of **Type-1**, i.e. $H_j > H_b$ and $Q_1 = Q_2 + Q_3$.

By the application of Energy equation between the junction J and the reservoirs A, B and C successively.

$$H_a = H_j + r_1 Q_1^2$$
$$H_b = H_j - r_2 Q_2^2$$
$$H_c = H_j - r_3 Q_3^2$$

$$\therefore \quad (H_a - H_b) = r_1 Q_1^2 + r_2 Q_2^2 = 126.00 - 109.00 = 17$$
$$(H_b - H_c) = r_3 Q_3^2 - r_2 Q_2^2 = 109.00 - 100.00 = 9$$

Substituting the values of r_1, r_2 and r_3

$$7617 Q_1^2 + 33051 Q_2^2 = 17$$
$$41313 Q_3^2 - 33051 Q_2^2 = 9$$

Put $Q_3 = mQ_2$ resulting in $Q_1 = (1 + m)Q_2$.

Substituting in the above equations

$$Q_2^2[7617\,(1 + m)^2 + 33051] = 17 \quad (1)$$
$$Q_2^2[41313\, m^2 - 33051] = 9 \quad (2)$$

Dividing (1) by (2)

$$\frac{7617(1+m)^2 + 33051}{41313m^2 - 33051} = 1.889$$

$$\frac{(1 + 2m + m^2) + 4.339}{5.424m^2 - 4.339} = 1.889$$

On simplifying $m^2 - 0.216\, m - 1.464 = 0$

$$\therefore \quad m = \frac{0.216 \pm \sqrt{(0.216)^2 + 4 \times 1.464}}{2} = 1.323$$

[*Note: only positive root is relevant.*]

Substituting in Eq. (2)

$$Q_2^2 = [41313\,(1.323)^2 - 33051] = 9$$

Hence Q_2 = **0.0151 m³/s**

$Q_3 = mQ_2 = 1.323\,(0.0151) =$ **0.200 m³/s**

$Q_1 = (1 + m)\,Q_2 = 2.323\,(0.151) =$ **0.0351 m³/s**

H_j = piezometric head at the junction $= H_b + r_2 Q_2^2$
$= 109.000 + 33051\,(0.0151)^2 =$ **116.54 m**

11.19 A water supply system consists of three reservoirs A, B and C connected to a common junction J as shown in Fig. 11.15.

Pipe	Dia	Length	f	Connectivity
1	0.30 m	200 m	0.020	AJ
2	0.30 m	125 m	0.016	BJ
3	0.30 m	250 m	0.016	CJ

Calculate the discharge in each pipe and the piezometric head at the junction.

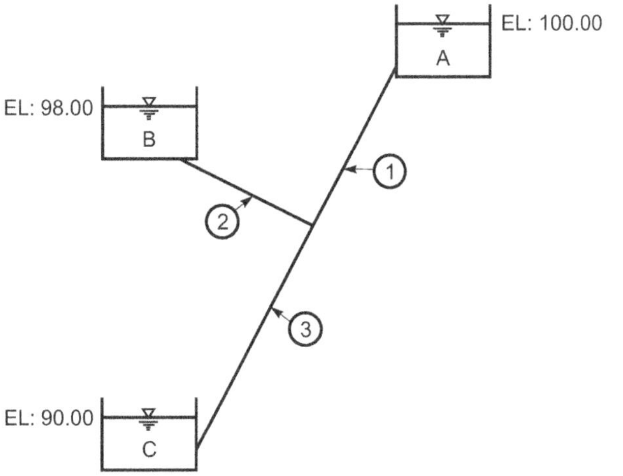

Fig. 11.15 *Example 11.19*

Solution:

Putting $h_f = r\,Q^2$

$$r = \frac{8fl}{\pi^2 g D^5} = \frac{8}{\pi^2 \times 9.81} \times \frac{fL}{(0.30)^5} = 34.0\,fL$$

Pipe	*L(m)*	*f*	*r*
1	200	0.02	136
2	125	0.016	68
3	250	0.016	136

Let H_j = piezometric head at the junction J. Similarly H_a, H_b and H_c are piezometric heads at reservoirs A, B and C respectively.

$$h' = \frac{H_a - H_b}{H_b - H_c} = \frac{100.00 - 98.00}{98.00 - 90.00} = 0.25$$

$$R_1 = \frac{r_1}{r_3} = \frac{136}{136} = 1.0 > h'$$

Since $h' < R_1$, the flow is of **Type-2**, i.e., $H_j < H_b$ and $Q_1 + Q_2 = Q_3$

The energy relationships for this type of flow are

$$H_a = H_j + r_1 Q_1^2$$
$$H_b = H_j + r_2 Q_2^2$$
$$H_c = H_j - r_3 Q_3^2$$

$\therefore \quad (H_a - H_b) = r_1Q_1^2 - r_2Q_2^2 = 100.00 - 98.00 = 2.0$

$(H_b = H_c) = r_2Q_2^2 + r_3Q_3^2 = 98.00 - 90.00 = 8$

Substituting the value of r_1, r_2 and r_3

$$136Q_1^2 - 68Q_2^2 = 2$$

$$68Q_2^2 + 136Q_3 = 8$$

Put $Q_3 = mQ_2$ resulting in $Q_1 = (m-1)Q_2$. Substituting in the above equations

$$Q_2^2\,[136\,(m-1]^2 - 68] = 2 \quad (1)$$

$$Q_2^2\,[68 + 136\,m^2] = 8 \quad (2)$$

Dividing (1) by (2)

$$\frac{136(m-1)^2 - 68}{68 + 136m^2} = 0.25$$

$$\frac{2(m-1)^2 - 1}{1 + 2m^2} = 0.25$$

On simplifying $m^2 - 2.667\,m + 0.5 = 0$

$$\therefore \quad m = \frac{2.667 \pm \sqrt{(2.667)^2 - 4 \times 0.5}}{2} = 2.46$$

[***Note:*** *only positive root is relevant as m < 1 gives negative values of* Q_1.]

Substituting in Eq. (2)

$$Q_2^2 = [68 - 138\,(2.46)^2] = 8$$

Hence $Q_2 = \mathbf{0.095\ m^3/s}$

$Q_3 = mQ_2 = 2.46\,(0.095) = \mathbf{0.234\ m^3/s}$

$Q_1 = (m-1)\,Q_2 = 1.46\,(0.095) = 0.139\ m^3/s$

H_f = piezometric head at the junction $= H_c + r_3Q_3^2$

$= 90.00 + 136\,(0.234)^2 = \mathbf{97.447\ m}$

11.20 For supplying water to a colony, it is intended to store water in a large tank whose water surface elevation is presently at 85.37 mRL. This tank is being supplied by a pipeline from a junction J, the friction loss in meters of water in this pipe being $0.6\,Q^{1.95}$, where Q is the discharge in m^3/s. Junction J receives water under pressure and without leakage from (i) a large reservoir I whose water surface elevation is constant at 100.00 mRL through a pipe characterised by $h_f = 0.35\,Q^2$; and (ii) a second large reservoir II with its constant water level at 96 mRL through a pipe characterised by $h_f = 0.46\,Q^{1.85}$. Neglecting losses other than due to pipe friction, determine the incoming flow into the tank in the colony.

Solution: Let the tank in the colony be designated as III and the pipelines from reservoirs to junctions be designated as 1, 2 and 3 as in Fig. 11.16.

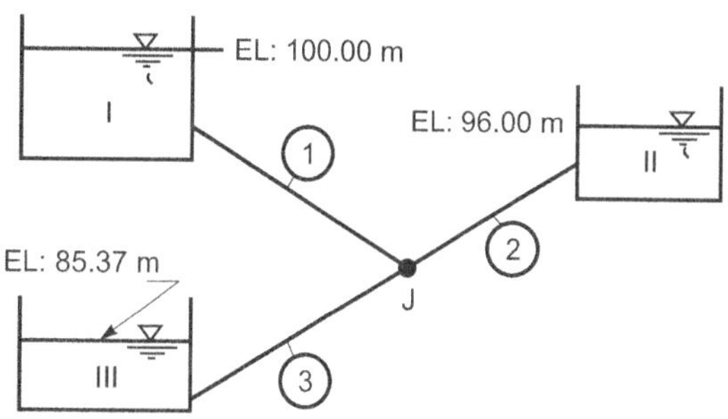

Fig. 11.16

Let H_j = Elevation of hydraulic gradeline at J.
For all pipes:

$$h_f = rQ^n \text{ with different values of } r \text{ and } n.$$

$$\text{Average value } \bar{n} = \frac{2.00 + 1.85 + 1.95}{3} = 1.933$$

A trial and error method is adopted.

[***Note:*** In this problem $h_f = rQ^n$. Hence $Q = (h_f/r)^{1/n}$. The correction ΔH_j is calculated as

$$\Delta H_j = (\bar{n} \times \Delta Q)/(\Sigma|Q/h_f|).]$$

First trial Assume H_j = 95.0 mRL

Pipe	*r*	*n*	*Estimated* h_f *(m)*	$Q = (h_f/r)^{1/n}$ *(m³/s)*	$\lvert Q/h_f \rvert$
1	0.35	2.0	100.00 – 95.00 = 5.00	+ 3.780	0.760
2	0.46	1.85	96.00 – 95.00 = 1.00	+ 1.522	1.522
3	0.60	1.95	95.00 – 85.37 = 9.63	– 4.151	0.431
				Σ = +1.151	Σ = 2.713

ΔQ = Inflow – outflow = + 1.151 m^3/s

ΔH_j = Correction to assumed $H_j = (\bar{n} \times \Delta Q)/(\Sigma|Q/h_f|)$

$$= +\frac{1.933 \times 1.151}{2.713} = 0.82 \text{ m}$$

New H_j (for 2nd trial) = 95.00 + 0.82 = 95.82 mRL

Second trial Assume H_j = 95.82 mRL

Pipe	*r*	*n*	*Estimated* h_f *(m)*	$Q = (h_f/r)^{1/n}$ *(m³/s)*	$\lvert Q/h_f \rvert$
1	0.35	2.0	100.00 – 95.82 = 4.18	+3.456	0.827
2	0.46	1.85	96.00 – 95.82 = 0.18	+0.602	3.346
3	0.60	1.95	95.82 – 85.37 = 10.45	–4.329	0.414
				Σ = –0.271	4.587

$$\Delta Q = -0.271\ m^3/s$$

$$\Delta H_j = \frac{1.933 \times 0.271}{4.587} = -0.11 \text{ m}$$

New H_j (for 3rd trial) = 95.82 – 0.11 = 95.71 mRL

Third trial Assume H_j = 95.71 mRL.

Pipe	r	n	Estimated h_f (m)	$Q = (h_f/r)^{1/n}$ (m^3/s)	$\lvert Q/h_f \rvert$
1	0.35	2.0	100.00 – 95.71 = 4.29	+3.501	0.002
2	0.46	1.85	96.00 – 95.71 = 0.29	+0.779	2.690
3	0.60	1.95	95.71 – 85.37 = 10.34	–4.306	0.416
				Σ = – 0.026	3.108

$$\Delta Q = -0.026 \text{ m}^3/\text{s}$$

As the ΔQ is less than 0.1% of discharge in pipe 3, no further corrections are deemed necessary. ΔQ is distributed equally to all the three pipes. Hence

Q_1 = discharge in pipe 1 = **3.51 m³/s**

Q_2 = discharge in pipe 2 = **0.79 m³/s**

Q_3 = discharge in pipe 3 = **4.30 m³/s**

$$\Delta H_j = \frac{1.933 \times 0.026}{3.108} = -0.01$$

Final H_j of the junction = **95.70 mRL**

11.21 A water tank discharges to the atmosphere through a 100 m long 20 cm diameter pipe branching into 60 m long 10 cm diameter and 120 m long 10 cm diameter pipes. The elevations above datum of the outlets are 100.00 m for 120 m branch and 105.00 m for the 60 m branch. The water surface in the tank is at an elevation of 125.00 m. Assuming $f = 0.02$ for all pipes, estimate the discharges in the two outlets.

Solution: Figure 11.17 shows the schematic layout of the tanks and pipelines. According to the given data:

Pipe	*Diameter*	*Length*	*Outlet*
1	20 cm	100 m	Junction J
2	10 cm	60 m	Atmosphere
3	10 cm	120 m	Atmosphere

[This problem is a variation of the three reservoir problem and the same technique is adopted for solution].

For pipe 1: $h_{f1} = r_1 Q_1^2 = \dfrac{f_1 L_1}{D_1} \cdot \dfrac{V_1^2}{2g} = \left(\dfrac{8fL}{\pi^2 g D^5}\right) Q_1^2$

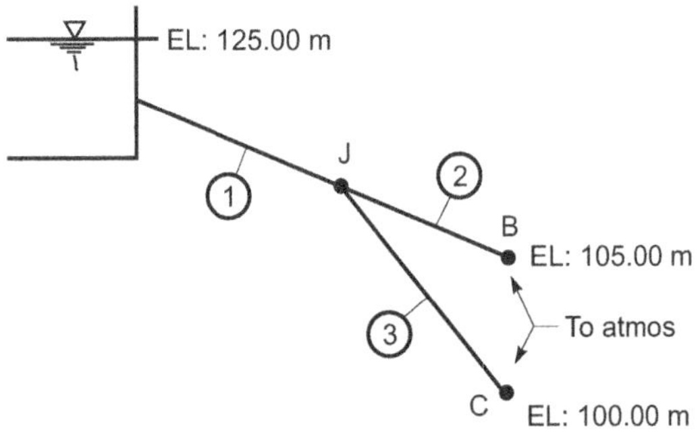

Fig. 11.17

$$r_1 = \left(\frac{8fL}{\pi^2 g D^5}\right) = \frac{8 \times 0.02 \times 100}{\pi^2 \times 9.81 \times (0.2)^5}$$

$$= 516.4$$

For pipe 2: $h_{f2} = r_2 Q_2^2 = \left(\dfrac{f_2 L_2}{D_2} + 1\right) \dfrac{V_2^2}{2g}$

$$r_2 = \frac{8\left(\dfrac{f_2 L_2}{D_2} + 1\right)}{\pi^2 g D^4} = \frac{8\left(\dfrac{0.02 \times 60}{0.1} + 1\right)}{\pi^2 \times g \times (0.1)^4}$$

$$r_2 = 10742$$

For pipe 3: $h_{f3} = r_3 Q_3^2 = \dfrac{8\left(\dfrac{f_3 L_3}{D_3} + 1\right)}{\pi^2 g D^4} Q_3^2$

$$r_3 = \frac{8\left(\dfrac{0.02 \times 120}{0.1} + 1\right)}{\pi^2 \times g \times (0.1)^4} = 20657$$

A trial and error procedure with an assumption of the energy head at the junction J is adopted.

First trial Let H_j = 121.0 m

Pipe No.	r	h_f (m)	$Q = \sqrt{h_f / r}$ (m^3/s)	Q (L/s)	$\lvert Q/h_f \rvert$ (Q in L/s)
1	516.4	125.0 – 121.0 = 4.0	+0.0880	+88.0	22.0
2	10742	121.0 – 105.0 = 16.0	–0.0386	–38.6	2.4
3	20657	121.0 – 100.0 = 21.0	–0.0319	–31.9	1.5
				Σ = +17.5	Σ = 25.9

$$\Delta H_j = \frac{2 \times \Sigma Q}{\Sigma \lvert Q / h_f \rvert} = \frac{2 \times 17.5}{25.9}$$

$$= 1.35 \text{ m}$$

New H_j = 121.0 + 1.35 = 122.35 m for the next trial.

Second trial H_j = 122.35 m

Pipe	r	h_f (m)	$Q = \sqrt{h_f / r}$ (m^3/s)	Q (L/s)	$\lvert Q/h_f \rvert$ (Q in L/s)
1	516.4	125.0 – 122.35 = 2.65	+0.0716	+71.6	27.0
2	10742	122.35 – 105.0 = 17.35	–0.0402	–40.2	2.3
3	20657	122.35 – 100.0 = 22.35	–0.0329	–32.9	1.5
				Σ = –1.5	Σ = 30.8

$$\Delta H_j = -\frac{2 \times 1.5}{30.8} = -0.1 \text{ m}$$

New H_j for the 3rd trial = 122.35 – 0.10 m

= 122.25 m

Third trial H_j = 122.25

Pipe	r	h_f (m)	$Q = \sqrt{h_f / r}$ (m^3/s)	Q (L/s)	$\lvert Q/h_f \rvert$ (Q in L/s)
1	516.4	125.0 – 122.25 = 2.75	+0.073	73.0	26.5
2	10742	122.25 – 105.0 = 17.25	–0.040	–40.0	2.3
3	20657	122.25 – 100.0 = 22.25	–0.033	–33.0	1.5
				Σ = – 0.0	Σ = 30.3

$$\Delta H_j = \frac{2 \times 0.0}{30.2} = -0.00$$

As the error in ΔQ is nil no further iteration is needed.

Hence $\mathbf{H_j = 122.25}$ m above datum.

and $\mathbf{Q_1 = 73.0}$ **L/s**

$\mathbf{Q_2 = 40.0}$ **L/s**

$\mathbf{Q_3 = 33.0}$ **L/s**

Alternative Precedure:

Exact Method Obviously the flow is the **Type-1**, i.e. $H_j > H_b$ and $Q_1 = Q_2 + Q_3$.

By the application of Energy equation between the junction J and the reservoir A, and outlets B and C successively

$$H_a = H_j + r_1 Q_1^2$$

$$H_b = H_j - r_2 Q_2^2$$

$$H_c = H_j - r_3 Q_3^2$$

$$\therefore \quad (H_a - H_b) = r_1 Q_1^2 + r_2 Q_2^2 = 125.00 - 105.00 = 20$$

$$(H_b - H_c) = r_3 Q_3^2 - r_2 Q_2^2 = 105.00 - 100.00 = 5$$

Substituting the values of r_1, r_2 and r_3

$$516.4 Q_1^2 + 10742 Q_2^2 = 20$$

$$20657 Q_3^2 - 10742 Q_2^2 = 5$$

Put $Q_3 = mQ_2$ resulting in $Q_1 = (1 + m)\,Q_2$.

Substituting in the above equations

$$Q_2^2\,[516.4\,(1 + m)^2 + 10742] = 20 \quad (1)$$

$$Q_2^2[20657\,m^2 - 10742] = 5 \quad (2)$$

Dividing (1) by (2)

$$\frac{561.4(1+m)^2 + 10742}{20657 m^2 - 10742} = 4.0$$

$$\frac{(1 + 2m + m^2) + 20.80}{40.0 m^2 - 20.80} = 4$$

On simplifying $m_2 - 0.01258\,m - 0.660 = 0$

$$\therefore \quad m = \frac{0.01258 \pm \sqrt{(0.01258)^2 + 4 \times 0.660}}{2} = 0.819$$

*[**Note:** Only positive root is relevant.]*

Substituting in Eq. (2)

$$Q_2^2 = [20657\,(0.819)^2 - 10742] = 5$$

Hence $Q_2 = 0.04\ m^3/s = \mathbf{40\ L/s}$

$Q_3 = m\,Q_2 = 0.819\,(0.04) = 0.0328\ m^3/s$

$= \mathbf{32.8\ L/s}$

$Q_1 = (1 + m)\,Q_2 = 1.819\,(0.0.04)$

$= 0.0728\ m^3/s = \mathbf{72.8\ L/s}$

11.22 For the reservoir system shown in Fig. 11.18 determine the flow in each pipe. Take $f = 0.018$ for all pipes. Neglect minor losses. At C the pipe discharges water to atmosphere at an elevation of 140.00 m.

Pipe	*Diameter*	*Length*
1	15 cm	800 m
2	20 cm	500 m
3	30 cm	600 m

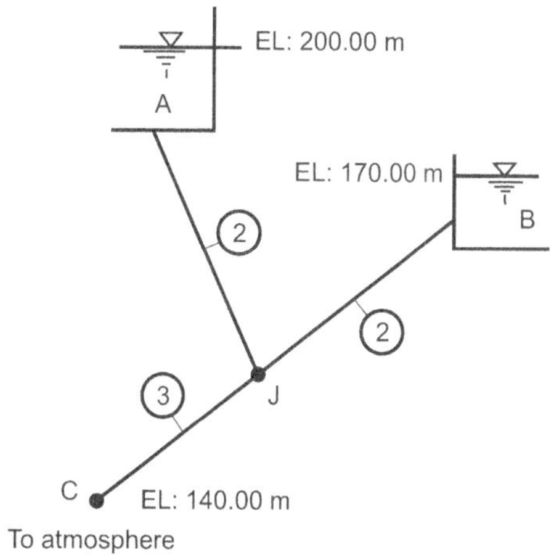

Fig. 11.18 *Example 11.22*

Solution: $h_f = r\,Q^2$ where $r = \dfrac{8fL}{\pi^2 g D^5}$

For pipe 1: $h_{f1} = r_1 Q_1^2$

$$r_1 = \frac{8fL}{\pi^2 g D^5} = \frac{8 \times 0.018 \times 800}{\pi^2 \times 9.81 \times (0.15)^5} = 15669$$

For pipe 2: $h_{f2} = r_2\,Q_2^2$

$$r_2 = \frac{8fL}{\pi^2 g D^5} = \frac{8 \times 0.018 \times 500}{\pi^2 \times 9.81 \times (0.20)^5} = 2324$$

For pipe 3: $h_{f3} = r_3 Q_3^2 = \left(\dfrac{f_3 L_3}{D_3} + 1\right)\dfrac{V_3^2}{2g}$

$$r_3 = \frac{8\left(\dfrac{f_3 L_3}{D_3} + 1\right)}{\pi^2 g D^4} = \frac{8\left(\dfrac{0.018 \times 600}{0.30} + 1\right)}{\pi^2 \times 9.81 \times (0.3)^4}$$

$= 377.4$

A trial and error procedure with an assumption of the energy head at the junction J is adopted.

First trial Let $H_j = 165.0$ m

Pipe No.	r	h_f (m)	$Q = \sqrt{h_f / r}$ (m^3/s)	Q (L/s)	$\lvert Q/h_f \rvert$ (Q in L/s)
1	15669	200.0 – 165.0 = 35.0	+0.0473	+47.3	1.35
2	2324	170.0 – 165.0 = 5.0	+0.0464	+46.4	9.28
3	377.4	165.0 – 140.0 = 25.0	–0.2574	–257.4	10.30
				Σ = –163.7	Σ = 20.93

$$\Delta H_j = \frac{2 \times \Sigma Q}{\Sigma \lvert Q / h_f \rvert} = -15.6 \text{ m}$$

New $H_j = 165.0 - 15.6 = 149.4$ m

Second trial $H_j = 149.4$ m

Pipe No.	r	h_f (m)	$Q = \sqrt{h_f / r}$ (m^3/s)	Q (L/s)	$\lvert Q/h_f \rvert$ (Q in L/s)
1	15669	200.0 – 149.4 = 50.6	+0.0568	+56.8	1.12
2	2324	170.0 – 149.4 = 20.6	+0.0941	+94.1	4.57
3	377.4	149.4 – 140.0 = 9.4	–0.1578	–157.8	16.79
				Σ = –6.9	Σ = 22.48

$$\Delta H_j = \frac{2 \times \Sigma Q}{\Sigma \lvert Q / h_f \rvert} = -\frac{2 \times 6.9}{22.48} = -0.6 \text{ m}$$

New H_j for third trial is $H_j = 149.4 - 0.6 = 148.8$ m

Third trial $H_j = 148.8$ m

Pipe No.	r	h_f (m)	$Q = \sqrt{h_f / r}$ (m^3/s)	Q (L/s)	$\lvert Q/h_f \rvert$ (Q in L/s)
1	15669	200.0 – 148.8 = 51.2	+0.0572	+57.2	1.12
2	2324	170.0 – 148.8 = 21.2	+0.0955	+95.5	4.51
3	377.4	148.8 – 140.0 = 8.8	–0.1527	–152.7	17.35
				Σ = 0	Σ = 22.98

As the continuity at the junction is satisfied, no further trials are needed. **Hence H_j = 148.8 m above datum.**

and **Q_1 = 57.2 L/s towards J**

Q_2 = 95.5 L/s, towards J

Q_3 = 152.7 L/s, away from J.

Alternative method:

Exact Method

$$h' = \frac{H_a - H_b}{H_b - H_c} = \frac{200.00 - 170.00}{120.00 - 140.00} = 1.0$$

$$R_1 = \frac{r_1}{r_3} = \frac{15669}{377.4} = 41.5 > h'$$

Since $h' < R_1$, the flow is of **Type-2,** i.e. $H_j < H_b$ and $Q_1 + Q_2 = Q_3$

The energy relationships for this type of flow are:

$$H_a = H_j + r_1 Q_1^2$$
$$H_b = H_j + r_2 Q_2^2$$
$$H_c = H_j - r_3 Q_3^2$$

$\therefore \quad (H_a - H_b) = r_1 Q_1^2 - r_2 Q_2^2 = 200.0 - 170.0 = 30.0$

$(H_b - H_c) = r_2 Q_2^2 + r_3 Q_3^2 = 170.0 - 140.0 = 30.0$

Substituting the values of r_1, r_2 and r_3

$$15669 Q_1^2 - 2324 Q_2^2 = 30$$
$$2324 Q_2^2 + 377.4 Q_3^2 = 30$$

Put $Q_3 = mQ_2$ resulting in $Q_1 = (m - 1)\, Q_2$. Substituting in the above equations

$$Q_2^2\, [15669\, (m-1)_2^2 - 2324] = 30 \qquad (1)$$
$$Q_2^2\, [22324 - 377.4 m^2] = 30 \qquad (2)$$

Dividing (1) by (2)

$$\frac{15669(m-1)^2 - 2324}{2324 + 377.4 m^2} = 1.0$$

$$\frac{m^2 - 2m + 1 - 0.1483}{0.0241 m^2 + 0.1483} = 1$$

On simplifying $m^2 - 2.0492\, m + 0.7207 = 0$

$$\therefore \quad m = \frac{2.0492 \pm \sqrt{(2.0492)^2 - 4 \times 0.7207}}{2} = 1.598$$

*[**Note:** only positive root is relevant as m < 1 gives negative values of Q_1.]*

Substituting in Eq. (2)

$$Q_2^2 = [2324 + 377.4\, (1.598)^2] = 30$$

Hence $Q_2 = 0.0955\ m^3/s$ = **95.5 L/s**

$Q_3 = m\, Q_2 = 1.598\, (0.0955) = 0.1526\ m^3/s$ = **152.6 L/s**

$Q_1 = (m - 1)\, Q_2 = 0.598\, (0.0955) = 0.0571\ m^3/s$ = **57.1 L/s**

11.23 A water supply main trifurcates at a junction point J into three branches each feeding a separate reservoir. The details of the pipes and the reservoir are as follows:

Pipe	*Diameter*	*Length*	*f*	*Feeding to reservoir elevation*
JA	20 cm	2.0 km	0.02	A = 80.00 m
JB	20 cm	2.5 km	0.02	B = 70.00 m
JC	20 cm	3.0 km	0.02	C = 60.00 m

If the inflow from the main at the junction is 0.250 m^3/s, determine the delivery into each reservoir. [Refer Fig. 11.19].

Solution:

$$h_f = rQ^2 \text{ where } r = \frac{8fL}{\pi^2 g D^5}$$

A trial and error solution is adopted.

Flow away from the junction is marked negative.

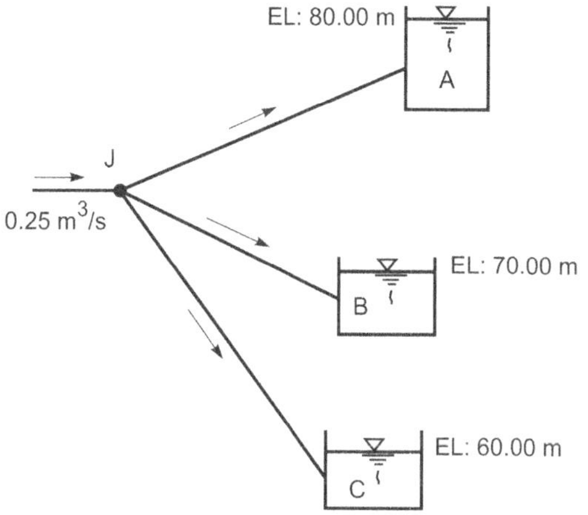

Fig. 11.19 *Example 11.23*

First trial

Let H_j = Elevation of hydraulic grade line at J
= 150.0 m

Pipe	r $\left(=\frac{8fL}{\pi^2 g D^5}\right)$	h_f (m)	$Q=\sqrt{h_f/r}$ (m³/s)	Q (L/s)	$\lvert Q/h_f\rvert$ (Q in L/s)
JA	10328	150 – 80 = 70	– 0.0823	–82.3	1.18
JB	12910	150 – 70 = 80	– 0.0787	–78.7	0.98
JC	15493	150 – 60 = 90	– 0.0762	–76.2	0.85
				Σ = –237.2	Σ = 3.01

ΔQ = Inflow + Σ Outflow = 250 – 237.2 = 12.8 L/s

$$\Delta H_j = \text{Correction to assumed } H_j = \frac{2\Delta Q}{\Sigma\lvert Q/h_f\rvert}$$

$$= \frac{2\times 12.8}{3.01} = 8.50 \text{ m}$$

New H_j = 158.5 m

Second trial

H_j = 158.5 m

Pipe	r	h_f (m)	$Q=\sqrt{h_f/r}$ (m³/s)	Q (L/s)	$\lvert Q/h_f\rvert$ (Q in L/s)
JA	10328	158.5 – 80 = 78.5	–0.0872	–87.2	1.11
JB	12910	158.5 – 70 = 88.5	–0.0828	–82.8	0.94
JC	15493	158.5 – 60 = 98.5	–0.0797	–79.7	0.81
				Σ = –249.7	Σ = 2.86

$$\Delta Q = 250 - 249.7 = +\,0.3 \text{ L/s}$$

$$\Delta H_j = +\frac{0.3\times 2}{2.86} = +\,0.2 \text{ m}$$

As ΔQ is very small no further iterations are necessary and ΔQ is distributed equally to the 3 pipes.

Hence

Q_{JA} = 87.3 L/s
Q_{JB} = 82.9 L/s
Q_{JC} = 79.8 L/s

The elevation of the hydraulic gradient line at the junction is 158.5 + ΔH_j = **158.7 m**.

11.24 For the network shown in Fig. 11.20(G) the head loss is given by $h_f = rQ^2$. The values of r for each pipe, and the discharge into or out of various nodes are shown in the sketch. The discharges are in an arbitrary unit. Obtain the distribution of discharge in the network.

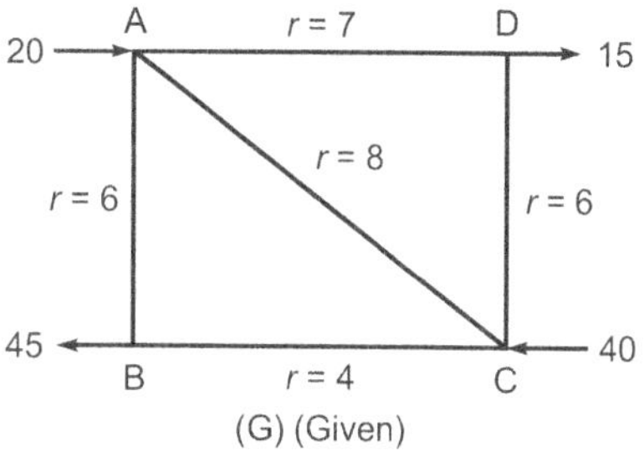

Fig. 11.20 *Example 11.24*

Solution: Flow direction is assumed positive clockwise for all loops. A first trial set of discharges is selected to satisfy continuity at each node [Fig. 11.20(a)), (i.e. flow into a node = flow out of the node). The Hardy–Cross method is used to find the corrections ΔQ.

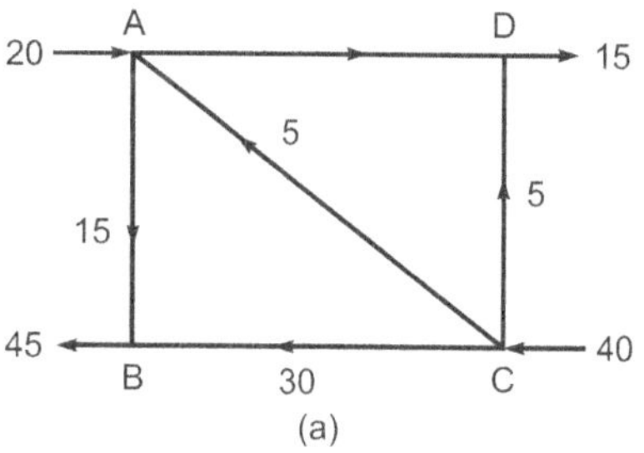

Fig. 11.20(a)

First Trial

Line	Loop ABC rQ^2	$\lvert 2rQ\rvert$
AB	$-6\times 15^2 = -1350$	$2\times 6\times 15 = 180$
CB	$4\times 30^2 = +3600$	$2\times 4\times 30 = 240$
CA	$-8\times 5^2 = -200$	$2\times 8\times 5 = 80$
	Σ = 2050	Σ = 500

Correction ΔQ for loop ABC

$$\Delta Q = -\frac{2050}{500}$$

$$\Delta Q = -4$$

	Loop ADC	
Line	rQ^2	$\|2rQ\|$
CD	$-6 \times 5^2 = -150$	$2 \times 6 \times 5 = 60$
DA	$7 \times 10^2 = +700$	$2 \times 7 \times 10 = 140$
AC	$8 \times 5^2 = +200$	$2 \times 8 \times 5 = 80$
	$\Sigma = 750$	$\Sigma = 280$

Correction ΔQ for loop ADC

$$\Delta Q = -\frac{750}{280}$$

$$\Delta Q = -3$$

The corrected flows are as in Fig. 11.20(b) Note that the line AC has two corrections.

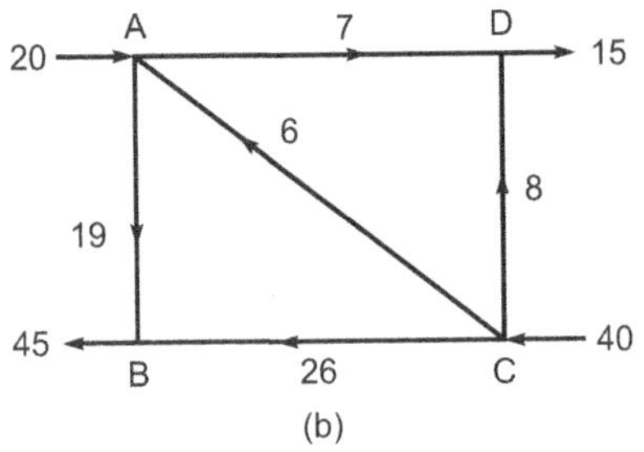

Fig. 11.20(b)

Second trial

	Loop ABC	
Line	rQ^2	$\|2rQ\|$
AB	$-6 \times 19^2 = -2166$	$2 \times 6 \times 19 = 228$
BC	$+4 \times 26^2 + 2704$	$2 \times 4 \times 26 = 208$
CA	$-8 \times 6^2 = -288$	$2 \times 8 \times 6 = 96$
	$\Sigma = 250$	$\Sigma = 532$

$$\Delta Q = -\frac{250}{532}$$

$$= -0.5$$

	Loop ADC	
Line	rQ^2	$\|2rQ\|$
CD	$-6 \times 8^2 = -384$	$2 \times 6 \times 8 = 96$
DA	$7 \times 7^2 = +343$	$2 \times 7 \times 7 = 98$
AC	$8 \times 6^2 = +288$	$2 \times 8 \times 6 = 96$
	$\Sigma = +247$	$\Sigma = +290$

$$\Delta Q = -\frac{247}{290} = -1.0$$

The corrections of the second trial are applied to get a distribution as shown in Fig. 11.20(c)

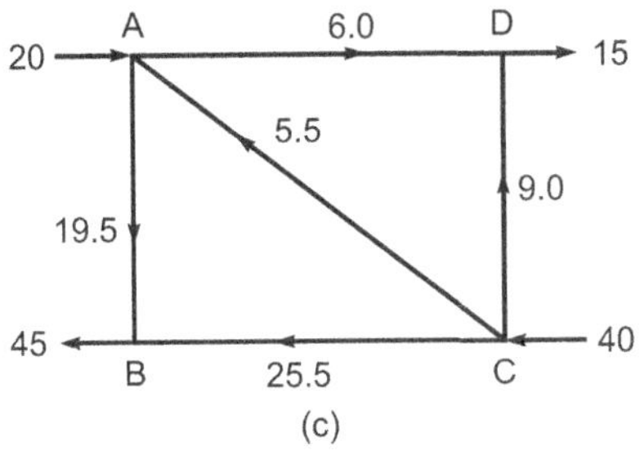

Fig. 11.20(c)

Third trial

	Loop ABC	
Line	rQ^2	$\|2rQ\|$
AB	$-6 \times 19.5^2 = -2281.5$	$2 \times 6 \times 19.5 = 234$
BC	$+4 \times 25.5^2 = 2601.0$	$2 \times 4 \times 25.5 = 204$
CA	$-8 \times 5.5^2 = -242.0$	$2 \times 8 \times 5.5 = 88$
	$\Sigma = 77.5$	$\Sigma = 526$

$$\Delta Q = -\frac{77.5}{526}$$

$$= -0.15$$

	Loop ADC	
Line	rQ^2	$\|2rQ\|$
CD	$-6 \times 9^2 = -486$	$2 \times 6 \times 9 = 108$
DA	$7 \times 6^2 = +252$	$2 \times 7 \times 6 = 84$
AC	$8 \times 5.5^2 = +242$	$2 \times 8 \times 5.5 = 88$
	$\Sigma = 8$	$\Sigma = 280$

$$\Delta Q = -\frac{8}{280}$$

$$= -0.03$$

The corrections are applied and further trials are discontinued by taking this level of distribution to be satisfactory. The calculations can be continued if further accuracy is needed. The final distribution is, therefore, as shown in Fig. 11.20(d).

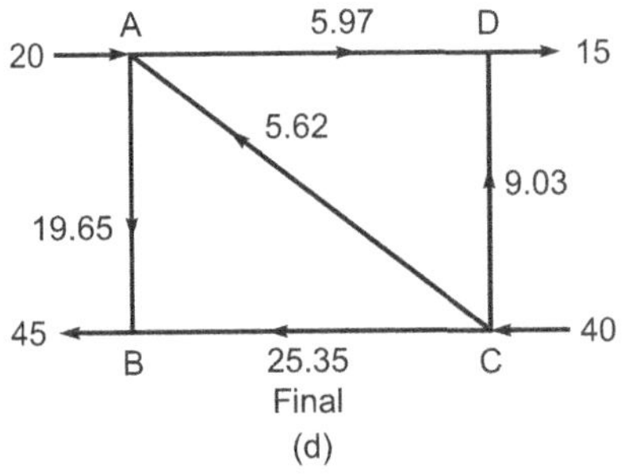

Fig. 11.20(d)

11.25 Calculate the discharge distribution in the network shown in Fig. 11.21(G). The head loss h_f in a pipe is given by $h_f = rQ^{1.85}$. The values of r for various pipes and also the inflows or outflows at nodes are shown in the figure. To start the calculations the following trial discharges are suggested.

Line	AB	BD	AD
Discharge Q (units)	56	3	44
Direction	A to B	B to D	A to D

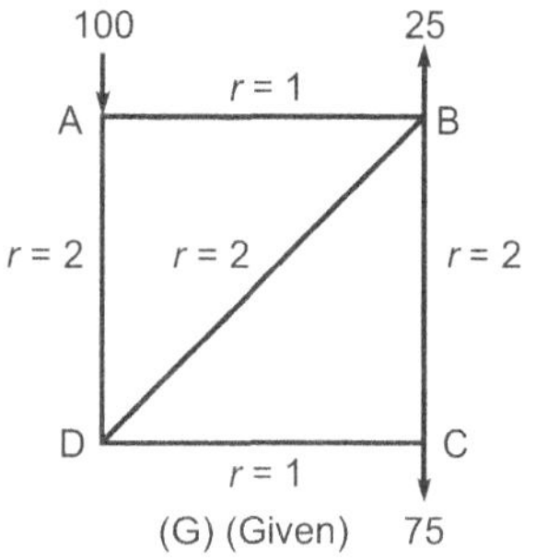

Fig. 11.21(G) *Example 11.25*

Solution: The Hardy–Cross method is used for the analysis of the network. The first trial discharges are as shown in Fig. 11.21(a).

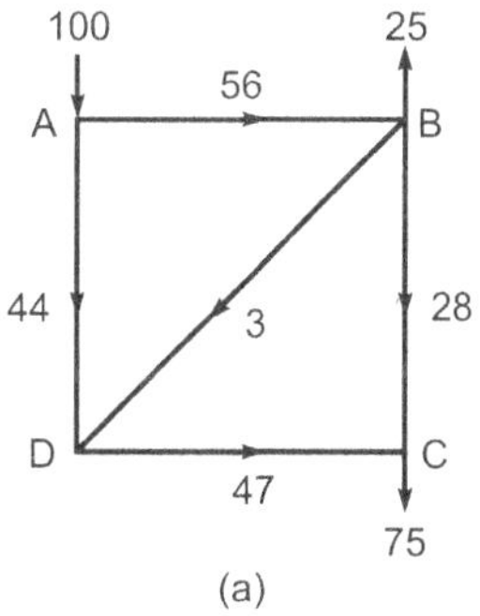

Fig. 11.21(a)

Calculations are shown in the following table. The quantities to be calculated are $(rQ^{1.85})$ and $(1.85\ rQ^{0.85})$ for each line. Clockwise flow in a loop is taken as positive.

First trial

Line	*Loop ABD* $rQ^{1.85}$	$\lvert 1.85\ rQ^{0.85}\rvert$
AB	$+1 \times (56)^{1.85} = 1714.5$	$1.85 \times 1 \times (56)^{0.85} = 56.64$
BD	$+2 \times (3)^{1.85} = 15.3$	$1.85 \times 2 \times (3)^{0.85} = 9.41$
DA	$-2 \times (44)^{1.85} = -2194.9$	$1.85 \times 2 \times (44)^{0.85} = 92.29$
	$\Sigma = -465.1$	$\Sigma = 158.34$
Correction	$\Delta Q = +\dfrac{465.1}{158.34} = 3.0$	

Line	*Loop BCD* $rQ^{1.85}$	$\lvert 1.85\ rQ^{0.85}\rvert$
BC	$+2 \times (28)^{1.85} = 951.2$	$1.85 \times 2 \times (28)^{0.85} = 62.85$
BD	$-2 \times (3)^{1.85} = 15.3$	$1.85 \times 2 \times (3)^{0.85} = 9.41$
DC	$-1 \times (47)^{1.85} = -1239.9$	$1.85 \times 1 \times (47)^{0.85} = 48.80$
	$\Sigma = -304.0$	$\Sigma = 121.06$
Correction	$\Delta Q = +\dfrac{304.0}{121.06} = 2.5$	

The corrected discharges for second trial are indicated in Fig. 11.21 (b).

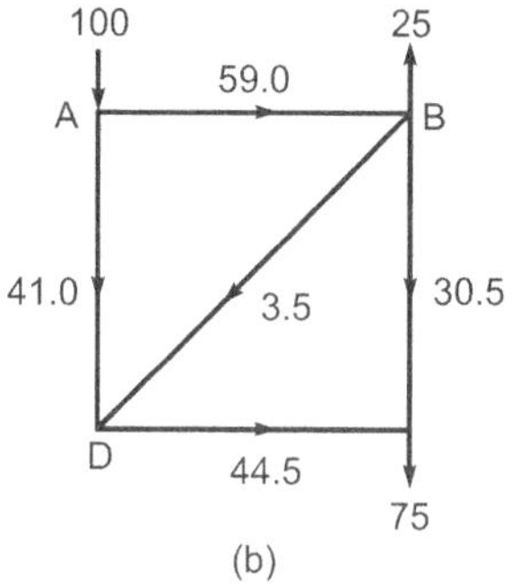

Fig. 11.21(b)

Note: The discharge in line AB = 56.0 + 3.0 = 59.0 from A to B.

AD = – 44.0 + 3 = – 41.0, i.e. 41.0 from A to D
BD = 3.0 + 3.0 – 2.5 = 3.5 from B to D
BC = 28.0 + 2.5 = 30.5 from B to C
DC = 47.0 – 2.5 = 44.5 from D to C

Second trial

Line	*Loop ABD* $rQ^{1.85}$	$\lvert 1.85\ rQ^{0.85}\rvert$
AB	$+1 \times (59.0)^{1.85} = 1888.3$	$1.85 \times 1 \times (59.0)^{0.85} = 59.2$
BD	$+2 \times (3.5)^{1.85} = 20.3$	$1.85 \times 2 \times (3.5)^{0.85} = 10.7$
DA	$-2 \times (41.0)^{1.85} = -1926.1$	$1.85 \times 2 \times (41.0)^{0.85} = 86.9$
	$\Sigma = -17.5$	$\Sigma = 156.8$
Correction	$\Delta Q = +\dfrac{17.5}{156.8} \approx +0.1$	

Line	*Loop BCD* $rQ^{1.85}$	$\lvert 1.85\ rQ^{0.85}\rvert$
BC	$+2 \times (30.5)^{1.85} = 1114.3$	$1.85 \times 2 \times (30.5)^{0.85} = 67.6$
BD	$-2 \times (3.5)^{1.85} = -20.3$	$1.85 \times 2 \times (3.5)^{0.85} = 10.7$
DC	$-1 \times (44.5)^{1.85} = -1120.6$	$1.85 \times 1 \times (44.5)^{0.85} = 46.6$
	$\Sigma = -26.6$	$\Sigma = 124.9$
Correction	$\Delta Q = +\dfrac{26.6}{124.9} \approx +0.2$	

As the correction are small, the corrections are applied and further trials are discontinued. The final distribution is obtained as below and is shown in Fig. 11.21(c).

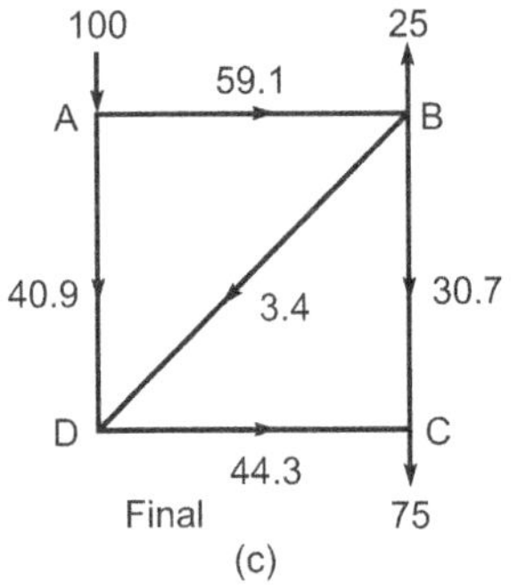

Fig. 11.21(c)

Discharge in line

AB = 59.0 + 0.1 = **59.1 from A to B**

AD = – 41.0 + 0.1 = – **40.9, i.e. 40.9 from A to D**

BD = + 3.5 + 0.1 – 0.2 = **3.4 from B to D**

BC = 30.5 + 0.2 = **30.7 from B to C**

DC = 44.5 – 0.2 = **44.3 from D to C**

11.26 Figure 11.22 shows a pipe network. The values of discharges (in arbitrary units) at the nodes and in each pipe are shown in the figure. If a conservative contaminant at a concentration of C_a enters the steady state network at node A, find the concentrations of the contaminant where the water leaves the system at points B, E and F. Water entering the network at D contains no contaminant. Assume perfect mixing at nodes.

Solution: Node A. From an inspection of Fig. 11.22, it is obvious that $C_{AB} = C_{AD} = C_a$

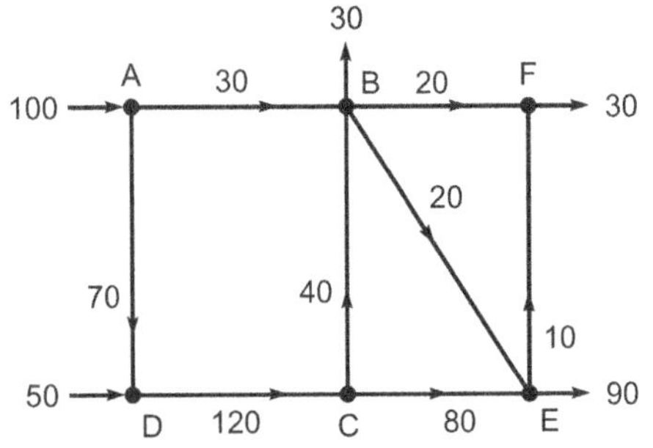

Fig. 11.22 *Example 11.26*

Node D: Concentration at D = $\dfrac{70 \times C_{AD}}{(70+50)}$

$$= \frac{70}{120} C_a = 0.583C_a$$

$$C_{DC} = C_{CB} = C_{CE} = 0.583C_a$$

Node B: Concentration at B = $\dfrac{30 \times C_{AB} + 40 \times C_{CB}}{(30+40)}$

$$= \frac{30C_a + 40(0.583)C_a}{70} = 0.762C_a$$

$$C_{BF} = C_{BE} = 0.762C_a$$

Contaminant concentration in flow leaving B = $\mathbf{0.762\ C_a}$

Node E: Concentration at E = $\dfrac{20 \times C_{BE} + 80 \times C_{CE}}{(20+80)}$

$$= \frac{20(0.762C_a) + 80(0.583)C_a}{100} = 0.619\ C_a$$

Contaminant concentration in flow leaving E = $\mathbf{0.619\ C_a}$

Node F: Concentration at F = $\dfrac{20 \times C_{BF} + 10 \times C_{EF}}{(20+10)}$

$$= \frac{20(0.762C_a) + 10(0.619C_a)}{30} = 0.7143\ C_a$$

Contaminant concentration in flow leaving

F = $\mathbf{0.7143\ C_a}$

Check:

Contaminant entering per second

= 100 C_a + 50 × (0) = 100 C_a units

Contaminant leaving per second

= 30 × (0.762 C_a) + 30 × (0.7143 C_a) + 90 × (0.619 C_a)

= 100 C_a units

11.27 A nozzle is fitted at the end of a pipe of diameter D carrying water. Show that for maximum kinetic energy to be supplied by the nozzle, the diameter of the nozzle d is given by

$$d = \left(\frac{D^5}{2fL}\right)^{1/4}$$

where f = friction factor and L = length of the pipe.

Solution: Let v = velocity in the nozzle and V = velocity in the pipe.

The total head H = velocity head at the nozzle + head lost in friction

$$H = \frac{v^2}{2g} + \frac{fLV^2}{2gD} \quad (1)$$

By continuity $\dfrac{\pi}{4} d^2 v = \dfrac{\pi}{4} D^2 V$

Hence $V = v\left(\dfrac{d^2}{D^2}\right)$

$$\therefore \quad \frac{v^2}{2g}\left(1 + \frac{fL}{D}\left(\frac{d}{D}\right)^4\right) = H$$

$$\frac{v^2}{2g} = \frac{H}{\left(1+\frac{fL}{D}\left(\frac{d}{D}\right)^4\right)} \quad (2)$$

Power of jet $\quad \mathbf{P} = \gamma Q \dfrac{v^2}{2g}$

From Eq. (1) $\quad \dfrac{v^2}{2g} = H - \dfrac{fL}{D}\dfrac{V^2}{2g}$

Thus $\quad \mathbf{P} = \gamma \dfrac{\pi}{4} d^2 v \left[H - \dfrac{fL}{D}\cdot\left(\dfrac{d}{D}\right)^4 \cdot \dfrac{v^2}{2g}\right]$

For maximum power $\quad \dfrac{d\mathbf{P}}{dv} = 0$

i.e.

$$H - 3\frac{fL}{D}\left(\frac{d}{D}\right)^4 \cdot \frac{v^2}{2g} = 0 \quad (3)$$

$$\frac{v^2}{2g} = \frac{H}{3\frac{fL}{D}\left(\frac{d}{D}\right)^4}$$

But by Eq. (2), $\quad \dfrac{v^2}{2g} = \dfrac{H}{1+\frac{fL}{D}\cdot\left(\frac{d}{D}\right)^4}$

Hence $\quad \dfrac{H}{3\frac{fL}{D}\left(\frac{d}{D}\right)^4} = H\Big/\left[1+\dfrac{fL}{D}\left(\dfrac{d}{D}\right)^4\right]$

or $\quad 1 + \dfrac{fL}{D}\left(\dfrac{d}{D}\right)^4 = 3\dfrac{fL}{D}\left(\dfrac{d}{D}\right)^4$

$$d^4 = \frac{D^5}{2fL}$$

$$d = \left(\frac{D^5}{2fL}\right)^{1/4}$$

[*Note:* From Eq. (3) at the condition of maximum power transmission, $h_f = 1/3\ H$.]

11.28 A 50 cm diameter pipe conveys water for purposes of power generation. The pipe is 1000 m long and has a head of 100 m at the inlet. If a nozzle, discharging into atmosphere is fitted at the outlet, estimate the diameter of the nozzle for maximum transmission of power. What is the magnitude of the maximum power? [Assume $f = 0.019$].

Solution: For the condition of maximum power, the diameter of the nozzle

$$d = \left(\frac{D^5}{2fL}\right)^{1/4} = \left(\frac{(0.5)^5}{2\times 0.019\times 1000}\right)^{1/4}$$

$= 16.93$ cm

Head loss $\quad h_f = \dfrac{H}{3} = \dfrac{100}{3} = 33.33$ m

$$h_f = \frac{fLV^2}{2gD} = \frac{0.019\times 1000}{2\times 9.81\times 0.5}\times V^2 = 33.33$$

V = Velocity in the pipe = 4.148 m/s

Discharge $\quad Q = \dfrac{\pi}{4}\times(0.5)^2\times 4.148$

$= 0.8146\ m^3/s$

Maximum Power

$$\mathbf{P}_m = \gamma Q(H - h_f)$$

$$= 9.79\times 0.8146\times \frac{2}{3}\times 100 \text{ kW}$$

$= \mathbf{531.6\ kW}$

11.29 A pipeline 60 cm in diameter takes off from a reservoir whose water surface elevation is 150 m above datum. The pipe is 5000 m long and is laid completely at the datum level. In the last 1200 m of the pipe, water is withdrawn by a series of pipes at a uniform rate of 0.088 m^3/s per 300 m. Find the pressure at the end of the pipeline. Assume $f = 0.02$ and the pipe to have a dead end.

Solution: First an expression for head loss in a pipe having a uniform withdrawal of q^* m^3/s per metre length is derived.

Consider a section at a distance x from the start of the uniform withdrawal at q^* per metre length (see Fig. 11.23).

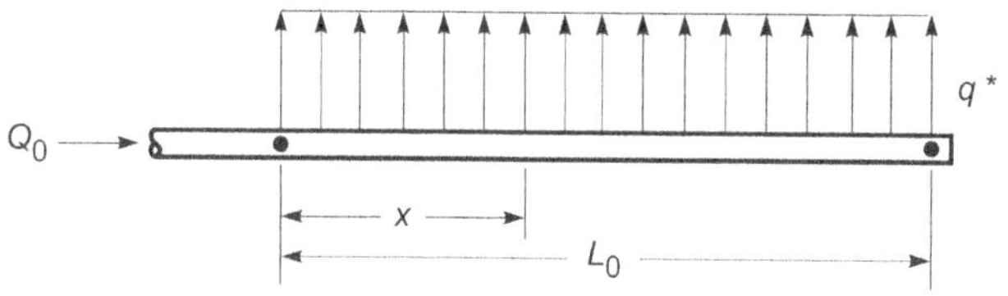

Fig. 11.23

Discharge $\quad Q_x = Q_0 - q^*x$

In a small distance dx,

$$dh_f = \frac{fLV^2}{2gD} = \frac{f(Q_0 - q^*x)^2}{2g\times\left(\frac{\pi}{4}\right)^2\times D^5}$$

$$dx = \frac{8f}{\pi^2 g D^5}(Q_0 - q^*x)^2\, dx$$

$$h_f = \int_0^{L_0} dh_f = \frac{-8f}{3\pi^2 g D^5}\frac{1}{q^*}\left[(Q_0 - q^*x)^3\right]_0^{L0}$$

$$= \frac{8f}{3\pi^2 g D^5}\frac{1}{q^*}\left[Q_0^3 - (Q_0 - q^*L_0)^3\right]$$

In the present problem,

$$Q_0 = \frac{1200}{300} \times (0.088) = 0.352 \text{ m}^3\text{/s}$$

$$q^* = \frac{0.088}{300} = 2.933 \times 10^{-4} \text{ m}^3\text{/s/m}$$

$$L_0 = 1200 \text{ m}$$

H_L = total head lost = (head lost in first (5000 – 1200) m with a discharge Q_0 = 0.352 m³/s) + head lost in 1200 with a uniform withdrawal of q^*.

$$= h_{f1} + h_{f2}$$

$$h_{f1} = \frac{0.02 \times 3800 \times (0.352)^2}{2 \times 9.81 \times \left(\frac{\pi}{4}\right)^2 \times (0.6)^5}$$

$$= 10.00 \text{ m}$$

$$h_{f2} = \frac{8f}{3\pi^2 g D^5} \frac{1}{q^*} [Q_0^3 - (Q_0 - q^* L_0)^3]$$

$$= \frac{8 \times 0.02}{3 \times \pi^2 \times 9.81 \times (0.6)^5} = \frac{1}{(2.933 \times 10^{-4})}$$

$$[(0.352)^3 - 0]$$

$$= 1.053 \text{ m}$$

Total head loss = 10.00 + 1.053 = 11.053 m

Residual head at the dead end = 150.0 – 11.053 = **138.947 m above datum.**

11.30 Water flows through a pipeline whose diameter varies from 20 cm to 10 cm in a length of 10 m. If the Darcy–Weisbach friction factor is assumed constant at 0.02 for the whole pipe, determine the head loss in friction when the pipe is flowing full with a discharge of 50 L/s.

Solution: Consider a stretch of length dx at a distance x from the 20 cm diameter end, (Fig. 11.24).

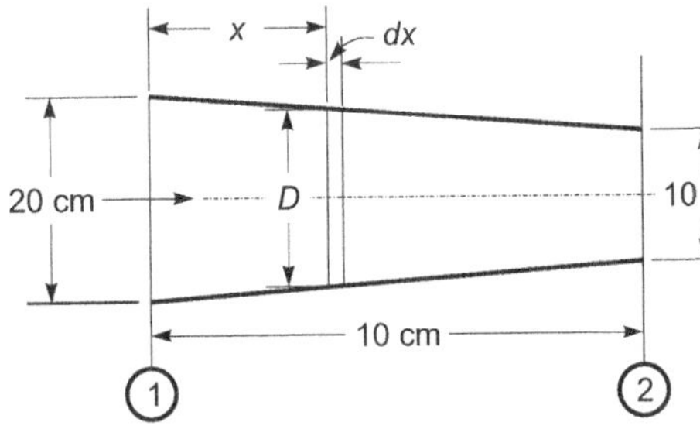

Fig. 11.24

$$dh_{fx} = \frac{f dx V^2}{2gD} = \frac{f Q^2 dx}{2g\left(\frac{\pi}{4}\right)^2 D^5}$$

where D = diameter at the section (x).

$$D = \left[20 - \frac{(20-10)}{10} x\right] \text{cm} = \frac{1}{100}(20 - x) \text{ m}$$

Hence

$$dh_{fx} = 0.008263 \, fQ^2 \frac{(10)^{10} dx}{(20-x)^5}$$

$$= 0.08263 \times 0.02 \times (0.050)^2 \times 10^{10} \frac{dx}{(20-x)^5}$$

$$= 41313 \frac{dx}{(20-x)^5}$$

Total head loss

$$h_f = \int_0^{10} dh_{fx}$$

$$= 41313 \int_0^{10} (20-x)^{-5} dx = 41313 \left[\frac{1}{4(20-x)^4}\right]_0^{10}$$

$$= \frac{41313}{4}\left[\frac{1}{(10)^4} - \frac{1}{(20)^4}\right] = \mathbf{0.968 \text{ m}}$$

11.31 For the pumping set-up shown in Fig. 11.25 estimate the (a) power required and (b) the pressure at the suction side of the pump. Include minor losses. Atmospheric pressure head = 10.0 m.

Pipe	*Diameter*	*Length*	*f*
AP	15 cm	20 cm	0.02
PC	12 cm	300 m	0.02

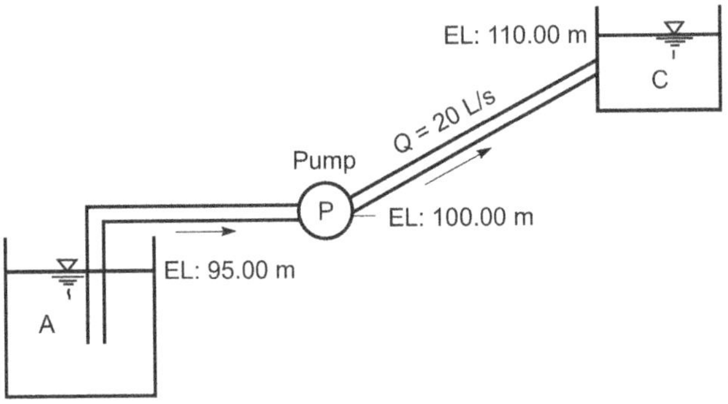

Fig. 11.25

Solution: Static head H_s = 110.00 – 95.00 = 15 m

$$V_1 = \text{Velocity in the suction pipe} = \frac{Q}{A_1}$$

$$V_1 = \frac{0.02}{\frac{\pi}{4} \times (0.15)^2} = 1.132 \text{ m/s}$$

$$V_2 = \text{Velocity in the delivery pipe} = V_1 \left(\frac{D_1}{D_2}\right)^2$$

$$= 1.132 \left(\frac{15}{12}\right)^2 = 1.769 \text{ m/s}$$

h_{f1} = head loss in suction pipe = $\frac{f_1 L_1}{D_1} \cdot \frac{V_1^2}{2g}$

$= \frac{0.02 \times 20}{0.15} \times \frac{(1.132)^2}{2 \times 9.81} = 2.667 \times 0.0653$

$= 0.174$ m

h_{L1} = Inlet loss $= 0.5 \frac{V_1^2}{2g} = 0.5 \times \frac{(1.132)^2}{2 \times 9.81}$

$= 0.5 \times 0.0653 = 0.033$ m

h_{f2} = friction loss in the delivery pipe = $\frac{f_2 L_2}{D_2} \frac{V_2^2}{2g}$

$= \frac{0.02 \times 300}{0.12} \times \frac{(1.769)^2}{2 \times 9.81} = 50 \times 0.1595$

$= 7.975$ m

h_{L2} = Loss at exit = $\frac{V_2^2}{2g} = \frac{(1.769)^2}{2 \times 9.81} = 0.160$ m

Total loss = 0.174 + 0.033 + 7.975 + 0.160 = 8.342 m

Head delivered by pump H_f = Static head + losses

= 15.0 + 8.342 = 23.342 m

Power delivered = $\gamma Q H_f$

= 9.79 × 0.02 × 23.342 kW

= **4.57 kW**

(b) By energy equation between reservoir A and the pump P:

Let p_s = Pressure at the suction side of the pump.

$95.00 + 0 + 0 = 100.00 + \frac{p_s}{\gamma} + \frac{V_1^2}{2g} +$ losses in the suction pipe

$= 100 + \frac{p_s}{\gamma} + 0.065 + 0.174 + 0.033$

$\frac{p_s}{\gamma} = -5.272$ m (gauge)

= 10.0 – 5.272 = 4.728 m (abs)

P_s = 4.728 × 9.79 = **46.29 kPa (abs)**

11.32 Two reservoirs A and B are connected through a piping system consisting of one 60 cm diameter pipe 500 m long branching into two pipes of 30 cm diameter and 40 cm diameter, each 700 m long. A pump situated at reservoir A pumps 0.40 m^3/s of water through this pipe system to reservoir B whose water surface elevation is 40 m above that of A. Assuming $f = 0.02$ determine the input power for the pump. Assume pump efficiency as 50%.

Solution: Refer to Fig. 11.26

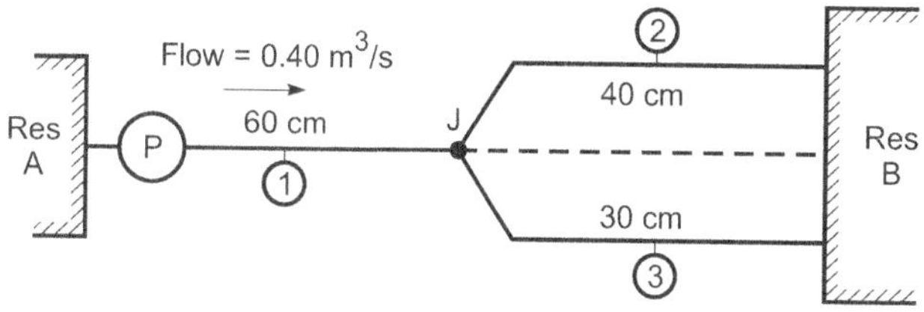

Fig. 11.26

Let the suffix 1 refer to the 60 cm pipe, 2 to the 40 cm pipe and 3 to the 30 cm pipe. Consider an equivalent pipe of diameter 60 cm to replace the two parallel pipes. The equivalent pipe (D_e, L_e, f_e) to replace a set of parallel pipes (D_2, L_2, f_2) and (D_3, L_3, f_3) is given by

$$\left(\frac{D_e^5}{f_e L_e}\right)^{1/2} = \left(\frac{D_2^5}{f_2 L_2}\right)^{1/2} + \left(\frac{D_3^5}{f_3 L_3}\right)^{1/2}$$

Here $f_e = f_2 = f_3$ and the other parameters are:

L_2 = 700 m L_3 = 700 m L_e = unknown

D_2 = 0.40 m D_3 = 0.30 m D_e = 0.60

Hence $\left(\frac{(0.60)^5}{L_e}\right)^{1/2} = \left(\frac{(0.4)^5}{700}\right)^{1/2} + \left(\frac{(0.3)^5}{700}\right)^{1/2}$

i.e., $\frac{0.2789}{L_e^{1/2}} = 3.8247 \times 10^{-3} + 1.8632 \times 10^{-3}$

L_e = 2403.5 m = equivalent length of 60 cm pipe to replace the parallel pipes. The total equivalent length of 60 cm pipe is now

= (500 + 2403.5) = 2903.5 m

$A = \frac{\pi}{4} \times (0.6)^2 = 0.2827$ m^2

$V = \frac{Q}{A} = \frac{0.40}{0.287} = 1.4147$ m/s

$V^2/2g = \frac{(1.4147)^2}{2 \times 9.81} = 0.102$ m

h_f = frictional head loss = $\frac{f_e L_e}{D} \frac{V^2}{2g}$

$h_f = \frac{0.02 \times 2903.5}{0.6} \times (0.102) = 9.87$ m

H_t = Total pumping head = 40.0 + 9.87

= 49.87 m

Power input for the pump

$$\mathbf{P}_1 = \frac{\gamma Q H_t}{\eta} = \frac{9.79 \times 0.4 \times 49.87}{0.5} = \mathbf{391\ kW}$$

[*Note:* This problem could also be solved without considering the equivalent pipe. First the discharge division in the parallel pipe is determined and then the total frictional loss. The calculation effort is definitely less switch the equivalent pipe method.]

11.33 A pump delivers water from a tank A (water surface elevation = 100.00 m) to tank B (water surface elevation = 150.00 m). The suction pipe is 50 m long ($f = 0.025$) and 30 cm in diameter. The delivery pipe is 900 m long ($f = 0.020$) and 20 cm in diameter. If the head discharge relationships for the pump is given by $H_p = 80 - 7000\,Q^2$, calculate the discharge in the pipeline and the power delivered by the pump. (Neglect minor losses). [In the pump equation H_p is in metres and Q in m^3/s].

Solution: Consider the schematic sketch, Fig. 11.27

Here,

$D_1 = 0.30$ m $\quad D_2 = 0.20$ m

$L_1 = 50.0$ m $\quad L_2 = 900$ m

$f_1 = 0.025 \quad f_2 = 0.020$

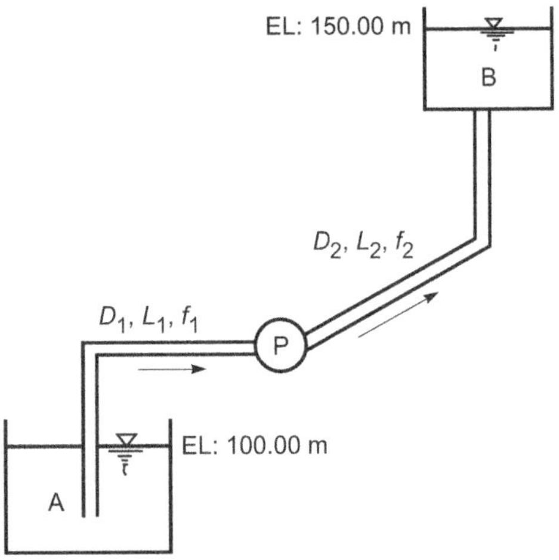

Fig. 11.27

Suction pipe:

$$\text{Head loss} = h_{L1} = \frac{f_1 L_1 V_1^2}{2g D_1} = \frac{0.025 \times 50}{0.30}\frac{V_1^2}{2g}$$

$$= 4.167 \frac{V_1^2}{2g} \text{ m}$$

Delivery pipe:

$$\text{Head loss} = h_{L2} = \frac{f_2 L_2 V_2^2}{2g D_2} = \frac{0.020 \times 900}{0.20}\frac{V_2^2}{2g}$$

$$= 90 \frac{V_2^2}{2g} \text{ m}$$

$$\text{Total head loss} = H_L = 41.167 \frac{V_1^2}{2g} + 90 \frac{V_2^2}{2g}$$

By continuity $V_1 (0.30)^2 = V_2 (0.20)^2$

$$V_1 = V_2 \left(\frac{0.20}{0.30}\right)^2 = 0.444\, V_2$$

$$\frac{V_1^2}{2g} = 0.1975 \frac{V_2^2}{2g}$$

$$H_L = [(4.167 \times 0.1975) + 90] \frac{V_2^2}{2g}$$

$$= 90.82 \frac{V_2^2}{2g} \text{ m}$$

Static head = 150.0 – 100 = 50.0 m

H_p = Head delivered by pump

= Static head + friction head

$$= 50 + 90.82 \frac{V_2^2}{2g}$$

$$= 50 + 90.82 \times \frac{Q^2}{\left[\frac{\pi}{4} \times (0.2)^2\right]^2} \times \frac{1}{2 \times 9.81}$$

$$H_p = 50 + 4690\, Q^2 \qquad (1)$$

But by the given pump performance relation

$$H_p = 80 - 7000\, Q^2$$

$$\therefore\ 80 - 7000\, Q^2 = 50 + 4690\, Q^2$$

$$Q^2 = \frac{30}{11690} = 2566 \times 10^{-3}$$

$Q = 0.0506\ m^3/s$ = **50.6 L/s**

$$H_p = 50 + 4690\,(0.0506)^2 = 62.04 \text{ m}$$

Power delivered by the pump

$$\mathbf{P} = \gamma Q H_p$$

$= 9.79 \times 0.0506 \times 62.04$ = **30.73 kW**

OBJECTIVE QUESTIONS

11.1 Hydraulic grade line for flow in a pipe of constant diameter is
(a) always above the centreline of the pipe
(b) always above the energy grade line
(c) always sloping downwards in the direction of the flow
(d) coincides with the pipe centreline

11.2 The head loss in a sudden expansion from 6 cm diameter pipe to 12 cm diameter pipe, in terms of velocity V_1 in the 6 cm pipe is
(a) $\frac{15}{16}\frac{V_1^2}{2g}$ (b) $\frac{3}{4}\frac{V_1^2}{2g}$
(c) $\frac{1}{4}\frac{V_1^2}{2g}$ (d) $\frac{9}{16}\frac{V_1^2}{2g}$

11.3 The head loss caused due to sudden expansion from area A_1 to area A_2 causing velocity to change from V_1 to V_2 is
(a) $\left(1-\frac{A_1}{A_2}\right)^2\frac{V_1^2}{2g}$ (b) $\left(1-\frac{A_1}{A_2}\right)^2\frac{V_2^2}{2g}$
(c) $\left(1-\frac{A_2}{A_1}\right)^2\frac{V_2^2}{2g}$ (d) $\left(1-\frac{A_1}{A_2}\right)\frac{V_1^2}{2g}$

11.4 A pipeline 5 cm diameter is reduced abruptly to 2.5 cm diameter at a section to enable measurement of flow. The loss head at the contraction is 0.5 m. The mean velocity at the reduced section will be about.
(a) 6.3 m/s (b) 3.0 m/s
(c) 4.0 m/s (d) 4.80 m/s

11.5 In a sudden contraction, the velocity head changes from 0.5 m to 1.25 m. The coefficient of contraction is 0.66. The head loss is this contraction is
(a) 0.133 m (b) 0.332 m
(c) 0.644 m (d) 0.750 m

11.6 The minor loss due to sudden contraction is due to
(a) flow contraction
(b) expansion of flow after sudden contraction
(c) boundary friction
(d) cavitation

11.7 A pipe has a well rounded entrance from a reservoir. If the head loss at the entrance is expressed as $K\frac{V^2}{2g}$, the value of K would be about
(a) 0.02 (b) 0.2
(c) 0.5 (d) 1.0

11.8 The loss at the exit of a submerged pipe in a reservoir is
(a) $0.5\frac{V^2}{2g}$ (b) $\frac{V^2}{2g}$
(c) $0.1\frac{V^2}{2g}$ (d) negligibly small

11.9 Minor losses in a pipe flow are those losses
(a) which are insignificantly small
(b) which can be neglected always
(c) caused by local disturbance due to pipe fittings
(d) caused by frictional resistance

11.10 Minor losses are usually neglected when they comprise 5% or less of the head loss due to friction. In a 10 cm diameter pipe ($f = 0.02$), the minimum length of the pipeline at which one can ignore minor losses due to sharp edged square entrance from a reservoir and a submerged square outlet to another reservoir, is
(a) 100 m (b) 15 m
(c) 150 m (d) 12000 m

11.11 In a 15 cm pipe line, the minor losses add up to $15\frac{V^2}{2g}$. The length of a pipe of 15 cm diameter ($f = 0.02$) equivalent to this loss is
(a) 75 m (b) 200 m
(c) 7.5 m (d) 150 m

11.12 Two identical pipes of length L, diameter D and friction factor f, are connected in *series* between two reservoirs. The size of a pipe of length L and of the same friction factor f, equivalent to the above pipeline, is
(a) 0.5 D (b) 0.87 D
(c) 1.15 D (d) 1.40 D

11.13 Two pipe systems in series are said to be equivalent when
(a) the average diameter in both systems is the same
(b) the average friction factor in both systems is the same
(c) total length of the pipe is the same in both the systems.
(d) the discharge under the same head is the same in both systems.

11.14 Two identical pipes of length L, diameter D and friction factor f, are connected in *parallel* between two reservoirs. The size of a pipe of length L and of same friction factor f, equivalent to the above pipes, is

(a) 0.5 D (b) 0.87 D
(c) 1.40 D (d) 2.0 D

11.15 Two identical pipes of length L, diameter D and friction factor f, are connected in parallel between two points. The length of a single pipe of diameter D and the same friction factor f, equivalent to the above pair, is

(a) $\sqrt{2L}$ (b) $\frac{L}{2}$
(c) $\frac{L}{\sqrt{2}}$ (d) $\frac{L}{4}$

11.16 In using Darcy–Weisbach equation for flow in a pipe, the friction factor is misjudged by + 25%. The resulting error in the estimated discharge Q is

(a) +25% (b) −16.67%
(c) −5% (d) −12.5%

11.17 A pipeline connecting two reservoirs has its diameter reduced by 10% over a length of time due to chemical deposit action. If the friction factor remains unaltered, for a given head difference in the reservoirs this would reflect in a reduction in discharge of

(a) 10% (b) 14.6%
(c) 23.2% (d) 31.6%

11.18 Two pipelines of equal length and diameter of 20 cm and 30 cm respectively are connected in parallel between two reservoirs. If the friction factor f is the same for both the pipes, the ratio of the discharges in the smaller to the larger size of the pipe is

(a) 0.363 (b) 0.444
(c) 0.667 (d) 0.137

11.19 Three pipes are connected in series. Then

(a) the head loss in each pipe is the same
(b) the total discharge is the sum of the discharge in the individual pipes
(c) the discharge through each pipe is the same
(d) the Reynolds number for each pipe is the same

11.20 For maximum transmission of power through a pipeline with a total head H, the head loss due to friction h_f is given by h_f =

(a) $H/3$ (b) $\frac{2}{3}H$
(c) $H/2$ (d) 0.1 H

11.21 A nozzle with a coefficient of velocity $C_v = 0.95$ is attached at the end of a pipe. If the head loss in the nozzle is $K(V^2/2g)$, where V = velocity of jet issuing from the nozzle, the value of K for this nozzle is

(a) 0.905 (b) 0.108
(c) 0.053 (d) 0.0028

11.22 Two reservoirs are connected by two pipes A and B of identical friction factor and length, in *series*. If the diameter of A is 30% larger than that of B the ratio of the head loss in A to that in B is

(a) 0.77 (b) 0.59
(c) 0.50 (d) 0.27

11.23 Two reservoirs connected by a pipe of friction factor f has a difference in water surface elevation of H. If H and f were to remain constant, to increase the flow by 100%, one would need to increase the cross sectional area of flow by

(a) 74% (b) 100%
(c) 50% (d) 32%

11.24 The discharge Q in a pipe of known f is estimated by using the head loss h_f in a length L and diameter D. If an error of 1% is involved in the measurement of D, the corresponding error in the estimation of Q is

(a) 2.5% (b) 1.0%
(c) 0.4% (d) 5%

11.25 In a pipeline the friction factor f is estimated by knowing the discharge Q in a pipe of diameter D and head loss h_f in a length L. If there is an error of + 1% in the measurement of Q, the corresponding error in f is

(a) 1/2% (b) −1/2%
(c) +2% (d) ±1%

11.26 In a pipeline design the usual practice is to assume that due to aging

(a) the effective roughness increases linearly with time
(b) the friction factor increases linearly with time
(c) the pipe becomes smoother with time
(d) the friction factor decreases linearly with time

11.27 A rectangular conduit 0.8 m × 0.4 m carries air (kinematic viscosity = 1.5×10^{-5} m^2/s) at a velocity of 3 m/s. The Reynolds number of the flow for calculation of friction factor f is

(a) 8×10^4 (b) 1.07×10^5
(c) 1.6×10^5 (d) 6×10^4

11.28 In a pipe network
(a) the algebraic sum of discharges around each elementary circuit must be zero
(b) the head at each node must be the same
(c) the algebraic sum of the piezometric head drops around each elementary circuit is zero
(d) the piezometric head loss in each line of a circuit is the same

PROBLEMS

11.1 A tank discharges water through a horizontal pipe into atmosphere. The pipe is 2.5 m long and contains a gate valve of $K = 0.2$ for fully open conditions. Calculate the discharge in the pipe for a head of 3.0 m in the tank when
(a) the pipe is of 8 cm diameter with rounded entrance ($K = 0.05$).
(b) the pipe is of 10 cm diameter with square entrance.
(Assume $f = 0.02$ for both cases and the valve to be fully open).

[*Ans.* (a) $Q_1 = 28.2$ L/s; (b) $Q_2 = 40.6$ L/s]

11.2 An 8 cm diameter pipe carrying water has an abrupt expansion 12 cm diameter at a section. If a differential mercury manometer connected to upstream and downstream sections of the expansion indicate a gauge reading of 2.0 cm, estimate the discharge in the pipe.

(*Ans.* $Q = 15.9$ L/s)

11.3 When a sudden contraction from 50 cm diameter to 25 cm diameter is introduced in a horizontal pipeline the pressure changes from 105 kPa to 69 kPa. Assuming a coefficient of contraction of 0.65, calculate the flowrate.
Following this contraction if there is a sudden enlargement to 50 cm and if the pressure in the 25 diameter section is 69 kPa, what is the pressure in the 50 cm section?

(*Ans.* $Q = 0.376$ m^3/s; $p_3 = 80$ kPa)

11.4 A 30 cm diameter pipe is required for a town's water supply. As pipes of this diameter were not available in the market, it was decided to lay two parallel pipes of equal diameter. Find he diameter of the parallel pipes. Assume f is same for all the pipes.

(*Ans.* $D_e = 22.74$ cm; the next higher standard size pipe would be used.)

11.5 Two pipes, each of 10 cm diameter and length 100 m, are connected in parallel, between two points. Calculate the
(a) equivalent length of a single pipe of 10 cm diameter.
(b) equivalent size (diameter) of a single pipe of length 100 m.
(Assume f is same for all the pipes).

(*Ans.* (a) $L_e = 25$ m; (b) $D_e = 0.132$ m)

11.6 Three pipes A, B and C with details as given in the following are connected in series:

Pipe	*Length*	*Diameter*	*f*
A	60 m	10.0 cm	0.018
B	80 m	8.0 cm	0.020
C	100 m	6.0 cm	0.020

Calculate (a) the size of a pipe of length 125 m and $f = 0.020$, equivalent to the pipeline ABC. (b) the length of an 8 cm diameter ($f = 0.015$) pipe equivalent to the pipeline ABC.

(*Ans.* (a) $D_e = 6.0$ cm; (b) $L_e = 692$ m)

11.7 If n number of pipes of different diameters but of same length L are connected in parallel between two points, obtain an expression for the size of an equivalent single pipe of length L. All the n pipes and the equivalent pipe can be assumed to have the same friction factor f.

$$\left(Ans.\ D_e = D_1 \left[\sum_1^n \left(\frac{D_i}{D_1} \right)^{5/2} \right]^{2/5} \right)$$

11.8 Two pipes A and B with details as below are connected in parallel between two points:

Pipe	*Length*	*Diameter*	*f*
A	150 m	10 cm	0.02
B	200 m	8 cm	0.015

Calculate the size of an equivalent pipe of length 175 m and having a friction factor of 0.015.

(*Ans.* $D_e = 11.67$ cm)

11.9 For the distribution main of a city water supply a 35 cm diameter pipe is required. As pipes above 30 cm diameter are not available, it is decided to lay two parallel mains of the same diameter. Assum-ing the friction factor f of all the pipes to be the same, determine the diameter of the parallel mains.

(*Ans.* D = 26.52 cm; the commercial pipe of next larger size would be needed)

11.10 Water from a tank A is siphoned over a summit to another tank B. The elevations of water levels in tanks A and B and of the pipe centreline at the summit are 100.00, 85.00 and 105.00 metres respectively. The pipe is of 15 cm diameter and is 800 m long from A up to tank B and 120 m long from A to the summit. Assume a friction factor of 0.025 and estimate the discharge and pressure at the summit. [Atmospheric pressure = 10.0 m (abs)].

(*Ans.* Q = 26.3 L/.s, p_2 = 47.85 kPa (abs)]

11.11 Two reservoirs are connected by a pipeline consisting of two pipes in series; one of 15 cm diameter and 6 m long and another of 22.5 cm diameter and 15 m long. If the difference in water levels of the reservoirs is 6.0 m, calculate the discharge by considering all losses. Assume f = 0.020 for both pipes.

(*Ans.* Q = 133.3 L/s)

11.12 Three pipes: 300 m long of 30 cm diameter, 150 m long of 20 cm diameter and 200 m long of 25 cm diameter are connected in series in the same order as indicated above between a high level reservoir and a low level reservoir. The friction factor f for the pipes are: 0.018, 0.02 and 0.019 respectively Determine the rate of flow for a difference in elevation of 15 m between the two reservoirs. Account for all losses. Contractions and expansions are sudden. (Assume k for contraction = 0.30).

(*Ans.* Q = 106.3 L/s)

11.13 Two pipes 1 and 2 are connected in parallel between points M and N. The details of the pipes are:

Pipe	*Length*	*Diameter*	f	Σk
1	75 m	8 cm	0.018	15
2	150 m	12 cm	0.020	7.5

(Σk = sum of minor loss coefficients in a pipe). The pipes are horizontal and the pressure difference between M and N is kN/m^2. Determine the discharge of water in each pipe.

(*Ans.* Q_1 = 6.3 L/s, Q_2 = 14.0 L/s)

11.14 A 30 cm pipeline is 750 m long and connects two reservoirs A and B. The elevation of the water surface in the upper reservoir A is 122.50 m. At a certain point, distance 500 m from A on the pipeline, the elevation of the centreline of the pipe is 122.00 m. If cavitation is expected at a pressure of 21 kPa (abs) determine the lowest elevation of the water surface in the reservoir B that is admissible. Neglect minor losses. [Assume f = 0.02 and atmospheric pressure = 10.3 m (abs)].

(*Ans.* Elevation of B = 109.895 m)

11.15 The reservoir M with its water surface at an elevation of 120.00 m above datum is connected to the reservoir N, whose water surface is at the elevation 100.00, through a pipeline system. Two pipes, AC (8 cm diameter and 100 m long) and BC (10 cm diameter and 800 m long) take off from the reservoir M and joint at a junction C. From C a pipe CD (12 cm diameter and 1200 m long) is connected a reservoir N. Assuming f = 0.02 for all the pipes and neglecting minor losses, estimate the discharges in all the pipes and the head at the junction.

(*Ans.* Q_1 = 7.976 L/s,
Q_2, = 4.084 L/s
and Q_3 = 12.060 L/s;
H_c = 111.59 m above datum)

11.16 Two reservoirs having a difference in water surface elevation of 12 m are connected by a pipeline system. This pipeline consists of a 30 cm diameter 1000 m long pipe leading from the higher reservoir to a junction from which two parallel pipes, each of 20 cm diameter and 800 m long, connect the lower reservoir. Assuming f = 0.020 for the 30 cm pipe and f = 0.015 for the 20 cm pipe, estimate the total discharge transferred to the lower reservoir?

(*Ans.* Q = 90.8 L/s)

11.17 Two reservoirs with 15 m difference in their water levels are connected by 300 mm diameter pipeline 3000 m long. Calculate the discharge. If a parallel pipeline of 300 mm diameter is attached to the last 1500 m length of the existing pipe, determine the modified discharge. [Assume f = 0.04]

(*Ans*: Q_o = 0.0606 m^3/s
and Q_{new} = 0.0767 m^3/s)

11.18 Three pipes whose data are given below are connected in parallel between two reservoirs A and B. If a total discharge of 50 L/s of water is transmitted from A to B, (a) estimate the discharge in each pipe. (b) What is the difference in the water surface elevations of reservoirs A and B?

Pipe	*Length*	*Diameter*	*f*
1	200 m	9 cm	0.020
2	250 m	10 cm	0.020
3	300 m	12 cm	0.018

(*Ans.* Q_1 = 12.7 L/s, Q_2 = 14.8 L/s and Q_3 = 22.5 L/s; h_f = 9.028 m)

11.19 Two pipes: L_1 = 400 m, D_1 = 30.0 cm and L_2 = 500 m and D_2 = 20 cm are connected in parallel between two reservoirs. If f_1 = 0.020 and f_2 = 0.015, what difference in reservoir water surface elevations will produce a total flow of 0.25 m^3/s from one reservoir to another?

(*Ans.* ΔH = 9.0 m)

11.20 Three pipes having the characteristics as given below are connected in parallel between two points A and B. The pressure drop between B, $(p_a - p_b)$ = 12 kPa and the elevation drop from A and B is 6 m. Estimate the discharge of water in these pipes. Pipe data:

Pipe	*Length*	*Diameter*	*f*
1	150 m	6 cm	0.023
2	120 m	8 cm	0.020
3	100 m	10 cm	0.018

(*Ans.* Q_1 = 4.44 L/s, Q_2 = 10.93 L/s and Q_3 = 22.0 L/s)

11.21 Three reservoirs A, B and C are interconnected as shown in Fig. 11.28. Determine the distribution of discharge in the pipes.

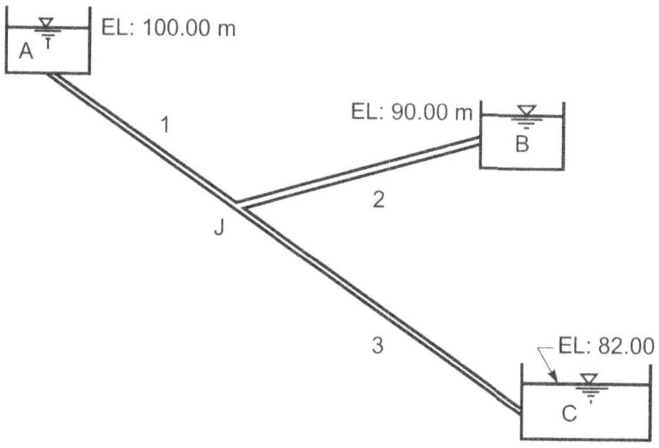

Fig. 11.28 *Problem 11.21*

Pipe	*Length*	*Diameter*	*f*
1	1500 m	30 cm	0.015
2	2000 m	25 cm	0.018
3	2000 m	30 cm	0019

(*Ans.* H_j = 91.36 m; Q_1 = 106.3 L/s; Q_2 = 21.1 L/s and Q_3 = 85.2 L/s)

11.22 The water levels in two reservoirs A and B are 65 m and 50 m above datum. The reservoirs are connected by two pipes to a common junction J, where the elevation of the hydraulic grade line is 45 m above datum. The junction J is also connected to another reservoir C. Estimate the discharge in the pipes and the elevation of the water surface in the reservoir C. The pipe data are:

Pipe	*Diameter*	*Length*	*f*
AJ	40 cm	1000 m	0.018
BJ	30 cm	750 m	0.020
JC	60 cm	850 m	0.022

(*Ans.* H_c = 41.11 m; Q_1 = 371 L/s from A; Q_2 = 99 L/s from B; Q_3 = 470 L/s to C)

11.23 Three reservoirs A, B and C are connected to a common junction J. The head loss is the connecting pipes is expressed as $h_f = rQ^2$. The following are the relevant reservoir and pipe data:

Pipe	*value of r*	*Reservoir*	*Water surface elevation (m)*
AJ	10	A	40.0
BJ	40	B	15.0
CJ	15	C	10.0

Estimate the elevation of the hydraulic grade line at the junction and the discharge in the pipes.

(*Ans.* H_j = 22.28 m, Q_1 = 1331 L/s, Q_2 = 427 L/s and Q_3 = 904 L/s)

11.24 A tank with its water level at 325.00 m above datum is discharging through a 15 cm diameter 200 m long pipe up to a junction J where it bifurcates into two branches each of 10 cm diameter. One branch JB is 55 mm long and its outlet B, at an elevation of 305.00 m, discharges freely to atmosphere. The other branch JC is 80 m long and its outlet C is at an elevation of 295.00 m and discharges freely to atmosphere. Assuming f = 0.02 for all the pipes, find the discharges in each pipe and the total discharge.

(*Ans.* H_j = 321.86 m, Q_1 = 26.9 L/s, Q_2 = 13.0 L/s and Q_3 = 13.9 L/s)

11.25 Water flows from the reservoir through a pipe of 0.15 m diameter and 180 m long to a point 13.5 m below the surface of the reservoir. Here it branches into two pipes, each of 0.1 m diameter, one of which is 48 m long discharging to atmosphere 18 m below the reservoir level and the other 60 m long discharging to atmosphere 24 m below the reservoir level. Assuming a constant friction factor $f = 0.032$, calculate the discharge from each pipe. Neglect any losses at the junction.

(*Ans.* $Q_1 = 0.0453$ m³/s. $Q_2 = 0.0193$ m³/s and $Q_3 = 0.0260$ m³/s)

11.26 A pipe network is shown in Fig. 11.29 in which Q and h_f refer to discharges and head losses respectively. Determine the head losses and discharges indicated by a question mark, for this pipe network.

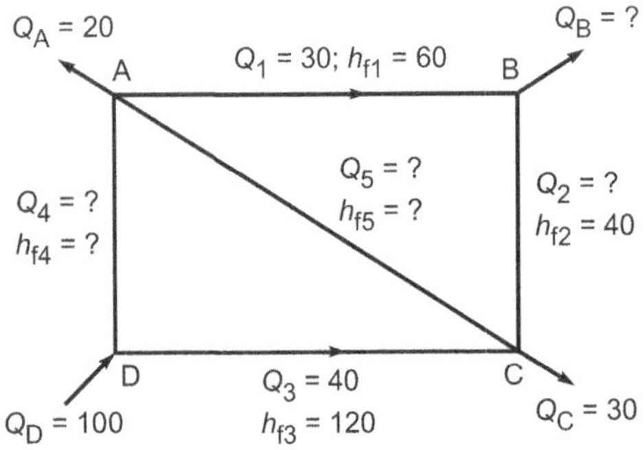

Fig. 11.29 *Related to Problem 11.26*

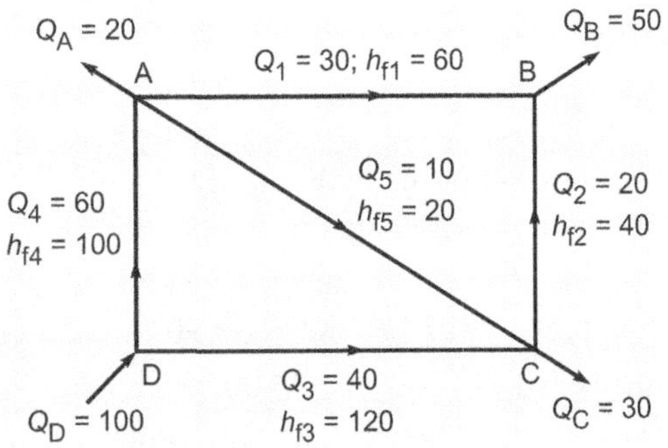

Fig. 11.29(a) *Answers to Problem 11.26*

11.27 Verify whether the following suggested distribution of discharges in the pipe network shown in Fig. 11.30 is satisfactory. If not, adjust the distribution. The head loss in a pipe is given by $h_f = rQ^2$.

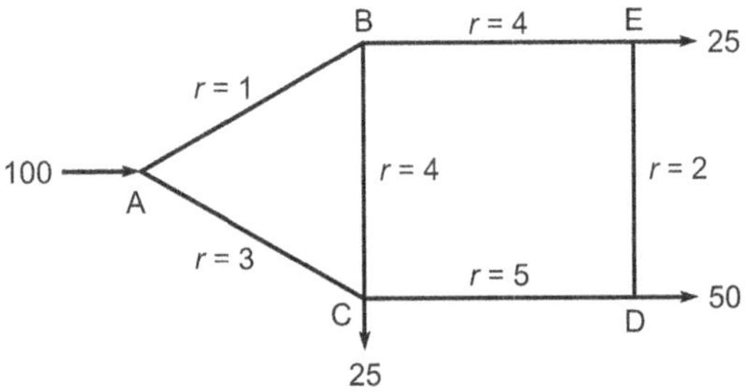

Fig. 11.30 *Related to Problem 11.27*

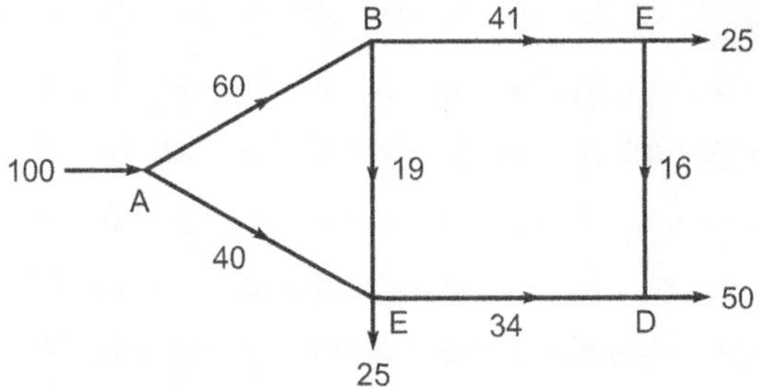

Fig. 11.30(a) *Answers to Problem 11.27*

Line	AB	BC	CA	BE	ED	DC
Suggested discharges (units)	60	19	40	41	16	34

An accuracy of 0.5 unit of discharge is adequate.

11.28 For the following pipe network (Fig. 11.31) obtain the discharge distribution. The head loss is given by $h_f = rQ^2$. An accuracy of 0.01 units of discharge is sufficient.

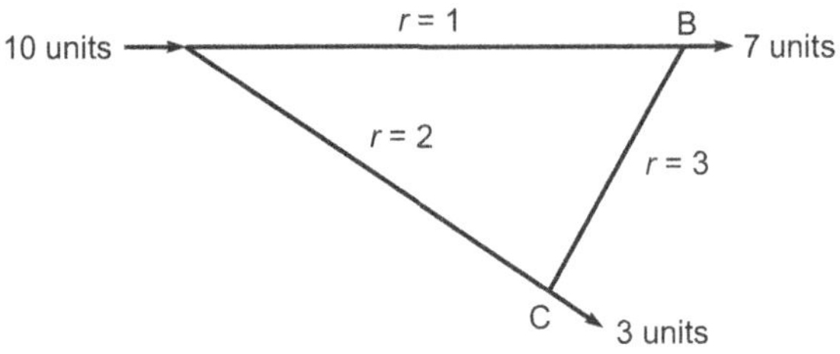

Fig. 11.31 *Related to Problem 11.28*

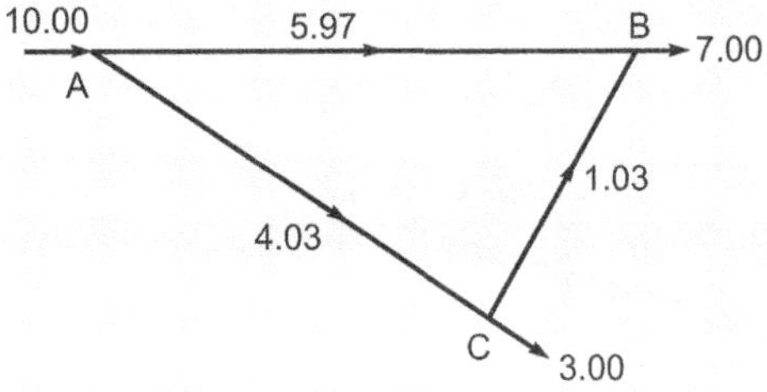

Fig. 11.31(a) *Answers to Problem 11.28*

11.29 Verify whether the suggested discharges in various pipelines in the following network (Fig. 11.32) are proper. If not, determine the proper distribution. The head loss in each pipe is given by $h_f = rQ^2$.

Line	AB	BC	CD	DA	AC
Suggested discharge (units)	58	42	32	18	20

An accuracy of 0.1 unit of discharge is adequate.

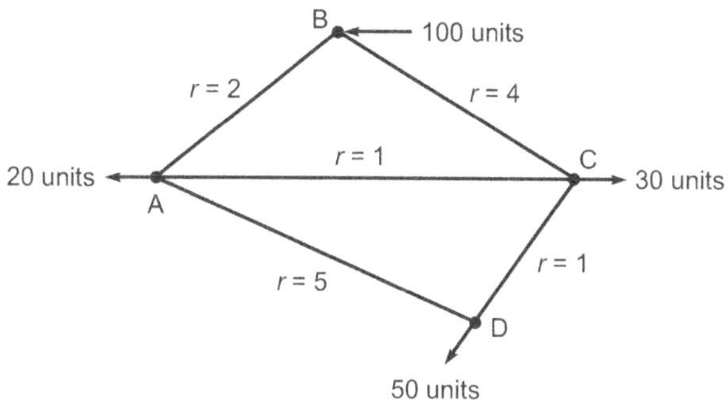

Fig. 11.32 *Related to Problem 11.29*

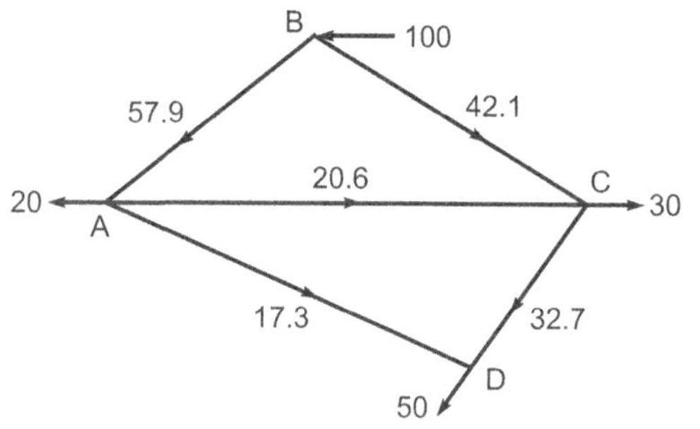

Fig. 11.32(a) *Answers to Problem 11.29*

11.30 If a conservative contaminant enters the steady state network of Fig. 11.33 at node D with a concentration C_d, and at the node E with a concentration 0.5 C_d, find the concentration of contaminant in the flows leaving the nodes E and F. Water entering the nodes A and B contains no contaminant. Assume perfect mixing at the nodes. The discharges entering the nodes and in the pipes are shown in the figure in arbitrary units.

(*Ans*: $C_f = 0.0893\ C_d$; $C_e = 0.2608\ C_d$)

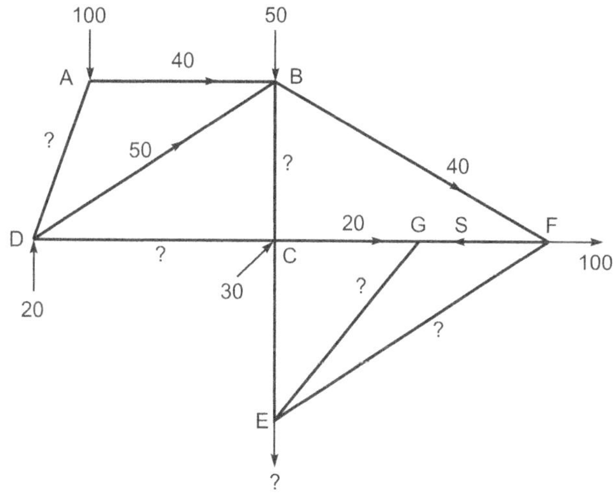

Fig. 11.33 *Problem 11.30*

[*Note:* The discharges in the pipes and at the nodes are indicated by numerals in the Figure. The arrows indicate the direction of flow.]

11.31 A pipe 10 cm in diameter and 400 m long has a nozzle fixed at the discharge end. The nozzle discharges at atmospheric pressure. If the pipe is horizontal and the head at the pipe inlet is 45 m, determine the maximum power that could be transmitted. What would be the corresponding size of the nozzle? (Assume $f = 0.02$).

(*Ans.* $\mathbf{P}_{max} = 4.42$ kW; $d = 28.1$ mm)

11.32 A 1000 m long pipe supplies water to a turbine which develops a brake power of 100 kW. The efficiency of the turbine is 90%. The head at the turbine inlet is 150 m and at the upstream and of the pipe it is 200 m. If $f = 0.02$, estimate the discharge and the velocity in the pipe. What is the diameter of the pipe?

(*Ans.* $Q = 75.66$ L/s, $V = 2.973$ m/s; $D = 18.0$ cm)

11.33 The supply of water from a reservoir to a nozzle situated 108 m below the reservoir free surface is through a 300 mm diameter pipe of 600 m length. The friction factor of the pipe is $f = 0.02$. Find the greatest possible power of the issuing jet.

(*Ans.* $\mathbf{P}_m = 209.4$ kW)

11.34 Kerosene ($\rho = 804$ kg/m^3) is pumped from one tank to another through a 100 m long 5 cm diameter pipe. The total minor loss coefficient in the system can be taken as $\Sigma k = 10.0$. If the total static lift is 5 m, calculate the power required to pump 0.3 m^3/min. Assume $f = 0.02$.

(*Ans.* $\mathbf{P} = 0.85$ kW)

11.35 Calculate the power required to pump oil (RD = 0.85) at a rate of 20 L/s from a storage sump to an elevated tank. The static lift is 20 m and the pipe is 10 cm in diameter and 250 m long. Assume $f = 0.02$. (Include entrance and exit losses).

(*Ans.* $\mathbf{P} = 6.16$ kW)

11.36 In a closed loop air drying system the conduit is 0.50 m × 0.30 m rectangular section and has a total length of 85 m. The minor losses can be accounted for by an equivalent length of 15 m. What power is required to circulate air through this loop at 25 m/s? Assume $f = 0.02$ and $\rho = 1.20$ kg/m^3.

(*Ans.* $\mathbf{P} = 8.91$ kW)

11.37 A fire brigade pump delivers water through a hose of 200 m length and 8 cm diameter to a nozzle which produces a 2 cm diameter jet. The C_v of the nozzle is 0.985 and the pipe friction factor $f = 0.025$. The end of the nozzle is 25 m above the water level in the supply tank. Calculate the power delivered by the pump when the nozzle velocity is 64 m/s.

(*Ans.* **P** = 57.3 kW)

11.38 A 25 cm pipe 2000 m long ($f = 0.002$) discharges freely into air at an elevation of 8 m below the water surfaces of the supply tank. It is required to increase the discharge by 75% by inserting a pump in the pipeline. If the pump has an efficiency of 60%, calculate the power to be supplied to the pump.

(*Ans.* **P** = 22.8 kW)

11.39 A pump with discharge characteristics

$$H_p = 80 - 2000\,Q^2$$

where H_p is in metres and Q is in m^3/s is used to pump gasoline (ρ = 680 kg/m^3) in a 500 m long pipe of 30 cm diameter ($f = 0.02$). The static lift is 10 m. What flow rate will result? What is the power expended on the fluid? (Neglect minor losses).

(*Ans.* Q = 173 L/s; **P** = 23.2 kW)

11.40 Two reservoirs are connected by 150 m of horizontal 25 cm diameter pipe with $f = 0.025$. Midway in the pipe is a turbine. If the difference in reservoir elevations is 25 m and the change in head at the turbine is 18 m, estimate the rate of flow. (Include minor losses).

(*Ans.* Q = 141.6 L/s).

12

Flow in Open Channels

CONCEPT REVIEW

12.1 OPEN CHANNEL FLOW

Flow in a conduit with a free surface is known as *open channel flow*. Flows in irrigation channels, streams and rivers, navigation channels, drainage channels and sewers are some examples of open channel flow. Basically, all open channels have a bottom slope and the component of the weight of the liquid along the slope acts as the driving fore. The boundary resistance at the perimeter acts as the resisting force. Water flow in open channels is largely in the turbulent flow regime with negligible surface tension effects. The pressure on the free surface is atmospheric at every point on the surface.

12.2 CLASSIFICATION

Open channel flows with no addition or withdrawal of flow along the channel are classified for identification and analysis as shown in the following chart.

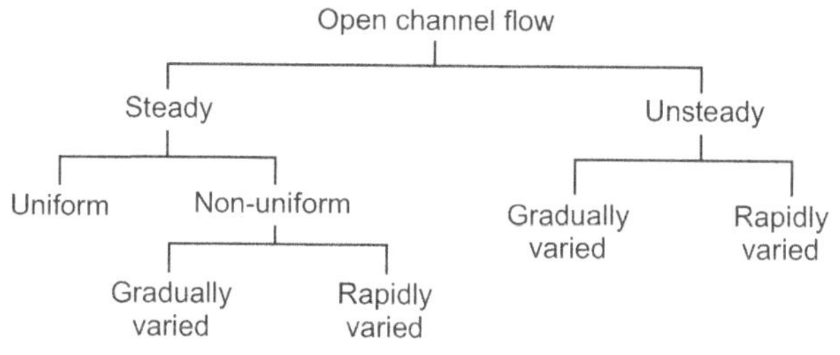

12.2.1 Uniform Flow

The depth, slope and velocity remain constant along the channel. The water surface slope S_w, the slope of the energy line S_f, and the bed slope S_0, will all be equal to each other.

12.2.2 Rapidly Varied Flow

The flow is non-uniform and the change in depth takes place rapidly. The frictional loses are relatively unimportant. The hydraulic jump is a typical example of the rapidly varied flow.

12.2.3 Gradually Varied Flow

The flow is non-uniform and the change in depth takes place gradually. The curvatures of the water surface are small and the pressure can be assumed to be hydrostatic. The frictional resistance at the boundary plays an important role in this phenomenon. The backwater curve produced by an obstruction to flow such as a weir is a typical example.

12.3 UNIFORM FLOW

12.3.1 Boundary Shear

The average shear stress τ_0 at the boundary of a channel in uniform flow (Fig. 12.1) is given by

$$\boxed{\tau_0 = \gamma R S_0}$$

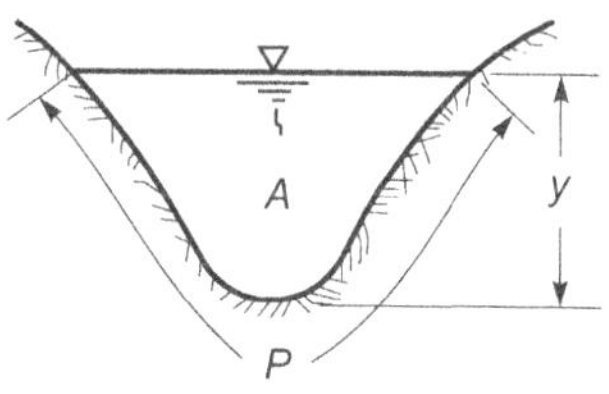

Fig. 12.1

where γ = unit weight of water

R = hydraulic radius = $\dfrac{\text{Area of flow}}{\text{wetted perimeter}} = A/P$

S_0 = bottom slope of the channel

12.3.2 Uniform Flow Resistance Formulas:

Manning's formula The most widely used resistance equation for uniform flow, known as *Manning's formula*, is

$$V = \frac{1}{n} R^{2/3} S_0^{1/2} \quad (12.1)$$

where V = average velocity of flow and n = a roughness coefficient known as Manning's roughness coefficient. This coefficient is essentially a function of the boundary surface and has the dimensions of $[L^{-1/3} T]$. It varies from 0.013 for a smooth float finished concrete surface to 0.025 for irregular channels in excavated rock. Natural streams and rivers may have a high value of n ranging from 0.020 to as high as 0.13.

Chezy formula Another popular resistance equation, known as *Chezy formula*, is

$$V = C\sqrt{RS_0} \quad (12.2)$$

where C = Chezy coefficient.

Darcy–Weisbach resistance equation The Darcy–Weisbach resistance equation, developed for pipe flow, could be used for representing open channel resistance as

$$V = \sqrt{8g/f}\,\sqrt{RS_0} \quad (12.3)$$

where f = Darcy–Weisbach friction factor the variation of which is given by the Moody Diagram as

$$f = \text{fn}\left[\left(Re = \frac{4RV}{\nu}\right), \left(\frac{4R}{\varepsilon_s}\right)\right] \quad (12.4)$$

where ε_s = equivalent sand grain roughness of the surface.

12.3.3 Relationship between *n*, *C* and *f*

The Manning's n, Chezy's C and Darcy–Weisbach's f are related as

$$C = \sqrt{8g/f} = \frac{1}{n} R^{1/6} \quad (12.5)$$

Thus $$f = \left(\frac{n^2}{R^{1/3}}\right)(8g) \quad (12.6)$$

and $$n = \frac{R^{1/6}}{C} \text{ and } n = \frac{R^{1/6}\sqrt{f}}{\sqrt{8g}} \quad (12.7)$$

It should be noted that Manning's formula is applicable to fully developed rough turbulent flow regimes only.

12.3.4 Uniform Flow Computation

The discharge $Q = AV$ and thus

$$Q = \frac{1}{n} AR^{2/3} S_0^{1/2} \quad (12.8)$$

Usually the uniform flow parameters are designated with a subscript 0. For channel cross sections whose top width is constant (e.g. rectangular) or increases with depth. (e.g. triangular, trapezoidal, parabolic) there is only one depth at which a given discharge will flow as uniform flow in a channel of known slope S_0. Such a depth is known as *normal depth*, y_0. This is a very important parameter in all open channel computations.

Uniform flow computations are relatively simple. The available relations are:

(a) Manning's formula (In this chapter Manning's formula is used in all cases unless specifically mentioned otherwise).
(b) Continuity equation and
(c) Geometry of the channel.

The following five types of basic problems can arise:

Problem type	*Given*	*Required*	*Method of solution*
1	y_0, n, S_0, Geometric elements (GE)	Q and V	Direct (explicit)
2	Q, y_0, n, GE	S_0	Direct
3	Q, y_0, S_0, GE	n	Direct
4	Q, n, S_0, GE	y_0	Trial and error
5	Q, y_0, n, S_0	GE	Trial and error

Typical geometric elements of common open channel cross sections are shown in Table 12.1.

12.3.6 Maximum Velocity and Discharge

Channels with closing top, i.e. those in which the top width decreases with depth in a certain range, do have an interesting property in which the velocity and discharge can reach a maximum value at finite values of depth.

For maximum discharge $$\frac{dQ}{dy} = 0 \quad (12.9)$$

i.e. $$\frac{d}{dy}(AR^{2/3}) = 0$$

Similarly for maximum velocity $$\frac{dV}{dy} = 0 \quad (12.10)$$

i.e. $$\frac{d}{dy}(R) = 0$$

Table 12.1 *Geometrical Elements of Common Open Channel Cross-sections*

Shape	*Rectangle* (*a*)	*Triangle* (*b*)	*Trapezoidal* (*c*)	*Circular* (*d*)
	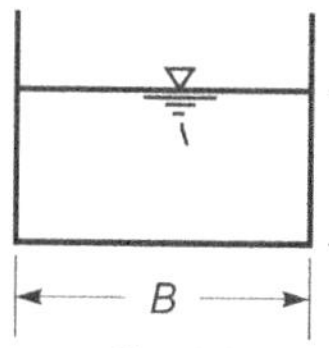 Fig. (a)	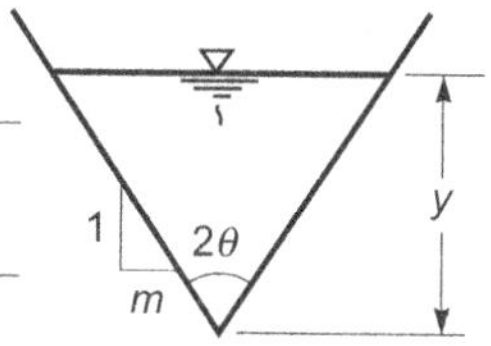 Fig. (b)	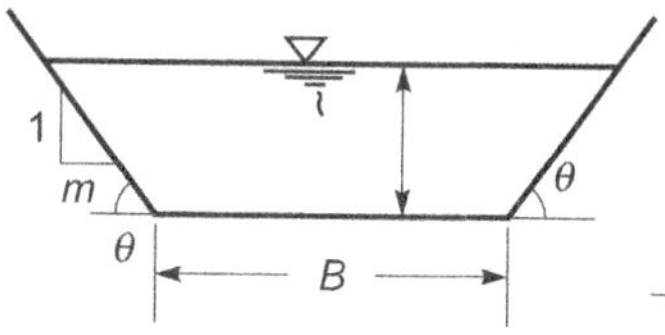 Fig. (c)	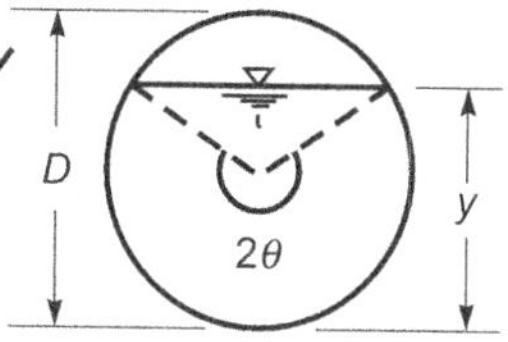 Fig. (d)
Area, A	By	my^2	$(B + my)y$	$\frac{D^2}{8}(2\theta - \sin 2\theta)$
Wetted perimeter, P	$B + 2y$	$2y\sqrt{m^2+1}$	$B + 2y\sqrt{m^2+1}$	$D\theta$
Top width, T	B	$2my$	$B + 2my$	$D \sin \theta$
Hydraulic radius, R	$\frac{By}{B+2y}$	$\frac{m}{2\sqrt{m^2+1}}\,y$	$\frac{(B+my)y}{B+2y\sqrt{m^2+1}}$	$\frac{D}{8}\;\frac{(2\theta - \text{Sin } 2\theta)}{\theta}$
Other relationships	For wide rectangular channel, i.e. ($y_0/B \le 0.02$), $R = y$	$m = \tan\theta$	$m = \cot\theta$	$\cos\theta = \left(1 - 2\frac{y}{D}\right)$

12.3.7 Hydraulically Efficient Channel Section

In uniform flow

$$Q = \frac{1}{n} AR^{2/3} S_0^{1/2} = \frac{1}{n} A^{5/3} \frac{1}{P^{2/3}} S_0^{1/2}$$

For a given roughness coefficient n, slope S_0 and area of channel A, a minimum perimeter channel section represents the hydraulically efficient section that will convey maximum discharge. Such a channel is called as the *efficient* or *best section*.

For an efficient section, A = constant and $dP/dy = 0$. Analysing various channel sections it can be shown that

(a) Of all the various possible open-channel sections, the semi-circular shape has the least amount of perimeter for a given area.

(b) For any other cross section shape, by using the subscript 'e' to denote the hydraulically efficient section value,

(i) For a rectangular section:

$$\boxed{B_e = 2y_e} \tag{12.11}$$

and

$$\boxed{R_e = y_e/2} \tag{12.12}$$

(ii) For a triangular section the vertex angle:

$$\boxed{2\theta_e = 90°} \tag{12.13}$$

(iii) For a trapezoidal section of side slope m horizontal: 1 vertical

$$\boxed{B_e = 2y_e(\sqrt{1+m^2} - m)} \tag{12.14}$$

$$\boxed{A_e = (2\sqrt{1+m^2} - m)\, y_e^2} \tag{12.15}$$

$$\boxed{R_e = y_e/2} \tag{12.16}$$

If the side slope can also be varied, the optimum value of $m = m_{cm} = \frac{1}{\sqrt{3}}$ gives the *most efficient section*.

12.3.8 Specific Energy and Critical Depth

(a) *Specific energy:* The total energy of a channel section referred to the channel bed as the datum is known as the specific energy, E. Thus

$$\boxed{E = y + \alpha \frac{V^2}{2g}} \tag{12.17}$$

where α = kinetic energy correction factor, usually taken as unity, when no other information on it is available.

(b) *Alternate depths:* For a given discharge Q in a channel, there will be two depths for a given specific energy E. These two depths are known as the *alternate depths*, (Fig. 12.2).

(c) *Critical depth:* The depth at which the specific energy is the minimum for a given discharge is

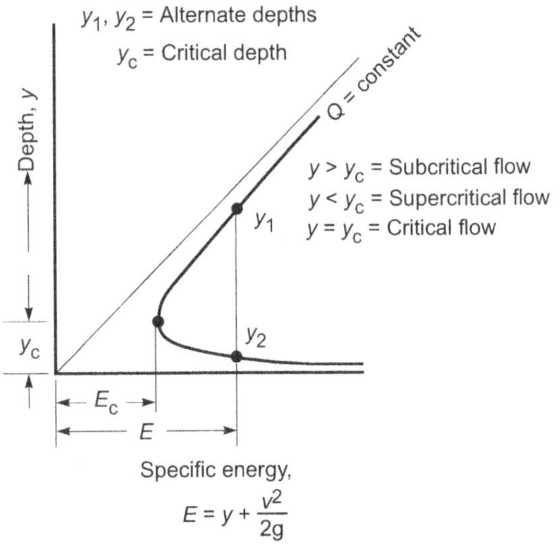

Fig. 12.2 *Specific Energy Diagram*

known as the *critical depth*. It can also be shown that the discharge for a given specific energy is maximum at the critical depth. Using subscript 'c' to denote the critical flow condition, at the critical depth y_c

$$\frac{Q^2}{g} = \frac{A_c^3}{T_c} \tag{12.18}$$

where T_c = top width at critical depth.

This relationship is used to define the Froude number F for open channels. At critical depth $F_c = 1$ and hence

$$\frac{(Q/A_c)^2}{g\,(A_c/T_c)} = F_c^2 = 1.$$

At any other depth y, the Froude number is given by

$$F = \frac{V}{\sqrt{g\,(A/T)}} \tag{12.19}$$

[*Note:* In this chapter the notation for the Froude number is F instead of Fr as used in Chapter 6. This is a commonly used notation in open cannel flow.]

12.3.9 Computation of Critical Depth

The critical depth y_c is calculated by the relation $\frac{Q^2}{g} = \frac{A_c^3}{T_c}$. For rectangular and triangular sections simple explicit expressions for y_c are obtained.

Rectangular channel:
$$y_c = \left(\frac{q^2}{g}\right)^{1/3} \tag{12.20}$$

where $q = Q/B$ = discharge intensity.

Triangular section:
$$y_c = \left(\frac{2Q^2}{gm^2}\right)^{1/5} \tag{12.21}$$

For other sections, a trial and error procedure is required for the evaluation of y_c. For details regarding the critical flow computation and use of specific energy relationship for solving problems relating to transition, the reader is advised to refer to Ref. 12.1.

12.4 RAPIDLY VARIED FLOW

12.4.1 Hydraulic Jump

Hydraulic jump is a case of rapidly varied flow. This phenomenon occurs when a super-critical flow stream tries to reach its alternate depth in sub-critical mode. In the process it loses substantial energy and falls short of the alternate depth. The depths on either side of the jump are known as *sequent depths* (Fig. 12.3).

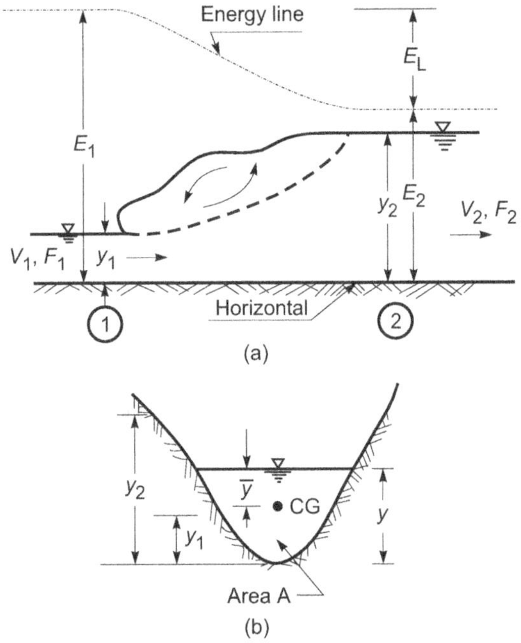

Fig. 12.3 *Hydraulic Jump*

Consider a horizontal, frictionless channel of any arbitrary shape (Fig. 12.3) By applying momentum equation in the flow direction to a control volume encompassing sections 1 and 2

$$P_1 - P_2 = M_2 - M_1 \tag{12.22}$$

where P = pressure force on a section = $\gamma A\,\bar{y}$

M = momentum flux passing a section = $\rho\, Q\, V = \rho\, Q^2/A$

In the above A = cross sectional area and $\bar{y}$ = depth of center of gravity of the area from the water surface.

Eq. 12.22 can be rearranged as

$$P + M = \gamma\left(A\bar{y} + \frac{Q^2}{gA}\right) = \text{constant}$$

$$A_1\bar{y}_1 + \frac{Q^2}{gA_1} = A_2\bar{y}_2 + \frac{Q^2}{gA_2} \qquad (12.23)$$

This is a general expression to determine the sequent depths, y_1 and y_2 in a hydraulic jump in channels of any shape. The energy loss E_L in the jump is

$$\boxed{E_L = E_1 - E_2} \qquad (12.24)$$

The computation usually involves trial and error procedure

Hydraulic jumps in rectangular channels

For a hydraulic jump in a horizontal, frictionless rectangular channel. Equation 12.23 can be simplified to obtain sequent depth ratio y_2/y_1 as

$$\frac{y_2}{y_1} = \frac{1}{2}[-1 + \sqrt{1 + 8F_1^2}] \qquad (12.25)$$

where $F_1 = \dfrac{V_1}{\sqrt{gy_1}}$ = Froude number of the approaching flow.

[*Note:* Since $\frac{y_2}{y_1} \to \sqrt{2}F_1$ for high value of F_1, for quick calculations $\frac{y_2}{y_1} \approx 1.4\,F_1$ can be used.]

The energy loss in a hydraulic jump occurring in a rectangular channel is

$$\boxed{E_L = E_1 - E_2 = \frac{(y_2 - y_1)^3}{4y_1y_2}} \qquad (12.26)$$

The length of the jump is $L_j = 6.1y_2$ for $F_1 > 4.5$. The power dissipated in the jump is

$$\boxed{\mathbf{P = \gamma QE_L}} \qquad (12.27)$$

12.5 GRADUALLY VARIED FLOW

12.5.1 Basic Equation

In gradually varied flow the energy slope S_f, the water surface slope S_w, and the bed slope S_o are all different. At any depth y the energy slope S_f is assumed to be given by Manning's formula.

$$V = \frac{1}{n} R^{2/3} S_f^{1/2}$$

or

$$\boxed{S_f = \frac{n^2V^2}{R^{4/3}} = \frac{n^2Q^2}{A^2R^{4/3}}} \qquad (12.28)$$

The basic differential equation governing the gradually varied flow is

$$\frac{dy}{dx} = \frac{S_o - S_f}{1 - \dfrac{Q^2T}{gA^3}} \qquad (12.29)$$

This could also be written in terms of specific energy E as

$$\frac{dE}{dx} = S_o - S_f \qquad (12.30)$$

Using Manning's formula Eq. 12.30 can be expressed in terms of y_o and y_c for a wide rectangular channel as

$$\boxed{\frac{dy}{dx} = \frac{1 - \left(\dfrac{y_o}{y_c}\right)^{3.33}}{1 - \left(\dfrac{y_c}{y}\right)^3}} \qquad (12.31)$$

12.5.2 Classification

For a given channel, when Q, n and S_0 are fixed, the normal depth y_0 and critical depth y_c are fixed depths. Depending upon the relative values of y_0 and y_c, the channels are divided into five categories as shown in Table 12.2. Further, based on the relative positions of the actual depth y, normal depth y_0 and critical depth y_c, the possible

Table 12.2 *Categories of Channel Flow*

Sl. No.	*Channel category*	*Symbol*	*Characteristic condition*	*Remarks*
1	Mild slope	M	$y_0 > y_c$	Subcritical flow at normal depth
2	Steep slope	S	$y_c > y_0$	Supercritical flow at normal depth
3	Critical slope	C	$y_c = y_0$	Critical flow at normal depth
4	Horizontal bed	H	$S_0 = 0$ $y_0 = \infty$	Cannot sustain uniform flow
5	Adverse slope	A	$S_0 < 0$	Cannot sustain uniform flow

gradually varied flow (GVF) profiles are grouped into twelve types as shown in Table 12.3, and also in Fig. 12.4.

Table 12.3 *Classification of GVF Profiles*

Channel	*Condition*	*Type*
Mild Slope	$y > y_0 > y_C$	M_1
	$y_0 > y > y_C$	M_2
	$y_0 > y_C > y$	M_3
Steep Slope	$y > y_C > y_0$	S_1
	$y_C > y > y_0$	S_2
	$y_C > y_0 > y$	S_3
Critical Slope	$y > (y_C = y_0)$	C_1
	$y < (y_C = y_0)$	C_3
Horizontal bed	$y > y_C$	H_2
	$y < y_C$	H_3
Adverse Slope	$y > y_C$	A_2
	$y < y_C$	A_3

12.5.3 Computation of GVF Profiles

There are a host of methods for computing the GVF profiles. The *direct step method* is a simple procedure suitable for use in prismatic channels. The basic equation

$$\frac{dE}{dx} = S_0 - S_f$$

is written in finite difference form as

$$\frac{\Delta E}{\Delta x} = S_0 - \bar{S}_f \tag{12.32}$$

to get

$$\Delta x = \frac{\Delta E}{S_0 - \bar{S}_f} \tag{12.33}$$

Where $\bar{S}_f = \dfrac{S_{f1} + S_{f2}}{2}$ = average friction slope for the reach.

Further details about the direct step method and other methods can be had from Ref. 12.1.

12.6 FLOW MEASUREMENT

Different kinds of structures, which obstruct the flow and create a unique head-discharge relationships, are employed for flow measurement in open channels. These include the weirs described in Chapter 13. A device which employs the occurrence of critical depth and finds

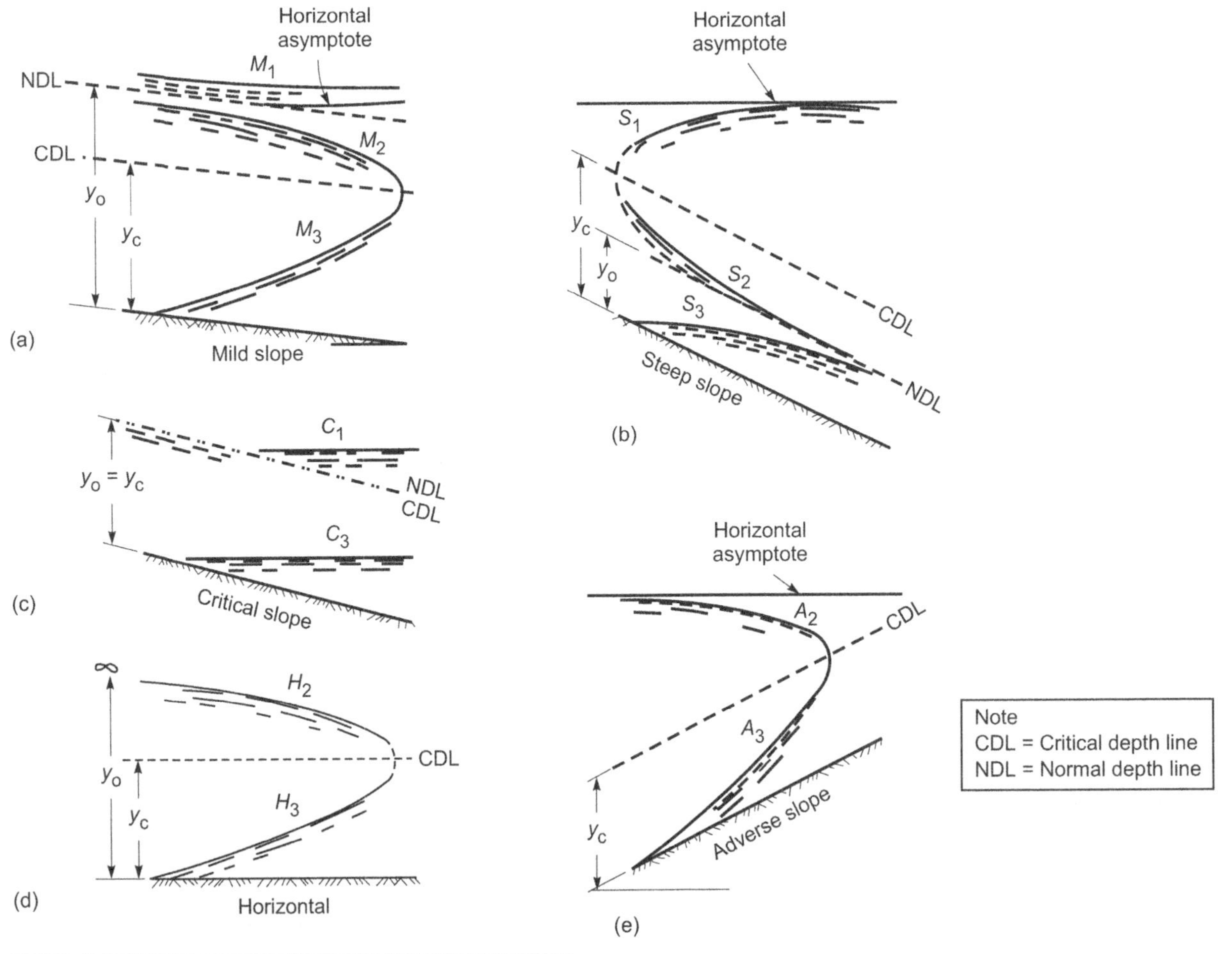

Fig. 12.4 *Gradually Varied Flow Profiles*

considerable use is the *broad-crested weir*. These weirs are sturdy structures with finite crest width in the direction of the flow (Fig. 12.5).

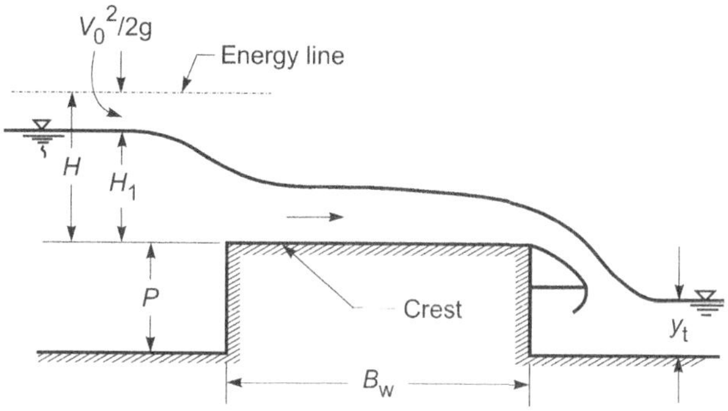

Fig. 12.5 *Broad Crested Weir*

The discharge over a broad crested weir is given by

$$Q = 1.705\ C_d\ LH^{3/2} \tag{12.34}$$

where L = length of the weir and H = energy head $= H_1 + \dfrac{V_0^2}{2g}$

The coefficient C_d depths on the geometry of the weir. For broad crested weirs with square entrance, C_d is about 0.85. For rounded entrance weirs C_d is about 0.98. It is usual to neglect the velocity of approach V_0.

[***Note:*** The topic of open-channel flows is indeed very vast and covers a very wide range of applications. The reader is reminded that the brief summary given above is but an introduction. There are books, treating in extensive detail, on open channel flow and the reader may refer to the following reference for further details.]

REFERENCE

12.1 Subramanya. K, *Flow in Open Channels*, Tata McGraw-Hill Pub. Co., New Delhi 2nd Ed., 11th Reprint, 2004.

WORKED EXAMPLES

A. Uniform Flow

12.1 A rectangular channel 2.5 m wide carries water at a depth of 1.2 m. The bed slope of the channel is 0.0036. Calculate the average shear stress on the boundary.

Solution:

Area $A = By = 2.5 \times 1.2 = 3.00\ \text{m}^2$

Wetted perimeter $P = B + 2y = 2.5 + (2 \times 1.2) = 4.9$ m

Hydraulic radius $R = A/P = \dfrac{3.00}{4.90} = 0.612$ m

Average boundary shear stress

$$\tau_0 = \gamma R S_0$$
$$= (998 \times 9.81) \times 0.612 \times 0.0036$$
$$= \mathbf{21.58\ Pa}$$

12.2 A trapezoidal channel has a bed width of 2.0 m and side slopes of 1.5 horizontal : 1 vertical. The channel has a longitudinal slope of 1/4000. If the Manning's coefficient of the channel boundary is 0.018, calculate the mean velocity and discharge in the channel for a depth of 1.4 m.

Solution: Referring to Fig. 12.6

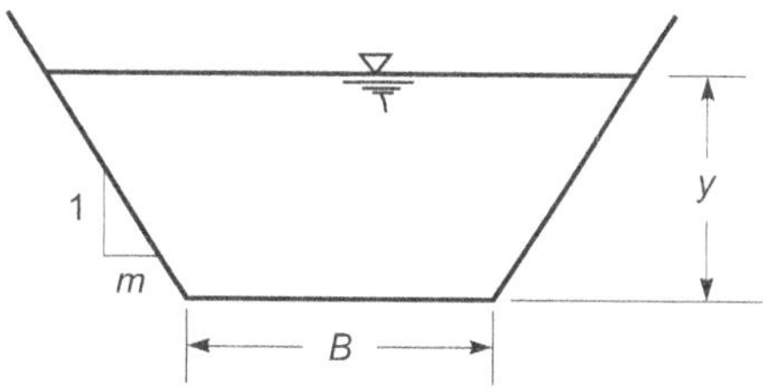

Fig. 12.6 *Trapezoidal channel*

Area $A = (B + my)y$

$$= (2.0 + 1.5 \times 1.4) \times 1.4 = 5.74\ \text{m}^2$$

Wetted perimeter $P = (B + 2\sqrt{m^2+1}\cdot y)$

$$= (2.0 + 2\sqrt{(1.5)^2+1} \times 1.4)$$
$$= 7.048\ \text{m}$$

Hydraulic radius $R = A/P = \dfrac{5.74}{7.048} = 0.814$ m

Mean velocity, by Manning's formula

$$V = \frac{1}{n} R^{2/3} S_0$$
$$= \frac{1}{0.018} \times (0.814)^{2/3} \times \left(\frac{1}{4000}\right)^{1/2}$$
$$= \mathbf{0.766\ m/s}$$

Discharge $Q = AV$

$$= 5.74 \times 0.766 = \mathbf{4.40\ m^3/s}$$

12.3 A trapezoidal channel has a bottom width of 10.0 m and a side slope of 1.5 horizontal: 1 vertical. The Manning's n can be taken as 0.015. What bottom slope is necessary to pass 100 m^3/s of discharge in this channel at a depth of 3.0 m?

Solution:

$$A = (10 + 1.5 \times 3.0) \times 3.0 = 43.5 \text{ m}^2$$

$$P = (10 + 2 \times 3.0 \times \sqrt{(1.5)^2 + 1}) = 20.817 \text{ m}$$

$$R = \frac{43.5}{20.817} = 2.09 \text{ m}$$

By Manning's formula

$$Q = \frac{1}{n} AR^{2/3} S_0^{1/2}$$

$$100 = \frac{1}{0.015} \times 43.5 \times (2.09)^{2/3} \times S_0^{1/2}$$

$$= 4740 \, S_0^{1/2}$$

$$S_0 = \mathbf{4.451 \times 10^{-4}}$$

12.4 For a flow in a rectangular channel of width 5 m and depth of flow of 2.3 m the Darcy–Weisbach friction factor f is estimated to be 0.02. Estimate the values of Chezy's C and Manning's n.

Solution: The coefficients C, n and the friction factor f are related as

$$C = \sqrt{\frac{8g}{f}} = \frac{1}{n} R^{1/6}$$

Area $A = 5.0 \times 2.3 = 11.5 \text{ m}^2$

Wetted perimeter $P = 5.0 + (2 \times 2.3) = 9.6 \text{ m}$

Hydraulic radius $R = A/P = \frac{11.5}{9.6} = 1.198 \text{ m}$

Hence $C = \sqrt{\frac{8 \times 9.81}{0.02}} = \mathbf{62.6}$

Also $\frac{1}{n} \cdot (1.198)^{1/6} = C = 62.6$

or $n = \frac{1.0306}{62.6} = \mathbf{0.0165}$

12.5 A trapezoidal channel has a bed width of 3.0 m and side slopes of 1 : 1. The bottom slope of the channel is 0.0036. If a discharge of 15 m^3/s passes in this channel at a depth of 1.25 m, estimate the value of Chezy coefficient C.

Solution:

Area $A = [3.0 + (1 \times 1.25)] \times 1.25$

$= 5.3125 \text{ m}^2$

Velocity $V = \frac{15.0}{5.3125} = 2.824 \text{ m/s}$

Wetted perimeter $P = B + 2y\sqrt{m^2 + 1}$

$= 3.0 + (2 \times 1.25 \times \sqrt{1+1})$

$= 6.5355 \text{ m}$

Hydraulic radius $R = A/P = \frac{5.3125}{6.5355} = 0.8129 \text{ m}$

By Chezy formula, $V = C\sqrt{RS_0}$

$$2.824 = C\sqrt{(0.8129 \times 0.0036)}$$

or $\mathbf{C = 52.2}$

12.6 What size of a circular drainage pipe is needed to carry 1.10 m^3/s of discharge when flowing half full? The pipe is laid at a slope of 0.0004 and the Manning's n for the pipe can be taken as 0.018.

Solution: Let D = diameter of the pipe

y = depth of flow = $D/2$

Area $A = \frac{1}{2} \times \frac{\pi D^2}{4} = \frac{\pi}{8} D^2$

Wetted perimeter $P = \frac{\pi}{2} D$

Hydraulic radius $R = A/P = \left(\frac{\pi}{8} D^2\right) \Big/ \left(\frac{\pi}{2} D\right) = D/4$

By Manning's formula, discharge

$$Q = \frac{1}{n} AR^{2/3} S_0^{1/2}$$

$$1.10 = \frac{1}{0.018} \times \left(\frac{\pi}{8} D^2\right) \times \left(\frac{D}{4}\right)^{2/3} (0.0004)^{1/2}$$

$$D^{8/3} = 6.353$$

$$\mathbf{D = 2.00 \text{ m}}$$

12.7 A rectangular channel 3.5 m wide is laid on a slope of 0.0005. Calculate the normal depth of flow for a discharge of 5.0 m^3/s in this channel. The Manning's coefficient n can be taken as 0.02.

Solution: Let y_0 = normal depth

Then area $A = 3.5\, y_0$

Wetted perimeter $P = 3.5 + 2y_0$

Hydraulic radius $R = \frac{3.5\, y_0}{(3.5 + 2\, y_0)}$

By Manning's formula

$$Q = \frac{1}{n} AR^{2/3} S_0^{2/3}$$

$$5.0 = \frac{1}{0.02} \times \frac{(3.5\, y_0)\,(3.5\, y_0)^{2/3}}{(3.5 + 2\, y_0)^{2/3}} \times (0.0005)^{1/2}$$

$$\frac{9.02\, y_0^{5/3}}{(3.5 + 2\, y_0)^{2/3}} = 5.0$$

or $(3.5 + 2y_0)^{2/3} = 1.804\ y_0^{5/3}$

By *trial and error,* $\mathbf{y_0 = 1.48}$ **m**

[*Note:* The determination of normal depth of flow involves trial and error solution in most of the cases).

12.8 A circular drainage pipe 0.80 m in diameter conveys a discharge at a depth of 0.30 m. If the pipe is laid on a slope of 1 in 900, estimate the discharge. Manning's $n = 0.015$.

Solution: Referring to Fig. 12.7

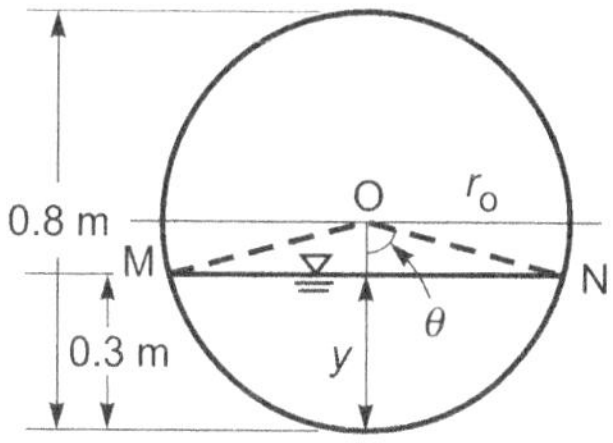

Fig. 12.7

$$D = 0.80 \text{ m}$$
$$y = 0.30 \text{ m}$$

Area of flow section

A = Area of sector OMN—Area of triangle OMN

$$= \frac{1}{2} r_0^2 \cdot 2\theta - \frac{1}{2} (2r_0 \sin \theta)\, r_0 \cos \theta$$

$$= \frac{D^2}{8} (2\theta - \sin 2\theta)$$

Also $\cos \theta = \left(\frac{D}{2} - y\right) \Big/ (D/2) = 1 - \frac{2y}{D}$

Hence $2\theta = 2 \cos^{-1} \left(1 - \frac{2y}{D}\right)$

$$= 2 \cos^{-1} \left(1 - \frac{2 \times 0.3}{0.8}\right) = 2.636 \text{ rad}$$

$$\sin 2\theta = 0.4841$$

Area $A = \frac{(0.8)^2}{8} (2.636 - 0.4841)$

$$= 0.1722 \text{ m}^2$$

Wetted perimeter $P = \frac{1}{2} D (2\theta)$

$$= \frac{0.8}{2} \times 2.636 = 1.055 \text{ m}$$

Hydraulic radius $R = A/P = \frac{0.1722}{1.055} = 0.1633$ m

By Manning's formula

$$Q = \frac{1}{n} AR^{2/3} S_0^{1/2}$$

$$= \frac{1}{0.015} \times (0.1722)\,(0.1633)^{2/3}\,(1/900)^{1/2}$$

$= 0.1143\ \text{m}^3/\text{s} =$ **114.3 L/s**

12.9 A triangular channel has a side slope of 1.5 horizontal: 1 vertical and is laid on a longitudinal slope of 1 in 1650. Assuming Manning's $n = 0.013$, estimate the normal depth required to pass a discharge of 0.30 m^3/s.

Solution: Referring to Fig. 12.8

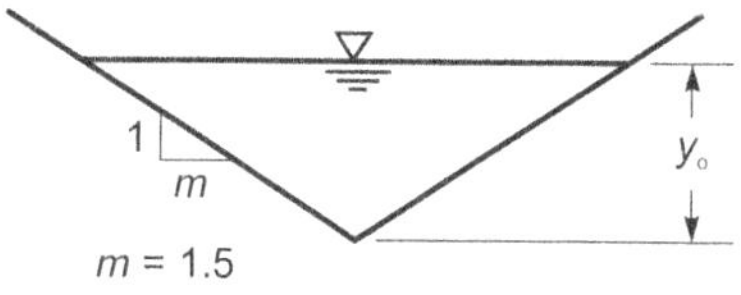

Fig. 12.8 *Triangular channel*

Let y_0 be the normal depth.

Here $m = 1.5$

$$A = my_0^2 = 1.5\ y_0^2$$

$$P = 2y_0\sqrt{m^2 + 1} = 2y_0\sqrt{(1.5)^2 + 1}$$

$$= 3.6056\ y_0$$

$$R = A/P = \frac{1.5\, y_0^2}{3.6056\, y_0} = 0.416\ y_0$$

By Manning's formula,

$$Q = \frac{1}{n} AR^{2/3} S_0^{1/2}$$

$$0.30 = \frac{1}{0.013} \times (1.5\ y_0^2)\,(0.416\ y_0)^{2/3}\,(1/1650)^{1/2}$$

$$0.30 = 1.583\ y_0^{8/3}$$

$\mathbf{y_0 = 0.54}$ **m**

12.10 A trapezoidal channel is to be designed to convey 50 m^3/s of water at a velocity of 2.0 m/s. The bed width to depth ratio is to be 8.0 and the side slopes are to be 1 horizontal : 1 vertical. It will be lined with a material whose $n = 0.02$. Calculate the bed width, depth of flow and slope of the channel.

Solution:

$$m = 1.0,\ Q = 50.0 \text{ m}^3/\text{s}$$
$$V = 2.0 \text{ m/s}$$

Area $A = \frac{Q}{V} = \frac{50.0}{2.0} = 25.0 \text{ m}^2$

But $A = (B + my)y$

$$= (8y + y)y = 9y^2 = 25.0$$

$y = \left(\frac{25}{9}\right)^{1/2} =$ **1.667 m**

$B = 8y =$ **13.33 m**

Wetted perimeter

$$P = B + 2y\sqrt{m^2+1}$$
$$= 13.33 + (2 \times 1.667 \times \sqrt{2})$$
$$= 18.047$$

$$R = A/P = \frac{25.0}{18.047} = 1.385 \text{ m}$$

By Manning's formula

$$V = \frac{1}{n} R^{2/3} S_0^{1/2}$$

$$2.0 = \frac{1}{0.02} \times (1.385)^{2/3} \times S_0^{1/2}$$

$$\mathbf{S_0 = 0.001036}$$

12.11 A canal is formed with side slopes of 2 horizontal: 1 vertical and a bottom width of 3.0 m. The longitudinal slope is 1 in 5000. Using Manning's formula and assuming Manning's n = 0.025, calculate the depth of water for a discharge of 3.1 m³/s in uniform flow.

Solution:

Let y_0 = normal depth

Area $A = (B + my_0)y_0$

$= (3.0 + 2y_0)y_0$

Wetted perimeter

$$P = B + 2y_0\sqrt{m^2+1}$$
$$= 3.0 + 2\sqrt{5}\, y_0$$

$$R = A/P = \frac{(3.0+2y_0)y_0}{3.0+4.472\,y_0}$$

By Manning's formula

$$Q = \frac{1}{n} AR^{2/3} S_0^{1/2}$$

$$3.10 = \frac{1}{0.025} \times (3.0 + 2y_0)\, y_0 \times \frac{[(3.0+2y_0)y_0]^{2/3}}{(3.0+4.472\,y_0)^{2/3}} \times \left(\frac{1}{5000}\right)^{1/2}$$

Simplifying

$$12.828\,(3.0 + 4.472\, y_0) = (3.0 + 2y_0)^{5/2} \cdot y_0^{5/2}$$

Solving by trial and error

$$\mathbf{y_0 = 1.20 \text{ m}}$$

B. Maximum Velocity and Discharge

12.12 Obtain an expression for the depth of flow in a circular channel which gives maximum velocity for a given longitudinal slope. The resistance to flow can be expressed by Manning's equation.

Solution: For a circular channel of diameter D

Area $$A = \frac{D^2}{8}(2\theta - \sin 2\theta)$$

Wetted perimeter $P = D\theta$

Velocity by Manning's formula is

$$V = \frac{1}{n} R^{2/3} S_0^{1/2}$$

For maximum velocity $\frac{d}{d\theta}(R) = 0$

i.e. $$\frac{d(A/P)}{d\theta} = P\frac{dA}{d\theta} - A\frac{dP}{d\theta} = 0$$

$$D\theta\left(\frac{D^2}{8}\cdot 2 - 2\frac{D^2}{8}\cos 2\theta\right) - \frac{D^2}{8}(2\theta - \sin 2\theta)\, D = 0$$

$$\theta - \theta\cos 2\theta - \theta + \frac{1}{2}\sin 2\theta = 0$$

On simplifying

$$\tan 2\theta = 2\theta$$

Solving by trial and error

$$2\theta = 4.4934 \text{ rad.} = 257.4528°$$
$$= 257°\ 27'\ 10''$$

$$\frac{y}{D} = \frac{1}{2}(1 - \cos\theta) = \mathbf{0.8128}$$

12.13 A triangular duct resting on a side carries water with a free surface (Fig. 12.9). Obtain the condition for maximum discharge in this channel when (a) m = 0.5 and (b) m = 1.0.

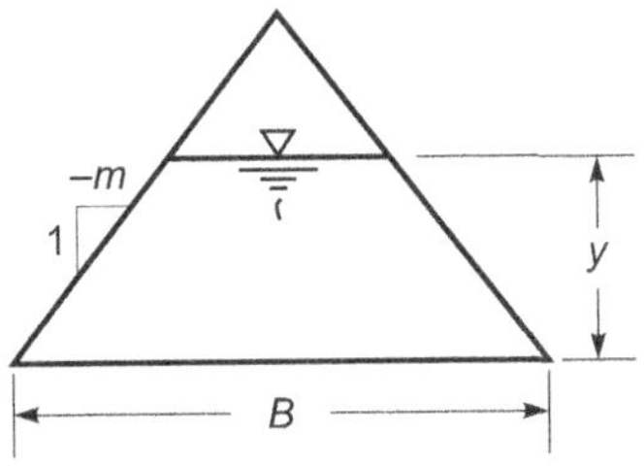

Fig. 12.9

Solution: For maximum discharge $\frac{dQ}{dy} = 0$

But $$Q = \frac{1}{n} AR^{2/3} S_0^{1/2}$$

Since n and S_0 are constants

$$\frac{dQ}{dy} = \frac{d}{dy}(AR^{2/3})$$
$$= \frac{d}{dy}(A^5/P^2) = 0$$

$$5P\frac{dA}{dy} - 2A\frac{dP}{dy} = 0$$

$$A = (B - my)\, y$$

$$P = B + 2y\sqrt{m^2+1}$$

Hence for maximum discharge,

$$5\,[(B + 2y\sqrt{m^2+1}\,]\,(B - 2my) - 2(B - my)\; y(2\sqrt{m^2+1}\,) = 0$$

$$\mathbf{5B^2 + (6By\sqrt{m^2+1} - 10\,B\,my] - 16y^2\,m\sqrt{m^2+1} = 0}$$

(a) When $m = 0.5$

$$5B^2 + 1.708\,By - 8.9443\,y^2 = 0$$

$$\mathbf{y = 0.849\,B}$$

(b) When $m = 1.0$

$$5B^2 - 1.5147\,By - 22.627\,y^2 = 0$$

$$\mathbf{y = 0.4378\,B}$$

12.14 Obtain an expression for the depth of flow in a circular channel which gives maximum discharge for a given longitudinal slope and
(a) constant value of Manning's n.
(b) constant value of Chezy C.

Solution: For a circular channel for diameter D,

Area $\quad A = \dfrac{D^2}{8}\,(2\theta - \sin 2\theta)$

Wetted perimeter $\quad P = D\theta$

(a) When Manning's formula is used:

$$Q = \frac{1}{n}\,AR^{2/3}\,S_0^{1/2} = \frac{1}{n}\,A^{5/3}\,P^{-2/3}\,S_0^{1/2}$$

For maximum discharge $\dfrac{d}{d\theta}\,(A^5/P^2) = 0$

i.e. $\quad 5P\,\dfrac{dA}{d\theta} - 2A\,\dfrac{dP}{d\theta} = 0$

$$5D\theta \cdot \frac{D^2}{8}\,(2 - 2\cos 2\theta) - 2\,\frac{D^2}{8}\,(2\theta - \sin 2\theta)\,D = 0$$

Simplifying $\quad 3\theta - 5\theta\cos 2\theta + \sin 2\theta = 0$

By trial and error the solution of the equation is obtained as

$$\boldsymbol{\theta} = 2.639 \text{ rad.} = \mathbf{151°\ 11'}$$

$$\boldsymbol{y/D} = \frac{1}{2}\,(1 - \cos\,\theta) = \mathbf{0.938}$$

(b) When Chezy formula with constant C is used:

$$Q = AC\sqrt{R}\,\sqrt{S_0} = C\sqrt{S_0}\;A^{3/2}\,P^{-1/2}$$

For maximum discharge $\dfrac{d}{d\theta}\,(A^3/P) = 0$

$$-\frac{A}{P}\frac{dP}{d\theta} + 3\frac{dA}{d\theta} = 0$$

$$-\frac{D^2\,(2\theta - \sin 2\theta)}{8\,D\theta}\cdot D + 3\,\frac{D^2}{8}\,(2 - 2\cos 2\theta) = 0$$

$$4\theta - 6\theta\cos 2\theta + \sin 2\theta = 0$$

Solving by trial and error

$$\theta = 2.689 \text{ rad.} = \mathbf{154°}$$

$$\frac{\boldsymbol{y}}{\boldsymbol{D}} = \frac{1}{2}\,(1 - \cos\,\theta) = \mathbf{0.95}$$

C. Hydraulically Efficient Channel

12.15 Show that for a hydraulically efficient triangular section the hydraulic radius

$$R = y/2\sqrt{2}.$$

Solution: Let the side slope of the channel be m horizontal : 1 vertical

Area $\quad A = my^2$ i.e. $m = A/y^2$

Perimeter $\quad P = 2y\sqrt{m^2+1}$

$$= 2y\sqrt{\frac{A^2}{y^4} + 1}$$

$$P = 2\left(\frac{A^2}{y^2} + y^2\right)^{1/2}$$

For an efficient section, $\dfrac{dP}{dy} = 0$

$$-2\,\frac{A_e^2}{y_e^3} + 2y_e = 0$$

$$y_e = \sqrt{A_e} \text{ or } m = 1.0$$

Hence $\quad P_e = 2\sqrt{2}\,y_e$ and $A_e = y_e^2$

Thus $\quad \boldsymbol{A_e/P_e = R_e = \dfrac{y_e}{2\sqrt{2}}}$

12.16 A lined channel (n = 0.014) is of trapezoidal section with one side vertical and the other on a side slope 1.5 horizontal : 1 vertical. If the channel is to deliver 9.0 m^3/s when laid on a slope of 0.0002, calculate the dimension of the efficient section which requires minimum lining. Also calculate the corresponding mean velocity.

Solution: Referring to Fig. 12.10

Area $\quad A = \left(B + \dfrac{my}{2}\right)y$

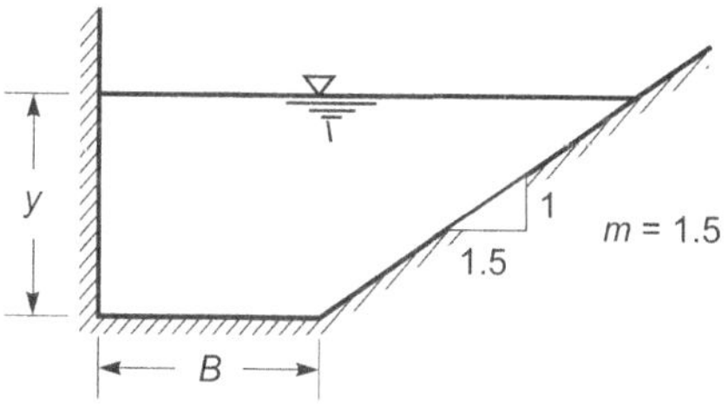

Fig. 12.10

$$B = \frac{A}{y} - \frac{my}{2}$$

Wetted perimeter

$$P = B + y + \sqrt{m^2+1}\,y$$

$$= \frac{A}{y} + \left[1 - \frac{m}{2} + \sqrt{m^2+1}\right] y$$

For minimum perimeter $\frac{dP}{dy} = 0$

$$-\frac{A}{y_e^2} + 1 - \frac{m}{2} + \sqrt{m^2+1} = 0$$

$$A = \left[1 - \frac{m}{2} + \sqrt{m^2+1}\right] y_e^2$$

Putting $m = 1.5$,

$$A = (1 - 0.75 + \sqrt{3.25})\, y_e^2$$

$$= 2.0528\, y_e^2$$

$$B_e = \frac{A}{y_e} - \frac{m}{2} y_e$$

$$= \left(2.0528 - \frac{1.5}{2}\right) y_e = 1.3028\, y_e$$

$$P_e = 1.3028\, y_e + y_e + \sqrt{3.25}\, y_e$$

$$= 4.1056\, y_e$$

$$R_e = A_e/p_e = \frac{2.0528}{4.1056} y_e = 0.5\, y_e$$

By Manning's formula

$$Q = \frac{1}{n} A R^{2/3} S_0^{1/2}$$

$$9.0 = \frac{1}{0.014} (2.0528\, y_e^2)(y_e/2)^{2/3} (0.0002)^{1/2}$$

$$= 1.3063\, y_e^{8/3}$$

$$y_e^{8/3} = 6.8896$$

$$\mathbf{y_e = 2.062\ m}$$

$$\mathbf{B_e = 1.3028\, y_e = 2.687\ m}$$

$$\mathbf{A_e = 2.0528\, y_e^2 = 8.7282\ m^2}$$

Mean velocity

$$V = \frac{Q}{A_e} = \frac{9.0}{8.7282} = \mathbf{1.031\ m/s}$$

12.17 A trapezoidal channel with side slopes of 2 horizontal : 1 vertical has to be designed to carry 15 m³/s at a slope of 1/5000. Determine the dimensions of the efficient section. Assume Manning's $n = 0.014$.

Solution: For efficient trapezoidal section

$$R = y_e/2$$

and $B + 2my_e = 2\sqrt{m^2+1} \cdot y_e$

Here $m = 2.0$ and hence

$$B_e + 4y_e = 2\sqrt{5}\, y_e$$

$$B_e = 0.4721\, y_e$$

Area $A = (B_e + my_e)\, y_e$

$$= (0.4721 + 2.0)\, y_e^2 = 2.4721\, y_e^2$$

$$R = y_e/2$$

Discharge by Manning's formula

$$Q = \frac{1}{n} A R^{2/3} S_0^{1/2}$$

$$15.0 = \frac{1}{0.014} \times (2.4721\, y_e^2)\,(y_e/2)^{2/3}\,(1/5000)^{1/2}$$

$$= 1.573\, y_e^{8/3}$$

$$y_e^{8/3} = 9.535 \text{ or } \mathbf{y_e = 2.329\ m}$$

Bottom width

$$\mathbf{B_e} = 2.329 \times 0.4721 = \mathbf{1.10\ m}$$

12.18 It is required to convey 10 m³/s of water at a mean velocity of 1.25 m/s. Calculate the dimensions of the most efficient section of the channel whose shape is (a) rectangular (b) triangular (c) trapezoidal and (d) circular. Which of these has (i) the least perimeter and (ii) the largest perimeter?

Solution:

Discharge $Q = 10.0$ m³/s

Velocity $V = 1.25$ m/s

Area $A = 10.0/1.25 = 8.0$ m²

(a) *For the most efficient rectangular channel:*

$$A = 2y_{em}^2$$

$$8.0 = 2y_{em}^2 \quad \therefore \mathbf{y_{em} = 2.0\ m}$$

Width $\mathbf{B_{em} = \frac{8.0}{2.0} = 4.0\ m}$

Perimeter $\mathbf{P_{em} = 4.0 + (2 \times 2.0) = 8.0\ m}$

(b) *For the most efficient triangular section:*

$$2\theta = 90°,$$

i.e. side slope = 1.0 Horizontal : 1 vertical

or $m = 1.0$

$$A = y_{em}^2$$

$$8.0 = y_{em}^2 \quad \text{i.e.} \quad \mathbf{y_{em} = \sqrt{8} = 2.828\ m}$$

Perimeter $P_{em} = 2\sqrt{2}\, y_{em}$

$$= 2\sqrt{2} \times 2.828 = \mathbf{8.0\ m}$$

(c) *For the most efficient trapezoidal section:* Side slopes are inclined at 60° to the horizontal, or $1/\sqrt{3}$ horizontal : 1 vertical, i.e. $m = 1/\sqrt{3}$.

$$A = \sqrt{3}\, y_{em}^2$$

$$8.0 = \sqrt{3}\, y_{em}^2$$

$$\mathbf{y_{em} = 2.149\ m}$$

Bottom width $B_{em} = \frac{2}{\sqrt{3}} y_{em}$

$= \frac{2}{\sqrt{3}} \times 2.149 = \mathbf{2.482\ m}$

Perimeter $P_{em} = 2\sqrt{3}\ y_{em}$

$= 2\sqrt{3} \times 2.149$

$= \mathbf{7.445\ m}$

(d) *Circular section:* The most efficient circular section is a semicircle. Hence

$$A = \frac{\pi}{2} y_{em}^2$$

$$8 = \frac{\pi}{2} y_{em}^2$$

$$\mathbf{y_{em} = 2.257\ m}$$

Perimeter $P_{em} = \pi\ y_{em} = \pi \times 2.257$

$= \mathbf{7.090\ m}$

The most efficient circular section has the least perimeter (P_{em} = 7.09 m) and the most efficient triangular and rectangular sections have the maximum perimeter of P_{em} = 8.00 m.

12.19 A lined rectangular channel with Manning's n = 0.015 is 5 m wide and has a flow depth of 2 m with bed slope of 1 in 1600. Retaining the rectangular shape of channel section and the same total area of lining, to what maximum extent can the discharge be increased without changing the slope?

Solution:

Existing: Area $A_1 = 5 \times 2 = 10\ m^2$

Wetted perimeter $P_1 = 5 + 2 + 2 = 9$ m

$$R_1 = A_1/P_1 = \frac{10}{9} = 1.111 \text{ m}$$

Discharge $Q_1 = \frac{1}{n} A_1 R_1^{2/3} S_0^{1/2}$

$$= \frac{1}{0.015} \times 10.0 \times (1.111)^{2/3} \times \left(\frac{1}{1600}\right)^{1/2}$$

$= 17.88\ m^3/s$

New channel: For maximum discharge the channel must be of efficient proportions. For rectangular shape

$$y = \frac{B}{2}$$

Assuming the free board h is same in both cases, the area of lining per metre length of channel is

$$P_1 + 2h = 2h + 2y + B$$

$$9.0 = 2 \times \frac{B}{2} + B = 2B$$

$B = 4.5$ m and $y = 2.25$ m

New area $A = 4.5 \times 2.25 = 10.125\ m^2$

$P = 9.0$ m

$$R = \frac{10.125}{9.0} = 1.125 \text{ m}$$

New discharge

$$Q = \frac{1}{0.015} \times 10.125 \times (1.125)^{2/3} \times \frac{1}{(1600)^{1/2}}$$

$= 18.25\ m^3/s$

Increase in discharge

$= 18.25 - 17.88 = \mathbf{0.37\ m^3/s}$

D. Standard Lined Canal Sections

12.20 A standard lined triangular section is shown in Fig. 12.11. The section consists of a triangular section of the side slope m horizontal : 1 vertical with its bottom being rounded off by a circular curve of radius equal to the full supply depth. For such a channel calculate the full supply depth corresponding to a full supply discharge of 25 m^3/s. The side slopes are 2 horizontal : 1 vertical, the longitudinal slope is 1 in 2000 and Manning's n = 0.017.

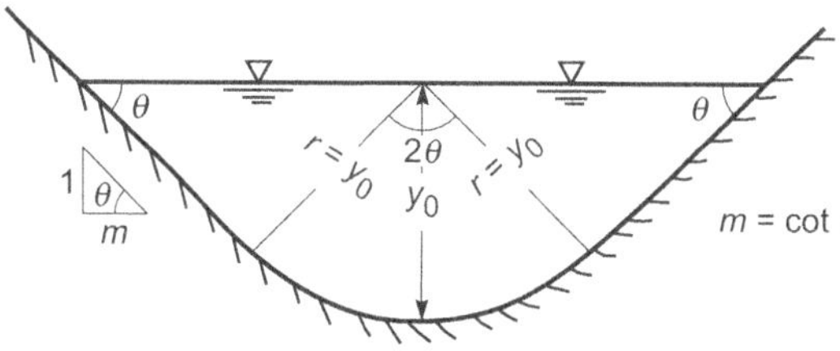

Fig. 12.11 *Standard Lined Triangular Canal Section*

Solution:

Let y_0 = normal depth = r_0

θ = inclination of the sides to the horizontal

$\cot\theta = m$

$$\theta = \tan^{-1}\frac{1}{m}$$

Area $A = 2\left(\frac{1}{2} y_0^2 \cot\theta\right) + \frac{1}{2} y_0^2\, 2\theta$

$= y_0^2(\theta + \cot\theta) = \varepsilon\, y_0^2$

where $\varepsilon = \theta + \cot\theta = \left(m + \tan^{-1}\frac{1}{m}\right)$

Wetted perimeter

$P = 2y_0 \cot\theta + 2y_0\,\theta$

$= 2y_0(\theta + \cot\theta) = 2\,\varepsilon\, y_0$

Hydraulic radius

$$R = A/P = \frac{\varepsilon\, y_0^2}{2\varepsilon\, y_0} = y_0/2$$

In the present case,

$$m = 2.0$$

$$\tan^{-1}\frac{1}{m} = \tan^{-1}\frac{1}{2} = 0.4636$$

$$\varepsilon = 2.4636$$

$$A = 2.4636\, y_0^2 \quad \text{and} \quad R = 0.5\, y_0$$

By Manning's equation

$$Q = \frac{1}{n} AR^{2/3} S_0^{1/2}$$

$$25 = \frac{1}{0.017} \times (2.4636\, y_0^2)\,(0.5\, y_0)^{2/3} \left(\frac{1}{2000}\right)^{1/2}$$

$$y_0^{8/3} = 12.2465$$

$$\mathbf{y_0 = 2.559\ m}$$

12.21 A standard lined trapezoidal section (Fig. 12.12) has a bottom width of 35 m and side slopes of 1.5 horizontal : 1 vertical. The longitudinal slope is 1 in 5000 and the Manning's n can be assumed to be 0.016. Estimate the discharge if the full supply depth y_0 is 3.5 m.

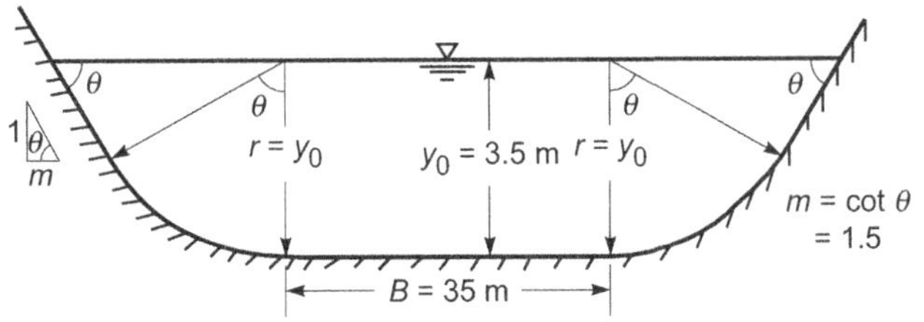

Fig. 12.12 *Standard Lined Trapezoidal Section*

Solution: Let the side slopes be m horizontal : 1 vertical. If θ = Inclination of the side to the horizontal

then $\cot\theta = m$ and $\theta = \tan^{-1}\dfrac{1}{m}$

If $\varepsilon = \cot\theta + \theta = m + \tan^{-1}\dfrac{1}{m}$

Area $A = By_0 + y_0^2(\cot\theta + \theta) = (B + \varepsilon\, y_0)\, y_0$

Perimeter $P = B + 2y_0(\cot\theta + \theta) = B + 2y_0\varepsilon$

In the present case, $m = 1.5$

$$\varepsilon = 1.5 + \tan^{-1}\frac{1}{1.5} = 2.088$$

$$A = [35.0 + (2.088)\times 3.5]\times 3.5 = 148.08\ \text{m}^2$$

$$P = 35.0 + 2 \times 2.088 \times 3.5 = 49.616\ \text{m}$$

$$R = A/P = 148.08/49.616 = 2.9845\ \text{m}$$

By Manning's formula,

$$V = \frac{1}{n} R^{2/3} S_0^{1/2}$$

$$= \frac{1}{0.016} \times (2.9845)^{2/3} \times \left(\frac{1}{5000}\right)^{1/2}$$

$$= 1.832\ \text{m/s}$$

Discharge $Q = AV = 148.08 \times 1.832$

$$= \mathbf{271.3\ m^3/s}$$

12.22 A standard lined trapezoidal channel section is shown in Fig. 12.13. For such a channel determine the bed width B and depth of flow y_0 to carry 90 m^3/s of discharge on a slope of 1 in 3000. The velocity of flow is to be 2.0 m/s and the side slopes are 1.25 horizontal : 1 vertical. Manning n can be taken as 0.015. Also $B > y_0$.

Solution: For the standard lined trapezoidal section,

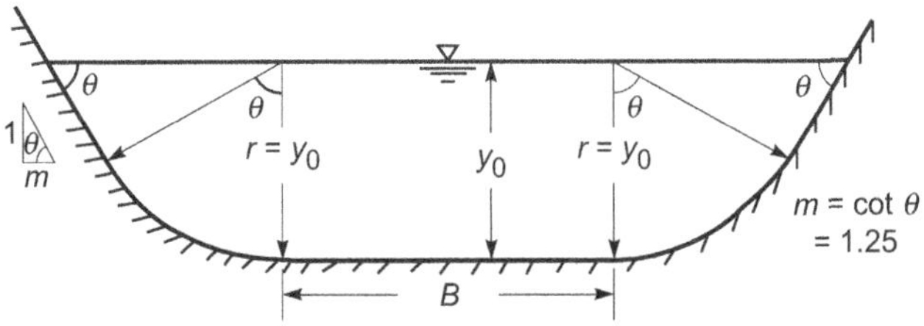

Fig. 12.13 *Standard Lined Trapezoidal Section*

Let θ = Inclination of sides to the horizontal, and side slope = m horizontal: 1 vertical.

$$\cot\theta = m \quad \text{or} \quad \theta = \tan^{-1}\frac{1}{m}$$

Let $\cot\theta + \theta = m + \tan^{-1}\dfrac{1}{m} = \varepsilon$

Area $A = By_0 + y_0^2(\cot\theta + \theta) = (B + \varepsilon y_0)\, y_0$

Perimeter $P = B + 2y_0(\cot\theta + \theta) = B + 2\varepsilon y_0$

In the present problem,

$$m = 1.25$$

$$\varepsilon = 1.25 + \tan^{-1}\frac{1}{1.25} = 1.9247$$

Area $A = Q/V = \dfrac{90.0}{2.0} = 45.0$

Also $A = By_0 + 1.925\, y_0^2 = 45$ (1)

By Manning's formula

$$V = \frac{1}{n} R^{2/3} S_0^{1/2}$$

$$2.0 = \frac{1}{0.015} \times R^{2/3} \times \frac{1}{(3000)^{1/2}}$$

$$R = 2.1063\ \text{m}$$

Also $R = A/P = \dfrac{45}{B + (2\times 1.9247)\, y_0}$

or $B = 21.364 - 3.8494\, y_0$ (2)

Substituting for B in Eq. (1)

$$21.364\, y_0 - 3.8494\, y_0^2 + 1.9247\, y_0^2 = 45$$

$$y_0^2 - 11.1\, y_0 + 23.38 = 0$$

Solving for y and noting that $B > y_0$

$$y_0 = \frac{1}{2}\,[11.1 - \sqrt{(11.1)^2 - 4 \times 23.38}\,] = 2.826 \text{ m}$$

$$B = 21.364 - (3.8494 \times 2.826) = \mathbf{10.488\ m}$$

E. Specific Energy and Critical Depth

12.23 A rectangular channel has a width of 1.8 m and carries a discharge of 1.8 m^3/s at a depth of 0.20 m. Calculate (a) the specific energy, (b) depth alternate to the existing depth and (c) Froude numbers at the alternate depths.

Solution:

Let $y_1 = 0.20$ m = Existing depth

Area $A_1 = By_1 = 1.8 \times 0.2 = 0.36\ m^2$

Velocity $V_1 = Q/A_1 = \dfrac{1.80}{0.36} = 5.0$ m/s

(a) Specific energy

$$E_1 = y_1 + \frac{V_1^2}{2g}$$

$$= 0.20 + \frac{(5.0)^2}{2 \times 9.81} = \mathbf{1.4742\ m}$$

(b) Let y_2 = depth alternate to y_1

Then $E_2 = E_1 = y_2 + \dfrac{V_2^2}{2g} = 1.4742$

$$y_2 + \frac{(1.8)^2}{(2 \times 9.81)(1.80)^2 \cdot y_2^2} = 1.4742$$

$$y_2 + \frac{0.05097}{y_2^2} = 1.4742$$

By trial and error, $\mathbf{y_2 = 1.45\ m}$

(c) Froude number for a rectangular channel is

$$F = V/\sqrt{g\,y}$$

For $y_1 = 0.2$ m, $F_1 = \dfrac{5.0}{\sqrt{9.81 \times 0.2}} = \mathbf{3.57}$

For $y_2 = 1.45$ m, $V_2 = \dfrac{Q}{(B \cdot y_2)} = \dfrac{1.80}{(1.80 \times 1.45)}$

$= 0.69$ m/s

$$F_2 = \frac{0.69}{\sqrt{9.81 \times 1.45}} = \mathbf{0.1829}$$

12.24 In a flow through a rectangular channel for a certain discharge the Froude numbers corresponding to the two alternate depths are F_1 and F_2. Show that

$$(F_2/F_1)^{2/3} = \frac{2 + F_2^2}{2 + F_1^2}$$

Solution: Let y_1 and y_2 be the alternative depths.

The specific energy $E = y_1 + \dfrac{V_1^2}{2g} = y_2 + \dfrac{V_2^2}{2g}$

$$y_1\left(1 + \frac{V_1^2}{2g\,y_1}\right) = y_2\left(1 + \frac{V_2^2}{2g\,y_2}\right)$$

Since $V/\sqrt{g\,y} = F$ = Froude number

$$\frac{y_1}{y_2} = \frac{(1 + F_2^2/2)}{(1 + F_1^2/2)} = \frac{(2 + F_2^2)}{(2 + F_1^2)}$$

Also $F_1^2 = \dfrac{Q^2}{B^2 g\, y_1^3}$ and $F_2^2 = \dfrac{Q^2}{B^2 g\, y_2^3}$

where Q = discharge in the channel and B = width of the channel, Hence,

$$\frac{y_1^3}{y_2^3} = \frac{F_2^2}{F_1^2} \quad \text{or} \quad \left(\frac{y_1}{y_2}\right) = \left(\frac{F_2}{F_1}\right)^{2/3}$$

Hence

$$\boldsymbol{\frac{y_1}{y_2} = \left(\frac{F_2}{F_1}\right)^{2/3} = \frac{(2 + F_2^2)}{(2 + F_1^2)}}$$

12.25 A rectangular channel 2.0 m wide carries a discharge of 6.0 m^3/s. Calculate the critical depth and specific energy at critical depth.

Solution:

At critical depth, $\dfrac{Q^2}{g} = \dfrac{A_c^3}{T_c}$

For a rectangular channel $A_c = B\,y_c$

Top width $T_c = B$

Hence $\dfrac{Q^2}{g} = \dfrac{B^3 y_c^3}{B}$

or $\dfrac{(Q^2/B^2)}{g} = \dfrac{q^2}{g} = y_c^3$

where q = discharge intensity = discharge per unit width

$$y_c = (q^2/g)^{1/3}$$

Here $q = \dfrac{6.0}{2.0} = 3.0\ m^3/s/m$

$$y_c = \left(\frac{(3.0)^2}{9.81}\right)^{1/3} = \mathbf{0.972\ m}$$

$$V_c = \frac{Q}{(B\,y_c)} = \frac{6.0}{(2.0 \times 0.972)} = 3.087 \text{ m/s}$$

E_c = Specific energy at critical depth = $y_c + \dfrac{V_c^2}{2g}$

$$= 0.972 + \frac{(3.087)^2}{2 \times 9.81} = \mathbf{1.458\ m}$$

12.26 If y_1 and y_2 are alternate depths in a rectangular channel show that

$$y_c^3 = \frac{2y_1^2 y_2^2}{(y_1 + y_2)}$$

and hence the specific energy $E = \frac{y_1^2 + y_1 y_2 + y_2^2}{(y_1 + y_2)}$.

Solution: Let q = discharge intensity. Then

$$q = V_1 y_1 = V_2 y_2$$

As y_1 and y_2 are alternative depths, the specific energy

$$E = y_1 + \frac{V_1^2}{2g} = y_2 + \frac{V_2^2}{2g}$$

$$= y_1 + \frac{q^2}{2g y_1^2} = y_2 + \frac{q^2}{2g y_2^2}$$

Hence $$(y_1 - y_2) = \frac{q^2}{2g}\left(\frac{1}{y_2^2} - \frac{1}{y_1^2}\right)$$

$$2(y_1 - y_2) = \frac{(y_1^2 - y_2^2)}{y_1^2 y_2^2} \cdot \frac{q^2}{g}$$

But in a rectangular channel $\frac{q^2}{g} = y_c^3$

Hence $$y_c^3 = \frac{2(y_1 - y_2)\, y_1^2 y_2^2}{(y_1^2 - y_2^2)}$$

$$\mathbf{y_c^3 = \frac{2y_1^2 y_2^2}{(y_1 + y_2)}}$$

Specific energy $$E = y_1 + \frac{1}{2}\cdot\left(\frac{q^2}{g}\right)\frac{1}{y_1^2}$$

$$= y_1 + \frac{1}{2y_1^2} \times \frac{2y_1^2 y_2^2}{(y_1 + y_2)}$$

$\therefore$ $$\mathbf{E = \frac{y_1^2 + y_1 y_2 + y_2^2}{(y_1 + y_2)}}$$

12.27 For a triangular channel having a vertex angle of 120°, calculate the critical depth for a discharge of 3.0 m^3/s.

Solution: Referring to Fig. 12.14

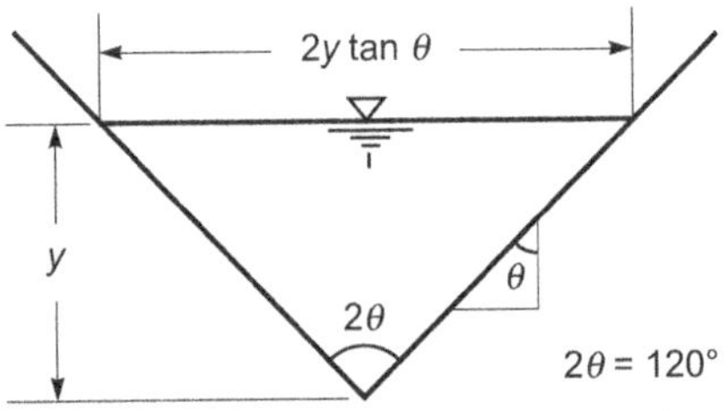

Fig. 12.14

Let y_c = Critical depth.

Top width $T_c = 2y_c \tan\theta$

Area $A_c = y_c^2 \tan\theta$

At critical depth $$\frac{Q^2}{g} = \frac{A_c^3}{T_c}$$

$$= \frac{y_c^6 \tan^3\theta}{2y_c \tan\theta} = \frac{1}{2} y_c^5 \tan^2\theta$$

$$y_c = \left(\frac{2Q^2}{g}\frac{1}{\tan^2\theta}\right)^{1/5}$$

$$= \left(\frac{2 \times (3.0)^2}{9.81 \times (\tan 60°)^2}\right)^{1/5}$$

$$= \mathbf{0.906\ m}$$

12.28 Calculate the critical depth corresponding to a discharge of 6.0 m^3/s in (a) rectangular channel of width 3.0 m, (b) triangular channel of side slope 1.5 horizontal : 1 vertical (c) trapezoidal channel of bottom width 2.0 m and side slope 1.0 horizontal : 1.0 vertical

Solution:

(a) Rectangular channel:

$$y_c = \left(\frac{q^2}{g}\right)^{1/3}$$

$$q = \frac{6.0}{3.0} = 2.0\ m^3/s/m$$

$$y_c = \left(\frac{(2.0)^2}{9.81}\right)^{1/3} = \mathbf{0.742\ m}$$

(b) In a triangular channel of side slope m horizontal : 1 vertical

$$A = my^2$$

$$T = 2my$$

At critical depth

$$\frac{Q^2}{g} = \frac{A_c^3}{T_c} = \frac{m^3 y_c^6}{2my_c}$$

$$y_c = \left(\frac{2Q^2}{gm^2}\right)^{1/5}$$

$$= \left[\frac{2 \times (6.0)^2}{9.81 \times (1.5)^2}\right]^{1/5} = \mathbf{1.267\ m}$$

(c) In a trapezoidal channel of side slope m horizontal : 1 vertical,

$$A = (B + my)y \quad \text{and} \quad T = (B + 2my)$$

At critical depth, $\frac{Q^2}{g} = \frac{A_c^3}{T_c}$

$$\frac{Q^2}{g} = \frac{(B+m y_c)^3 y_c^3}{(B+2m y_c)}$$

$$\frac{(6.0)^2}{9.81} = \frac{(2.0+y_c)^3 y_c^3}{(2.0+2y_c)}$$

$$(2.0 + y_c^3)\, y_c^3 = 7.34\,(1 + y_c)$$

By trial and error, $\mathbf{y_c = 1.504\ m}$

12.29 Calculate the discharge corresponding to a critical depth of 1.20 m in (a) a rectangular channel of width 3.0 m, (b) a triangular channel of side slope 2 horizontal : 1 vertical, and (c) a trapezoidal channel of side slope 1.5 horizontal : 1 vertical and bottom width 2.5 m.

Solution:

(a) Rectangular channel

$$y_c = (q^2/g)^{1/3} \quad \text{or} \quad q = (g y_c^3)^{1/2}$$

$$q = [9.81 \times (1.20)^3]^{1/2} = 4.117\ \text{m}^3/\text{s}$$

$$Q = Bq = 3.0 \times 4.117 = \mathbf{12.352\ m^3/s}$$

(b) Triangular channel

$$\frac{Q^2}{g} = \frac{m^2 y_c^5}{2}$$

$$Q = \left[\frac{g m^2 y_c^5}{2}\right]^{1/2} \quad \text{Here } m = 2 \text{ and } y_c = 1.2\ \text{m}$$

$$\text{Hence } Q = \left[\frac{9.81 \times (2.0)^2 \times (1.2)^5}{2}\right]^{1/2} = \mathbf{6.987\ m^3/s}$$

(c) Trapezoidal section

$$\frac{Q^2}{g} = \frac{(B+m y_c)^3 y_c^3}{(B+2m y_c)}; \qquad m = 1.5$$

$$\frac{Q^2}{9.81} = \frac{[2.5+(1.5\times 1.2)]^3 (1.2)^3}{[2.5+(2\times 1.5\times 1.2)]}$$

$$Q^2 = 9.81 \times 22.5226 = 220.947$$

$$\mathbf{Q = 14.864\ m^3/s}$$

12.30 In a trapezoidal channel of bottom width 3.5 m and side slopes 2.0 horizontal : 1 vertical, the critical depth is found to be 0.9 m. What is the minimum specific energy corresponding to this critical depth?

Solution: At critical depth,

$$\frac{Q^2}{g} = \frac{A_c^3}{T_c}$$

As $\quad V_c = Q/A_c, \ \frac{V_c^2}{2g} = \frac{Q^2}{2g \cdot A_c^2} = \frac{A_c^3}{2T_c A_c^2} = \frac{A_c}{2T_c}$

Minimum specific energy

$$E_c = y_c + \frac{V_c^2}{2g} = y_c + \frac{A_c}{2T_c}$$

Thus $\quad E_c = y_c + \frac{(B+m y_c)\, y_c}{2(B+2m y_c)}$

$$= 0.90 + \frac{[3.5+(2.0\times 0.9)]\times 0.9}{2\times[3.5+(2\times 2\times 0.9)]}$$

$$= 0.90 + 0.336 = \mathbf{1.236\ m}$$

12.31 On what slope should one construct a 3 m wide rectangular channel (n = 0.014) so that critical flow will occur at a normal depth of 1.2 m?

Solution: At critical flow

$$\frac{Q^2}{g} = \frac{A_c^3}{T_c}$$

If $\quad y_c = 1.2$ m = critical depth

$$A_c = 3 \times 1.2 = 3.6\ \text{m}^2$$

$$T_c = 3.0\ \text{m}$$

$$Q^2 = g \cdot \left(\frac{A_c^3}{T_c}\right) = \frac{9.81 \times (3.6)^3}{3.0} = 151.565$$

$$Q = 12.35\ \text{m}^3/\text{s}$$

If this discharge is to flow at a normal depth of 1.2 m, by Manning's formula

$$Q = \frac{1}{n} A R^{2/3} S_0^{1/2}$$

$$A = 3.6\ \text{m}^2,\ P = (3.0 + 2 \times 1.2) = 5.4\ \text{m}$$

$$R = \frac{3.6}{5.4} = 0.6667\ \text{m}$$

$$12.35 = \frac{1}{0.014} \times (3.6)\,(0.6667)^{2/3}\, S_0^{1/2}$$

$$= 196.236\, S_0^{1/2}$$

$$\mathbf{S_0 = 3.96 \times 10^{-3}}$$

[***Note*:** The slope of channel which carries a given discharge as a uniform flow at the critical depth is called the *critical slope*, S_c. In the above problem the required slope of the channel is the critical slope.]

12.32 A circular pipe 0.9 m diameter is flowing half full and the flow is in critical state (Fig. 12.15). Estimate the discharge. If n = 0.015 calculate the critical slope S_c.

Solution:

$$A = \frac{1}{2} \times \frac{\pi D^2}{4} = \frac{\pi}{8} \times (0.9)^2 = 0.3181\ \text{m}$$

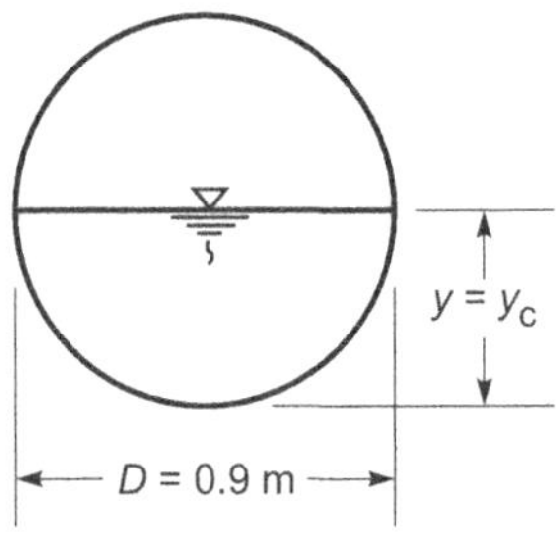

Fig. 12.15

Top width $T = 0.9$ m

$$\frac{A^3}{T} = \frac{(0.3181)^3}{0.9} = 0.03576$$

At critical flow

$$\frac{Q^2}{g} = \frac{A_c^3}{T_c} = 0.03576$$

$$Q = (9.81 \times 0.03576)^{1/2} = 0.592 \text{ m}^3/\text{s}$$

Wetted perimeter

$$P = \frac{\pi D}{2} = \frac{\pi \times 0.9}{2} = 1.4137 \text{ m}$$

Hydraulic radius

$$R = D/4 = \frac{0.9}{2} = 0.225 \text{ m}$$

By Manning's formula

$$Q = \frac{1}{n} AR^{2/3} S_0^{1/2}$$

$$0.592 = \frac{1}{0.015} (0.3181)(0.225)^{2/3} S_0^{1/2}$$

$$S_0 = 5.695 \times 10^{-3}$$

As $y = y_c$, $S_0 = S_c$ = **Critical slope = 5.695 × 10⁻³**

F. Transitions

12.33 A 3.6 m wide rectangular channel carries water at a depth of 1.8 m. In order to measure the discharge, the channel width is reduced to 2.4 m and a hump of 0.3 m is provided in the bottom. Calculate the discharge if the water surface in the contracted section drops by 0.15 m. Assume no losses.

Solution: Refer to Fig. 12.16. By applying energy equation to sections 1 and 2 with channel bed at section 1 as the datum,

$$y_1 + \frac{V_1^2}{2g} = \Delta z + y_2 + \frac{V_2^2}{2g}$$

Since there is no loss of energy between sections (1) and (2).

$$\frac{V_1^2}{2g} + 0.15 = \frac{V_2^2}{2g} \quad \text{(see Fig. 12.16)}$$

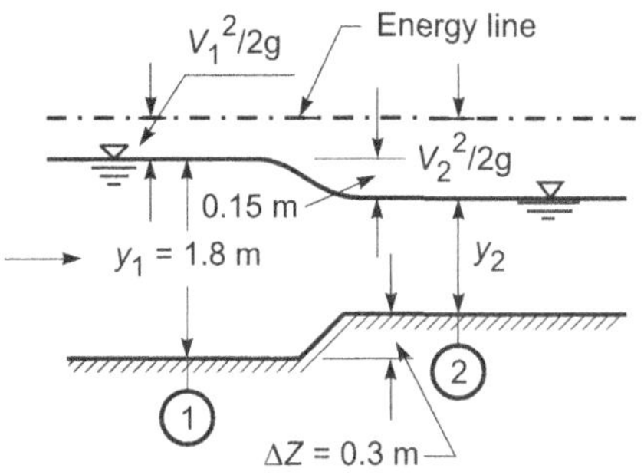

Fig. 12.16

Hence $y_1 + \dfrac{V_1^2}{2g} = \Delta z + y_2 + 0.15 + \dfrac{V_2^2}{2g}$

or

$$y_2 = y_1 - \Delta z - 0.15$$
$$= 1.80 - 0.30 - 0.15 = 1.35 \text{ m}$$

Further

$$\frac{V_1^2}{2g} = \frac{Q^2}{2 \times 9.81\,(3.6 \times 1.8)^2} = \frac{Q^2}{823.85}$$

$$\frac{V_2^2}{2g} = \frac{Q^2}{2 \times 9.81 \times (2.4 \times 1.35)^2} = \frac{Q^2}{205.96}$$

Further

$$\frac{V_2^2}{2g} - \frac{V_1^2}{2g} = 0.15$$

Thus

$$Q^2 \cdot \left[\frac{1}{205.96} - \frac{1}{823.85}\right] = 0.15$$

$$\mathbf{Q = 6.418 \text{ m}^3/\text{s}}$$

12.34 A wide rectangular channel carries a flow of 2.76 m³/s per metre width, the depth of flow being 1.524 m. (a) Calculate the minimum rise in the floor at a section required to produce critical flow conditions. (b) What is the corresponding fall in the water level?

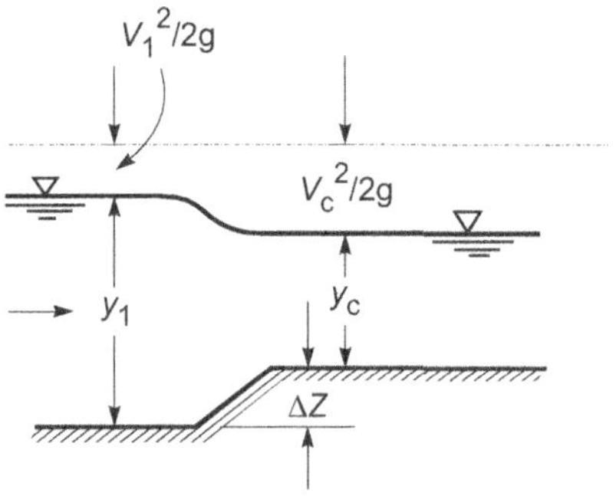

Fig. 12.17

Solution:

(a) $$y_c = (q^2/g)^{1/3} = \left(\frac{(2.76)^2}{9.81}\right)^{1/3} = 0.919 \text{ m}$$

$$V_c^2/2g = \frac{y_c}{2} = 0.460 \text{ m}$$

$$y_1 = 1.524 \text{ m},$$
$$V_1 = q/y_1 = 2.76/1.524 = 1.811 \text{ m/s}$$

$$\frac{V_1^2}{2g} = \frac{(1.811)^2}{2 \times 9.81} = 0.167 \text{ m}$$

By energy equation

$$y_1 + \frac{V_1^2}{2g} = \Delta z + y_c + \frac{V_c^2}{2g}$$

$$1.524 + 0.167 = \Delta z + 0.919 + 0.460$$

$$\mathbf{\Delta z = 0.312 \text{ m}}$$

(b) Change in the water surface elevation

$$= \Delta h = \frac{V_c^2}{2g} - \frac{V_1^2}{2g}$$

$$= 0.460 - 0.167 = \mathbf{0.293 \text{ m}}$$

12.35 In a rectangular channel 3.5 m wide, laid at a slope of 0.0036, uniform flow occurs at a depth of 2.0 m. Find how high can a hump be raised on the channel bed without causing a change in the upstream depth. If the upstream depth is to be raised to 2.4 m what should be the height of the hump? Assume Manning's $n = 0.015$.

Solution:

$$A_1 = 3.5 \times 2.0 = 7.0 \text{ m}^2$$
$$P_1 = 2 \times 2.0 + 3.5 = 7.5 \text{ m}$$
$$R_1 = \frac{7.0}{7.5} = 0.933 \text{ m}$$

Now, $$Q = \frac{1}{n} A R^{2/3} S_0^{1/2}$$

$$= \frac{1}{0.015} \times (7.0)\,(0.933)^{2/3}\,(0.0036)^{1/2}$$

$$= 26.74 \text{ m}^3\text{/s}$$

$$V_1 = \frac{26.74}{7.0} = 3.820 \text{ m}$$

$$F_1 = \frac{3.820}{\sqrt{9.81 \times 2}}$$

= 0.8625, hence the flow is subcritical.

Case 1: At the maximum height of the hump Δz_m,
Upstream depth = y_1
At the contracted section $y_2 = y_c$

$$y_c = \left(\frac{q^2}{g}\right)^{1/3} = \left[\frac{(26.74/3.5)^2}{9.81}\right]^{1/3} = 1.812 \text{ m}$$

$$\frac{V_c^2}{2g} = \frac{y_c}{2} = 0.906 \text{ m}$$

By the energy equation, assuming no loss between sections 1 and 2.

$$y_1 = \frac{V_1^2}{2g} = \Delta z_m + y_c + \frac{V_c^2}{2g}$$

$$E_1 = 2.0 + \frac{(3.820)^2}{2 \times 9.81} = 2.744$$

$$\Delta z_m = 2.744 - 1.812 - 0.906 = \mathbf{0.026 \text{ m}}$$

Case 2: The discharge remains at 26.74 m^3/s. When the upstream depth is 2.4 m due to the hump, the depth of flow at the contracted section y_2 will still be the critical depth.

Hence $$y_1' + \frac{V_1'^2}{2g} = \Delta z + y_c + \frac{V_c^2}{2g}$$

$$= \Delta z + 2.718$$

$$V_1' = \frac{26.74}{(3.5 \times 2.4)} = 3.1833 \text{ m/s}$$

$$\frac{V_1'^2}{2g} = \frac{(3.1833)^2}{2 \times 9.81} = 0.5165$$

Hence $$2.40 + 0.517 = \Delta z + 2.718$$

$$\mathbf{\Delta z = 0.199 \text{ m}}$$

12.36 A rectangular channel is 2.5 m wide and carries a flow of 2.75 m^3/s at a depth of 0.9 m. A contraction of the channel width is proposed at a certain section (Fig. 12.18).

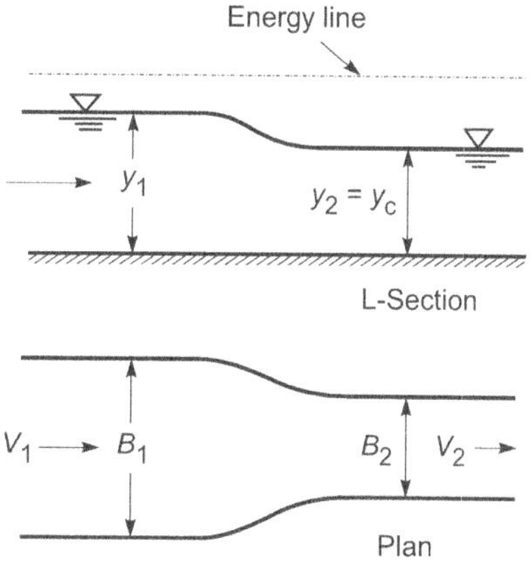

Fig. 12.18

Find the smallest allowable contracted width that will not affect the upstream flow conditions.

Solution:

$$q_1 = \frac{Q}{B_1} = \frac{2.75}{2.50} = 1.10 \text{ m}^3\text{/s/m}$$

Upstream conditions: $$V_1 = \frac{q_1}{y_1} = \frac{1.10}{0.9} = 1.222 \text{ m/s}$$

$$F_1 = V_1/\sqrt{g y_1} = \frac{1.222}{\sqrt{9.81 \times 0.9}} = 0.411$$

(∴ subcritical flow)

$$E_1 = y_1 + \frac{V_1^2}{2g} = 0.90 + \frac{(1.222)^2}{2 \times 9.81} = 0.9761 \text{ m}$$

At the maximum contraction, critical depth will occur at the contracted section. Thus $y_2 = y_{c2}$.

Thus $E_1 = E_2 = E_c = 0.9761$ m.

$$y_{c2} = \text{Critical depth at section 2} = \frac{2}{3} E_c$$

$$= \frac{0.9761 \times 2}{3} = 0.6508 \text{ m}$$

$$y_{c2} = (q_2^2/g)^{1/3} = 0.6508$$

$$q_2 = [(0.6508)^3 \times 9.81]^{1/2} = 1.644 \text{ m}^3/\text{s/m}$$

As $q_2 = Q/B_2$

$$B_2 = Q/q_2 = \frac{2.75}{1.644} = \mathbf{1.672 \text{ m}}$$

= Minimum permissible width at contraction

12.37 A rectangular channel 5.2 m wide has a discharge of 10.0 m^3/s at a velocity of 1.25 m/s. At a certain section the bed width is reduced to 3.0 m through a smooth transition. A smooth flat hump is to be built in this contracted section to cause critical flow for flow measurement purposes. Estimate the height of the hump necessary for this purpose. [Assume no loss of energy at the transition].

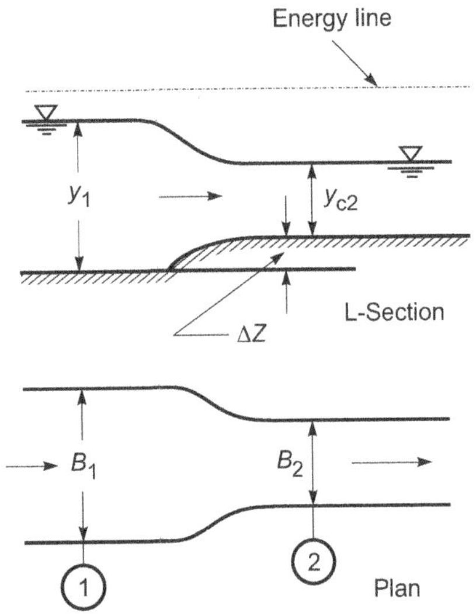

Fig. 12.19

Solution:

Discharge $Q = 10.0$ m^3/s

Upstream section

$$A_1 = \frac{10.0}{1.25} = 8.0 \text{ m}^2$$

$$y_1 = \frac{8.0}{5.2} = 1.5385 \text{ m}$$

$$q_1 = y_1 V_1 = 1.5385 \times 1.25$$
$$= 1.9231 \text{ m}^3/\text{s/m}$$

$$V_1 = 1.25 \text{ m/s}, \quad \frac{V_1^2}{2g} = \frac{(1.25)^2}{2 \times 9.81} = 0.0796$$

Specific energy

$$E_1 = y_1 + \frac{V_1^2}{2g}$$
$$= 1.5385 + 0.0796 = 1.6181 \text{ m}$$

Froude number

$$F_1 = V_1/\sqrt{g y_1} = \frac{1.25}{\sqrt{9.81 \times 1.5385}}$$

$= 0.3217$ (∴ The flow is subcritical flow.)

Downstream section:

$$q_2 = \frac{Q}{B_2} = \frac{10.0}{3.0} = 3.333 \text{ m}^3/\text{s/m}$$

Critical depth at contracted section

$$y_{c2} = (q_2^2/g)^{1/3}$$
$$= \left[\frac{(3.333)^2}{9.81}\right]^{1/3} = 1.0424 \text{ m}$$

If Δz = height of the hump needed to cause critical flow at section 2, by energy equation between sections 1 and 2.

$$E_1 = \Delta z + y_{c2} + \frac{V_{c2}^2}{2g} = \Delta z + 1.5\, y_{c2}$$

$$1.6181 = \Delta z + (1.5 \times 1.0424)$$

$$\mathbf{\Delta z = 0.0545 \text{ m}}$$

Required height of the hump is 0.055 m.

12.38 Water flows at a depth of 1.6 m and velocity of 1.10 m/s in an open channel of rectangular cross section of width 4.0 m. At a certain section the width is reduced to 3.5 m and the bed is raised by 0.35 m through a smooth flat top hump (Fig. 12.20). Calculate the water surface elevations at the contracted section as well as at a section upstream of it. Assume energy losses to be negligible.

Solution: Refer to Fig. 12.20

Upstream section (Section 1)

$$A_1 = 1.6 \times 4.0 = 6.4 \text{ m}^2$$

$$Q = 1.10 \times 6.4 = 7.04 \text{ m}^3/\text{s}$$

Discharge intensity

$$q_1 = \frac{Q}{B_1} = \frac{7.04}{4.0} = 1.76 \text{ m}^3/\text{s/m}$$

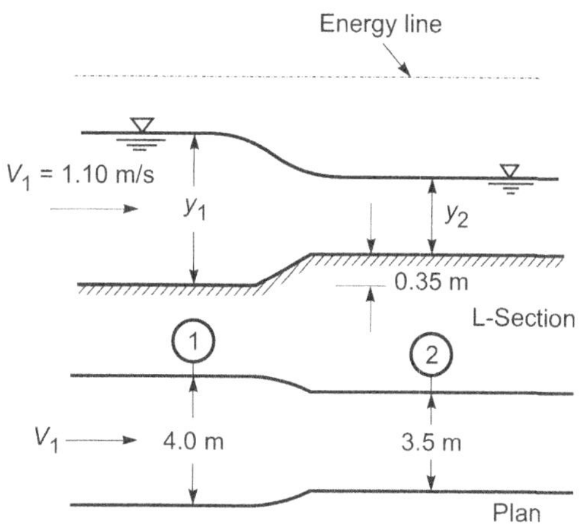

Fig. 12.20

$$V_1 = 1.10 \text{ m/s}, \frac{V_1^2}{2g} = \frac{(1.10)^2}{2 \times 9.81} = 0.06167 \text{ m}$$

Specific energy

$$E_1 = y_1 + \frac{V_1^2}{2g}$$

$$= 1.60 + 0.06167 = 1.66167 \text{ m}$$

Froude number

$$F_1 = \frac{V_1}{\sqrt{g y_1}} = \frac{1.10}{\sqrt{9.81 \times 1.6}} = 0.2776$$

As $F_1 < 1.0$, the upstream flow is subcritical. The water surface will drop down at the contracted section.

Contracted section: (Section 2)

$$q_2 = \frac{Q}{B_2} = \frac{7.04}{3.50} = 2.0114 \text{ m}^3\text{/s/m}$$

Critical depth

$$y_{c2} = (q_2^2/g)^{1/3} = \left(\frac{(2.0114)^2}{9.81}\right)^{1/2} = 0.7444 \text{ m}$$

Minimum specific energy at section 2 = E_{c2}

$$E_{c2} = y_{c2} + \frac{V_{c2}^2}{2g} = 1.5\, y_{c2} = 1.5 \times 0.7444$$

$$= 1.1165 \text{ m}$$

Let H_m = minimum energy at section 2 with upstream bed as datum.

$$H_m = \Delta z + E_{c2} = 0.350 + 1.1165 = 1.4665 \text{ m}$$

As $H_m < E_1$ the depth of flow in the critical section will be larger than the critical depth. Thus

$$y_2 > y_{c2}$$

By energy equation

$$E_1 = 1.66167 = y_2 + \frac{V_2^2}{2g} + \Delta z = y_2 + \frac{q_2^2}{2g\, y_2^2} + \Delta z$$

$$= y_2 + \frac{(2.0114)^2}{2 \times 9.81 \times y_2^2} + 0.35$$

$$y_2 + \frac{0.2062}{y_2^2} = 1.3117$$

By trial and error and noting that $y_1 > y_2 > y_{c2}$ (i.e. $1.6 > y_2 > 0.7444$) the value of y_2 is found as

$$y_2 = 1.158 \text{ m}$$

The upstream depth will remain unaffected at

$$y_1 = 1.60 \text{ m}$$

Hence, with the bed level of the section 1 as datum

Elevation of upstream water surface = 1.60 m

Elevation of water section at the contracted section

$$= y_2 + \Delta z = 1.158 + 0.35 = \mathbf{1.508 \text{ m}}$$

G. Hydraulic Jump

12.39 In hydraulic jump occurring in a rectangular horizontal channel, the discharge per unit width is 2.5 $m^3/s/m$ and the depth before the jump is 0.25 m. Estimate (i) the sequent depth and (ii) the energy loss.

Solution: Referring to Fig. 12.21,

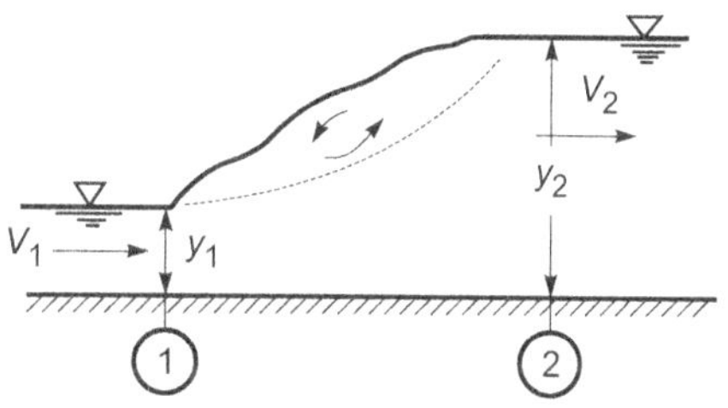

Fig. 12.21 *Hydraulic Jump*

$$q = 2.5 \text{ m}^3\text{/s/m} \quad \text{and} \quad y_1 = 0.25 \text{ m}$$

$$V_1 = \frac{q}{y_1} = \frac{2.5}{0.25} = 10.0 \text{ m/s}$$

Initial Froude number

$$F_1 = \frac{V_1}{\sqrt{g y_1}} = \frac{10.0}{\sqrt{9.81 \times 0.25}} = 6.386$$

(i) The sequent depth ratio y_2/y_1 is given by

$$\frac{y_2}{y_1} = \frac{1}{2}\left[-1 + \sqrt{1 + 8F_1^2}\,\right]$$

$$\frac{y_2}{0.25} = \frac{1}{2}\left[-1 + \sqrt{1 + 8 \times (6.386)^2}\,\right]$$

$$y_2 = 2.136 \text{ m} = \text{Sequent depth}$$

(ii) The energy loss E_L is given by

$$E_L = \frac{(y_2 - y_1)^3}{4 y_1 y_2}$$

$$= \frac{(2.136 - 0.250)^3}{4 \times 2.136 \times 0.25} = 3.141 \text{ m}$$

12.40 A stationary hydraulic jump occurs in a rectangular channel with the initial and sequent depths being equal to 0.20 m and 1.20 m respectively. Estimate
(i) the discharge per unit width and
(ii) the energy loss

Solution:

(i) $$\frac{y_2}{y_1} = \frac{1}{2}\,[-1 + \sqrt{1 + 8F_1^2}\,]$$

$$\frac{1.20}{0.20} = \frac{1}{2}\,[-1 + \sqrt{1 + 8F_1^2}\,]$$

$$F_1^2 = 21 \quad \text{and} \quad F_1 = 4.583$$

$$F_1 = \frac{V_1}{\sqrt{9.81 \times 0.2}} = 4.583$$

$$V_1 = 6.419 \text{ m/s}$$

Discharge per unit width

$$q = V_1 y_1 = 6.419 \times 0.2 = \mathbf{1.284\ m^3/s/m}$$

(ii) The energy loss E_L is given by

$$E_L = \frac{(y_2 - y_1)^3}{4\,y_1 y_2} = \frac{(1.20 - 0.20)^3}{4 \times 0.20 \times 1.20}$$

$$= \mathbf{1.042\ m}$$

12.41 If, in a hydraulic jump occurring in a horizontal rectangular channel, the Froude number before the jump is 10.0 and the energy loss is 3.20 m estimate the (i) sequent depths (ii) the discharge intensity and (iii) the Froude number after the jump.

Solution:

$$F_1 = 10.0 \quad \text{and} \quad E_L = 3.20 \text{ m}$$

The sequent depth ratio

$$\frac{y_2}{y_1} = \frac{1}{2}\,[-1 + \sqrt{1 + 8F_1^2}\,]$$

$$= \frac{1}{2}\,(-1 + \sqrt{1 + 8 \times (10)^2}\,) = 13.651$$

Energy loss $$E_L = \frac{(y_2 - y_1)^3}{4\,y_1 y_2}$$

$$\frac{E_L}{y_1} = \frac{(y_2/y_1 - 1)^3}{4\,(y_2/y_1)}$$

$$\therefore \quad \frac{3.20}{y_1} = \frac{(13.651 - 1)^3}{4 \times 13.651} = 37.08$$

(i) y_1 = depth before the jump = $\frac{3.20}{37.08}$ = **0.0863 m**

y_2 = depth after the jump = 13.651 × 0.0863 = **1.178 m**

(ii) $$F_1 = \frac{V_1}{\sqrt{g\,y_1}}$$

$$10.0 = \frac{V_1}{\sqrt{9.81 \times 0.0863}}$$

$$V_1 = 9.201 \text{ m/s}$$

Discharge intensity

$$q = V_1 y_1 = 9.201 \times 0.0863$$

$$= \mathbf{0.7941\ m^3/s/m}$$

(iii) Froude number after the jump = F_2

$$F_2 = \frac{V_2}{\sqrt{g\,y_2}} = \frac{q}{y_2\sqrt{g\,y_2}} = \frac{0.7941}{1.178\sqrt{9.81 \times 1.178}}$$

$$= \mathbf{0.1983}$$

12.42 For a hydraulic jump in a rectangular channel the velocity and depth after the jump are known to be 0.80 m/s and 1.75 m respectively. Calculate the depth before the jump, the energy loss and the power dissipated per metre width.

Solution:

$$F_2 = \frac{V_2}{\sqrt{g\,y_2}} = \text{Froude number after the jump}$$

$$= \frac{0.80}{\sqrt{9.81 \times 1.75}} = 0.1931$$

The sequent depth ratio is given in terms of F_2 as

$$\frac{y_1}{y_2} = \frac{1}{2}\,(-1 + \sqrt{1 + 8F_2^2}\,)$$

$$= \frac{1}{2}\,(-1 + \sqrt{1 + 8 \times (0.1931)^2}\,) = 0.0697$$

$$y_1 = 0.0697 \times 1.75 = \mathbf{0.122\ m}$$

Energy loss $$E_L = \frac{(y_2 - y_1)^3}{4\,y_1 y_2}$$

$$= \frac{(1.750 - 0.122)^3}{4 \times 0.122 \times 1.750} = \mathbf{5.054\ m}$$

Power dissipated

$$\mathbf{P} = \gamma Q E_L$$

$$= 9790 \times (0.80 \times 1.75) \times 5.054 \text{ W/m}$$

$$= 69270 \text{ W/metre width}$$

$$= \mathbf{69.27\ kW/metre\ width}$$

12.43 Given the energy loss and Froude number after the jump as 9.00 m and 0.12 respectively estimate the discharge intensity and initial depth of the hydraulic jump in a rectangular channel.

Solution:

$F_2 = 0.12$

$$\frac{y_1}{y_2} = \frac{1}{2}\left(-1 + \sqrt{1+8F_2^2}\right) = \frac{1}{2}\left(-1+\sqrt{1+8\times(0.12)^2}\right)$$
$$= 0.0280$$

Energy loss $E_L = \dfrac{(y_2-y_1)^3}{4y_1y_2}$

$$\frac{E_L}{y_2} = \frac{\left(1-\dfrac{y_1}{y_2}\right)^3}{4(y_1/y_2)} = \frac{(1-0.028)^3}{4\times 0.028} = 8.1994$$

$$y_2 = \frac{E_L}{8.1994} = \frac{9.0}{8.1994} = 1.0976 \text{ m}$$

$$y_1 = 0.028 \times 1.0976 = 0.0307 \text{ m}$$

$$F_2 = \frac{V_2}{\sqrt{g y_2}} = \frac{V_2}{\sqrt{9.81\times 1.0976}} = 0.12$$

$$V_2 = 0.3938 \text{ m/s}$$

Discharge intensity

$$q = V_2 y_2 = 0.3938 \times 1.0976 = \mathbf{0.4322\ m^3/s/m}$$

12.44 In a hydraulic jump occurring in a rectangular channel the discharge per unit width is 2.0 m^3/s/m and the energy loss is 1.75 m. Calculate the initial depth, initial Froude number and the sequent depths of the jump.

Solution: For the solution of this kind of problem, a trial and error procedure is adopted.

First trial Assume $F_1 = 4.5$.

$$F_1 = \frac{V_1}{\sqrt{g y_1}} = \frac{q}{y_1\sqrt{g y_1}}$$

$$\therefore \quad y_1^3 = \frac{q^2}{gF_1^2} = \frac{(2.0)^2}{9.81\times(4.5)^2} = 0.02014$$

$$y_1 = 0.272 \text{ m}$$

$$\frac{y_2}{y_1} = \frac{1}{2}\left(-1 + \sqrt{1+8F_1^2}\right)$$
$$= \frac{1}{2}\left(-1+\sqrt{1+8\times(4.5)^2}\right) = 5.884$$

$$E_L = \frac{(y_2-y_1)^3}{4y_1y_2} = y_1\left[\frac{(y_2/y_1-1)^3}{4(y_2/y_1)}\right]$$

$$= 0.272 \times \frac{(5.884-1)^3}{4\times 5.884} = 1.346 \text{ m}$$

Actual $E_L = 1.75$ hence assumed F_1 is not correct. Next assumption of F_1 should be higher than 4.5.

Second trial Assume $F_1 = 5.10$.

$$y_1 = \left(\frac{q^2}{gF_1^2}\right)^{1/3} = \left(\frac{(2.0)^2}{9.81\times(5.10)^2}\right)^{1/3} = 0.25 \text{ m}$$

$$\frac{y_2}{y_1} = \frac{1}{2}\left(-1 + \sqrt{1+8\times(5.10)^2}\right) = 6.73$$

$$E_L = 0.25\left[\frac{(6.73-1)^3}{4\times 6.73}\right] = 1.75 \text{ m}$$

= same as per given data

Hence the assumption $\mathbf{F_1 = 5.10}$ is all right.

$\mathbf{y_1 = 0.25\ m}$

$y_2 = 0.25 \times 6.73 = \mathbf{1.683\ m}$

12.45 A hydraulic jump occurs in a rectangular channel and the depths of flow before and after the jump are 0.5 m and 3.0 m. Calculate (i) the critical depth and (ii) power lost per unit width of channel.

Solution: $y_1 = 0.5$ m and $y_2 = 3.0$ m

The sequent depths in a jump are related to the critical depth as

(i) $$y_1y_2(y_1+y_2) = 2y_c^3$$
$$0.5\times 3.0\times(0.5+3.0) = 2y_c^3$$
$$y_c^3 = 2.625 \quad \text{and} \quad \mathbf{y_c = 1.379\ m}$$

(ii) $$\frac{q^2}{g} = y_c^3 = (1.379)^3 = 2.625$$
$$q = (2.625\times 9.81)^{1/2} = 5.075 \text{ m}^3\text{/s/m}$$

Head loss $E_L = \dfrac{(y_2-y_1)^3}{4y_1y_2}$

$$= \frac{(3.0-0.5)^3}{4\times 3\times 0.5} = 2.604 \text{ m}$$

Power lost per metre width

$$\mathbf{P} = \gamma q E_L$$
$$= 9790\times 5.075\times 2.604 \text{ W}$$
$$= 129386 \text{ W} = \mathbf{129.4\ kW}$$

12.46 Show that the Froude numbers F_1 and F_2 before and after the jump respectively, in a hydraulic jump taking place in a rectangular channel are related as

$$F_2^2 = \frac{8F_1^2}{\left(-1+\sqrt{1+8F_1^2}\right)^3}$$

Solution: $$F_2^2 = \frac{V_2^2}{g y_2} = \frac{q^2}{g y_2^3}$$

$$= \frac{q^2}{g y_1^3 (y_2/y_1)^3} = \frac{F_1^2}{(y_2/y_1)^3}$$

But the sequent depth ratio y_2/y_1 is related to F_1 as

$$\frac{y_2}{y_1} = \frac{1}{2}(-1 + \sqrt{1+8F_1^2})$$

Hence $$F_2^2 = \frac{8F_1^2}{(-1+\sqrt{1+8F_1^2})^3}$$

12.47 At the toe of a hydraulic jump the Froude number and depth of flow are 10.0 and 0.4 m respectively. Estimate the specific energy head at the heel of the jump.

Solution:

$$F_1 = 10.0 \text{ and } y_1 = 0.4$$

Let y_2 = sequent depth to y_1. Then

$$y_2/y_1 = \frac{1}{2}(-1 + \sqrt{1+8F_1^2})$$

$$= \frac{1}{2}(-1 + \sqrt{1+8(10)^2})$$

$$\frac{y_2}{0.4} = 13.651$$

$$y_2 = 5.46 \text{ m}$$

Energy loss $$E_L = \frac{(y_2 - y_1)^3}{4y_1y_2} = \frac{(5.46-0.4)^3}{4\times 0.4 \times 5.46}$$

$$= 14.83 \text{ m}$$

Initial specific energy $$= E_1 = y_1 + \frac{V_1^2}{2g} = y_1\left(1 + \frac{F_1^2}{2}\right)$$

$$E_1 = 0.4\left(1 + \frac{(10)^2}{2}\right) = 20.40 \text{ m}$$

Specific energy at the heel = $E_2 = E_1 - E_L$

$E_2 = 20.40 - 14.83 =$ **5.57 m**

12.48 An overflow spillway has its crest at elevation 136.00 m and a horizontal apron at an elevation of 102.00 m on the downstream side. Estimate the tail water elevation required to form a hydraulic jump when the elevation of the energy line just upstream of the spillway crest is 138.00 m. Assume $C_d = 0.735$ for the spillway. Neglect energy loss due to flow over the spillway.

Solution: Refer to Fig. 12.22,
The discharge per unit width of the spillway q is

$$q = \frac{2}{3} C_d \sqrt{2g}\, H^{3/2}$$

$$H = 138.00 - 136.00 = 2.00 \text{ m}$$

$$q = \frac{2}{3} \times 0.735 \times \sqrt{2\times 9.81} \times (2.0)^{3/2}$$

$$= 6.139 \text{ m}^3\text{/s/m}$$

$$E_1 = 138.00 - 102.00 = 36.00 \text{ m}$$

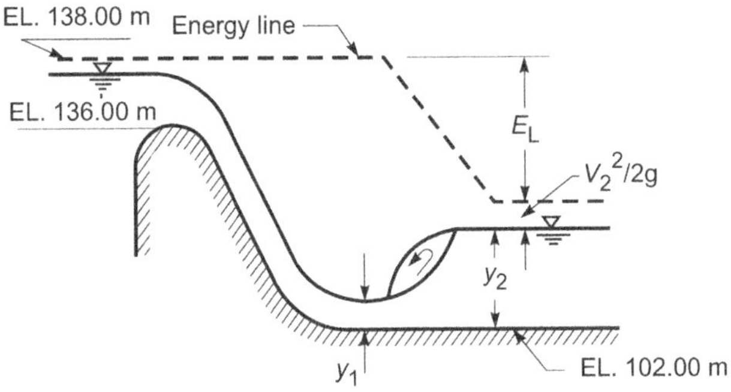

Fig. 12.22 *Overflow spillway*

$$E_1 = y_1 + \frac{V_1^2}{2g} = y_1 + \frac{q^2}{2gy_1^2}$$

$$= y_1 + \frac{(6.139)^2}{19.62\, y_1^2}$$

Hence $$y_1 + \frac{1.9208}{y_1^2} = 36.00$$

By trial and error $y_1 = 0.2317$ m

$$V_1 = \frac{q}{y_1} = \frac{6.139}{0.2317} = 26.495 \text{ m/s}$$

$$F_1 = \frac{V_1}{\sqrt{gy_1}} = \frac{26.495}{\sqrt{9.81\times 0.2317}} = 17.574$$

$$\frac{y_2}{y_1} = \frac{1}{2}(-1 + \sqrt{1+8F_1^2})$$

$$= \frac{1}{2}(-1 + \sqrt{1+8\times(17.574)^2}) = 24.36$$

$$y_2 = 0.2317 \times 24.36 = 5.644 \text{ m}$$

Required tailwater elevation = 102.000 + 5.644
= **107.644 m**

12.49 A hydraulic jump occurs in a horizontal 90° triangular channel. If the sequent depths in this channel are 0.60 m and 1.20 m respectively, estimate (i) the flow rate, (ii) Froude number at the beginning and end of the jump and (iii) energy loss in the jump.

Solution:

(i) Consider a triangular channel of side slope m horizontal : 1 vertical as in Fig. 12.23. (In the present case $m = 1$)

$$P = \text{Pressure force} = \gamma A\bar{y} = \gamma(my^2)\frac{y}{3} + \gamma my^3/3$$

$$M = \text{Momentum flux} = \frac{\rho Q^2}{A} = \frac{\rho Q^2}{my^2}$$

For a hydraulic jump in horizontal, frictionless channel

$$P_1 + M_1 = P_2 + M_2$$

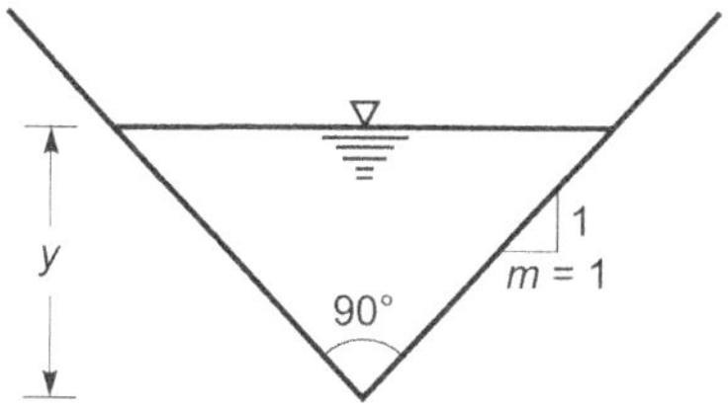

Fig. 12.23

$$\frac{\gamma m y_1^3}{3} + \frac{\rho Q_1^2}{m y_1^2} = \frac{\gamma m y_2^3}{3} + \frac{\rho Q_2^2}{m y_2^2}$$

$$\frac{Q^2}{m}\left[\frac{1}{y_1^2} - \frac{1}{y_2^2}\right] = \frac{gm}{3}\,(y_2^3 - y_1^3)$$

On simplifying $\quad \dfrac{Q^2}{g} = \dfrac{m^2}{3}\left[\dfrac{y_1^3(\eta^3-1)\,\eta^2 y_1^4}{(\eta^2-1)\,y_1^2}\right]$

where $\quad \eta = \dfrac{y_2}{y_1}$

In the present problem $m = 1$,

$$\eta = \frac{y_2}{y_1} = 1.2/0.6 = 2.0$$

$$\frac{Q^2}{g} = \frac{1}{3}\,(0.6)^5\,\frac{(2^3-1)}{(2^2-1)} = 0.24192$$

Hence $\quad$ $\mathbf{Q = 1.541\ m^3/s}$

(ii) For triangular channel $F = \dfrac{Q}{A\sqrt{gA/T}}$. As such

$$F^2 = \frac{Q^2 T}{gA^3} = \frac{Q^2(2my)}{gm^2y^6} = \frac{2Q^2}{gm^2y^5}$$

$$F_1^2 = \frac{2(1.541)^2}{9.81\times 1\times (0.6)^5} = 6.222$$

$$\mathbf{F_1 = 2.494}$$

Froude number at the end of the jump:

Since $\quad F^2 = \dfrac{2Q^2}{gm^2y^5},\quad \dfrac{F_1}{F_2} = \left(\dfrac{y_2}{y_1}\right)^{5/2}$

$$\frac{F_1}{F_2} = \left(\frac{1.20}{0.60}\right)^{5/2} = 5.657$$

$$F_2 = 2.494/5.657 = 0.441$$

(iii) Energy loss

$$E_L = E_1 - E_2 = \left(y_1 + \frac{V_1^2}{2g}\right) - \left(y_2 + \frac{V_2^2}{2g}\right)$$

$$A_1 = 1\times(0.6)^2 = 0.36\ \text{m}^2,$$

$$V_1 = 1.541/0.36 = 4.281\ \text{m/s}$$

$$A_2 = 1\times(1.2)^2 = 1.44\ \text{m}^2$$

$$V_2 = 1.541/1.44 = 1.070\ \text{m/s}$$

$$E_L = \left[0.6 + \frac{(4.281)^2}{2\times 9.81}\right] - \left[1.2 + \frac{(1.070)^2}{2\times 9.81}\right]$$

$$= 1.534 - 1.258 = \mathbf{0.276\ m}$$

H. Gradually Varied Flow

12.50 Water flows in a triangular channel of side slope 1 horizontal : 1 vertical and longitudinal slope of 0.001. Determine whether the channel is mild, steep or critical when a discharge of 0.2 m^3/s flows through it. Assume Manning's $n = 0.015$. For what range of depths will the flow be on a type 1, 2 or 3 curve?

Solution: For a depth of flow of y in the channel,

$$A = y^2$$

$$T = 2y$$

$$R = y^2/2\sqrt{2}y = \frac{1}{2\sqrt{2}}\,y$$

Critical depth y_c:

$$\frac{Q^2}{g} = \frac{A_c^3}{T_c} = \frac{y_c^6}{2y_c} = \frac{y_c^5}{2}$$

$$y_c = \left(\frac{2Q^2}{g}\right)^{1/5} = \left(\frac{2\times(0.2)^2}{9.81}\right)^{1/5} = 0.382\ \text{m}$$

Normal depth y_0:

$$Q = \frac{1}{n}\,AR^{2/3}\,S_0^{1/2}$$

$$= \frac{1}{n}\,(y_0)^2\left(\frac{y_0}{2\sqrt{2}}\right)^{2/3} S_0^{1/2}$$

$$= \frac{0.5}{n}\,y_0^{8/3}\,S_0^{1/2}$$

$$y_0^{8/3} = \frac{nQ}{0.5\,S_0^{1/2}} = \frac{0.015\times 0.2}{0.5\times(0.001)^{1/2}} = 0.18974$$

$$y_0 = 0.536\ \text{m}$$

Since $y_0 > y_c$, the channel is a mild slope channel for this discharge. If y is the depth of flow:

For $\quad$ **M_1 curve** $\quad$ $\mathbf{y > 0.536\ m}$

M_2 curve $\quad$ $\mathbf{0.536\ m > y > 0.382\ m}$

M_3 curve $\quad$ $\mathbf{y < 0.382\ m}$

12.51 A wide rectangular channel has a Manning's coefficient of 0.018. For a discharge intensity of 1.5 $m^3/s/m$, identify the possible types of gradually varied flow (GVF) profiles produced in the following break in the grade of the channel:

(a) $S_{01} = 0.0004$ and $S_{02} = 0.016$

(b) $S_{01} = 0.015$ and $S_{02} = 0.0003$

Solution: Discharge intensity $q = 1.5$ m^3/s/m

Critical depth $y_c = (q^2/g)^{1/3} = \left(\dfrac{(1.5)^2}{9.81}\right)^{1/3} = 0.612$ m

Normal depth y_0: For a wide rectangular channel $R = y_0$

$$q = \frac{1}{n} y_0 y_0^{2/3} S_0^{1/2}$$

$$y_0 = \left[\frac{nq}{\sqrt{S_0}}\right]^{3/5} = \left[\frac{0.018 \times 1.5}{\sqrt{S_0}}\right]^{3/5}$$

Slope	y_0
0.0003	1.305 m
0.0004	1.197 m
0.015	0.404 m
0.016	0.396 m

Case	y_c *(m)*	y_{01} *(m)*	y_{02} *(m)*	*Type of grade change*
(a)	0.612	1.197	0.396	Mild to Steep
(b)	0.612	0.404	1.305	Steep to Mild

The resulting water surface profiles are:

Case (a): M_2 curve on Mild slope and S_2 curve on Steep slope (Fig. 12.24 (a))

Case (b): (i) M_3 curve and a jump or
(ii) Jump and S_1 curve (Fig. 12.24 (b))
(iii) in a limiting case of (i) and (ii) the jump may occur at the break in grade and there will be no GVF profiles in such a case.

The resulting water surface profile will be any one of the above three cases depending upon the sequent depths of the resulting jump.

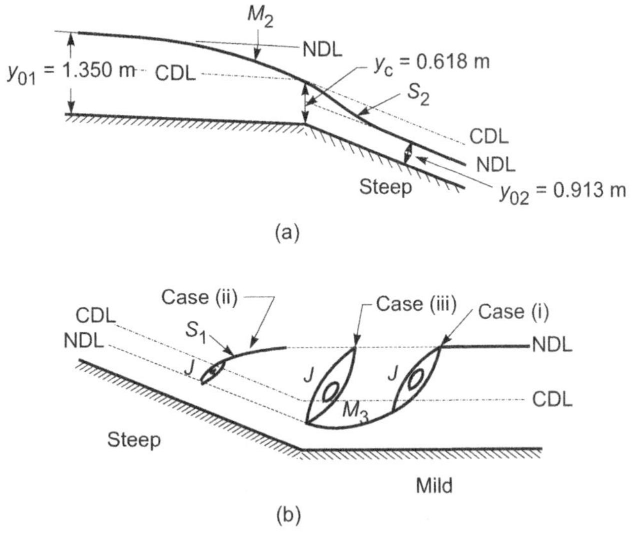

Fig. 12.24 *Example 12.51*

12.52 For a wide horizontal channel, by assuming Chezy coefficient C = constant, show that the gradually varied flow profiles can be expressed as

$$x = \frac{C^2}{g}\left[y - \frac{y^4}{4y_c^3}\right] + \text{constant}$$

Solution: The basic differential equation of GVF is

$$\frac{dy}{dx} = \frac{S_0 - S_f}{1 - \dfrac{Q^2T}{gA^3}} \qquad (1)$$

For a wide rectangular channel, by considering unit width,

$$T = 1,\ A = y \quad \text{and} \quad Q = q = Vy,\ R = y.$$

Also by Chezy formula

$$V = C\sqrt{RS_f} = C\sqrt{yS_f}$$

$$S_f = \frac{V^2}{C^2 y} = \frac{q^2}{C^2 y^3}$$

For a horizontal channel $S_0 = 0$

Hence, with these for a wide horizontal channel, Eq. (1) becomes

$$\frac{dy}{dx} = \frac{-S_f}{1 - \dfrac{Q^2T}{gA^3}} = \frac{-q^2/C^2y^3}{1 - \dfrac{q^2}{g}\cdot\dfrac{1}{y^3}}$$

But $q^2/g = y_c^3$

Hence
$$\frac{dy}{dx} = \frac{-\dfrac{g}{C^2}\left(\dfrac{y_c}{y}\right)^3}{1 - (y_c/y)^3}$$

$$dx = \left[\frac{1}{-\dfrac{g}{C^2}(y_c/y)^3} + \frac{\left(\dfrac{y_c}{y}\right)^3}{\dfrac{g}{C^2}(y_c/y)^3}\right] dy$$

$$x = \left[-\frac{C^2}{4g}\frac{y^4}{y_c^3} + \frac{C^2}{g}y\right] + \text{constant}$$

$$\mathbf{x = \frac{C^2}{g}\left[y - \frac{y^4}{4y_c^3}\right] + constant}$$

12.53 In a gradually varied flow in a rectangular channel of bottom width 3.0 m the discharge is 8 m^3/s and the depth of flow changes from 1.4 m at section M to 1.05 m at section N. Calculate the average energy slope between these two sections. Assume $n = 0.018$.

Solution: In gradually varied flow the Manning's formula is written for any section as

$$V = \frac{1}{n} R^{2/3} S_f^{1/2}$$

where S_f = energy slope at that section.

Hence $$S_f = \frac{n^2 V^2}{R^{4/3}}$$

Average energy slope between two sections

$$\bar{S}_f = \frac{S_{f1} + S_{f2}}{2}$$

The calculations are shown in Table 12.3.

Table 12.3 *Energy Slope Calculation*

Property	*Section M*	*Section N*
A	$3.0 \times 1.4 = 4.2$ m^2	$3.0 \times 1.05 = 3.15$ m^2
P	$3.0 + (2 \times 1.4) = 5.8$ m	$3.0 + (2 \times 1.05) = 5.10$ m
R	$4.2/5.8 = 0.724$ m	$3.15/5.10 = 0.6176$ m
V	$8.0/4.2 = 1.9048$ m/s	$8.0/3.15 = 2.5397$ m/s
S_f	$\frac{(0.018)^2 \times (1.9048)^2}{(0.724)^{4/3}} = 1.8077 \times 10^{-3}$	$\frac{(0.018)^2 \times (2.5397)^2}{(0.6176)^{4/3}} = 3.973 \times 10^{-3}$

$$\bar{S}_f = \frac{1.8077 \times 10^{-3} + 3.9730 \times 10^{-3}}{2} = \mathbf{2.890 \times 10^{-3}}$$

12.54 A rectangular channel has a bed width = 4.0 m, bottom slope = 0.0004 and Manning's n = 0.02. The normal depth of flow in this channel is 2.0 m. If the channel empties into a pool at the down stream end and the pool elevation is 0.60 m higher than the canal bed elevation at the downstream end, calculate the coordinates of the resulting GVF profile.

Solution: For uniform flow,

$$y_0 = 2.0 \text{ m}$$

$$A_0 = 4.0 \times 2.0 = 8.0 \text{ m}^2$$

$$R_0 = \frac{8.0}{(4.0 + 2 \times 2.0)} = 1.0 \text{ m}$$

$$Q_1 = Q_0 = \frac{1}{n} A_0 R_0^{2/3} S_0^{1/2}$$

$$= \frac{1}{0.02} \times 8.0 \times (1.0)^{2/3} \times (0.0004)^{1/2}$$

$$= 8.0 \text{ m}^3/\text{s}$$

For critical depth:

$$q = \frac{Q}{B} = \frac{8.0}{4.0} = 2.0 \text{ m}^3/\text{s/m}$$

Critical depth

$$y_c = (q^2/g)^{1/3} = (2^2/9.81)^{1/3} = \mathbf{0.742 \text{ m}}$$

Since $y_0 > y_c$, the channel slope is *mild*. Since the downstream pool elevation is 0.6 m above the canal invert at that section, and y_c = 0.742 m, the critical depth will be the downstream control. The water surface profile will be an M_2 curve extending from y_0 = 2.0 m to y_c = 0.742 m at the downstream end.

The direct step method with 4 steps are used. The calculations start at $y = y_c = 0.742$ m and end at $y = 0.99 y_0 = 1.98$ m. The calculations are shown in Table 12.4.

12.55 A sluice gate discharges 2 m^3/s/m into a wide horizontal rectangular channel. The depth at the *vena contracta* is 0.15 m. The tailwater depth is 1.80 m. Assuming the channel to have a Manning's n = 0.015, determine the location of the hydraulic jump.

Solution: In Fig. 12.25,

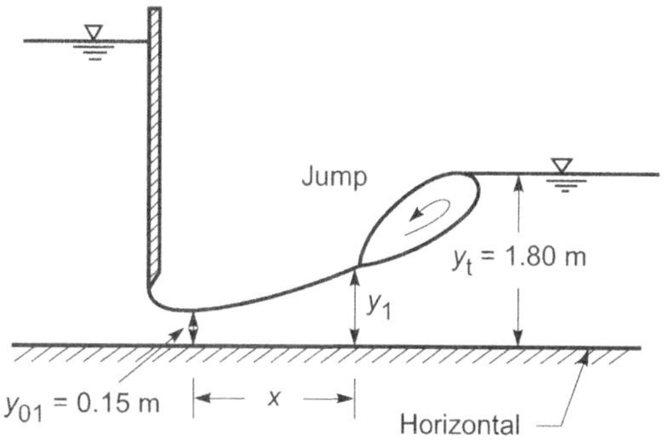

Fig. 12.25 *Example 12.55*

y_t = tailwater = 1.80 m

y_1 = beginning of the jump.
= depth sequent to y_t.

y_{01} = depth at *vena contracta*.

x = distance of the beginning of the jump from *vena contracta*.

To find y_1: At *vena contracta*,

$$V_{01} = \frac{q}{y_{01}} = \frac{2.0}{0.15} = 13.33 \text{ m/s}$$

$$F_{01} = \frac{V_{01}}{\sqrt{g\, y_{01}}} = \frac{13.33}{\sqrt{9.81 \times 0.15}} = 11.0$$

If a jump is to take place at *vena contracta*, the sequent depth required y_{02} is

$$\frac{y_{02}}{y_{01}} = \frac{1}{2} [-1 + \sqrt{1 + 8 F_{01}^2}\,]$$

$$= \frac{1}{2} [-1 + \sqrt{1 + 8 (11)^2}\,] = 15.05$$

$$y_{02} = 2.258 \text{ m}$$

Table 12.4 *Computation of M_2 Profile of Example 12.54*

$Q = 8.0\ m^3/s$ $\quad n = 0.02 \quad y_0 = 2.0\ m$

$B = 4.0\ m \quad S_0 = 0.0004 \quad y_c = 0.742\ m$

1	2	3	4	5	6	7	8	9	10	11
y (m)	A (m²)	R (m)	V (m/s)	E (m)	ΔE (m)	S_f	$\bar{S}_f$	$S_0 - \bar{S}_f$	Δx (m)	x (m)
0.742	2.968	0.5412	2.695	1.1129		6.5872×10^{-3}				0
					0.0917		46.673×10^{-4}	-42.673×10^{-4}	−21.5	
1.00	4.000	0.6666	2.000	1.2039		2.7473×10^{-3}				21.5
					0.3001		19.024×10^{-4}	-15.024×10^{-4}	−199.7	
1.40	5.600	0.8235	1.4285	1.5040		1.0575×10^{-3}				221.2
					0.3589		7.9411×10^{-4}	-3.9411×10^{-4}	−910.7	
1.80	7.200	0.9474	1.1111	1.8629		5.3071×10^{-4}				1132
					0.1691		4.7079×10^{-4}	-0.7079×10^{-4}	−2389	
1.98	7.920	0.9950	1.0101	2.0320		4.1086×10^{-4}				3521

Note:
- Col. 2: $A = By = 4.0 \times \text{Col. 1}$
- Col. 3: $R = \dfrac{A}{P} = \dfrac{\text{Col. 2}}{(4 + 2 \times (\text{Col. 1}))}$
- Col. 4: $V = \dfrac{Q}{A} = \dfrac{8.0}{\text{Col. 2}}$
- Col. 5: $E = \dfrac{V^2}{2g} + y = \dfrac{(\text{Col. 4})^2}{2 \times 9.81} + \text{Col. 1}$
- Col. 6: $\Delta E = E_2 - E_1$
- Col. 7: $S_f = \dfrac{n^2V^2}{R^{4/3}} = \dfrac{(0.02)^2 \times (\text{Col. 4})^2}{(\text{Col. 3})^{4/3}}$
- Col. 8: $\bar{S}_f = \dfrac{1}{2}(S_{f2} + S_{f1})$
- Col. 9: $S_0 - \bar{S}_f = 0.0004 - \text{Col. 8}$
- Col. 10: $\Delta x = \dfrac{\Delta E}{S_0 - \bar{S}_f} = \dfrac{\text{Col. 6}}{\text{Col. 9}}$
- Col. 11: $x = \Sigma \Delta x$

As the actual tailwater depth is 1.8 m, the jump will be repelled (i.e. will take place downstream of the *vena contracta* at a distance x. The depth at the toe of the jump will by y_1 and will be related to y_t = tailwater depth by

$$\frac{y_1}{y_t} = \frac{1}{2}\left[-1 + \sqrt{1 + 8F_t^2}\right]$$

$$y_t = 1.80\ \text{m}$$

$$V_t = \frac{2.00}{1.80} = 1.111\ \text{m/s}$$

F_t = tailwater Froude number

$$= \frac{V_t}{\sqrt{g y_t}} = \frac{1.111}{\sqrt{9.81 \times 1.80}} = 0.2644$$

$$\frac{y_1}{1.80} = \frac{1}{2}\left[-1 + \sqrt{1 + 8 \times (0.2644)^2}\right] = 0.12436$$

$$y_1 = 0.224\ \text{m}$$

To find distance x: Since the channel is horizontal, and the flow from the *vena contracta* is supercritical, the water surface profile between y_{01} and y_1 will be an H_3 curve. The distance x in this curve is calculated by using the direct step method. Calculations are shown in Table 12.5. Two steps are used in this calculation. Larger number of steps will give greater accuracy.

In Table 12.5

$$V = q/y = \frac{2.0}{\text{Col. 1}}$$

$$E = y + \frac{V^2}{2g} = \text{Col. 1} + \frac{(\text{Col. 2})^2}{2 \times 9.81}$$

$$\Delta E = E_2 - E_1$$

$$S_f = \frac{n^2V^2}{y^{4/3}} = \frac{n^2q^2}{y^{10/3}} = \frac{(0.015)^2(2)^2}{y^{10/3}} = \frac{0.0009}{(\text{Col. 1})^{10/3}}$$

$$\bar{S}_f = (S_{f2} + S_{f1})/2$$

Table 12.5 *Calculation of H_3 Profile–Example 12.55*

$q = 2.0\ m^3/s/m$ $n = 0.015$

1	2	3	4	5	6	7	8
y (m)	V (m/s)	E (m)	ΔE (m)	S_f	$-\bar{S}_f$	Δx (m)	x (m)
0.15	11.11	6.4724		0.27332			0
			−1.17556		−0.2328	5.05	
0.20	10.00	5.2968		0.19237			5.05
			−1.00963		−0.16211	6.23	
0.224	8.929	4.2872		0.13185			11.28

Since $S_0 = 0,\ \Delta x = -\dfrac{\Delta E}{\bar{S}_f} = \dfrac{\text{Col. 4}}{\text{Col. 6}}$

The toe of the jump will be at a distance of **11.28 m** from the *vena contracta*.

1. Broad Crested Weir

12.56 Derive an expression for discharge over a broad crested weir.

Solution: Referring to Fig. 12.26,

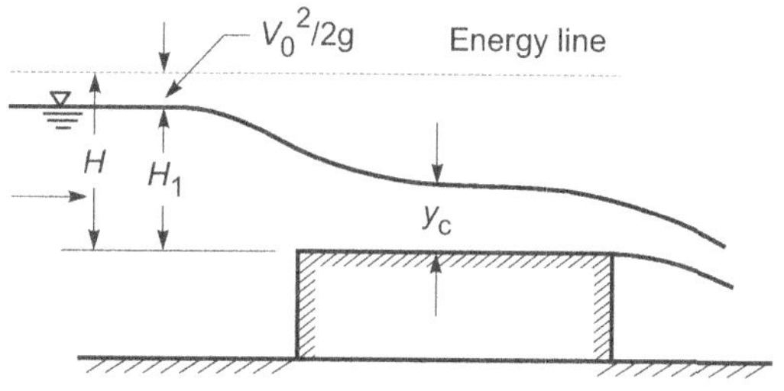

Fig. 12.26 *Broad Crested Weir*

The critical depth y_c occurs over the weir crest. Assuming no loss of energy and considering the elevation of the weir crest as datum the energy equation is

$$H_1 + \frac{V_0^2}{2g} = H = y_c + \frac{V_c^2}{2g}$$

$$= \frac{3}{2} y_c \qquad \left(\because\ y_c = \frac{V_c^2}{g}\right)$$

Thus $\quad y_c = \dfrac{2}{3} H$ and $V_c = \sqrt{g y_c}$

Ideal discharge $Q_t = L V_c y_c$ where L = length of weir.

$$= \frac{2}{3}\sqrt{(2/3\,g)}\ LH^{3/2}$$

Putting $g = 9.81$, $Q_t = 1.705\ LH^{3/2}$

When the weir occupies the full width of the channel $L = B$ = width of channel. To account for energy losses a coefficient of discharge C_d is introduced as

$$Q = Q_t C_d = \mathbf{1.705\ C_d\ LH^{3/2}}$$

It is to be noted that $H = H_1 + \dfrac{V_0^2}{2g}$ where V_0 is the velocity of approach. It is usual to neglect the velocity of approach and in that case

$$\mathbf{Q = 1.705\ C_d\ LH_1^{3/2}}$$

Usually C_d is about 0.85 for a weir with square entrance. For weirs with rounded entrance C_d is about 0.98.

12.57 A broad crested weir spanning the full width of a 2.0 m wide channel is 1.5 m high and has a square entrance. What head is required to pass a discharge of 3.0 m^3/s?

Solution: For a broad-crested weir,

$$Q = 1.705\ C_d\ LH^{3/2}$$

where $\quad H = H_1 + V_0^2/2g$

V_0 = Velocity of approach

Assume $C_d = 0.85$ as the weir has a square entrance.

$$B = L = 2.0\ \text{m}$$

$$Q = 3.0 = 1.705 \times 0.85 \times 2.0 \times H^{3/2}$$

$$H = 1.023\ \text{m} = \left(H_1 + \frac{V_0^2}{2g}\right)$$

To find V_0: Assume $H_1 = H = 1.023$ m. Then

$$V_0 = \frac{Q}{B(H_1 + P)} = \frac{3.0}{2.0\,(1.023 + 1.5)} = 0.594\ \text{m/s}$$

$$\frac{V_0^2}{2g} = \frac{(0.594)^2}{2 \times 9.81} = 0.018\ \text{m}$$

$$H_1 = 1.023 - 0.018 = 1.005\ \text{m}$$

New $\quad V_0 = \dfrac{Q}{B(H_1 + P)} = \dfrac{3.0}{2.0\,(1.005 + 1.5)}$

$$= 0.599\ \text{m/s}$$

$$\frac{V_0^2}{2g} = 0.0183 \approx 0.018 \text{ as obtained earlier}$$

Hence H_1 = head over the broad-crested weir = **1.005 m**

OBJECTIVE QUESTIONS

Uniform Flow—Definitions

12.1 Uniform flow in a channel is characterized by the following statement:
(a) Total energy remains constant along the channel.
(b) Gradient of the total energy is parallel to the channel bed
(c) Specific energy decreases along the channel.
(d) Total energy line either rises or falls depending upon the Froude number.

12.2 Uniform flow in an open channel exists when the flow is steady and the
(a) channel is frictionless
(b) channel is non-prismatic
(c) channel is prismatic
(d) channel is prismatic and the depth of flow is constant along the channel.

12.3 In defining a Froude number applicable to channels of any shape, the length parameter used is the
(a) depth of flow
(b) hydraulic radius
(c) ratio of area to top width
(d) wetted perimeter

12.4 The flow can be uniform in
(a) a non-prismatic channel
(b) a wide rectangular channel
(c) a horizontal trapezoidal channel
(d) a frictionless rectangular channel

Specific Energy and Critical Depth

12.5 A rectangular channel has its width reduced from 6.0 m to 4.0 m at a transition. If the depth of flow upstream of the contraction is 1.2 m, the change in the bottom elevation at the transition required to cause zero change in the water surface elevation is
(a) 0.60 m drop (b) 0.60 m rise
(c) 0.30 m drop (d) 0.30 m rise

12.6 The term *alternate depths* in open channel flow is used to designate the depths
(a) at the beginning and end of a hydraulic jump
(b) having the same kinetic energy for a given discharge
(c) having the same specific energy for a given discharge
(d) at the beginning and end of a gradually varied flow profile

12.7 Which of the following conditions is the chief characteristic of critical flow?
(a) $\frac{Q^2 T}{g A^3} = 1$ (b) $\frac{Q T^2}{g A^2} = 1$
(c) $\frac{Q^2 R}{g A^3} = 1$ (d) $\frac{Q^2 T^2}{g A^3} = 1$

12.8 If the alternate depths for certain flow in a rectangular channel are 0.5 m and 3.0 m respectively, the critical depth for this channel is
(a) 1.087 m (b) 1.333 m
(c) 1.500 m (d) 3.500 m

12.9 While determining the Froude number applicable to channels of any shape, the length parameter used is the
(a) ratio of area to wetted perimeter
(b) wetted perimeter
(c) depth of flow
(d) ratio of area to top width

12.10 For a triangular channel having side slopes of 2 horizontal : 1 vertical, the Froude number is given by $F =$
(a) $V/\sqrt{gy}$ (b) $2V/\sqrt{gy}$
(c) $V/\sqrt{2gy}$ (d) $V/\sqrt{g(y/2)}$

12.11 In a rectangular channel if the critical depth is 2.0 m, the specific energy at critical depth is
(a) 3.0 m (b) 1.5 m
(c) 2.0 m (d) 2.5 m

12.12 In a rectangular channel the depth of flow is 1.6 m and the specific energy at that section is 2.8 m. The flow is
(a) sub-critical (b) supercritical
(c) critical (d) not possible

12.13 For a uniform flow with a depth of 0.6 m and Froude number of 2.0 in a rectangular channel, the specific energy will be
(a) 2.4 m (b) 0.8 m
(c) 2.6 m (d) 1.8 m

12.14 A rectangular channel carries a uniform flow with a Froude number of 2.83. The ratio of critical depth to normal depth of this flow is
(a) 1.68 (b) 2.83
(c) 2.00 (d) 4.75

12.15 In a triangular channel with side slopes of 2.0 horizontal : 1 vertical, the critical depth is 2.8 m. The specific energy at critical depth is
(a) 3.5 m (b) 3.0 m
(c) 4.2 m (d) 3.72 m

12.16 For a given discharge in a channel at critical depth,
(a) the total energy is minimum
(b) the total energy is maximum
(c) the specific energy is maximum
(d) the specific energy is minimum.

12.17 At critical depth,
(a) the discharge is minimum for a given specific energy
(b) the discharge is maximum for a given specific force
(c) the discharge is minimum for a given specific force
(d) the discharge is maximum for a given specific energy

12.18 The specific energy E_c in a critical flow at a depth y_c occurring in a triangular channel is given by E_c =
(a) 1.25 y_c (b) 1.50 y_c
(c) 1.75 y_c (d) 2.5 y_c

12.19 For a given discharge in a channel the critical depth is a function of
(a) slope of the channel
(b) roughness of the channel
(c) geometry of the channel
(d) viscosity of the liquid

12.20 If the Froude number characterising the flow in an open channel is less than unity, an increase in the channel width at a transition causes the water surface elevation to
(a) remain unchanged
(b) decrease
(c) increase
(d) form ripples

12.21 In a supercritical flow in a rectangular channel, a smooth expansion changes the width from B_1 to B_2. This causes the water surface elevation after the expansion to
(a) increase
(b) decrease
(c) remain unchanged
(d) increase or decrease depending upon the channel roughness

12.22 In subcritical flow in a channel, Δz_m is the minimum height of a smooth hump that can be installed to cause critical flow over the hump. If the hump of height $\Delta z > \Delta z_m$ is installed, then the flow over the hump will be
(a) subcritical
(b) supercritical
(c) critical and the upstream water surface will rise
(d) critical and a lowering of the upstream water surface will occur

12.23 In subcritical flow in a channel Δz_m represents the maximum height of a hump that can be built without causing any change in the upstream water surface. If a hump of height $\Delta z < \Delta z_m$ is built, then
(a) the upstream water surface will be lowered
(b) the flow over the hump will be critical
(c) the flow over the hump will be supercritical
(d) the flow over the hump will be subcritical

12.24 For a given discharge in a horizontal frictionless channel two depths may have the same specific force. These two depths are known as
(a) specific depths
(b) sequent depths
(c) alternate depths
(d) normal and critical depths

12.25 For flow under a sluice gate where the upstream depth is 1.2 m and the depth at the *vena contracta* is 0.3 m, the discharge per metre width would be nearly.
(a) 0.36 m^3/s (b) 1.25 m^3/s
(c) 1.45 m^3/s (d) 4.0 m^3/s

Uniform Flow: Resistance and Computation

12.26 A rectangular channel 3 m wide is laid on a slope of 0.0002. The average boundary shear stress for depth of flow of 1.5 m is nearly
(a) 0.90 N/m^2 (b) 0.45 N/m^2
(c) 0.30 N/m^2 (d) 0.15 N/m^2

12.27 Normal depth of flow computed using Manning's formula in a rectangular open channel of given slope and roughness and carrying a discharge of q m^3/s per metre width is proportional to
(a) $q^{-0.67}$ (b) $q^{0.5}$
(c) $q^{0.6}$ (d) $q^{0.67}$

12.28 The dimensions of Manning's roughness coefficient n are
(a) $L^{1/2} T^{-1}$ (b) $L^{-1/3}T$
(c) $M^0L^0T^0$ (d) L

12.29 The dimensions of Chezy coefficient C are
(a) $L^{-1/3}\,T$ (b) $M^0L^0T^0$
(c) $L^{1/2}\,T^{-1}$ (d) LT^{-1}

12.30 Manning's roughness coefficient n is related to Darcy–Weisbach friction factor f as
(a) $n = \left[\dfrac{fR^{1/3}}{8g}\right]$ (b) $n = R^{2/3}/\sqrt{8gf}$
(c) $n = [8fR^{1/3}/g)^{1/2}$ (d) $n = \sqrt{8g/f}\;R^{1/6}$

12.31 The Chezy coefficient C and Manning's n are related as
(a) $C = n^{1/3}\,R^{1/6}$ (b) $C = \dfrac{1}{n}\,R^{1/6}$
(c) $C = \dfrac{n^{1/6}}{R}$ (d) $n = C\,R^{1/6}$

12.32 A rectangular channel has a velocity of 1.5 m/s and an average shear stress of 6 N/m^2. The Chezy coefficient for this flow is
(a) 70.7 (b) 40.4
(c) 50.5 (d) 60.6

12.33 In a wide rectangular channel the full supply depth is 1.52 m. If 50% of the full supply discharge is flowing in this channel, the depth of flow will be
(a) 0.76 m (b) 0.90 m
(c) 1.00 m (d) 0.43 m

12.34 A channel of bed slope 0.0009 carries a discharge of 30 m^3/s when the depth of flow is 1.0 m. What discharge does an exactly similar channel at the same depth of flow carry if the slope is decreased to 0.0001?
(a) 10 m^3/s (b) 15 m^3/s
(c) 60 m^3/s (d) 90 m^3/s

12.35 In a wide rectangular channel, an increase in a normal depth by 20% correspond to an increase in discharge by about
(a) 13% (b) 25%
(c) 36% (d) 48%

12.36 For a hydraulically efficient rectangular channel of bed width 4.0 m, the depth of flow is
(a) 4.0 m (b) 8.0 m
(c) 1.0 m (d) 2.0 m

12.37 In a wide rectangular channel the mean velocity is given by the Chezy formula with Chezy coefficient C = constant. In this channel as the normal depth increase, the Froude number of the flow
(a) increases
(b) decreases
(c) remains constant
(d) increases initially and then becomes constant.

12.38 For a hydraulically efficient triangular section the hydraulic radius R =
(a) $2\sqrt{2}\,y$ (b) $y/2\sqrt{2}$
(c) $y/2$ (d) y

12.39 In a hydraulically efficient circular channel the ratio of the hydraulic radius to the diameter of the channel is
(a) 1.0 (b) 0.5
(c) 0.25 (d) 0.125

12.40 In a hydraulically most efficient trapezoidal channel section the hydraulic radius R =
(a) $y/2$ (b) y
(c) $y/2\sqrt{2}$ (d) $\dfrac{4}{\sqrt{3}}\,y$

12.41 In a hydraulically most efficient trapezoidal channel section the ratio of the bed width to depth is
(a) 0.50 (b) 0.707
(c) 0.866 (d) 1.155

12.42 At the same mean velocity, the ratio of head loss per unit length for a sewer pipe running full to that for the same pipe flowing half full would be
(a) 2.0 (b) 1.67
(c) 1.0 (d) 0.67

Hydraulic Jump

12.43 The sequent depth ratio of a hydraulic jump in a rectangular channel is 16.48. The Froude number at the beginning of the jump is
(a) 5.0 (b) 8.0
(c) 10.0 (d) 12.0

12.44 The Froude number at the end of a hydraulic jump in a rectangular channel is 0.25. The sequent depth ratio of this jump is
(a) 2.5 (b) 5.2
(c) 8.9 (d) 9.8

12.45 In a horizontal rectangular channel a hydraulic jump with a sequent depth ratio of 5.0 is formed. This jump can be classified as
(a) weak jump
(b) oscillating jump
(c) strong jump
(d) steady jump

12.46 The sequent depths in a hydraulic jump formed in a 4.0 m wide rectangular channel are 0.2 m and 1.0 m. The discharge in the channel, in m^3/s, is
(a) 5.00 (b) 1.12
(c) 2.17 (d) 4.34

12.47 The sequent depths in a hydraulic jump formed in a horizontal rectangular channel are 0.2 m and 2.0 m. The length of the jump is about

(a) 50 m (b) 12 m
(c) 8 m (d) 2 m

12.48 In a hydraulic jump occurring in a horizontal rectangular channel the sequent depths are 0.25 m and 1.25 m.
The energy loss in this jump is

(a) 0.8 m (b) 1.0 m
(c) 1.25 m (d) 1.50 m

12.49 The discharge per metre width at the foot of a spillway is 10 m^3/s at a velocity of 20 m/s. A perfect free jump will occur at the foot of the spillway when the tail water depth is nearly

(a) 4.5 m (b) 5.00 m
(c) 6.50 m (d) 8.50 m

12.50 A hydraulic jump occurs near the junction of

(a) mild slope and milder slope
(b) steep slope and mild slope
(c) mild slope and steep slope
(d) steep slope and steeper slope

[The sequence of slopes is in the order mentioned]

Gradually Varied Flow

12.51 The differential equation of the gradually varied flow can be written by using Manning's formula for the case of a wide rectangular channel as $\frac{dy}{dx}$ =

(a) $S_0 \frac{1-(y_0/y)^{3.33}}{1-(y_c/y)^3}$ (b) $S_0 \frac{1-(y_c/y)^{3.33}}{1-(y_0/y)^3}$

(c) $S_0 \frac{1-(y_0/y)^3}{1-(y_c/y)^3}$ (d) $S_0 \frac{1-(y_0/y_c)^3}{1-(y_c/y)^{3.33}}$

12.52 If E = specific energy at a section in a gradually varied flow, then dE/dx =

(a) $S_0 + S_f$ (b) $S_0 - S_f$
(c) $S_f - S_0$ (d) $S_f/S_0 - 1$

where S_f = energy slope and S_0 = bed slope.

12.53 A long prismatic channel ends in an abrupt drop. For an initial subcritical flow the water surface profile

(a) starts from the critical depth line and joins the normal depth line asymptotically
(b) starts from the critical depth at the drop and decreases in the upstream direction.
(c) lies wholly above the normal depth line
(d) lies wholly below the critical depth line

12.54 If in a gradually varied flow dy/dx is positive, then dE/dx

(a) is always negative
(b) is always positive
(c) is positive if $y/y_c > 1$
(d) is negative if $y > y_c$

12.55 A 3 m wide rectangular channel flowing at its normal depth of 0.8 m carries a discharge of 9.5 m^3/s. The channel slope is

(a) steep (b) critical
(c) mild (d) none of the above

12.56 In an M_1 type of gradually varied flow profile

(a) $y_0 > y > y_c$ (b) $y_0 > y_c > y$
(c) $y > y_0 > y_c$ (d) $y_c > y_0 > y$

12.57 In an M_2 type of gradually varied flow profile

(a) $y_0 > y > y_c$ (b) $y > y_0 > y_c$
(c) $y_0 > y_c > y$ (d) $y_c > y > y_0$

12.58 In an S_1 type of gradually varied flow

(a) $y_c > y > y_0$ (b) $y > y_c > y_0$
(c) $y_c > y_0 > y$ (d) $y > y_0 > y_c$

12.59 The flow will be in supercritical state in the following profiles

(a) M_3, S_3 and M_1 (b) M_2, S_1 and M_3
(c) S_2, S_3 and M_3 (d) S_1, S_2 and S_3

12.60 The following types of gradually varied flow profiles do not exist.

(a) C_2, H_1 and A_1 (b) C_3, H_2 and A_2
(c) C_2, S_1 and A_3 (d) C_2, H_2 and A_1

12.61 Fig 12.27 shows a gradually varied flow situation in an open channel with a break in bed slope. Types of water surface profiles occurring from left to right are

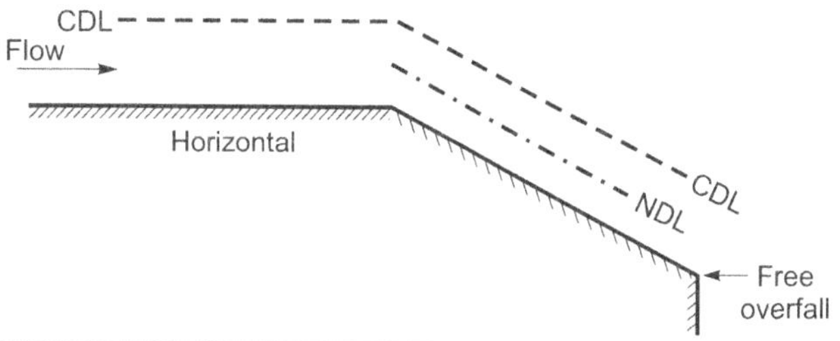

Fig. 12.27 *Question 12.61*

(a) H_3, S_3 (b) H_2, S_2
(c) H_2, M_3 (d) H_3, M_2

12.62 A sluice gate is operated in a long open channel with a steep slope as shown in Fig. 12.28 in such a manner that its opening is smaller than the normal depth.

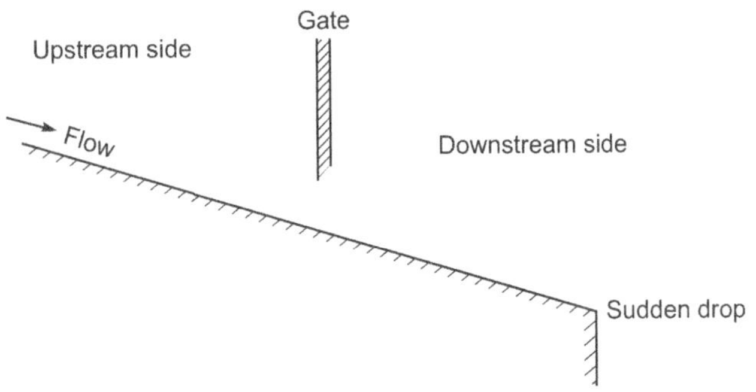

Fig. 12.28 *Question 12.62*

In this situation, the hydraulic jump
(a) will occur on the downstream side
(b) will occur on the upstream side
(c) will occur on both upstream and downstream sides
(d) will not occur

12.63 Which of the following is the correct representation of sequence of surface profiles if the channel slope changes from Mild to Steep?
(a) M_1, S_1 (b) M_1, S_2
(c) M_2, S_3 (d) M_2, S_2

12.64 A wide rectangular channel carries a flow of 2.96 m^3/s per metre width. The bed slope of the channel is 1.0×10^{-4} and Manning's $n = 0.021$. If at a section the depth of flow is 1.5 m the energy slope at that section is
(a) 0.01 (b) 0.00228
(c) 0.0009 (d) 0.001

12.65 A 2 m wide rectangular channel flowing at its normal depth of 1.2 m carries a discharge of 6.0 m^3/s. If at a section, the depth of flow is 1.10 m, the water surface at that location is a part of the gradually varied flow of type
(a) S_2 (b) M_2
(c) M_3 (d) S_3

12.66 The flow in a long 4.0 m wide rectangular channel is 8.0 m^3/s. The normal depth of flow is 1.5 m. If, at a certain section A, the depth of flow in the channel is 1.0 m, the depth of flow at a section downstream of A would be
(a) > 1.0 m (b) < 1.0 m
(c) = 1.0 m (d) ≤ 1.0 m

Broad-crested Weir

12.67 The discharge Q over a broad-crested weir of length L is often expressed as $Q = 1.705\ C_d\ LH^{3/2}$. In this expression H is the difference in elevation between
(a) the upstream energy line and the crest
(b) the upstream water surface and the crest
(c) the upstream water surface and the upstream bed
(d) the upstream energy line and the downstream energy line

12.68 The modular limit of a broad-crested weir is about
(a) 15% (b) 35%
(c) 67% (d) 90%

12.69 If the C_d and length L of a rectangular notch and a broad-crested weir are the same, then for the same head on both of these
(a) the broad-crested weir passed 73% more discharge than the rectangular notch
(b) the rectangular notch passes 57.7% less discharge
(c) the rectangular notch passes 73% more discharge than the broad-crested weir
(d) both notch and the broad-crested weir pass equal discharges.

PROBLEMS

A. Uniform Flow

12.1 A trapezoidal channel has a bottom width of 3.5 m and side slopes of 1.5 horizontal : 1 vertical. The channel has a longitudinal slope of 1/4000 and carries a certain discharge at a depth of 1.70 m. Calculate the average shear stress on the boundary.
(*Ans.* $\tau_0 = 2.614$ Pa)

12.2 Show by using Manning's formula that the average boundary shear stress is given by

$$\tau_0 = \frac{\gamma n^2 V^2}{R^{1/3}}$$

12.3 Show that the Chezy coefficient C, Manning's coefficient n and Darcy–Weisbach friction factor f are related as

$$f = \frac{8g}{C^2} = \frac{8gn^2}{R^{1/3}}$$

12.4 A trapezoidal channel has a bed width of 2.0 m, side slope of 1.25 horizontal : 1 vertical and carries a discharge of 9.00 m^3/s at a depth of 2.0 m. Calculate the average velocity and bed slope of the channel. [Assume Manning's coefficient $n = 0.015$].
(*Ans.* $V = 1.0$ m/s; $S_0 = 2.0534 \times 10^{-4}$)

12.5 In a trapezoidal channel of bottom width 3.0 m and side slopes 2 horizontal : 1 vertical, the depth of flow is 1.4 m. If the channel has a Manning's coefficient $n = 0.015$, estimate the values of Chezy coefficient C and Darcy–Weisbach friction factor f.

(*Ans.* $C = 65.2$; $f = 0.0184$)

12.6 An open channel of trapezoidal section has a base width of 2.5 m and sides inclined at 60° to the horizontal. The bed slope is 1 in 500. It is found that when the discharge is 1.5 m^3/s the normal depth is 0.5 m. Using Manning's formula calculate the discharge when the normal depth is 0.70 m.

(*Ans.* $Q = 2.594$ m^3/s)

12.7 A trapezoidal channel with cross section as in Fig. 12.29 carries a discharge of 10 m^3/s at a depth of 1.5 m under uniform flow conditions. The longitudinal slope of the channel bed is 0.001. Compute the average shear stress in N/m^2 on the boundary. Also compute the value of Manning's n.

(*Ans.* $\tau_0 = 9.134$ Pa; $n = 0.025$)

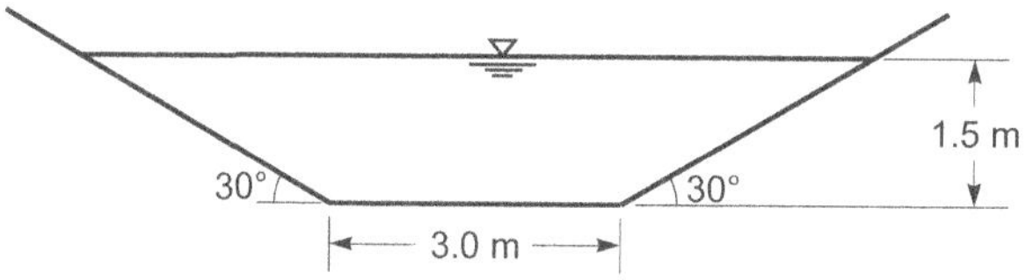

Fig. 12.29

12.8 Figure 12.30 shows the situation in a uniform flow in a wide rectangular channel. Calculate the longitudinal slope if $n = 0.02$.

(*Ans.* $S_0 = 3.96 \times 10^{-3}$)

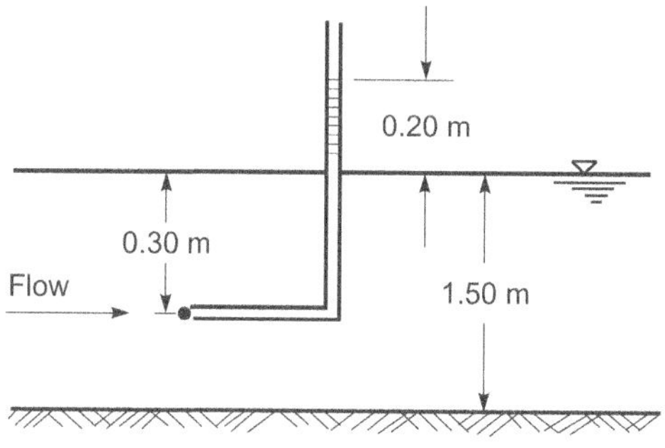

Fig. 12.30

12.9 In a wide open channel of depth 0.6 m the velocity of 0.12 m below water surface is 2.35 m/s and at a depth of 0.48 m below the water surface it is 0.95 m/s. Estimate the discharge per metre width and the shear stress on the bed if the bed slope of the channel is 0.0004.

(*Ans.* $q = 0.99$ $m^3/s/m$; $\tau_0 = 23.5$ Pa)

12.10 A triangular channel has one side vertical and the other side has a slope of 1.5 horizontal : 1 vertical. Calculate the discharge in this channel for a depth of flow of 0.90 m. The channel has a longitudinal slope of 1/3000 and Manning's $n = 0.02$.

(*Ans.* $Q = 0.2146$ m^3/s)

12.11 A rectangular channel 3.0 m wide had a badly damaged lining whose Manning's n was estimated as 0.025. The lining was repaired and it now has an n value of 0.014. If the depth of flow remains the same at 1.30 m as before the repair, estimate the new discharge and percentage increase in discharge as a result of repair.

(*Ans.* $Q_2 = 4.894$ m^3/s; 78.5% increase)

12.12 A rectangular channel 2.5 m wide and laid on a slope of 0.005 has a discharge of 10.0 m^3/s at a depth of 1.2 m. What would be the discharge when the depth is 0.80 m?

(*Ans.* $Q_2 = 5.728$ m^3/s)

12.13 A circular concrete pipe 0.60 m in diameter is laid on a slope of 0.001. Manning's n for the pipe can be taken as 0.014. Estimate the discharge when the depth of flow is (a) 42 cm, (b) 30 cm and (c) 12 cm.

[*Ans.* (a) $Q = 151$ L/s; (b) $Q = 90$ L/s; $Q = 15.8$ L/s]

12.14 What diameter of a semicircular channel will have the same discharge as a rectangular channel of width 2.0 m and depth 1.2 m? Assume the bed slope and Manning's n are the same for both the channels.

(*Ans.* $D = 2.396$ m)

12.15 For a channel shown in Fig. 12.31 the bed slope is 1 in 1000 and Manning's $n = 0.014$. Calculate the discharge.

(*Ans.* $Q = 1.898$ m^3/s)

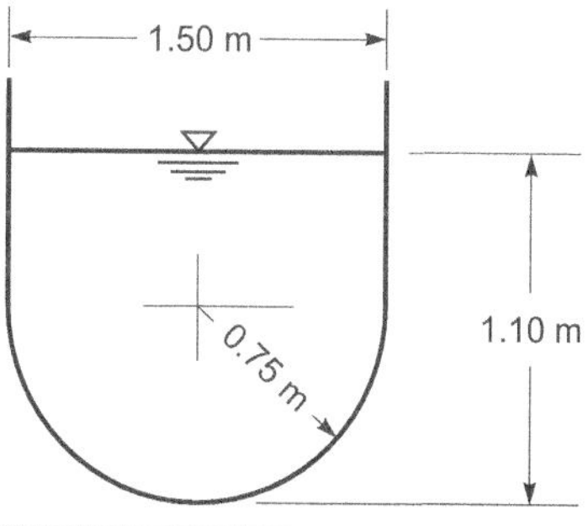

Fig. 12.31 *Problem 12.15*

12.16 Calculate the quantity of water that will be discharged at a uniform depth of 0.9 m in a 1.2 m diameter pipe which is laid at a slope of 1 in 1000. Use Chezy formula with $C = 58$.

(*Ans.* $Q = 1.0\ m^3/s$)

12.17 A commonly used lined canal section is shown in Fig. 12.32. It consists of a triangular section with side slopes of m horizontal : 1 vertical. Further, it is rounded off at the bottom by a radius equal to full supply depth.
For one such channel, commonly known as the *standard lined triangular section*, the full supply depth is 2.0 m, $m = 1.5$ and bed slope = 1/4000. Manning's $n = 0.015$. Determine the full supply discharge.

(*Ans.* $Q = 8.804\ m^3/s$)

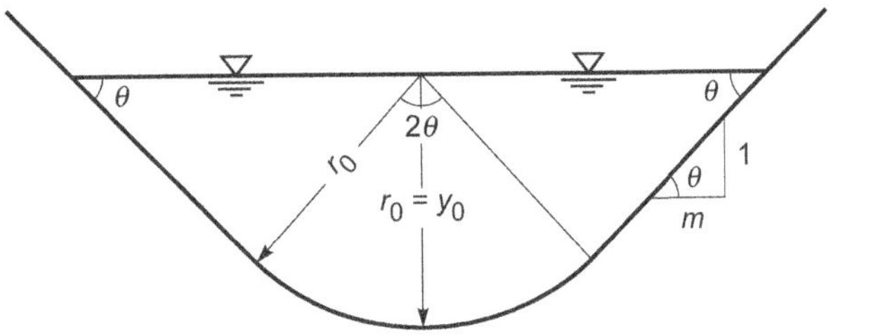

Fig. 12.32 *Standard Lined Triangular Channel*

12.18 For a standard lined triangular section (Fig. 12.32) the full supply depth is 1.5 m. The side slopes are 2.0 horizontal : 1 vertical and Manning's $n = 0.014$. Determine the bed slope necessary to pass a full supply discharge of 10.0 m^3/s.

(*Ans.* $S_0 = 9.361 \times 10^{-4}$)

12.19 Show that for a standard lined triangular section (Fig. 12.32), the hydraulic radius is equal to $y_0/2$, where y_0 = full supply depth.

B. Normal Depth

12.20 A wide rectangular channel has a slope of 0.0004 and its Manning's roughness coefficient is 0.02. If a discharge intensity of 2.54 m^3/s per metre width is to be passed in this channel, estimate the normal depth.

(*Ans.* $y_0 = 1.75$ m)

12.21 A triangular channel has a vertex angle of 75° and a longitudinal slope of 0.001. If Manning's $n = 0.015$, estimate the normal depth for a discharge of 250 L/s in this channel.

(*Ans.* $y_0 = 0.668$ m)

12.22 A 1.5 m wide rectangular channel carries a discharge of 300 L/s. This longitudinal slope of the channel is 0.00016. If Manning's $n = 0.018$, estimate the normal depth.

(*Ans.* $y_0 = 0.595$ m)

12.23 A channel with a trapezoidal cross section is to carry 30.0 m^3/s. In this channel the bottom width = 6.0 m, side slopes = 2 horizontal : 1 vertical, $n = 0.014$ and longitudinal bed slope = 0.0002. Determine the normal depth.
[***Hint:*** A trial and error procedure will be needed to estimate the normal depth in this case.]

(*Ans.* $y_0 = 2.218$ m)

12.24 A wide rectangular channel is to carry a discharge of 2.5 m^3/s/metre width at a Froude number of 0.5. Assuming Manning's $n = 0.015$, calculate the (a) normal depth, (b) required bed slope and (c) average bed shear stress.

(*Ans.* (a) $y_0 = 1.6815$ m;
(b) $S_0 = 4.973 \times 10^{-4}$;
(c) $\tau_0 = 8.19$ Pa)

12.25 Show that the normal depth in a triangular channel of side slopes m horizontal: 1 vertical is given by

$$y_0 = 1.1892 \left[\frac{Qn}{\sqrt{S_0}}\right]^{3/8} \left[\frac{m^2+1}{m^5}\right]^{1/8}$$

12.26 For a standard lined triangular section (Fig. 12.32) the side slopes are 1.25 horizontal : 1 vertical and the bed slope is 1 in 1600. Calculate the normal depth for a discharge of 10.0 m^3/s. (Take Manning's $n = 0.016$).

(*Ans.* $y_0 = 1.866$ m)

C. Maximum Velocity and Discharge

12.27 Water flows in a triangular duct resting on one of its sides. The duct is in the shape of an isosceles triangle of bed width B and sides making an angle 45° with the bed. The water surface is below the vertex and the bed is horizontal laterally. Determine the relationship between the depth of flow y and the bed width B for maximum velocity condition.

(*Ans.* $y = 0.338\ B$)

12.28 A square conduit of sides b, placed with its diagonal vertical, carries water with a free surface. Show that the channel can convey maximum discharge when $y = 1.259\ b$.

12.29 A circular channel of diameter 0.6 m is laid on a slope of 1 in 3000. Calculate the maximum discharge it can convey as an open channel. (Manning's $n = 0.018$).

(*Ans.* $Q_m = 0.0871\ m^3/s$)

D. Efficient Channel Sections

12.30 A rectangular channel (Manning's n = 0.020) is 5.0 m wide, 0.9 m deep and has a slope of 1 in 1600. If the channel had been designed to be of efficient rectangular section, for the same wetted perimeter what additional discharge would it carry?

(*Ans.* ΔQ = 2.211 m^3/s; 51.76% increase)

12.31 A trapezoidal channel has one side vertical and the other side has a slope of 1.5 horizontal : 1 vertical. This channel is to carry a discharge of 15 m^3/s with a velocity of 1.5 m/s. Calculate the dimensions of an efficient section of this shape and also the bottom slope necessary to achieve this discharge. (Manning's n = 0.013).

(*Ans.* y_e = 2.207 m; B_e = 2.876 m; $S_0 = 3.335 \times 10^{-4}$)

12.32 Determine the dimensions of a concrete lined (n = 0.014) trapezoidal channel of *most efficient proportions* to carry a discharge of 10.0 m^3/s. The bed slope of the channel is 0.005.

(*Ans.* y_{em} = 1.250 m; B_{em} = 1.444 m; m = 0.5773)

12.33 Determine the efficient section and bed slope of a trapezoidal channel (n = 0.025) designed to carry 15 m^3/s of flow. To prevent scouring the velocity is to be 1.0 m/s and the side slopes of the channel are 1 vertical to 2 horizontal.

(*Ans.* y_e = 2.463 m; B_e = 1.163 m; $S_0 = 4.735 \times 10^{-4}$)

12.34 What should be the dimensions of the efficient trapezoidal section of side slopes 1.5 horizontal to 1 vertical, if the design discharge is 12 m^3/s and the channel slope is 0.0005? (Take n = 0.02).

(*Ans.* y_e 2.19 m; B_e = 1.326 m)

12.35 What are the dimensions of an efficient rectangular brick channel (n = 0.015) designed to carry 6.0 m^3/s of water with a bed slope of 0.002? How much additional discharge is carried by a semicircular channel of the same area?

(*Ans.* y_s = 1.192 m; B_e = 2.384 m; Q_s = 6.5033 m^3/s; ΔQ = 0.5033 m^3/s)

12.36 A rectangular channel 8 m wide and 1.5 m deep has a slope of 1 in 1000 and is lined with smooth concrete plaster. It is desired to enhance the discharge to a maximum by changing the dimensions of the channel but keeping the same quantity of lining. Work out the new dimensions and the percentage increase in discharge. Assume Manning's n = 0.015 and the bed slope to remain unchanged.

(*Ans.* y_e = 2.75 m; B_e = 5.50 m; Q = 39.43 m^3/s, 47.06% increase)

12.37 A trapezoidal channel is 6.0 m wide and has a side slope of 0.5 horizontal : 1 vertical. If the bed slope of the channel is 0.0004 and Manning's n = 0.02, find the discharge which can make the channel a hydraulically efficient section.

(*Ans.* Q = 73.9 m^3/s)

12.38 For a triangular cross sectional canal, show that the cross section is most efficient when the sides make an angle of 45° with the vertical. Determine the hydraulic radius in terms of the depth of flow for such a channel.

(*Ans.* $R = y/2\sqrt{2}$)

12.39 A trapezoidal channel has side slopes of 1.5 horizontal : 1 vertical and is required to carry 15 m^3/s with a slope of 1 in 4000. If the channel is lined, the Manning's coefficient is 0.014 and it is 0.028 if unlined. What percentage of earthwork is saved in a lined section relative to an unlined section when hydraulically efficient section is used in both cases? The free board can be assumed to be 0.75 m in both cases and the lining can be assumed to be up to the top of the section.

(*Ans.* Earth work saved = 33.4%)

12.40 Show that the normal depth in a hydraulically *most efficient* trapezoidal section is given by

$$y_{em} = 0.9678\left(\frac{Qn}{\sqrt{S_0}}\right)^{3/8}$$

12.41 Show that the flow rate in a *most efficient* trapezoidal section is given by

$$Q = \frac{0.7435}{n}\,B^{8/3}\,\sqrt{S_0}$$

12.42 The following Table 12.6 lists some typical problems relating to efficient trapezoidal channels. Out of the nine variables listed in the Table, four are given in each problem. You are to determine the others and fill in the blanks in the table. Note that when m = 0 the channel is rectangular and when B_e = 0 the channel is triangular.

Table 12.6 *Dimensions of Efficient Channels–Problem 12.42*

Prob.	Q	B_e	y_e	m	V	P_e	A_e	n	S_0
	(m^3/s)	(m)	(m)		(m/s)	(m)	(m^2)		
(i)	15			2.0				0.015	0.0004
(ii)				1.5	1.5			0.020	0.0009
(iii)				1.5		12.0		0.020	0.0006
(iv)				1.0			5.0	0.015	0.0004
(v)			1.2	1.5				0.018	0.001
(vi)		1.5		1.0				0.014	0.0004
(vii)	6.0			0				0.015	0.0008
(viii)	3.0	0						0.015	0.0008

Answers to Problem 12.42

Prob.	Q	B_e	y_e	m	V	P_e	A_e	n	S_0
	(m^3/s)	(m)	(m)		(m/s)	(m)	(m^2)		
(i)	15.0	0.991	2.10	2.0	1.376	10.383	10.90	0.015	0.0004
(ii)	1.256	0.84	0.20	1.5	1.50	0.842	0.837	0.020	0.0009
(iii)	30.94	1.726	2.85	1.5	1.551	12.0	19.952	0.020	0.0006
(iv)	5.87	1.370	1.654	1.0	1.175	6.047	5.0	0.015	0.0004
(v)	3.79	0.727	1.20	1.5	1.250	5.053	3.032	0.018	0.001
(vi)	8.01	1.50	1.811	1.0	1.337	6.621	5.994	0.014	0.0004
(vii)	6.0	1.190	2.380	0	2.118	9.522	2.833	0.015	0.0008
(viii)	3.0	0	1.544	1.0	1.059	4.366	2.832	0.015	0.0008

E. Specific Energy and Critical Depth

12.43 A circular channel 1.20 m in diameter is laid on a slope of 0.008. The Manning's n for the channel can be taken as 0.018. Calculate the specific energy when the depth of flow is 0.60 m.

(*Ans.* $E = 0.853$ m)

12.44 A rectangular channel 1.5 m wide carries a discharge of 9.0 m^3/s at a depth of 2.0 m. Calculate the (a) specific energy and (b) depth alternate to the given depth of 2.0 m.

(*Ans.* $E = 2.4587$ m; $y_2 = 1.214$ m)

12.45 A rectangular channel 2.0 m wide carries a discharge of 6.0 m^3/s. If the specific energy is 1.8 m calculate the alternate depths.

(*Ans.* $y_1 = 0.625$ m; $y_2 = 1.626$ m)

12.46 In a rectangular channel 2.0 m wide the two alternate depths to a given specific energy are 0.30 m and 1.2 m respectively. Calculate the discharge.

(*Ans.* $Q = 2.446$ m^3/s)

12.47 For a constant specific energy of 3.0 m, what maximum flow may occur in a rectangular channel of 4.5 m bed width?

(*Ans.* $Q_m = 39.87$ m^3/s)

12.48 What is the lowest possible specific energy for a water discharge of 12 m^3/s to flow through a 3.0 m wide rectangular channel?

(*Ans.* $E_{min} = 1.766$ m)

12.49 A rectangular channel 12.0 m wide carries 200 m^3/s. Find the critical depth and critical velocity. What slope will produce this critical velocity in the channel if $n = 0.02$?

(*Ans.* $S_0 = 4.682 \times 10^{-3}$; $y_c = 3.048$ m; $V_c = 5.468$ m/s)

12.50 What is the maximum discharge that may be carried by a 3.2 m wide rectangular channel at a specific energy of 1.8 m?

(*Ans.* $Q_m = 13.18$ m^3/s)

12.51 Fill in the missing data in Table 12.7, connected with critical depth computation in a rectangular channel.

Table 12.7 *Data on Calculation of y_c in a Rectangular Channel–Problem 12.51*

Prob.	Q (m^3/s)	B (m)	y_c (m)	E_c (m)
(i)		2.6		0.40
(ii)	10.48		0.70	
(iii)	6.0	2.0		
(iv)		5.0		1.2

Answers to Problem 12.51

Prob.	Q (m^3/s)	B (m)	y_c (m)	E_c (m)
(i)	5.15	2.6	0.40	0.60
(ii)	10.48	4.0	0.70	1.05
(iii)	6.0	2.0	0.917	1.376
(iv)	14.0	5.0	0.80	1.2

12.52 In a rectangular channel the specific energy is 2.0 m and one of the alternate depths is 0.5 m. Calculate the (a) other alternate depth, (b) critical depth and (c) discharge intensity.

(*Ans.* (a) $y_1 = 1.896$ m; (b) $y_c = 1.333$ m; (c) $q = 4.822$ m^3/s/m)

12.53 Corresponding to a discharge of 9.0 m^3/s, calculate the critical depth in

(i) a rectangular channel of bed width 2.5 m;
(ii) triangular channel of vertex angle 72°;
(iii) trapezoidal channel of bottom width 3.0 m and side slopes 0.5 horizontal : 1 vertical

[***Note:*** Prob (iii) requires trial and error method.]

(*Ans.* (a) $y_c = 1.097$ m; (b) $y_c = 1.99$ m; (c) $y_c = 0.921$ m)

12.54 Calculate the discharge corresponding to a critical depth of 1.5 m in a

(a) rectangular channel of width 1.5 m;
(b) triangular channel of vertex angle 120°;
(c) trapezoidal channel of bed width = 6.0 m and side slopes 1.5 horizontal : 1 vertical.

(*Ans.* (a) $Q = 8.631$ m^3/s; (b) $Q = 10.571$ m^3/s; (c) $Q = 42.08$ m^3/s)

12.55 A 1.0 m diameter circular culvert is flowing half full and the flow is in critical state. Estimate the discharge.

(*Ans.* $Q = 0.771$ m^3/s)

12.56 Show that for a triangular channel the Froude numbers corresponding to alternate depths are given by

$$\frac{F_1}{F_2} = \frac{(4 + F_1^2)^{5/2}}{(4 + F_2^2)^{5/2}}$$

12.57 If y_1 and y_2 are alternate depths in a triangular channel of side slope m horizontal: 1 vertical show that

$$\frac{4 y_1^4 y_2^4}{(y_1^2 + y_2^2)(y_1 + y_2)} = y_c^5$$

where y_c = critical depth. Further, show that the specific energy E is given by

$$\frac{E}{y_1} = \frac{\eta^4 + \eta^3 + \eta^2 + \eta}{(1 + \eta^2)(1 + \eta)} \qquad \text{where} \quad \eta = y_2/y_1$$

12.58 A triangular channel has an apex angle of 60° and carries a flow with a velocity of 2.0 m/s and depth of 1.25 m (a) Is the flow sub-critical or super critical? (b) What is the critical depth? (c) What is the specific energy? (d) What is the alternate depth possible for this specific energy?

[*Ans*: (a) Sub-critical, (b) $y_c = 1.148$ m, (c) E = 1.454 m and (d) $y_2 = 1.08$ m]

12.59 Water is flowing at critical depth of 1.4 m in a channel of cross section shown in Fig. 12.33 Estimate the discharge and specific energy in the channel.

[*Ans*: $Q = 11.50$ m^3/.s; E = 2.59 m]

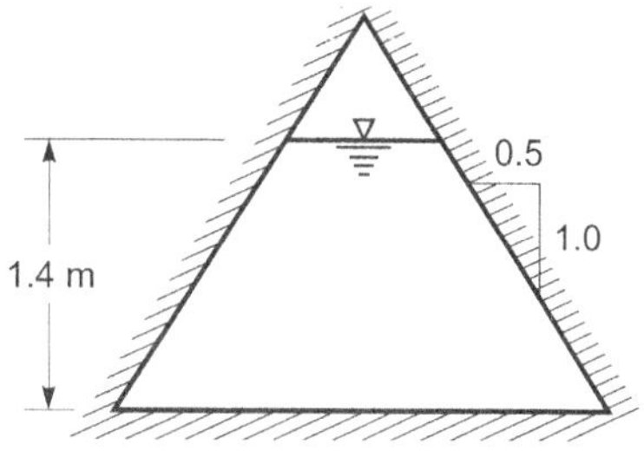

Fig. 12.33

12.60 Water is flowing at critical depth in a channel of shape shown in Fig. 12.34. If the depth of flow is 1.2 m, estimate the discharge and specific energy in the channel.

(*Ans.* $Q = 12.84$ m^3/s; $E_c = 3.00$ m)

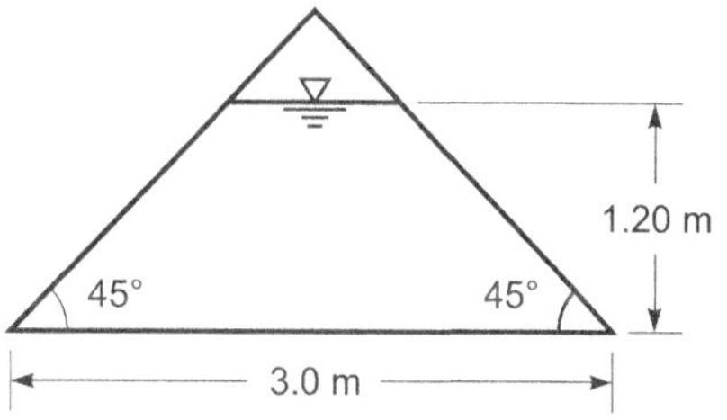

Fig. 12.34

12.61 Calculate the bottom width of a trapezoidal channel required to carry a discharge of 15 m^3/s as a critical flow at a depth of 1.2 m. The side slope of the section is 1.5 m horizontal : 1 vertical.

[***Hint:*** A trial and error method will be needed.]

(*Ans.* $B = 2.535$ m)

12.62 A concrete triangular channel ($n = 0.015$) with side slopes of 1.5 horizontal : 1 vertical is laid on a slope of 0.005. For a depth of flow of 1.5 m, estimate (i) the average shear stress on the boundary and (ii) the critical depth.

[*Ans.* (i) $\tau_0 = 432$ Pa (ii) $y_c = 1.65$ m)

12.63 Show that for a wide rectangular channel the critical slope S_c and critical depth y_c are related as

$$S_c = gn^2/y_c^{1/3}$$

[*Note:* The slope of a channel which carries a given discharge as a uniform flow at the critical depth is known as the critical slope, S_c.]

12.64 A triangular channel with a vertex angle of 120° carries a discharge of 2.0 m^3/s. Find the critical depth and the specific energy corresponding to critical flow. If the slope of the channel is 0.001 and Manning's n = 0.015, what is the normal depth?

(*Ans.* $y_c = 0.771$ m; $y_0 = 1.515$ m; $E_c = 0.964$ m)

F. Transitions

12.65 A horizontal rectangular channel 3.0 m wide is narrowed to 1.5 m width to cause critical flow in the contracted section. If the depth in the contracted section is 0.8 m, calculate the discharge in the channel and the possible depths of flow and corresponding Froude numbers in the 3.0 m wide section. Neglect energy losses in the transition.

(*Ans.* $Q = 3.362$ m^3/s; $y_1 = 1.153$ m; $y_2 = 0.261$ m; $F_1 = 0.289$; $F_2 = 2.683$)

12.66 A 2.0 m wide rectangular channel has a flow with a velocity of 1.3 m/s and depth of 1.2 m. A smooth hump is to be built at a section to cause critical flow. Calculate the minimum height of the hump required to achieve this.

(*Ans.* $\Delta z = 0.344$ m)

12.67 A discharge of 10.0 m^3/s flows at a depth of 2.0 m in a rectangular channel 4.0 m wide. At a section the width is reduced to 3.5 m. It is desired to provide a smooth hump at this contracted section to achieve critical depth without affecting the upstream depth. Estimate the height of the hump required. Neglect frictional losses.

(*Ans.* $\Delta z = 0.669$ m)

12.68 A 3.0 m wide rectangular channel carries a discharge of 10.0 m^3/s at a depth of 1.35 m. If at a section a smooth hump of height 0.25 m is built, estimate the depth of flow over the hump and upstream of it by neglecting frictional losses.

(*Ans.* $y_2 = y_c = 1.042$ m; $y_1' = 1.59$ m)

12.69 A rectangular channel 5.0 m wide carries water with a velocity of 2.0 m/s and a depth of 1.5 m. At a section a smooth hump of height 0.10 m is provided. What are the depths of flow over the hump and upstream of it?

(*Ans.* $y_2 = 1.354$ m; $y_1 = 1.50$ m)

12.70 A discharge of 15 m^3/s flows through a rectangular channel 3.0 m wide. The depth of flow is 2.0 m. A smooth hump of 0.10 m is built at a section. Also, at that section the bottom width is reduced to 2.8 m. What are the depths of water at the contracted section and upstream of it? Neglect frictional losses.

(*Ans.* $y_2 = 1.73$ m; $y_1 = 2.00$ m)

12.71 Water flows at a velocity of 2.40 m/s and a depth of 1.80 m in an open channel of rectangular cross section. If the channel width is reduced from 3.00 m to 2.10 m and the bottom elevation is increased by 0.30 m at a section, to what extent the water surface elevations are effected by this boundary contraction? Assume zero energy loss.

(*Ans.* $y_2 = y_{c2} = 1.5717$ m; $y_1' = 2.50$ m)

12.72 A rectangular channel 3.0 m wide has a smooth contraction which reduces the width to 2.5 m. If the depths of flow at the contraction and at an upstream section are 1.2 m and 2.0 m respectively, estimate the discharge in the channel. (Neglect frictional losses).

(*Ans.* $Q = 13.72$ m^3/s)

12.73 Water flows in a rectangular channel at a depth of 1.5 m. A smooth hump 30 cm high in the bed produces a drop of 15 cm in the water surface elevation without affecting the upstream depth. Neglecting losses, calculate the rate of flow per metre width.

(*Ans.* $q = 2.52$ m^3/s/m)

G. Hydraulic Jump

12.74 Water is being discharged from under a sluice gate at a rate of 18 m^3/s in a 3.0 m wide rectangular channel. A hydraulic jump is found to occur at a section where the depth of flow is 0.50 m. Determine the (a) depth after the jump, (b) energy dissipated in the jump and (c) length of the jump.

[*Ans.* (a) $y_2 = 3.59$ m; (b) $E_L = 4.11$ m; (c) $L_j = 21.9$ m]

12.75 A hydraulic jump occurs in a horizontal rectangular channel with sequent depths of 0.70 m and 4.2 m. Calculate the rate of flow per unit width, energy loss and the initial Froude number.

(*Ans.* $q = 8.4$ m^3/s/m; $E_L = 3.65$ m; $F_1 = 4.58$)

12.76 A hydraulic jump occurs in a horizontal rectangular channel at an initial Froude number of 10.0. What percentage of initial energy is lost in this jump?

(*Ans.* $E_L/E_1 = 72.7\%$)

12.77 A hydraulic jump formed below a sluice gate in a horizontal rectangular channel has a depth of 0.60 m before the jump and an initial specific energy of 10.0 m. Find the sequent depth and percentage of initial energy lost in the jump.

(*Ans.* $y_2 = 4.459$ m; $E_L/E_1 = 53.7\%$)

12.78 The Froude number of the sub-critical flow after a hydraulic jump in a horizontal rectangular channel is 0.3. Estimate the Froude number of the supercritical approach flow.

(*Ans.* $F_1 = 4.883$)

12.79 A hydraulic jump in a rectangular channel has the Froude number at the beginning of the jump $F_1 = 5.0$. Find the Froude number F_2 at the end of the jump.

(*Ans.* $F_2 = 0.2956$)

12.80 At what initial Froude number F_1 and would the energy loss in a hydraulic jump in a horizontal rectangular channel be 50% of the initial specific head?

(*Ans.* $F_1 = 5.11$)

12.81 Complete the following Table 12.8 relating to elements of a hydraulic jump in a horizontal rectangular channel.

12.83 At the foot of a 30 m wide spillway in a dam where the discharge velocity is 28.2 m/s and the depth is 0.96 m, a hydraulic jump is formed on a horizontal apron. Calculate the height of the jump and the total power dissipated in the jump.

(*Ans.* $H_j = 11.04$ m; **P** = 232.17 MW)

12.84 Water is being discharged from a reservoir under a sluice gate at a rate of 18 m^3/s in a horizontal rectangular channel of width 2 m. At a point where the depth is 0.6 m a hydraulic jump is known to occur. Determine the

(a) velocities before and after the jump.
(b) energy loss in the jump and
(c) the power dissipated.

(Ans. $V_1 = 15.0$ m/s; $V_2 = 1.816$ m/s; $E_L = 6.946$ m; **P** = 1224 kW)

12.85 A hydraulic jump in a 1.5 m wide horizontal rectangular channel was used as a flow measuring device. If the flow depths before and after the jump are 0.3 m and 2.1 m respectively in that channel, estimate the discharge.

(*Ans.* $Q = 4.085$ m^3/s)

12.86 In a hydraulic jump on a horizontal rectangular channel the depth and Froude number before the jump are 0.20 m and 9.0 respectively. Estimate the energy loss and specific head at the end of the jump.

(*Ans.* $E_L = 5.798$ m; $E_2 = 2.5$ m)

Table 12.8 *Elements of Hydraulic Jumps–Problem 12.81*

Prob.	V_1	y_1	q	F_1	y_2	V_2	F_2	E_L	E_L/E_1
	(m/s)	(m)	(m^3/s/m)		(m)	(m/s)		(m)	
(i)				12.0				9.0	
(ii)					2.5	0.6			
(iii)							0.20	8.0	
(iv)			3.87					6.3	

(Note: Prob. (iv) requires a trial and error procedure).

Answers to Problem 12.81

Prob.	V_1	y_1	q	F_1	y_2	V_2	F_2	E_L	E_L/E_1
No.	(m/s)	(m)	(m^3/s/m)		(m)	(m/s)		(m)	%
(i)	15.03	0.16	2.405	12.0	2.636	0.9126	0.1794	9.0	77.1
(ii)	21.02	0.0714	1.50	25.12	2.5	0.6	0.1212	20.06	88.8
(iii)	14.55	0.2237	3.255	9.82	3.00	1.085	0.2	8.0	72.63
(iv)	13.44	0.288	3.87	8.00	3.116	1.242	0.2246	6.3	66.3

12.82 A rectangular channel carrying a supercritical stream is to be provided with a hydraulic jump type of energy dissipator. If it is desired to have an energy loss of 5 m in the jump when the inlet Froude number is 8.5, determine the sequent depths.

(*Ans.* $y_1 = 0.198$ m; $y_2 = 2.277$ m)

12.87 An overflow spillway is 40 m high. At the design head of 2.5 m over the spillway crest, calculate (i) the water depth at the toe of the spillway

(ii) sequent depth required for the formation of the jump and (iii) loss of head in the jump. (Assume $C_d = 0.738$. Neglect energy loss due to flow over the face of the spillway).

[*Ans.* $y_1 = 0.30$ m, $y_2 = 6.954$ m, $E_L = 35.30$ m]

12.88 A hydraulic jump takes place in a horizontal, triangular channel having side slopes of 1.5 H : IV. The depths before and after the jumps are 0.30 m and 1.20 m respectively. Estimate (i) the flow rate, (ii) Froude numbers at the beginning and end of the jump and (iii) energy loss in the jump.

[*Ans*: $Q = 1.096$ m^3/s, $F_1 = 6.693$, $F_2 = 0.209$ and $E_L = 2.447$ m]

12.89 For a hydraulic jump taking place in a horizontal, frictionless, triangular channel, show that the sequent depths y_1 and y_2 are related to the prejump Froude number F_1 as

$$F_1^2 = \frac{2\eta^2(\eta^3 - 1)}{3(\eta^2 - 1)} \quad \text{where } \eta = y_2/y_1$$

H. Gradually Varied Flow

12.90 A wide rectangular channel ($n = 0.015$) carries a flow of 2.5 m^3/s per metre width. The bed slope of the channel is 0.005. Determine whether the channel slope is mild, steep or critical.

(*Ans.* Steep)

12.91 A wide rectangular channel has a longitudinal slope of 0.008 and a discharge intensity of 10.0 m^3/s per metre width. If Manning's $n = 0.02$, determine whether the channel slope is mild, steep or critical. For what range of depths will the flow be on a type 1, 2 or 3 curve?

(*Ans.* Steep = S_1 curve for $y > 2.168$ m; S_2 curve for 2.168 m > y > 1.621 m; S_3 curve for $y < 1.621$ m)

12.92 Sketch the possible GVF profiles in the following break in grades. The flow is from left to right:
(a) Horizontal channel to steep.
(b) Mild to milder.
(c) Steep to steeper.

[Ans. (a) H_2, S_2; (b) M_1 on mild channel; (c) S_2 on step channel).

12.93 Sketch the possible gradually varied flow profiles in the following serial arrangement of channels and controls. The flow is from left to right.
(a) Mild-sluice gate-steep
(b) Free intake-steep-sluice gate-mild-sudden drop.

[*Ans.* (a) M_1; S_3; (b) S_2, Jump, S_3; M_3, Jump, M_2]

12.94 Establish that the gradually varied flow profiles in a frictionless rectangular channel are given by

$$x = \left(\frac{y}{S_0}\right)\left[1 + \frac{1}{2}\left(\frac{y_c}{y}\right)^3\right] + \text{constant}$$

12.95 Show that for a horizontal wide rectangular channel, by using Manning's formula, the gradually varied flow profiles are given by

$$x = \left(\frac{y_c^{4/3}}{n^2 g}\right)\left[\frac{3}{4}\left(\frac{y}{y_c}\right)^{4/3} - \frac{3}{13}\left(\frac{y}{y_c}\right)^{13/3}\right] + \text{const.}$$

12.96 Water is ponded up to a depth of 5 m just upstream of a weir in a wide channel. Estimate the depth of flow 1.5 km upstream of the weir, given $q = 2$ m^3/s/m, Manning's $n = 0.018$ and $S_0 = 0.001$. Take two steps.

(*Ans.* 3.51 m)

12.97 At a certain section M in rectangular channel of bed width 2 m the depth of flow is 1.20 m. When the flow rate is 6.0 m^3/s, estimate the distance from M to another section N where the depth is 1.40 m. The bed slope is 0.0020 and Manning's $n = 0.015$. Take two steps.

(*Ans.* 245 m)

12.98 A rectangular brick-lined channel ($n = 0.016$) has a bed width of 4.0 m and a longitudinal slope of 0.0009. At a certain discharge the normal depth was 2.0 m. In a reach where the flow was non-uniform the depth of flow at a section A was 2.6 m. Calculate the depth at a section B, 500 m downstream of A, by using (a) one step only and (b) two steps.

[*Ans.* (a) 2.85 m; (b) 2.87 m]

12.99 A wide rectangular channel carries a discharge intensity of 4.0 m^3/s per metre width. The longitudinal slope of the channel is 0.0005. Calculate the GVF profile produced by a sudden drop in the bed of the channel. Assume Manning's $n = 0.025$.

(*Ans.* An M_2 curve from 2.456 m to 1.177 m at the brink)

12.100 Water emerges from a spillway with a velocity of 20 m/s and a depth of 0.5 m at the toe. Estimate the distance of the hydraulic jump from the spillway toe if no specific measures are taken for the formation of the jump at the toe. Manning's $n = 0.02$ and $S_o = 0.001$. Assume the channel to be a wide rectangular channel. Calculations may be performed by Direct step method taking only one step.

[*Ans:* $x = 83$ m]

[***Hint:*** Depth after the jump (y_2) = Normal depth in the channel (y_o)]

12.101 A rectangular channel is 3.0 m wide and has a bed slope of 0.001. A sluice gate in this channel issues out a flow with a depth of 0.3 m at the *vena contracta*. If a hydraulic jump takes place downstream of the gate to the normal depth of 2.0 m in the channel, estimate the location of the jump. Assume Manning's $n = 0.015$.

(*Ans.* $x = 56.5$ from *vena contracta*)

12.102 A weir is 3.0 m long and has a head of 1.2 m over the weir crest. The height of the weir crest above the channel bed is 0.8 m. Neglecting the velocity of approach, estimate the discharge if the weir is

(a) a broad crested weir with well rounded entrance.

(b) a broad crested weir with square entrance.

(c) a sharp crested weir.

[*Ans.* (i) $Q = 6.59$ m^3/s; (ii) $Q = 5.715$ m^3/s; (iii) $Q = 8.513$ m^3/s]

12.103 A broad crested weir is 2.5 m long and passes a discharge of 3.5 m^3/s under a head of 0.9 m. Neglecting the velocity of approach, estimate its coefficient of discharge.

(*Ans.* $C_d = 0.962$ m)

13

Flow Measurement

CONCEPT REVIEW

13.1 INTRODUCTION

All engineering applications involving fluids require measurement of flow properties in one way or other. These can be measurement of the geometry of flow (depths, areas of cross sections of flow, etc.), local properties (velocities, pressures, etc.) or integrated properties (discharge, power, etc.). The measurement methods employed are many and depend on whether the flow is confined as in a pipeflow or open to atmosphere as in open channel flows. Also different techniques are employed depending on whether the flow is incompressible or compressible.

In this chapter basic properties of orifices which find considerable applications in flow measurement are dealt with in detail. Also, the theories of pitot tube method of velocity measurement, venturimeter for pipe flow measurement and weirs for discharge measurement in open channels are covered in sufficient detail. Detailed information on various flow-measuring devices may be found from specialised treatises on flow measurement available in literature.

13.2 ORIFICES

An orifice is an opening in a fluid container. It is an important flow element which finds application in diverse fluid flow situations including fluid flow measurement and control. Figure 13.1 shows the trajectory from a circular orifice of diameter d and area a in a tank containing a liquid to a height H above the centreline of the orifice. If $d \ll H$, it is called a small orifice. Figure 13.1(b) shows the flow in the immediate vicinity of the orifice. It is seen that the jet from the orifice attains a minimum area at a small distance from the plane of the orifice. This section is called the *vena contracta*.

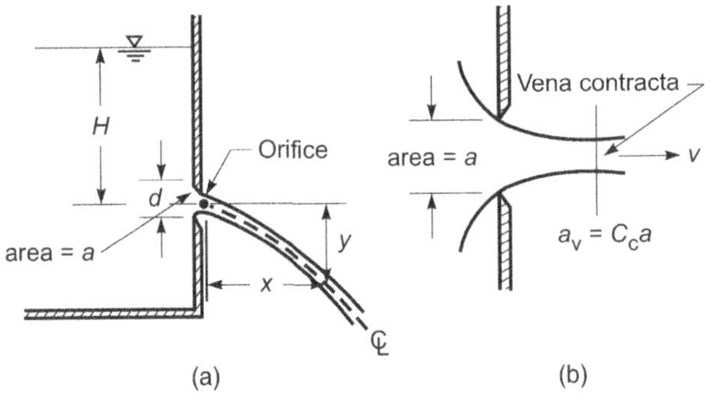

Fig. 13.1

13.2.1 Definitions

The following definitions are used in orifice flows:

(i) Coefficient of contraction

$$C_c = \frac{\text{area of jet at } vena\ contracta}{\text{area of orifice opening}}$$

or $$C_c = \frac{a_v}{a} \tag{13.1}$$

(ii) Coefficient of velocity

$$C_v = \frac{\text{actual mean velocity at } vena\ contracta}{\text{ideal mean velocity of the jet}}$$

or $$C_v = \frac{V_a}{\sqrt{2gH}} \tag{13.2}$$

(iii) Coefficient of discharge

$$C_d = \frac{\text{actual discharge}}{\text{ideal flow}}$$

$$C_d = \frac{C_c a C_v \sqrt{2gH}}{a\sqrt{2gH}} = C_c \cdot C_v \quad (13.3)$$

Usually $C_c = 0.61$ for very small orifices and C_v will be about 0.97 and C_d will have a value around 0.60.

The trajectory of the jet from the *vena contracta* of the orifice [Fig. 13.1(a)] will be that of a projectile under the action of gravity and is given by

$$x = V_a t = C_v \sqrt{2gH} \cdot t$$

and $$y = \frac{1}{2} g t^2$$

Thus $$\frac{x}{\sqrt{4yH}} = C_v \quad (13.4)$$

13.2.2 Loss of Head

The loss of head H_L between a section upstream of the orifice and the *vena contracta* is

$$H_L = H - \frac{V_a^2}{2g} = H(1 - C_v^2)$$

$$\therefore \quad H_L = \left(\frac{1}{C_v^2} - 1\right)\frac{V_a^2}{2g} \quad (13.5)$$

where V_a = velocity at the contracted section.

13.2.3 Time of Emptying of a Tank through an Orifice

If a tank of surface area (Fig. 13.2) A is being drained by an orifice at area a, then at any instant when the water surface is at a height h above the centre of the orifice, in an elemental time interval dt

$$-A\,dh = C_d a \sqrt{2gh}\,dt$$

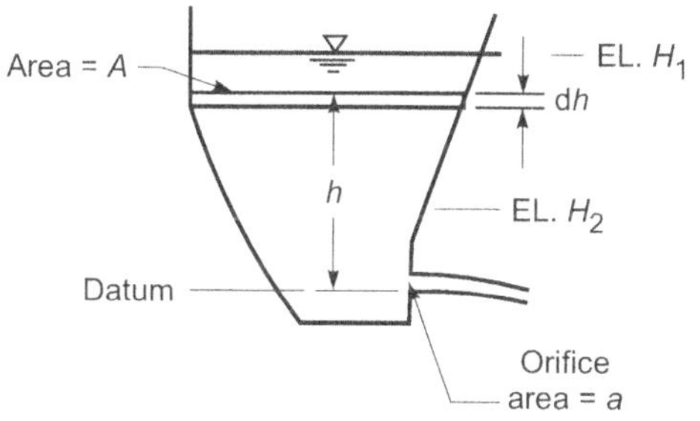

Fig. 13.2

The time required to lower the water surface from elevation H_1 to H_2 measured above the centreline of the orifice is

$$T = \int_{H_2}^{H_1} \frac{A}{C_d a \sqrt{2g}} \frac{dh}{\sqrt{h}} \quad (13.6)$$

It is important to know that, in general, A is a function of h and the appropriate function must be used to replace A before integration. Usually C_d, a and $\sqrt{2g}$ are constants.

13.3 ORIFICE METER

A plate with an orifice of diameter D_2 (area A_2) inserted axially in a pipe of diameter D_1 causes a pressure difference between an upstream section and the downstream section (Fig. 13.3). By Bernoulli theorem the discharge in the pipe is expressed in terms of the difference in pressure heads.

$$Q = C_d A_2 \frac{\sqrt{2g\Delta H}}{\sqrt{1 - C_c^2 (D_2/D_1)^4}} \quad (13.7)$$

where $\Delta H = \dfrac{(p_1 - p_2)}{\gamma}$ = difference in pressure heads upstream and downstream of the orifice plate.

C_c = Coefficient of contraction of the orifice

C_d = Coefficient of discharge

A_2 = area of the orifice

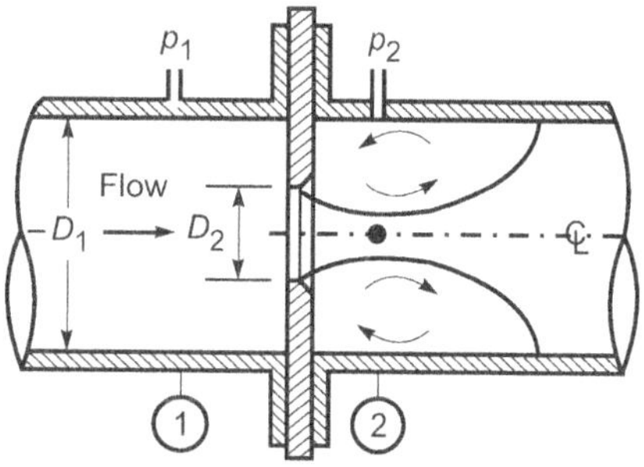

Fig. 13.3 *Oritfice Meter*

Equation (13.7) is usually written in terms of an instrument constant K_0, known as *flow coefficient*, as

$$Q = K_0 A_2 \sqrt{2g\Delta H} \quad (13.8)$$

where K_0 = flow coefficient of the orifice meter = fn (Reynolds number, A_2/A_1) and varies between 0.60 and 0.80. As the coefficient of discharge for a given orifice meter depends upon the location of the pressure trappings, it is usual to specify the location of the pressure tappings in the description of an orifice meter.

13.3.1 Flow Nozzle

A flow nozzle is essentially an orifice meter is which the jet contraction is eliminated by smooth entrance boundary, i.e., C_c = 1.0 (Fig. 13.4). The discharge is given by a formula similar to Eq. 13.8 as

$$\boxed{Q = K_f A_2 \sqrt{2 g \Delta H}} \tag{13.9}$$

where K_f = flow coefficient of the flow nozzle.

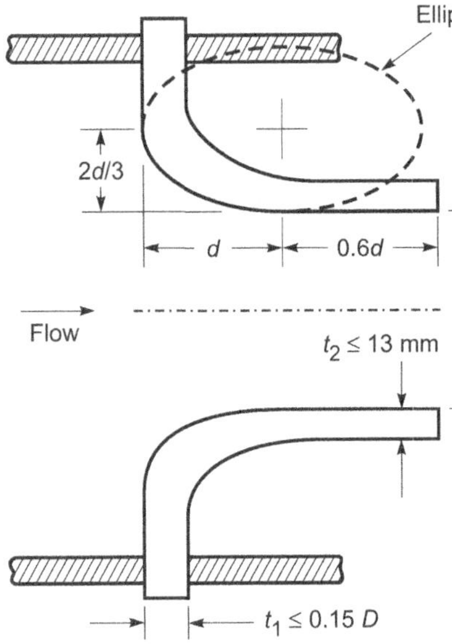

Fig. 13.4 *Flow Nozzle*

The flow coefficient K_f is practically independent of the Reynolds number in the operating range and is essentially a function of A_2/A_1. Its value varies from 0.985 for A_2/A_1 = 0.1 to 1.08 for A_2/A_1 = 0.50 for short radius ISA nozzles. The flow nozzles are costlier than orifice meters but the overall losses are much smaller than in an orifice meter.

13.4 VENTURIMETER

Venturimeter is one of the popular devices for measuring flow in pipes. It consists of a converging tube, a small throat section and an expansion tube (Fig. 13.5). The inlet and outlet diameters are the same as the diameter of the pipe in which it is to be installed. The expansion angle is kept very small to reduce the possibility of flow separation. Due to the convergence of the inlet section the velocity in the throat section is greater than in pipe and consequently, by Bernoulli principle, the piezometric head at the throat will be smaller than at the entrance. The difference in the piezometric heads between the inlet and the throat sections is a measure of the discharge in the pipe. From Fig. 13.5.

$$Q = A_1 V_1 = A_2 V_2$$

and by Bernoulli equation, for no-loss condition

$$\frac{p_1}{\gamma} + Z_1 + \frac{V_1^2}{2g} = \frac{p_2}{\gamma} + Z_2 + \frac{V_2^2}{2g}$$

Hence
$$\boxed{\left(\frac{V_2^2 - V_1^2}{2g}\right) = h_2 - h_1 = \Delta h}$$

where h = piezometric head = $\left(\dfrac{p}{\gamma} + Z\right)$

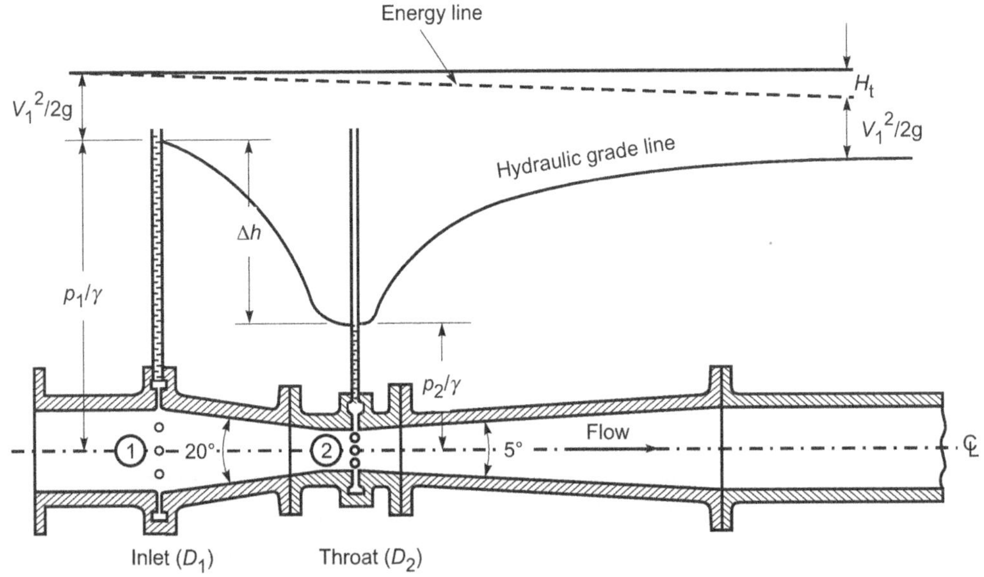

Fig. 13.5 *Venturimeter*

Substituting $V_1 = (A_2/A_1)V_2$ and simplifying, the ideal discharge (i.e. without any loss of head) is given by

$$Q_t = \frac{A_2}{\sqrt{1-(D_2/D_1)^4}}\sqrt{2g\Delta h}$$

Introducing the coefficient of discharge C_d to account for the losses and hence for the variation of Q_t from the actual discharge Q,

$$Q = C_d \frac{A_2}{\sqrt{1-(D_2/D_1)^4}}\sqrt{2g\Delta h} \qquad (13.10)$$

As the coefficient of contraction C_c for a venturimeter is 1.0, $C_d = C_v$.

The inlet head loss H_{Li} between the inlet and the throat is

$$H_{Li} = \frac{V_2^2}{2g}\left(\frac{1}{C_d^2}-1\right)[1-(D_2/D_1)^4] \qquad (13.11)$$

Also, it can be shown that

$$H_{Li} = (1-C_d^2)\,\Delta h \qquad (13.12)$$

The total head loss H_t due to the introduction of a venturimeter in a pipe is

$$H_t = H_{Li} + \text{(Loss in the expanding portion of the tube)} \qquad (13.13)$$

Usually D_2/D_1 is between 1/4 to 3/4. The coefficient of discharge C_d for a given geometry is

$$C_d = \text{fn (Reynolds number, } D_2/D_1)$$

However, in the normal designed operating range of the meter the effect of Reynolds number is very little and thus for a given tube C_d is essentially a constant:

The venturimeter is used for measuring rates of flow in both incompressible and compressible fluids.

13.4.1 Venturimeter with a Differential U-tube Manometer

Very often the change in the piezometric head between the inlet and throat of a venturimeter is measured by a differential U-tube manometer. For the general situation shown in Fig. 13.6,

$$\left(\frac{p_1}{\gamma}+Z_1\right) + x + y = y\,\frac{S_m}{S_p} + x + \left(Z_2 + \frac{p_2}{\gamma}\right)$$

$$\left(\frac{p_1}{\gamma}+Z_1\right) - \left(\frac{p_2}{\gamma}+Z_2\right) = \Delta h = y\left(\frac{S_m}{S_p}-1\right) \qquad (13.14)$$

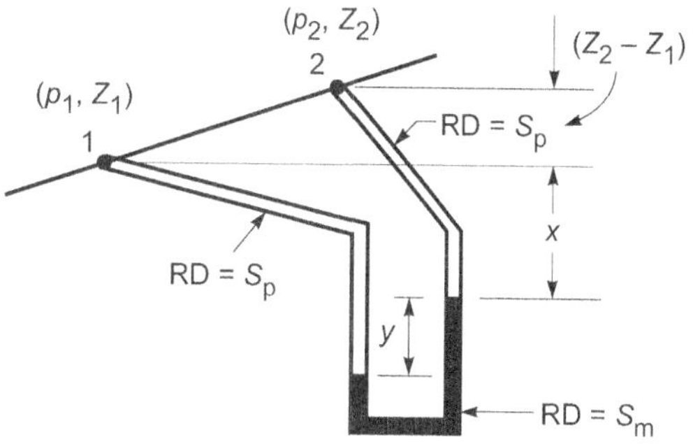

Fig. 13.6 *Differential Manometer*

In Eq. 13.14

S_m = relative density of manometric liquid

S_p = relative density of the fluid flowing in the pipe

y = reading of the differential manometer.

It is to be noted that the piezometric head difference Δh depends upon the gauge reading y regardless of the orientation of the venturimeter, whether it is horizontal, vertical or inclined, the same relation of Eq. 13.14, viz.

$$\Delta h = y\left(\frac{S_m}{S_p}-1\right)$$

holds goods.

[***Note:*** If an inverted U-tube differential manometer with $S_m < S_p$ is used then Δh is given by

$$\Delta h = y\left(1-\frac{S_m}{S_p}\right) \qquad (13.14(a)]$$

13.5 PITOT TUBE

If a small obstruction, in the form of a tube shown in Fig. 13.7, is inserted into a fluid so as to squarely face the flow, then at a point near the nose of the tube the velocity will be zero. This point is called a *stagnation*

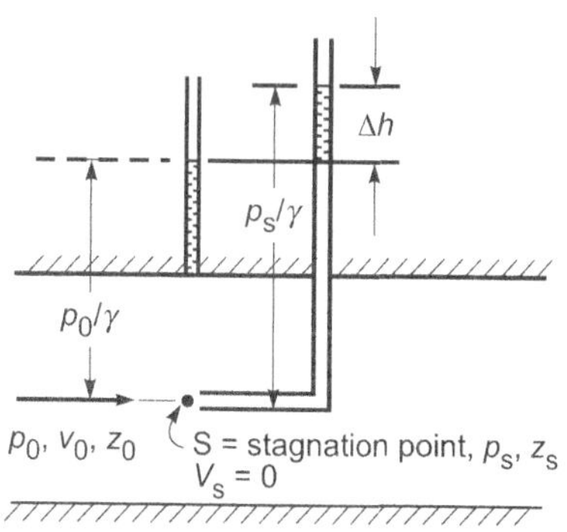

Fig. 13.7

point. The pressure at the stagnation point p_s is obtained by Bernoulli principle as

$$\frac{p_s}{\gamma} + Z_s + 0 = \frac{p_0}{\gamma} + Z_0 + \frac{V_0^2}{2g} \quad (13.15)$$

where p_0, Z_0 and V_0 are the pressure, elevation and velocity of the approaching flow. Then, by assuming the elevation difference $(Z_s - Z_0)$ to be negligible

$$\frac{V_0^2}{2g} = \left(\frac{p_s}{\gamma} - \frac{p_0}{\gamma}\right) = \frac{\Delta p}{\gamma} = \Delta h \text{ where } \Delta p = (p_s - p_0)$$

$$\boxed{V_0 = \sqrt{2g\left(\frac{p_s - p_0}{\gamma}\right)} = \sqrt{2g\,\Delta h}} \quad (13.16)$$

A pitot tube is a device to measure the velocity of fluid flow and is based on this relationship (Eq. 13.16) between the stagnation pressure and the static pressure. It consists essentially of a cylindrical tube bent into L shape. The pressure at the nose (stagnation pressure) is measured by a manometer. The static pressure in the flow can be measured separately by a piezometer. For a free surface the elevation of the free surface enables the static pressure to be determined easily.

For compressible fluids Eq. 13.16 is applicable to low Mach numbers (say < 0.2) only. The relationship between pressure coefficient $\left(\frac{p_s - p_0}{\rho V^2/2}\right)$ and Mach number M_0 for aircraft flight measurements using pitot tube is

$$\boxed{\left(\frac{p_s - p_0}{\rho V^2/2}\right) = 1 + \frac{1}{4}M_0^2} \quad (13.16\text{-a})$$

A pitot tube which combines a static pressure hole also, is known as a *pitot-static tube* (Fig. 13.8).

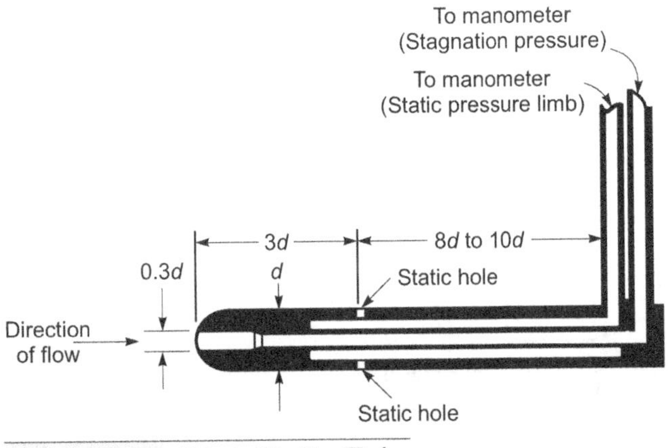

Fig. 13.8 *Pitot-Static Tube*

In a pitot tube the velocity of the stream is given by

$$\boxed{V_0 = C\sqrt{2g\,(p_s - p_0)/\gamma} = C\sqrt{2g\,\Delta h}} \quad (13.17)$$

where C = coefficient of the instrument with usual values between 0.98 to 1.0.

13.6 WEIRS

Weirs are the most popular and standard devices for measuring stream flow in open channels. Basically they are obstructions across a flow to cause a unique head-discharge relationship. When the flow from a weir is independent of the downstream water level, the flow is called a *free flow*. If it is affected by the downstream water level then it is a *drawned* or *submerged flow*. The weir is *sharp crested* if the flow springs off the upstream edge not to come in contact with the crest anywhere else. Thin plate weirs are known as *notches*.

13.6.1 Rectangular Weirs Flowing Free

(i) *Suppressed weir* When the weir is spanning the full width of a rectangular channel it is called a *suppressed rectangular weir*, (Fig. 13.9). Such weirs require aeration on the downstream side. For a fully ventilated, sharp crested, suppressed, rectangular weir flowing free, the discharge Q is related to the head over the weir H_1 as

$$\boxed{Q = \frac{2}{3} C_d \sqrt{2g}\, L H_1^{3/2}} \quad (13.18)$$

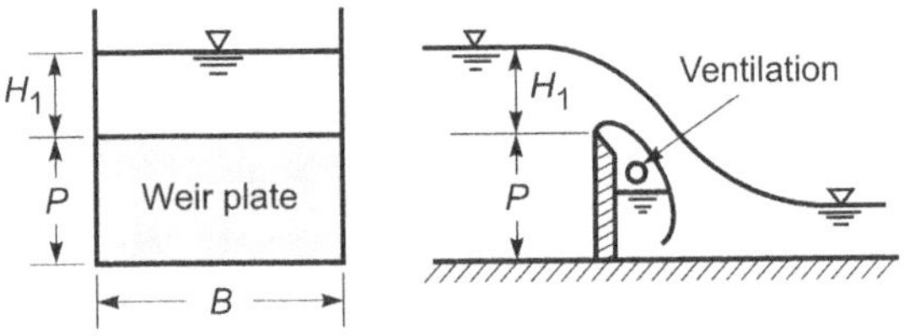

Fig. 13.9 *Suppressed Weir*

where L = length of the weir

and C_d = coefficient of discharge which takes into account the velocity of approach. C_d is given by the *Rehbock formula*

$$\boxed{C_d = 0.611 + 0.075\,\frac{H_1}{P}} \quad (13.19)$$

where P = height of the weir crest above the channel bottom.

This relationship for C_d is valid for $H_1/P \le 5$.

(ii) *Contracted weirs* When the length of the weir L is less than the width of the channel B, (Fig. 13.10) the weir is known as a contracted weir. Due to the pressure of the

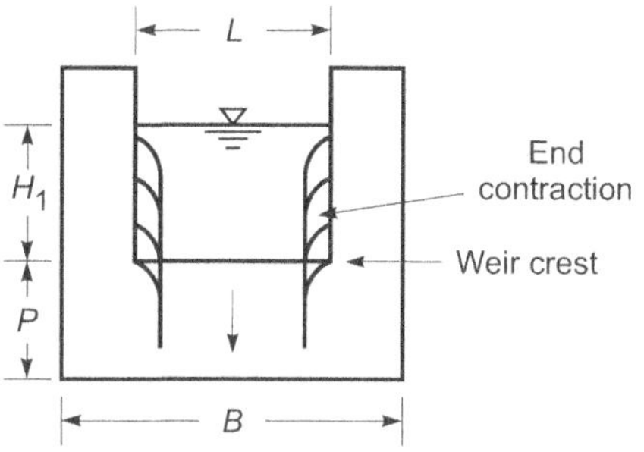

Fig. 13.10 *Contracted Weir*

end contractions the effective length of the weir will be smaller than L. The discharge equation for contracted rectangular weirs flowing free is given by *Francis formula*

$$Q = \frac{2}{3} C_c \sqrt{2g}\,(L - 0.1 n H_1)\left[\left(H_1 + \frac{V_0^2}{2g}\right)^{3/2} - \left(\frac{V_0^2}{2g}\right)^{3/2}\right] \tag{13.20}$$

In this formula V_0 = velocity of approach, n = number of end contractions and C_c = coefficient of contraction whose value is usually taken as 0.622. This formula is valid for $L > 3H_1$, $H_1/P < 1.0$.

13.6.2 Triangular Weir

For a triangular weir with a central angle θ (Fig. 13.11), the discharge under a head H_1 is given by

$$Q = C_d \frac{8}{15} \sqrt{2g} \tan\frac{\theta}{2} \cdot H_1^{5/2} \tag{13.21}$$

The coefficient of discharge C_d is, in general, a function of θ and has a value around 0.58.

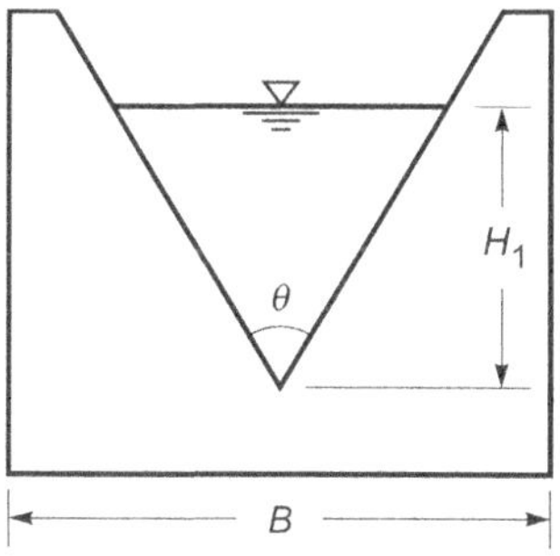

Fig. 13.11 *Triangular Weir*

13.6.3 Trapezoidal Weirs

The flow from a trapezoidal weir (Fig. 13.12) of side slope m horizontal : 1 vertical is considered as a combination of flow from a suppressed rectangular weir of length $L = B$, and from a triangular weir with a central angle of 2θ, where $\tan\theta = m$.

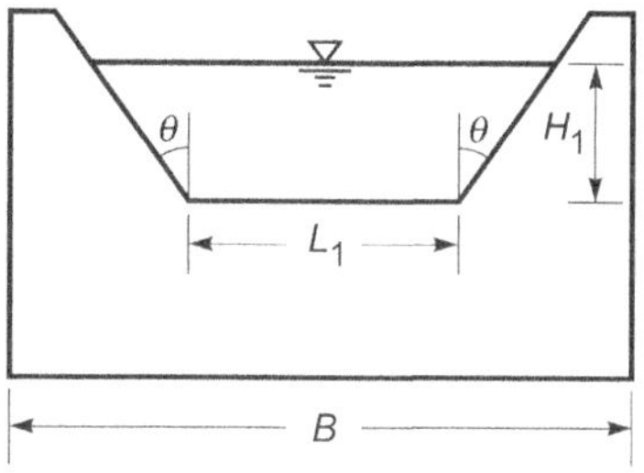

Fig. 13.12 *Trapezoidal Weir*

However, an exception to the above rule is for a weir with side slope value of 1 horizontal: 4 vertical; the discharge is calculated by using the suppressed weir formula, viz.

$$Q = \frac{2}{3} C_{dc} \sqrt{2g}\, B H_1^{3/2} \tag{13.22}$$

where the coefficient of discharge C_{dc} is a constant and has a value of about 0.63. This weir is known as *Cipolletti weir.*

13.6.4 Submerged Flow

If the tailwater level in a weir is above the weir crest, the weir will be functioning under submerged flow mode (Fig. 13.13).

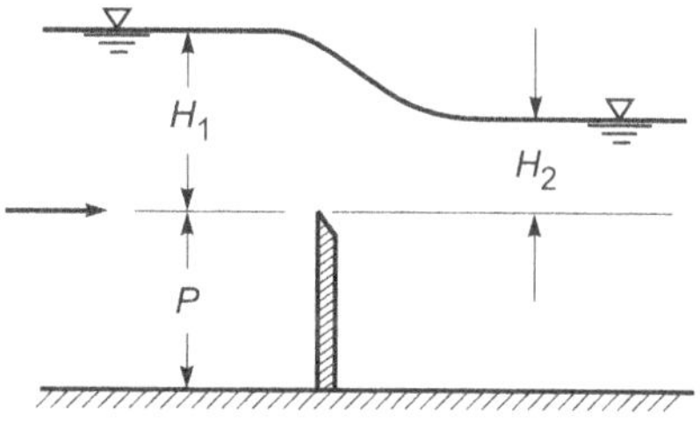

Fig. 13.13 *Submerged Weir Flow*

The discharge over the weir Q_s is estimated by the *Villemonte formula*

$$Q_s = Q_1\left[1 - \left(\frac{H_1}{H_2}\right)^n\right]^{0.385} \tag{13.23}$$

where Q_1 = free-flow discharge under the head H_1, n = exponent in the head-discharge relationship for the weir (For rectangular weir $n = 1.5$, and for triangular weir $n = 2.5$), H_2 = downstream water surface elevation measured above the weir crest, and H_1 = upstream head.

13.6.5 Current Meter

A current meter measures point velocity in the cross section of an open channel. It is a mechanical device, consisting of a rotating element (cup assembly/propeller) the rotational speed of which, when immersed in the flow, is a measure of the velocity at the meter. Figure 13.14

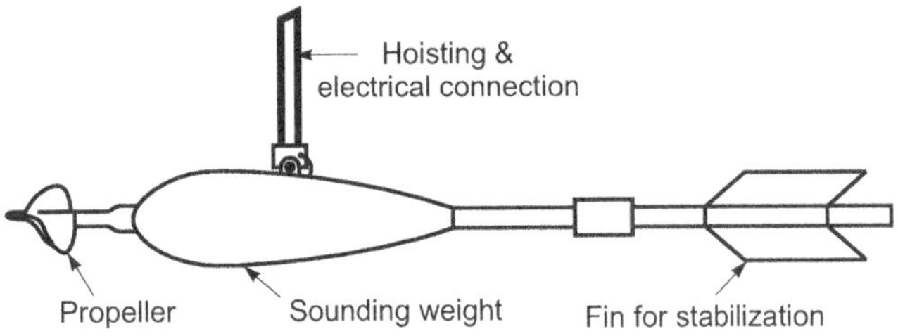

Fig. 13.14 *Horizontal Axis Current Meter*

shows a horizontal axis propeller type current meter. This type of meter comes in a wide variety of size with propellers diameters in the range 6 cm – 15 cm to register velocities in the range 0.15 – 4.0 m/s. A current meter is so designed that its rotational speed varies linearly with the stream velocity.

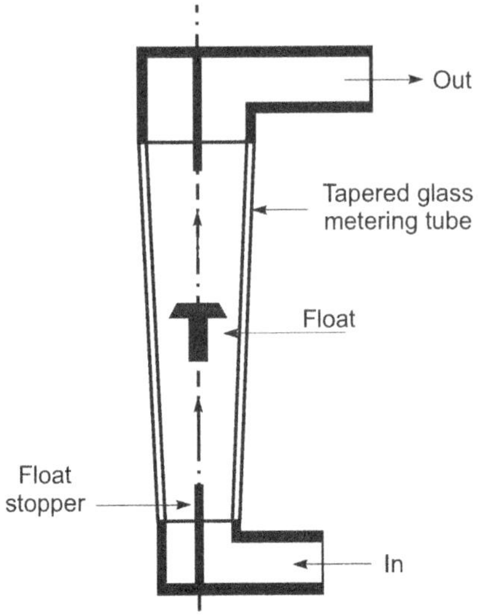

Fig. 13.15 *Rotameter*

13.7 ROTAMETER

A rotameter is a device to measure flow rate of fluid in a pipe. A rotameter consists of a transparent vertical tapering tube with a float in it (Fig. 13.15). Fluid entering from the bottom of the tube will raise the float increasing the annular area between the tube and the float. An equilibrium position is reached at which the upward force of the fluid on the float is balanced by the weight of the float. The position of the float is a measure of the discharge, the greater the flow the higher the position of the float in the tube. The float is so shaped that it will rotate in the flow (hence the name) and maintains its position on the axis of the tube. The shapes of the tube and float are so adjusted to get a linear discharge scale, which is etched on the tube. In practice, it is necessary to install the rotameter in a vertical position with the flow entering from bottom position.

WORKED EXAMPLES

A. Orifices

13.1 A 4 cm diameter orifice in the vertical side of a tank discharges water. The water surface in the tank is at a constant level of 2.0 m above the centre of the orifice. If the head loss in the orifice is 0.20 m and the coefficient of contraction can be assumed to be 0.63, estimate (i) the values of the coefficient of velocity and the coefficient of discharge; (ii) the discharge through the orifice and (iii) the location of the point of impact of the jet on a horizontal plane located 0.5 m below the centre of the orifice.

Solution:

Ideal velocity $V = \sqrt{2gH} = \sqrt{2g \times 2.0}$

$= 6.264$ m/s

Head loss $= h_L = H - \dfrac{V_a^2}{2g}$

$$\therefore \quad 2.0 - \frac{V_a^2}{2g} = 0.2$$

$$V_a = \sqrt{2 \times 9.81 \times (2.0 - 0.2)} = 5.943 \text{ m/s}$$

(i) Coefficient of velocity $C_v = \dfrac{V_a}{V}$

$= \dfrac{5.943}{6.264} = \mathbf{0.949}$

Coefficient of discharge $C_d = C_c \cdot C_v$

$= 0.63 \times 0.949$

$= \mathbf{0.598}$

(ii) Discharge through the orifice

$$Q = C_d a \sqrt{2gH}$$

$$= 0.598 \times \left(\frac{\pi}{4} \times (0.04)^2\right) \times (2 \times 9.81 \times 2.0)^{1/2}$$

$= 4.707 \times 10^{-3}$ m³/s = **4.707 L/s**

(iii) The trajectory of the jet from the *vena contracta* is

$$C_v = \frac{x}{\sqrt{4yH}}$$

or $x = C_v\sqrt{4yH}$

for $y = 0.50$ m, $x = 0.949\ \sqrt{4 \times 0.50 \times 2.0}$

$= \mathbf{1.898\ m}$

13.2 Oil flows through a 25 mm diameter orifice under a head of 5.5 m at a rate of 3 L/s. The jet strikes a wall 1.5 away and 120 mm vertically below the centreline of the contracted jet. Calculate the coefficients of velocity, contraction and discharge.

Solution:

Here $H = 5.5$ m, $x = 1.5$ m and $y = 0.12$ m.

$$C_v = \frac{x}{\sqrt{4yH}} = \frac{1.5}{\sqrt{4 \times 0.12 \times 5.5}}$$

$= \mathbf{0.923}$

$$Q = C_d a\sqrt{2gH}$$

$$3 \times 10^{-3} = C_d\left(\frac{\pi}{4} \times (0.025)^2\right) \times \sqrt{2 \times 9.81 \times 5.5}$$

$C_d = \mathbf{0.588}$

Since $C_d = C_v\ C_c$

$$C_c = \frac{0.588}{0.923} = \mathbf{0.637}$$

13.3 A vessel has two identical orifices A and B with identical values of C_v provided in one of its vertical sides at depths H_a and H_b, below the liquid surface in the tank, respectively. Show that the point of intersection of the two jets is H_b below orifice A and H_a below orifice B. Further find the distance of the point of intersection from the plane of the orifices.

Solution: Refer to Fig. 13.16.

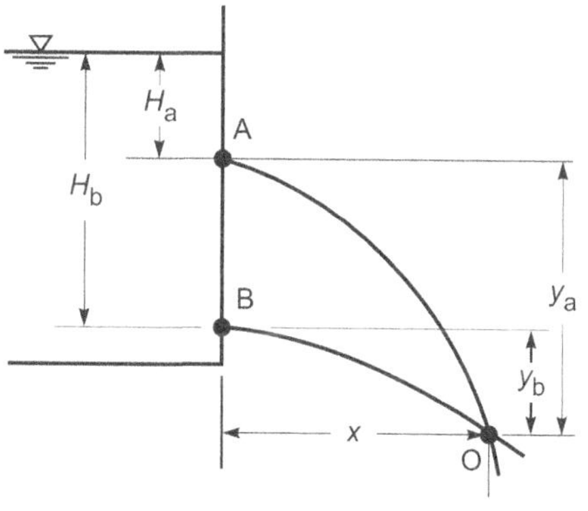

Fig. 13.16

The trajectory is given by

$$x^2 = 4yHC_v^2 \quad (1)$$

At the point of intersection,

$$x_a^2 = 4y_a\ H_a C_v^2 = x_b^2 = 4y_b H_b C_v^2$$

Hence $$\frac{y_a}{y_b} = \frac{H_b}{H_a} \quad (2)$$

Also $H_a + y_a = H_b + y_b$

$$ya - y_b = H_b - H_a \quad (3)$$

Solving for y_a from Eqs. (2) and (3)

$\mathbf{y_a = H_b}$ and $\mathbf{y_b = H_a}$

Substituting in Eq. (1),

$$x^2 = 4y_a\ H_a\ C_v^2 = 4y_b\ H_b C_v^2$$

$$\therefore \quad x = \sqrt{4H_a H_b C_v^2} = 2C_v\sqrt{H_a H_b}$$

13.4 A closed cylindrical tank is 3.5 m high and contains an oil of relative density 0.85 to a height of 3.0 m above the bottom. The space above the oil surface contains air under a pressure of 50 kPa. If an 8 cm diameter orifice is provided on the side of the tank with its centre 25 cm above the bottom, estimate the weight of fluid discharged in one minute. (Take $C_d = 0.60$).

Solution: Referring to Fig. 13.17,

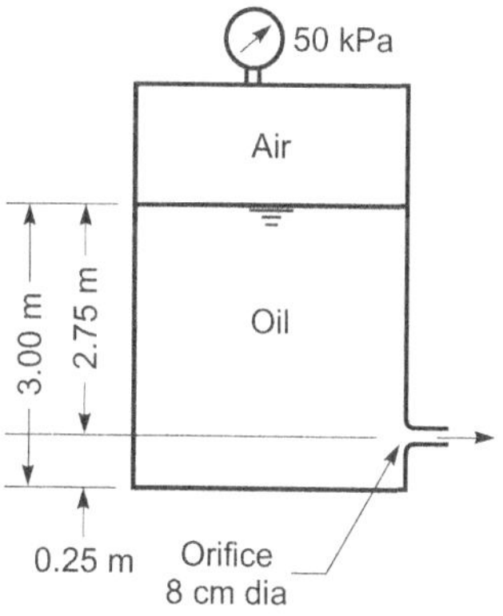

Fig. 13.17

Head over the orifice,

$$H = \frac{p}{\gamma} + (3.0 - 0.25)$$

$$= \frac{50.0}{9.79 \times 0.85} + 2.75 = 8.759 \text{ m of oil}$$

$$a = \text{area of orifice} = \frac{\pi}{4} \times d^2$$

$$= 0.7854 \times (0.08)^2 = 5.027 \times 10^{-3}\ \text{m}^2$$

Discharge $Q = C_d a\sqrt{2gH}$

$$= 0.6 \times 5.027 \times 10^{-3} \times \sqrt{2 \times 9.81 \times 8.759}$$

$$= 0.03954 \text{ m}^3\text{/s}$$

Weight of oil discharged per minute $W = 60\ \gamma Q$

$$= 60 \times (9.79 \times 0.85) \times 0.03954 \text{ kN/min}$$

$$= \mathbf{19.74\ kN/min}$$

13.5 A 10 cm diameter pipe has a nozzle at its end. If the velocity in the pipe is 2.5 m/s and pressure 50 kPa, calculate the velocity, the diameter of the jet and the head loss in the jet. The jet is discharged into atmosphere and the coefficient of velocity in the jet can be taken as 0.98.

Solution:

Total head $H = \dfrac{p_1}{\gamma} + \dfrac{V_1^2}{2g}$

$$= \frac{50}{9.79} + \frac{(2.5)^2}{2 \times 9.81} = 5.426 \text{ m}$$

$$V_j = \text{velocity of jet} = C_v \sqrt{2gH}$$

$$= 0.98\sqrt{2 \times 9.81 \times 5.426} = \mathbf{10.11\ m/s}$$

Discharge $Q = A_1 V_1 = a_j V_j$

$$\frac{\pi}{4} \times (0.10)^2 \times 2.5 = a_j \times 10.11$$

$$a_j = 1.942 \times 10^{-3} \text{ m}^2 = \frac{\pi}{4}(D_j)^2$$

$$D_j = 0.0497 \text{ m} = \mathbf{4.97\ cm}$$

Head loss $H_L = \left(\dfrac{1}{C_v^2} - 1\right)\dfrac{V_j^2}{2g}$

$$= \left(\frac{1}{(0.98)^2} - 1\right) \times \frac{(10.11)^2}{2 \times 9.81} = \mathbf{0.215\ m}$$

13.6 A nozzle attached to the end of a 10 cm diameter pipe has 3 cm diameter at the other end. If the nozzle issues out a jet of water at 25 m/s into atmosphere, calculate the pressure at the base of the nozzle. The coefficients of velocity and contraction for the nozzle can be taken as 0.96 and 0.80 respectively.

Solution: In Fig. 13.18, let the suffix j refer to the actual jet at the *vena contracta*. Thus a_j = area of the jet at the *vena contracta* and v_j = velocity of the jet at that section. It is given that

$$v_j = 25 \text{ m/s}$$

$$a_j = C_c \times (\text{area of the nozzle end})$$

$$= 0.80 \times \frac{\pi}{4} \times (0.03)^2 = 5.655 \times 10^{-4} \text{ m}^2$$

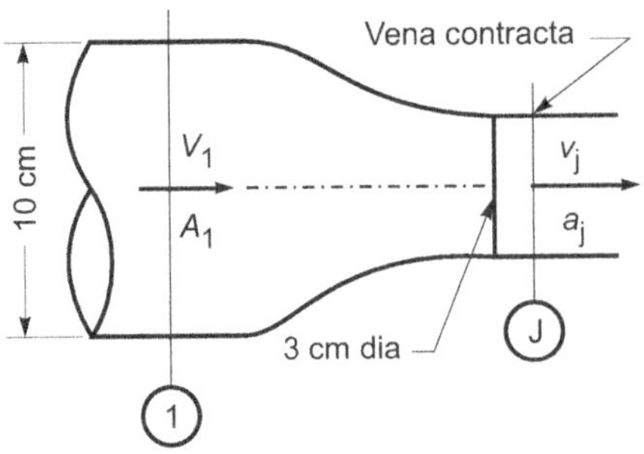

Fig. 13.18

Discharge $Q = a_j v_j = 5.655 \times 10^{-4} \times 25$

$$= 0.01414 \text{ m}^3\text{/s}$$

By continuity,

$$Q = a_j v_j = C_c a v_j = A_1 V_1$$

$$\therefore \quad V_1 = C_c\left(\frac{a}{A_1}\right) v_j = 0.8 \times \left(\frac{3}{10}\right)^2 \times 25$$

$$= 1.8 \text{ m/s}$$

Energy loss in the nozzle

$$= H_L = \left(\frac{1}{C_v^2} - 1\right)\frac{v_j^2}{2g}$$

By applying the Bernoulli theorem to section 1 and J

$$\left(\frac{p_1}{\gamma} + Z_1 + \frac{V_1^2}{2g}\right) - H_L = \frac{p_0}{\gamma} + Z_0 + \frac{v_j^2}{2g}$$

$$\frac{p_0}{\gamma} = 0 = \text{atmospheric pressure head,}$$

$$Z_1 = Z_0$$

$$\left(\frac{p_1}{\gamma} + \frac{V_1^2}{2g}\right) - \left(\frac{1}{C_v^2} - 1\right)\frac{v_j^2}{2g} = \frac{v_j^2}{2g}$$

$$\frac{p_1}{\gamma} = \frac{v_j^2}{2g}\frac{1}{C_v^2} - \frac{V_1^2}{2g}$$

$$\frac{1}{2 \times 9.81} = \left[\frac{(25)^2}{(0.96)^2} - (1.8)^2\right]$$

$$= 34.40 \text{ m}$$

$$p_1 = 34.40 \times 9.79 = \mathbf{336.8\ kPa}$$

[*Check:* You can check your calculations by noting that

$$v_j = C_v\sqrt{2gH} \quad \text{and} \quad H = \frac{p_1}{\gamma} + \frac{V_1^2}{2g}$$

$$= 34.40 + \frac{(1.8)^2}{2 \times 9.81}$$

$$\therefore \quad H = 34.565 \text{ m,}$$

and $v_j = 0.96 \times \sqrt{2 \times 9.81 \times 34.565} = 25.0$ m/s]

13.7 Water at 20°C flows in a 20 cm diameter pipe in which a standard 10 cm diameter orifice meter is installed. The meter is provided with *vena contracta* taps. Calculate the discharge corresponding to a pressure head differential of 0.5 m. For water at 20°C, $\rho = 998$ kg/m^3 and $\mu = 0.001$ Pa.s. The flow coefficient K_0 of the meter for $D_1/D_2 = 0.5$ has the following variation with the pipe Reynolds number:

$Re = \frac{VD_1}{\nu}$	10^4	2×10^4	4×10^4	6×10^4	10^5	10^6
K_0	0.64	0.63	0.63	0.62	0.62	0.62

Solution:

Discharge $\quad Q = K_0 \cdot A_2\sqrt{2g\Delta H}$

$$A_2 = \text{Area of the orifice} = \frac{\pi}{4} \times (0.10)^2$$

$$= 7.854 \times 10^{-3} \text{ m}^2$$

$$\Delta H = 0.5 \text{ m}, \sqrt{2g\Delta H} = \sqrt{2 \times 9.81 \times 0.5}$$

$$= 3.132 \text{ m/s}$$

As K_0 is not known to start with, a trial and error method is adopted. Let $K_0 = 0.62$

$$Q = 0.62 \times (7.854 \times 10^{-3}) \times 3.132$$

$$= \mathbf{0.01525\ m^3/s}$$

Velocity $\quad V = \frac{Q}{(\pi D_1^2/4)} = \frac{0.01525}{\left(\frac{\pi}{4} \times 0.20^2\right)} = 0.485$ m/s

Reynolds number

$$Re = \frac{VD_1}{\nu} = \frac{0.485 \times 0.20}{(0.001/998)} = 96806 = 10^5$$

For this value of Re, the flow coefficient $K_0 = 0.62$ which is the same as the assumed value. Hence no more iterations are required. The discharge $Q = 0.01525$ m^3/s = 15.25 L/s.

13.8 A liquid with a relative density of 0.70 flows through a 7.5 cm diameter pipe A 3.0 cm diameter ISA flow nozzle is inserted at a certain location to measure the flow. If the flow coefficient of the nozzle is 0.99, calculate the piezometric head difference across the nozzle for a discharge of 20 L/s. If the instrument is horizontal, what is the pressure difference across the nozzle?

Solution:

Discharge $\quad Q = K_f A_2\sqrt{2g\Delta H}$

$$Q = 0.002 \text{ m}^3\text{/s}, K_f = 0.99$$

$$A_2 = \text{area of the nozzle} = \frac{\pi}{4} \times (0.03)^2$$

$$= 7.0686 \times 10^{-4}$$

$$0.002 = 0.99 \times 7.0686 \times 10^{-4}\sqrt{2 \times 9.81 \times \Delta H}$$

ΔH = piezometric head difference across the nozzle = 0.4163 m

$$\Delta H = \left(\frac{p_1}{\gamma} - \frac{p_2}{\gamma}\right) + (Z_1 - Z_2)$$

For a horizontal nozzle $Z_1 - Z_2 = 0$. Hence

$$\Delta p = (p_1 - p_2) = \gamma \Delta H$$

$$= 9.79 \times 0.70 \times 0.4163 = \mathbf{2.853\ kPa}$$

13.9 An external cylindrical mouthpiece is fitted to the side of a tank containing water. Obtain an expression for the minimum pressure developed in the mouthpiece if the coefficient of contraction is 0.62. At what head over the mouthpiece would cavitation occur in the mouthpiece?
Assume local atmospheric pressure = 96.48 kPa (abs) and vapour pressure = 4.0 kPa (abs)

Solution: A mouthpiece is a short tube fitted in place of an orifice. The flow in a mouthpiece first contracts to a *vena contracta* as in an orifice (Section C-C in Fig. 13.19 and then it expands to fill the tube of area a. The loss of head in the tube is essentially due to sudden expansion from the contracted section C-C to full pipe section (as at 2-2). Hence,

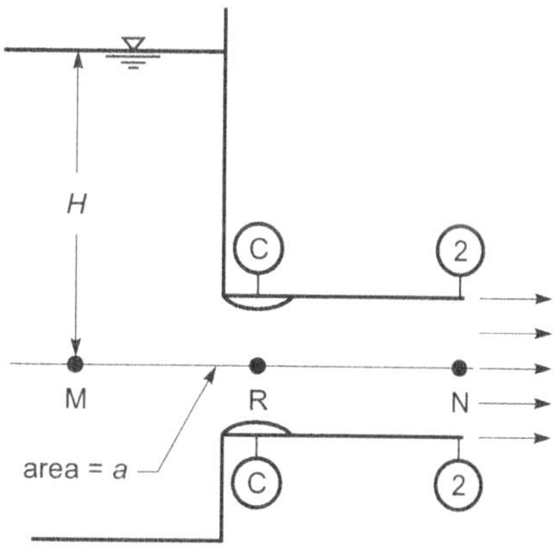

Fig. 13.19

Head loss $\quad H_L = (V_c - V_2)^2/2g$

But $\quad Q = V_c a_c = V_2 a$

Also $\quad a_c = C_c\, a$

Thus $\quad V_c = \frac{V_2}{C_c}$

$$H_L = \frac{V_2^2}{2g}\left(\frac{1}{C_c} - 1\right)^2 = \frac{V_2^2}{2g}\left(\frac{1}{0.62} - 1\right)^2$$

$$= 0.376\,\frac{V_2^2}{2g}$$

Applying Bernoulli equation between points M and N and, if

H_a = atmospheric pressure head,

$$H_a + H = H_a + \frac{V_2^2}{2g} + H_L$$

$$1.376\,\frac{V_2^2}{2g} = H \quad \text{or} \quad V_2 = 0.853\sqrt{2gH}$$

$$V_c = \frac{V_2}{C_2} = \frac{0.853}{0.62}\sqrt{2gH} = 1.376\sqrt{2gH}$$

Since the velocity in the tube is the highest at the *vena contracta*, the minimum pressure occurs at section c-c. By applying Bernoulli equation to N and R.

$$H_a + \frac{V_2^2}{2g} + H_L = \left(\frac{p_c}{\gamma}\right)_{abc} + \frac{V_c^2}{2g}$$

or $$H_a - \frac{p_c}{\gamma} = \frac{V_2^2}{2g}\left(\frac{1}{0.62}\right)^2 - 1.376\frac{V_2^2}{2g}$$

$$= 1.225\,\frac{V_2^2}{2g} = 1.225 \times \left(\frac{H}{1.376}\right)$$

$$= 0.890\,H$$

Hence $$\left(\frac{p_c}{\gamma}\right)_{abs} = (H_a - 0.890\,H) \qquad \text{(absolute)}$$

[The gauge $\left(\frac{p_c}{\gamma}\right)$ will be negative by an extent of 0.89 H.]

(ii) For cavitation to take place the local pressure must reach the vapour pressure. Hence, the cavitation conditions.

$$p_c = p_v = 4.0 \text{ kPa (abs)}$$

Also, $$\frac{p_a}{\gamma} = H_a = \frac{96.48}{9.79} = 9.855 \text{ m}$$

$$\left(\frac{p_c}{\gamma}\right)_{abs} = \frac{4.0}{9.79} = 0.409 \text{ m}$$

∴ $$0.409 = 9.855 - 0.890\,H$$

or $$H = \frac{(9.855 - 0.409)}{0.890} = 10.61 \text{ m}$$

Thus cavitation will take place in the tube at a head H = **10.61 m.**

13.10 A circular tank 2 m diameter has its axis vertical and contains a 10 cm diameter orifice at its bottom. If the coefficient of discharge of the orifice is 0.63, estimate the time required to lower the water surface from 3 m to 2 m height above the bottom.

Solution: Refer to Fig. 13.20.

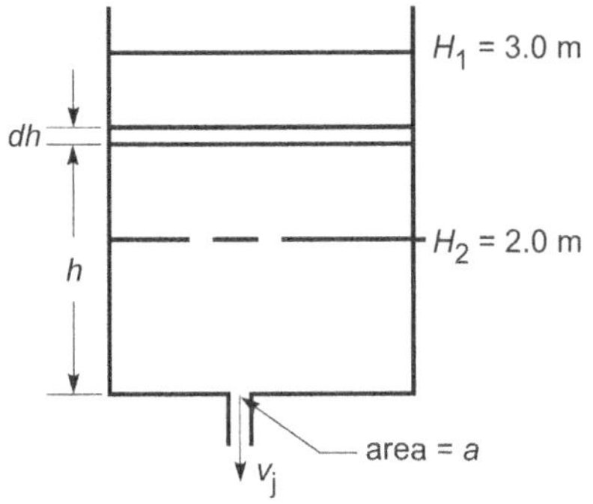

Fig. 13.20

Let A = area of the tank at height h.
Let the water level fall by dh in time dt. Then

$$-A\,dh = C_d a\sqrt{2gH}\,dt$$

$$T = \int dt = \int_{H_1}^{H_2} \frac{A\,dh}{C_d a\sqrt{2g}\,h^{1/2}}$$

Here A is constant with respect to h and as such

$$T = \frac{2A}{C_d a\sqrt{2g}}(\sqrt{H_1} - \sqrt{H_2})$$

$$A = \text{area of tank} = \frac{\pi}{4} \times (2)^2$$

$$a = \text{area of orifice} = \frac{\pi}{4} \times (0.1)^2$$

∴ $$\frac{A}{a} = \left(\frac{2}{0.1}\right)^2 = 400$$

C_d = 0.63, H_1 = 3.0 m, and H_2 = 2.0
T = time to lower the water surface from 3.0 m to 2.0 m

$$= \frac{2 \times 400}{0.63 \times \sqrt{2 \times 9.81}}(\sqrt{3} - \sqrt{2}) = \mathbf{91.1\ s}$$

13.11 A cylindrical tank of radius R and length L laid with its axis horizontal contains water. An orifice of area a and coefficient of discharge C_d is provided at the bottom of the tank. Find the time taken to lower the water surface in the tank from an elevation of H_1 to H_2 measured with respect to the centre of the orifice.

Solution: Figure 13.21 shows a cross section of the tank. At any instant t, let the water surface be at a height h above the orifice. The top width of water surface

$$b = 2x = 2 \times \sqrt{(OB)^2 - (OA)^2}$$

$$= 2\sqrt{(R^2 - (R-h)^2} = 2\sqrt{(2Rh - h^2)}$$

Area of the water surface $A = L \times b = 2L\sqrt{(2Rh - h^2)}$

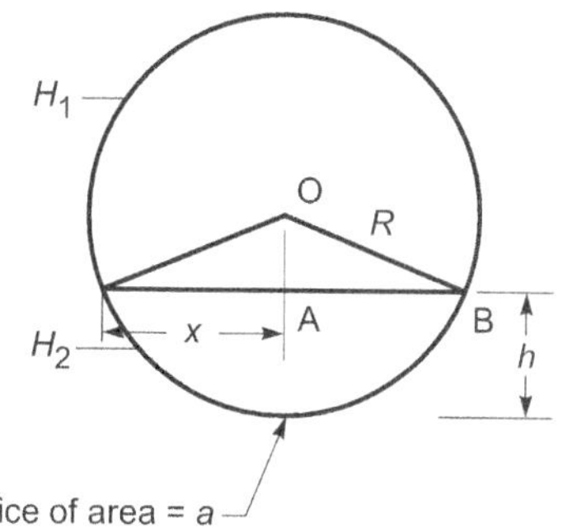

Fig. 13.21

If the water surface drops by an amount dh in time dt,

$$-A\,dh = C_d a\sqrt{2gh}\,dt$$

$$T = \int_0^T dt = -\int_{H_1}^{H_2} \frac{1}{C_d a\sqrt{2g}}\cdot\frac{2L\sqrt{2Rh-h^2}}{\sqrt{h}}\,dh$$

$$= -\frac{2L}{C_d a\sqrt{2g}}\int_{H_1}^{H_2}\sqrt{2R-h}\,dh$$

$$= \frac{2L}{C_d a\sqrt{2g}}\times\frac{2}{3}\,[(2R-H_2)^{3/2}-(2R-H_1)^{3/2}]$$

$$\mathbf{T = \frac{4}{3}\cdot\frac{L}{C_d a\sqrt{2g}}\,[(2R-H_2)^{3/2}-(2R-H_1)^{3/2}]}$$

13.12 A conical tank has an apex angle of 60° and is held with its vertex downwards and axis vertical. A 12 cm diameter orifice plate with a 10 cm diameter orifice (C_d = 0.8) is provided at the bottom and forms the base. If the tank contains water to a height of 2.0 m above the orifice (Fig. 13.22) find the (i) time taken to lower it by 1.0 m and (ii) velocity and accelerations at the beginning and at the end of the event.

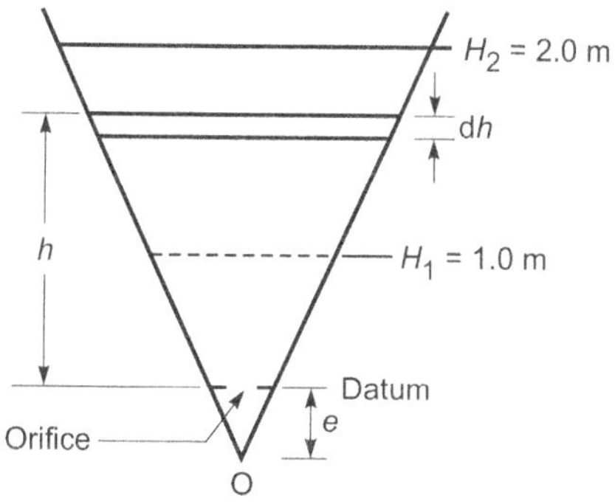

Fig. 13.22

Solution:

Let e = height of orifice above the vertex O.

$$\frac{0.12}{2}/e = \tan 30°$$

$$e = \frac{0.06}{\tan 30°} = 0.1039 \text{ m}$$

The water surface elevation is measured from the orifice. Let at any instant the water surface be at an elevation of h above the orifice. It drops by dh in time dt.

A = Water surface area at elevation h above orifice.

$= \pi\,[(h+e)\tan 30°)^2 = 1.0472\,(h+e)^2$

(i) By continuity

$$-A\,dh = C_d a\sqrt{2gH}\,dt$$

$$dt = -\frac{A}{C_d a\sqrt{2g}}\frac{dh}{\sqrt{h}}$$

$$= -\frac{1.0472(h^{3/2}+2\sqrt{h}\,e+e^2/\sqrt{h})\,dh}{0.8\times\left(\frac{\pi}{4}\times(0.1)^2\right)\times\sqrt{19.62}}$$

$$= -37.63\,(h^{3/2}+2\sqrt{h}\,e+e^2h^{-1/2})\,dh$$

Integrating between H_1 and H_2

$$T = 37.63\left[\frac{2}{5}(H_1^{5/2}-H_2^{5/2})+2\times\frac{2}{3}(H_1^{3/2}-H_2^{3/2})e+2e^2\,(H_1^{1/2}-H_2^{1/2})\right]$$

Substituting $e = 0.1039$, $H_1 = 2.00$ m and $H_2 = 1.0$ m,

$$T = 37.63\left[\frac{2}{5}(2^{5/2}-1^{5/2})+\frac{4}{3}(2^{3/2}-1^{3/2})\times 0.1039+2\,(0.1039)^2\,(2.0^{1/2}-1.0^{1/2})\right]$$

$= \mathbf{79.96\ s}$

(ii) (a) Velocity of water surface = V_s

$$V_s = \frac{dh}{dt} = \frac{C_d a\sqrt{2g}\sqrt{h}}{A}$$

$$= -\frac{C_d a\sqrt{2g}}{\pi}\frac{\sqrt{h}}{(h+e)^2}$$

$$= -\frac{0.8\times\frac{\pi}{4}\times(0.1)^2\times\sqrt{19.62}}{\pi}\left(\frac{\sqrt{h}}{(h+0.1039)^2}\right)$$

$$= -8.859\times10^{-3}\left(\frac{\sqrt{h}}{(h+0.1039)^2}\right)$$

When h = 2.0 m,

$$V_{s1} = -8.859\times10^{-3}\left(\frac{\sqrt{2}}{(2.1039)^2}\right) = \mathbf{-0.00283\ m/s}$$

$$h = 1.0 \text{ m},\ V_{s2} = -8.859\times10^{-3}\,\frac{\sqrt{1}}{(1.1039)^2}$$

$= \mathbf{-0.00727\ m/s}$

(b) Accelerations

Acceleration $a_s = \dfrac{d^2h}{dt^2}$

$$\frac{dh}{dt} = -K\left(\frac{\sqrt{h}}{(h+e)^2}\right)$$

where $K = 8.859 \times 10^{-3}$ and $e = 0.1039$

$$\frac{d^2h}{dt^2} = -K\left[-\frac{2\sqrt{h}}{(h+e)^3} + \frac{1}{2\sqrt{h}\,(h+e)^2}\right]$$

when $h = 2.0$ m,

$$a_{s1} = -8.859 \times 10^{-3}\left[\frac{-2\sqrt{2}}{(2.1039)^3} + \frac{1}{2\sqrt{2}\,(2.1039)^2}\right]$$

$$= \mathbf{1.983 \times 10^{-3}\ m^2/s}$$

when $h = 1.0$ m,

$$a_{s2} = -8.859 \times 10^{-3}\left[-\frac{2}{(1.1039)^3} + \frac{1}{2(1.1039)^2}\right]$$

$$= \mathbf{0.00954\ m^2/s}$$

13.13 A prismatic tank 3 m long and 2 m wide is divided by a vertical partition wall into two parts so that the area of one is twice the areas of the other. The partition contains a 6 cm dia orifice through which the liquids in the tanks are connected. The liquid levels in the tanks are always above the orifice. If at an instant the liquid level in the larger tank is 2 m above that of the smaller find the time taken for the difference to become 1.5 m. Assume $C_d = 0.63$ for the orifice.

Solution: Referring to Fig. 13.23,

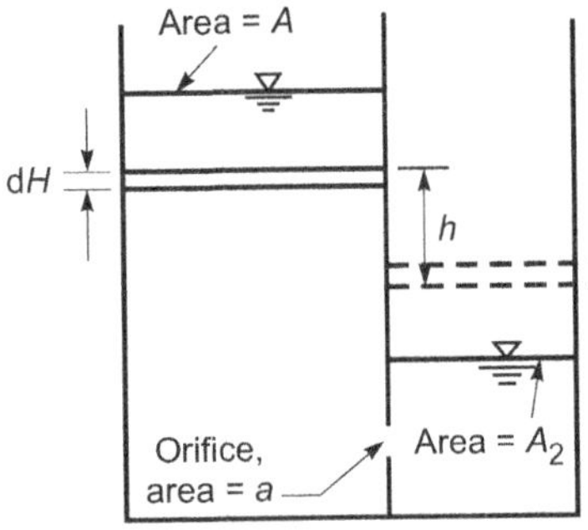

Fig. 13.23

Let at any instant of time the difference in head is h, and let dH be the drop in tank A_1.

New head difference

$$h - dh = h - dH - dH \cdot \frac{A_1}{A_2}$$

$$dh = dH\left(1 + \frac{A_1}{A_2}\right) \quad \text{or} \quad dH = \frac{dh}{(1 + A_1/A_2)}$$

$$-A_1\, dH = C_d a\sqrt{2gH}\, dt$$

$$T = \int_0^T dt = -\int_{H_1}^{H_2} \frac{A_1 h^{-1/2} dh}{C_d a\,(1 + A_1/A_2)\sqrt{2g}}$$

$$T = \frac{2A_1}{C_d a\,(1 + A_1/A_2)\sqrt{2g}}\,(H_1^{1/2} - H_2^{1/2})$$

Here $A = A_1 + A_2 = 6\ m^2$

$A_1/A_2 = 2 \quad \therefore A_1 = 4\ m^2$ and $A_2 = 2\ m^2$

$$a = \frac{\pi}{4} \times (0.06)^2 = 2.827 \times 10^{-3}\ m^2$$

$H_1 = 2.0$ m, $H_2 = 1.5$ m

Substituting in the expression for T

$$T = \frac{2 \times 4 \times (\sqrt{2} - \sqrt{1.5})}{0.63 \times (2.827 \times 10^{-3}) \times (1+2) \times \sqrt{2 \times 9.81}} = \mathbf{64\ s}$$

13.14 A cylindrical tank is placed with its axis vertical and is provided with a circular orifice of 4 cm diameter at the bottom. A steady inflow and free discharge at the bottom orifice causes the depth of water in the tank to rise from 0.59 m to 0.75 m in 106 s. Further, it is observed that the depth rises from 1.20 m to 1.29 m in 129 s. Determine the inflow rate and the diameter of the tank. Assume $C_d = 0.62$.

Solution: Referring to Fig. 13.24.

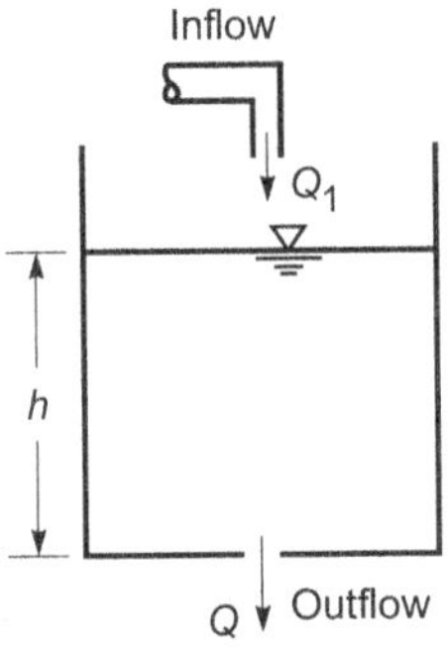

Fig. 13.24

Let Q_1 = Steady inflow rate

At any instant, let h be the head over the orifice. In a time dt, by continuity

$$A\, dh = Q_1 dt - C_d a\sqrt{2gH}\, dt$$

$$= (Q_1 - K\sqrt{h})\, dt$$

where $K = C_d a\sqrt{2g}$

$$A = \text{Area of the tank}$$

and a = area of the orifice.

$$\frac{dh}{dt} = \frac{Q_1}{A} - \frac{K}{A}\sqrt{h}$$

in the present case

$$a = \frac{\pi}{4} \times (0.04)^2 = 1.2566 \times 10^{-3}\ \text{m}^2$$

$$K = 0.62 \times 1.2566 \times 10^{-3} \times \sqrt{2 \times 9.81}$$

$$= 3.451 \times 10^{-3}$$

Also,

(i) $$\frac{dh}{dt} = \frac{(0.75 - 0.59)}{106}$$

$= 1.5094 \times 10^{-3}$ at an average head

$$h_1 = \frac{0.75 + 0.59}{2} = 0.67\ \text{m}$$

(ii) $$\frac{dh}{dt} = \frac{(1.29 - 1.20)}{129}$$

$= 6.9767 \times 10^{-4}$ at an average head

$$h_2 = \frac{1.29 + 1.20}{2} = 1.245\ \text{m}$$

Substituting in Eq. (1)

$$1.5094 \times 10^{-3} = \frac{Q_1}{A} - \frac{K}{A}\sqrt{0.67}$$

$$6.9767 \times 10^{-4} = \frac{Q_1}{A} - \frac{K}{A}\sqrt{1.245}$$

Solving

$$\frac{K}{A}(1.1158 - 0.8185) = 8.1173 \times 10^{-4}$$

$$\frac{K}{A} = 2.7307 \times 10^{-3}$$

or $$A = \frac{2.7307 \times 10^{-3}}{3.451 \times 10^{-3}} = 0.79128\ \text{m}^2$$

As, $$A = \frac{\pi D^2}{4},$$

$$D = \left(\frac{0.79128}{\pi/4}\right)^{1/2} = \mathbf{1.004\ m}$$

$$\frac{Q_1}{A} = 1.5094 \times 10^{-3} + (2.7307 \times 10^{-3}) \times \sqrt{0.67}$$

$$= 3.745 \times 10^{-3}\ \text{m}^3/\text{s}$$

$$Q_1 = (0.79128)(3.745 \times 10^{-3})$$

$$= 2.963 \times 10^{-3}\ \text{m}^3/\text{s}$$

$$= \mathbf{2.963\ L/s}$$

B. Venturimeter

13.15 A vertical venturimeter 40 cm × 20 cm is provided in a vertical pipe to measure the flow of oil of relative density 0.80. The different in elevations of the throat section and the entrance section is 1 m, the direction of flow of oil being vertically upwards. The oil-mercury differential gauge shows deflection of mercury (RD = 13.6) equal to 40 cm (Fig. 13.25). Determine the quantity of oil flowing in the pipe. Neglect losses.

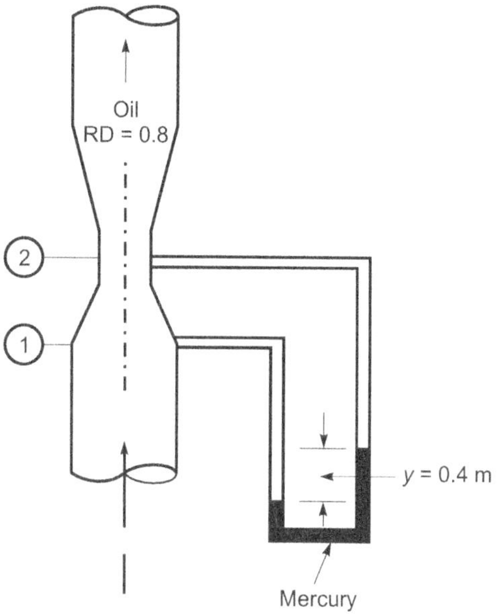

Fig. 13.25

Solution: The discharge in the venturimeter

$$Q = \frac{C_d A_2}{\sqrt{1 - (D_2/D_1)^4}} \sqrt{2g\Delta h}$$

The differential manometer reading y will give directly the difference in piezometric head Δh as

$$\Delta h = y\left(\frac{S_m}{S_p} - 1\right)$$

Here $S_m = 13.6$, $S_p = 0.80$, $y = 0.4$ m

$$\Delta h = 0.4\left(\frac{13.6}{0.80} - 1\right) = 6.4\ \text{m of oil}$$

$C_d = 1.0$ as the losses are neglected.

$$A_2 = \frac{\pi}{4}(0.2)^2 = 0.031416$$

$$\frac{D_2}{D_1} = \frac{20}{40} = 0.5, \quad \sqrt{1 - (0.5)^4} = 0.96825$$

Substituting the various values,

$$Q = \frac{0.031416}{0.96825} \times \sqrt{2 \times 9.81 \times 6.4}$$

∴ $Q = 0.3636\ \text{m}^3/\text{s}$ = **363.6 L/s of oil**

13.16 Show that the head loss in the inlet section of a venturimeter is $H_L = (1 - C_d^2)\,\Delta h$ where C_d = coefficient of discharge of the meter and Δh =

piezometric head difference between the inlet and the throat.

Solution: If suffixes 1 and 2 refer to the inlet and the throat respectively,

$$\frac{p_1}{\gamma} + \frac{V_1^2}{2g} + Z_1 = \frac{p_2}{\gamma} + \frac{V_2^2}{2g} + Z_2 + H_L$$

where H_L = head loss between 1 and 2.

But $$\left(\frac{p_1}{\gamma} + Z_1\right) - \left(\frac{p_2}{\gamma} + Z_2\right) = \Delta h$$

Also $V_1 = \dfrac{Q}{A_1}$ and $V_2 = \dfrac{Q}{A_2}$

Hence $$\frac{Q^2}{2g}\left(\frac{1}{A_2^2} - \frac{1}{A_1^2}\right) = \Delta h - H_L$$

$$\frac{Q^2}{2g}\frac{1}{A_2^2}\left(1 - \left(\frac{A_2}{A_1}\right)^2\right) = \Delta h - H_L$$

i.e. $$\frac{Q^2}{2g}\frac{1}{A_2^2}\left(1 - \left(\frac{D_2}{D_1}\right)^4\right) = \Delta h - H_L \qquad (1)$$

But the venturimeter flow equation

$$Q = \frac{C_d A_2}{\sqrt{1-(D_2/D_1)^4}}\sqrt{2g\Delta h}$$

or $$\frac{Q^2}{2gA_2^2} = \frac{C_d^2}{(1-(D_2/D_1)^4}\Delta h \qquad (2)$$

Substituting from (2) in (1)

$$C_d^2\,\Delta h = \Delta h - H_L$$

or $$\mathbf{H_L = (1 - C_d^2)\,\Delta h}$$

13.17 A venturimeter of throat diameter 5 cm is fitted into a 12.5 cm diameter water pipeline. The coefficient of discharge is 0.96. Calculate the flow in the pipeline when the reading on a mercury-water differential U-tube manometer connected to the upstream and throat sections shows a reading of 20 cm. If the energy loss in the downstream divergent cone of the meter is 10 times the velocity head in the pipe, calculate the total head loss of the meter.

Solution:

(i) The venturimeter discharge is given by

$$Q = \frac{C_d A_2}{\sqrt{1-(D_2/D_1)^4}}\sqrt{2g\Delta h}$$

The differential manometer reading y is related to Δh as

$$\Delta h = y\left(\frac{S_m}{S_p} - 1\right)$$

In this case,

$$\Delta h = 0.20\left(\frac{13.6}{1.0} - 1\right) = 2.52 \text{ m}$$

$$A_2 = \frac{\pi}{4} \times (0.05)^2 = 1.9635 \times 10^{-3} \text{ m}^2$$

$$\frac{D_2}{D_1} = \frac{5}{12.5} = 0.4, \quad \sqrt{1-\left(\frac{D_2}{D_1}\right)^4} = 0.9871$$

$$Q = \frac{0.96 \times 0.019635}{0.9871}\sqrt{2 \times 9.81 \times 2.52}$$

$$= 0.01343 \text{ m}^3\text{/s} = 13.43 \text{ L/s}$$

(ii) V_1 = Velocity in the pipe = $\dfrac{Q}{A_1}$

$$= \frac{0.01343}{\frac{\pi}{4} \times (0.125)^2} = 1.094 \text{ m/s}$$

$$\frac{V_1^2}{2g} = 0.061 \text{ m}$$

H_{Li} = head loss in the converging cone.

$= (1 - C_d^2)\,\Delta h$

$= [1 - (0.96)^2] \times 2.52 = 0.1976$ m of water

H_{Ld} = head loss in the diverging cone

$$= 10\frac{V_1^2}{2g} = 10 \times 0.061 = 0.61 \text{ m of water}$$

Total head loss in the meter

$H_L = H_{Li} + H_{Ld}$

$= 0.1976 + 0.61$

$=$ **0.8076 m of water**

13.18 An oil of relative density 0.90 flows through a vertical pipe of diameter 20 cm. The flow in measured by a 20 cm × 10 cm venturimeter. The throat is 30 cm above the inlet section. A differential U-tube manometer containing mercury is connected to the throat and the inlet. If $C_d = 0.99$, what is (a) the flow for a manometer reading of 9 cm and (b) the manometer reading for a flow of 50 L/s?

Solution: The discharge in the venturimeter is

$$Q = \frac{C_d A_2}{\sqrt{1-(D_2/D_1)^4}}\sqrt{2g\Delta h}$$

For a differential manometer,

$$\Delta h = y\left(\frac{S_m}{S_p} - 1\right)$$

In the present case, $S_m = 13.6$, $S_p = 0.9$

$$\therefore \quad \Delta h = y\left(\frac{13.6}{0.9} - 1\right) = 14.11\, y$$

$$A_2 = \frac{\pi}{4} \times (0.1)^2 = 7.854 \times 10^{-3} \text{ m}^2$$

$$\frac{D_2}{D_1} = 0.5, \quad \sqrt{1-(D_2/D_1)^4} = 0.96825$$

$$Q = \frac{0.99 \times 7.854 \times 10^{-3}}{0.96825} \times \sqrt{2 \times 9.81 \times 14.11 y}$$

$$Q = 0.1336\sqrt{y}$$

(i) When $y = 9$ cm $= 0.09$ m, $Q = 0.1336\sqrt{0.09}$
$= 0.040$ m^3/s $=$ **40.0 L/s**

(ii) When $Q = 50$ L/s $= 0.050$ m^3/s

$$0.050 = 0.1336\sqrt{y}$$

$y = 0.14$ m $=$ **14 cm**

13.19 Determine the maximum discharge of water that can be carried without cavitation by a horizontal 100 mm × 50 mm venturimeter, which has a coefficient of discharge of 0.95. The inlet pressure is 10 kPa (gauge), the vapour pressure of water is 4 kPa (abs) and the local atmospheric pressure is 96.0 kPa (abs).

Solution: Cavitation occurs when the local pressure reaches the vapour pressure. Hence, for maximum discharge the pressure at throat $p_2 = p_v = 4.0$ kPa (abs)

Inlet pressure (abs)

= 10 kPa (gauge) + atmospheric pressure

$$p_1 = 10.0 + 96.0 = 106.0 \text{ kPa (abs)}$$

For a horizontal venturimeter $Z_1 = Z_2$

$$\therefore \quad \Delta h = \left(\frac{p_1}{\gamma} + Z_1\right) - \left(\frac{p_2}{\gamma} + Z_2\right) = \frac{p_1 - p_2}{\gamma}$$

$$= \frac{106.0 - 4.0}{9.79} = 10.419 \text{ m}$$

The venturimeter discharge equation is

$$Q = \frac{C_d A_2}{\sqrt{1-(D_2/D_1)^4}} \times \sqrt{2 g \Delta h}$$

Here $C_d = 0.95$, $A_2 = \frac{\pi}{4} \times (0.05)^2 = 1.9635 \times 10^{-3}$ m^2

$$\frac{D_2}{D_1} = \frac{50}{100} = 0.5,$$

$$\sqrt{1-(D_2/D_1)^4} = \sqrt{1-(0.5)^4} = 0.96825$$

$$\therefore \quad Q = \frac{0.95 \times 1.9635 \times 10^{-3}}{0.96825} \times \sqrt{2 \times 9.81 \times 10.419}$$

$= 0.0275$ m^3/s

= **27.5 L/s (maximum discharge)**

13.20 Show that the head loss between inlet and throat of a venturimeter can be expressed in terms of the throat velocity head as

$$H_{Li} = \left(\frac{1}{C_d^2} - 1\right)\left(1 - \left(\frac{D_2}{D_1}\right)^4\right)\frac{V_2^2}{2g}$$

Solution: First consider the situation without losses

$$\frac{p_1}{\gamma} + \frac{V_1^2}{2g} + Z_1 = \frac{p_2}{\gamma} + \frac{V_{2t}^2}{2g} + Z_2$$

where V_{2t} = theoretical velocity at the throat

$$\text{But } \left(\frac{p_1}{\gamma} + Z_1 - \frac{p_2}{\gamma} - Z_2\right) = \Delta h$$

= difference in piezometric heads.

Also $\quad V_1 D_1^2 = V_2 D_2^2$

$$\frac{V_{2t}^2}{2g}\left(1 - \left(\frac{D_2}{D_1}\right)^4\right) = \Delta h$$

or

$$V_{2t} = \frac{1}{\sqrt{1-(D_2/D_1)^4}} \times \sqrt{2 g \Delta h}$$

Now let us consider the actual flow:
The actual discharge

$$Q = \frac{C_d A_2}{\sqrt{1-(D_2/D_1)^4}} \times \sqrt{2 g \Delta h} = V_2 A_2$$

$\therefore \quad V_2$ = actual velocity at section 2

$$= \frac{C_d\sqrt{2 g \Delta h}}{\sqrt{1-(D_2/D_1)^4}} = C_d V_{2t}$$

Hence here $C_d = C_v$ = coefficient of velocity

$$\text{Head loss } H_{Li} = \frac{p_1}{\gamma} + Z_1 + \frac{V_1^2}{2g} - \frac{p_2}{\gamma} - Z_2 - \frac{V_2^2}{2g}$$

$$= \Delta h + \frac{V_1^2}{2g} - \frac{V_2^2}{2g}$$

Substituting $\Delta h = [1 - (D_2/D_1)^4]\, \frac{1}{C_d^2}\frac{V_2^2}{2g}$

and

$$\frac{V_1^2}{2g} = \frac{V_2^2}{2g}\left(\frac{D_2}{D_1}\right)^4$$

$$H_{Li} = \frac{V_2^2}{2g}\left[\frac{1}{C_d^2}\left\{1 - \left(\frac{D_2}{D_1}\right)^4\right\} + \left(\frac{D_2}{D_1}\right)^4 - 1\right]$$

$$H_{Li} = \left(\frac{1}{C_d^2} - 1\right)\left\{1 - \left(\frac{D_2}{D_1}\right)^4\right\}\frac{V_2^2}{2g}$$

13.21 A pipe carrying water has a 30 cm × 15 cm venturimeter which is positioned inclined at 30° to the horizontal. The flow is upwards. The converging cone is 45 cm in length and the C_d of the meter is 0.98. A differential U-tube manometer with mercury as indicating fluid is connected to the inlet and to the throat and shows a differential column height of 30 cm (Fig. 13.26).

(i) Calculate the discharge in the pipe.

(ii) If the pressure in the inlet section is 50 kPa determine the pressure at the throat.

(iii) Find the head loss in the converging section of the venturimeter.

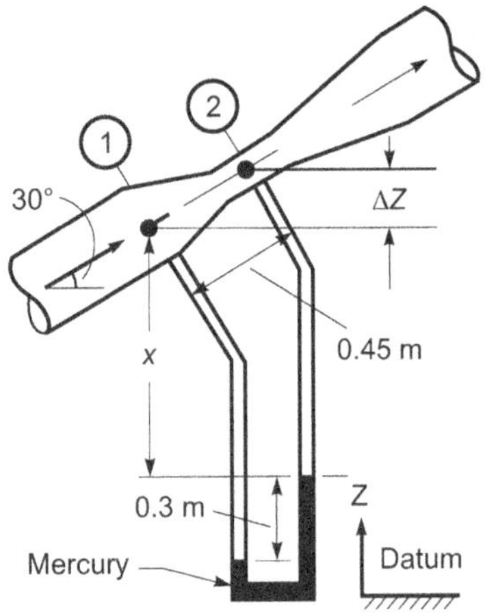

Fig. 13.26

Solution:

(i) $$\left(\frac{p_1}{\gamma} + Z_1\right) - \left(\frac{p_2}{\gamma} + Z_2\right) = \Delta h$$

From differential manometer

$$\Delta h = y\left(\frac{S_m}{S_p} - 1\right)$$

$$= 0.3\left(\frac{13.6}{1} - 1\right) = 3.78 \text{ m}$$

The discharge Q by the venturimeter equation is

$$Q = \frac{C_d A_2}{\sqrt{1 - (D_2/D_1)^4}} \times \sqrt{2g\Delta h}$$

$$C_d = 0.98,\ A_2 = \frac{\pi}{4} \times (0.15)^2 = 0.01767 \text{ m}^2$$

$$\frac{D_2}{D_1} = \frac{15}{30} = 0.5,\ \sqrt{1 - (D_2/D_1)^4} = 0.96825$$

$$\therefore \quad Q = \frac{0.98 \times 0.01767}{0.96825}\sqrt{19.62 \times 3.78}$$

$$= 0.154 \text{ m}^3/\text{s} = \mathbf{154\ L/s}$$

(ii) $\Delta Z = Z_2 - Z_1$

= difference in elevation between the throat and the inlet

$= 0.45 \sin 30° = 0.225$ m

Since $$\left(\frac{p_1}{\gamma} + Z_1\right) - \left(\frac{p_2}{\gamma} + Z_2\right) = \Delta h = 3.78 \text{ m}$$

$$\frac{p_2}{\gamma} = \frac{p_1}{\gamma} - (Z_2 - Z_1) - \Delta h$$

$$= \frac{50}{9.79} - 0.225 - 3.78 = 1.102 \text{ m}$$

$$p_2 = 1.102 \times 9.79 = \mathbf{10.79\ kPa}$$

(iii) Head loss in the converging section

$$H_{Li} = \left(\frac{1}{C_d^2} - 1\right)\left(1 - \left(\frac{D_2}{D_1}\right)^4\right)\frac{V_2^2}{2g}$$

$$= \left(\frac{1}{0.98^2} - 1\right)\left\{1 - \left(\frac{15}{30}\right)^4\right\}\left(\frac{0.154}{0.01767}\right)^2 \frac{1}{2 \times 9.81}$$

$$= 0.0387 \times 3.871 = \mathbf{0.15\ m}$$

Alternatively: $H_L = (1 - C_d^2)\,\Delta h$

$= (1 - 0.98^2)(3.78) = \mathbf{0.15\ m}$

13.22 A 20 cm diameter pipe carrying water is fitted with a venturimeter of throat diameter 10 cm. The venturimeter is vertical and the throat is 20 cm above the inlet section. The difference in pressure between the inlet and the throat is measured by an inverted U-tube manometer containing a liquid of relative density 0.60. It is known that the inlet head loss is 3.0% of the piezometric head difference between the inlet and the throat. (a) Calculate the discharge in the tube when the U-tube reading is 15 cm. (b) What is the head loss in the inlet section?

Solution:

(a) H_{Li} = Head loss at inlet = $(1 - C_d^2)\,\Delta h$

$= 0.03\,\Delta h$

$\therefore \quad C_d = 0.985$

From the inverted differential manometer

$$\Delta h = y\left(1 - \frac{S_m}{S_p}\right)$$

$$= 0.15\left(1 - \frac{0.6}{1.0}\right) = 0.06 \text{ m}$$

By the venturimeter equation

$$Q = \frac{C_d A_2}{\sqrt{1-(D_2/D_1)^4}}\sqrt{2g\Delta h}$$

$$C_d = 0.985,\quad A_2 = \frac{\pi}{4}\times(0.1)^2 = 7.854\times10^{-3}\ \text{m}^2$$

$$\frac{D_2}{D_1} = \frac{10}{20} = 0.5,\ \sqrt{1-(D_2/D_1)^4} = 0.96825$$

Hence

$$Q = \frac{0.985\times7.854\times10^{-3}}{0.96825}\times\sqrt{2\times9.81\times0.06}$$

$$= 8.667\times10^{-3}\ \text{m}^3/\text{s} = \mathbf{8.67\ L/s}$$

(b) Head loss in the inlet section $H_{Li} = 0.03 \times \Delta h$

$$= 0.03\times0.06 = 1.8\times10^{-3}\ \text{m}$$

$$= \mathbf{1.8\ mm\ of\ water}$$

13.23 A venturimeter with a throat diameter of 10 cm is connected to a 20 cm diameter main carrying water. The head loss between the inlet and the throat is known to be $0.1\,\frac{V_2^2}{2g}$ where V_2 = velocity at the throat. Estimate the discharge when the inverted differential U-tube manometer connected to the inlet and throat records a reading of 30 cm. The manometer fluid has a relative density of 0.75. What is the coefficient of discharge of the meter?

Solution: For the inverted differential *U*-tube manometer,

Δh = difference in piezometric heads

$$= y\left(1-\frac{S_m}{S_p}\right) = 30\left(1-\frac{0.75}{1.0}\right) = 7.5\ \text{cm}$$

$$= 0.075\ \text{m}$$

The loss of head between inlet and throat:

$$H_{Li} = \left(\frac{1}{C_d^2}-1\right)\left(1-\left(\frac{D_2}{D_1}\right)^4\right)\frac{V_2^2}{2g}$$

$$0.1\,\frac{V_2^2}{2g} = \left(\frac{1}{C_d^2}-1\right)\left[1-\left(\frac{0.1}{0.2}\right)^4\right]\frac{V_2^2}{2g}$$

$$\left(\frac{1}{C_d^2}-1\right) = 0.10667$$

$$\mathbf{C_d = 0.95}$$

Also, $\quad H_{Li} = (1 - C_d^2)\,\Delta h$

$$\therefore\quad 0.1\frac{V_2^2}{2g} = [1-(0.95)^2]\,(0.075)$$

$$= 7.313\times10^{-3}$$

$$\frac{V_2^2}{2g} = 0.07313 \quad\text{and}\quad V_2 = 1.198\ \text{m/s}$$

Discharge $\quad Q = A_2V_2$

$$= \frac{\pi}{4}\times(0.1)^2\times1.198$$

$$= 9.41\times10^{-3}\ \text{m}^3/\text{s} = \mathbf{9.41\ L/s}$$

C. Pitot Tube

13.24 A pipe carries a flow of an oil of RD = 0.85. A pitot-static tube is inserted into the pipe to measure the velocity at a point M. If a differential mercury-oil gauge connected to the pitot-static tube indicates a reading of 4 cm, calculate the velocity at M. Assume the coefficient of the pitot tube as 0.99.

Solution: For the differential manometer,

$$\Delta h = y\left(\frac{S_m}{S_p}-1\right) = 0.04\left(\frac{13.6}{0.85}-1\right) = 0.6\ \text{m of oil}$$

For the pitot tube

$$V_0 = C\sqrt{2g\Delta h} = 0.99\sqrt{2\times9.81\times0.6}$$

Velocity at M = **3.397 m/s**

13.25 A pitot tube is mounted on an air plane to indicate the speed of the plane relative to the prevailing wind. What differential pressure intensity in kPa will the instrument register when the plane is travelling at a speed of 200 km/h in a wind of 60 km/h blowing against the direction of the plane (Fig. 13.27). Use $\rho_{air} = 1.20$ kg/m^3.

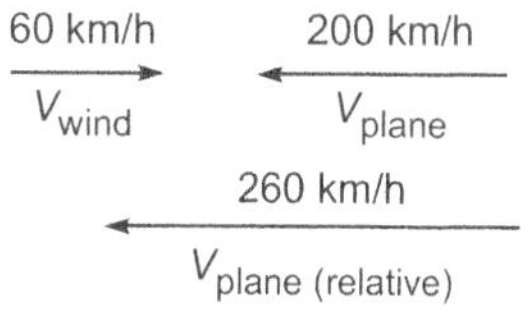

Fig. 13.27

Solution: Relative velocity of plane with respect to wind.

$$V_{PR} = 200 - (-60)$$

$$= 260\ \text{km/h} = 72.2\ \text{m/s}$$

In a pitot tube (assuming $C = 1$)

$$\frac{\Delta p}{\gamma} = \frac{V_{PR}^2}{2g} \quad\text{or}\quad \Delta p = \frac{\rho V_{PR}^2}{2}$$

$$\Delta p = \frac{1.20\times(72.2)^2}{2} = 3127.7\ \text{Pa}$$

$\therefore\quad \Delta p = P_s - p_0$ = Differential pressure intensity in the instrument = **3.128 kPa**.

13.26 In a pitot static tube the stagnation pressure is 3.0 kPa and the static pressure is –3.0 kPa (gauge). The fluid is air (ρ = 1.20 kg/m^3). Calculate the velocity of flow by taking the instrument coefficient as 0.98.

Solution: In a pitot-static tube,

$$\Delta h = \frac{p_s}{\gamma} - \frac{p_0}{\gamma} = \left(\frac{p_s - p_0}{\gamma}\right)$$

$$V_0 = C\sqrt{2g\left(\frac{p_s - p_0}{\gamma}\right)} = C\sqrt{2\left(\frac{p_s - p_0}{\rho}\right)}$$

where p_s = stagnation pressure and p_0 = static pressure.

$$V_0 = 0.98\sqrt{\frac{2\,[3.0 - (-3.0)]}{1.20} \times 1000}$$

$$= \mathbf{98\ m/s}$$

13.27 A pitot tube is inserted in a pipe of 30 cm diameter. The static pressure of the tube is 10 cm of mercury, vacuum. The stagnation pressure at the centre of the pipe recorded by the pitot tube is 1.0 N/cm^2. Calculate the rate of flow of water through the pipe, if the mean velocity of flow is 0.85 times the central velocity. Assume coefficient of the tube to be 0.98.

Solution:

$$\frac{p_0}{\gamma} = \frac{p_{static}}{\gamma} = -\frac{10}{100} \times 13.6 = -1.36 \text{ m}$$

$$\frac{p_s}{\gamma} = \frac{p_{stagnation}}{\gamma} = \frac{1 \times 10^4}{9790} = 1.021 \text{ m}$$

$$\Delta h = \left(\frac{p_s - p_0}{\gamma}\right) = [1.021 - (-1.36)] = 2.381 \text{ m}$$

$$v_m = \text{Centreline velocity} = C\sqrt{2g\Delta h}$$

$$= 0.98\sqrt{2 \times 9.81 \times 2.381} = 6.698 \text{ m/s}$$

Mean velocity in the pipe = $0.85 \times v_m$

$$V = 0.85 \times 6.698 = 5.693 \text{ m/s}$$

Discharge $Q = \frac{\pi}{4} \times (0.3)^2 \times (5.693) = \mathbf{0.402\ m^3/s}$

13.28 To calibrate a repaired pitot tube, an air jet discharging from an orifice is used. A chamber containing air under a pressure of 150 Pa (ρ_{air} = 1.2 kg/m^3) discharges the same through a sharp edged orifice of C_v = 0.98. The pitot tube is placed on the centreline at the *vena contracta* and the stagnation pressure measured by an inclined manometer. The manometer is inclined at 30° to the horizontal, contains an oil of RD = 0.85 and records a reading of 38 mm. Calculate the coefficient of the pitot tube.

Solution: For the orifice

$$V = C_v\sqrt{2g\Delta h_0} = C_v\sqrt{2\frac{(\Delta p)}{\rho}}$$

Here $\Delta p = P_{chamber} - P_{atmospheric} = 150$ Pa

$$V = 0.98\sqrt{2 \times \frac{150}{1.2}} = 15.495 \text{ m/s}$$

For the inclined manometer

$$\Delta h_m = y \sin 30° = \frac{38}{1000} \times \sin 30° = 0.019 \text{ m of liquid}$$

$$\Delta p_{manometer} = p_{stagnation} - p_{atmospheric} = \Delta p_s$$

$$= 0.019 \times 0.85 \times 9790 = 158.1 \text{ Pa}$$

For the pitot tube $V = C\sqrt{2(\Delta p_s/\rho)}$

$$15.495 = C \cdot \sqrt{2 \times \frac{158.1}{1.2}}$$

Coefficient of the pitot tube C = **0.955**

13.29 For the flow situation shown in Fig. 13.28 determine the ratio h_1/h_2 if the area ratio $A_1/A_2 = 2.0$. Neglect frictional losses.

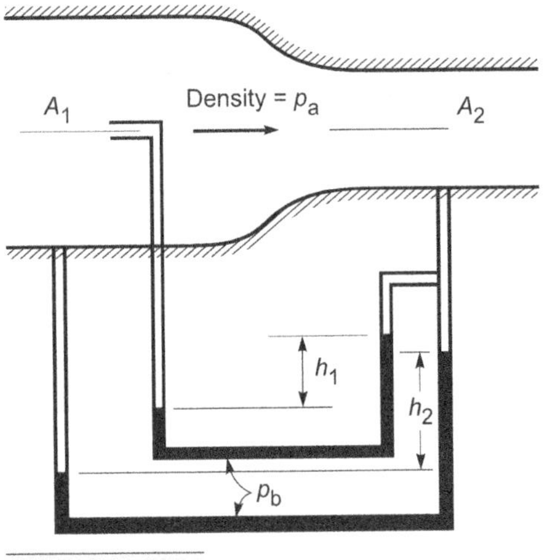

Fig. 13.28

Solution: For pitot tube:

$$p_s + h_1(\rho_a g) = p_2 + h_1(\rho_b g)$$

For piezometric tubes

$$p_1 + h_2(\rho_a g) = p_2 + h_2\,\rho_b g$$

$$\therefore\ (p_s - p_1) + \rho_a g\,(h_1 - h_2) = (h_1 - h_2)\,\rho_b g$$

$$(p_s - p_1) = (h_1 - h_2)(\rho_b - \rho_a)g$$

$$\left(\frac{p_s - p_1}{\rho_a g}\right) = \frac{V_1^2}{2g} = (h_1 - h_2)\left(\frac{\rho_b}{\rho_a} - 1\right) \quad (1)$$

From piezometer tappings

$$(p_1 - p_2) = h_2 g\,(\rho_b - \rho_a)$$

$$\left(\frac{p_1 - p_2}{\rho_a g}\right) = h_2\left(\frac{\rho_b}{\rho_a} - 1\right)$$

By Bernoulli equation:

$$\frac{p_1}{\rho_a g} + \frac{V_1^2}{2g} = \frac{p_2}{\rho_a g} + \frac{V_2^2}{2g}$$

$$\left(\frac{p_1 - p_2}{\rho_a g}\right) = \frac{V_2^2 - V_1^2}{2g} = \frac{V_1^2}{2g}\left[\left(\frac{V_2}{V_1}\right)^2 - 1\right]$$

$$= \frac{V_1^2}{2g}\left[\left(\frac{A_1}{A_2}\right)^2 - 1\right]$$

$$\therefore\ h_2\left(\frac{\rho_b}{\rho_a} - 1\right) = (h_1 - h_2)\left(\frac{\rho_b}{\rho_a} - 1\right)(4 - 1)$$

$$\left(\frac{h_1}{h_2} - 1\right)3 = 1$$

$$\frac{h_1}{h_2} = \frac{1}{3} + 1 = \frac{4}{3}$$

$$\boldsymbol{\frac{h_1}{h_2} = \frac{4}{3}}$$

D. Weirs

13.30 A suppressed rectangular weir in a rectangular channel of width 2.5 m carries a discharge with a head of 0.35 m over the crest. If the crest is 0.70 m above the channel bed, calculate the discharge over the weir. The flow can be assumed to be free.

Solution:

Here $H_1 = 0.35$ m, $P = 0.70$ m, $L = 2.5$ m

By Rehbock formula

$$C_d = 0.611 + 0.075\, H_1/P$$

$$= 0.611 + \left(0.075 \times \frac{0.35}{0.70}\right) = 0.649$$

The discharge

$$Q = \frac{2}{3} C_d \sqrt{2g}\; L H_1^{3/2}$$

$$= \frac{2}{3} \times 0.649 \times \sqrt{2 \times 9.81} \times 2.5 \times (0.35)^{3/2}$$

$$= 0.9925 \text{ m}^3/\text{s}$$

$$\boldsymbol{Q = 992.5 \text{ L/s}}$$

13.31 During the calibration of a rectangular suppressed weir in a 0.40 m laboratory channel the discharge passing over the weir was measured volumetrically. What is the coefficient of discharge of this weir if it was found to pass a discharge of 25 L/s under a head of 10 cm?

Solution: By the weir formula

$$Q = \frac{2}{3} C_d \sqrt{2g}\; L H_1^{3/2}$$

$$0.025 = \frac{2}{3} C_d \sqrt{2 \times 9.81} \times 0.40 \times (0.10)^{3/2}$$

$$= 0.03735\, C_d$$

$$\boldsymbol{C_d = 0.669}$$

13.32 A 10 cm high, sharp crested, rectangular weir spans the full width of a 0.80 m rectangular channel and is installed at the end of the channel. What maximum discharge can be passed in this channel if the side walls are 0.75 m high and the minimum specified free board is 15 cm?

Solution:

Here $P = 10$ cm $= 0.1$ m and $L = 0.80$ m

Also $H_1 + P$ = Channel height – free broad

$= 0.75 - 0.15 = 0.60$ m

$\therefore\ H_1 = 0.60 - 0.10 = 0.50$ m

$$H_1/P = \frac{0.50}{0.10} = 5.0$$

This is at the limit of the applicability of Rehbock's formula for C_d.

$$\therefore\ C_d = 0.611 + 0.075\,(H_1/P)$$

$$= 0.611 + 0.075\,(5.0) = 0.986$$

The discharge over the weir, by the weir formula, is

$$Q = \frac{2}{3} C_d \sqrt{2g}\; L H_1^{3/2}$$

$$= \frac{2}{3} \times 0.986 \times \sqrt{2 \times 9.81} \times 0.80 \times (0.50)^{3/2}$$

$$= \mathbf{0.8235\ m^3/s}$$

13.33 A rectangular channel 3.0 wide carries a discharge of 1.5 m³/s. Calculate the height of a rectangular suppressed weir in this channel which can pass this discharge while maintaining an upstream depth of flow of 1.5 m.

Solution: Refer to Fig. 13.29.

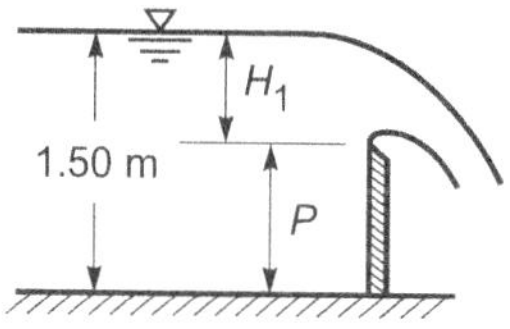

Fig. 13.29

In this problem since both H_1 and P are unknowns, a trial and error procedure is used.

First assume a value of C_d.

I trial: Assume $C_d = 0.64$

$$Q = \frac{2}{3} C_d \sqrt{2g}\ LH_1^{3/2}$$

$$1.5 = \frac{2}{3} \times 0.64 \times \sqrt{2 \times 9.81} \times 3.0 \times H_1^{3/2}$$

$$H_1^{3/2} = 0.2646, \quad H_1 = 0.412 \text{ m}$$

$$P = 1.50 - 0.412 = 1.088 \text{ m}$$

By Rehbock formula $C_d = 0.611 + 0.075 \dfrac{H_1}{P}$

$$= 0.611 + 0.075 \times \left(\frac{0.412}{1.088}\right) = 0.639$$

2nd trial: Use $C_d = 0.639$ in the 2nd trial for the calculation of discharge by the weir formula.

$$Q = 1.50 = \frac{2}{3} \times 0.639 \times \sqrt{2 \times 9.81} \times 3.0 \times (H_1)^{3/2}$$

$$H_1^{3/2} = 0.2650, \ H_1 = 0.4125 \text{ m}$$

$$P = 1.50 - 0.4125 = 1.0875 \text{ m}$$

$H_1/P = 0.3793$ and substituting this in the Rehbock formula,

$$C_d = 0.611 + 0.075 \times 0.3793 = 0.639$$

No further trials are necessary. Thus the weir height P = **1.0875 m**.

13.34 A rectangular weir 0.75 m high and 1.5 m long is to be used for discharging water from a tank under a head of 0.5 m. Estimate the discharge (a) When it is used as a suppressed weir. (b) When it is used as a contracted weir.
(Assume C_d is given by the Rehbock equation in both cases and neglect velocity of approach).

Solution: By Rehbock formula,

$$C_d = 0.611 + 0.075 \frac{H_1}{P}$$

$$= 0.611 + 0.075 \left(\frac{0.5}{0.75}\right)$$

$$= 0.661$$

(a) *For a suppressed weir,*

$$Q = \frac{2}{3} C_d \sqrt{2g}\ LH_1^{3/2}$$

$$= \frac{2}{3} \times 0.661 \times \sqrt{2 \times 9.81} \times 1.5 \times (0.5)^{3/2}$$

$$= \mathbf{1.035\ m^3/s}$$

(b) *For a contracted weir,*

L_e = effective length of weir = $L - 0.1 \times 2 \times H$

$= 1.5 - (0.1 \times 2 \times 0.5) = 1.4$ m

$$Q = \frac{2}{3} \times C_d \sqrt{2g}\ L_e H_1^{3/2}$$

$$= \frac{2}{3} \times 0.661 \times \sqrt{2 \times 9.81} \times 1.4 \times (0.5)^{3/2}$$

$$= \mathbf{0.966\ m^3/s}$$

13.35 A rectangular contracted weir has a length of 2.5 m between the abutments. There are two 0.15 m thick rectangular pier-like obstructions on the crest. Estimate the discharge under a head of 0.70 m. Assume $C_d = 0.62$ and the velocity of approach to be negligible.

Solution: Effective crest length = $L_e = L - 0.1\ nH_1$

Here crest length $L = 2.5 - 2 \times 0.15 = 2.20$ m

n = number of end contractions = $2 + (2 \times 2) = 6$

$L_e = 2.20 - (0.1 \times 6 \times 0.7) = 1.78$ m

The discharge, from Francis formula by neglecting the velocity of approach,

$$Q = \frac{2}{3} C_d \sqrt{2g}\ L_e H_1^{3/2}$$

$$= \frac{2}{3} \times 0.62 \times \sqrt{2 \times 9.81} \times 1.78 \times (0.7)^{3/2}$$

$$= \mathbf{1.91\ m^3/s}$$

13.36 The flow in a 2.0 m wide rectangular channel is measured by a rectangular weir 1.0 m long and 0.60 m high. Find the discharge in the channel when the head over the weir is 0.30 m. Take C_d = 0.62.

Solution: This is the case of a contracted weir.

Here $L = 1.0$ m, $P = 0.60$ m, $H_1 = 0.30$ m.

n = number of end contractions = 2.

The discharge is estimated by the Francis formula

$$Q = \frac{2}{3} C_d \sqrt{2g}\ (L - 0.1\ nH_1) \left[\left(H_1 + \frac{V_0^2}{2g}\right)^{3/2} - \left(\frac{V_0^2}{2g}\right)^{3/2}\right]$$

Since the discharge is involved in both sides of this equation, a trial and error method is adopted. Since $(V_0^2/2g)$ is usually a very small quantity as a first trial, Q is calculated by assuming $V_0 = 0$.

First trial.

$$Q_1 = \left\{\frac{2}{3} \times 0.62 \times \sqrt{2 \times 9.81} \times (1.0 - 0.2 \times 0.3)\right\} (0.30)^{3/2}$$

$$= 1.721\ (0.30)^{3/2} = 0.2828 \text{ m}^3/\text{s}$$

Second trial using the above Q, the velocity of approach

$$V_0 = \frac{Q}{A} = \frac{Q}{B(H+P)} = \frac{0.2828}{2.0 \times 0.9} = 0.157 \text{ m/s}$$

$$\frac{V_0^2}{2g} = 0.001258$$

Revised discharge

$$Q = Q_2$$
$$= 1.721\ [(0.30 + 0.001258)^{3/2} - (0.001258)^{3/2}]$$
$$= 0.2845\ \text{m}^3/\text{s}$$

Third trial: Using the above Q,

revised $$V_0 = \frac{0.2845}{2 \times 0.9} = 0.15805\ \text{m/s}$$

$$\frac{V_0^2}{2g} = 0.001273$$

Revised discharge

$$Q_3 = 1.721 \times [(0.3 + 0.001273)^{3/2} - (0.001273)^{3/2}]$$
$$= 0.2845\ \text{m}^3/\text{s}$$

This is the same as in the previous trial. Thus no more trials are needed.

∴ Discharge in channel $\mathbf{Q = 0.2845\ m^3/s}$.

13.37 The flow of water in a channel varies from 100 L/s to 500 L/s. It is desired to measure the discharge in the canal with an accuracy of 0.5 per cent throughout the entire range. The water surface elevation could be measured to an accuracy of 0.5 mm. If a sharp crested weir is used for discharge measurement, what is the maximum length of the weir to satisfy the above requirements? Assume $C_d = 0.62$ and neglect the velocity of approach.

Solution: For a contracted weir, by neglecting the velocity of approach,

$$Q = \frac{2}{3} C_d \sqrt{2g}\ L_e H_1^{3/2}$$

where L_e = effective length. Considering L_e = constant

$$\frac{dQ}{Q} = 1.5 \frac{dH_1}{H_1}$$

Considering the accuracies desired

$$0.005 = 1.5 \times \frac{0.0005}{H_m}$$

H_m = minimum head desired = 0.15 m = 15 cm.

Thus H_1 must be greater than or equal to 0.15 m. By the discharge equation, for the smallest discharge at the maximum L_e,

$$Q = 0.100 = \frac{2}{3} \times 0.62 \times \sqrt{2 \times 9.81}\ L_e\ (0.15)^{3/2}$$

$$L_e = \frac{0.100}{0.106} = 0.94\ \text{m}$$

L_e should be less than or equal to 0.94 m

Since $L_e = L - 0.2\ H_1$

L = length of the weir $\le (0.94 + (0.2 \times 0.15))$

Maximum length of weir = **0.97 m**.

13.38 A discharge of 100 L/s is to be measured by triangular notch of vertex angle 60°. What would be the head over the vertex of the notch? If the accuracy of reading the head is 1 mm what error in discharge can be expected at this level? Assume $C_d = 0.58$.

Solution: For a triangular notch, the discharge Q is given by

$$Q = \frac{8}{15} C_d \sqrt{2g}\ \tan\frac{\theta}{2}\ H_1^{5/2}$$

$$0.100 = \left\{\frac{8}{15} \times 0.58 \times \sqrt{19.62} \times \tan 30°\right\} H_1^{5/2}$$

$$0.100 = 0.79107\ H_1^{5/2}$$

$$\mathbf{H_1 = 0.437\ m}$$

Writing the discharge equation as $Q = KH_1^{5/2}$

$$dQ = \frac{5}{2} KH_1^{3/2}\ dH_1$$

$$\therefore \quad \frac{dQ}{Q} = \frac{5}{2}\frac{dH_1}{H_1}$$

Here $dH = 0.001$ m and $H_1 = 0.437$ m

$$\frac{dQ}{Q} = \frac{5}{2} \times \frac{0.001}{0.437} = 5.72 \times 10^{-3}$$
$$= 0.572\%$$

Since $Q = 100$ L/s, $dQ = 0.572$ L/s

Possible error in discharge = **± 0.572 L/s**

13.39 For a head of 25 cm on a 90° V-notch, what error in the measured head will produce the same percentage error in the computed discharge as an error of 1° in the vertex angle θ?

Solution: For a V-notch,

$$Q = \frac{8}{15} C_d \sqrt{2g}\ \tan\frac{\theta}{2}\ H_1^{5/2}$$

$$= K \tan\frac{\theta}{2}\ H_1^{5/2} \quad \text{where } K = \text{a constant.}$$

$$\frac{dQ}{Q} = \frac{5}{2}\frac{dH_1}{H_1}$$

At $H = 0.25$ cm,

$$\frac{dQ}{Q} = \frac{5}{2} \times \frac{1}{0.25}\ dH_1 = 10\ dH_1$$

$$\frac{dQ}{d\theta} = KH_1^{5/2} \cdot \frac{1}{2} \cdot \left(\sec^2\frac{\theta}{2}\right)$$

$$\frac{dQ}{Q} = \frac{\sec^2(\theta/2)\,d\theta}{2\tan(\theta/2)}$$

At $\theta = 90°$ and $d\theta = 1° = \frac{1}{57.296}$ rad

$$\frac{dQ}{Q} = \frac{1}{2}\frac{1}{\cos^2(45°)}\frac{d\theta}{\tan(45°)}$$

$$= \frac{1}{57.296}$$

If $\frac{dQ}{Q}$ is to be the same in both cases

$$10\,dH_1 = \frac{1}{57.296}$$

$dH_1 = 1.745 \times 10^{-3}$ m = **1.745 mm**

= required error in measurement of H_1.

13.40 A trapezoidal notch has a base width of 0.75 m and a side slope of 1 horizontal : 2 vertical. Calculate the discharge over the notch for a head of 0.50 m by assuming $C_d = 0.63$.

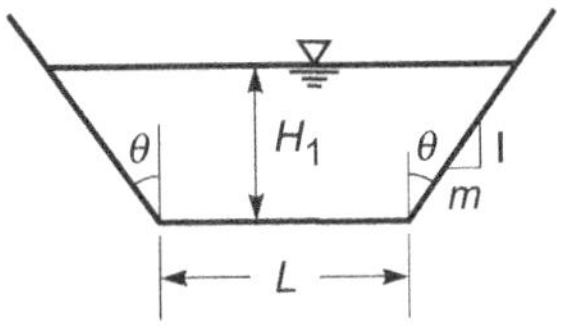

Fig. 13.30 *Trapezoidal Notch*

Solution:

$\tan\theta = m = 0.5$

The discharge Q for a head H_1 is

$$Q = \frac{2}{3} C_d \sqrt{2g}\, H_1^{3/2}\left(L + \frac{4}{5}H_1 \tan\theta\right)$$

$$= \frac{2}{3} \times 0.63 \times \sqrt{2g} \times (0.50)^{3/2}$$

$$\left\{0.75 + \left(\frac{4}{5} \times 0.5 \times 0.5\right)\right\}$$

$= 0.65774\ (0.95) =$ **0.625 m³/s**

13.41 A flow from a channel is controlled by a trapezoidal notch so that the full supply discharge of 2.0 m³/s flows over the notch at a head of 1.2 m measured over the crest. At half this head, a discharge of 0.6 m³/s passes over the notch. Assuming $C_d = 0.62$, calculate the base width and side slopes of the notch.

Solution: For a trapezoidal notch,

$$Q = \frac{2}{3} C_d \sqrt{2g}\, H_1^{3/2}\left(L + \frac{4}{5}H_1 \tan\theta\right)$$

$$= \frac{2}{3} \times 0.62 \times \sqrt{2 \times 9.81}\, H_1^{3/2}\,(L + 08\, H_1 \tan\theta)$$

$$= 1.8308\, H_1^{3/2}\,(L + 0.8\, H_1 \tan\theta)$$

when $H_1 = 1.2$ m,

$$2.0 = 1.8308 \times (1.2)^{3/2}\,(L + (0.8 \times 1.2 \times \tan\theta)]$$

Simplifying $L + 0.96 \tan\theta = 0.8310$ (1)

When $H_1 = 1.2/2 = 0.6$ m,

$$0.6 = 1.8308 \times (0.6)^{3/2}\,(L + (0.8 \times 0.6 \times \tan\theta)]$$

Simplifying $L + 0.48 \tan\theta = 0.7051$

Subtracting Eq. (2) from Equation (1) (2)

$0.48 \tan\theta = (0.8310 - 0.7051) = 0.1259$

or $\tan\theta = 0.2623$

$\boldsymbol{\theta = 14.7°}$

From Eq. (2)

$L = 0.7051 - 0.1259 =$ **0.5792 m**

13.42 A plate having a cutout in the form of a triangle with apex angle θ and base length L is provided at the end of a channel to act as a weir. The apex is above the water surface and the base which is horizontal acts as the crest of the notch. Obtain an expression for the discharge over the notch.
If the base is 1.2 m long and the apex angle is 60°, estimate the discharge for a head of 0.30 m over the crest. (Assume $C_d = 0.60$).

Solution: The notch in question is a trapezoidal notch with inward sloping sides (Fig. 13.31). This can be considered equivalent to be suppressed rectangular notch of length L minus a triangular notch of apex angle θ, as shown in Fig. 13.31 (b), both having the same C_d. Hence the discharge for a head H_1 is

$$Q = \frac{2}{3} C_d \sqrt{2g}\, L H_1^{3/2} - \frac{8}{15} C_d \sqrt{2g} \tan\frac{\theta}{2} H_1^{5/2}$$

$$Q = \frac{2}{3} C_d \sqrt{2g}\, H_1^{3/2}\left(L - \frac{4}{5}H_1 \tan\frac{\theta}{2}\right)$$

In the problem:

$\theta = 60°$

$H_1 = 0.30$ m

$C_d = 0.60$

$L = 1.20$ m

$$Q = \frac{2}{3} \times 0.60 \times \sqrt{19.62} \times (0.30)^{3/2}$$

$$\left\{1.2 - \left(\frac{4}{5} \times 0.3 \times \tan 30°\right)\right\}$$

= **0.309 m³/s**

13.43 A channel has a Cipolletti weir of 50 cm base installed at a section. Estimate the discharge in the

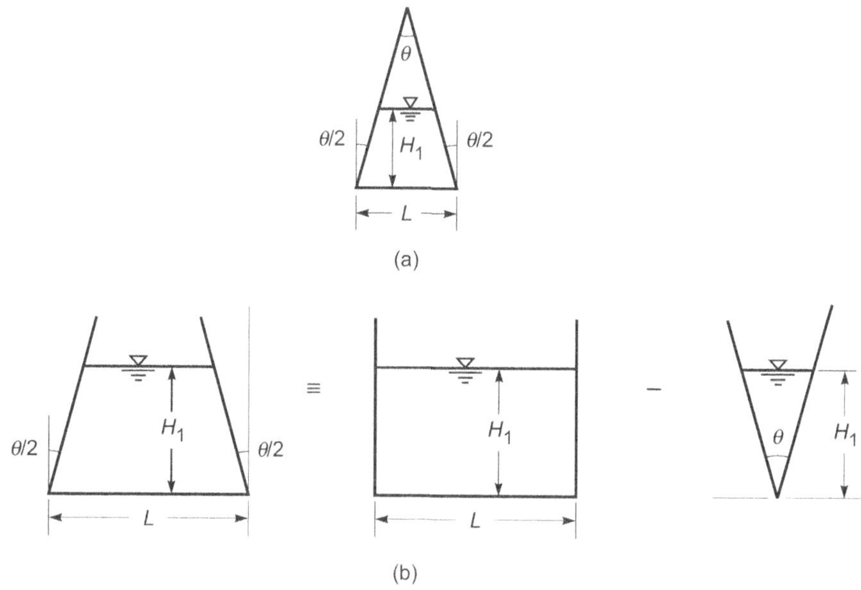

Fig. 13.31 *Inward Sloping Trapezoidal Notch*

channel when the head over the weir crest is 0.25 cm. Assume $C_d = 0.63$.

Solution: The discharge over a Cipolletti weir is calculated by using the suppressed weir formula.

$$Q = \frac{2}{3} C_d \sqrt{2g}\, LH_1^{3/2}$$

$$= \frac{2}{3} \times 0.63 \times \sqrt{2 \times 9.81} \times 0.50 \times (0.25)^{3/2}$$

$= 0.1163$ m³/s = **116.3 L/s**

13.44 A 75° triangular notch is discharging under submerged conditions. The vertex of the notch is at a height of 30 cm from the channel bed. The elevations of water surfaces upstream and downstream of the notch, measured from the channel bed are 75 cm and 50 cm respectively. Assuming $C_d = 0.60$, estimate the discharge over the notch.

Solution: This is a case of submerged flow.

$$H_1 = 75 - 30 = 45 \text{ cm} = 0.45 \text{ m}$$

$$H_2 = 50 - 30 = 20 \text{ cm} = 0.20 \text{ m}$$

Q_1 = free flow mode discharge under H_1

$$= \frac{8}{15} \times C_d \sqrt{2g} \tan \frac{\theta}{2} H_1^{5/2}$$

$$= \frac{8}{15} \times 0.6 \times \sqrt{2 \times 9.81} \tan (37.5°) \times (0.45)^{5/2}$$

$= 0.1477$ m³/s

By the Villemonte equation

$$Q_s = Q_1 \left[1 - \left(\frac{H_2}{H_1} \right)^n \right]^{0.385}$$

For a triangular notch, $n = 5/2$.

$$Q_s = 0.1477 \left[1 - \left(\frac{0.20}{0.45} \right)^{2.5} \right]^{0.385}$$

$$= 0.1477\, [1 - 0.1317)^{0.385} = 0.140 \text{ m}^3/\text{s}$$

The discharge over the notch is **140 L/s**.

13.45 A tank of constant cross sectional area A has a V notch to control its water surface. If the water surface in the tank is at an elevation H_1 above the vertex of the notch, obtain an expression for the time required to lower it to H_2 through an outflow over the V-notch.

Solution: Let the water surface be at an elevation h above the vertex at any instant t, and in time dt let it drop by dh. By continuity, the volume of outflow

$$-A\, dh = Q \cdot dt = \frac{8}{15} C_d \sqrt{2g} \tan \frac{\theta}{2} h^{5/2}\, dt$$

where θ = angle of the V-notch.

$$dt = \frac{-A}{\frac{8}{15} C_d \sqrt{2g} \tan \frac{\theta}{2}} \frac{dh}{h^{5/2}}$$

$$\int_0^T dt = T = \int_{H_1}^{H_2} \frac{-A}{\frac{8}{15} C_d \sqrt{2g} \tan \frac{\theta}{2}} (h^{-5/2}\, dh)$$

$$T = \frac{2}{3} \times \frac{-A}{8/15\, C_d \sqrt{2g} \tan \frac{\theta}{2}} (H_2^{-3/2} - H_1^{-3/2})$$

or $$T = \frac{5}{4} \cdot \frac{A}{C_d \sqrt{2g} \tan \frac{\theta}{2}} \left(\frac{1}{H_2^{3/2}} - \frac{1}{H_1^{3/2}} \right)$$

13.46 A prismatic reservoir 1.4×10^5 m^2 in area is to be controlled by a rectangular weir without end contractions, with its crest level at elevation 100.00 m. It is required to lower the water level from 101.60 m to 101.00 m in half an hour. Estimate the length of the weir required. The weir can be assumed to have a constant $C_d = 0.735$.

Solution: Let, at any instant of time, the water level be at a height h above the weir crest. In time dt, volume of water outflow is

$$-A\,dh = Q\,dt = \frac{2}{3} C_d \sqrt{2g}\; Lh^{3/2}$$

$$dt = -\frac{A\,dh}{\frac{2}{3} C_d \sqrt{2g}\, Lh^{3/2}}$$

$$T = \int_0^T dt = \int_{H_2}^{H_1} \frac{A}{\frac{2}{3} C_d \sqrt{2g}\, L} (h^{-3/2})\,dh$$

$$= \frac{2A}{\frac{2}{3} C_d \sqrt{2g}\, L} \left(\frac{1}{\sqrt{H_2}} - \frac{1}{\sqrt{H_1}} \right)$$

Here $T = 30$ min $= 1800$ s

$H_1 = 1.60$ m, $H_2 = 10$ m

$$L = \frac{2 \times 1.4 \times 10^5}{\left(\frac{2}{3}\right) \times (0.735)\sqrt{19.62} \times 1800} \left(1 - \frac{1}{\sqrt{1.6}} \right)$$

$= \mathbf{15.0\ m}$

OBJECTIVE QUESTIONS

13.1 A 20 cm diameter orifice discharging from a tank issues out a jet of 15.75 cm diameter at the *vena contracta*. The coefficient of contraction is

(a) 0.520 (b) 0.620
(c) 0.790 (d) 0.887

13.2 The loss of head H_L in an orifice discharging under a head H is

(a) $\sqrt{H}\,(C_v - 1)$ (b) $H\,(1 - C_v)$
(c) $H\,(1 - C_v)^2$ (d) $\left[\frac{1}{C_v^2} - 1\right] H$

13.3 If a tank discharges water from an orifice under variable head h, the water surface will be lowered at constant velocity, if the surface area of the tank varies as

(a) $\sqrt{h}$ (b) $\frac{1}{\sqrt{h}}$
(c) $\frac{1}{h}$ (d) h

13.4 A 10 cm diameter Borda's internal mouthpiece running free, discharges 24 L/s under a head of 2.0 m. The coefficient of velocity C_v is

(a) 0.5 (b) 0.948
(c) 0.976 (d) 0.995

13.5 The C_d of an orifice is always

(a) greater than C_c (b) equal to C_v
(c) equal to C_c (d) less than C_c

13.6 An orifice is discharging under a head of 1.25 m of water. A pitot tube kept at its centre line at the *vena contracta* indicates a head of 1.20 m of water. The coefficient of velocity of the orifice is

(a) 0.990 (b) 0.980
(c) 0.965 (d) 0.960

13.7 The head loss at an orifice ($C_v = 0.98$) discharging under a head of 2.0 m is

(a) 0.02 m (b) 0.04 m
(c) 0.06 m (d) 0.08 m

13.8 The velocity of efflux from an orifice is observed to be 3.1 m/s. If the C_v of the orifice is 0.98, the head loss at the orifice is

(a) 0.441 m (b) 0.009 m
(c) 0.015 m (d) 0.020 m

13.9 The water level in a cylindrical tank with its axis vertical was lowered form an elevation of 4 m to 2.0 m in 200 s by discharging the contents through a orifice at the bottom of the tank. If the diameter of the orifice is doubled, the time required for the above lowering of the water surface would be

(a) 50 s (b) 100 s
(c) 12.5 s (d) 200 s

13.10 In a prismatic tank of area A the water surface is lowered by efflux from an orifice at the bottom. If the head of water above the orifice at any time t is h, the velocity of water surface in the tank is

(a) $-dA/dt$ (b) $-dh/dt$

(c) $\sqrt{2gh}$ (d) $C_v\sqrt{2gh}$

13.11 In submerged orifice flow, the discharge is proportional to
(a) square root of the upstream head H_1
(b) square root of the downstream head H_2
(c) square root of the difference between upstream and downstream heads, $(H_1 - H_2)$.
(d) square of the upstream head H_1.

13.12 An orifice meter consists of an orifice of diameter d in a pipe of diameter D. In general, the C_d of the orifice meter is
(a) a function of d/D only
(b) a function of Reynolds number only
(c) independent of d/D and Reynolds number.
(d) a function of d/D and Reynolds number

13.13 For a given Reynolds number, the C_d of an orifice meter of orifice diameter d in a pipe of diameter D
(a) increase with an increase in d/D
(b) decreases with increase in d/D
(c) independent of d/D
(d) increases with d/D up to $d/D = 0.5$.

13.14 An orifice meter with $d/D = 0.5$ and Reynolds number $= 10^6$ is expected to have C_d of about
(a) 0.45 (b) 0.61
(c) 0.80 (d) 0.95

13.15 The percentage error in the estimation of the discharge due to an error of 2% in the measurement of the reading of a differential manometer connected to an orifice meter is
(a) 4% (b) 2%
(c) 1% (d) 0.5%

13.16 Identify the *incorrect statement*:
A flow nozzle
(a) has a contraction coefficient of unity
(b) is less costly than a venturimeter
(c) is more efficient than an orifice meter
(d) has overall losses much smaller than in a venturimeter

13.17 A standard, long radius, flow nozzle has a discharge coefficient C_d larger than that of a corresponding standard venturimeter. The non-recoverable energy loss under identical condition is
(a) larger in the flow nozzle
(b) larger in the venturimeter
(c) essentially same in both the meters
(d) larger in the flow nozzle upto a critical Reynolds number and then onwards it is smaller.

13.18 In a standard orifice meter.
(a) the bevel of the plate is on the upstream
(b) the bevel angle is 45° to 60°
(c) the non-recoverable energy loss is independent of the location of the pressure taps
(d) the C_d value is independent of the location of the pressure taps.

13.19 A standard venturimeter of diameter ratio 0.5, at a Reynolds number of 10^6 has a coefficient of discharge of about
(a) 0.65 (b) 0.75
(c) 0.87 (d) 0.97

13.20 The discharge coefficient of a standard venturimeter can be expressed, in genial as $C_d =$
(a) fn (Re)
(b) fn (β)
(c) fn (Re, β)
(d) a constant for all Re and β.
where Re = Reynolds number and β = ratio of throat to inlet diameter.

13.21 A venturimeter has a differential mercury water manometer connected to its inlet and throat. The gauge reading y of the manometer for a given discharge in the pipe
(a) depends on the orientation of the venturimeter
(b) is independent of the orientation of the venturimeter
(c) varies as the slope of the venturimeter with respect to the horizontal
(d) depends on whether the manometer is above or below the pipe centreline.

13.22 A venturimeter has a $C_d = 0.95$. For a differential head of 2.8 m across the inlet and the throat, the loss of head between the inlet and throat is
(a) 0.273 m (b) 0.140 m
(c) 0.302 m (d) 0.95 m

13.23 A venturimeter with a throat diameter of 6 cm is connected to a pipe of 10 cm diameter. The C_d of the instrument is 0.95. For this meter, the head loss from the inlet to throat section can be expressed as $KV^2/2g$ where V = velocity at the throat and K =
(a) 0.0975 (b) 0.0105
(c) 0.014 (d) 0.108

13.24 The value of the coefficient of discharge of venturimeter lies in the range
(a) 0.60 to 0.75 (b) 0.70 to 0.80
(c) 0.81 to 0.94 (d) 0.95 to 0.98

13.25 The stagnation pressure in front of an object in a fluid flow is equal to

(a) static pressure
(b) dynamic pressure
(c) sum of the static and dynamic pressures
(d) piezometric head

13.26 A static tube is used to measure
(a) the velocity
(b) undisturbed fluid pressure
(c) the total head
(d) datum head

13.27 The pitot-static tube measures
(a) the dynamic pressure
(b) the static pressure
(c) the total head
(d) the difference in static and dynamic pressures

13.28 To measure static pressure in a pipe, one uses a pressure gauge connected to a
(a) pitot tube
(b) venturimeter
(c) orifice meter
(d) piezometer tapping.

13.29 Preston tube is a device to measure
(a) boundary pressure
(b) mean velocity of flow
(c) boundary shear stress
(d) turbulent velocity fluctuation

13.30 A Pitot-static tube indicates a differential head of 0.75 m of water between its two openings when inserted in a stream of water. If the coefficient of the tube is 0.99, the velocity in m/s, at the location of the tube is
(a) 4.43 (b) 0.78
(c) 3.84 (d) 3.80

13.31 A Pitot tube (coefficient = 1.0) is used to measure the velocity of air mass density 1.2 kg/m^3. If the head difference in a vertical U-tube fitted with water is 12 mm, then the velocity of air in m/s is
(a) 10 (b) 14
(c) 17 (d) 20

13.32 A hot-wire anemometer is used to measure essentially the
(a) wind speed over ground surfaces
(b) turbulent velocity fluctuations in a flow
(c) shear stress on a boundary
(d) drag force on a body

13.33 A laser-doppler anemometer (LDA) is a device to measure
(a) the turbulent velocity fluctuations in a flow
(b) shear stress at a boundary
(c) drag force on an airfoil
(d) surface tension of a fluid

13.34 The discharge per metre length over a suppressed rectangular weir of a crest height 10 cm above the bed under a head of 10 cm, in L/s/m, is
(a) 73 (b) 59
(c) 106 (d) 64

13.35 The discharge, in L/min, in a 90° V-notch having a C_d = 0.58 under a head of 0.10 m, is
(a) 260 (b) 310
(c) 130 (d) 173

13.36 The coefficient of discharge by Rehbock formula of a sharp crested suppressed rectangular weir of crest height P and passing a free flow with a head H_1 is C_d =

(a) $0.611 + 0.8 \dfrac{H_1}{P}$

(b) $0.611 + 0.075 \dfrac{H_1}{P}$

(c) $0.602 + 0.075 \dfrac{P}{H_1}$

(d) $0.675 + 0.0075 \dfrac{H_1}{P}$

13.37 The coefficient of discharge of a suppressed rectangular weir at the limit of application of Rehbock formula is
(a) 0.786 (c) 0.886
(c) 0.986 (d) 1.06

13.38 A suppressed sharp crested weir is 0.6 m high and discharges water at a head of 1.2 m. The coefficient of discharge of this weir is
(a) 0.611 (b) 0.701
(c) 0.736 (d) 0.761

13.39 The Rehbock formula for estimating the C_d of a suppressed rectangular weir is valid up to H_1/P
(a) ≤ 0.50 (b) ≤ 5.0
(c) ≤ 10.0 (d) > 20

13.40 The discharge over a suppressed sharp crested rectangular weir is given by

$Q = \frac{2}{3} C_d \sqrt{2g}\ L H_1^{3/2}$. Here H_1 is

(a) the difference in elevation between the upstream water surface and the energy line
(b) the difference in elevation between the energy line and the weir crest
(c) the difference in elevation of the water surface and the weir crest
(d) the difference in elevation between the water surface and the bottom of the channel.

13.41 For a suppressed rectangular weir an arrangement for aeration of nappe is necessary.
(a) to maintain water quality
(b) to prevent submergence of the weir
(c) to have the highest value of C_d
(d) to have a constant head-discharge relationship which is independent of time.

13.42 Identify the *incorrect* statement:
The important requirements in the installation of a rectangular suppressed weir are:
(a) the weir crest must be level
(b) the underside of the nappe on the downstream must be ventilated
(c) the upstream top edge of the plate must be bevelled
(d) the approach channel shall be straight and prismatic

13.43 In a triangular notch there is an error of 4% in observing the head. The error in the computed discharge is
(a) 4% (b) 10%
(c) 6% (d) 2.5%

13.44 In a suppressed rectangular weir the computed discharge was found to be 3% in excess of the actual discharge. If this discrepancy was due to an error in reading the head, the measured head was
(a) 3% excess (b) 2% less
(c) 2% excess (d) 1.2% excess

13.45 In a 90° triangular notch, for a given head, the error in the estimated discharge due to a 2% error in the measurement of the vertex angle is
(a) π% (b) 5.0%
(c) 3.0% (d) $\pi/2$%

13.46 A 1.5 m long, suppressed rectangular weir had a head of water of 62 cm. However, it was measured as 60.0 cm and used in computation. The percentage error in computed flow is
(a) 3.3 (b) 5.0
(c) 6.7 (d) 2.0

13.47 The discharge in a triangular notch and a rectangular suppressed weir both having the same head and C_d are identical when the ratio of the water surface width in the V-notch to the length of the rectangular weir is
(a) 1 (b) $1\frac{1}{2}$
(c) $1\frac{7}{8}$ (d) $2\frac{1}{2}$

13.48 An overflow spillway is 10.0 m long between two square abutments and has two piers of 0.25 m width on its crest when the head over the weir is 0.60 cm. The effective length of the spillway for calculation of the discharge by the weir formula is
(a) 9.14 m (b) 9.50 m
(c) 9.26 m (d) 9.40 m

13.49 A separate arrangement for aeration of nappe is desired in the following:
(a) triangular weir
(b) trapezoidal weir
(c) contracted rectangular weir
(d) none of the above

13.50 A suppressed rectangular weir and a triangular V-notch are installed in a large tank such that the vertex of the notch and the weir are at the same level. If C_d of both the weir and the notch is same.
(a) the discharge from the V-notch will be larger for all heads.
(b) the discharge from the rectangular weir will be larger for all heads
(c) the rectangular weir will have higher discharge from zero value to a critical head and then onwards it will be smaller than that of the V-notch.
(d) the triangular notch will have higher discharge from zero value to a critical head beyond which it will be smaller than that of the rectangular weir.

13.51 The discharge over a 90° V-notch is written as $Q = 1.37\ H^{5/2}$, where Q is in m^3/s and H is in metres. The C_d of this notch is
(a) 0.611 (b) 0.580
(c) 0.464 (d) 0.710

13.52 The discharge over a Cipolletti weir of base width B is written as $Q = KH^{3/2}$, where Q is in m^3/s and H in metres. If the $C_d = 0.62$, the coefficient K =
(a) 1.86 B (b) 1.83 B
(c) 1.46 B (d) 1.52 B

13.53 A Cipolletti weir and a suppressed rectangular weir both have the same crest length, C_d and head over the weir. Then
(a) the discharge in the Cipolletti weir is larger
(b) the discharge in the rectangular weir is larger
(c) both the weirs will have the same discharge
(d) Rectangular weir will have smaller discharge at small heads and both the discharges will be same after a critical head is exceeded.

13.54 A Cipolletti weir has a side slope of
(a) 1 vertical : 4 horizontal
(b) 1 vertical : 2 horizontal
(c) 1 horizontal : 4 vertical
(d) 1 horizontal : 2 vertical

13.55 A submerged weir is one in which the water level on the downstream of the weir is
(a) just at the crest level
(b) below the crest level
(c) is above the crest level
(d) at the same elevation as the upstream water surface.

13.56 In submerged flows over triangular notches, the ratio of submerged flow to the corresponding free flow under the same head, Q_s/Q_1 is given by the Villemonte equation as, Q_s/Q_1 =

(a) $\left[1-\left[\frac{H_2}{H_1}\right]^{3/2}\right]^{0.385}$

(b) $\left[1-\left[\frac{H_2}{H_1}\right]^{5/2}\right]^{0.385}$

(c) $\left[1-\left[\frac{H_2}{H_1}\right]\right]^{0.385}$

(d) $\left[1-\left[\frac{H_2}{H_1}\right]^{0.385}\right]^{3/2}$

where H_1 and H_2 are heads above the vertex on the upstream and downstream of the notch respectively.

13.57 In a rectangular suppressed weir the tailwater head is 30% of the upstream head both measured above the crest. The submerged flow discharge is x per cent of the free flow discharge at the same upstream head, where x is
(a) 98.1% (b) 70%
(c) 93.3% (d) 87.7%

13.58 While conducting flow measurement using a rectangular notch, an error of 2% in head over the notch and an error of –3% in the length of the notch occurred. The percentage error in the computed discharge would be
(a) +6% (b) –1%
(c) –2.5% (d) zero

PROBLEMS

A. Orifices

13.1 A closed tank A contains 3.0 m depth of water and an air space at 15 kPa pressure. A 5 cm diameter orifice at the bottom of the tank discharges the water to a tank B containing pressurised air at 25 kPa. If the coefficient of discharge of the orifice is 0.61, calculate the discharge of water from tank A.

(*Ans.* Q = 7.46 L/s)

13.2 A 5 cm diameter orifice discharges 7.75 L/s of water under a head of 2.0 m. A flat plate held normal to the jet just downstream of the *vena contracta* experiences a force of 4.5 N. Find he values of C_c, C_v and C_d of the orifice.

(*Ans.* $C_c = 0.678$, $C_v = 0.929$ and $C_d = 0.630$)

13.3 A 25 mm diameter nozzle discharges 0.76 m^3/min of water when the head is 60 m. The diameter of the jet is 22.5 mm. Determine (i) the values of the coefficients C_c, C_v and C_d and (ii) the loss of head due to fluid resistance in the nozzle.

(*Ans.* $C_c = 0.810$, $C_v = 0.930$ and $C_d = 0.750$; $H_L = 8.106$ m)

13.4 A jet issues in an upward trajectory out of an orifice located on the side of a tank which has an inclination of α with the horizontal. If the head of the liquid in the tank over the centre of the orifice is H, show that the maximum elevation of the jet Y and its horizontal distance X from the *vena contracta* are given by

$$X = C_v^2\, H \sin^2 \alpha \quad \text{and} \quad Y = C_v^2\, H \cos^2\alpha$$

13.5 Water discharges at the rate of 100 L/s through a 12 cm diameter vertical sharp edged orifice placed under a constant head of 10 m. A point on the jet, issuing into atmosphere, has coordinates measured from the *vena contracta* of 4.5 m horizontal and 0.55 m vertical. Find the coefficients C_v, C_c and C_d.

(*Ans.* $C_v = 0.959$, $C_c = 0.658$ and $C_d = 0.631$)

13.6 A 12 cm × 3 cm nozzle is attached to the end of a 12 cm diameter water pipe. The pressure at the base of the nozzle is 200 kPa. If the coefficient, of velocity is 0.96 and the coefficient of contraction is 0.90, determine the discharge of the jet.

(*Ans.* $Q = 12.246$ L/s)

13.7 A closed tank contains kerosene (RD = 0.8) to a depth of 2.5 m. The top portion of the tank contains air under a pressure of 25 kPa. If a sharp edged circular orifice of diameter 3 cm ($C_d = 0.61$) is provided at the bottom of the tank, estimate the discharge through the orifice.

(*Ans.* $Q = 4.294$ L/s)

13.8 A 9 cm diameter base and 4.5 cm diameter tip nozzle has a pressure of 60 kPa at the base. If the coefficients of velocity and contraction of the nozzle are 0.95 and 0.85 respectively, determine the jet velocity and power of the jet of water.

(*Ans.* $V_j = 10.636$ m/s and $\mathbf{P} = 0.813$ kW)

13.9 A nozzle 5 cm in diameter is attached to a 10 cm diameter pipe. The pressure at the base of the nozzle is 8 N/cm^2. If the coefficient of contraction = 0.99 and the coefficient of velocity = 0.98, find the following: (a) the velocity of the jet, (b) discharge, (c) power available from the jet, (d) efficiency of the nozzle and (e) height to which the jet will rise if directed vertically upwards.

(*Ans.* (a) $V_j = 12.8$ m/s; (b) $Q = 24.88$ L/s (c) $\mathbf{P} = 2.034$ kW; (d) $\eta = 96.04\%$; (e) $h_m = 8.35$ m)

13.10 Determine the discharge from a 10 cm diameter mouthpiece fitted to the side of a tank containing 3 m of water above the centreline of the mouthpiece. The coefficient of contraction for the mouthpiece can be taken as 0.65.

(*Ans.* $Q = 53.1$ L/s)

13.11 A rectangular tank of cross section 0.9 m × 1.2 m is 3.0 m high. At 20 cm from the bottom, an 8 cm orifice with $C_d = 0.65$ is provided. If the tank is full when the orifice is opened find the time taken to lower the water surface by 1.0 m.

(*Ans.* $T = 49.5$ s)

13.12 A vertical prismatic tank of cross section area 1.2 m^2 has a 4 cm diameter orifice at the bottom. If it takes 60 s to lower the water surface elevation from 1.2 m to 1.0 m above the orifice, find the coefficient of discharge of the orifice.

(*Ans.* $C_d = 0.686$)

13.13 A cylindrical water tank 0.6 m in diameter has its axis vertical and is to be provided with an orifice at its bottom. If the C_d of the orifice is 0.65, what size of orifice is needed to lower the water surface elevation, measured above the base of the tank, from 1.5 m to 0.8 m in 30 s?

(*Ans.* $d = 5.25$ cm)

13.14 A tank is in the form of a frustum of a cone with its axis vertical. It has a diameter of 1.5 m at the top and 0.75 m diameter at the bottom which is 1.5 m below the top. Estimate the time required to lower the water surface from the top of the tank by a depth of 0.75 m through a 5 cm sharp edged orifice (with $C_d = 0.63$) situated in the bottom.

(*Ans.* $T = 138.2$ s)

13.15 A cylindrical tank with its axis horizontal is 2.5 m in diameter and is 4.0 m long and contains oil. Estimate the time required to lower the oil surface in the tank from 2.0 m to 1.20 m above the bottom of the tank, through a 12 cm sharp edged orifice of discharge coefficient 0.63 situated in the bottom of the tank.

(*Ans.* $T = 191$ s)

13.16 Two water tanks of area 2 m^2 and 0.8 m^2 have a common partition wall. A circular orifice of diameter 10 cm allows flow of water between the tanks. If initially the level of the water in the larger tank is 1.5 m above that in the other, determine the time required for the difference in water surface elevations to reduce to 0.5 m. The coefficient of discharge of the orifice can be taken as 0.7.

(*Ans.* $T = 24.3$ s)

13.17 A hemispherical tank 3 m in diameter has an orifice 15 cm diameter at the bottom. Assuming $C_d = 0.62$, find the time required to lower the level of the water surface from 2.0 m to 1.2 m above the orifice.

(*Ans.* $T = 90.37$ s)

13.18 A vertical prismatic tank of cross sectional area A has a steady inflow of Q into the tank, while an orifice of area a at the bottom is discharging the flow freely. Show that the time taken to lower or raise the water surface between heights H_1 and H_2 above the orifice is

$$T = -\frac{2A}{K^2}\left[Q \ln\left(\frac{Q - K\sqrt{H_2}}{Q - K\sqrt{H_1}}\right) + K(\sqrt{H_2} - \sqrt{H_1})\right]$$

where $K = C_d\, a\sqrt{2g}$.

B. Venturimeter

13.19 A vertical venturimeter 15 cm × 10 cm installed in a pipe carrying water downwards shows the same pressure at the inlet and at the throat. The throat is 25 cm below the inlet. If C_d of the meter is 0.95, calculate the discharge in the pipe.

(*Ans.* Q = 18.4 L/s)

13.20 A horizontal venturimeter 9 cm × 4 cm is installed in a 9 cm water pipe. A differential water-mercury manometer reads 30 cm. If the coefficient C_d of the meter is 0.96, estimate the (i) discharge in the pipe and (ii) head loss in the converging section of the meter.

(*Ans.* Q = 10.6 L/s, H_L = 0.296 m)

13.21 A venturimeter is used for the measurement of discharge of water in a horizontal pipeline. The upstream diameter is 300 mm, the throat is of 150 mm diameter and the difference in pressure between the inlet and throat is 3 m head of water. If the loss of head through the converging section of the meter is one-eighth of the throat velocity head, calculate the discharge in the pipe.

(*Ans.* Q = 131.5 L/s)

13.22 Oil of relative density 0.9 flows in a 10 cm pipe. A 10 cm × 5 cm venturimeter (C_d = 0.97) in the pipe exhibits a reading of 12 cm in a mercury-oil differential manometer. Calculate the head loss in the inlet to throat region and estimate the discharge.

(*Ans.* H_L = 0.10 m, Q = 11.33 L/s)

13.23 Crude oil (relative density = 0.85) flows upwards at a volumetric rate of 60 L/s through a vertical venturimeter with an inlet diameter of 200 mm and a throat diameter of 100 mm. The coefficient, of discharge of the venturimeter is 0.98. The vertical distance between the pressure tappings is 300 mm.

(i) Determine the difference in the readings of pressure gauges connected to the inlet and throat sections.

(ii) If a differential U-tube mercury-oil manometer is used to connect the two tappings determine the manometer reading.

[*Ans.* (i) Δp = 26.66 kPa; (ii) y = 19.36 cm]

13.24 A 20 cm × 10 cm venturimeter with C_d = 0.96 carries 30 L/s of water. A differential gauge has an indicator liquid M and the manometer reading is 1.16 m. What is the relative density of the manometer liquid M?

(*Ans.* S_m = 1.75)

13.25 A 15 cm × 7.5 cm venturimeter has C_d = 0.96. If the loss of head between inlet and throat is 0.78 cm calculate the discharge

(*Ans.* Q = 6.12 L/s)

13.26 A 20 cm × 10 cm venturimeter (C_d = 0.96) carries water. A differential mercury water manometer shows a reading of 10 cm. (i) Calculate the discharge. (ii) If the loss of head in the expansion part of the meter is taken as eight times the velocity head in the pipe, calculate the total head loss due to the venturimeter.

[*Ans.* (i) Q = 38.7 L/s; (ii) H_L = 0.7178 m]

13.27 A 20 cm × 15 cm venturimeter is fixed vertically in a pipe carrying water. The throat is 45 cm above the inlet. If the inlet pressure is 103 kPa (abs), what is the maximum discharge that can be passed through the pipe without cavitation occurring in the meter? The vapour pressure of water = 3.5 kPa (abs). The C_d of the meter is 0.95.

(*Ans.* Q_{max} = 280 L/s)

13.28 A 15 cm diameter pipe has a discharge of 60 L/s of water and the pressure head at a section is 5 m (gauge). If a horizontal venturimeter is to be installed at this section, what is the minimum diameter to be adopted to ensure that the pressure head at the throat does not go below 2.3 m (abs)? The atmospheric pressure is 10.3 m. Assume C_d = 0.95.

(*Ans.* D_2 = 7.19 cm)

13.29 A venturimeter with a throat diameter of 12 cm is inserted in a pipe of 20 cm diameter. The pipe carries a liquid of relative density 0.65. The differential mercury-liquid U-tube manometer connected to the throat and the inlet records a readings of 4.0 cm. If the head loss between the inlet and the throat is 8% of the velocity head in the throat, estimate (a) discharge and (b) the coefficient of discharge of the venturimeter.

(*Ans.* (a) Q = 45.9 L/s; (b) C_d = 0.957)

C. Pitot Tube

13.30 A pitot tube is inserted into an air stream at 95.0 kPa (abs). If a mercury-air manometer indicates a positive gauge pressure with a reading of 20 mm, calculate the velocity of air. ρ_{air} = 1.22 kg/m^3 and P_{atmos} = 100 kPa. Assume the pitot tube coefficient C = 0.98.

(*Ans.* V = 109.8 m/s)

13.31 The velocity of an oil flow (RD = 0.90) was measured by a pitot-static tube. The instrument had a coefficient of 0.98. Calculate the velocity corresponding to a mercury-oil differential manometer, connected to the pitot-static tube, reading of 6 cm.

(*Ans.* $V_0 = 4.0$ m/s).

13.32 For the pitot tube shown in Fig. 13.32 determine the velocity at point A. Assume the instrument coefficient to be 0.99.

(*Ans.* $V_0 = 2.426$ m/s)

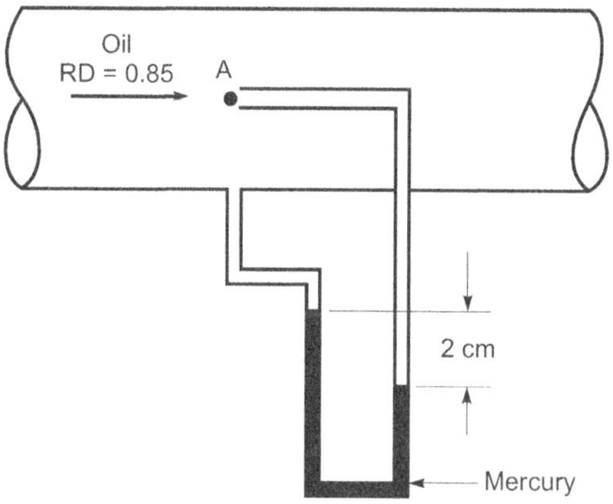

Fig. 13.32

13.33 For the flow of water in a frictionless uniform pipe, a pitot tube was arranged on the centreline as shown in Fig. 13.33. Calculate he centreline velocity in the pipe by assuming the instrument coefficient to be 1.0.

(*Ans.* $V_0 = 3.851$ m/s)

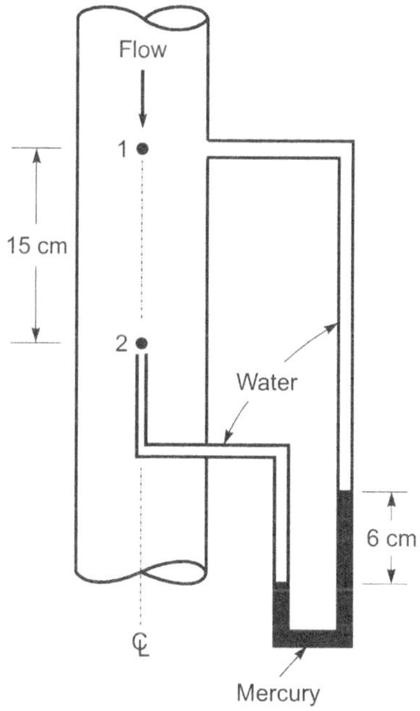

Fig. 13.33

13.34 A pitot tube is placed on the centreline of a pipe carrying kerosene (RD = 0.8) as shown in Fig. 13.34. Neglecting frictional losses in the pipe, estimate the centreline velocity. [Assume $C = 0.99$].

(*Ans.* $V_0 = 3.07$ m/s)

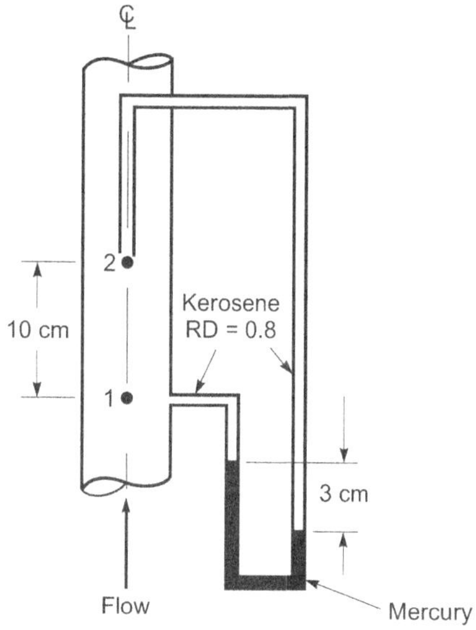

Fig. 13.34

13.35 Kerosene (RD = 0.81) is flowing in a pipe and the static pressure at a station is 3 kPa. If the stagnation pressure of a pitot tube ($C = 0.99$) inserted at the centreline at that section indicates 4.0 kPa, what is the centreline velocity at that section?

(*Ans.* $V_0 = 1.56$ m/s)

13.36 A pitot-static tube ($C = 0.97$) is connected to an inverted U-tube manometer containing an oil of relative density 0.82. If the pitot tube is to measure the velocity of water up to 0.6 m/s, what would be the largest manometer reading?

(*Ans.* $y = 10.8$ cm)

13.37 A pipe is known to be conveying water at a centreline velocity of 2.52 m/s. A pitot-static tube is inserted at the centreline and is connected to a differential carbon tetrachloride (RD = 1.60)-water differential manometer. If the manometer reading is 55 cm, what is the value of the instrument coefficient?

(*Ans.* $C = 0.99$)

13.38 At the summit of a siphon in a water pipe the centreline velocity of flow is 2.5 m/s and the pressure is 3 m of water (vacuum). If a pitot tube ($C = 0.99$) is inserted into the pipe centreline at this section, what will be the stagnation pressure in kPa (abs)? Atmospheric pressure = 101 kPa.

(*Ans.* P_{at} = 74.81 kPa (abs))

D. Weirs

13.39 A suppressed rectangular weir in a 1.5 m wide rectangular channel is located so that its crest is 0.30 m above the bed. Estimate the discharge over the weir for a head of 0.30 m over the weir.

(*Ans.* Q = 0.499 m^3/s)

13.40 A sharp crested rectangular weir is 2.0 m long. Calculate the height of the weir to pass a flow of 1.10 m^3/s while maintaining an upstream depth of 1.20 m.

(*Ans.* P = 0.767 m)

13.41 The head on a sharp crested rectangular suppressed weir 1.2 m long and 0.9 m high is 10 cm. Calculate the discharge and the velocity of approach.

(*Ans.* Q = 69.4 L/s, V_a = 0.0578 m/s)

13.42 What is the coefficient of discharge of a suppressed rectangular weir of 1.5 m length which passes 1.12 m^3/s under a head of 0.50 m?

(*Ans.* C_d = 0.715)

13.43 A 20 cm high sharp crested rectangular weir plate is installed at the end of a 1.2 m wide rectangular channel. The weir plate spans the full width of the channel. What maximum discharge can be passed if the side walls of the channel are 0.90 m high and the minimum specified free board is 0.20 m?

(*Ans.* Q_m = 1.0 m^3/s)

13.44 A sharp crested suppressed rectangular weir of height 0.6 m is used to measure the discharge in a 1.8 m wide rectangular channel. At a certain discharge the head over the weir was recorded as 0.6 m by a point gauge. It was however found later that the point gauge had a zero error and was recording heads 2 cm too small. Determine the percentage error in the estimated discharge corresponding to an observed head of 0.6 m.

(*Ans.* Error = 5.14%)

13.45 A partially suppressed rectangular weir consists of a 0.85 m long weir with only one end contraction built in a 2 m wide channel. For a head of 0.40 m over the crest, estimate the discharge by assuming C_d = 0.62 and the velocity of approach to be negligible.

(*Ans.* Q = 0.375 m^3/s)

13.46 An overflow weir is 16 m long between the abutments. There are three piers of 0.30 m thick on the weir. Estimate the discharge for a head of 0.5 m over the weir. Assume C_d = 0.63 and the velocity of approach to be zero.

(*Ans.* Q = 9.67 m^3/s)

13.47 A rectangular channel 1.8 m wide has, at its end, a sharp crested rectangular weir 1.2 m long and 0.50 m high. Calculate the discharge in the channel when the head recorded over the weir is 0.25 m, C_d of the weir = 0.62.

(*Ans.* Q = 0.266 m^3/s)

13.48 Find the discharge in a triangular notch of vertex angle 30° corresponding to a head of 35 cm. What is the head corresponding to a discharge of 50 L/s? Take C_d = 0.60.

(*Ans.* Q = 27.5 L/s; H_1 = 44.4 cm)

13.49 A right angled triangular notch is used for measuring the discharge in a laboratory flume. The coefficient of discharge of the notch is 0.59. If the heads can be measured with an accuracy of 2 mm, find the head measured above the level of the vertex of the notch and the likely error in the calculated discharge of 60 L/s.

(*Ans.* H_1 = 28.4 cm; error in Q = ± 1.056 L/s)

13.50 It is desired to measure a discharge varying from 50 L/s to 150 L/s with an accuracy of 1% throughout the range. The depth can be measured with an accuracy of 1 mm. What is the maximum permissible vertex angle of a V-notch that will satisfy this condition? Take C_d = 0.60.

(*Ans.* $Q \le 96° \ 55' \ 12''$)

13.51 A 60° triangular weir has a coefficient of discharge of 0.59. If the accuracy of measuring the angle is 1°, estimate the likely error in the estimated discharge when the head over the vertex of the weir is 40 cm.

(*Ans.* Likely error = 1.64 L/s)

13.52 A trapezoidal sharp-crested weir has a base width of 1.2 m and side slopes of 1.5 horizontal : 1 vertical. Calculate the discharge over the weir for a head of 35 cm, assuming C_d = 0.62.

(*Ans.* Q = 0.614 m^3/s)

13.53 Estimate the head over a Cipolletti weir of base width 0.90 m required to pass a discharge of 600 L/s. Assume C_d = 0.63.

(*Ans.* H_1 = 50.45 cm)

13.54 A trapezoidal notch is to be designed to pass a discharge of 1.0 m^3/s at a head of 0.8 m over the crest and 0.50 m^3/s at a head of 0.51 m. Assuming $C_d = 0.7$, calculate the base width and side slope of the notch.

(*Ans.* $L = 0.643$ m, $\theta = 2.96°$)

13.55 Calculate the bottom width and side slope of a trapezoidal notch to discharge 4.0 m^3/s at a head of 1.5 m and 1.20 m^3/s at a head of 0.75 m. Assume $C_d = 0.63$.

(*Ans.* $L = 0.816$ m, $\theta = 16.46°$)

13.56 Find the discharge through a trapezoidal weir with inward sloping sides as shown in Fig. 13.35 $C_d = 0.60$.

(*Ans.* $Q = 210$ L/s)

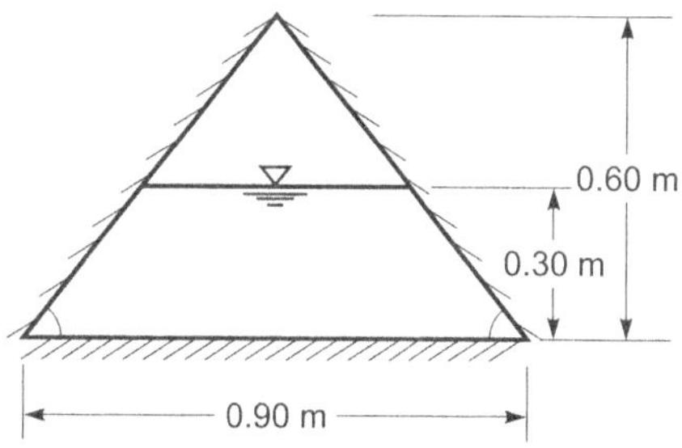

Fig. 13.35

13.57 A suppressed rectangular weir is fitted in a 2.5 m channel. Calculate the discharge over the weir when the water surface elevations on the upstream and downstream of the weir, measured from the weir crest, are respectively 0.60 m and 0.30 m. A constant value of $C_d = 0.62$ can be assumed.

(*Ans.* $Q_s = 1.798$ m^3/s)

13.58 A triangular notch in a channel is known to have carried 60 L/s under a head of 30 cm when flowing free. If it is submerged with water surface elevations at 50 cm and 30 cm above the vertex on the upstream and downstream of the notch respectively, estimate the discharge in the channel.

(*Ans.* $Q_s = 190$ L/s)

13.59 The discharge over a weir could be expressed as $Q = KH^n$ where H is the head over the weir and K and n are constants. It was found that the discharge was 7.48 m^3/s and 3.84 m^3/s when the head H was 2.0 m and 1.25 m, respectively. Calculate (a) the discharge over the weir for a head of 1.50 m and (b) head required to pass a discharge of 2.5 m^3/s.

(*Ans.* (a) $Q = 4.973$ m^3/s; (b) $H = 0.9237$ m)

13.60 A suppressed rectangular weir of length L and a V-notch of angle θ are installed in a large tank such that the crest of the weir and vertex of the notch are at the same level. For what range of head H would the triangular notch have a higher discharge than the weir? Assume that both the weir and the notch have the same C_d.

$$\left(Ans.\ H > \frac{5}{4}\frac{L}{(\tan \theta/2} \right)$$

13.61 A tank with vertical sides and a horizontal cross sectional area of 2 m^2 is provided with notch cut at the top of one of its sides. Water flowing into the tank at constant rate was discharged over the notch and the head over the notch crest was observed to be 0.27 m. The supply of water was suddenly stopped and it was observed that the head over the notch started falling at a rate of 5.04 mm/s. When the head had fallen to 0.12 m, it was observed that the head over the notch was falling at the rate of 1.8 mm/s. Estimate the rate of inflow when there is a steady head of 0.15 m over the notch.
[assume $Q = KH^n$ for the notch.]

(*Ans*: $Q = 4.545$ L/s
($K = 0.0594$ and $n = 1.3548$))

13.62 A sharp crested rectangular suppressed weir 90 cm long and a 90° V-notch are placed in the same vertical face of a tank. The vertex of the notch is 20 cm below the crest of the weir. Assuming C_d to be the same for both the notch and the weir, find: (a) the head over the V-notch when the discharges in the notch and the weir are identical, and (b) the head above the vertex of the V-notch when the rectangular weir discharges its greatest excess amount over and above the discharge of the V-notch.

(*Ans.* (a) $H = 0.645$ m;
(b) $H = 0.535$ m)

13.63 Estimate the time required to lower the water level from an elevation of 13.00 m to an elevation of 12.00 m in a reservoir through a flow over a 75° V-notch whose vertex is at elevation 10.00 m. The tank has a constant cross sectional area of 6000 m^2. (The C_d of the notch = 0.60).

(*Ans.* T = 9 min 52.5 s)

13.64 Find the time required to lower the water surface of a rectangular tank of cross sectional area 5000 m^2 from an elevation of 10.20 m to 9.20 m.

A rectangular suppressed weir of length 2.0 m with its crest at elevation 7.7 m above the datum is used for lowering the water surface (Assume $C_d = 0.62$).

(*Ans.* T = 8 min 22.6 s)

13.65 A tank with a constant cross sectional area of 2.0 m^2 was emptied by a rectangular 30 cm long suppressed weir. If the rate of lowing of the head in the tank was 4.50 cm/s when the head was 30 cm, estimate the discharge coefficient C_d.

(*Ans.* $C_d = 0.618$)

14

Unsteady Flow

CONCEPT REVIEW

14.1 INTRODUCTION

A flow is unsteady when flow parameters such as discharge, velocity, pressure vary with time. In unsteady flows additional forces due to time variation of velocity and momentum come into play and the problem becomes more difficult to solve. Typical examples of unsteady flow are: (i) emptying of a tank containing liquid through an orifice (Chapter 13), (ii) mass oscillation of a liquid column, (iii) gradual establishment of flow in a long pipe due to sudden opening of a valve, (iv) water hammer pressure phenomenon due to sudden closure of a valve in a pipeline, (v) surges in open channels and, (vi) passage of flood waves in rivers. In this chapter only three kinds of unsteady phenomenon are dealt with in detail. These are: (a) Surges in prismatic channels, (b) Water hammer phenomenon in pipes and (c) Establishment of flow in pipes due to sudden opening of a valve. The aspects relating to emptying of vessels through orifices or weirs is already dealt with in Chapter 13 and is not repeated here. For details on other types of unsteady flows the reader is referred to standard texts on the subject.

14.2 SURGES IN OPEN CHANNELS

Whenever there is a sudden change in the discharge or depth or both in an open channel, a rapidly varied unsteady phenomenon, known as surge, develops. Such situations occur during sudden operation of a control gate. A surge producing an increase in depth is known as *positive surge* and the one which causes a decrease in the depth is known as *negative surge*. Positive surges have steep fronts, more like a hydraulic jump, and the shape of the wave does not change during its translation. They are also known as *moving hydraulic jumps*. These are relatively easy to analysis than negative surges. In this section only positive surges are considered.

14.2.1 Positive Surge Moving Downstream

Figure 14.1 shows a horizontal, frictionless rectangular channel in which a positive surge in moving downstream. Suffix 1 refers to the conditions before the arrival of the surge and suffix 2 refers to a section after the passage of the surge. The absolute velocity of the surge is V_w and is assumed constant. The unsteady flow situation can be simulated to an equivalent steady state flow by superimposing a velocity $(-V_w)$, (directed to the left in Fig. 14.1 (a)) at all sections [Fig. 14.1(b)]. The conditions are now

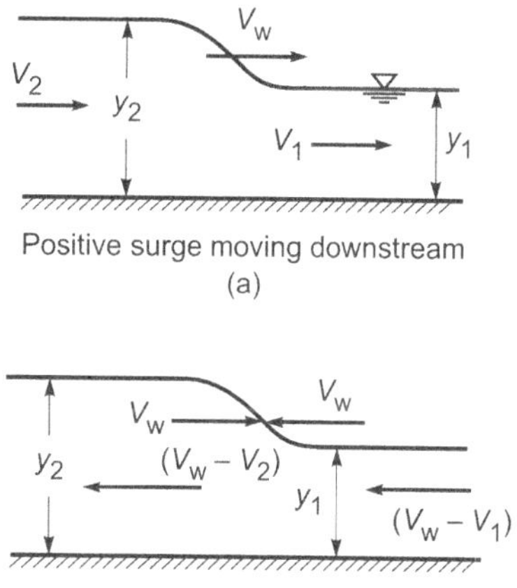

Positive surge moving downstream
(a)

Simulated steady flow
(b)

Fig. 14.1

similar to that of a hydraulic jump with approach velocity of $(V_w - V_1)$ and approach depth y_1. The depth after the surge (jump) is y_2 and the corresponding velocity is $(V_w - V_2)$.

The continuity equation is,

$$A_1 (V_w - V_1) = A_2 (V_w - V_2) \quad (14.1)$$

By considering unit width of the rectangular channel

$$y_1 (V_w - V_1) = y_2 (V_w - V_2) \quad (14.2)$$

By application of momentum equation

$$\boxed{\frac{(V_w - V_1)^2}{g y_1} = \frac{1}{2}\frac{y_2}{y_1}\left(\frac{y_2}{y_1} + 1\right)} \quad (14.3)$$

From Eqs. 14.2 and 14.3 two out of the five variables V_1, V_2, y_1, y_2 and V_w can be evaluated if the other three are given.

14.2.2 Positive Surge Moving Upstream

Figure 14.2(a) shows a positive surge moving upstream. This kind of surge occurs on the upstream of a sluice gate when the gate is closed suddenly and in the phenomenon of *tidal bores*. The unsteady flow situation is converted to a simulated steady flow by superposition of a velocity V_w directed downstream [to the right in Fig. 14.2(b)]. As before, suffixes 1 and 2 refer to the conditions at sections of the channel before and after the passage of the surge. The continuity equation is

$$\boxed{A_1 (V_w + V_1) = A_2 (V_w + V_2)} \quad (14.4)$$

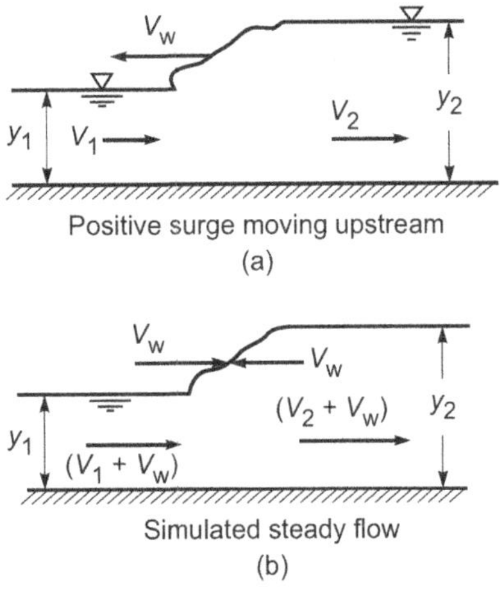

Fig. 14.2

For unit width of the rectangular channel, Eq. 14.4 becomes

$$y_1 (V_w + V_1) = y_2 (V_w + V_2) \quad (14.4)$$

From Fig. 14.2(b) it is easy to see that the flow is similar to that of a hydraulic jump with initial velocity of $(V_w + V_1)$ and initial depth y_1. The final velocity is $(V_w + V_2)$ and the depth after the surge is y_2. By momentum equation

$$\boxed{\frac{(V_w + V_1)^2}{g y_1} = \frac{1}{2}\left(\frac{y_2}{y_1}\right)\left(1 + \frac{y_2}{y_1}\right)} \quad (14.5)$$

From Eqs. 14.4 and 14.5 two of the five variables y_1, y_2, V_1, V_2 and V_w can be determined if three other variables are given. It is to be remembered that, in the real flow, V_w is directed upstream (to the left in the figure).

14.2.3 Other Forms of Equations

Equations 14.3 and 14.5, being symmetrical, could also be expressed as follows to suit the problem at hand:

Alternative form of Eq. 14.3:

$$\frac{(V_w - V_2)^2}{g y_2} = \frac{1}{2}\frac{y_1}{y_2}\left(1 + \frac{y_1}{y_2}\right) \quad (14.3a)$$

Alternative form of Eq. 14.5

$$\frac{(V_w + V_2)^2}{g y_2} = \frac{1}{2}\frac{y_1}{y_2}\left(1 + \frac{y_1}{y_2}\right) \quad (14.5a)$$

14.3 WATER HAMMER

When a liquid flow in a long pipeline is reduced suddenly, due to compressibility of the liquid, the sudden change in momentum would cause a pressure surge to develop. This pressure moves through the pipe at high speed and undergoes reflection at the ends. The phenomenon is known as *Water hammer* and is of importance in all major pipeline designs.

14.3.1 Velocity of Pressure Wave

If the pipe is rigid, the pressure wave will travel in the medium at a velocity C equal to the velocity of sound in an infinite expansion of the medium, and is given by

$$\boxed{C = \sqrt{K/\rho}} \quad (14.6)$$

where K = bulk modulus of the liquid in Pa.

and ρ = mass density of the liquid in kg/m^3.

If the pipe material is elastic, the velocity of propagation will be less than that given by Eq. 14.6 and depends upon the diameter D, thickness t of the pipe and the modulus of elasticity E of the pipe material. The velocity of the pressure wave in an elastic pipe is given by

$$\boxed{C = \sqrt{K/\rho}\left[\frac{1}{1 + (D/t)(K/E)}\right]^{1/2}} \quad (14.7)$$

For water, normally $\sqrt{K/\rho}$ is about 1400 m/s and velocity of pressure wave C by Eq. 14.7 for normal dimensions would be between 900 and 1200 m/s.

14.3.2 Rapid Closure

Consider a pipe of length L leading from a reservoir and terminating in a valve at its downstream end. When the valve is instantaneously closed a pressure of magnitude p_h is formed and moves up with a velocity C. The wave undergoes reflections at the reservoir end as well as at the valve. For the case of a frictionless flow, the various stages of the pressure wave are shown in Fig. 14.3 (a and b). Figure 14.4 shows the water hammer pressure p_h as a function of time at various locations of the pipe. It is seen that at all locations the time period for a complete cycle is $4L/C$. The time $T_0 = 2L/c$ is called as *critical time*. If the time of closure T of a valve is such that $T < T_0$, the pressure head at the valve will be same as that for instantaneous closure. As such, the time of closure $T \le 2L/C$ is known as *rapid closure*.

If x_0 = length of the pipe having peak pressure in a closure time $T < 2\,L/C$, the distance x_0 is given by

$$\boxed{x_0 = L - \frac{CT}{2}} \qquad (14.8)$$

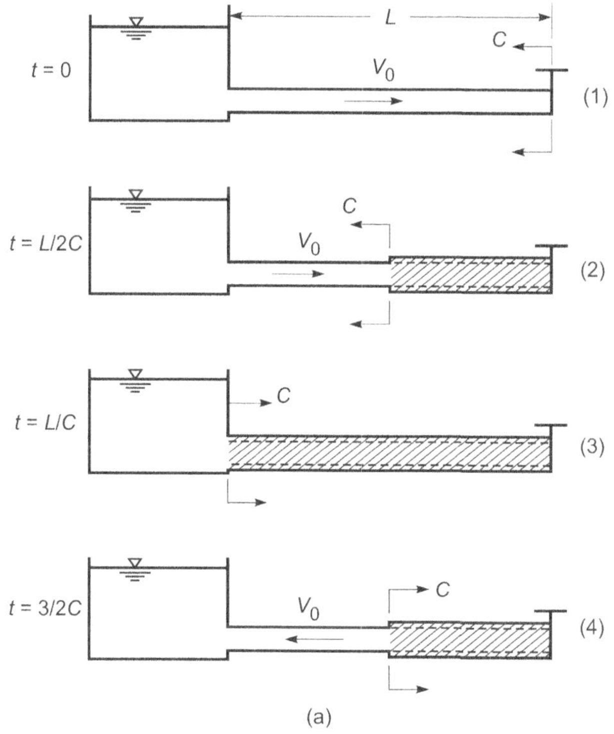

Fig. 14.3(a) *Various Stages of a Water Hammer Pressure Wave*

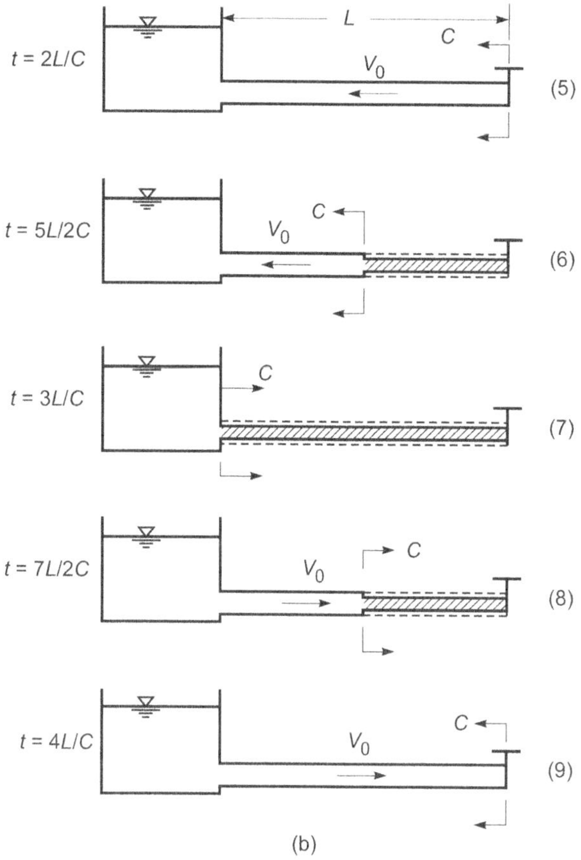

Fig. 14.3(b) *Various Stages of a Water Hammer Pressure Wave*

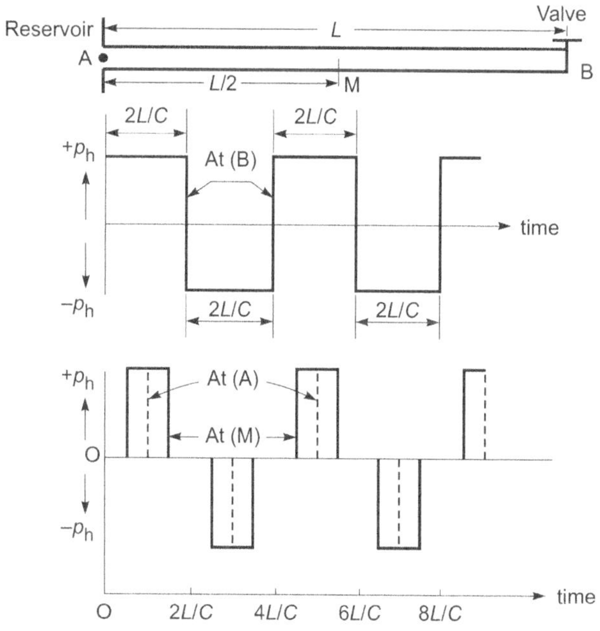

Fig. 14.4 *Variation of Water Hammer Pressure with Time*

Thus it is seen that in instantaneous closure ($T = 0$), $x_0 = L$, i.e. the entire pipe will have peak pressure.

Water Hammer Pressure Let the velocity of flow in a pipe be changed from V_1 to V_2 *rapidly*. The resulting water hammer pressure p_h is given in pascals by

$$p_h = -\rho C \,\Delta V \quad \text{where} \quad \Delta V = (V_2 - V_1) \qquad (14.9)$$

Thus, if the flow in a pipe is completely stopped, $\Delta V = -V_1$ and the water hammer pressure is, in pascals,

$$\boxed{p_h = \rho C V_1} \quad \text{(for complete closure)} \qquad (14.10)$$

14.3.3 (a) Slow Closure

If the time of closure $T > T_o$ (critical time) then the closure is known as *slow closure*. If T is slightly larger than T_o (say $1 < T/T_o \le 1.5$), compressibility effects are important but the peak water hammer pressure will be less than that in rapid closure. An approximate value of peak water hammer pressure can be obtained by

$$\boxed{\frac{p_{hs}}{p_{hr}} = \frac{T_o}{T}} \quad (\text{for } 1 < T/T_o \le 1.5) \qquad (14.11\text{-a})$$

where $T_0 = 2L/C$ = critical time.
T = actual time of slow closure, ($T > T_0$)
P_{hs} = peak water hammer pressure in slow closure
P_{hr} = peak water hammer pressure in rapid closure [Eq. 14.9 or Eq. 14.10 as is the case].

(b) Very Slow Closure

If the actual valve closure time is many times greater than T_o, the compressibility effects are no longer important and the pressure rise is due to change in momentum and is given by

$$\boxed{p_{vsc} = \frac{\rho L V}{T}} \qquad (14.11\text{-b})$$

14.3.4 Design of Pipe Thickness

The water hammer pressure p_h will be over and above the steady state pressure in the pipe, p_s (referred commonly as static pressure). Hence, the pipe will have to withstand a total pressure p_t given by

$$\boxed{p_t = p_s + p_h} \qquad (14.12)$$

The stress σ in the pipe wall is given by the thin cylinder formula

$$\boxed{\sigma = \frac{p_t D}{2t}} \qquad (14.13)$$

where t = thickness of the pipe wall and D = diameter of the pipe. For design, σ should be less than the working stress σ_w of the pipe material. Thus the minimum thickness of a pipewall is

$$\boxed{t_m = \frac{(p_s + p_h)\,D}{2\sigma_w} = \frac{p_t D}{2\sigma_w}} \qquad (14.14)$$

For steel, the normal working stress is 0.1 kN/mm^2 (= 100 MPa).

14.4 ESTABLISHMENT OF FLOW

Consider a pipe leading from a large tank. A valve at the downstream end controls the flow (Fig. 14.5). If the valve is opened suddenly, it takes a while for the flow to be fully established. By considering the fluid as *incompressible* and the pipe to be *rigid*, the time for flow establishment is obtained by integration of the Euler equation.

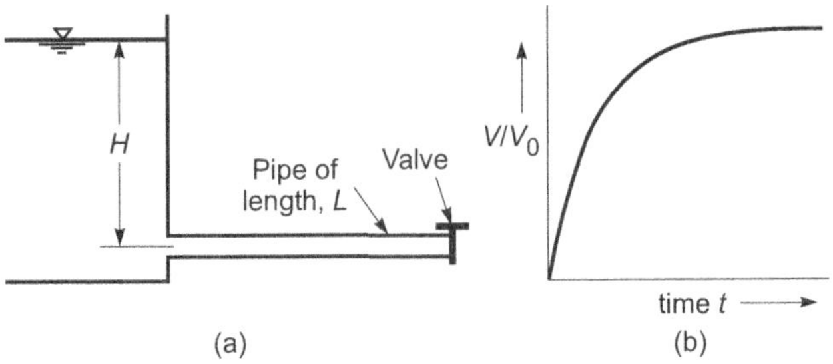

Fig. 14.5 *Establishment of Flow*

Let V = velocity at any time t
V_0 = final steady state velocity.
f = Darcy–Weisbach friction factor
k_1 = sum of minor loss coefficients for the pipe
H = static head at the outlet.

$$\text{Total head loss} = \left(\frac{fL}{D} + k_1\right)\frac{V^2}{2g}$$

$$= k\frac{V^2}{2g} \quad \text{where} \quad k = \frac{fL}{D} + k_1$$

Then $$V_0 = \sqrt{\frac{2gH}{(1+k)}}$$

If $V = 0$ at $t = 0$,

the time taken to attain a velocity V after sudden opening of the valve, is given by

$$\boxed{t = \frac{L}{(1+k)\,V_0}\, ln \frac{(1 + V/V_0)}{(1 - V/V_0)}} \qquad (14.15)$$

It is seen that V approaches V_0 asymptotically (Fig. 14.5 (b)]. For a frictionless pipe, $k = 0$.

WORKED EXAMPLES

A. Surges in Canals

14.1 A 2.0 m wide rectangular channel has a flow with a velocity of 2.0 m/s and depth of 1.3 m. The rate of inflow at the upstream end is suddenly increased to an extent that the depth is doubled in magnitude. Estimate the absolute velocity of the resulting surge and the new discharge.

Solution: Conditions of relative steady flow are simulated by adding the velocity $-V_w$ vectorially, i.e. to the left in Fig. 14.6.

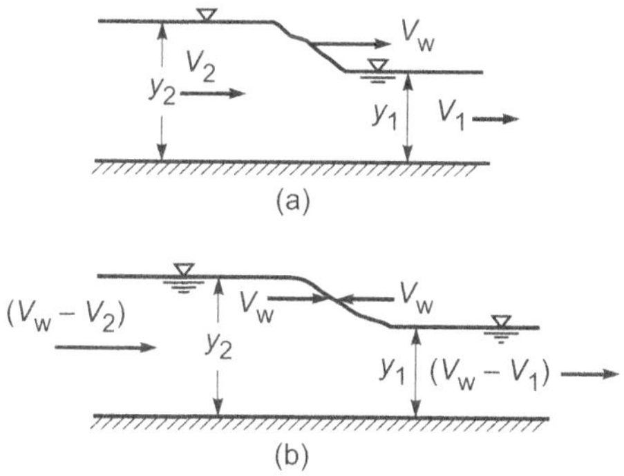

Fig. 14.6 *Surge Moving Downstream*

Here $y_1 = 1.30$ m, $V_1 = 2.0$ m/s

$y_2 = 2.60$ m

By the surge equation obtained from a combination of momentum and continuity equations:

$$\frac{V_w - V_1}{\sqrt{g y_1}} = \left[\frac{1}{2}\frac{y_2}{y_1}\left(\frac{y_2}{y_1}+1\right)\right]^{1/2}$$

$$V_w - 2.0 = \sqrt{9.81\times 1.3}\left[\frac{1}{2}\times 2\times(2+1)\right]^{1/2}$$

$$= 6.185$$

$$\mathbf{V_w = 8.185\ m/s\ (in\ simulated\ flow)}$$

The surge moves *downstream* with a velocity of 8.185 m/s.

By continuity equation:

$$y_1 (V_w - V_1) = y_2 (V_w - V_2)$$

$$1.3\,(8.185 - 2.0) = 2.6\,(8.185 - V_2)$$

$$V_2 = 8.185 - 3.0925 = 5.093 \text{ m/s}$$

New discharge $Q_2 = By_2 V_2 = 2.0 \times 2.6 \times 5.093$

$$\mathbf{= 26.484\ m^3/s}$$

14.2 A rectangular channel is discharging 1.60 m³/s per metre width at a depth of 0.8 m. If a sudden release of flow at the upstream end doubles the discharge intensity, determine the absolute velocity of the resulting surge and the new depth.

Solution: The absolute velocity of the surge is V_w in the downstream direction. By superposing a velocity V_w in the opposite direction (i.e. to the left) a steady flow is simulated as in Fig. 14.7.

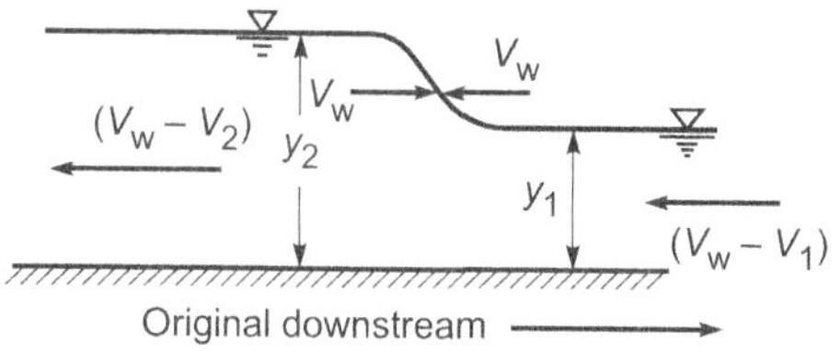

Fig. 14.7

Here $y_1 = 0.8$ m

$$V_1 = \frac{1.6}{0.8} = 2.0 \text{ m/s}$$

$$V_2 y_2 = 2 \times 1.6 = 3.2 \quad \text{and} \quad V_2 = \frac{3.2}{y_2}$$

By continuity equation:

$$y_1 (V_w - V_1) = y_2 (V_w - V_2)$$

$$0.8\,(V_w - 2.0) = V_w y_2 - 3.2$$

$$V_w (y_2 - 0.8) = 1.6$$

$$V_w = 1.6/(y_2 - 0.8)$$

From the surge equation, (Eq. 14.3).

$$\frac{(V_w - V_1)^2}{g y_1} = \frac{1}{2}\frac{y_2}{y_1}\left(1+\frac{y_2}{y_1}\right)$$

$$\frac{\left(\dfrac{1.6}{y_2 - 0.8} - 2.0\right)^2}{9.81\times 0.8} = \frac{1}{2}\frac{y_2}{0.8}\left(1+\frac{y_2}{0.8}\right)$$

$$\left(\frac{3.2 - 2.0\,y_2}{y_2 - 0.8}\right)^2 = 6.131\, y_2(0.8 + y_2)$$

By trial and error, $\mathbf{y_2 = 1.088\ m}$

$$V_2 = \frac{3.2}{y_2} = \frac{3.2}{1.088} = 2.941 \text{ m/s}$$

$$V_w = \frac{1.6}{(1.088 - 0.8)} = 5.556 \text{ m/s}$$

$$\mathbf{V_w = +\,5.556\ m/s\ in\ simulated\ flow.}$$

Hence the surge moves *downstream* with a velocity of 5.556 m/s.

14.3 A 10 m wide rectangular channel supplies 35 m^3/s of water at a normal depth of 2.0 m to a turbine installation. Due to a major reduction in load, the supply to the turbine is cut down to 5 m^3/s only Determine the velocity of propagation of the surge wave and also the height of the surge.

Solution: A relative steady flow situation is simulated by superimposing a velocity V_w to the right on the system (Fig. 14.8).

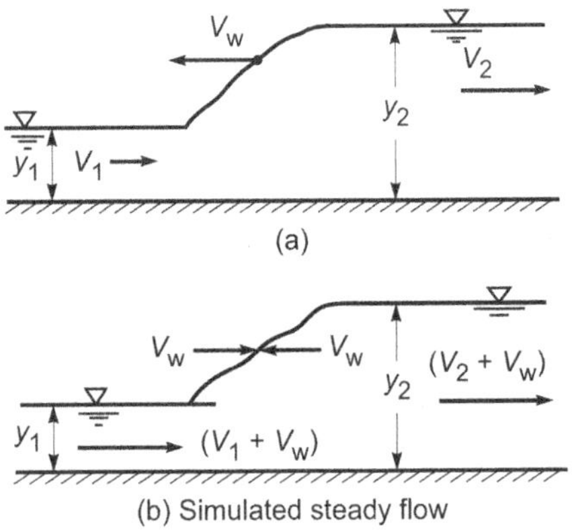

Fig. 14.8 *Positive Surge Moving Upstream*

Here $y_1 = 2.0$ m

$$V_1 = \frac{35.0}{2.0 \times 10.0} \ 1.75 \text{ m/s}$$

Also $Q_2 = 10 \times y_2 \ V_2 = 5.0 \ m^3/s$

$$V_2 \ y_2 = 0.5 \ m^3/s/m$$

$$V_2 = 0.5/y_2$$

Also, by the continuity equation

$$y_1 (V_w + V_1) = y_2 (V_w + V_2)$$

$$2.0 (V_w + 1.75) = y_2 V_w + 0.5$$

$$V_w (2.0 - y_2) = -3.0$$

$$V_w = \left(\frac{3.0}{y_2 - 2.0}\right)$$

From the surge equation, (Eq. 14.5),

$$\frac{(V_w + V_1)^2}{g y_1} = \frac{1}{2} \frac{y_2}{y_1} \left(\frac{y_2}{y_1} + 1\right)$$

Substituting for V_w,

$$\frac{\left(\frac{3.0}{y_2 - 2.0} + 1.75\right)^2}{9.81 \times 2.0} = \frac{1}{2} \frac{y_2}{2.0} \left(\frac{y_2}{2.0} + 1\right)$$

$$\left(\frac{3.0}{y_2 - 2.0} + 1.75\right)^2 = 2.4525 \ y_2 (y_2 + 2.0)$$

By trial and error, $y_2 = 2.763$ m.

$$V_2 = \frac{0.5}{2.763} = 0.181 \text{ m/s}$$

$$V_w = \frac{3.0}{y_2 - 2.0} = \frac{3.0}{2.763 - 2.0}$$

$= 3.932$ m/s in simulated flow.

The surge moves *upstream* with a velocity.

$$\mathbf{V_w = 3.932 \ m/s}$$

Height of surge $= y_2 - y_1 = 2.763 - 2.000$

$= \mathbf{0.763 \ m.}$

14.4 Water flows at a depth of 0.8 m at a velocity of 0.9 m/s in a wide rectangular channel in a reach immediately upstream of a control gate. The gate is suddenly closed completely. Determine the resulting surge conditions.

Solution: Let V_w = velocity of the surge. Superimpose a velocity V_w to the right to simulate the steady flow conditions as in Fig. 14.9.

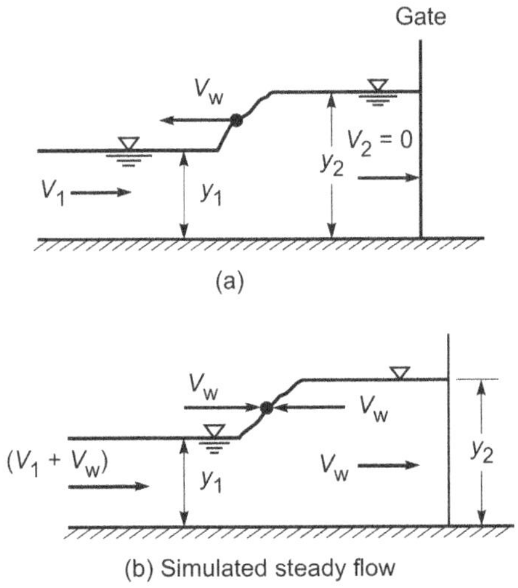

Fig. 14.9

Here $V_2 = 0$

$y_1 = 0.8$ m

$V_1 = 0.9$ m/s

By continuity equation,

$$y_1 (V_1 + V_w) = y_2 (V_2 + V_w)$$

$$0.8 (0.9 + V_w) = y_2 (0 + V_w)$$

$$V_w = \frac{0.72}{(y_2 - 0.8)}$$

By the surge equation, (Eq. 14.5),

$$\frac{(V_1 + V_w)^2}{g y_1} = \frac{1}{2} \frac{y_2}{y_1} \left(1 + \frac{y_2}{y_1}\right)$$

$$\left\{0.9+\left(\frac{0.72}{y_2-0.8}\right)\right\}^2 = 9.81\times 0.8\times\frac{1}{2}\times\frac{y_2}{0.8}\left(1+\frac{y_2}{0.8}\right)$$

$$\left(\frac{0.9\,y_2}{y_2-0.8}\right)^2 = 6.13125\,y_2(0.8+y_2)$$

By trial and error $\mathbf{y_2 = 1.075\ m.}$

Velocity of surge $V_w = \dfrac{0.72}{(1.075-0.8)}$

$= \mathbf{2.618\ m/s\ in\ simulated\ flow.}$

The surge moves *upstream* with a velocity of 2.618 m/s.

14.5 In a wide tidal river the velocity is 0.75 m/s and the depth of flow is 1.30 m. If a tidal bore is observed to move upstream with a velocity of 4.0 m/s in this river, determine the velocity and depth of flow after the bore has passed.

Solution: Superimpose a velocity of V_w to the right as shown in Fig. 14.10 to simulate steady flow situation.

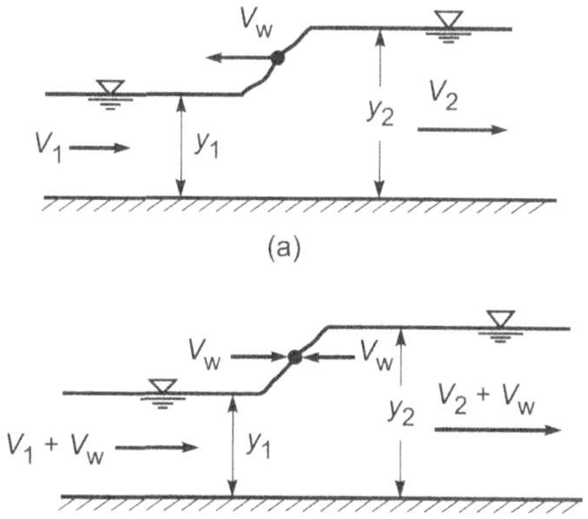

(b) Simulated steady flow

Fig. 14.10

Here $y_1 = 1.30$ m,

$V_1 = 0.75$ m/s

$V_w = 4.0$ m/s

By continuity,

$$y_1(V_1+V_w) = y_2(V_2+V_w)$$

$$1.3\times(0.75+4.0) = y_2(V_2+4.0)$$

$$V_2 = \frac{6.175}{y_2} - 4.0$$

By the surge equation,

$$\frac{V_1+V_w}{\sqrt{g y_1}} = \left[\frac{1}{2}\frac{y_2}{y_1}\left(1+\frac{y_2}{y_1}\right)\right]^{1/2}$$

$$\frac{(0.75+4.0)^2}{9.81\times 1.30} = \frac{1}{2}\frac{y_2}{y_1}\left(1+\frac{y_2}{y_1}\right)$$

$$\left(\frac{y_2}{y_1}\right)^2 + \left(\frac{y_2}{y_1}\right) - 3.5384 = 0$$

$$\frac{y_2}{y_1} = \frac{-1\pm\sqrt{1+4\times 3.5384}}{2} = 1.446$$

$y_2 = 1.30\times 1.446 = \mathbf{1.88\ m}$

$$V_2 = \frac{6.175}{1.88} - 4.0 = -0.715\ \text{m/s}$$

$= \mathbf{0.715\ m/s\ in\ upstream\ direction.}$

14.6 A river flowing toward the sea at a velocity of 5 km/h and having a depth of 2.0 m encounters an incoming tide which increases the depth to 5.0 m and a tidal bore travels upstream. Estimate the speed of the bore and speed and direction of flow after the passage of the bore.

Solution: Let V_w (directed upstream) to be velocity of the bore. Superimpose a velocity V_w directed to the right, to get simulated steady flow as shown in Fig. 14.11.

Here $V_1 = \dfrac{5.0\times 1000}{60\times 60} = 1.389$ m/s

$y_1 = 2.0$ m, $y_2 = 5.0$ m

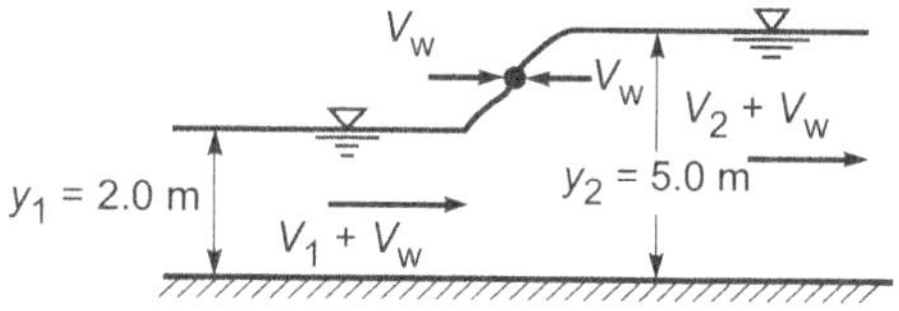

Fig. 14.11 *Simulated Steady Flow*

By continuity equation:

$$y_1(V_1+V_w) = y_2(V_2+V_w)$$

$$2.0\times(1.389+V_w) = 5.0(V_2+V_w)$$

$$V_2 = \frac{1}{5}(2.778-3V_w)$$

By the surge equation

$$\frac{(V_w+V_1)^2}{g y_1} = \frac{1}{2}\frac{y_2}{y_1}\left(\frac{y_2}{y_1}+1\right)$$

$$\frac{(V_w+1.389)^2}{9.81\times 2.0} = \frac{1}{2}\times\frac{5.0}{2.0}\left(\frac{5.0}{2.0}+1\right)$$

$$(V_w+1.389)^2 = 85.838$$

$$\mathbf{V_w = 7.876\ m/s}$$

The bore travels upstream with a velocity of 7.876 m/s.

Velocity $V_2 = \dfrac{1}{5}(2.778 - 3\times 7.876)$

$= \mathbf{-4.17\ m/s}$

After the passage of the tidal bore the flow in the river is *upstream* with a velocity of 4.17 m/s.

14.7 A tidal bore is observed to move at a velocity of 6.0 m/s upstream in a wide river of depth 2.0 m. If the depth of flow after the passing of the bore is 4.25 m, estimate the velocities in the river before and after the bore.

Solution: By superimposing a velocity V_w to the right, a steady flow situation is simulated, as in Fig. 14.12.

Here $V_w = 6.0$ m, $y_1 = 2.0$ and $y_2 = 4.25$ m

By continuity equation

$$y_1 (V_1 + V_w) = y_2 (V_2 + V_w)$$

$$2.0 (V_1 + 6.0) = 4.25 (V_2 + 6.0)$$

$$V_2 = 0.4706 V_1 - 3.1765$$

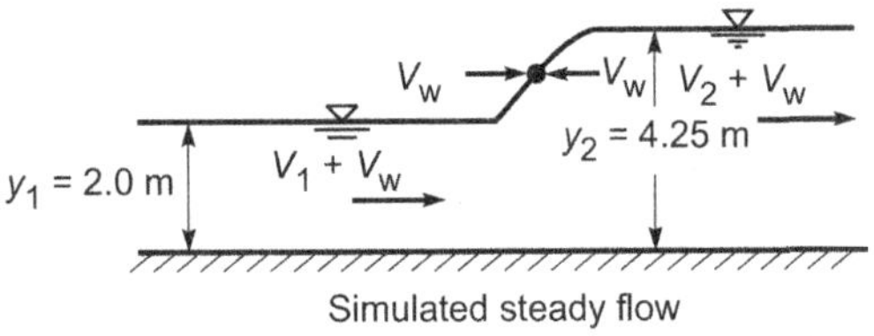

Fig. 14.12

By the surge equation, (Eq. 14.5),

$$\frac{(V_1 + V_w)^2}{g y_1} = \frac{1}{2}\frac{y_2}{y_1}\left(1 + \frac{y_2}{y_1}\right)$$

$$\frac{(V_1 + 6.0)^2}{9.81 \times 2.0} = \frac{1}{2} \times \frac{4.25}{2.0}\left(1 + \frac{4.25}{2.0}\right)$$

$$(V_1 + 6)^2 = 65.145$$

By taking the positive root

$$V_1 = \mathbf{2.07\ m/s}$$

$$V_2 = 0.4706 \times 2.07 - 3.1765$$

$$= \mathbf{-\ 2.202\ m/s}$$

The flow in the river after the bore will be in the *upstream* direction with a velocity of 2.202 m/s.

14.8 A rectangular channel had a positive surge of velocity 8.0 m/s moving down the channel. If the depth of flow and velocity after the passing of the surge are 3.5 m and 4.5 m/s respectively, estimate the depth and velocity before the passage of the surge.

Solution: Let V_w = velocity of surge wave.

Superimpose a velocity V_w to the left (towards upstream) to simulate steady flow as in Fig. 14.13.

Here $y_2 = 3.5$ m, $V_2 = 4.5$ m/s, and $V_w = 8.0$ m/s

By the continuity equation,

$$y_1 (V_w - V_1) = y_2 (V_w - V_2)$$

$$y_1 (8.0 - V_1) = 3.5 (8.0 - 4.5)$$

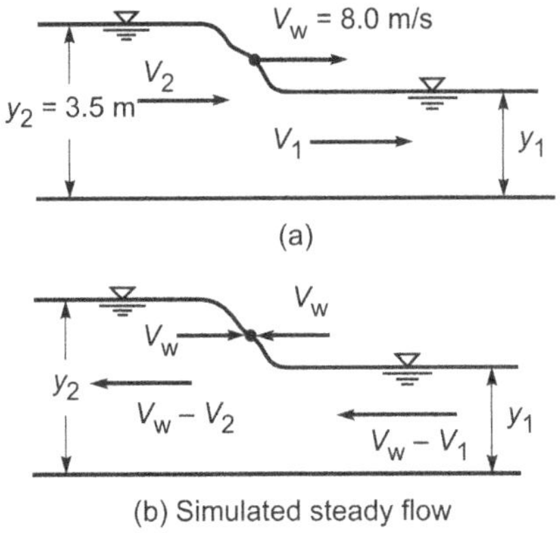

Fig. 14.13

$$V_1 = 8.0 - \frac{12.25}{y_1}$$

By the surge equation, (Eq. 14.3),

$$\frac{(V_w - V_1)^2}{g y_1} = \frac{1}{2}\frac{y_2}{y_1}\left(1 + \frac{y_2}{y_1}\right)$$

$$\frac{\left(8.0 - 8.0 + \frac{12.25}{y_1}\right)^2}{9.81\, y_1} = \frac{1}{2}\frac{3.5}{y_1}\left(1 + \frac{3.5}{y_1}\right)$$

$$\frac{15.297}{y_1^3} = \frac{1.75}{y_1}\left(1 + \frac{3.5}{y_1}\right)$$

$$y_1^2 = 3.5\, y_1 - 8.741 = 0$$

$$y_1 = \frac{-3.5 \pm \sqrt{(3.5)^2 + 4 \times 8.741}}{2}$$

$$y_1 = \mathbf{1.686\ m}$$

$$V_1 = 8.0 - \frac{12.25}{1.686} = \mathbf{0.734\ m/s}$$

Alternative method: Writing the alternative form of the surge equation Eq. (14.3a)

$$\frac{(V_w - V_2)^2}{g y_2} = \frac{1}{2}\left(\frac{y_1}{y_2}\right)\left(1 + \frac{y_1}{y_2}\right)$$

$$\frac{(8.0 - 4.5)^2}{9.81 \times 3.5} = \frac{1}{2}\left(\frac{y_1}{3.5}\right)\left(1 + \frac{y_1}{3.5}\right)$$

$$y_1 + \frac{y_1^2}{3.5} - 2.4975 = 0$$

$$y_1^2 + 3.5\, y_1 - 8.741 = 0$$

Solving $\quad$ $\mathbf{y_1 = 1.686\ m}$

14.9 A 2.0 m wide rectangular channel was carrying a flow with a velocity of 2.0 m/s. Due to sudden operation of a downstream gate a positive surge of velocity 4.5 m/s and travelling upstream was generated. If the depth of flow after the passage of the surge is 3.6 m, calculate the velocity of flow after the passage of the surge. What was the discharge before the passage of the surge?

Solution: Let V_w (directed to the left) be the velocity of the surge wave. Superimpose a velocity V_w directed to the right, to obtain a simulated steady flow, as in Fig. 14.14.

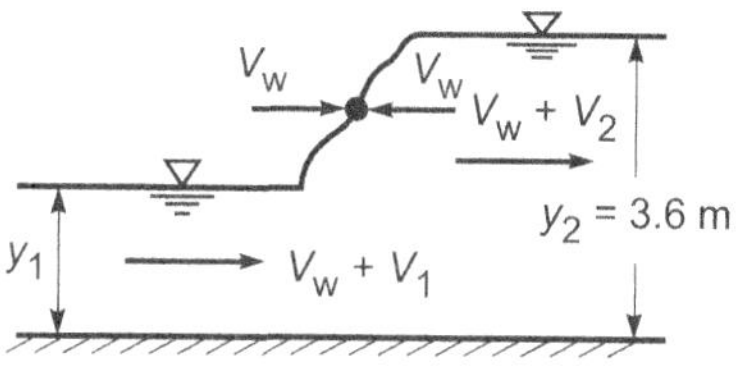

Fig. 14.14 *Simulated Steady Flow*

Here
$$V_1 = 2.0 \text{ m/s}$$
$$V_w = 4.5 \text{ m/s}$$
$$y_2 = 3.6 \text{ m}$$

By continuity equation:
$$y_1 (V_w + V_1) = y_2 (V_w + V_2)$$
$$y_1 (4.5 + 2.0) = 3.6 (4.5 + V_2)$$
$$6.5\, y_1 = 16.2 + 3.6\, V_2$$
$$V_2 = 1.806\, y_1 - 4.5$$

By the surge equation:
$$\frac{(V_w + V_1)^2}{g y_1} = \frac{1}{2}\frac{y_2}{y_1}\left(1 + \frac{y_2}{y_1}\right)$$
$$\frac{(4.5 + 2.0)^2}{9.81\, y_1} = \frac{1}{2}\left(\frac{3.6}{y_1}\right)\left(1 + \frac{3.6}{y_1}\right)$$
$$\left(1 + \frac{3.6}{y_1}\right) = 2.3927$$
$$y_1 = 2.585 \text{ m}$$

Discharge before the passage of the surge
$$= Q_1 = B y_1 V_1$$
$$= 2.0 \times 2.585 \times 2.0 = \mathbf{10.34\ m^3/s}$$
V_2 = Velocity after the passage of the surge
$$= 1.806\, y_1 - 4.5$$
$$= (1.806 \times 2.585) - 4.5 = \mathbf{0.168\ m/s}$$

14.10 Show that in a positive surge moving down a rectangular channel with absolute velocity V_w, the depths before the passage of the surge y_1 and after the passage of the surge y_2 are related as
$$\frac{y_2}{y_1} = \frac{1}{2}\left[-1 + \sqrt{1 + 8F_a^2}\right]$$
where
$$F_a^2 = \frac{(V_w - V_1)^2}{g y_1}$$
and V_1 = velocity in the channel before the passage of the surge.

Solution: The unsteady flow due to passage of the surge is as shown in Fig. 14.15(a). By superimposing a velocity V_w directed upstream (to the left) a steady flow is simulated, [Fig. 14.15(b)].

Here by continuity:
$$y_1 (V_w - V_1) = y_2 (V_w - V_2) \qquad (1)$$

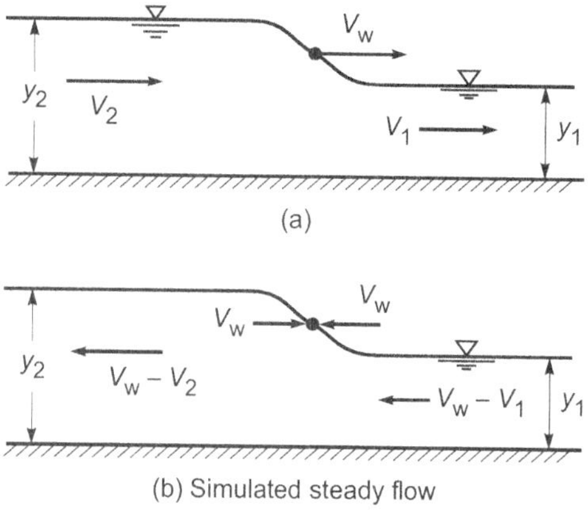

Fig. 14.15 *Positive Surge Moving Downstream*

By momentum equation:
$$\frac{\gamma y_1^2}{2} - \frac{\gamma y_2^2}{2} = \rho (V_w - V_1)\, y_1 \{(V_w - V_2) - (V_w - V_1)\}$$
$$y_1^2 - y_2^2 = \frac{2}{g}(V_w - V_1)\, y_1 \times \left\{\frac{y_1}{y_2}(V_w - V_1) - (V_w - V_1)\right\}$$
$$y_2^2 - y_1^2 = \frac{2}{g}(V_w - V_1)^2\, y_1 \left(1 - \frac{y_1}{y_2}\right)$$
$$(y_2 + y_1)(y_2 - y_1) = \frac{2}{g}(V_w - V_1)^2 \frac{y_1}{y_2}(y_2 - y_1)$$
$$\left(\frac{y_2}{y_1} + 1\right)\frac{y_2}{y_1} = \frac{2}{g}\frac{(V_w - V_1)^2}{y_1} = 2F_a^2$$
$$\left(\frac{y_2}{y_1}\right)^2 + \left(\frac{y_2}{y_1}\right) - 2F_a^2 = 0$$

Considering the positive root
$$\frac{y_2}{y_1} = \frac{1}{2}\left[-1 + \sqrt{1 + 8F_a^2}\right]$$

B. Water Hammer

14.11 Calculate the velocity of propagation of a pressure surge in a circular conduit with rigid walls in the following cases:

Case	*Liquid*	*Bulk modulus of elasticity K in MPa*
(a)	Water	2.22×10^3
(b)	Crude oil (RD = 0.80)	1.50×10^3
(c)	Gasolene (RD = 0.68)	9.58×10^2

Solution: Velocity of pressure surge = sonic velocity =

$$C = \sqrt{K/\rho}$$

(a) for water

$$C = \sqrt{\frac{2.22 \times 10^9}{998}} = \mathbf{1491.5\ m/s}$$

(b) for crude oil (RD = 0.8)

$$C = \left(\frac{1.50 \times 10^9}{0.8 \times 998}\right)^{1/2} = \mathbf{1370.7\ m/s}$$

(c) for gasolene (RD = 0.68)

$$C = \left(\frac{9.58 \times 10^8}{0.68 \times 998}\right)^{1/2} = \mathbf{1188\ m/s}$$

14.12 Calculate the velocity of propagation of a pressure wave in a steel pipe ($E = 2.07 \times 10^5$ MPa) of 2.5 m diameter carrying kerosene (RD = 0.80 and $K = 1.43 \times 10^3$ MPa). The pipe thickness is 2 cm.

Solution: The velocity of pressure wave C in a elastic pipe carrying a liquid is

$$C = \sqrt{K/\rho}\left(\frac{1}{1+(D/t)(K/E)}\right)^{1/2}$$

$$= \left(\frac{1.43 \times 10^9}{0.8 \times 998}\right)\left[\frac{1}{1+\left(\frac{250}{2}\right)\left(\frac{1.43 \times 10^9}{2.07 \times 10^{11}}\right)}\right]^{1/2}$$

$= 1338.3 \times 0.7325 =$ **980.4 m/s**

14.13 A valve is suddenly closed at the downstream end of a 0.90 m diameter pipeline carrying water in such a manner that the velocity is decreased from 4.0 m/s to 1.0 m/s instantaneously. Estimate the maximum pressure rise at the valve. Assume the pipe to the rigid and K for water $= 2.20 \times 10^3$ MPa.

Solution: Change in pressure

$$\Delta p_h = -\rho C \Delta V$$

Here C = Velocity of pressure surge

$= \sqrt{K/\rho}$ (as the pipe is rigid)

$$C = \left(\frac{2.2 \times 10^9}{998}\right)^{1/2} = 1484.7 \text{ m/s}$$

Also $V_2 - V_1 = \Delta V = 1.0 - 40 = -3.0$ m/s

$\Delta p_h = 998 \times 1484.7 \times 3 = 4.445 \times 10^6$ Pa

$=$ **4.445 MPa**

14.14 A cast iron pipe ($E = 1.0 \times 10^{11}$ Pa) is 0.9 m in diameter and carries water ($K = 2.0 \times 10^9$ Pa) at a velocity of 2.60 m/s. A valve in this pipe is instantaneously closed bringing the flow to a sudden stop at the valve end. Estimate the water hammer head produced due to this action. The pipe thickness is 1.25 cm and the pipe can be treated as elastic.

Solution: The velocity of pressure wave

$$C = \sqrt{K/\rho}\left(\frac{1}{1+(D/t)(K/E)}\right)^{1/2}$$

$$= \left(\frac{2.0 \times 10^9}{998}\right)^{1/2} \times \left[\frac{1}{1+(90/1.25)(2.0 \times 10^9/1.0 \times 10^{11})}\right]^{1/2}$$

$= 1415.6 \times 0.64 = 906.2$ m/s

Pressure rise $\Delta p_h = -\rho C \Delta V$

Water hammer head $= h_h = \dfrac{\Delta p_h}{\gamma} = -\dfrac{C \Delta V}{g}$

Here $\Delta V = 0 - V = -2.60$ m/s.

Hence, $h_w = \dfrac{906.2 \times 2.60}{9.81} =$ **240.2 m**

14.15 A steel penstock 60 cm in diameter has a shell thickness of 1.2 cm. The modulus of elasticity of the shell material is 2.1×10^5 N/mm^2 and the volume modulus of elasticity of water is 2.1×10^3 N/mm^2. The pipe is designed to discharge water at a mean velocity of 2.1 m/s. Determine the water hammer pressure rise caused by the sudden closure of a valve at the downstream end (i) by neglecting the elasticity of the pipe material and also (ii) by considering the elasticity of the pipe material.

Solution:

(i) By neglecting the elasticity of the pipe material, velocity of pressure wave

$$C = \sqrt{K/\rho}$$

$$= \left(\frac{2.1 \times 10^3 \times 10^6}{998}\right)^{1/2} = 1450.6 \text{ m/s}$$

Water hammer pressure rise

$$\Delta p_h = -\rho C \Delta V$$
$$= -998 \times 1450.6 \times (0 - 2.1)$$
$$= \mathbf{3.04\ MPa}$$

(ii) By considering the elasticity of the pipe material:

$$C = \sqrt{K/\rho}\left[\frac{1}{1+(D/t)(K/E)}\right]^{1/2}$$

$$= 1450.6\left[\frac{1}{1+(60/1.2)(2.1\times10^9/2.1\times10^{11})}\right]^{1/2}$$

$$= 1184.4 \text{ m/s}$$

Water hammer pressure rise,

$$\Delta p_h = -\rho C \Delta V$$
$$\Delta p_h = -998 \times 1184.4 \times (0 - 2.1)$$
$$= \mathbf{2.48\ MPa}$$

14.16 Estimate the maximum time for a sudden closure of a valve in a pipe of 80 cm diameter, 1000 m long carrying kerosene at 1.25 m^3/s. For kerosene: Relative density = 0.8, bulk modulus = 1.43×10^9 Pa. For the pipe material: Modulus of elasticity = 2.10×10^{11} Pa. Wall thickness of the pipe is 80 mm.

Solution: Velocity of pressure wave,

$$C = \sqrt{K/\rho}\left[\frac{1}{1+(D/t)(K/E)}\right]^{1/2}$$

$$= \left(\frac{1.43\times10^9}{0.8\times998}\right)^{1/2} \times \left[\frac{1}{1+\left(\frac{800}{80}\right)(1.43\times10^9/2.10\times10^{11})}\right]^{1/2}$$

$$= 1295 \text{ m/s}$$

Critical time $T_0 = 2L/C$

$$= (2 \times 1000)/1295 = \mathbf{1.544\ s}$$

Hence, *maximum* time for a sudden closure is 1.544 s.

14.17 A 0.5 m diameter and 100 m long pipeline carrying 0.5 m^3/s of water is a fitted with a valve at the downstream end. Calculate the rise of pressure caused within the pipe due to valve closure (i) instantaneously and (ii) in 1 second. [Take the sonic velocity as 1430 m/s]

Solution:

Critical time $T_o = 2\text{ L/C} = \dfrac{2\times100}{1430} = 0.14$ s.

(i) For instantaneous closure, water hammer pressure $p_h = -\rho C \Delta V$

Here $-\Delta V = V = \dfrac{0.5}{\left(\frac{\pi}{4}\right)\times(0.5)^2} = 0.2546$ m/s

Hence $p_h = 998 \times 1430 \times 0.2546 = 363350$ Pa

$$= \mathbf{363.35\ kPa}$$

(ii) Closure in T = 1s is a very slow closure as T/T_o = (1/0.14) ≈ 7 Hence, at the valve, the pressure rise is approximately

$$P_{vsc} = \frac{\rho LV}{T} = \frac{998\times100\times0.2546}{1} = 25409 \text{ Pa}$$
$$= \mathbf{25.41\ kPa}$$

14.18 The velocity of water in a 60 cm diameter and 15 mm thick cast iron pipe ($E = 1.04 \times 10^{11}$ N/m^2) is changed from 3.0 m/s to zero in 1.25 s by closure of a valve (a). If the pipe length is 800 m what will be the water hammer pressure at the valve? What will be the corresponding pressure rise if the closure taken place in (b) 2.0 s (c) 0.8 s respectively. K for water = 2.11×10^9 N/m^2.

Solution: Velocity of pressure wave

$$C = \sqrt{K/\rho}\left[\frac{1}{1+(D/t)(K/E)}\right]^{1/2}$$

$$C = \left(\frac{2.11\times10^9}{998}\right)^{1/2} \times \left[\frac{1}{1+\left(\frac{60}{1.5}\right)(2.11\times10^9/1.04\times10^{11})}\right]^{1/2}$$

$$= 1454 \times 0.743 = 1080.4 \text{ m/s}$$

Critical time $T_0 = 2L/C = 2 \times 800/1080.4$

$$= 1.48 \text{ s}$$

(a) Closure in T = 1.25 s is rapid closure as $T < T_0$. Hence, water hammer pressure

$$p_h = -\rho C \Delta V = -998 \times 1080.4 \times (-3.0) \text{ Pa}$$
$$= \mathbf{3.235\ MPa}$$

(b) When T = 3.0 s, as T/T_0 = (2.0/1.48) = 1.35 it is a slow closure. The pressure rise at the valve is approximately

$$p_{hs} = \frac{T_o}{T}(p_h)_{rapid} = \frac{1.48}{2.00} \times 3.235 = \mathbf{2.394\ MPa}$$

(c) When T = 0.8 s, as $T < T_0$, it is a rapid closure and p_h is the same as in case (a) viz

$$\mathbf{p_h = 3.235\ MPa}$$

14.19 A 40 cm diameter steel wipe is 2500 m long and conveys water with a head of 250 m at its downstream end. The pipe thickness is 10 mm. If the valve at the downstream end of a pipe gradually reduces the velocity from 3.0 m/s to 0.5 m/s in 5 s, estimate the total pressure on the pipe walls at the valve.

For water:

$K = 2.10 \times 10^9$ Pa; for steel; $E = 2.11 \times 10^{11}$ Pa.

Solution: Velocity of pressure wave,

$$C = \sqrt{K/\rho}\left[\frac{1}{1+(D/t)(K/E)}\right]^{1/2}$$

$$= \left(\frac{2.10\times10^9}{998}\right)^{1/2}$$

$$\left[\frac{1}{1+\left(\frac{400}{10}\right)(2.10\times10^9/2.11\times10^{11})}\right]^{1/2}$$

$$= 1450.6 \times 0.8457 = 1227 \text{ m/s}$$

Critical time for rapid closure

$$T_0 = 2L/C = (2 \times 2500)/1227 = 4.076 \text{ s}$$

Actual time of closure $T = 5.0$ and $T/T_0 = 5.0/4.076 = 1.227$ it is a slow closure case.

For a rapid closure:

$$p_{hr} = \text{water hammer pressure} = -\rho C \Delta V$$

$$= -998 \times 1227 \times (0.5 - 3.0)$$

$$p_{hr} = 3.061 \times 10^6 \text{ Pa} = 3061 \text{ kPa}$$

At slow closure, pressure rise:

$$p_{hs} = \frac{T_0}{T}\, p_{hr} = \frac{4.076}{5.0} \times 3061 = 2495 \text{ kPa}$$

Total pressure at the valve end

$$p_t = p_{static} + p_{hs}$$

$$= (9.79 \times 250) + 2495 = \mathbf{4943\ kPa}$$

14.20 A valve at the downstream end of a 3500 m long pipeline is closed in 4.0 s. The pipe carries water flowing at a velocity of 0.8 m/s. Assuming the pipe to be rigid, calculate (i) the peak water hammer pressure produced by this closure and (ii) the length of the pipe subjected to the above peak pressure at the end of the closure time. [$K_{water} = 2.20 \times 10^3$ MPa].

Solution: As the pipe is rigid, velocity of pressure wave

$$C = \sqrt{K/\rho} = \left(\frac{2.2\times10^9}{998}\right)^{1/2} = 1484.7 \text{ m/s}$$

Critical time

$$T_0 = \frac{2L}{C} = \frac{2\times3500}{1484.7} = 4.715 \text{ s}$$

As T = time of closure = 4.0 s, the closure is rapid. Hence the water hammer pressure

$$p_h = \rho C V_0 = 998 \times 1484.7 \times 0.8 \text{ Pa}$$

$$= \mathbf{1.185\ MPa}$$

Length of the pipe, from the valve end, affected by this peak pressure, during the closure time.

$$x_0 = L - \frac{CT}{2}$$

$$= 3500 - \frac{1484.7\times4.0}{2} = \mathbf{530.6\ m}$$

14.21 A 2800 m long pipeline conveys water at a velocity of 2.0 m/s. If a valve at the downstream end is closed suddenly, calculate the maximum pressure rise. Sketch the variation of the water hammer pressure with time at (i) the valve end (ii) mid-point B of the pipe and (iii) at point C 2100 m upstream of the valve. The pipe starts from a lake at the upstream end. Assume the pipe to be rigid. For water, take $K = 1.956 \times 10^9$ Pa.

Solution:
$$C = \sqrt{K/\rho} = \left(\frac{1.956\times10^9}{998}\right)^{1/2} = 1400 \text{ m/s}$$

Critical time $T_0 = \dfrac{2L}{C} = \dfrac{2\times2800}{1400} = 4$ s

Period of pressure fluctuation: $2T_0 = 8$ s

Water hammer pressure

$$p_h = \rho C V_0 = 998 \times 1400 \times 2.0 \text{ Pa}$$

$$= \mathbf{2.794\ MPa}$$

(i) Figure 14.16 (a) shows the variation of the water hammer pressure p_h, at the valve V with the static pressure at V as datum.

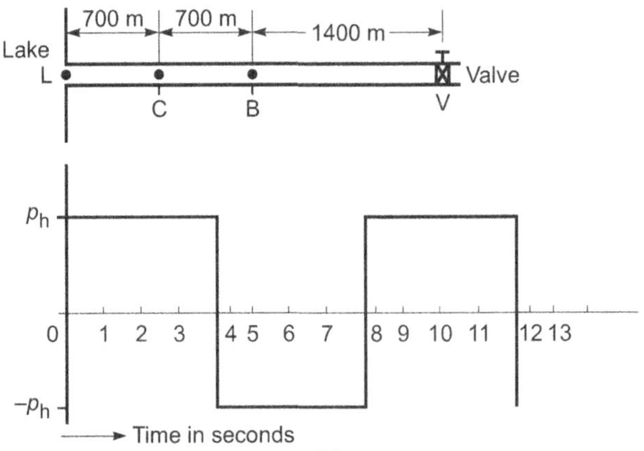

(a)

Fig. 14.16(a) *Variation of p_h at Valve V*

Time taken for p_h to reach the lake = 2800/1400 = 2 s

Time taken for the reflected wave to reach V = 2 s

Hence from 0 to 4 s, the pressure at V will be $+p_h$ and from 4 to 8 s, it will be $-p_h$

and from 8 to 12 s, it will be $+p_h$

and so on.

Note the period of the wave is 8 s.

(ii) Figure 14.16(b) shows the variation of the water hammer pressure p_h, at mid point B. The static pressure at B is taken as datum.

Time for the pressure wave to reach

$$B = 1400/1400 = 1 \text{ s}$$

Time for p_h to reach the lake = 1 s

Time for the reflected wave to reach = 1 s

Hence, duration of max pressure p_h at B is 2 s, i.e. from t = 1.0 s to 3.0 s.

Time for the reflected wave to reach the valve = 1 s

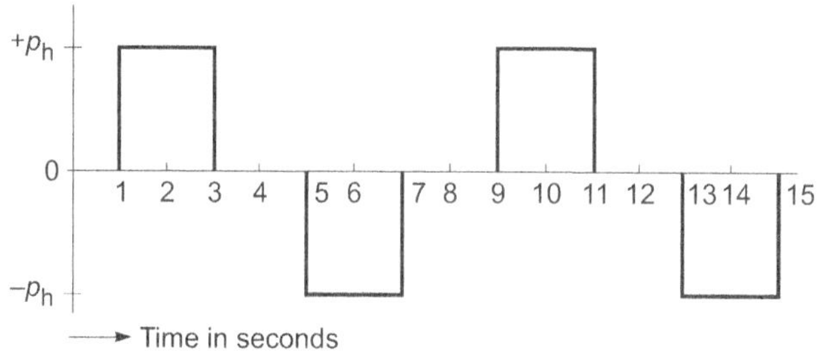

Fig. 14.16(b) *Variation of p_h of mid-point B*

Time for the $-p_h$ wave to reach point B = 1 s

Hence from 3.0 s to 5.0 s the pressure at B will remain at zero (relative to the static pressure).

From 5.0 s to 7.0 s it will register – p_h.

From 7.0 s to 9.0 s the pressure B will be 0.

From 9.0 s to 11.0 s it will be + p_h and so on.

The period of pressure fluctuation is **8 s**.

(iii) Fig. 14.16(c) shows the variation of the water hammer pressure p_h, at point C. The static pressure at C is taken at datum.

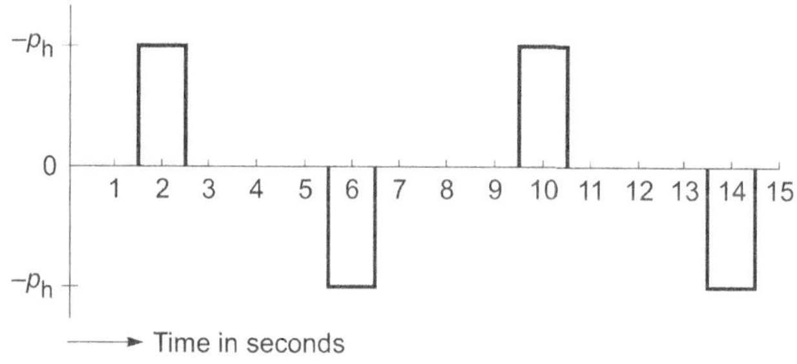

Fig. 14.16(c) *Variation of p_h at point C*

Time for pressure p_h to reach C = 2100/1400 = 1.5 s

Time for p_h to reach the lake = 0.5 s.

Time for reflected pressure to reach C = 0.5 s.

Hence, pressure at C will remain zero (relative to static pressure at C) from 2.5 s to 5.5 s.

From 5.5 s to 6.5 s the pressure at C will be $-p_h$

From 6.5 to 9.5 s the pressure at C will be zero

From 9.5 s to 10.5 s, it will be + p_h, and so on.

Note the period of pressure fluctuation at C = **8 s.**

14.22 A steel pipe 1.20 m in diameter conveys 1.40 m^3/s of water under a head of 300 m. A valve at the downstream end can be expected to close suddenly. Estimate the water hammer pressure due to this closure. Also determine the minimum thickness of the wall to the nearest millimeter needed to withstand the pressures involved. For steel; E = 210 kN/mm^2, and safe working stress = 0.1 kN/mm^2. For water: K = 2.10 kN/mm^2.

Solution: Assume the pipe to be rigid. The velocity of pressure wave for this case

$$C = \sqrt{K/\rho} = \left(\frac{2.10 \times 10^9}{998}\right)^{1/2} = 1450.6 \text{ m/s}$$

Velocity of flow

$$= V_0 = \frac{Q}{A} = \frac{1.40}{\left(\frac{\pi}{4}\right)(1.20)^2} = 1.238 \text{ m/s}$$

Water hammer pressure

$$p_h = \rho C V_0$$
$$= 998 \times 1450 \times 1.238 \times 10^{-3} \text{ kPa}$$
$$= 1792 \text{ kPa}$$

Total pressure

$$p_t = \text{Static pressure + water hammer pressure}$$
$$= (9.79 \times 300) + 1792$$
$$p_t = \mathbf{4729\ kPa}$$

Minimum thickness of pipe required

$$t_m = \frac{p_t D}{2\sigma_w} \quad \text{where} \quad \sigma_w = \text{working stress}$$
$$= (4729 \times 1.2)/(2 \times 0.1 \times 10^6)$$
$$= 0.02837 \text{ m} = 28.3 \text{ mm}$$

When the elasticity of the pipe is considered, the thickness required will be slightly less than 28 mm. A trial and error approach is adopted.

Trial 1 : Assume t = 26 mm

$$D/t = \frac{1200}{26} = 46.15 \quad \text{and} \quad \frac{K}{E} = \frac{2.1}{210} = \frac{1}{100}$$

Velocity of pressure, wave,

$$C = 1450.6 \times \left[\frac{1}{1 + (D/t)(K/E)}\right]^{1/2}$$

$$= 1450.6 \times \left[\frac{1}{1 + (46.15)(1/100)}\right]^{1/2} = 1200 \text{ m/s}$$

Water hammer pressure

$$p_h = \rho C V_0$$

$p_h = 998 \times 1200 \times 1.238 \times 10^{-3}$ kPa = 1483 kPa

p_s = Static pressure = 9.79 × 300 = 2937 kPa

p_t = total pressure = $p_s + p_h$ = 4420 kPa

$$t_m = \text{minimum thickness needed} = \frac{p_t D}{2\sigma_w}$$

$$= \frac{4420 \times 1.2}{2 \times 0.1 \times 10^6} = 0.02652 \text{ m} = 26.5 \text{ mm}$$

Hence adopt **27 mm** thickness pipe.

As a check:

D/t = 1200/27 = 44.44

$$C = 1450.6 \times \left[\frac{1}{1+(44.44)(1/100)}\right]^{1/2} = 1207 \text{ m/s}$$

$p_h = 998 \times 1207 \times 1.238 \times 10^{-3}$ kPa = 1491 kPa

p_s = 2937 kPa

p_t = 4428 kPa

$$\sigma = \text{stress in the pipe wall} = \frac{p_t D}{2t}$$

$$= \frac{4428 \times 1.20}{2 \times 0.027} = 98405 \text{ kPa} < \sigma_w$$

Hence the adopted thickness t = 27 mm is satisfactory.

14.23 A steel pipe 20 cm in diameter and 1000 m long conveys a discharge of 40 L/s of crude oil. A valve at the downstream end of the pipe is completely closed in 1.25 s. If the thickness of the pipe is 10 mm, calculate the additional stress in the pipe due to the valve operation.

For crude oil: Relative density = 0.80

Bulk modulus = 1520 MPa

For steel pipe:

Modulus of elasticity $= 2.07 \times 10^5$ MPa

Solution: Velocity of pressure wave,

$$C = \sqrt{K/\rho}\left[\frac{1}{1+(D/t)(K/E)}\right]^{1/2}$$

$$= \left(\frac{1.52 \times 10^9}{0.8 \times 998}\right)\left[\frac{1}{1+\left(\frac{200}{10}\right)(1520/207000)}\right]^{1/2}$$

= 1380 × 0.9338 = 1289 m/s

$$\text{Velocity of flow} = V_0 = \frac{0.040}{\left(\frac{\pi}{4}\right) \times (0.2)^2}$$

= 1.273 m/s.

Critical closure time $T_0 = 2L/C$

= (2 × 1000)/1289 = 1.55 s

Actual closure time $T = 1.25 \text{ s} < T_0$

Hence the closure is rapid.

Water hammer pressure

$p_h = \rho C V_0$

$p_h = (998 \times 0.8) \times 1289 \times 1.273$ Pa

= 1310 kPa

Additional stress due to water hammer:

$$\sigma_a = \frac{p_h D}{2t}$$

$$\sigma_a = \frac{1310 \times 0.20}{2 \times 0.01} = 13100 \text{ kPa}$$

= **0.0131 kN/mm²**

14.24 A steel pipeline is 30 cm in diameter and has a wall thickness of 3 mm. The pipe is 1000 m long and conveys a flow of 100 L/s of oil (RD = 0.82). The static head at the outlet is 160 m of oil. If the working stress of steel is 0.1 kN/mm², calculate the minimum time of closure of a downstream valve. For oil: $K = 1 \times 10^9$ Pa and for steel: $E = 2.14 \times 10^{11}$ Pa.

Solution: Velocity of pressure wave,

$$C = \sqrt{K/\rho}\left[\frac{1}{1+(D/t)(K/E)}\right]^{1/2}$$

$$= \left(\frac{1 \times 10^9}{0.82 \times 998}\right)^{1/2} \times \left[\frac{1}{1+\left(\frac{300}{3}\right)(1 \times 10^9/2.14 \times 10^{11})}\right]^{1/2}$$

= 1105.4 × 0.8255 = 912.6 m/s

$$\text{Velocity of flow } V_0 = \frac{0.100}{\left(\frac{\pi}{4}\right)(0.3)^2} = 1.415 \text{ m/s}$$

Critical time of closure

$T_0 = 2L/C$

= (2 × 1000)/912.6 = 2.19 s

At rapid closure, at $T \le T_0$,

Water hammer pressure

$p_h = \rho C V_0$

$p_h = (0.82 \times 998) \times 912.6 \times 1.415$ Pa

= 1057 kPa

Static pressure $p_s = 160 \times (0.82 \times 9.79)$

= 1284 kPa

Total pressure p = 1284 + 1057 = 2341 kPa

$$\text{Stress in the pipe } \sigma = \frac{pD}{2t}$$

$$= \frac{2341 \times 300}{2 \times 3} = 1.17 \times 10^5 \text{ kPa}$$

As the working stress $\sigma_w = 1.0 \times 10^5$ kPa is lower than this value, the valve must be closed with a time $T > T_0$ to have slow closure. Maximum allowable total pressure is therefore

$$p_t = \frac{\sigma_w 2t}{D} = \frac{1.0 \times 10^5 \times 2 \times 3}{300} = 2000 \text{ kPa}$$

Allowable water hammer pressure in slow closure

$$p_{hw} = 2000 - p_{static} = 2000 - 1284 = 716 \text{ kPa}$$

In slow closure with time T,

$$\frac{p_{hw}}{p_h} = \frac{T_0}{T} \quad \left(\text{for } 1 < \frac{T}{T_0} \lesssim 1.5\right)$$

$\therefore$ T = Time of slow closure to attain p_{hw}

$$= T_0 (p_h/p_{hw}) = \frac{2.19 \times 1057}{716} = 3.233$$

Since $T/T_0 = 3.233/2.19 = 1.476 < 1.5$ the assumption of slow closure is O.K.

Hence, minimum time of closure T_m = **3.23 s**.

C. Establishment of Flow

14.25 A 15 cm diameter, 1500 m long pipe leads from a large reservoir to an outlet which is 20 m below the water level in the reservoir. If a valve at the pipe outlet is suddenly opened estimate the time required to reach (i) 50% and (ii) 95% of steady state discharge. Assume $f = 0.02$ and minor losses as $k_1 \frac{V^2}{2g}$ with $k_1 = 5.0$.

Solution:

Friction loss $\quad h_f = \frac{fLV^2}{2gD} = k_f \frac{V^2}{2g}$

$$k_f = \frac{fL}{D} = \frac{0.02 \times 1500}{0.15} = 200$$

Minor loss coefficient = $k_1 = 5.0$

Total has coefficient = $k = k_1 + k_f = 205$

For steady flow: $\quad V_0 = \sqrt{2gH/(1+k)}$

$$= \left[\frac{2 \times 9.81 \times 20}{(1+205)}\right]^{1/2} = 1.380 \text{ m/s}$$

During the establishment of flow,

$$t = \frac{L}{(1+k)V_0} \ln \frac{V_0 + V}{V_0 - V}$$

$$= \frac{1500}{(1+205) \times 1.380} \ln \frac{1+(V/V_0)}{1-(V/V_0)}$$

$$= 5.276 \ln \frac{1+(V/V_0)}{1-(V/V_0)}$$

(i) When $Q/Q_0 = 0.50$, $V/V_0 = 0.5$

$$\ln \frac{1+(V/V_0)}{1-(V/V_0)} = \ln \left(\frac{1+0.5}{1-0.5}\right) = 1.0986$$

$$t = 5.276 \times 1.0986 = \mathbf{5.8 \text{ s}}$$

(ii) When $Q/Q_0 = 0.95$, $V/V_0 = 0.95$

$$\ln \frac{1+(V/V_0)}{1-(V/V_0)} = \ln \left(\frac{1+0.95}{1-0.95}\right) = 3.6636$$

$$t = 5.276 \times 3.6636 = \mathbf{19.33 \text{ s}}$$

14.26 Solve the previous example (Example 14.25) by neglecting all losses.

Solution: If friction is neglected, $k = 0$ and the time of flow establishment

$$t = \frac{L}{V_0} \ln \frac{V_0 + V}{V_0 - V}$$

Here $\quad V_0 = \sqrt{2gH} = (2 \times 9.81 \times 20)^{1/2} = 19.81$ m/s

$$t = \frac{1500}{19.81} \ln \frac{1+(V/V_0)}{1-(V/V_0)}$$

$$= 75.72 \ln \frac{1+(V/V_0)}{1-(V/V_0)}$$

(i) When $Q/Q_0 = 0.5$, $V/V_0 = 0.5$

$$t = 75.72 \ln \frac{1+0.5}{1-0.5} = \mathbf{83.19 \text{ s}}$$

(ii) When $Q/Q_0 = 0.95$, $V/V_0 = 95$

$$t = 75.72 \ln \frac{1+0.95}{1-0.95} = \mathbf{277.4 \text{ s}}$$

14.27 A 40 cm diameter pipe is 2 km in length and ends in a 10 cm nozzle. The nozzle is used to drive a turbine. If the ultimate velocity head at the jet exit is 40 m, estimate the time required for the turbine to reach 98 per cent full speed after the control valve is opened. Assume $f = 0.02$ for the pipe and all other losses can be ignored.

Solution: Refer to the schematic layout shown in Fig. 14.17.

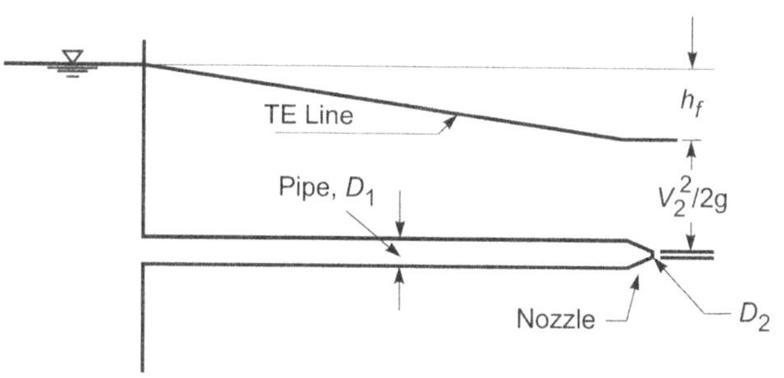

Fig. 14.17

Let $\quad V_1$ = ultimate velocity in the pipe.

V_2 = ultimate velocity in the nozzle.

By continuity $\quad V_1 \frac{\pi}{4} D_1^2 = V_2 \frac{\pi}{4} D_2^2$

$$V_1 = V_2 (D_2/D_1)^2$$

$$\frac{V_2^2}{2g} = 40.0 \text{ m},$$

hence $V_2 = \sqrt{2 \times 9.81 \times 40} = 28.0$ m/s

$$V_1 = 28.0 \times \left(\frac{10}{40}\right)^2 = 1.75 \text{ m/s}$$

Head loss $h_{f1} = \frac{fL}{D_1}(V_1^2/2g) = k\,\frac{V_1^2}{2g}$

$$k = \frac{fL}{D_1} = \frac{0.02 \times 2000}{0.40} = 100$$

For 98% ultimate flow establishment, time required

$$t = \frac{L}{(1+k)\,V_1} \ln \frac{1 + V/V_1}{1 - V/V_1}$$

$$= \frac{2000}{(1+100) \times 1.75} \ln \frac{1+0.98}{1-0.98}$$

$= 11.315 \ln 99 =$ **52 s**

OBJECTIVE QUESTIONS

Surges in Canals

14.1 In a rectangular channel of depth 1.2 m and velocity 2.0 m/s, an elementary wave travelling upstream will have an absolute velocity of
(a) 5.43 m/s (b) 3.43 m/s
(c) 1.43 m/s (d) 2.0 m/s

14.2 A sluice gate controlling flow in a canal is suddenly lowered by an amount to cause partial closure. This will produce.
(a) a negative wave on the upstream
(b) a positive surge on the downstream
(c) a positive wave on the upstream
(d) a standing wave on the downstream

14.3 A positive surge travels upstream in a canal with an absolute velocity V_w. With suffixes 1 and 2 referring to sections upstream and downstream of the surge respectively, the continuity equation is written as
(a) $A_1 V_1 = A_2 V_2$
(b) $A_1 (V_1 + V_w) = A_2 (V_2 - V_w)$
(c) $A_2 (V_w - V_2) = A_1 (V_w + V_1)$
(d) $A_1 (V_1 + V_w) = A_2 (V_2 + V_w)$

14.4 In analyzing the elements of a positive surge in a prismatic canal the following equations are used:
(a) continuity and energy equations
(b) continuity and momentum equations
(c) energy equation with method of superposition
(d) only the continuity equation

14.5 In a rectangular channel carrying a flow with a depth of 1.2 m and velocity of 2.0 m/s, a gate on the downstream is suddenly closed. If a positive surge of speed 3.75 m/s travelling upstream is produced, the height of the surge is
(a) 1.5 m (b) 0.25 m
(c) 2.3 m (d) 0.8 m

14.6 A tidal bore is a phenomenon in which
(a) a positive surge travels down a tidal river with the outgoing tide.
(b) a positive surge travels upstream in a tidal river with the incoming tide.
(c) a positive surge moves upstream or downstream in a tidal river depending upon the tide.
(d) a negative surge is associated with an incoming tide.

Water Hammer

14.7 If the bulk modulus of water is 1.96×10^9 N/m^2 the water hammer wave velocity through a rigid pipe is
(a) 448 m/s (b) 996 m/s
(c) 4390 m/s (d) 1401 m/s

14.8 The velocity of pressure wave in a rigid pipe carrying a fluid of density ρ and viscosity μ varies as
(a) ρ (b) $\sqrt{\rho}$
(c) ρ/μ (d) $1/\sqrt{\rho}$

14.9 In a pipe flow the ratio of the bulk modulus of the fluid and the coefficient of elasticity of the pipe material is 1/100. The thickness of the pipe wall is 1/50 of the diameter of the pipe. The ratio of the velocity of propagation of a pressure wave C in this pipe to the velocity of sound in an infinite medium of the fluid C_1 is given by $C/C_1 =$
(a) 0.667 (b) 0.816
(c) 0.500 (d) 0.732

14.10 The velocity of a pressure wave in water of infinite extent is 1440 m/s, For a pipe with diameter = 40 cm, thickness = 4 mm, with E of the pipe material = 2.1×10^{11} Pa and K for water = 2.1×10^9

Pa, the velocity of propagation of water hammer pressure wave, in m/s, is

(a) 1440 (b) 720
(c) 2036 (d) 1018

14.11 A pipe 1000 m long conveys a fluid whose velocity of propagation of pressure wave is 1000 m/s. After a sudden closure of the downstream end valve, the peak water hammer pressure will exist for a duration of

(a) 2s (b) 1s
(c) 4s (d) 0.5s

14.12 The velocity of pressure wave in water hammer is equal to the velocity of sound in water. Is this statement correct?

(a) No, because the elasticity of the pipe reduces the velocity of pressure wave.
(b) No, because the elasticity of the pipe increases the velocity of pressure wave.
(c) Yes, the pipe is more rigid than water.
(d) Yes, the pressure disturbances always propagate with the velocity of sound.

14.13 A penstock is 3000 m long. Pressure wave travels in it with a velocity of 1500 m/s. If the gates of the turbine are closed uniformly and completely in 4 s, then the closure is called

(a) rapid (b) slow
(c) sudden (d) ultra-rapid

14.14 A penstock is 2000 m long and the velocity of pressure wave in it is 1000 m/s. Water hammer pressure head for instantaneous closure of valve at the downstream end of the pipe is 60 m. If the valve is closed in 4 s, then the peak water hammer pressure is equal to

(a) 15 m (b) 30 m
(c) 60 m (d) 120 m

14.15 In a pipe 2000 m long carrying oil, the velocity of propagation of the pressure wave is 1000 m/s. A valve at the down-stream end is closed suddenly. At the midpoint of the pipeline, the peak water hammer pressure will exist for a duration of

(a) 1.0 s (b) 4.0 s
(c) 3.0 s (d) 2.0 s

14.16 A valve at the downstream end of 2500 m long pipe is closed in 4 s. If the velocity of propagation of the pressure wave in this pipe is 1000 m/s, the length of the pipe subjected to peak water hammer pressure at the end of the closure time is

(a) 1000 m (b) 500 m
(c) 2000 m (d) 2500 m

14.17 In an 800 m long pipe the velocity of propagation of pressure wave $C = 960$ m/s. If the peak water hammer pressure due to sudden closure of the flow at the downstream end is 900 kPa, the peak water hammer pressure due to closure of the same valve in 2.0 s is

(a) 900 kPa (b) 750 kPa
(c) 1080 kPa (d) 821 kPa

14.18 A long pipeline 3840 m in length carries water. It has a velocity of propagation of pressure wave $C = 960$ m/s. When a valve at the downstream end of the pipe was closed in T seconds, 960 m of the pipeline from the valve end was subjected to peak pressure at the end of the closure time. The time of closure T is

(a) 5s (b) 6s
(c) 8s (d) 12s

14.19 In a pipeline 1920 m long the velocity of propagation of pressure wave is 960 m/s. If a rapid closure of a downstream end valve is desired the largest time of closure is

(a) 2s (b) 4s
(c) 6s (d) 8s

14.20 In a 2000 m long pipeline the velocity of pressure wave is 1000 m/s. If a flow of water with a velocity of 0.8 m/s in this pipe is suddenly stopped completely in 1.5 s the resulting peak water hammer pressure in kPa is

(a) 798.4 (b) 1197.6
(c) 532.2 (d) 81.5

14.21 A 3150 m long pipeline has a velocity of propagation of pressure wave $C = 1050$ m/s. If a flow of water with a velocity of 1.2 m/s is stopped completely by a downstream valve in 8 s, the approximate peak water hammer pressure, in kPa, is

(a) 1257 (b) 1656
(c) 943 (d) 1050

14.22 In a long pipeline at a downstream valve the normal static pressure head is 140 m. If the valve is suddenly closed, the expected water hammer pressure head is 80 m. If the pipe thickness is to be designed for this contingency, it will have to be designed to withstand a pressure head of

(a) 60 m (b) 80 m
(c) 140 m (d) 220 m

14.23 Indicate the *incorrect* statement:

A pipe can be protected from high water hammer pressure through

(a) use of slow closing valves

(b) use of longer pipelines
(c) use of automatic pressure regulating valves
(d) use of surge chambers

14.24 A surge tank is provided in hydropower schemes to
(a) strengthen the penstocks
(b) reduce water hammer pressure
(c) reduce frictional losses in the system
(d) increase the net head

14.25 A surge tank is provided in a hydropower scheme to
(a) provide additional storage close to the penstock
(b) take care of change of slope, alignment of size of water conductor system
(c) reduce the pressure under transient condition.
(d) provide convenient overspills.

Flow Establishment

14.26 A long pipe is connected to a water reservoir and a downstream control valve is suddenly opened. If Q is the discharge at any instant during the establishment of flow and Q_0 is the ultimate discharge then Q/Q_0 varies
(a) linearly with time t
(b) as square root of time t
(c) as a function of terms involving e^t
(d) as a function of terms involving $ln\ t$

14.27 A downstream end valve in a long pipe connected to a water tank is suddenly opened. If t_0 is the time to reach 95% of the ultimate flow by neglecting friction and other losses and t_1 is the corresponding time obtained by including friction and other losses, then
(a) $t_1 > t_0$ (b) $t_1 = t_0$
(c) $t_1 \geq t_0$ (d) $t_1 < t_0$

14.28 Indicate the *incorrect* statement:
The time of establishment of 95% of the ultimate flow due to sudden opening of a downstream valve in a pipeline
(a) will increase if the pipe length is increased
(b) will increase if the pipe friction is reduced
(c) will decrease if the head of the reservoir is increased
(d) will decrease if the pipe diameter is increased

14.29 A pipe of length L leads from a large reservoir. When a downstream valve is suddenly opened, the time to attain 25% of the ultimate flow is estimated as 2.18 s. Then, the time to attain 75% of the flow would be.
(a) 6.54 s (b) 3.78 s
(c) 15.15 s (d) 8.30 s

14.30 In a pipe leading from a reservoir the time for the establishment of flow due to sudden opening of a downstream valve is calculated. If the friction and other minor losses are neglected the time to attain 50% of the ultimate flow is 1215 s. The corresponding time when there is a head loss, expressed as equal to k times the velocity head is 15 s. The value of the loss coefficient k is
(a) 81 (b) 80
(c) 9 (d) 10

PROBLEMS

Open Channels

Surges

14.1 The depth and velocity of flow in a rectangular channel are 0.9 m and 1.5 m/s respectively. If a gate at the downstream end of the channel is completely closed abruptly, what will be the height and absolute velocity of the resulting surge?
(*Ans.* $y_2 - y_1 = 0.50$ m, $V_w = 2.70$ m/s)

14.2 A rectangular channel carries a flow with a velocity of 0.65 m/s and depth of 1.40 m. If the discharge in the channel is abruptly increased three-fold by a sudden lifting of a gate on the upstream, estimate the velocity and height of the resulting surge.
(*Ans.* $y_2 - y_1 = 0.36$ m, $V_w = 5.06$ m/s in downstream direction)

14.3 A rectangular channel 3.0 m wide is conveying 15.0 m^3/s of discharge at a depth of 1.50 m. If a downstream sluice gate is abruptly lowered to reduce the discharge by 60%, estimate the characteristics of the resulting surge.
(*Ans.* $y_2 = 2.64$ m, $V_w = 2.632$ m/s (travels upstream))

14.4 A tidal bore moves up a wide river with a velocity of 6.0 m/s. If the river had a depth of 1.8 m and velocity of 0.85 m/s before the passing of the bore, estimate the height of the bore and velocity of flow in the river after the passage of the bore.
(*Ans.* Height of bore = 1.546 m, $V_2 = -2.315$ m/s))

14.5 In a tidal river the depth and velocity of flow are 0.90 m and 1.25 m/s respectively. Due to tidal action a tidal bore of height 1.2 m is observed to travel upstream. Estimate the speed of the bore and the speed of flow after the passage of the bore.

(*Ans.* V_w = 4.61 m/s (travels upstream) V_2 = –2.1 m/s)

14.6 In a rectangular channel a positive surge of velocity 6.0 m/s was seen moving down the stream. If the depth and velocity after the passage of the surge are 2.5 m and 5.0 m/s respectively, estimate the height of the surge.

(*Ans.* Height of surge = 2.31 m)

14.7 For a positive surge in a horizontal rectangular channel fill in the blanks in Table 14.1. The normal downstream direction is taken as positive.

Table 14.1 *Data on Positive Surge in Rectangular Channel*

Sl. No.	y_1 (m)	y_2 (m)	V_1 (m/s)	V_2 (m/s)	V_w (m/s)
(a)	1.2	3.6	0.8	...	...
(b)	...	4.0	...	5.3	9.2
(c)	...	3.4	...	–0.25	– 5.0
(d)	...	0.6	1.50	0	...
(e)	1.75	...	0.70	...	–5.50
(f)	...	1.80	2.80	...	–3.50

Answers to Question 14.7

Sl. No.	y_1 (m)	y_2 (m)	V_1 (m/s)	V_2 (m/s)	V_w (m/s)
(a)	1.2	3.6	0.8	6.403 –4.803	+9.204 –7.604
(b)	2.05	4.0	1.59	5.3	9.2
(c)	2.60	3.4	1.211	–0.25	–5.00
(d)	0.30	0.6	1.50	0	–1.50
(e)	1.75	4.388	0.70	–2.607	–5.50
(f)	0.515	1.8	2.8	–1.698	–3.50

Water Hammer

14.8 Calculate the velocity of propagation of pressure wave in the following cases of flow through pipes:

Case	Liquid	Density kg/m³	Bulk modulus of elasticity (K) MPa	Pipe dia mm	Pipe thickness mm	Material	Modulus of elasticity (E) MPa
(i)	Water	998	2190	300	8.0	Steel	2.1×10^5
(ii)	Water	998	2190	800	10.0	Cast iron	1.0×10^5
(iii)	water	998	2190	450	———	Rigid pipe	———
(iv)	Sea water	1025	2280	300	———	Rigid pipe	———
(v)	Mercury	13550	25500	10	———	Rigid pipe	———

(*Ans.* (i) 1256 m/s (ii) 893 m/s (iii) 1481.3 m/s (iv) 1491.4 m/s (v) 1371.8 m/s)

14.9 A valve at the downstream end of a 1.20 m diameter steel penstock carrying water is operated suddenly so as to reduce the flow from 3.0 m³/s to 0.5 m³/s. Estimate the maximum water hammer pressure rise at the valve if the pipe thickness is (i) 7.5 mm and (ii) 12 mm. [For water: K = 2200 MPa; for steel: E : 2.10×10^5 MPa].

(*Ans.* (i) Δp_h = 2.264 MPa; (ii) Δp_h = 2.589 MPa)

14.10 A brass pipe ($E = 0.8 \times 10^{11}$ Pa) is 5 cm in diameter and 600 m long. A liquid of density 800 kg/m³ and bulk modulus of elasticity 9.0×10^8 Pa is conveyed through this pipe at a rate of 450 L/min. The thickness of the pipe wall is 3 mm. Estimate the hoop stress in the pipe wall due to water hammer pressure when the flow is suddenly stopped by a valve at the downstream end of the pipe.

(*Ans.* σ = 24.77 MPa)

14.11 A valve is closed in 5.5 s at the end of a 4000 m long pipe carrying oil of density 917 kg/m³ at a velocity of 2.0 m/s. Assuming the pipe to be rigid, estimate the peak pressure developed by this closure and the length of the pipe subjected to this peak pressure at the end of the closure time. For the oil $K = 1.38 \times 10^9$ Pa.

(*Ans.* x_0 = 626.4 m; p_h = 2250 kPa)

14.12 A cast iron pipe 30 cm in diameter and 8 mm thick is 1500 m long. The pipe is to convey 200 L/s of water. (a) Estimate the maximum time of closure of a valve at the downstream end that would be reckoned as rapid closure. (b) What is the peak water hammer pressure produced by rapid closure. (c) What is the length of the pipe subjected to peak water hammer pressure at the and of the valve closure time if the time of closure is 2.0 s? [For Water : K = 2200 MPa; for cast iron: $E = 80 \times 10^9$ Pa]

(*Ans.* (a) 2.88 s, (b) p_h = 2.942 MPa, (c) x_0 = 458 m)

14.13 A 20 cm steel pipe is 1500 m long and conveys 50 L/s of water with a static head of 200 m at the downstream end of the pipe. If a valve at the downstream end is closed in 3 s, estimate the stress in the pipe wall at the valve. The pipe thickness is 6 mm.
[For water: $K = 2.20 \times 10^9$ Pa; for steel: $E = 2.11 \times 10^{11}$ Pa]

(*Ans.* $\sigma = 0.059$ kN/mm^2)

14.14 A 2300 m long pipeline leading from a large tank has a diameter of 15 cm and thickness of 2.8 mm. When a discharge of 2200 L/min of water was flowing the valve was suddenly closed completely. Sketch the variation of the water hammer pressure with time at (i) the valve end, (ii) a distance of 575 m from the valve and (iii) 57.5 m from the upstream tank.
[Take $K = 2.0 \times 10^9$ Pa for water and $E = 2.08 \times 10^{11}$ Pa for steel.]

(*Ans.* Fig. 14.18)

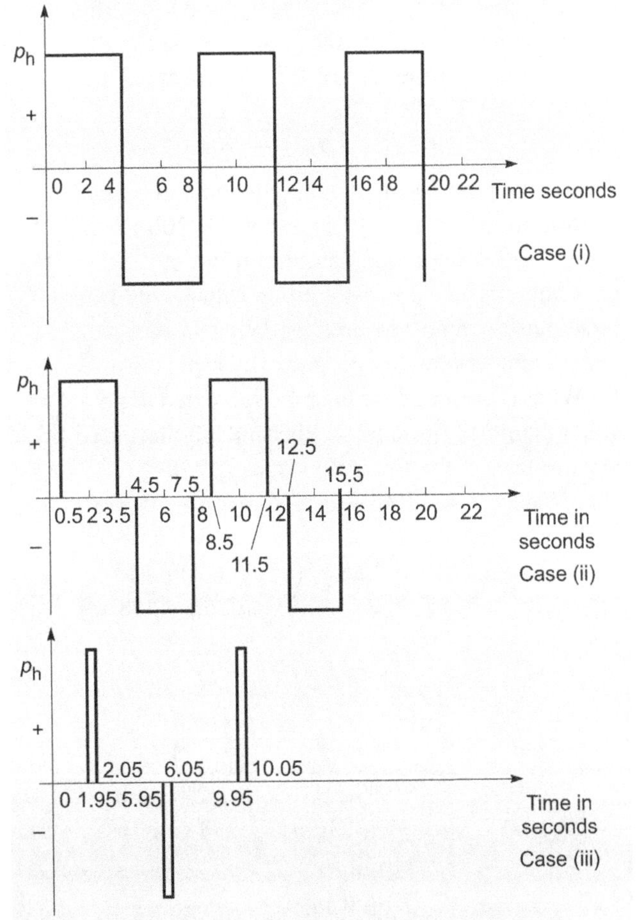

Fig. 14.18 *Answer to Problem 14.14*

14.15 A steel pipeline 0.50 m in diameter and 3.0 km long discharge 250 L/s of water freely at its lower end under a head of 160 m. The thickness of the pipewall is 6 mm. A valve at the downstream end is used to regulate the flow. If the working stress in the steel is 0.11 kN/mm^2, calculate the minimum time of complete closure of the valve.
[For water, $K = 2.2 \times 10^9$ Pa and for steel, $E = 2.1 \times 10^{11}$ Pa.]

(*Ans.* $T_{min} = 7.09$ s)

14.16 A steel pipe 45 cm in diameter conveys water at a velocity of 0.90 m/s. The pipe is 2000 m long and has a static head of 150 m at a valve provided at tits downstream end. The valve can be expected to be closed rapidly and the pipe thickness has to be designed to include the contingency. Estimate the minimum thickness of the pipewall to the nearest millimetre.
[For steel: $E = 2.21 \times 10^{11}$ Pa and safe working stress $\sigma_w = 0.12$ kN/mm^2. For water: $K = 2.05 \times 10^9$ Pa.]

(*Ans.* $t = 5$ mm)

14.17 A steel pipe 3000 m long has a diameter of 0.90 m. It is a carry 0.6 m^3/s of oil of density 800 kg/m^3 with a static pressure of 900 kPa at the outlet valve. A valve at the outlet is designed to close completely in 6 s. Estimate the minimum thickness of pipe, to the nearest mm, required at the valve.
[For steel $E = 2.05 \times 10^{11}$ Pa and working stress $\sigma_w = 0.125$ kN/mm^2; for oil $K = 1500$ MPa.]

(*Ans.* $t = 6$ mm)

Establishment of Flow

14.18 Two reservoirs with a constant difference of 10 m in their water surface elevation are connected by a 15 cm diameter pipe of length 400 m and $f = 0.025$. The minor loses in the pipe can be taken as 15 times the velocity head in the pipe. If a valve controlling the flow is suddenly opened, (a) estimate the time for 95% of ultimate flow to be established and (b) find the flow at the end of 10 s from the start of the valve operation.

(*Ans.* (a) 11.51 s
(b) 92% of ultimate flow)

14.19 A turbine is supplied with water through a 30 cm diameter pipe leading 1500 m from reservoir. The friction factor f for the pipe can be taken as 0.015 and all other losses can be neglected. If the rate of flow during normal operation is 140 L/s, (a) estimate the minimum time required for the turbine the reach 98% of its capacity from fully shut off condition. (b) What is the corresponding time if the pipe can be assumed to be frictionless?

(*Ans.* (a) 45.8 s (b) 3479 s)

14.20 A pipe 3000 m long and of 0.45 m diameter leads from a reservoir of water surface elevation of 120.00 m. The outlet of the valve is at an elevation of 100.00 m and discharges to atmosphere. If minor losses are 16 $V^2/2g$ where V = velocity in the pipe and f = 0.018, (a) estimate the time, after sudden opening of the outlet valve, for the flow to attain 90% of the ultimate value and (b) what is the velocity in the pipe after 20 s?

(*Ans.* (a) 38.1 s (b) V = 1.097 m/s)

14.21 A 7.5 m diameter pipeline is 350 m long and discharges water from a large tank through a 3.0 cm nozzle into atmosphere at the outlet. If a valve at the outlet is suddenly opened and if the ultimate velocity head at the nozzle is 12 m, (i) estimate the time required for the flow to attain 95% of its ultimate value. Assume f = 0.020 and minor losses = 15 $V^2/2g$ where V = velocity in the pipe. (ii) What would be the corresponding time if the friction and minor losses are neglected?

(*Ans.* (i) 4.82 s (ii) 522.3 s)

15

Compressible Flow

CONCEPT REVIEW

15.1 INTRODUCTION

In the previous chapters the fluid was considered incompressible and the mechanics of fluids under various situations was studied. However, the assumption of incompressibility is applicable when the compressibility effects are very small. It is well known that gases are highly compressible. If the Mach number represented by the ratio of flow velocity to sonic velocity ($M = V/C$) is small (say < 0.3) the flow of gases can be considered as incompressible. For other cases, the thermodynamic properties of the gas play an important role and the flow must be analysed by taking due recognition of these properties. In this chapter, certain basic thermodynamic relationships and basic compressibility flow situations are dealt with. It is emphasised that the subject of flow of compressible fluids is indeed a very vast subject and for detailed information on various aspects the reader should refer to a textbook on Fluid Mechanics or to a book on Gas Dynamics (See References at page 376).

The Mach number M is the most important parameter in compressible flow and based on its value the fluid flow is usually classified as follows for purposes of analysis in aerospace related problems:

$M < 0.3$	Incompressible	Density effects are negligible
$0.3 < M < 0.8$	Subsonic	Density effects are important. No shock wave.
$0.8 < M < 1.2$	Transonic flow	Shock waves divide flow into subsonic and supersonic regions.
$1.2 < M < 3.0$	Supersonic	Shock waves are present. No subsonic region.
$3.0 < M$	Hypersonic	Shock waves and other flow changes are very strong.

However, for purpose of illustrating the compressiblity effects, in this chapter only two classes viz. subsonic flow and supersonic flow are adopted.

15.2 THERMODYNAMIC PRINCIPLES

The basic thermodynamic principles and properties of gases used in the brief treatment of this chapter are as follows:

(1) The gas is assumed to be a *perfect gas*.

Equation of State: $$\frac{p}{\rho} = pv = RT \qquad (15.1)$$

where p = *absolute* pressure

ρ = density (mass per unit volume)

v = specific volume (= $1/\rho$ = volume per unit mass)

R = gas constant

T = *absolute* temperature in Kelvin.

[***Note:*** It is important to note that only absolute values are used in compressible fluid flow. Absolute values are measured above absolute zero. Gauge pressures and vacuum pressures must be first converted to absolute values. For temperatures t°C = 273 + t Kelvin = TK. It is a convention not to indicate degree (°) in Kelvin. It is simply T kelvin or TK. (like for example 293 K).]

The gas constant R is given by

$$\boxed{R = \frac{8314}{M_g}\ \text{J/(kg}\cdot\text{K)}} \qquad (15.2)$$

where M_g = molecular weight of gas. For air $M_g = 28.97$ and $R = 287$ J/(kg·K).

Further $\boxed{R = c_p - c_v}$ (15.3)

where c_p = specific heat at constant pressure
and c_v = specific heat at constant volume.
The ratio $c_p/c_v = k$ is an important thermodynamic property of a gas. For air and other diatomic gases $k = 1.4$.
From Eq. 15.3

$$\boxed{c_v = \frac{R}{k-1}} \qquad (15.4a)$$

and

$$\boxed{c_p = \frac{kR}{k-1}} \qquad (15.4b)$$

Note that the units of R, c_p and c_v are J/(kg·K).
Since 1 Joule (J) = N·m = (kg·m/s^2) (m) the unit J/(kg·K) = m^2/(s^2·K).
If R is to be expressed in heat units, the relevant conversion is 1 kcal = 4812 J.
Thus for air

$$R = 287\ \text{J/(kg}\cdot\text{K)} = \frac{287}{4812} = 0.0596\ \text{kcal/(kg}\cdot\text{K)}$$

Table 15.1 gives the values of M_g, ρ, R, c_v, c_p and k for some common gases.
(2) Another fundamental equation for a perfect gas is

$$\boxed{\frac{p}{\rho^k} = pv^k = \text{constant}} \qquad (15.4)$$

Thus by Eqs 15.4 and 15.1, between two sections 1 and 2,

$$\frac{p_1}{\rho_1^k} = \frac{p_2}{\rho_2^k} \quad \text{and} \quad \left(\frac{T_2}{T_1}\right) = \left(\frac{p_2}{p_1}\right)^{(k-1)/k} \qquad (15.5)$$

15.3 BASIC DEFINITIONS

Internal energy, u, is the energy of unit mass of fluid due to molecular activity. Change of internal energy

$$\boxed{u_2 - u_1 = c_v\,(T_2 - T_1)} \qquad (15.6)$$

Thus

$$\boxed{c_v = \left(\frac{\partial u}{\partial T}\right)_{v\,=\,\text{constant}}} \qquad (15.7)$$

Enthalpy, h

The energy possessed by a unit mass of a gas by virtue of its absolute temperature under which it exists is known as enthalpy. It is represented as a sum of pressure per unit mass (p/ρ) and internal energy per unit mass (u) as

$$\boxed{h = u + p/\rho} \qquad (15.8)$$

For a perfect gas the change in enthalpy

$$\Delta h = c_p \Delta T (15.9)$$

Thus

$$\boxed{c_p = \left(\frac{\partial h}{\partial T}\right)_{p\,=\,\text{constant}}} \qquad (15.10)$$

Entropy, s

Entropy is defined as a measure of the availability of energy for conversion into mechanical work. The entropy change ds for a perfect gas is

$$T\text{d}s = \text{d}h - \frac{\text{d}p}{\rho} \qquad (15.11)$$

Table 15.1 *Properties of Common Gases at 1 atm. and 20°C*

Gas	*Formula*	*Mol. Weight* M_g	ρ *kg/m*^{3}	*R* *J/(kg·K)*	c_p *J/(kg·K)*	c_V *J/(kg·K)*	*k*
Air		29	1.205	287	1003	716	1.40
Carbon dioxide	CO_2	44	1.84	188	858	670	1.28
Carbon monoxide	CO	28	1.16	297	1040	743	1.40
Oxygen	O_2	32	1.33	260	909	649	1.40
Nitrogen	N_2	28	1.16	297	1040	743	1.40
Chlorine	Cl_2	71	2.946	117	461	344	1.34
Methane	CH_4	16	0.668	520	2250	1730	1.30
Helium	He	4	0.166	2077	5220	3143	1.66
Hydrogen	H_2	2	0.0839	4120	14450	10330	1.40

Putting $dh = c_p\, dT$ and $\rho T = p/R$

$$\int_1^2 ds = \int_1^2 c_p \frac{dT}{T} - R\int_1^2 \frac{dp}{p}$$

$$s_2 - s_1 = c_p \ln\frac{T_2}{T_1} - R\ln\frac{p_2}{p_1} \qquad (15.12)$$

A flow in which entropy does not change is known as *isentropic flow*, i.e. $s_2 - s_1 = 0$

For an isentropic flow

$$\ln\left(\frac{T_2}{T_1}\right) = \frac{R}{c_p}\ln\frac{p_2}{p_1}$$

$$\frac{p_2}{p_1} = \left(\frac{T_2}{T_1}\right)^{k/(k-1)} = \left(\frac{\rho_2}{\rho_1}\right)^k \qquad (15.13)$$

Isothermal process is a process in which the temperature is constant

$$\frac{p}{\rho} = \text{constant} \qquad (15.14)$$

Adiabatic process is a process in which there is no heat transfer to or from a gas. A *frictionless adiabatic process* is *isentropic*, and for such a process [from Eq. (15.13)],

$$\frac{p}{\rho^k} = \text{constant} \qquad (15.15)$$

15.4 BASIC EQUATIONS FOR COMPRESSIBLE FLUID FLOW

(i) *Continuity*: Mass rate of flow

$$\dot{m} = \rho A V = \text{constant} \qquad (15.16)$$

In differential form $$\frac{dA}{A} + \frac{d\rho}{\rho} + \frac{dV}{V} = 0 \qquad (15.16a)$$

(ii) *Energy equation for isentropic flow*: If *no heat* is added and no mechanical work is done,

$$\frac{p}{\rho} + u + \frac{V^2}{2} + gZ = \text{constant} \qquad (15.17)$$

In gas flow, the Z terms are negligible and hence,

$$h + \frac{V^2}{2} = \text{constant} \qquad (15.18)$$

i.e. $$c_p T + \frac{V^2}{2} = \text{constant} \qquad (15.19)$$

i.e. $$\left(\frac{k}{k-1}\right)\frac{p}{\rho} + \frac{V^2}{2} = \text{constant} \qquad (15.20)$$

15.5 APPLICATION OF ENERGY EQUATION

It a gas is expanded to perform work, then per unit mass, by energy equation

Heat absorbed by gas = Work done by gas + Increase in internal energy.

$$q = w + \Delta e \qquad (15.21)$$

Isothermal process If the process is from state 1 to state 2 then in isothermal conditions, $T_1 = T_2$. Hence there will be no change in internal energy, $\Delta e = 0$. Thus

Work done by gas per unit mass, w = Heat absorbed by gas unit mass, q

Also the work done/kg by gas = $\int_1^2 p\,dv$

$$w = (p_1/\rho_1)\cdot \ln\frac{\rho_1}{\rho_2} = RT_1 \ln\left(\frac{\rho_1}{\rho_2}\right) \qquad (15.22)$$

In expansion, $\rho_2 < \rho_1$ and work done by gas is positive. If w is negative, as in compression, it means that work is done on the gas and heat is rejected.

Isentropic process: In an isentropic process q = heat absorbed by unit mass of gas = 0.

Hence,

$$w = -\Delta u = (e_1 - e_2)$$

showing that the work done in expansion is at the expense of the internal energy of the gas.

Work done/kg by gas = $\int_1^2 p\,dv$

$$w = \frac{(p_1/\rho_1) - (p_2/\rho_2)}{(k-1)} = \frac{R(T_1 - T_2)}{(k-1)}$$

$$= c_v (T_1 - T_2) \qquad (15.23)$$

Note that in isentropic expansion there will be a decrease in internal energy and in isentropic compression there will be an increase in internal energy.

15.6 SONIC VELOCITY

The sonic velocity, i.e. the velocity of sound is the speed of propagation of a pressure wave in the medium and is given for an isentropic flow as

$$C = \sqrt{kp/\rho} = \sqrt{kRT} \qquad (15.24)$$

The ratio of the velocity of flow to the sonic speed is known as Mach number M.

Thus $$M = V/C \quad (15.25)$$

In supersonic flow, if a point source of disturbance is present, it creates a conical space in which all the disturbances are piled up. The boundary of the cone is a *shock* where there is a sudden change in the fluid properties like density and pressure. Figure 15.1 is a schematic diagram of a supersonic flow moving past a disturbance.

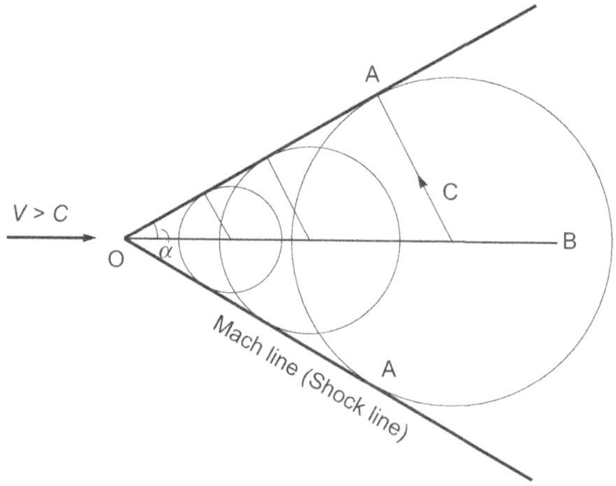

Fig. 15.1

OAA′ is the trace of a cone of disturbance, and this conical zone is known as *Mach cone*. The semi vertex angle AOB = α is the Mach angle. The Mach angle α is given by

$$\sin\alpha = \frac{1}{M} \quad (15.26)$$

15.6.1 Stagnation Values

When the flow velocity $V = 0$, the corresponding values of other parameters, viz. pressure, temperature and density are known as *stagnation values* and are designated with suffix 0. Thus stagnation pressure = p_0, stagnation temperature = T_0 and stagnation density is ρ_0. Stagnation values are important reference parameters in gas dynamics.

The energy equation [Eq. 15.19] becomes

$$c_p T_1 + \frac{V_1^2}{2} = c_p T_0 \quad (\text{as } V_0 = 0)$$

and hence $$V_1^2 = 2c_p (T_0 - T_1) \quad (15.27)$$

The ratios of pressure p_1, density ρ_1, and temperature T_1 with their corresponding stagnation values p_0, ρ_0 and T_0 are expressed in terms of Mach number M_1 (= V_1/C) and k as follows:

$$\frac{T_0}{T_1} = \left(1 + \frac{k-1}{2} M_1^2\right) \quad (15.28)$$

$$\frac{p_0}{p_1} = \left(1 + \frac{k-1}{2} M_1^2\right)^{k/(k-1)} = \left(\frac{T_0}{T_1}\right)^{k/(k-1)} \quad (15.29)$$

$$\frac{\rho_0}{\rho_1} = \left(1 + \frac{k-1}{2} M_1^2\right)^{1/(k-1)} = \left(\frac{T_0}{T_1}\right)^{1/(k-1)} \quad (15.30)$$

15.6.2 Effect of Area Variation

For Steady, one-dimensional, isentropic flow the Euler equation is

$$V\mathrm{d}V + \frac{\mathrm{d}p}{\rho} = 0 \quad (15.31)$$

Combining with the continuity equation Eq. 15.16 (a)]

$$\frac{\mathrm{d}A}{A} + \frac{\mathrm{d}\rho}{\rho} + \frac{\mathrm{d}V}{V} = 0$$

and noting that $C^2 = \mathrm{d}p/\mathrm{d}\rho$ and $M = V/C$

the following important relation between variation of area and Mach number is obtained.

$$\frac{\mathrm{d}V}{V} = \frac{\mathrm{d}A}{A}\left(\frac{1}{M^2 - 1}\right) = -\frac{\mathrm{d}p}{\rho V^2} \quad (15.32)$$

An analysis of this equation indicates the qualitative relationship between M, duct geometry and other flow parameters as in Table 15.2.

15.7 FLOW IN A NOZZLE DISCHARGING FROM A TANK

Consider a tank containing a gas at pressure p_0, temperature T_0 and density ρ_0 discharging through a converging nozzle. Since the velocity in the tank is zero for all practical purposes, the pressure p_0 temperature T_0 and density ρ_0 are all stagnation values.

Let the pressure outside the nozzle (ambient pressure) be p_2 (Fig. 15.3).

Let p_1, V_1, ρ_1 and T_1 are the values of pressure, velocity, density and temperature at the throat (exit) of the nozzle. These values, which depend on the nature of flow are controlled by a critical pressure ratio p_1^*/p_0 given by the following relationship.

$$\frac{p_1^*}{p_0} = \left[\frac{2}{(k+1)}\right]^{k/(k-1)} = 0.528 \quad \text{when } k = 1.4 \quad (15.33)$$

Table 15.2 *Compressible Flow in Converging and Diverging Ducts*

Geometry	*Subsonic* *M<1*	*Supersonic* *M>1*
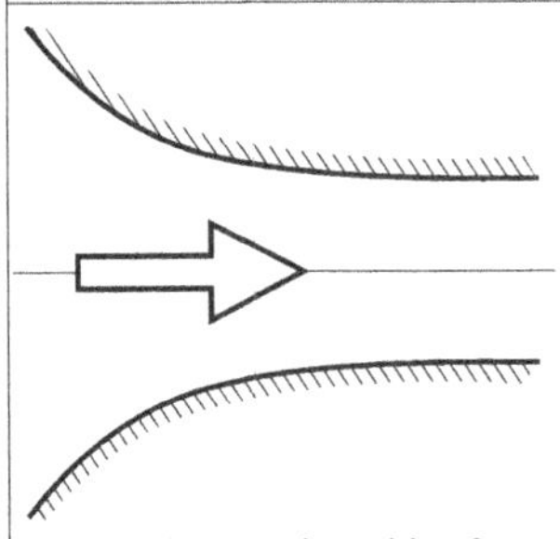 Converging, $dA < 0$ **Fig. 15.2(a)** *Coverging Duct*	$dV > 0$, $dp < 0$ Along the flow: • Velocity increases • Pressure decreases • Density decreases • Temperature decreases • Boundary is a *subsonic nozzle.*	$dV < 0$, $dp > 0$ Along the flow: • Velocity decreases • Pressure increases • Density increases • Temperature increases • Boundary is a *supersonic diffuser.*
	Subsonic *M<1*	*Supersonic* *M>1*
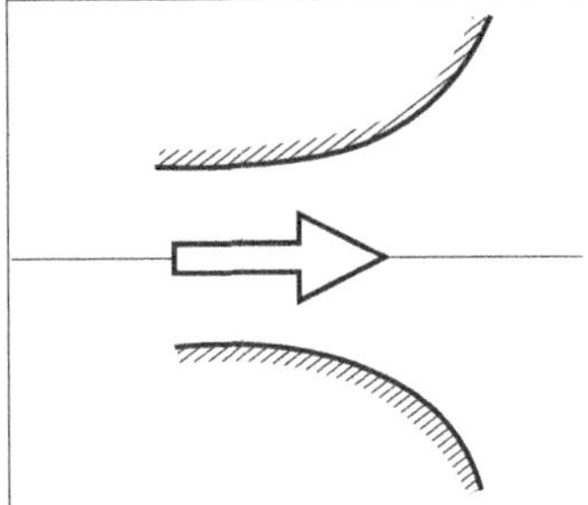 Diverging, $dA < 0$ **Fig. 15.2(b)** *Diverging Duct*	$dV > 0$, and $dp < 0$ Along the flow: • Velocity decreases • Pressure increases • Density increases • Temperature increases • Boundary is a *subsonic diffuser*	$dV < 0$, $dp > 0$ Along the flow: • Velocity increases • Pressure decreases • Density decreases • Temperature decreases • Boundary is a *supersonic nozzle.*

(Note the opposing behaviour of subsonic and supersonic flows in a given geometry.)

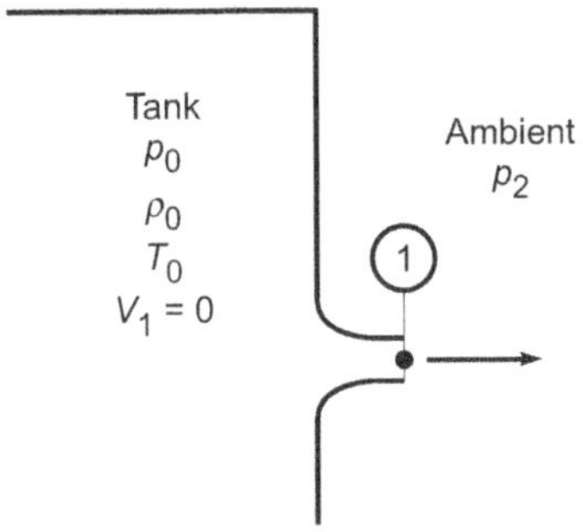

Fig. 15.3 *Nozzle Discharging from a Tank*

This critical pressure p_1^* corresponds to occurrence of sonic flow at the throat. Two types of flow are possible.

(1) When $p_2 > p_1^*$: Subsonic flow prevails through the nozzle and the nozzle exit pressure $p_1 = p_2$. The relationship of the various flow parameters will be as follows:

When $p_2 > p_1^*$, $\quad p_1 = p_2$

and
$$\frac{p_0}{p_1} = \left(1 + \frac{k-1}{2} M_1^2\right)^{1/(k-1)} \tag{15.34}$$

$$\frac{T_0}{T_1} = \left(1 + \frac{k-1}{2} M_1^2\right) \tag{15.35}$$

$$\frac{\rho_0}{\rho_1} = \left(1 + \frac{k-1}{2} M_1^2\right)^{k/(k-1)} \tag{15.36}$$

The exit velocity

$$V_1 = \sqrt{2\left(\frac{k}{k-1}\right)\frac{p_0}{\rho_0}\left[1 - \left(\frac{p_1}{p_0}\right)^{(k-1)/k}\right]} \tag{15.37}$$

Mass rate of flow

$$\dot{m} = A_1 \rho_1 V_1 \tag{15.38}$$

$$\boxed{\dot{m} = A_1 \rho_1 \sqrt{2\left(\frac{k}{k-1}\right)\frac{p_0}{\rho_0}\left[1 - \left(\frac{p_1}{p_0}\right)^{(k-1)/k}\right]}} \tag{15.38a}$$

(2) When $p_1^* \geq p_2$: Sonic conditions ($M_1 = 1.0$) prevail at the nozzle exit and $p_1 = p_1^*$. The mass rate of flow

$$\dot{m}^* = A_1 \rho_1^* V^* \quad (15.39)$$

where the parameters with superscript* represent critical condition. This mass rate $\dot{m}^*$ is the maximum mass rate of flow that the nozzle can discharge under the given upstream conditions. The values critical flow parameters are

$$\frac{p_1^*}{p} = \left(\frac{2}{k+1}\right)^{k/(k-1)} = 0.528 \quad \text{for } k = 1.4 \quad (15.40)$$

$$\frac{T_1^*}{T_0} = \left(\frac{2}{k+1}\right) = 0.833 \quad \text{for } k = 1.4 \quad (15.41)$$

$$\frac{\rho_1^*}{\rho_0} = \left(\frac{2}{k+1}\right)^{1/(k-1)} = 0.634 \quad \text{for } k = 1.4 \quad (15.42)$$

$$V_1^* = C_1^* = \sqrt{kRT_1^*} \quad (15.43)$$

Thus for air, when $\dfrac{p_1}{p_2} < 0.528$ critical flow prevails in the nozzle and the maximum mass rate will pass through the nozzle. Also

$$\boxed{\dot{m} = A_2 \rho_1^* \sqrt{kRT^*}} \quad (15.44)$$

The condition of occurrence of critical flow at the nozzle exit is known as *choking* condition.

15.8 NORMAL SHOCK WAVE

In a supersonic stream, under certain conditions, a shock wave normal to the flow direction may occur and the flow may change into subsonic state after the shock. Such a shock wave is known as *normal shock* wave. This shock wave will have very little thickness and is analogous to the hydraulic jump that takes place in an open channel flow. Considering a section 1 on the upstream and a section 2 on the downstream of the shock and noting that $A_1 = A_2 = A$ (as the thickness of the shock is very small) the controlling equations for adiabatic flow are

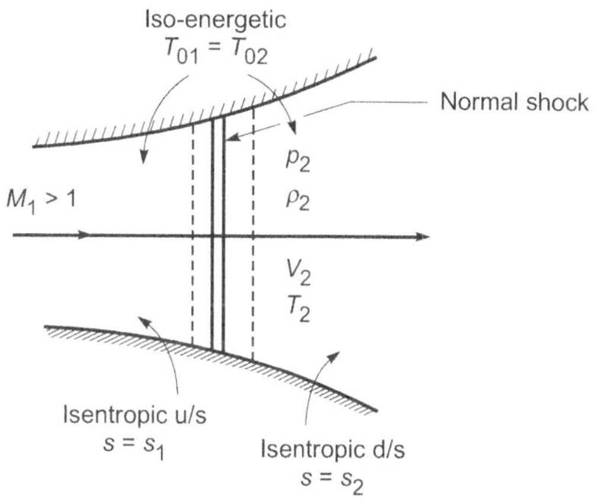

Fig. 15.4 *Normal Shock Wave*

Continuity: $\dot{m} = \rho_1 V_1 = \rho_2 V_2$

Momentum: $p_2 - p_1 = \rho_1 V_1^2 - \rho_2 V_2^2$

Energy: $\dfrac{V_1^2}{2} + h_1 = \dfrac{V_2^2}{2} + h_2 = h_0 = \text{constant}$

$= c_p T_0$

(by considering no heat transfer and no work done)

Perfect gas relationship: $\dfrac{p_1}{\rho_1 T_1} = \dfrac{p_2}{\rho_2 T_2}$

It is found that the entropy changes across the shock and the change is given by

$$\boxed{s_2 - s_1 = c_v \ln\left[\left(\frac{p_2}{p_1}\right)\left(\frac{\rho_1}{\rho_2}\right)^k\right]} \quad (15.45)$$

The relationships between the various fluid and flow parameters on either side of the shock are obtained as

(a) Mach number relation,

$$\boxed{M_2^2 = \frac{2 + (k-1) M_1^2}{2kM_1^2 - (k-1)}} \quad (15.46)$$

and

$$\boxed{M_1^2 = \frac{2 + (k-1) M_2^2}{2kM_2^2 - (k-1)}} \quad (15.47)$$

(b) Pressure ratio,

$$\boxed{\frac{p_2}{p_1} = \frac{1 + kM_1^2}{1 + kM_2^2} = \frac{2kM_1^2 - (k-1)}{(k+1)}} \quad (15.48)$$

(c) Density ratio,

$$\boxed{\frac{\rho_2}{\rho_1} = \frac{M_1^2 (k+1)}{2 + M_1^2 (k-1)}} \quad (15.49)$$

(d) Velocity ratio,

$$\boxed{\frac{V_2}{V_1} = \frac{1}{(\rho_2/\rho_1)} = \frac{2 + M_1^2 (k-1)}{M_1^2 (k+1)}} \quad (15.50)$$

(e) Temperature: Stagnation temperature is constant across the shock. Thus,

$$\boxed{T_{01} = T_{02}} \quad (15.51)$$

(f) Stagnation pressure ratio and density ratio,

$$\frac{p_{02}}{p_{01}} = \frac{\rho_{02}}{\rho_{01}} = \left[\frac{(k+1)\,M_1^2}{2+(k-1)\,M_1^2}\right]^{k/(k-1)} \times \left[\frac{k+1}{2kM_1^2-(k-1)}\right]^{1/(k-1)} \qquad (15.52)$$

Note the following characteristics of a normal shock:

- The upstream flow is supersonic and the downstream flow is subsonic.
- The stagnation temperature remains the same across the shock and hence all over the flow.
- Stagnation pressure and stagnation density decrease with M_1 in the same ratio, viz.

$$\frac{p_{02}}{p_{01}} = \frac{\rho_{02}}{\rho_{01}}$$

- Entropy increases across a shock with consequent decrease in stagnation pressure and stagnation density across the shock.

REFERENCES

15.1 A.Y. Pope, *Aerodynamics of Subsonic Flight*, 2nd ed., Pitman, New York, 1958.
15.2 S. M. Yahya, *Fundamentals of Compressible Flow*, Wiley Eastern, New Delhi, 1982.
15.3 A.H. Shapiro, *The Dynamics and Thermodynamics of Compressible Fluid Flow*, Vols. I & II, Ronald Press, New York, 1954.
15.4 F. M., White, *Fluid Mechanics*, 3rd ed., McGraw-Hill, Singapore, 1994.

WORKED EXAMPLES

A. Thermodynamic Properties

15.1 Calculate the value of the gas constant R for carbon dioxide and oxygen.

Solution: Gas constant $R = \dfrac{8314}{\text{Molecular weight of gas}}$

$$= \frac{8314}{M_g}$$

For carbon dioxide, $M_g = 44$

Hence $R = \dfrac{8314}{44} = 189$ N·m/kg·K)

For oxygen, $M_g = 32$

Hence $R = \dfrac{8314}{32}$

$$= \mathbf{260\ N{\cdot}m/(kg{\cdot}K)}$$

15.2 If the molecular weight of air is 28.96 and $k = 1.4$, calculate c_p and c_v for air in J/(kg·K) and kcal/kg·K) units.

Solution: Gas constant

$$R = \frac{8314}{M_g} = \frac{8314}{28.96}$$

$$= 287 \text{ J/(kg · K)}$$

$$c_p = \frac{k}{k-1}\,R = \frac{1.4}{(1.4-1.0)} \times 287$$

$$= \mathbf{1004.5\ J/(kg \cdot K)}$$

$$c_v = \frac{R}{k-1} = \frac{287}{(1.4-1.0)} = \mathbf{717.5\ J/\,(kg \cdot K)}$$

In heat units,

1 kcal = 4187 J

$$\therefore \quad c_p = \frac{1004.5}{4187} = \mathbf{0.240\ Kcal/(kg \cdot K)}$$

$$c_v = \frac{717.5}{4187} = \mathbf{0.171\ kcal/(kg \cdot K)}$$

15.3 Compute the change in internal energy and enthalpy of 1000 N of carbon dioxide if its temperature is increased from 15°C to 65°C. The initial pressure is 100 kPa (abs). Take $c_p = 858$ J/(kg·K) and $c_v = 670$ J/(kg·K)

Solution: m = Mass of 1000 N of CO_2 $= \dfrac{1000}{9.81}$

$$= 101.94 \text{ kg}$$

$$T_1 = 273 + 15 = 288 \text{ K}$$

$$T_2 = 273 + 65 = 338 \text{ K}$$

$$(T_2 - T_1) = (338 - 288) = 50 \text{ K}$$

(1) Change in internal energy per unit mass,

$$\Delta u = c_v\,(T_2 - T_1)$$

$$= 670 \times 50 = 33{,}500 \text{ J/kg}$$

Total change in internal energy = $m\Delta u$

$$= 101.91 \times 33,\ 500 = 3414{,}880 \text{ J}$$

$$= \mathbf{3415\ kJ}$$

(2) Change in enthalpy per unit mass,

$$\Delta h = c_p(T_2 - T_1)$$
$$= 858 \times 50 = 42{,}900 \text{ J/kg}$$

Total change in enthalpy $= m\Delta h$

$$= 101.91 \times 42{,}900$$
$$= 4373{,}226 \text{ J}$$
$$= \mathbf{4373 \text{ kJ}}$$

15.4 Oxygen of mass 5 kg at 25°C is compressed isentropically to 50% its original volume. Find the final temperature, pressure, the work required and changes in internal energy and enthalpy. Take initial pressure as 100 kPa (abs); $c_v = 649$ J(kg·K) and $c_p = 909$ J/(kg·K).

Solution:

$$k = \frac{c_p}{c_v} = \frac{909}{649} = 1.4$$

m = mass of gas = 5.0 kg

For an isentropic process,

$$pv^k = \text{constant}$$

For a perfect gas,

$$pv = RT$$

Hence $pv^k = (pv)\,(v^{k-1}) = (RT)\,v^{k-1}$

Since R = constant, $T\,v^{k-1}$ = constant

$$T_1 v_1^{k-1} = T_2 v_2^{k-1}$$
$$T_1 = 25 + 273 = 298 \text{ K}$$
$$v_1/v_2 = 1.0/(0.5) = 2.0$$

(1)
$$T_2 = T_1\left(\frac{v_1}{v_2}\right)^{k-1} = 298 \times (2)^{(1.4-1.0)}$$
$$= 393.2 \text{ K} = \mathbf{120.2°C}$$

(2) Since
$$\frac{p_1 v_1}{T_1} = \frac{p_2 v_2}{T_2}$$
$$p_2 = p_1\,(v_1/v_2)\,(T_2/T_1)$$
$$= 100\,(2.0)\left(\frac{393.2}{298}\right)$$
$$= \mathbf{263.9 \text{ kPa (abs)}}$$

(3) Change in internal energy per unit mass,

$$\Delta u = c_v\,(T_2 - T_1) = 649\,(393.2 - 298)$$
$$= \mathbf{61784.8 \text{ J/kg}}$$

Total change of internal energy = Work required

$$= m\Delta u = 5.0 \times 61784.8 \times \frac{1}{1000} \text{ kJ}$$

$= \mathbf{308.9 \text{ kJ}}$ = Work done on the gas

(4) Change in enthalpy per kg:

$$\Delta h = c_p\,(T_2 - T_1) = 909 \times (393.2 - 298)$$
$$= \mathbf{86537 \text{ J/kg}}$$

Total change of enthalpy $= m\Delta h$

$$= 5.0 \times 86537 \times \frac{1}{1000} \text{ kJ}$$
$$= \mathbf{432.7 \text{ kJ}}$$

15.5 Determine the change in entropy for 7 kg of oxygen when initial pressure and temperature are 150 kPa (abs) and 27°C respectively and final pressure and temperature are 450 kPa (abs) and 85°C respectively. Take R = 260 J/(kg·K) and k = 1.40.

Solution:

$$c_v = \frac{R}{k-1} = \frac{260}{(1.4 - 1.0)} = 650 \text{ J/(kg · K)}$$

$T_1 = 27 + 273 = 300$ K and $T_2 = 85 + 273 = 358$ K

$p_1 = 150$ kPa (abs) and $p_2 = 450$ kPa (abs)

Change in entropy per kg:

$$s_2 - s_1 = c_v \ln\left[\left(\frac{T_2}{T_1}\right)^k \left(\frac{p_1}{p_2}\right)^{k-1}\right]$$
$$= 650 \ln\left[\left(\frac{358}{300}\right)^{1.4}\left(\frac{150}{450}\right)^{(1.4-1.0)}\right]$$
$$= -124.8 \text{ J/K per kg}$$

Total change in entropy

$$= m(s_2 - s_1) = 7 \times (-124.8) = \mathbf{-873.6 \text{ J/K}}$$

15.6 Air of mass 3 kg expands from an initial state $p_1 = 1000$ kPa (abs) and $T_1 = 250$°C to a final state of 200 kPa (abs). Find the initial volume, final volume, temperature and the work done in (i) isothermal and (ii) isentropic expansions. Take $k = 1.4$ and R = 287 J/(kg·K).

Solution: $T_1 = 273 + 250 = 523 \text{ K} = T_2$

Initial density
$$\rho_1 = \frac{p_1}{RT} = \frac{1000 \times 1000}{287 \times 523}$$
$$= 6.662 \text{ kg/m}^3$$

Initial volume
$$V_1 = m/\rho = \frac{3.0}{6.662} = \mathbf{0.450 \text{ m}^3}$$

(i) In an isothermal process

$$\frac{p_1}{\rho_1} = \frac{p_2}{\rho_2} \quad \text{as} \quad T_1 = T_2$$
$$\rho_2 = \frac{p_2}{p_1}\rho_1 = \frac{200}{1000} \times 6.662$$
$$= 1.3324 \text{ kg/m}^3$$

Final volume

$$V_2 = \frac{3.0}{1.3324} = \mathbf{2.252 \text{ m}^3}$$

Work done $W = m\,RT_1 \ln \dfrac{\rho_1}{\rho_2}$

$$= 3.0 \times 287 \times 523 \ln \frac{6.662}{1.3324}$$

$$= 724700 \text{ J} = \mathbf{724.7 \ kN \cdot m}$$

Final temperature

$$T_2 = 532 \text{ K} = \mathbf{250°C}$$

(ii) In isentropic expansion

$$\frac{p_1}{\rho_1^k} = \frac{p_2}{\rho_2^k}$$

or $$\rho_2 = \left(\frac{p_2}{p_1}\right)^{1/k} \rho_1 = \left(\frac{200}{1000}\right)^{1/1.4} \times 6.662$$

$$= 2.1103 \text{ kg/m}^3$$

Final volume

$$V_2 = \frac{3.0}{2.1103} = \mathbf{1.42 \ m^3}$$

Final temperature

$$T_2 = \left(\frac{p_2}{p_1}\right)^{(k-1)/k} T_1$$

$$= \left(\frac{200}{1000}\right)^{0.4/1.4} \times 523 = 330.2 \text{ K}$$

$$= \mathbf{57.2°C}$$

Work done per unit mass

$$= c_p (T_1 - T_2)$$

$$= \frac{R}{k-1}(T_1 - T_2)$$

$$= \frac{287}{0.4}(523 - 330.2) = 138334 \text{ J/kg}$$

W = Total work done = $m\,c_p(T_1 - T_2)$

$$= 3.0 \times 138334 = 415002 \text{ J}$$

$$= 415 \text{ kJ} = \mathbf{415 \ kN \cdot m}$$

Speed of Sound

15.7 Find the speed of sound in oxygen at a pressure of 100 kPa (abs) and 25°C. Take R = 260 J/kg·K) and k = 1.40.

Solution: $C = \sqrt{kRT}$

$k = 1.40$, $R = 260$ J/kg·K

and $T = 273 + 25 = 298$ K

$$C = \sqrt{1.4 \times 260 \times 298} = \mathbf{329.4 \ m/s}$$

15.8 An aeroplane is to move at Mach number of 1.5 at altitudes of 1000 m and 10,000 m. The atmospheric pressure and densities at these elevation are:

Elevation (m)	*Pressure [kPa (abs)]*	*Density (kg/m³)*
1000	89.89	1.112
10,000	26.42	0.4125

Calculate the speed of the plane, in km/h, at these altitudes. Assume k = 1.4.

Solution:

(1) At 1000 m altitude,

$$C = \sqrt{kp/\rho} = \sqrt{\frac{1.4 \times 89890}{1.112}} = 336.4 \text{ m/s}$$

Mach number

$$M = \frac{V}{C}$$

$$V = C\,M = 336.4 \times 1.5 = 504.6 \text{ m/s}$$

$$= \frac{504.6 \times 60 \times 60}{1000} = \mathbf{1817 \ km/h}$$

(2) At 10,000 m altitude

$$C = \sqrt{kp/\rho} = \sqrt{(1.4 \times 26420)/0.4125}$$

$$= 299.5 \text{ m/s}$$

$$V = C\,M = 299.5 \times 1.5 = 449.2 \text{ m/s}$$

$$= \frac{449.2 \times 60 \times 60}{1000} = \mathbf{1617 \ km/h}$$

[***Note:*** If the value of k is not specified it is usual to assume k = 1.4.]

15.9 A supersonic plane moves with a Mach number of 2.2 at an elevation of 1000 m above a certain point F. It the atmospheric temperature is 22°C, what is the time lapse at which the sonic boom reaches an observer at F after the plane is directly overhead ? How far will the plane be form F at that instant of the boom reaching F? (R = 287 J/kg·K) and k = 1.4).

Solution: $C = \sqrt{kRT} = \sqrt{1.4 \times 287 \times (273 + 22)}$

$$= 344.3 \text{ m/s}$$

$$V = \text{Velocity of plane} = CM = 344.3 \times 2.2$$

$$= 757.4 \text{ m/s}$$

Considering the Fig. 15.5, when the plane is at B the sonic boom reaches F. If angle ABF = α then

$$\sin \alpha = \frac{1}{M} = \frac{1}{2.2}$$

$$\alpha = 27.036°$$

AB = Distance traversed by plane = $V\,t$

$$AB = \frac{AF}{\tan \alpha} = \frac{1000}{\tan(27.036°)}$$

$$= \mathbf{1959.6 \ m}$$

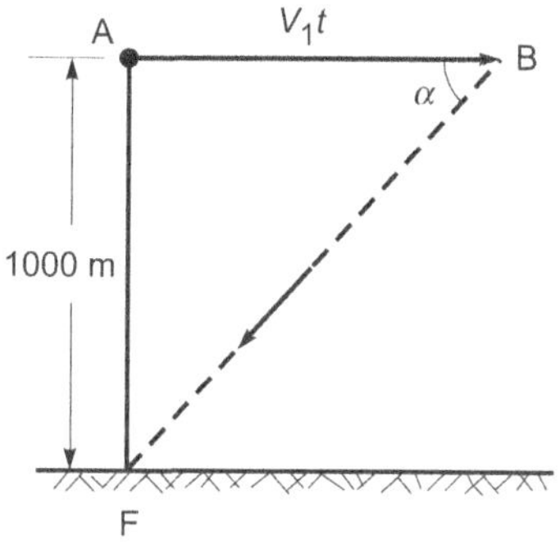

Fig. 15.5

t = time elapsed in the boom reaching the

$$\text{point } F = \frac{AB}{V} = \frac{1959.6}{757.4} = \mathbf{2.59\ s}$$

15.10 A supersonic fighter plane flies at an altitude of 3000 m. An observer on the ground hears the sonic boom 7.5 s after the passing of the plane over his head. Estimate the speed of the plane in km/h and the Mach number. (Assume R = 287 J/(kg·K) and the average temperature to be 11°C).

Solution: As the value of k is not given, k = 1.4 for air is assumed. Sonic velocity

$$C = \sqrt{kRT} = \sqrt{1.4 \times 287 \times (273 + 11)} = 337.8 \text{ m/s}$$

In Fig 15.6, B is the location of the plane when the sonic boom is heard at 0.

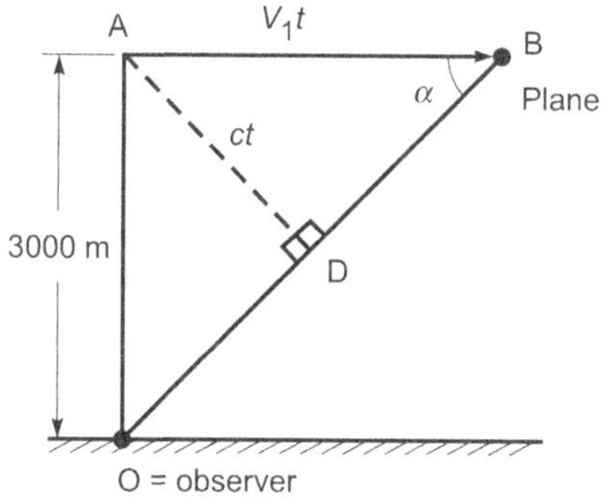

Fig. 15.6

$$A\hat{B}O = \alpha \quad \text{where} \quad \sin\alpha = \frac{1}{M}$$

If a perpendicular AD is drawn at A, then AD = Ct.

In triangle AOD, AOD = $(90 - \alpha)$

$$\therefore \quad \frac{AD}{AO} = \frac{Ct}{(AO)} = \sin(90 - \alpha)$$

$$= \cos\alpha = (1 - \sin^2\alpha)^{1/2}$$

$$= \left(1 - \frac{1}{M^2}\right)^{1/2}$$

$$\frac{337.8 \times 7.5}{3000} = \left(1 - \frac{1}{M^2}\right)^{1/2}$$

$$1 - \frac{1}{M^2} = (0.8445)^2$$

$$M = \mathbf{1.867}$$

$$\text{Mach number} \quad M = \frac{V}{C}$$

$$1.867 = \frac{V}{337.8}$$

$$\text{Speed of plane} \quad V = 1.867 \times 337.8 = 630.7 \text{ m/s}$$

$$= 630.7 \times \frac{60 \times 60}{1000}$$

$$= \mathbf{2271\ km/h}$$

B. Stagnation Properties

15.11 An air plane is moving in an atmosphere with pressure p = 44 kPa (abs) and density ρ = 0.63 kg/m³. A pitot tube on the plane records the stagnation pressure as 70 kPa (abs). Estimate the speed of the airplane and the stagnation temperature. (k = 1.4 and R = 287 J/(kg·K).

Solution:

(1) Stagnation pressure p_0 is given by

$$\frac{p_0}{p} = \left[1 + \frac{k-1}{2} M_1^2\right]^{k/(k-1)}$$

$$\frac{70}{40} = \left[1 + \frac{(1.4 - 1.0)}{2} M_1^2\right]^{1.4/(1.4-1)}$$

$$1.5909 = [1 + 0.2\, M_1^2]^{3.5}$$

$$M_1^2 = 0.7093 \quad \text{and} \quad M_1 = 0.8422$$

Sonic speed

$$C = \sqrt{kp/\rho} = \sqrt{(1.4 \times 44000)/0.63} = 312.7 \text{ m/s}$$

Mach number

$$M_1 = V_1/C$$

Velocity of plane

$$V_1 = C\,M_1 = 312.7 \times 0.8422 = \mathbf{263.4\ m/s}$$

(2) Temperature of the atmosphere,

$$T_1 = p/\rho R$$

$$T_1 = \left(\frac{44000}{0.63 \times 287}\right) = 243.35 \text{ K}$$

Stagnation temperature T_0 is given by

$$\frac{T_0}{T_1} = \left[1 + \frac{k-1}{2} M_1^2\right]$$

$$\frac{T_0}{243.35} = \left[1 + \left(\frac{0.4}{2}\right)(0.8422)^2\right] = 1.1419$$

$$T_0 = 277.87 \text{ K} = \mathbf{4.87°C}$$

15.12 A supersonic plane flies at 1900 km/h in air having a pressure of 28.5 kPa (abs) and density of 0.439 kg/m^3. Calculate the (1) temperature, (2) pressure and (3) density of air at the stagnation point on the nose of the plane. (Take $k = 1.4$ and $R = 287$ J/(kg·K)).

Solution:

Sonic speed $C = \sqrt{kp/\rho} = \sqrt{\dfrac{1.4 \times 28500}{0.439}}$

$= 301.5$ m/s

Speed of plane $V = \dfrac{1900 \times 1000}{3600} = 527.8$ m/s

Mach number $M = V/C = \dfrac{527.8}{301.5} = 1.7505$

For the atmosphere:

$$p_1 = 28500 \text{ Pa (abs)}$$

$$\rho_1 = 0.439 \text{ kg/m}^3$$

Temperature $T_1 = p_1/\rho_1 R = \dfrac{28500}{0.439 \times 287} = 226.2$ K

(1) Stagnation temperature T_0

$$\frac{T_0}{T_1} = \left[1 + \frac{k-1}{2} M_1^2\right]$$

$$\frac{T_0}{226.2} = \left[1 + \frac{(1.4 - 1.0)}{2} \times (1.7505)^2\right] = 1.61285$$

$$T_0 = 364.8 \text{ K} = \mathbf{91.8°C}$$

(2) Stagnation pressure p_0

$$\frac{p_0}{p_1} = \left[1 + \frac{k-1}{2} M_1^2\right]^{k/(k-1)}$$

$$\frac{p_0}{28.5} = \left[1 + \frac{(1.4 - 1.0)}{2}(1.7505)^2\right]^{1.4/(1.4-1.0)}$$

$$= (1.61285)^{3.5} = 5.328$$

$$p_0 = \mathbf{151.85 \text{ kPa (abs)}}$$

(3) Stagnation density ρ_0

$$\rho_0 = \frac{p_0}{RT_0} = \frac{151850}{287 \times 364.8} = \mathbf{1.450 \text{ kg/m}^3}$$

15.13 Find the stagnation temperature and pressure for carbon dioxide flowing at 150 m/s if the pressure and temperature in undisturbed flow are 500 kPa (abs) and 30°C respectively. (For CO_2: $k = 1.28$ and $R = 188$ J/(kg·K)).

Solution: Speed of sound in CO_2

$$= C = \sqrt{kRT}$$

$$= \sqrt{1.28 \times 188 \times (273 + 30)}$$

$$= 270 \text{ m/s}$$

Velocity $V_1 = 150$ m/s

Mach number

$$M_1 = V_1/C = \frac{150}{270} = 0.5555$$

Stagnation temperature T_0:

$$\frac{T_0}{T_1} = \left[1 + \left(\frac{k-1}{2}\right) M_1^2\right]$$

$$\frac{T_0}{(273 + 30)} = \left[1 + \frac{(1.28 - 1)}{2}(0.5555)^2\right] = 1.0432$$

$$T_0 = 316 \text{ K} = \mathbf{43°C}$$

Stagnation pressure p_0:

$$\frac{p_0}{p_1} = \left[1 + \frac{k-1}{2} M_1^2\right]^{k/(k-1)}$$

$$\frac{p_0}{500} = \left[1 + \frac{(1.28 - 1.00)}{2}(0.5555)^2\right]^{1.28/(1.28-1.0)}$$

$$= (1.0432)^{4.57143} = 1.2133$$

$$p_0 = \mathbf{606.65 \text{ kPa (abs)}}$$

C. Pitot-Static Tube

15.14 In the flight of an aircraft at an altitude of 5500 m where the ambient pressure is known to be 50 kPa (abs) and the temperature is –17°C, a pitot tube was used to measure the speed of the plane. The indicated stagnation pressure was 75 kPa (abs). Calculate the speed of the aircraft. ($k = 1.4$ and $R = 287$ J/(kg·K)).

Solution: Referring to Fig. 15.7.

$$T_1 = 273 - 17 = 256 \text{ K}$$

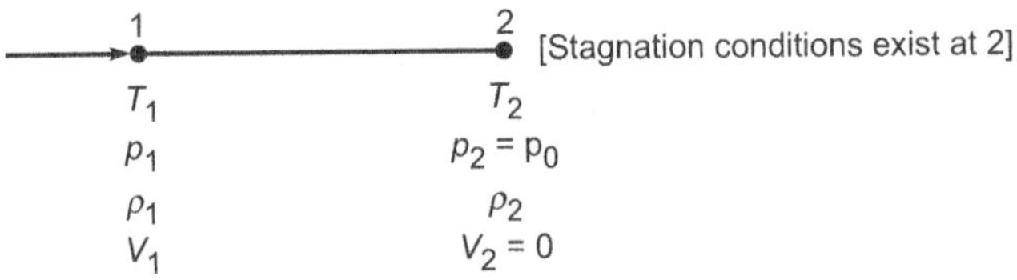

Fig 15.7

Sonic speed $C_1 = \sqrt{kRT_1}$

$$= \sqrt{1.4 \times 287 \times 256} = 320.72 \text{ m/s}$$

The stagnation pressure p_0 is related to free stream pressure p_1 as

$$\frac{p_0}{p_1} = \left(1 + \frac{k-1}{2} M_1^2\right)^{k/(k-1)}$$

$$= \left(1 + \frac{1.4 - 1.0}{2} M_1^2\right)^{1.4/(1.4-1)}$$

$$\frac{75}{50} = (1 + 0.2\, M_1^2)^{3.5}$$

$$M_1^2 = 0.6141$$

and $\quad M_1 = 0.7837$

Since $\quad M_1 = \dfrac{V_1}{C_1}$, the speed of plane V_1 is

$$V_1 = C_1 M_1 = 320.72 \times 0.7837 = \mathbf{251.3 \text{ m/s}}$$

15.15 Air flow in a duct is measured by a pitot-static tube. The stagnation pressure was found to be 95 kPa (abs) and the static pressure was recorded as 50 kPa (abs). The stagnation temperature was measured to be 30°C. Calculate the velocity of air (i) by considering compressibility effects and (ii) by ignoring compressibility effects.
[Take $k = 1.4$ and $R = 287$ J/(kg·K)]

Solution:

(i) $\quad T_0 = 273 + 30 = 303$ K

For isentropic flow with subscript zero denoting stagnation values

$$T_1 = T_0 \left(\frac{p_1}{p_0}\right)^{(k-1)/k}$$

$$= 303 \left(\frac{50}{95}\right)^{(1.4-1)/1.4} = 252.2 \text{ K}$$

Also $\quad \dfrac{V_1^2}{2} = \dfrac{kR}{(k-1)} (T_0 - T_1) = c_p (T_0 - T_1)$

$$V_1^2 = \frac{2kR}{(k-1)} (T_0 - T_1)$$

$$= \frac{2 \times 1.4 \times 287}{(1.4 - 1.0)} (303 - 252.2)$$

$$= 102057.2$$

$$V_1 = \mathbf{319.5 \text{ m/s}}$$

(ii) When compressibility effects are ignored:

$$\rho_1 = \rho_0 = p_0/RT_0 = \frac{95000}{287 \times 303}$$

$$= 1.092 \text{ kg/m}^3$$

$$\frac{(p_0 - p)}{\frac{1}{2}\rho_1 V_1^2} = 1$$

$$\therefore \quad V_1^2 = \frac{2(p_0 - p)}{\rho_1} = \frac{2 \times (95 - 50) \times 1000}{1.092}$$

$$= 82418$$

$$V_1 = \mathbf{287.1 \text{ m/s}}$$

15.16 A tank containing air at a pressure of 250 kPa (abs) and 35°C is discharging it though a convergent nozzle. If the velocity of flow at the nozzle throat is 200 m/s, calculate the Mach number at that section.
[Take $k = 1.4$ and $R = 287$ J/(kg·K)].

Solution: Inside the tank stagnation conditions prevail.

Hence $\quad V_0 = 0$, and $T_0 = 273 + 35 = 308$ K

$$p_0 = 250 \text{ kPa (abs)}$$

At the nozzle throat (exit): $V_1 = 200$ m/s

$$V_1^2 = 2c_p(T_0 - T_1)$$

$$c_p = \frac{kR}{k-1} = \frac{1.4}{0.4} \times 287 = 1004.5$$

$$200^2 = 2009\,(308 - T_1)$$

$$T_1 = 288 \text{ K}$$

The Mach number M_1 at the exit is given by

$$\frac{T_0}{T_1} = \left(1 + \frac{k-1}{2} M_1^2\right)$$

$$\frac{308}{288} = 1 + 0.2\, M_1^2$$

$$M_1^2 = 0.3472 \text{ or } M_1 = \mathbf{0.589}$$

D. Flow Through a Nozzle

15.17 Air at 40°C flows a large tank through a converging nozzle of 40 mm diameter. The tank contains air at 150 kPa (abs) and the discharge is to atmosphere of pressure 95 kPa (abs). Calculate the mass rate of flow through the nozzle.
[Take $k = 1.4$ and $R = 287$ J/(kg·K)].

Solution: Referring to Fig. 15.8,
Temperature inside the tank $= T_0 = 273 + 40 = 313$ K

$$\frac{p_1}{p_0} = \frac{95}{150} = 0.6333$$

As this is larger than the *critical value* of 0.528 the flow in the nozzle throat will be subsonic.

$$\rho_0 = p_0/RT_0 = \frac{150000}{287 \times 313} = 1.6698 \text{ kg/m}^3$$

$$\frac{\rho_0}{\rho_1} = \left(\frac{p_0}{p_1}\right)^{1/k}$$

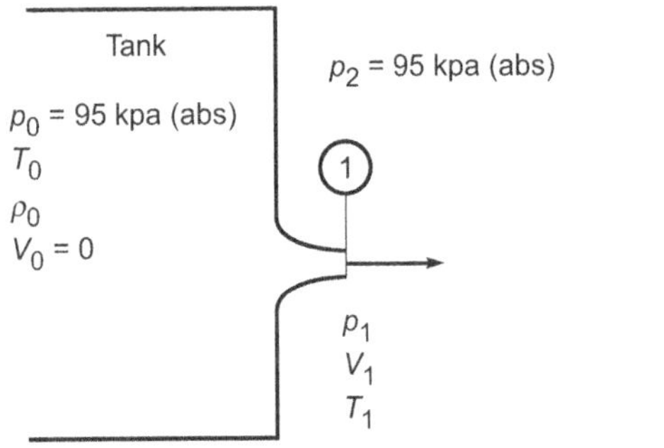

Fig. 15.8 *Flow through a nozzle*

$$\therefore \quad \rho_1 = \rho_0\left(\frac{p_1}{p_0}\right)^{1/k} = 1.6698\ (0.6333)^{1/1.4}$$

$$= 1.205 \text{ kg/m}^3$$

Velocity of flow at nozzle throat, V_1

$$V_1 = \left\{2\left(\frac{k}{k-1}\right)\frac{p_0}{\rho_0}\left[1-\left(\frac{p_1}{p_0}\right)^{(k-1)/k}\right]\right\}^{1/2}$$

$$= \left\{2\left(\frac{1.4}{0.4}\right)\left(\frac{150{,}000}{1.6698}\right)[1-(0.6333)^{0.4/1.4}]\right\}^{1/2}$$

$$= \{628.818\ (1 - 0.87765)\}^{1/2} = 2774 \text{ m/s}$$

$$A_1 = \text{area of nozzle exit} = \frac{\pi}{4} \times (0.04)^2$$

$$= 1.2566 \times 10^{-3} \text{ m}^2$$

Mass rate of flow

$$\dot{m} = \rho_1 A_1 V_1$$

$$= 1.205 \times 1.2566 \times 10^{-3} \times 277.4$$

$$= \mathbf{0.42\ kg/s}$$

15.18 Air at 60°C is stored in a large tank under a pressure of 200 kPa (abs). A 30 mm diameter converging nozzle is provided to discharge the nozzle to the atmosphere. Assuming isentropic conditions, estimate the mass rate of flow from the nozzle when it is discharging to environment at a pressure of 98 kPa (abs). What is the pressure at the nozzle throat?
[Take $k = 1.4$ and $R = 287$ J/(kg·K)]

Solution:

Inside the tank $T_0 = 273 + 60 = 333$ K

$p_0 = 200{,}000$ Pa (abs)

$$\rho_0 = \frac{p_0}{RT_0} = \frac{200{,}000}{287 \times 333} = 2.093 \text{ kg/m}^3$$

Pressure ratio $p_1/p_0 = 98/200 = 0.49$

As this is less than the *critical* value of $\dfrac{p_1^*}{p_0} = 0.528$ chocking condition will prevail. The flow at the nozzle throat (exit) will be critical, i.e. $M_1 = 1.0$. At critical condition

$$\frac{T_1^*}{T_0} = 0.833 \text{ and hence } T_1^* = 0.833 \times 333 = 277.4 \text{ K}$$

$$C_1^* = \text{Sonic velocity at throat} = \sqrt{kRT_1^*}$$

$$= \sqrt{1.4 \times 287 \times 277.4} = 333.8 \text{ m/s}$$

$$A_1 = \text{area of throat} = \frac{\pi}{4} \times (0.03)^2 = 7.0686 \times 10^{-4} \text{ m}^2$$

$$\frac{\rho_1^*}{\rho_0} = 0.634$$

$$\therefore \quad \rho_1^* = 0.634 \times 2.093 = 1.327 \text{ kg/m}^3$$

Mass rate of flow under choked condition

$$= m = \rho_1^* A_1 C_1^*$$

$$= 1.327 \times 7.0686 \times 10^{-4} \times 333.8$$

$$= 0.313 \text{ kg/s}$$

Pressure at the nozzle throat $= p_1^*$

$$p_1^* = 0.528\ p_0 = 0.528 \times 200 = \mathbf{105.6\ kPa\ (abs)}$$

[***Note:*** When the chocking condition prevails, in a converging nozzle the pressure at the throat will be p_1^*. The receiver pressure p_2 will be either equal to or less than p_1.]

15.19 It is proposed to discharge helium [$k = 1.66$ and $R = 2077$ J/(kg·K)] from a container tank to a receiver through a 2.0 cm convergent nozzle. The container has a pressure of 300 kPa (abs) and temperature of – 10°C. What is the maximum rate of mass flow possible under this arrangement? What is the corresponding maximum pressure in the receiver tank?

Solution: Referring to Fig 15.9,

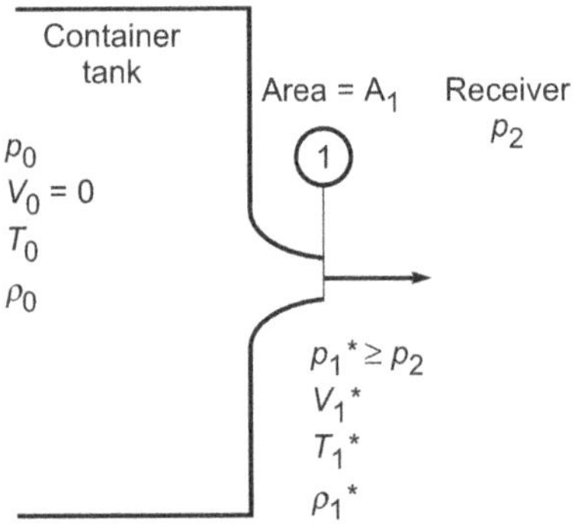

Fig 15.9

$$T_0 = 273 - 10 = 263 \text{ K}$$

$$p_0 = 300{,}000 \text{ Pa (abs)}$$

$$\rho_0 = \frac{p_0}{RT_0} = \frac{300{,}000}{2077 \times 263}$$
$$= 0.549 \text{ kg/m}^3$$

Under conditions of maximum flow, critical conditions will prevail at the nozzle throat. At critical condition

$$M_1 = M_1^* = 1.0 \qquad \text{Hence } C_1^* = V_1^*$$

$$\frac{p_1^*}{p_0} = \left(\frac{2}{k+1}\right)^{k/(k-1)} = \left(\frac{2}{1+1.66}\right)^{1.66/(1.66-1)}$$
$$= (0.7519)^{2.515} = 0.488$$
$$p_1^* = 0.488 \times 300 = 146.43 \text{ kPa (abs)}$$

The receiver pressure p_2 should be less than or equal to p_1^* = 146.43 kPa (abs) for maximum rate of flow condition. Hence maximum downstream receiver pressure possible = p_1^* = 146.43 kPa (abs)

$$\frac{T_1^*}{T_0} = \left(\frac{2}{k+1}\right) = (0.7519)$$
$$T_1^* = 263 \times 0.7519 = 197.7 \text{ K}$$
$$\frac{\rho_1^*}{\rho_0} = \left(\frac{2}{k+1}\right)^{1/(k-1)} = (0.7519)^{1/0.66} = 0.649$$
$$\rho_1^* = 0.549 \times 0.649 = 0.3565 \text{ kg/m}^3$$
$$V_1^* = C_1^* = \text{Speed of sound} = \sqrt{kRT_1^*}$$
$$= \sqrt{1.66 \times 2077 \times 197.7} = 825.6 \text{ m/s}$$

Area of nozzle

$$A_1 = \frac{\pi}{4} \times (0.02)^2 = 3.1416 \times 10^{-4} \text{ m}^2$$

Maximum mass rate of flow under the flow conditions prevailing

$$\dot{m}_{max} = \rho_1^* A_1 V_1^*$$
$$= 0.3565 \times 3.1416 \times 10^{-4} \times 825.6$$
$$= \mathbf{0.0925 \text{ kg/s}}$$

E. Energy Equation in Isentropic flow

15.20 A duct conveys an isentropic flow of air. At section 1, velocity, pressure and temperature are 100 m/s, 200 kPa (abs) and 70°C respectively. At another downstream section (Section 2) the velocity is 250 m/s. Calculate (i) the temperature at section 2, (ii) Mach number and pressure at section 2 and (iii) the densities at sections 1 and 2 [Take k = 1.4 and R = 287 J/(kg·K)]

Solution:

$$T_1 = 273 + 70 = 343 \text{ K}$$
$$V_1 = 100 \text{ m/s and } V_2 = 250 \text{ m/s}$$
$$p_1 = 200 \text{ kPa (abs)}$$

$$c_p = \frac{k}{k-1} R = \frac{1.4}{0.4} \times 287 = 1004.5$$

(i) $V_2^2 - V_1^2 = 2c_p(T_1 - T_2)$

$$(250)^2 - (100)^2 = 2 \times 1004.5\,(343 - T_2)$$
$$T_2 = 316.87 \text{ K} = \mathbf{43.87°C}$$

Sonic speed at section $2 = C_2 = \sqrt{kRT_2}$

$$C_2 = \sqrt{1.4 \times 287 \times 316.87}$$
$$= 356.82 \text{ m/s}$$

(ii) Mach number at Section 2 $= M_2 = \dfrac{V_2}{C_2}$

$$= \frac{250}{356.82} = \mathbf{0.70}$$

(iii) At Section 1, $V_1 = 100$ m/s

Sonic speed = $C_1 = \sqrt{kRT_1} = \sqrt{1.4 \times 287 \times 343}$

$$= 371.24 \text{ m/s}$$

Mach number $M_1 = V_1/C_1 = \dfrac{100}{371.24} = 0.2694$

Stagnation pressure = p_{01}

$$p_{01} = p_1\left(1 + \frac{k-1}{2} M_1^2\right)^{k/(k-1)}$$
$$= 200\,[1 + 0.2 \times (0.2694)^2]^{3.5} \text{ kPa (abs)}$$
$$= 210.3 \text{ kPa (abs)} = p_{02}$$

At Section 2:

$$V_2 = 250 \text{ m/s}$$
$$M_2 = 0.70$$

Stagnation pressure = p_{02}

$$\frac{p_{02}}{p_2} = \left(1 + \frac{k-1}{2} M_2^2\right)^{k/(k-1)}$$
$$\frac{p_{02}}{p_2} = [1 + 0.2 \times (0.7)^2\,]^{3.5} \text{ kPa (abs)}$$
$$= 1.3871$$
$$p_2 = \frac{210.3}{1.3871} = \mathbf{151.6 \text{ kPa (abs)}}$$

Densities:

$$\rho_1 = \frac{p_1}{RT_1} = \frac{200{,}000}{287 \times 343} = \mathbf{2.0317 \text{ kg/m}^3}$$
$$\rho_2 = \frac{p_2}{RT_2} = \frac{151600}{287 \times 316.87} = \mathbf{1.667 \text{ kg/m}^3}$$

[*Note:*

1. Check on calculation —

$$\rho_{01} = \text{Stagnation density at 1}$$
$$= \rho_1\left(1 + \frac{k-1}{2} M_1^2\right)^{1/(k-1)}$$

$= 2.0317[1 + 0.2 \times (0.2694)^2]^{2.5} = 2.106$

$$p_{02} = p_2\left(1 + \frac{2}{k-1}M_2^2\right)^{1/(k-1)}$$

$= 1.667[1 + 0.2 \times (0.700)^2]^{2.5} = 2.106$

$p_{02} = p_{01} = 2.106$

2. In isentropic flow the stagnation pressure p_0, stagnation temperature T_0 and stagnation density ρ_0 remain constant throughout the flow.]

F. Normal Shock

15.21 In a normal shock wave taking place in methane ($k = 1.32$) the downstream Mach number is 0.40. Estimate the upstream Mach number M_1.

Solution:

$$M_2^2 = \frac{2 + (k-1)M_1^2}{2kM_1^2 - (k-1)}$$

$$(0.4)^2 = \frac{2.0 + (1.32 - 1.0)M_1^2}{2 \times 1.32 \times M_1^2 - (1.32 - 1.0)}$$

$0.4224\ M_1^2 - 0.0512 = 2.0 + 0.32\ M_1^2$

$M_1^2 = 20.03$

and $\boldsymbol{M_1 = 4.476}$

15.22 In a normal shock wave occurring in helium ($k = 1.66$) the density downstream of the shock is three times that on the upstream. Calculate the corresponding pressure ratio and velocity ratio. What are the Mach numbers upstream and downstream of the shock?

Solution:

Given $\rho_2/\rho_1 = 3.0$

$$\frac{\rho_2}{\rho_1} = \frac{1 + \beta(p_2/p_1)}{\beta + (p_2/p_1)}$$

where $\beta = \dfrac{k+1}{k-1} = \dfrac{1.66 + 1.0}{1.66 - 1.0} = 4.03$

$$3.0 = \frac{1 + 4.03(p_2/p_1)}{4.03 + (p_2/p_1)}$$

$(4.03 - 3.0)(p_2/p_1) = 12.09 - 1.0$

$(p_2/p_1) = 10.77$ = pressure ratio

Also the pressure ratio

$$= \left(\frac{p_2}{p_1}\right) = \frac{2kM_1^2 - (k-1)}{(k+1)}$$

$$10.77 = \frac{2 \times 1.66 \times M_1^2 - (0.66)}{2.66}$$

$M_1^2 = 8.8278$ and $M_1 = 2.971$

Upstream Mach number $\boldsymbol{M_1 = 2.971}$

Velocity ratio $\dfrac{V_2}{V_1} = 1/(\rho_2/\rho_1) = \dfrac{1}{3} = 0.333$

Downstream Mach number,

$$M_2^2 = \frac{2 + (k-1)M_1^2}{2kM_1^2 - (k-1)}$$

$$= \frac{2 + (1.66 - 1.0)(2.971)^2}{2 \times 1.66 \times (2.971)^2 - (1.66 - 1.0)}$$

$= 7.826/28.645 = 0.2732$

$\boldsymbol{M_2 = 0.523}$

15.23 The pressure, velocity and temperature just upstream of a normal shock wave in air are 100 kPa (abs), 660 m/s and –20°C respectively. Calculate the pressure, velocity and temperature just downstream of the shock wave.
[Take $k = 1.4$ and $R = 287$ J/(kg·K)]

Solution:

$$\rho_1 = \frac{p_1}{RT_1} = \frac{100{,}000}{287 \times (273 - 20)} = 1.3772 \text{ kg/m}^3$$

Sonic speed $C_1 = \sqrt{kRT_1}$

$= \sqrt{1.4 \times 287 \times 253} = 318.83$ m/s

Mach number $M_1 = \dfrac{V_1}{C_1} = \dfrac{660}{318.83} = 2.07$

Pressure ratio $\dfrac{p_2}{p_1} = \dfrac{2kM_1^2 - (k-1)}{k+1}$

$$= \frac{2 \times 1.4 \times (2.07)^2 - (1.4 - 1.0)}{(1.4 + 1.0)}$$

$= 4.832$

$p_2 = 100 \times 4.832 =$ **483.2 kPa (abs)**

Velocity ratio $\dfrac{V_2}{V_1} = \dfrac{(k-1)M_1^2 + 2.0}{(k+1)M_1^2}$

$$= \frac{(1.4 - 1.0)(2.07)^2 + 2.0}{(1.4 + 1.0)(2.07)^2} = 0.361$$

$V_2 = 660 \times 0.361 =$ **238.4 m/s**

$\rho_1 A_1 V_1 = \rho_2 A_2 V_2$ and since $A_2 = A_1$

$\rho_1 V_1 = \rho_2 V_2$

$$\rho_2 = \rho_1 \frac{1}{(V_2/V_1)}$$

$= 1.3772/0.361 = 3.815$ kg/m^3

Temperature after the shock,

$$T_2 = \frac{p_2}{\rho_2 R}$$

$$T_2 = \frac{483.2 \times 1000}{3.815 \times 287}$$

$= 441.3$ K = **168.3°C**

15.24 Just downstream of a normal shock in air, the pressure, temperature and velocity are 250 kPa (abs), 100°C and 180 m/s respectively. Calculate the pressure, temperature and velocity prior to the shock wave. [Take $k = 1.4$ and $R = 287$ J/(Kg·K)]

Solution:

Density $$\rho_2 = \frac{p_2}{RT_2} = \frac{250{,}000}{287 \times (273 + 100)}$$

$= 2.335$ kg/m^3

Sonic speed $$C_2 = \sqrt{kRT_2} = \sqrt{1.4 \times 287 \times 373}$$

$= 387.1$ m/s

Mach number after the shock

$$M_2 = V_2/C_2$$

$$M_2 = \frac{180.0}{387.1} = 0.465$$

From the Mach number relation

$$M_1^2 = \frac{2 + (k-1)M_2^2}{2kM_2^2 - (k-1)}$$

$$= \frac{2 + 0.4 \times (0.465)^2}{2 \times 1.4 \times (0.465)^2 - 0.4}$$

$$M_1^2 = 10.1567$$

$$M_1 = 3.187$$

Pressure ratio $$\frac{p_2}{p_1} = \frac{1 + kM_1^2}{1 + kM_2^2} = \frac{1 + 1.4\,(3.187)^2}{1 + 1.4\,(0.465)^2}$$

$= 11.684$

$p_1 = 250/11.684 =$ **21.4 kPa (abs)**

Velocity ratio $$\frac{V_2}{V_1} = \frac{(k-1)\,M_1^2 + 2}{(k+1)\,M_1^2}$$

$$= \frac{0.4 \times (3.187)^2 + 2}{(2.4)\,(3.187)^2} = 0.2487$$

$V_1 = 180/0.2487 =$ **723.8 m/s**

Density ρ_1: $$\rho_1 = \frac{\rho_2 V_2}{V_1} = 2.335 \times 0.2487$$

$= 0.5807$ kg/m^3

Temperature: $$T_1 = \frac{p_1}{R\rho_1} = \frac{21.4 \times 1000}{287 \times 0.5807}$$

$= 128.4$ K

= **– 144.6°C**

15.25 When a pitot tube is used in supersonic flow a normal shock will stand in front of the probe. If in such a pitot-static tube placement, the stagnation pressure measured by the probe is 220 kPa (abs), the static pressure is 170 kPa (abs) and the stagnation temperature is 350 K estimate the velocity upstream of the shock and the corresponding Mach number.
[Take $k = 1.4$ and $R = 287$ J/(kg·K).]

Solution: Referring to Fig 15.10 for the pitot tube,

$$p_{02} = 220 \text{ kPa (abs)}$$
$$p_2 = 170 \text{ kPa (abs)}$$

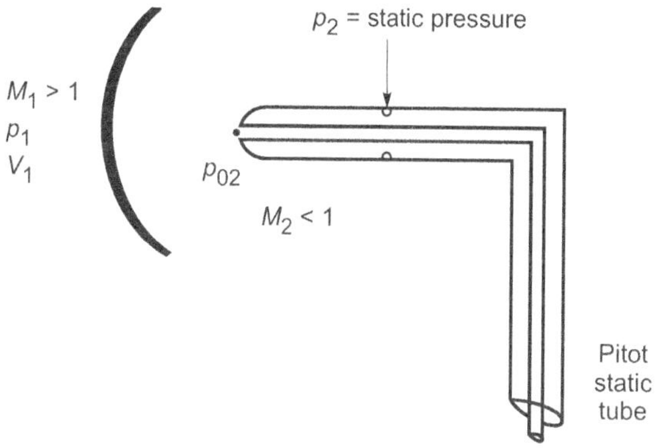

Fig. 15.10

Stagnation pressure ratio

$$\frac{p_{02}}{p_2} = \left(1 + \frac{k-1}{2} M_2^2\right)^{k/(k-1)}$$

As $k = 1.4$,

$$\frac{p_{02}}{p_2} = \frac{220}{170} = (1 + 0.2\,M_2^2)^{3.5}$$

$M_2^2 = 0.382$ and $M_2 = 0.618$

For the normal shock from the Mach number relation

$$M_1^2 = \frac{2 + (k-1)M_2^2}{2kM_2^2 - (k-1)}$$

$$= \frac{2 + 0.4 \times (0.618)^2}{2 \times 1.4 \times (0.618)^2 - 0.4} = 3.216$$

$$\mathbf{M_1 = 1.793}$$

Since stagnation temperature $T_0 = 350$ K

$$\frac{T_0}{T_1} = 1 + \frac{k-1}{2} M_1^2$$

$= 1 + 0.2 \times (1.793)^2 = 1.643$

$T_1 = 350/1.643 = 213$ K

Sonic speed at $T_1 = C_1 = \sqrt{kRT_1}$

$$C_1 = \sqrt{1.4 \times 287 \times 213} = 292.55 \text{ m/s}$$

$V_1 = C_1 M_1 = 292.55 \times 1.793 =$ **524.5 m/s**

OBJECTIVE QUESTIONS

15.1 A gas has a molecular weight of 44. The gas constant R for the gas, in J/(kg·K),
(a) 0.045 (b) 189
(c) 1130 (d) 1854

15.2 A gas has a molecular weight of 16 and has a c_v = 1730 J/(kg·K). The value of the specific heat ratio k for this gas is
(a) 1.30 (b) 1.40
(c) 1.65 (d) 1.21

15.3 The specific heat ratio k is given by the following expression
(a) $1 + (c_p/R)$ (b) $\dfrac{1}{1-(R/c_v)}$
(c) $1 + (R/c_v)$ (d) $1 + (c_v/R)$

15.4 A gas of 2 kg with c_p = 5220 J/(kg·K) and c_v = 3143 J/(kg·K) has its temperature raised by 20°C isentropically. The change in internal energy is
(a) 62.86 kJ (b) 313.2 kJ
(c) 156.60 kJ (d) 125.72 kJ

15.5 A gas of 3 kg with c_v = 745 J/(kg·K) and k = 1.40 has its temperature raised by 30°C isentropically. The change in enthalpy is
(a) 67.0 kJ (b) 94.5 kJ
(c) 31.5 kJ (d) 22.3 kJ

15.6 An *isentropic process* is
(a) adiabatic and irreversible
(b) adiabatic and frictionless
(c) reversible and isothermal
(d) any adiabatic process

15.7 The speed of sound in air varies as
(a) $\sqrt{T}$ (b) $\sqrt{\rho}$
(c) $1/\sqrt{p}$ (d) p
where T = absolute temperature, p = density and p = absolute pressure.

15.8 In an atmosphere where pressure p = 16.5 kN/m^2 (abs), density ρ = 0.265 kg/m^3 and k = 1.4, the speed of sound in the medium is
(a) 9.34 m/s (b) 78.2 m/s
(c) 295 m/s (d) 334 m/s

15.9 An aircraft moves at 1580 km/h in an atmosphere where the temperature is –60°C. If k = 1.4 and R = 287 J/(kg·K) the Mach number of the plane is
(a) 0.67 (b) 1.50
(c) 2.10 (d) 5.4

15.10 In an atmosphere the speed of sound is 300 m/s. If a plane travels at 1620 km/h in this atmosphere the Mach angle is
(a) 30.5° (b) 56.3°
(c) 10.7° (d) 41.8°

15.11 In a standard atmosphere the temperature is 15°C at sea level and –56°C at an altitude of 20 km. A supersonic plane has the same speed at the sea level as well as at an altitude of 20 km. If its Mach number is 1.5 at sea level, its Mach number at an altitude of 20 km in standard atmosphere is
(a) 2.80 (b) 1.30
(c) 1.99 (d) 1.73

15.12 The differential equation for energy for reversible adiabatic flow may take the form
(a) $d\rho + d(\rho V^2) = 0$ (b) $VdV + C^2\dfrac{d\rho}{\rho} = 0$
(c) $VdV + \dfrac{d\rho}{\rho} = 0$ (d) $2VdV + \dfrac{d\rho}{\rho} = 0$

15.13 Considering that p/ρ = constant, expresses an isentropic process, which one of the following is **NOT** a representation of speed of sound?
(a) $\sqrt{kgRT}$ (b) $\sqrt{\dfrac{dp}{d\rho}}$
(c) $\sqrt{k/\rho}$ (d) $\sqrt{p/\rho}$

15.14 The maximum Mach number for which the flow of air can be considered incompressible within 1% error is
(a) 0.1 (b) 0.2
(c) 0.4 (d) 0.6

15.15 An airplane is cruising at a speed of 800 km/h at an altitude where the air temperature is 0°C. The flight Mach number at this speed is nearly
(a) 1.33 (b) 0.67
(c) 0.25 (d) 2.4

15.16 In the flow of air (k = 1.4) in a duct the ambient temperature is 30°C and the stagnant temperature is measured as 59.7°C. The Mach number of the flow is
(a) 0.50 (b) 2.22
(c) 1.59 (d) 0.70

15.17 Air (k = 1.4) flows at a Mach number of 1.5 in a duct. If the ambient temperature is 7°C, the stagnation temperature is

(a) 133°C (b) 10.5°C
(c) 406°C (d) 256°C

15.18 If air [k = 1.4 and R = 287 J/(kg·K)] at 19°C is expanded isentropically, the maximum velocity that can be achieved is

(a) 342.5 m/s (b) 766 m/s
(c) 1000 m/s (d) 518 m/s

15.19 For isentropic flow of air (k = 1.4), the stagnation temperature T_0 and the temperature T at any Mach number are related as T_0/T =

(a) $(1 + 0.2\ M^2)$ (b) $(1 + 0.2\ M^2)^{2.5}$
(c) $(1 + 0.2\ M^2)^{3.5}$ (d) $(1 + 0.2\ M^2)^{-1/2}$

15.20 For air (k = 1.4) the critical pressure ratio P_1^*/P_0 for isentropic flow is

(a) 0.833 (b) 0.728
(c) 0.628 (d) 0.528

15.21 For helium (k = 1.66) the critical pressure ratio p_1^*/p_0 is

(a) 0.728 (b) 0.528
(c) 0.488 (d) 0.833

15.22 For N_2O the critical temperature ratio is 0.866 and the critical density ratio si 0.628. Its critical pressure ratio is

(a) 0.544 (b) 0.725
(c) 0.737 (d) 0.528

15.23 In isentropic flow of a perfect gas

(a) the velocity always decreases in conduits of increasing area.
(b) the velocity is always critical at the throat of a nozzle.
(c) if the flow is subsonic flow through a convergent nozzle, the maximum velocity is always at the throat.
(d) if the flow is in a convergent-divergent nozzle, the maximum velocity is always at the throat.

15.24 A gas is being discharged isentropically from a tank of pressure p_1(abs) through a converging nozzle in a receiving chamber of pressure p_2 (abs) with critical flow condition in the nozzle exit. If p_2 is kept constant and p_1 is doubled, the mass rate of flow through the nozzle will

(a) not change
(b) be doubled
(c) be halved
(d) increase by $\sqrt{2}$ times

15.25 In isentropic flow between two points

(a) stagnation pressure and stagnation temperature may vary
(b) the stagnation pressure decreases in the direction of the flow
(c) the stagnation temperature and stagnation pressure decrease with increase in the velocity
(d) the stagnation temperature and stagnation pressure remain constant

15.26 Identify the *incorrect* statement
For isentropic flow through a duct, the maximum mass flow possible is

(a) proportional to the stagnation pressure
(b) proportional to the stagnation temperature
(c) inversely proportional to the square root of stagnation temperature
(d) proportional to the area of the throat

15.27 In an isentropic flow of air (k = 1.4) the stagnation temperature T_0 = 300 K. If at a section the temperature is 166.7 K, the Mach number of the flow is

(a) 0.56 (b) 1.80
(c) 2.0 (d) 4.0

15.28 Which of the following is analogous to normal shock wave?

(a) An elementary wave in a still liquid
(b) Hydraulic jump
(c) Flow of liquid in an expanding nozzle
(d) Subcritical flow in a rough channel

15.29 In a normal shock taking place in a gas

(a) the velocity, pressure and density increase across the shock
(b) the entropy remains constant
(c) the entropy decreases across the shock
(d) the entropy increases across the shock

15.30 In a normal shock in a gas

(a) the upstream flow is supersonic
(b) the upstream flow is subsonic
(c) the downstream flow is sonic
(d) the downstream flow as well as the upstream flow is supersonic

15.31 In a normal shock in a gas

(a) the stagnation pressure remains the same on both sides of the shock
(b) the stagnation density remains the same on both sides of the shock
(c) the stagnation temperature remains the same on both sides of the shock
(d) the Mach number remains the same on both sides of the shock

15.32 A normal shock wave in a gas
(a) is reversible
(b) is isentropic throughout
(c) is irreversible
(d) causes an increase in Mach number

15.33 In a normal shock occurring in air ($k = 1.4$), if the upstream Mach number is 3.52, the Mach number after the shock is
(a) 0.61 (b) 0.45
(c) 0.13 (d) 0.28

15.34 In a normal shock wave in a gas with $k = 1.4$, one of the Mach numbers is 0.5. The other Mach number is
(a) 2.65 upstream of the shock
(b) 0.06 downstream of the shock
(c) 0.02 upstream of the shock
(d) 3.75 upstream of the shock

15.35 In a normal shock wave in air ($k = 1.4$), the density ratio across the shock ρ_2/ρ_1 is 3.0. The corresponding pressure ratio p_2/p_1 is
(a) 3.0 (b) 0.61
(c) 5.70 (d) 1.5

15.36 In a compressible flow the area of flow = A, velocity V and mass density = ρ. At a particular section, the differential form of continuity equation is
(a) $\frac{dA}{A} = \frac{dV}{V} + \frac{d\rho}{\rho}$ (b) $\frac{dA}{A} = -\frac{dV}{V} + \frac{d\rho}{\rho}$
(c) $\frac{dA}{A} = \frac{dV}{V} - \frac{d\rho}{\rho}$ (d) $\frac{dA}{A} = \frac{dV}{V} - \frac{d\rho}{\rho}$

15.37 In a supersonic flow, a diverging passage results in
(a) increase in velocity and pressure
(b) pressure, density and temperature increase
(c) velocity, pressure and density increase
(d) pressure, density and momentum flux increase

PROBLEMS

15.1 Calculate the value of R for chlorine, helium and hydrogen. Express R in both (J/(kg·K)) and (kcal/kg·K) units.

(Ans.)

Gas	R (J/kg·K)	R (kcal/kg·K)
Cl_2	117	0.0243
He	2079	0.4319
H_2	4157	0.8639

15.2 A mass of 4 kg of oxygen ($k = 1.4$) has its temperature decreased from 85°C to 10°C at constant volume conditions. Calculate c_p, c_v and work involved.

(Ans. c_p = 909.3 J/(kg·K); c_v = 649.5 J/(kg·K); Work W = 194.85 kJ (work is done by the gas)

15.3 Show that for a perfect gas $c_p = c_v + R$.

15.4 A mass of 5 kg of a gas of molecular weight 44 has a work of 100.5 kJ done on it at constant volume. This causes the temperature of the gas to rise by 30°C. Calculate R, c_v, c_p and k of the gas.

(Ans. R = 189 J/(kg·K); c_v = 670; c_p = 859; k 1.282)

15.5 Six kilograms of oxygen at 200 kPa (abs) and 10°C, in a container, is expanded isentropically to 120 kPa (abs). Find the final temperature and the work involved. ($k = 1.4$).

(Ans. T_2 = – 28.4°C; W = work done by the gas = 149.6 kJ)

15.6 Carbon dioxide of mass 2 kg is expanded isothermally from 400 kPa (abs) to 100 kPa (abs) at –10°C. Determine the work done by the gas, the initial volume and final volume, Take k = 1.30 and R = 189 J/(kg·K)

(Ans. W = 137.8 kJ, $\forall_1$ = 0.2485 m^3; $\forall_2$ = 0.994 m^3)

15.7 Calculate the speed of sound wave in the following fluids:

No.	*Gas*	*R (J/kg·K)*	*Temp. (°C)*	*k*
1.	Air	287	20	1.4
2.	CO_2	189	20	1.3
3.	H_2	4124	20	1.41
4.	N_2O	189	20	1.31

No.	*Liquid*	*Bulk modulus K (N/m^2)*	*Density (kg/m^3)*
5.	Water	2.19×10^9	998
6.	Gasoline	9.58×10^8	680
7.	Mercury	2.55×10^{10}	13550

Answers to Problem 15.7

No.	Fluid	C(M/s)
1.	Air	343.1
2.	CO_2	268.3
3.	H_2	1305
4.	N_2O	269.3
5.	Water	1481
6.	Gasoline	1187
7.	Mercury	1372

15.8 An airplane is flying at a Mach number of 1.8 in an atmosphere where the pressure is 14 kPa (abs) and density is 0.225 kg/m^3. Calculate the speed of the plane.

(*Ans.* $V = 531$ m/s)

15.9 A rocket travels at 1800 km/h in air of pressure 35.6 kPa (abs) and temperature –37°C. Find the Mach number and Mach angle. [Take $k = 1.5$ and $R = 287$ J/(kg·K)].

(*Ans.* $M = 1.624$, $\alpha = 38°$)

15.10 A supersonic plane flies at an altitude of 2500 m and 6.5 s after it has passed over the head of an observer on the ground, the sonic boom is heard. Calculate the speed of the plane and its Mach number. The average temperature of the atmosphere can be assumed to be 5°C. Take $R = 287$ J/(kg·K).

(*Ans.* $M = 2.02$ and $V = 675.2$ m/s)

15.11 Consider the general case of Problem 15.10 with h = elevation of the plane above the observer, t = time elapsed between passing of the plane over the observer and the hearing of the sonic boom. Show that the Mach number of the plane is given by

$$\frac{1}{M^2} = \left(1 - \frac{C^2 t^2}{h^2}\right)$$

where C = velocity of sound.

15.12 The Concorde airplane flies at a Mach number of 2.2. If it flies in a standard atmosphere at 15000 m altitude where the pressure is 12 kPa (abs) and density is 0.1935 kg/m^3, calculate the pressure, temperature and density at a stagnation point. [Take $R = 287$ J/(kg·K) and $k = 1.4$]

(*Ans.* $T_0 = 152$°C, $p_0 = 128.3$ kPa (abs); $\rho_0 = 1.052$ kg/m^3)

15.13 A conduit conveys air at a Mach number of 0.70. At a certain section the static pressure is 30 kPa (abs) and the temperature is 25°C. (i) Calculate the stagnation temperature and pressure (ii) If the stagnation temperature is 90°C, what would be the Mach number of the flow? (Take $k = 1.4$.)

(*Ans.* (i) $T_0 = 59.7$°C, $p_0 = 41.61$ kPa (abs); (ii) $M_1 = 0.995$)

15.14 Estimate the maximum velocity of air at 35°C so that the stagnation point temperature is less than 40°C. [Take $k = 1.4$ and $R = 287$ J/(kg·K).]

(*Ans.* $V_1 = 100.2$ m/s)

15.15 A pitot-static tube mounted on an airplane in flight records a stagnation pressure of 90 kPa (abs) and a static pressure of 70 kPa (abs). If the density of air in the atmosphere at the level is 0.80 kg/m^3, calculate the velocity of the airplane. (Take $k = 1.4$).

(*Ans.* $V_1 = 213.5$ m/s)

15.16 A pitot-static tube to be used in an airplane for on-flight measurement was calibrated on ground by using air of density 1.20 kg/m^3. A measurement taken at an altitude of 3000 m where the ambient pressure and density were 70 kPa (abs) and 0.91 kg/m^3 respectively, indicated a velocity of 200 m/s by using the above ground level calibration. Estimate the true speed of the airplane [Take $k = 1.4$ and $R = 287$ J/(kg·K)].

(*Ans.* $V_1 = 217.5$ m/s)

15.17 Show that in an isentropic flow of a gas in a duct, if a pitot-static tube measures the stagnation pressure p_0, stagnation temperature T_0 and static pressure p_1, the velocity of flow V_1 can be calculated by the relation.

$$V_1 = \{2c_p T_0[1 - (p_1/p_0)^{(k-1)/k}]\}^{1/2}$$

15.18 Show that, in an isentropic flow of a gas, the velocity of flow V_1 at a section is given by

$$V_1^2 = 2c_p(T_0 - T_1)$$

where T_1 = the temperature of the flow at the section and T_0 = the stagnation temperature of the flow at that section.

15.19 Air flow in a duct can be considered to be isentropic. At section 1, the velocity, pressure and temperature are 125 m/s, 200 kPa (abs) and 300 K respectively. If at a downstream section the velocity is 220 m/s, calculate the (i) Mach number, temperature and pressure at section 2 and (ii) the densities at sections 1 and 2.

(*Ans.* (i) $M_1 = 0.360$, $T_2 = 283.7$ K, $p_2 = 164.4$ kPa (abs) (ii) $\rho_1 = 2.3229$ kg/m^3, $\rho_2 = 2.0196$ kg/m^3)

15.20 Air at 200 kPa (abs) and 27°C is expanded isentropically. What is the maximum possible attainable speed? Take $k = 1.4$ and $R = 287$ J/(kg·K).

(*Ans.* $V_m = 776.3$ m/s)

15.21 Oxygen [$k = 1.4$ and $R = 260$ J/(kg·K)] is contained in a tank at 150 kPa (abs) and 20°C. If it is expanded isentropically to attain a Mach number of unity, what is the sonic speed and temperature at that section?

(*Ans.* $C_1 = 298.6$ m/s, $T_1^* = -29$°C)

15.22 Air at 40°C and pressure of 300 kPa (abs) flows from a large tank through a converging nozzle. If the Mach number at the outlet of the nozzle is 0.5, calculate the velocity, pressure, temperature and density at the nozzle exit. (Take $k = 1.4$ and $R = 287$ J/(kg·K).

(*Ans.* $V_1 = 173.04$ m/s; $p_1 = 252.9$ kPa (abs); $T_1 = 298.1$ K; $\rho_1 = 2.956$ kg/m^3)

15.23 A tank contains nitrogen [$k = 1.4$ and $R = 297$ J/(kg·K)] at 225 kPa (abs) and 50°C. A convergent nozzle of exit area 0.05 m^2 is used to exhaust this gas to an ambient pressure of 100 kPa (abs). Calculate the mass rate of flow and the pressure at the nozzle throat.

(*Ans.* $\dot{m} = 26.81$ kg/s, $p_1^* = 118.8$ kPa (abs))

15.24 Oxygen flows steadily from a reservoir at –10°C and 265 kPa (abs) through a convergent nozzle of exit diameter of 10 cm into another large tank where the pressure is 180 kPa (abs). Calculate the mass rate of flow and Mach number at the nozzle exit. Assume isentropic flow with $k = 1.4$ and $R = 260$ J/(kg·K).

(*Ans.* $\dot{m} = 5.17$ kg/s, $M_1 = 0.915$)

15.25 Chlorine [$k = 1.34$ and $R = 117$ J/(kg·K)] is stored in a tank at 300 kPa (abs) and 5°C. If a convergent nozzle of exit area 0.08 m^2 discharges the gas to another tank having a pressure of 150 kPa (abs), calculate the mass rate of flow. What is the pressure at the nozzle throat?

(*Ans.* $\dot{m} = 89.72$ kg/s, $p_1^* = 161.6$ kPa (abs))

15.26 Air form tank A at 80°C flows through a convergent nozzle of 8 cm diameter isentropically into a receiving atmosphere of pressure 100 kPa (abs). Estimate the mass flow rate when the pressure inside the tank A is (a) 200 kPa (abs), (b) 300 kPa (abs) and (c) 400 kPa (abs). Take $k = 1.4$ and $R = 287$ J/(kg·K).

(*Ans.* (a) $\dot{m} = 0.2162$ kg/s; (b) $\dot{m} = 0.3243$ kg/s; (c) $\dot{m} = 0.4324$ kg/s)

15.27 Air with pressure p_0 and temperature T_0 in a tank is discharged through a convergent nozzle into a receiver with ambient pressure p_1. If p_1 is kept constant, show that the mass rate of flow increases linearly with p_0 when $p_0 > 1.893\, p_1$.

15.28 In a supersonic stream of air a pitot-static tube creates a stagnation pressure of 50 kPa (abs) and a stagnation temperature of 410 K. A normal shock occurs in front of the tube. If the static pressure is 35 kPa (abs), estimate the velocity of the supersonic stream. Take $k = 1.4$ and $R = 287$ J/(kg·K).

(*Ans.* $V_1 = 485.9$ m/s)

15.29 In a supersonic stream of air at 600 m/s and a temperature of 250 K, a normal shock wave occur in front of a body. If the static pressure of the supersonic stream is 20 kPa (abs), calculate (i) the pressure after the shock, (ii) the temperature and velocity after the shock. Take $k = 1.4$ and $R = 287$ J/(kg·K).

(*Ans.* (i) $p_2 = 80.3$ kPa (abs); (ii) $T_2 = 127.5$°C, $V_2 = 239.5$ m/s)

15.30 Air at 5°C is moving at Mach number 2.5. If a normal shock wave occurs, estimate the ratios of pressure, velocity, temperature and density across the shock. Assume $k = 1.4$ and $R = 287$ J/(kg·K).

(*Ans.* $p_2/p_1 = 7.125$, $V_2/V_1 = 0.3$, $T_2/T_1 = 2.1375$, $\rho_2/\rho_1 = 3.333$)

15.31 If for a normal shock occurring in a supersonic stream of air at Mach number of 1.5, the upstream stagnation pressure is 210 kPa (abs), calculate the stagnation pressure after the shock. ($k = 1.5$).

(*Ans.* $p_{02} = 195.24$ kPa (abs))

15.32 In a normal shock wave in air the Mach number after the shock is 0.60. If the temperature and pressure after the shock are 150°C and 360 kPa (abs) respectively, determine the Mach number, pressure, temperature and density of the supersonic stream before the shock . Take $k = 1.4$ and $R = 287$ J/(kg·K).

(*Ans.* $M_1 = 1.878$; $\rho_1 = 91.19$ kPa (abs); $V_1 = 614$ m/s; $\rho_1 = 1.195$ kg/m^3; $T_1 = -7.1$°C)

15.33 The velocity ratio V_2/V_1 across a normal shock wave in air ($k = 1.4$) is 0.5. Estimate corresponding pressure ratio. What are the Mach numbers upstream and downstream the shock?

(*Ans.* $p_2/p_1 = 2.75$, $M_1 = 1.581$, $M_2 = 0.674$)

16

Fluid Flow Machines

CONCEPT REVIEW

16.1 FLUID FLOW MACHINES

Fluid flow machines can be very broadly classified as Turbo-machine and positive displacement machines.

A turbo machine is a device which adds energy to a fluid or extracts energy from it by virtue of a rotating system of blades. If the machine adds energy it is called a pump; if it extracts energy it is called a turbine. There are a wide variety of pumps and turbines and in this chapter only the basic fluid mechanics characteristics of turbo machines are dealt with in Sections 16.2 and 16.3. Reciprocating pumps, which form the most widely used type of positive displacement pumps, are described in Section 16.4.

16.2 TURBINES

16.2.1 Francis Turbine

Basic Equations Figure 16.1(a) shows the schematic sketch of a Francis turbine and Fig. 16.1(b) shows the inward flow through a radial turbine runner, typically a Francis turbine.

The following notations are employed:

Suffix 1 is used for inflow and suffix 2 for outflow conditions. Further,

V = absolute velocity of the fluid.

v = relative velocity of fluid with respect to the blade.

u = peripheral velocity of the blade $= \dfrac{2\pi r N}{60}$.

r = radius

N = revolutions per minute.

α = angle made by the absolute velocity vector $\mathbf{V}$ with the positive direction of the peripheral velocity u.

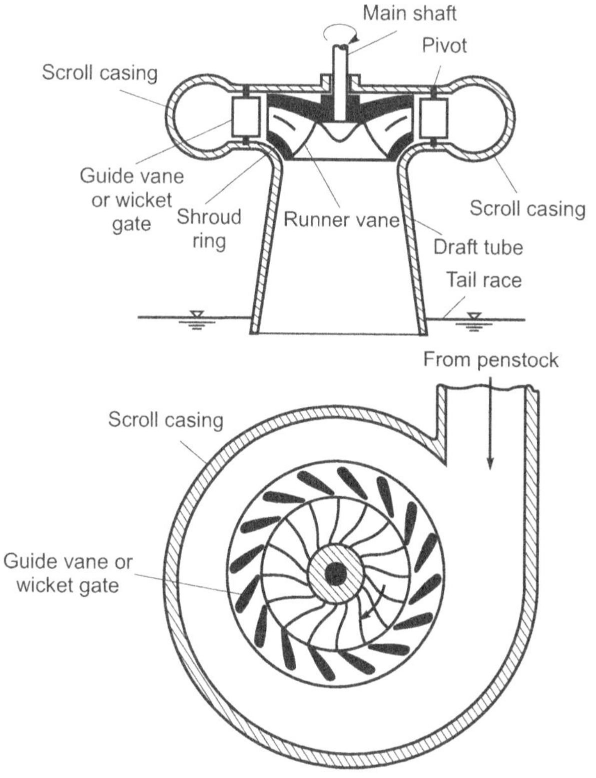

Fig. 16.1(a) *Schematic Sketch of a Francis Turbines*

β = angle made by the relative velocity vector $\mathbf{v}$ with the negative direction of the peripheral velocity u; known as blade angle.

$\beta' = 180 - \beta$.

V_u = tangential component of absolute velocity V $= V \cos \alpha$, known as swirl velocity.

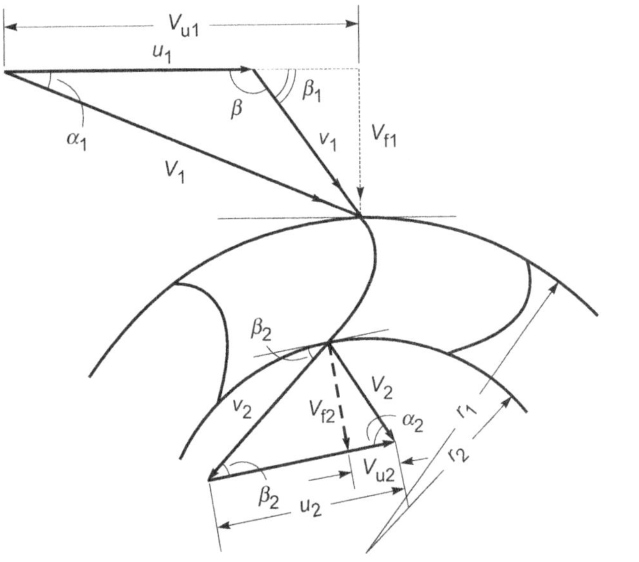

Fig. 16.1(b) *Inward Flow through a Turbine*

V_f = radial component of absolute velocity V
= $V \sin \alpha$, known as flow velocity.
b = width of flow passage.

Discharge: By continuity

$$Q = 2\pi r_1 b_1 V_1 \sin\alpha_1 = 2\pi r_2 b_2 V_2 \sin\alpha_2$$

or
$$\boxed{Q = 2\pi r_1 b_1 V_{f1} = 2\pi r_2 b_2 V_{f2}} \quad (16.1)$$

Torque: Torque on the axis of the runner,

$$T = \rho Q(r_1 V_1 \cos\alpha_1 - r_2 V_2 \cos\alpha_2)$$

$$\boxed{T = \rho Q(r_1 V_{u1} - r_2 V_{u2})} \quad (16.2)$$

Power: Power delivered to the runner,

$$\boxed{\mathbf{P} = T\omega = \rho Q(u_1 V_{u1} - u_2 V_{u2})} \quad (16.3)$$

Head: If H_e = head utilised by the turbine

$$\mathbf{P} = \rho Q H_e = \rho Q(u_1 V_{u1} - u_2 V_{u2})$$

$$\boxed{H_e = \frac{u_1 V_{u1} - u_2 V_{u2}}{g}} \quad (16.4)$$

Net available head, H: In a reaction turbine the net available head H is the difference between the energy levels just upstream of the turbine and that at the tail race.

Hydraulic efficiency, η_H:

$$\eta_H = \frac{\text{Head extracted}}{\text{Net available head}} = \frac{H_e}{H}$$

Thus
$$\boxed{\eta_H = \frac{(u_1 V_{u1} - u_2 V_{u2})}{gH}} \quad (16.5)$$

Mechanical efficiency, η_m:

$$\eta_m = \frac{\text{power available at the shaft}}{\text{power exerted by the water on the rotor}}$$

$$= \frac{\text{brake power}}{(\text{brake power} + \text{power used up in mechanical friction})} \quad (16.6)$$

Overall efficiency, η_0:

$$\eta_0 = \frac{\text{Power delivered to the shaft}}{\text{power available in water}}$$

$$= (\text{brake power})/\gamma Q H \quad (16.7)$$

Thus
$$\boxed{\eta_0 = \eta_m \cdot \gamma_H} \quad (16.8)$$

Specific speed, N_s: The specific speed N_s, in SI units, is defined for turbines as

$$\boxed{N_s = \frac{N\sqrt{\mathbf{P}}}{H^{5/4}}} \quad (16.9)$$

where **P** = power in kW, H = net available head in metres and N = speed of rotation in rpm. The specific speed of Francis turbine runners ranges from 40 to 420 but more commonly it is in the range 75 to 305.
Note that the dimensions of N_s are $[F^{1/2} L^{-3/4} T^{-3/2}]$ but, as a convention, the units are not written. As the specific speed is a dimensional quantity its value for a given machine depends on the system of units adopted. In this book the SI units are used. A dimensionless form of specific speed, known as *shape number,* is sometimes used. It is written as

Shape number = Dimensionless specific speed

Symbolically
$$\boxed{S_p = \frac{N\sqrt{\mathbf{P}}}{\rho^{1/2}(gH)^{5/4}}} \quad (16.10)$$

General Features of Francis Runner The relative velocity must be tangential to the blade at the entry to cause the least amount of disturbance to the flow. Also, for maximum efficiency the fluid should leave the outlet with zero swirl velocity (i.e. $V_{u2} = 0$ which means $\alpha_2 = 90°$). In solving problems, if no mention is made about the discharge at outlet, the radial discharge (i.e. $\alpha_2 = 90°$) can be assumed as this is a common practice in design.

Speed factor, ϕ: The speed factor ϕ is defined as

$$\boxed{\phi = \frac{\text{peripheral velocity of a rotating element}}{\sqrt{2gH}} = \frac{u_1}{\sqrt{2gH}}} \quad (16.11)$$

where $u_1 = \dfrac{\pi D_1 N}{60}$ in which N = rpm and D_1 = diameter at the inlet.

The value of ϕ for Francis turbine ranges from 0.70 to 0.85.

Flow ratio, ψ: The flow ratio ψ is defined as

$$\psi = \frac{\text{flow velocity at inlet}}{\sqrt{2gH}} = \frac{V_{f1}}{\sqrt{2gH}} \tag{16.12}$$

For Francis turbines ψ ranges from 0.15 to 0.35.

16.2.2 Propeller and Kaplan Turbines

These turbines are of the axial flow type. A propeller turbine with an adjustable blade is known Kaplan turbine. Figure 16.2 shows a propeller turbine in a schematic form. The following are some special features of these turbines:

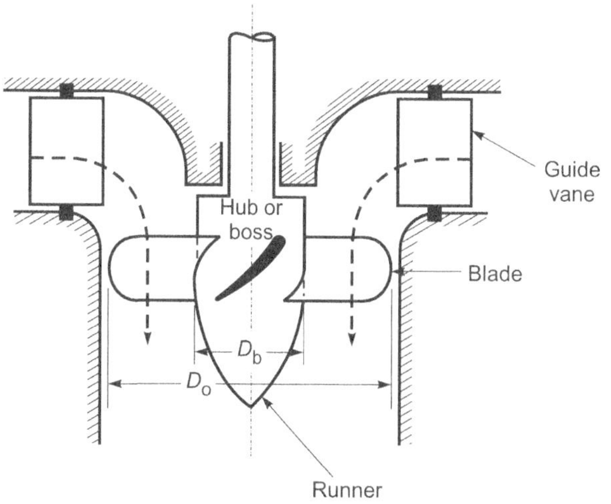

Fig. 16.2 *Schematic Sketch of a Propeller Turbine*

(1) The peripheral velocity at inlet and outlet are the same

Hence $$u_1 = u_2 = \frac{\pi D_0 N}{60} \tag{16.13}$$

(2) The flow velocity remains unchanged from inlet to outlet.

Hence, $$V_{f1} = V_{f2} \tag{16.14}$$

(3) It is usual to assume the relative velocities to remain unchanged, i.e.

$$v_1 = v_2 \tag{16.15}$$

(4) The inlet area = outlet area = $\dfrac{\pi}{4}(D_o^2 - D_b^2)$ (16.16)

where D_o = outside diameter of the runner and D_b = diameter of the hub (or boss).

(5) The flow leaves the runner radially, i.e. $\alpha_2 = 90°$.

(6) The specific speed $N_s = \dfrac{N\sqrt{\mathbf{P}}}{H^{5/4}}$ for propeller type turbines ranges from 380 to 950. The speed factor ϕ for Kaplan turbines ranges from 1.40 to 2.0.

16.2.3 Impulse Turbine

Figure 16.3 shows the schematic definition sketch of an impulse turbine. The net available head H for an impulse turbine is the energy head available at the base of the nozzle. As in a reaction turbine

$$\text{Power } \mathbf{P} = \gamma Q H \eta_0 \tag{16.17}$$

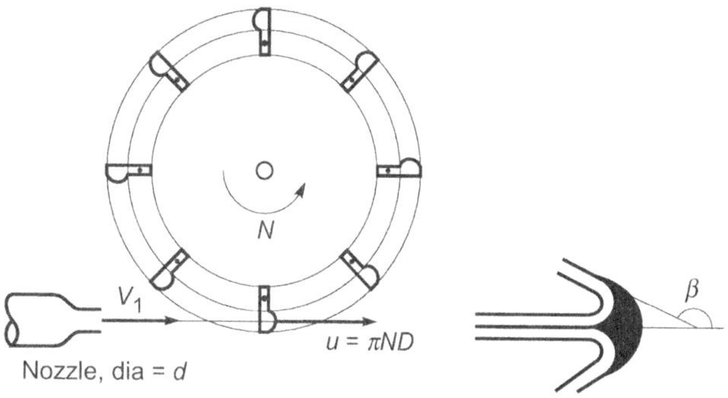

Fig. 16.3 *Definition Sketch of an Impulse Turbine*

where η_o= overall efficiency = $\eta_H\ \eta_m$. A small part of the head H is lost in friction in the nozzle, a portion is expended in bucket friction and in the kinetic energy carried away by the water leaving the buckets. A part of the energy developed is lost in mechanical friction and windage losses.

Figure 16.4 shows the inlet and outlet velocity vectors in an impulse turbine bucket. Here, the suffixes 1 and 2 denote the inlet and outlet conditions respectively.

$u = u_1 = u_2$ = peripheral velocity of the wheel.

V_1 = jet velocity = absolute velocity of impingement.

$= C_v\sqrt{2gH}$ (16.18)

where C_v = coefficient of velocity with a value of about 0.95 to 0.98

$v_1 = (V_1 - u)$ = relative velocity at inlet. (16.19)

v_2 = relative velocity at outlet = $\mathbf{V}_2 - \mathbf{u}_2 = kv_1$

where k = a coefficient to account for losses in the wheel. When $k = 1$, $v_2 = v_1$.

β = bucket angle = deflection angle of the jet and is usually an obtuse angle.

$\beta' = (180 - \beta)$.

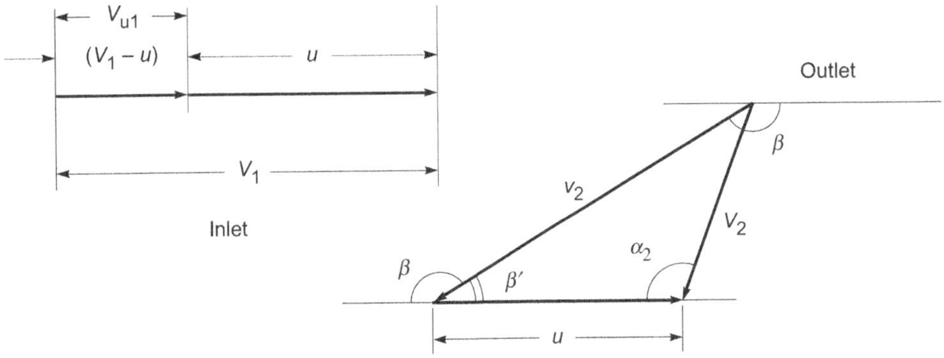

Fig. 16.4 *Velocity Vectors in Impulse Turbine*

By momentum equation, the force exerted on the bucket in x-direction (i.e. in the initial direction of the jet).

$$F_x = \rho\, Q(V_1 - u)\,(1 + k\cos\beta') \qquad (16.20)$$

Power transmitted to the bucket = power extracted.

$$\boxed{\mathbf{P} = \rho\, Qu\,(V_1 - u)\,(1 + k\cos\beta')} \qquad (16.21)$$

Head extracted,

$$\boxed{H_e = \frac{1}{g}u(V_1 - u)\,(1 + k\cos\beta')} \qquad (16.22)$$

Hydraulic efficiency

$$\boxed{\eta_H = \frac{H_e}{H} = \frac{u(V_1 - u)\,(1 + k\cos\beta')}{gH}}$$

Mechanical efficiency,

$$\boxed{\eta_H = \frac{\text{power available at shaft}}{\rho Qu(V_1 - u)\,(1 + k\cos\beta')}}$$

Speed ratio, ϕ: The ratio of peripheral velocity u to ideal velocity $\sqrt{2gH}$ is known as speed ratio or speed factor, ϕ. Thus

$$\boxed{\phi = u/\sqrt{2gH}} \qquad (16.23)$$

For most efficient operations ϕ depends upon the specific speed to some extent and is found to be in the range 0.43 to 0.47.

Specific speed, $\boxed{N_s = \dfrac{N\sqrt{\mathbf{P}}}{H^{5/4}}}$

For multiple jet impulse turbines, the specific speed is based on the brake power per jet. The impulse turbines have specific speeds ranging from 8 to 28 and attain best efficiency at a value of N_s around 17.

16.2.4 Similitude in Turbines

Scale models are often used in designing and other studies relating to turbines. Geometric similarity is a basic requirement. Kinematic similarity is assured by having geometrically similar velocity vector diagrams. It is usual to neglect viscous effects in the model studies. The model and prototype characteristic relationships are usually expressed in terms of the following relationships between the variables:

$$\frac{N_m D_m}{\sqrt{H_m}} = \frac{N_p D_p}{\sqrt{H_p}} \qquad (16.24)$$

$$\frac{Q_m}{N_m D_m^3} = \frac{Q_p}{N_p D_p^3} \qquad (16.25)$$

$$\frac{P_m}{N_m^3 D_m^5} = \frac{P_p}{N_p^3 D_p^5} \qquad (16.26)$$

It is seen from the above that the specific speed N_s is same for both the model and prototype. Also for a given diameter ratio

$$N \propto H^{1/2},$$
$$Q \propto H^{1/2}$$
$$\mathbf{P} \propto H^{3/2}$$

This fact is expressed in terms of *unit quantities*. The *unit speed* N_u is defined as the speed of a geometrically similar turbine working under a head of 1 m.

$$\boxed{N_u = N/\sqrt{H}} \qquad (16.27)$$

The *unit discharge* Q_u = flow rate in a geometrically similar turbine working under a head of 1 m

$$\boxed{Q_u = Q/\sqrt{H}} \qquad (16.28)$$

Unit Power P_u = Power generated in a geometrically similar turbine working under a head of 1 m.

$$\boxed{\mathbf{P}_u = \mathbf{P}/H^{3/2}} \qquad (16.29)$$

These relationships are useful in studying the performance of a turbine under varying heads.

16.2.5 Typical Characteristics of Common Turbines

Table 16.1 gives the ranges of various characteristics of commonly used turbines.

Table 16.1 *Ranges of Characteristics of Common Turbines*

Characteristics	*Pelton*	*Francis*	*Kaplan & Propeller*
Head H(m)	100–1760	30–450	1.5–75
Speed N(rpm)	75–1000	70–1000	70–600
Specific speed N_s	8–30	40–420	380–950
Max. power $\mathbf{P}_m$ (MW)	105	600	125

Typical inlet and outlet velocity triangles of the different kinds of turbines are shown in Fig. 16.5.

16.3 ROTODYNAMIC PUMPS

16.3.1 Definitions

A rotodynamic pump consists essentially of a rotating element inside a casing. The rotating element is called an *impeller*. The fluid enters the casing at its centre and flows outward by the action of the rotating impeller and is discharged around the circumference of the casing. During this process the fluid receives energy. Basically three kinds of rotodynamic pumps are recognised. They

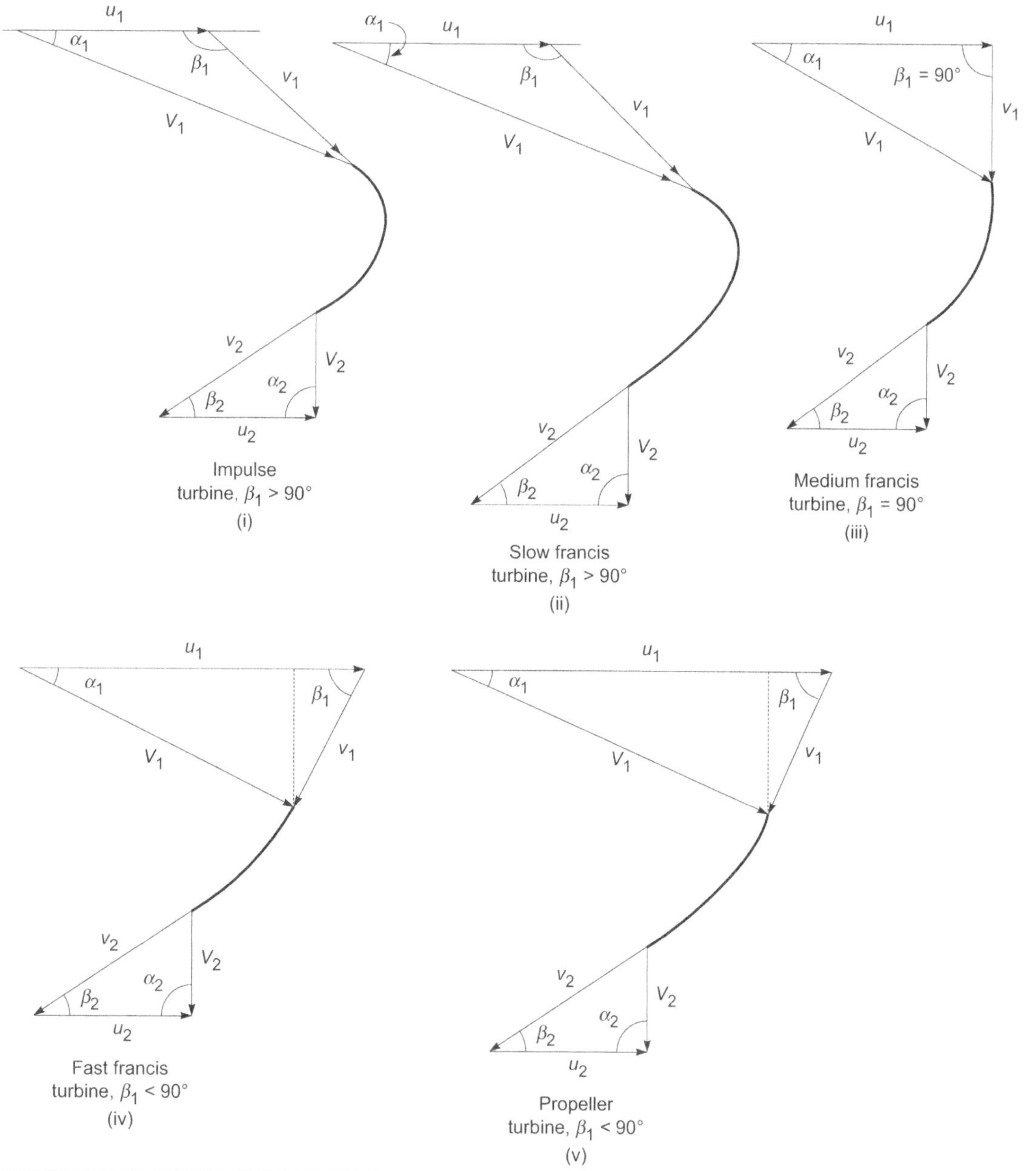

Fig. 16.5 *Different Kinds of Turbines*

are: (i) centrifugal pumps, (ii) mixed flow pumps, and (iii) axial flow pumps. This classification is based on the direction of flow of the fluid in the impeller. Figure 16.6 shows a typical centrifugal pump and a typical casing.

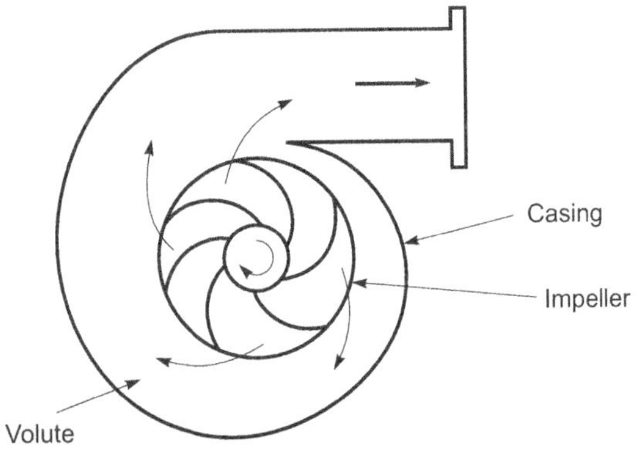

Fig. 16.6 *Centrifugal Pump*

Net Head Developed, *H* In a pump the mechanical energy through the shaft and impeller is converted to fluid energy. The difference between the total energy heads at the intake and discharge flanges of the pump is denoted as net head H developed in the pump. The intake end is commonly called the *suction end* and the discharge end as the *delivery end*. Denoting these by suffixes s and d respectively (see Fig. 16.7) the net head $H = H_d - H_s$

or $$H = \left(\frac{p_d}{\gamma} + Z_d + \frac{V_d^2}{2g}\right) - \left(\frac{p_s}{\gamma} + Z_s + \frac{V_s^2}{2g}\right) \quad (16.30)$$

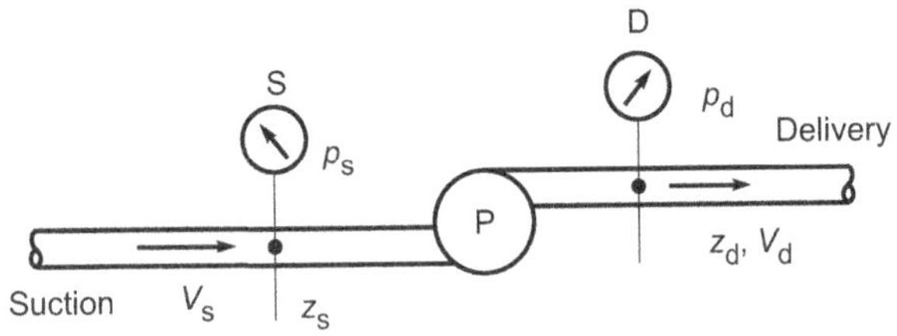

Fig. 16.7

Velocity Triangles Velocity triangles at inlet and outlet of an impeller are shown in Fig. 16.8 (a), (b) and (c). Three kinds of vane configurations viz. (i) backward curved vanes (ii) radial vane and (iii) forward curved vanes are shown. The backward curved vane pumps are the most common. Unless the data are explicitly clear about any other types, the backward curved blade with (β_1 and β_2) < 90° is assumed in all problems.

Notations The following notations are used in connection with flow in pump impellers:

Inlet	*Outlet*
V_1 = absolute velocity	V_2 = absolute velocity
v_1 = relative velocity	v_2 = relative velocity
u_1 = peripheral velocity	u_2 = peripheral velocity
α_1 = inclination of V_1 with u_1 direction = direction of absolute velocity at inlet	α_2 = inclination of V_2 with u_2 direction = direction of absolute velocity at exit
β_1 = inclination of v_1 with u_1 direction = blade angle at inlet	β_2 = inclination of v_2 with u_2 direction = blade angle at exit

Note that all angles are measured with respect to *positive direction* of *u*. If some angles are obtuse their complementary angles are used with a prime. For example: $\beta_1' = 180 - \beta_1$ and $\beta_2' = 180 - \beta_2$. The suffixes 1 and 2 are used for inlet and outlet conditions respectively.

Torque The torque acting on the fluid is

$$T = \rho\, Q\, (r_2 V_2 \cos \alpha_2 - r_1 V_1 \cos \alpha_1)$$
$$= \rho\, Q\, (r_2 V_{u2} - r_1 V_{u1}) \quad (16.31)$$

where Q = discharge

and $V_u = V \cos \alpha$ = whirl velocity.

The discharge Q itself is given in terms of flow velocity $V_f = V \sin \alpha$

as $$Q = \pi D_2 b_2 V_{f2} = \pi D_1 b_1 V_{f1} \quad (16.32)$$

where b = width of the impeller at the given radius.

Power P The power transmitted by the impeller to water is

$$\mathbf{P} = T\omega = \rho\, Q\, (u_2 V_{u2} - u_1 V_{u1})$$
$$= \gamma Q H' \quad (16.33)$$

In Eq. 16.33 H' = theoretical head $= \frac{1}{g}(u_2 V_{u2} - u_1 V_{u1})$ (16.34)

This is also known as *manometric head*. Now H' = head transferred from the shaft = head supplied to machine. If H = actual head delivered to water (= net head developed), then $H' = H + h_L$ where h_L = losses. The losses are composed of shock losses at entrance and exit to the blades, frictional losses in blade passages and circulation in the passages.

Hydraulic Efficiency, η_H

$$\eta_H = \frac{H}{H'} = \frac{gH}{u_2 V_{u2} - u_1 V_{u1}} = 1 - \frac{h_l}{H'} \quad (16.35)$$

For radial entry, i.e. $V_{u1} = 0$, $\boxed{\eta_H = \frac{gH}{u_2 V_{u2}}}$ (16.36)

The hydraulic efficiency is also known as *manometric efficiency*.

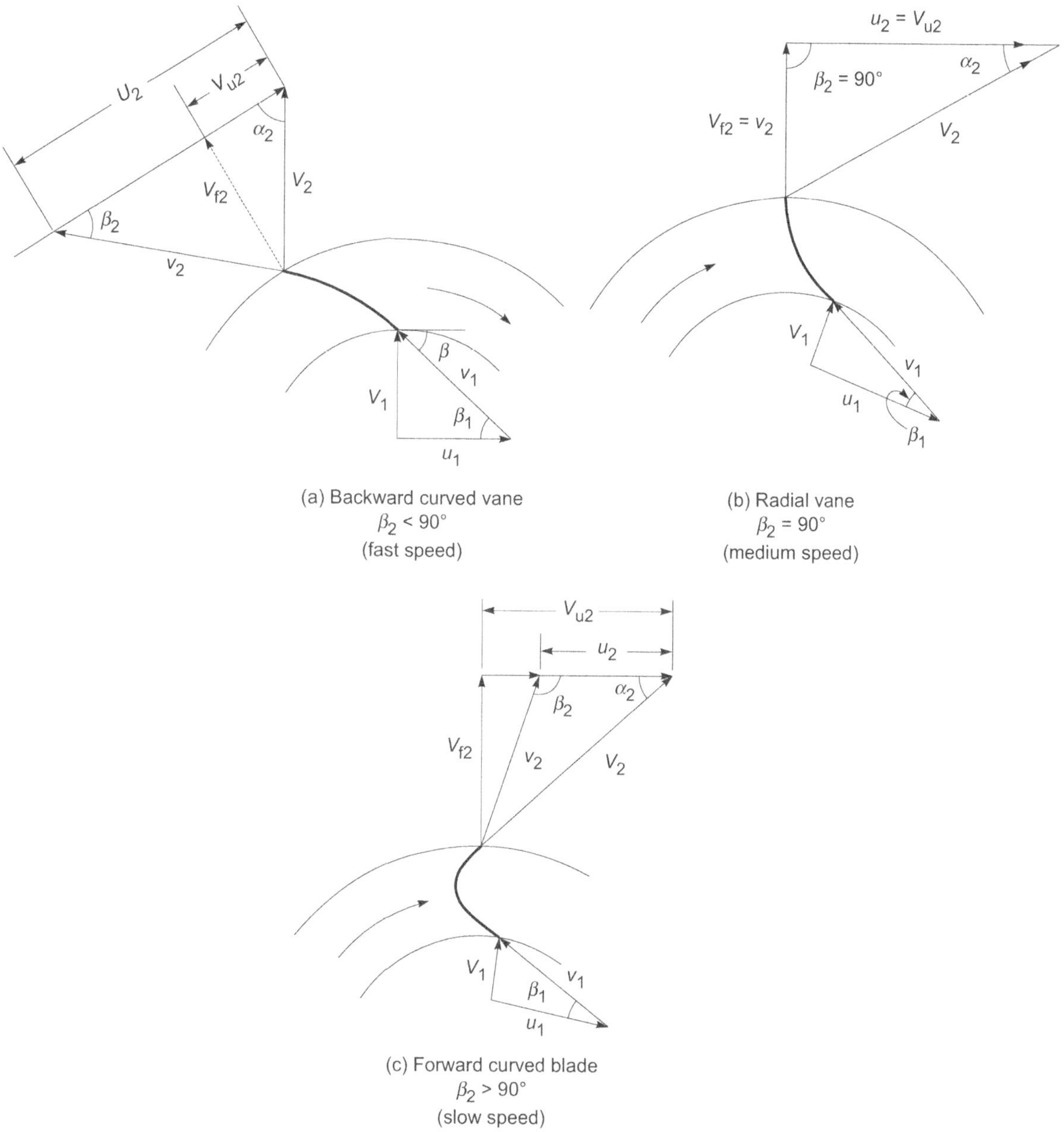

Fig. 16.8 *Different Vane Configurations of Impellers*

Volumetric Efficiency, η_v

$$\boxed{\eta_v = \frac{Q}{Q + Q_L}} \tag{16.37}$$

where Q = discharge actually delivered and Q_L = leakage of discharge.

Mechanical Efficiency, η_m

$$\eta_m = \frac{\text{brake power} - \text{power loss due to friction}}{\text{brake power}}$$

$$= \boxed{\eta_m \ \frac{(\mathbf{BP}) - \mathbf{P_f}}{(\mathbf{BP})}} \tag{16.38}$$

Overall efficiency, η_0

$$\eta_0 = \frac{\text{power delivered to the fluid}}{\text{power put into the shaft (BP)}}$$

$$\boxed{\eta = \frac{\gamma QH}{(\mathbf{BP})} = \eta_v\, \eta_m\, \eta_H} \tag{16.39}$$

16.3.2 Basic Features of Rotodynamic Pumps

Radial Entry At the design value of discharge, the flow is assumed to enter the impeller radially. This implies $\alpha_1 = 90°$ and $V_{f1} = V_1$. If this information is not explicitly specified, it can be assumed in the solution of problems. The inlet velocity diagram will be as in Fig. 16.9. The manometric head

$$H' = \frac{u_2 V_{u2}}{g}$$

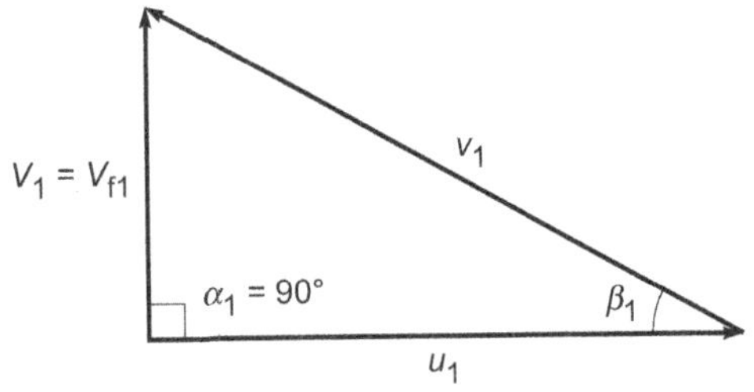

Fig. 16.9 *Radial Entry*

Specific Speed, N_s For a pump, the specific speed N_s is defined in SI units as

$$N_s = \frac{N\sqrt{Q}}{H^{3/4}} \tag{16.40}$$

where N = rotational speed in rpm for maximum efficiency, Q = discharge in m^3/s, and H = net head developed in m. Note that the specific speed N_s has the dimensions of $[L^{3/4}\,T^{-3/2}]$. The non-dimensional form of specific speed for pumps, known as *shape number* is

$$S_q = \frac{N\sqrt{Q}}{(gH)^{3/4}} \tag{16.41}$$

For double suction pumps the specific speed N_s is based on one half of the total capacity of the pump. The typical ranges of N_s of commonly used pump types are as follows:

Radial flow	Slow speed	10 – 30
	Median speed	30 – 50
	High speed	50 – 80
Mixed flow pumps		80 – 200
Axial flow pumps		200 – 300

Minimum Speed When the pump is switched on, the flow will take place only when the rise in pressure due to impeller action is large enough to overcome manometer head. The pressure head created by centrifugal action on the rotating liquid is $(u_2^2 - u_1^2)/2g$. Thus the flow will commence only if

$$(u_2^2 - u_1^2)/2g \geq H' \tag{16.42}$$

where H' = manometric head given by Eq. 16.34. Noting that $u_2 = \pi D_2 N/60$ and $u_1 = \pi D_1 N/60$ the minimum speed (in rpm) to start the pump is

$$N_{min} = \frac{60}{\pi\,[D_2^2 - D_1^2]^{1/2}}\sqrt{2gH'} \tag{16.43}$$

16.3.3 Similarity Laws

In the testing of pumps one is interested in the operation of pumps at a certain speed as determined by the motor to which it is coupled. Hence, the rotative speed N, diameter D are the basic repeating variables in the similarity relations. The similitude laws for discharge, head and power are expressed in terms of N and D. For homologous pumps, by using the suffixes m and p to designate the model and prototype respectively, we have the following relations:

$$\frac{H_m}{D_m^2 N_m^2} = \frac{H_p}{D_p^2 N_p^2} \tag{16.44-a}$$

$$\frac{Q_m}{N_m D_m^3} = \frac{Q_p}{N_p D_p^3} \tag{16.44-b}$$

$$\frac{\mathbf{P}_m}{\gamma_m D_m^5 N_m^3} = \frac{\mathbf{P}_p}{\gamma_m D_p^5 N_p^3} \tag{16.44-c}$$

Note that only the power relation contains the fluid property term, γ.

From the above relationships, it follows that for two homologous pumps the specific speed N_s will be the same. Thus between a geometric model and its prototype,

$$N_{sm} = \left(\frac{N_m\sqrt{Q_m}}{H_m^{3/4}}\right) = N_{sp} = \left(\frac{N_p\sqrt{Q_p}}{H_p^{3/4}}\right) \tag{16.45}$$

16.3.4 Pumps in Parallel

If two similar pumps A and B are connected in parallel (Fig. 16.10) the combined discharge Q_t will be the sum of the individual discharges Q_a and Q_b, i.e.

$$Q_t = Q_a + Q_b \tag{16.46}$$

The head H will however be the same in both the pumps and will also be the net head of the combined discharge. Thus,

$$H = H_a = H_b \tag{16.47}$$

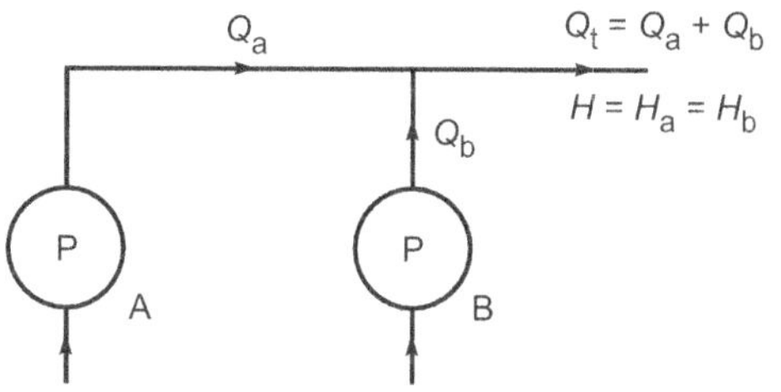

Fig. 16.10 *Pumps Connected in Parallel*

16.3.5 Pumps in Series

If two similar pumps 1 and 2 are connected in series (Fig. 16.11) the discharge will not change and the heads will be added up. Thus the total net head

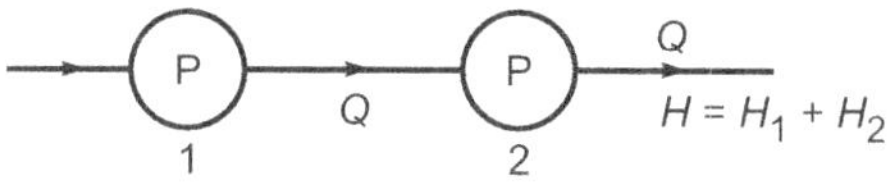

Fig. 16.11 *Pumps in Series*

$$H_t = H_1 + H_2 \tag{16.48}$$

and the discharge $Q = Q_1 = Q_2$ (16.49)

16.3.6 Cavitation

When the absolute local pressure at any point in a conduit carrying a liquid approaches the vapour p_v of the liquid, the dissolved gases and liquid vapour emerge out of the liquid as bubbles. These bubbles may travel to regions of higher pressure and collapse. At the point of bubble collapse the boundary gets damaged. The phenomenon of formation, travel and collapse of vapour bubbles is known as cavitation.

In pumps the suction end of the pump and blade passages are susceptible to cavitation. Cavitation in pumps causes reduction of the efficiency and often causes damage to the material of the pump assembly. An important factor in the pump operation is the avoidance of cavitation. A cavitation parameter σ (also known as *Thoma number*) is defined as

$$\boxed{\sigma = \frac{\dfrac{(p_s)_{abs}}{\gamma} + \dfrac{V_s^2}{2g} - \dfrac{p_v}{\gamma}}{H} = \frac{(\text{NPSH})}{H}} \tag{16.50}$$

in which suffix s denotes suction side and

$(p_s)_{abs}$ = absolute pressure at pump intake
V_s = Velocity at pump intake
p_v = vapour pressure (absolute)
NPSH = net positive suction head

$$= \left\{\frac{(p_s)_{abs}}{\gamma} + \frac{V_s^2}{2g} - \frac{p_v}{\gamma}\right\} \tag{16.51}$$

Depending upon the pump, a critical cavitation number (= critical Thoma number) σ_c is identified. Any value of $\sigma < \sigma_c$ will result in cavitation and cause severe reduction in the pump efficiency. Hence for operational purposes it should be seen that

$$\sigma \geq \sigma_c \tag{16.52}$$

i.e., $(\text{NPSH}) \geq \sigma_c H$

Hence, the minimum NPSH = $\sigma_c H$ (16.53)

16.4 RECIPROCATING PUMP

16.4.1 Introduction

A reciprocating pump is a positive displacement pump. In this a piston moving inside a cylinder draws in the fluid by suction and discharges it to the delivery pipe by pushing it bodily by the action of the piston. The piston gets its reciprocating motion through a crank and connecting mechanism that are connected to a prime mover. One-way valves provided in the suction and delivery sides of the piston help in pumping action. Figure 16.12(a) shows schematically a single acting reciprocating pump. In the single acting pump shown in Fig. 16.12(a) the suction stroke (crank angle 0° – 180°) gets the liquid from the sump to the cylinder and the delivery stroke (crank angle 180° – 360°) drives the liquid out of the cylinder. The suction and delivery strokes take place alternatively. The variation of the discharge with the crank angle is shown in Fig. 16.12(b).

In a double acting reciprocating pump, the suction and delivery strokes occur simultaneously. Figures 16.13(a) and (b) show a schematic sketch of a double acting reciprocating pump and the variation of discharge with crank angle in such a pump.

Multi-cylinder pumping arrangements are employed to get more steady flow in the delivery pipe. In these the cranks in a common drive provide the necessary phase shift and the outlets from a set of multi-cylinders are connected to a common delivery pipe. Figures 16.14(a, b and c) show two-throw and three-throw pumps and the discharge performance of a three–throw pump respectively.

16.4.2 Notations

A = Cross sectional area of piston
d_p = Diameter of piston
L = Length of stroke = twice the crank radius
r = crank radius
N = Revolutions per minute of the crank
H_d = Static delivery head
d_d = diameter of delivery pipe
L_d = Length of delivery pipe
H_s = Static suction head

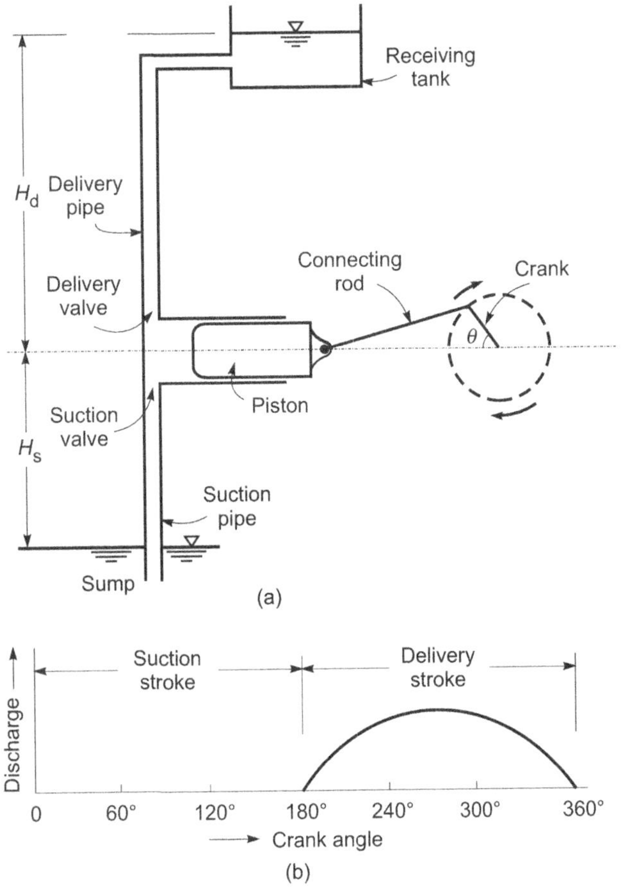

Fig. 16.12 *Single Acting Reciprocating Pump (a) Schematic Sketch (b) Variation of Discharge with Crank Angle*

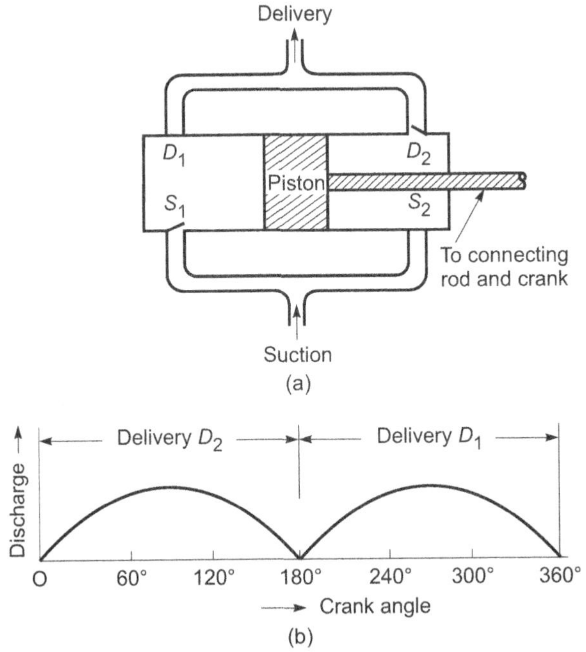

Fig. 16.13 *Double Acting Reciprocating Pump (c) Schematic Sketch (d) Variation of Discharge with Crank Angle*

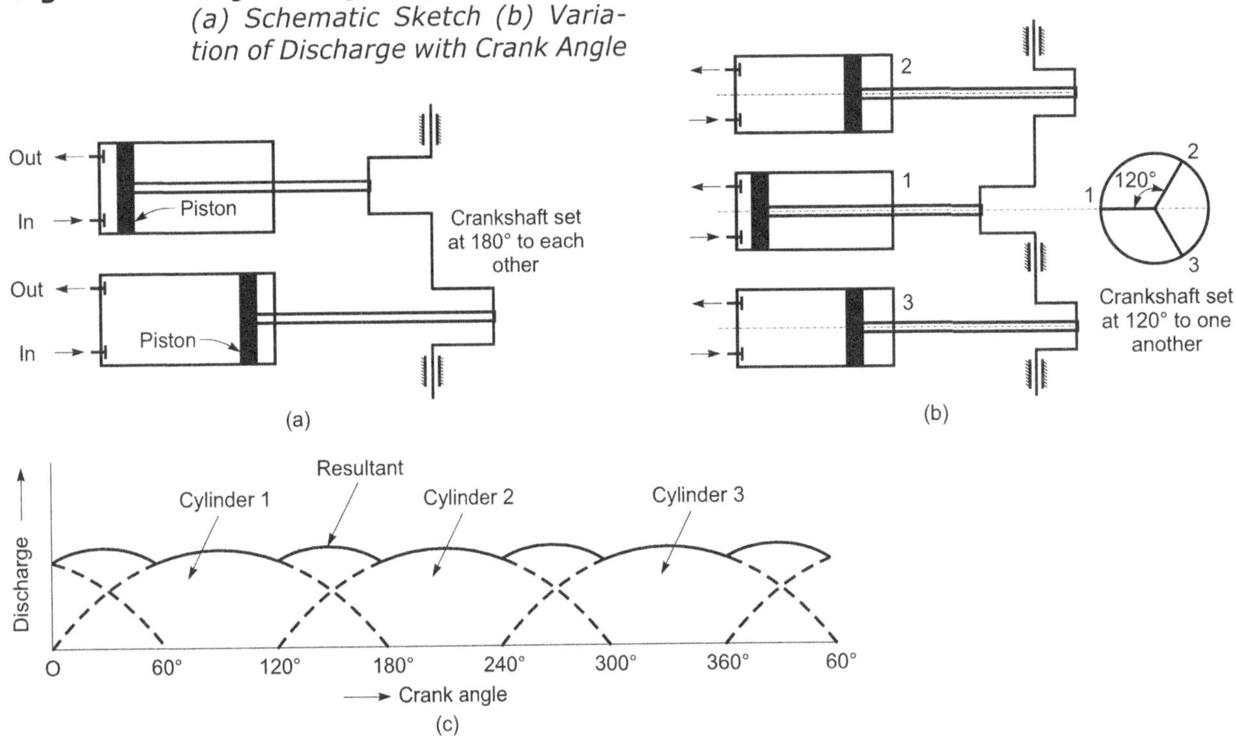

Fig. 16.14 *Multi Throw Reciprocating Pump (a) Two-throw Pump (b) Three-throw Pump (c) Variation of Discharge in a Three-throw Pump*

d_s = Diameter of suction pipe
L_s = Length of suction pipe
Q_t = Theoretical discharge
P = Power

16.4.3 Discharge

Theoretical discharge for single acting pump

$$Q_t = \frac{ALN}{60} \tag{16.54}$$

For a double acting pump

$$Q_t = \frac{2ALN}{60} \tag{16.55}$$

Volumetric efficiency (percentage)

$$\eta_e = \frac{\text{Actual discharge}}{\text{Theoretical discharge}} \times 100$$

$$= \frac{Q_t}{Q_a} \times 100 \tag{16.56}$$

$$\text{Slip} = \left(\frac{Q_t - Q_a}{Q_t}\right) \times 100 = 100 - \eta_e \tag{16.56a}$$

Coefficient of discharge

$$= \frac{Q_a}{Q_t} \tag{16.57}$$

16.4.4 Simple Harmonic Motion

The motion of the piston in the cylinder is treated as a simple harmonic motion. The crank rotates with an angular velocity ω radians/second. Then in a time t reckoned from inner dead centre

$$\theta = \omega t = \frac{2\pi N}{60} t$$

where N = rotational speed in rpm,
r = crank radius = stroke/2 = $L/2$

Displacement $x = r(1 - \cos\theta)$

Velocity of piston

$$v = \frac{dx}{dt} = \omega r \sin\theta = \omega r \sin \omega t$$

If A = area of piston and A_1 = area of a pipe (suction or delivery pipe)

Velocity of water in the pipe at any instant

$$v_1 = \frac{A}{A_1} \omega r \sin\theta \tag{16.58}$$

Maximum velocity in the pipe

$$v_{1m} = \frac{A}{A_1} \omega r$$

Time averaged velocity per cycle

$$V_1 = \frac{A}{A_1} \frac{\omega r}{\pi} \tag{16.59}$$

Acceleration of piston

$$= a = \frac{dv_p}{dt} = \omega^2 r \cos \omega t = \omega^2 r \cos\theta$$

Acceleration of fluid in the pipe (suction or delivery) at any instant

$$= a_1 = \frac{A}{A_1} \omega^2 r \cos\theta \tag{16.60}$$

$$\text{Maximum acceleration in the pipe} = a_{1m} = \frac{A}{A_1} \omega^2 r$$

16.4.5 Acceleration Head

The acceleration of fluid in the pipe requires a force F given by

$$F = \rho A_1 L_1 \left(\frac{A}{A_1}\right) \omega^2 r \cos\theta$$

The pressure head caused by this force F is

$$H_{a1} = \frac{F}{A_1 \rho g} = \frac{L_1}{g} \frac{A}{A_1} \omega^2 r \cos\theta \tag{16.61}$$

H_a = is known as acceleration or inertial head. This is an additional head that is required to be developed by the pump.

By replacing the suffix 1 denoting any pipe by the specific suffixes s to denote suction and d to denote delivery pipes

$$H_{as} = \frac{F}{A_s \rho g} = \frac{L_s}{g} \frac{A}{A_s} \omega^2 r \cos\theta$$

Maximum Acceleration head in suction pipe is at $\theta = 0$ and is

$$H_{asm} = \frac{L_s}{g} \frac{A}{A_s} \omega^2 r \tag{16.62}$$

For delivery pipe:

$$H_{ad} = \frac{F}{A_d \rho g} = \frac{L_d}{g} \frac{A}{A_d} \omega^2 r \cos\theta$$

Maximum acceleration head in delivery pipe is at $\theta = 0$ and is

$$H_{adm} = \frac{L_d}{g} \frac{A}{A_d} \omega^2 r \tag{16.63}$$

16.4.6 Friction Head

Darcy–Weisbach formula with friction factor f is used to estimate the friction losses h_{fs} and h_{fd} in suction pipe and delivery pipe respectively.

(i) *For suction pipe*

$$h_{fs} = \frac{flV_s^2}{2gd_s} = \frac{fL_s}{2gd_s}\left(\frac{A}{A_s} r\omega \sin\theta\right)^2 \quad (16.64)$$

Maximum h_{fs} is at ($\theta = \pi/2$) and hence h_{fsm}

$$= \frac{fL_s}{2gd_s}\left(\frac{A}{A_s} r\omega\right)^2$$

$$h_{fsa} = \text{Average } h_{fs} = (2/3)\, h_{fsm}$$

$$= \frac{1}{3}\frac{fL_s}{gd_s}\left(\frac{A}{A_s} r\omega\right)^2 \quad (16.65)$$

(ii) *Similarly for delivery pipe*

$$h_{fd} = \frac{flV_d^2}{2gd_d} = \frac{fL_d}{2gd_d}\left(\frac{A}{A_d} r\omega \sin\theta\right)^2 \quad (16.66)$$

Maximum h_{fd} is at ($\theta = \pi/2$) and hence

$$h_{fdm} = \frac{fL_d}{2gd_s}\left(\frac{A}{A_d} r\omega\right)^2$$

$$h_{fda} = \text{Average } h_{fd} = (2/3)\, h_{fdm}$$

$$= \frac{1}{3}\frac{fL_d}{gd_s}\left(\frac{A}{A_d} r\omega\right)^2 \quad (16.67)$$

Table 16.1 *Combined Effect of Acceleration and Friction*

Item	*Total head in Suction Pipe* $H_{ts} = H_{atmos} - (H_{as} + H_s + h_{fs})$ *(absolute)*	*Total head in Delivery Pipe* $H_{td} = H_{atmos} + (H_{ad} + H_d + h_{fd})$ *(absolute)*
Beginning of Storke ($\theta = 0$)	$h_{fs} = 0$	$h_{fd} = 0$
Mid-stroke ($\theta = 90°$)	$H_{as} = 0$	$H_{ad} = 0$
End of stroke ($\theta = 180°$)	$h_{fs} = 0$	$h_{fd} = 0$

16.4.7 Indicator Diagram

An indicator diagram is a plot showing the variation of pressure in the cylinder at various stages of the strokes. The area of the indicator diagram represents the work done by the pump. Figure 16.15 is an indicator diagram showing effect of acceleration and friction in a single stage reciprocating pump.

- The parabola 1-2-3 represents work done against friction in suction pipe
- The parabola 4-5-6 represents work done against friction in delivery pipe

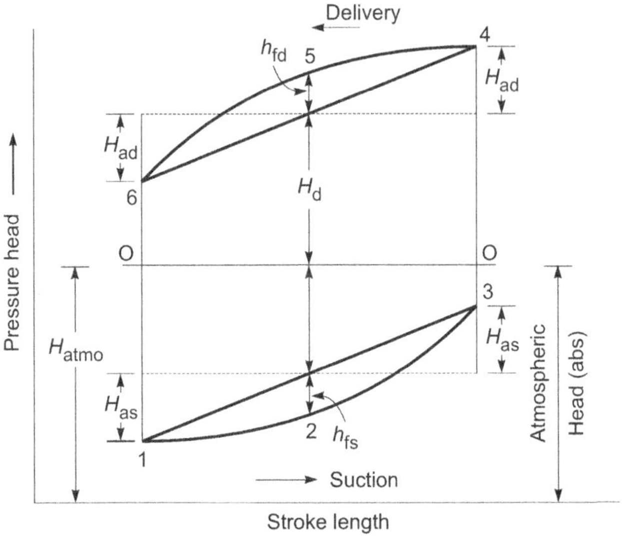

Fig. 16.15 *Indicator diagram*

If L = length of the stroke, and noting that the mean ordinate of a parabola is equal to 2/3 of the maximum ordinate

Total work done in one complete cycle of the crank = area of the indicator diagram

$$= (H_sL + (2/3)\, h_{fs}L + H_dL + (2/3)\, h_{fd}L)$$

$$= L\,(H_s + h_{fsa} + H_d + h_{fda})$$

Total work done per second = Power expended

$$\mathbf{P} = \frac{\gamma AN}{60} \text{ (Area of indicator diagram)}$$

$$= \frac{\gamma ALN}{60}(H_s + h_{fsa} + H_d + h_{fda}) \quad (16.68a)$$

For a double acting pump

$$\mathbf{P} = \frac{2\gamma ALN}{60}(H_s + h_{fsa} + H_d + h_{fda}) \quad (16.68b)$$

The safe speed of the pump is decided of the following considerations:

- At the beginning of the suction stroke the net head should be more than the minimum head at which cavitation (Separation of vapour) can occur

  i.e, At ($\theta = 0°$)

  $$H_{ts} = H_{atmo} - (H_{as} + H_s) > H_v \text{ (abs)} \quad (16.69a)$$

- At the end of the delivery stroke the net head should be more than the minimum head at which cavitation (Separation of vapour) can occur

  At ($\theta = 180°$)

  $$H_{td} = H_{atmo} + (H_{ad} + H_d) > H_v \text{ (abs)} \quad (16.69b)$$

16.4.8 Air Vessel

An air vessel is a large closed chamber fitted to a reciprocating pump to eliminate pulsations of pressure and discharge in suction and delivery pipes. These are fitted in either/both delivery and suction sides and as close to the pump cylinder as possible. Figure 16.16 shows a schematic arrangement of two air vessels in a pump system.

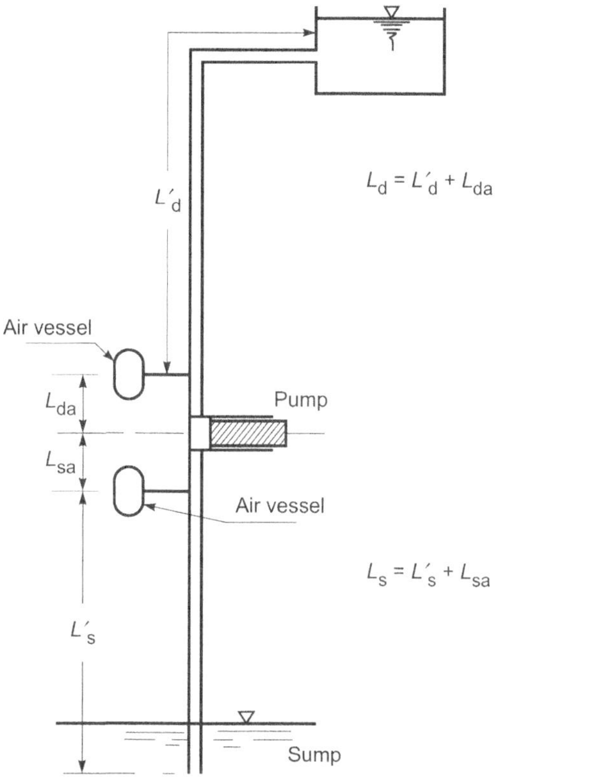

Fig. 16.16 *Schematic Arrangement of Air Vessels*

The chief advantages of air vessels in a reciprocating pump are

- Air vessel on suction side:
  - (i) Reduces cavitation possibility
  - (ii) For a given minimum pressure head, the pump can run at higher speed
  - (iii) Suction pipe length can be increased
- Air Vessel on delivery side:
  - (i) Almost constant delivery discharge is obtained
  - (ii) Reduction in friction loss and hence saving in power

Referring to Fig. 16.16, the acceleration heads are confined to the lenghts L_{sa} and L_{da} only. Beyond the air vessels, the velocity is constant at V_s and V_d in the suction pipe of length L'_a and delivery pipe of length L'_d respectively.

In the suction pipe

$$V_s = \frac{Q_t}{A_s} = \left(\frac{A}{A_s}\right)\left(\frac{LN}{60}\right) = \left(\frac{A}{A_s}\right)\frac{r\omega}{\pi}$$

Similarly in delivery pipe

$$V_d = \frac{Q_t}{A_d} = \left(\frac{A}{A_d}\right)\left(\frac{LN}{60}\right) = \left(\frac{A}{A_d}\right)\frac{r\omega}{\pi}$$

If Q_i = instantaneous discharge, then in any pipe (suction or delivery),

$$Q_i = A_1 v_1 = A_1 \omega r \sin\theta$$

Discharge in air vessel

$$Q_{av} = Q_t - Q_i = A_1 r\omega\left(\sin\theta - \frac{1}{\pi}\right) \quad (16.70a)$$

For a double acting pump,

$$Q_{av} = Q_t - Q_i = A_1 r\omega\left(\sin\theta - \frac{2}{\pi}\right) \quad (16.70b)$$

(In the above equations, the suffix 1 should be replaced by suffix s for suction pipe and by suffix d for delivery pipe).

- If the discharge Q_{av} is positive, then the flow is into the air vessel.
- If the discharge Q_{av} is negative, then the flow is out of the air vessel.
- For zero flow in the an air vessel in a single acting pump, $\sin\theta = 1/\pi$ or $\theta = 18.56°$ or $161.44°$.
- Similarly for zero into an air vessel in a double acting pump, $\sin\theta = 2/\pi$ or $\theta = 39.54°$ or $140.46°$.

16.4.9 Work Done Per Second (Power Consumed)

Work done per second (Power consumed) in kW with air vessels in the suction and delivery pipes is

$$\mathbf{P}_t = \frac{\gamma Q_t}{1000}\left[H_s + \frac{f(L_s - L_{sa})^2}{2gd_s}V_s^2 + \frac{1}{3}\frac{fL_{sa}}{gd_s}\left(\frac{A}{A_s}r\omega\right)^2 + \right.$$

$$\left. H_d + \frac{f(L_d - L_{da})^2}{gd_d}V_d^2 + \frac{1}{3}\frac{fL_{da}}{2gd_d}\left(\frac{A}{A_d}r\omega\right)^2\right] \quad (16.71)$$

In this Q_t = Theoretical discharge = $\frac{ALN}{60}$ and $\omega = \frac{2\pi N}{60}$.

If η is the efficiency of the pumping system, then the power (in kW) to be supplied is

$$\mathbf{P} = \eta\,\mathbf{P}_t \quad (16.72)$$

WORKED EXAMPLES

A. Francis Turbine

16.1 An inward flow reaction turbine has inlet and outlet diameters of 1.2 m and 0.6 m respectively. The breadth at inlet is 0.25 m and at outlet it is 0.35 m. At a speed of rotation of 250 rpm, the relative velocity at entrance is 3.5 m/s and is radial. Calculate the (i) absolute velocity at entrance and the inclination to the tangent of the runner, (ii) discharge and (iii) the velocity of flow at the outlet.

Solution: Referring to Fig. 16.17,

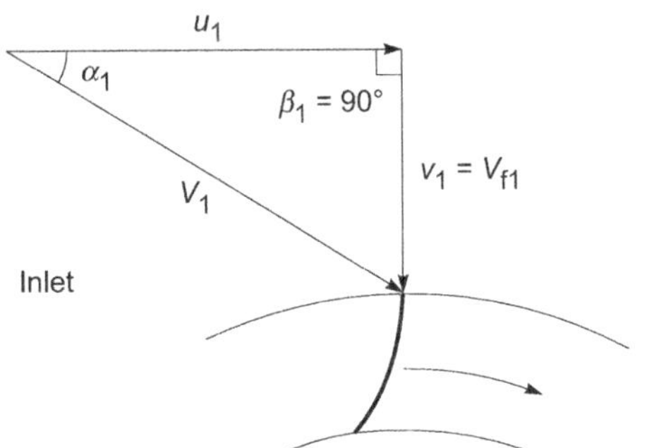

Fig. 16.17

$$u_1 = \frac{\pi D_1 N}{60} = \frac{\pi \times 1.2 \times 250}{60} = 15.7 \text{ m/s}$$

As $\beta_1 = 90°$, and $v_1 = V_{f1} = 3.5$ m/s

$$\tan \alpha_1 = \frac{v_1}{u_1} = \frac{3.5}{15.7} = 0.2228$$

(i) $\alpha = 12.56°$

$$V_1 = \text{absolute velocity at entry} = \frac{v_1}{\sin \alpha}$$

$$= \frac{3.5}{\sin (12.56°)} = \mathbf{16.093 \text{ m/s}}$$

(ii) Discharge

$$Q = \pi D_1 b_1 V_{f1}$$
$$= \pi \times 1.2 \times 0.25 \times 3.5 = \mathbf{3.299 \text{ m}^2\text{/s}}$$

(iii) At outlet

$$Q = \pi D_2 b_2 V_{f2}$$
$$3.299 = \pi \times 0.6 \times 0.35 \times V_{f2}$$
$$V_{f2} = \text{velocity of flow at outlet} = \mathbf{5.0 \text{ m/s}}$$

16.2 A Francis turbine has an inlet diameter of 2.0 m and an outlet diameter of 1.2 m. The breadth of the blades is constant at 0.2 m. The runner rotates at a speed of 250 rpm with a discharge of 8.0 m^3/s. The vanes are radial at the inlet and the discharge is radially outwards at the outlet. Calculate the angle of guide vane at the inlet and blade angle at the outlet.

Solution: Referring to Fig. 16.18,

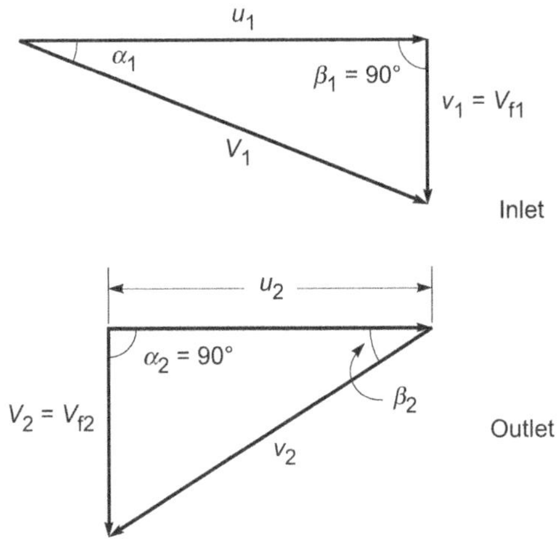

Fig. 16.18

Peripheral velocity

$$u_1 = \frac{\pi D_1 N}{60} = \frac{\pi \times 2.0 \times 250}{60} = 26.18 \text{ m/s}$$

At the outlet $u_2 = \frac{D_2}{D_1} u_1 = \frac{1.2}{2.0} \times 26.18 = 15.708$ m/s

Discharge $Q = \pi D_1 b V_{f1}$

$$8.0 = \pi \times 2.0 \times 0.2 \times V_{f1}$$

Flow velocity at inlet

$$V_{f1} = \frac{8.0}{\pi \times 0.4} = 6.366 \text{ m/s}$$

Velocity of flow at outlet

$$V_{f2} = \frac{D_1}{D_2} \times 6.366$$
$$= \frac{2.0}{1.2} \times 6.366$$
$$= 10.61 \text{ m/s}$$

As the vanes are radial at the inlet, $v_1 = V_{f1} = 6.366$ m/s. From the inlet velocity triangle,

$$\tan \alpha_1 = \frac{V_{f1}}{u_1} = \frac{6.366}{26.18} = 0.243$$

$\therefore \quad$ **$\alpha_1 = 13.67°$ = Inlet guide vane angle**

At the outlet,

$$\alpha_2 = 90°$$

$$\tan \beta_2 = \frac{V_{f2}}{u_2} = \frac{10.61}{15.708} = 0.675$$

$\therefore \quad$ **$\beta_2 = 34.04°$ = outlet blade angle**

16.3 At what angles the wicket gates of a Francis turbine be set to extract 8000 kW of power from a flow of 30 m^3/s when running at a speed of 200 rpm? The diameter of the runner at inlet is 3.0 m and the breadth of openings at inlet is 0.9 m (Fig. 16.19). The flow can be assumed to leave the runner radially, and the blade angle β is obtuse.

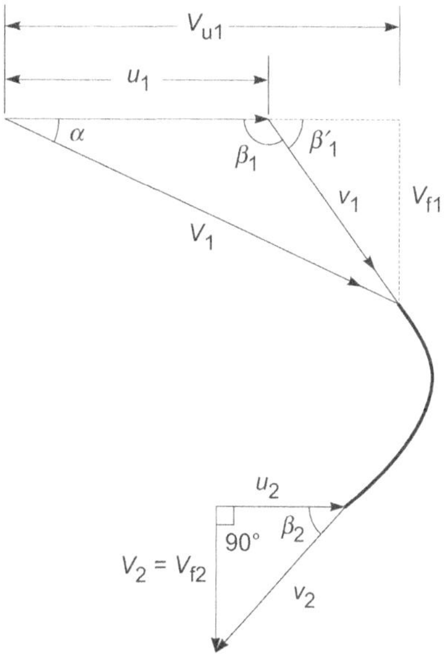

Fig. 16.19

Solution:

$$u_1 = \frac{\pi D_1 N}{60} = \frac{\pi \times 3.0 \times 200}{60} = 31.42 \text{ m/s}$$

Discharge $\quad Q = \pi D_1 b V_{f1}$

$$30 = \pi \times 3.0 \times 0.9 \times V_{f1}$$

V_{f1} = velocity of flow at inlet = 3.537 m/s

Power $\quad \mathbf{P} = \rho Q(u_1 V_{u1} - u_2 V_{u2})$

Since the outlet flow is radial

$$\alpha_2 = 90° \quad \text{and} \quad V_{u2} = 0$$

$$\mathbf{P} = \rho Q u_1 V_{u1}$$

$$8000 \times 10^3 = 998 \times 30.0 \times 31.42 \times V_{u1}$$

$$V_{u1} = 8.505 \text{ m/s}$$

From the inlet velocity triangle

$$\tan \alpha = \frac{V_{f1}}{V_{u1}} = \frac{3.537}{8.505} = 0.4159$$

$\alpha = 22.58°$

16.4 A reaction turbine works at 450 rpm under a head of 115 m. The diameter of the inlet is 1.20 m and the flow area is 0.4 m^2. At the inlet the absolute and relative velocities make angles of 20° and 60° respectively with the tangential velocity. Determine (i) the power developed and (ii) hydraulic efficiency. Assume the velocity of whirl at the outlet to be zero.

Solution: Refer to Fig. 16.20.

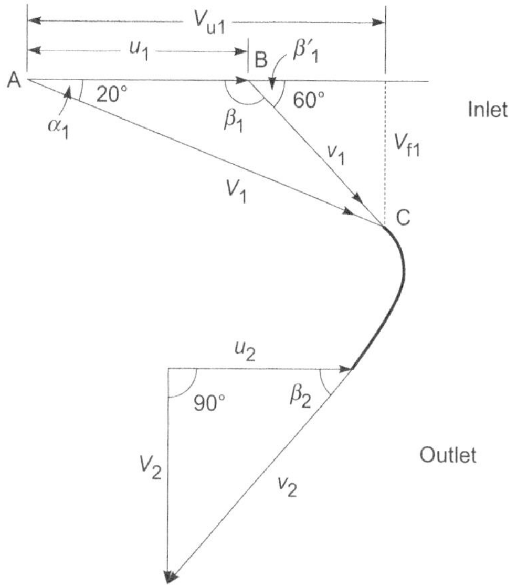

Fig. 16.20

At the inlet $\quad \alpha_1 = 20°, \quad \beta'_1 = 60°, \quad \beta = 120°$

$$u_1 = \frac{\pi D_1 N}{60} = \frac{\pi \times 1.2 \times 450}{60}$$

$$= 28.27 \text{ m/s}$$

Form Δ ABC, $\quad A\hat{C}B = 40°$

$$\frac{V_1}{\sin 120°} = \frac{u_1}{\sin 40°}$$

$$V_1 = 28.27 \times \frac{\sin 120°}{\sin 40°} = 38.09 \text{ m/s}$$

Velocity of flow $\quad V_{f1} = V_1 \sin 20° = 38.09 \sin 20°$

$$= 13.03 \text{ m/s}$$

Velocity of whirl $\quad V_{u1} = 38.09 \cos 20° = 35.79$ m/s

Discharge $\quad Q = \text{area} \times V_{f1} = 0.4 \times 13.03$

$$= 5.212 \text{ m}^3/\text{s}$$

Power developed $\quad \mathbf{P} = \rho\, Q(u_1 V_{u1} - u_2 V_{u2})$

But $\quad V_{u2}$ = velocity of whirl at outlet = 0

(i) Hence $\quad \mathbf{P} = \rho\, Q u_1 V_{u1}$

$$= 998 \times 5.212 \times 28.27 \times 35.79$$

$$= 5263 \times 10^3 \text{ W}$$

$$= \mathbf{5263 \text{ kW}}$$

(ii) Efficiency $\eta_H = \dfrac{u_1 V_{u1}}{gH} = \dfrac{28.27 \times 35.79}{9.81 \times 115} = 0.897$

$= \mathbf{89.7\%}$

16.5 An inward flow reaction turbine has an inlet guide vane angle of 30° and inlet edges of runner blade at 120° to the direction of rotation. The breadth of the runner at inlet is a quarter of the diameter and there is no whirl velocity at outlet. The overall head is 15 m and the speed is 1000 rpm (Fig. 16.21). The hydraulic and overall efficiencies may be assumed to be 88% and 85% respectively. Estimate the runner diameter at inlet and the power developed.

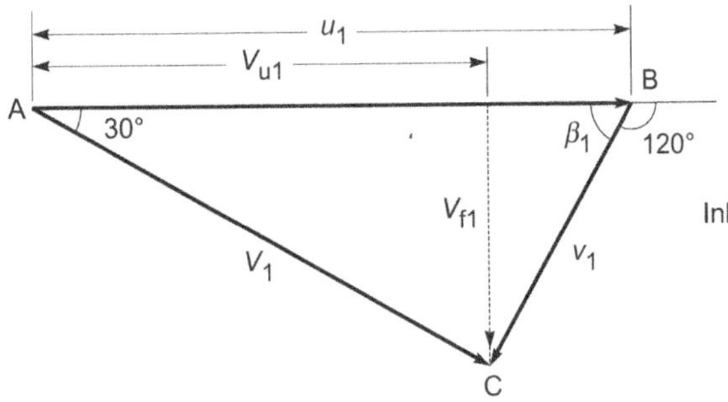

Fig. 16.21

Solution:

$$\alpha_1 = 30°,\ \beta_1 = 120°,\ \beta_1' = 60°$$

In triangle ABC, $A\hat{C}B = 90°$

$$V_1 = u_1 \cos 30° = 0.866\ u_1$$

$$V_{u1} = V_1 \cos 30° = (0.866)^2\ u_1 = 0.75\ u_1$$

As the outflow is radial, $V_{u2} = 0$

Head extracted $H_e = \dfrac{u_1 V_{u1}}{g} = \eta_H \cdot H$

$$= \frac{0.75\ u_1^2}{9.81} = 0.88 \times 15.0 = 13.2 \text{ m}$$

$$u_1 = 13.14 \text{ m/s}$$

If D_1 = Diameter of the runner at the inlet

$$u_1 = \frac{\pi D_1 N}{60}$$

$$13.14 = \frac{\pi \times D_1 \times 1000}{60}$$

(i) $D_1 = 0.25 \text{ m} = \mathbf{25\ cm}$

(ii) Discharge $Q = \pi D_1 b V_{f1}$

$$b = \frac{D_1}{4} = \frac{0.25}{4} = 0.0625 \text{ m}$$

$$V_{f1} = V_1 \sin 30° = 0.866 \times 13.14 \times 0.5$$
$$= 5.69 \text{ m/s}$$

$$Q = \pi \times 0.25 \times 0.0625 \times 5.69$$
$$= 0.279 \text{ m}^3/\text{s}$$

Available power $= \gamma Q H_e$

Developed power $= \eta_o \gamma Q H$

where η_o = overall efficiency

$$\mathbf{P} = \eta_o \gamma Q H = 0.85 \times (9.81 \times 998) \times 0.279 \times 15$$
$$= 34864 \text{ W} = \mathbf{34.864\ kW}$$

16.6 An inward flow reaction turbine has a runner of 60 cm diameter and the breadth at inlet is 6 cm. The inner diameter is 45 cm and the breadth at outlet is 8 cm. The vanes occupy 5% of the peripheral area. The guide vane angle at the inlet is 22° and the vane angles, at inlet and outlet are 80° and 25° respectively. The head on the turbine is 60 m and the hydraulic efficiency is known to be 90%. Assuming a mechanical efficiency of 95% calculate the power delivered to the turbine. Also determine the rotational speed of the turbine.

Solution: Refer to Fig. 16.22.

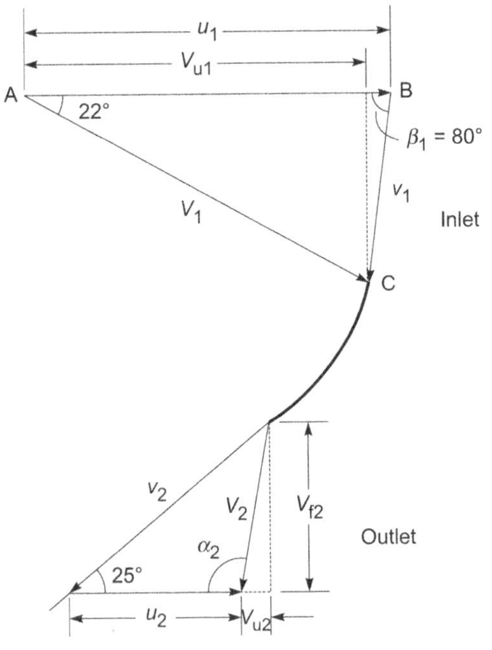

Fig. 16.22

Given data: $\alpha_1 = 22°,\ \beta_1 = 80°,\ D_1 = 0.60$ m

$\beta_2 = 25°,\ D_2 = 0.45$ m, $H_t = 60$ m

Discharge $Q = \pi D_1 b_1 (1 - 0.05)\ V_{f1}$

$= \pi D_2 b_2 (1 - 0.05)\ V_{f2}$

$$\frac{V_{f1}}{V_{f2}} = \frac{D_1 b_1}{D_2 b_2} = \frac{0.60}{0.45} \times \frac{0.06}{0.08} = 1.0$$

Hence $V_{f1} = V_{f2}$

At the inlet from Δ ABC, $A\hat{C}B = 78°$

$$\frac{u_1}{\sin 78°} = \frac{V_1}{\sin 80°}$$

$$V_1 = 1.0068\ u_1$$

$$\tan \alpha_1 = \frac{V_{f1}}{V_{u1}} = \tan 22° = 0.404$$

$$V_{f1} = 0.404\ V_{u1} \quad \text{i.e.} \quad V_{u1} = 2.4751\ V_{f1}$$

Also $$\frac{V_{f1}}{u_1 - V_{u1}} = \tan \beta_1 = \tan 80° = 5.671$$

$$u_1 - V_{u1} = 0.1763\ V_{f1}$$

$$u_1 = V_{u1} + 0.1763\ V_{f1}$$

$$= (2.4751 + 0.1763)\ V_{f1}$$

$$u_1 = 2.6514\ V_{f1}$$

On the outlet side:

$$u_2 = \frac{D_2}{D_1} u_1 = \frac{0.45}{0.60} u_1 = 0.75\ u_1$$

$$= 0.75 \times 2.6514\ V_{f1} = 1.9885\ V_{f1}$$

Since $$\tan 25° = \frac{V_{f2}}{u_2 + V_{u2}}$$

$$0.4663 = \frac{V_{f1}}{1.9885\ V_{f1} + V_{u2}} \qquad (\because \quad V_{f1} = V_{f2})$$

$$V_{u2} = 0.156\ V_{f1}$$

$$\tan \alpha_2' = \frac{V_{f2}}{V_{u2}} = \frac{V_{f1}}{0.156\ V_{f1}}$$

$$\tan \alpha_2' = 6.41, \quad \alpha_2' = 81.13°$$

or $$\alpha_2 = 98.87°$$

Actual head extracted:

$$H_e = \frac{u_1 V_{u1} - u_2 V_{u2}}{g}$$

$$= \left[\frac{(2.6514 \times 2.4751) - (1.9885 \times 0.156)}{9.81}\right] V_{f1}^2$$

$$= 0.6373\ V_{f1}^2$$

Since the hydraulic efficiency

$$\eta_H = \frac{H_e}{H} = 0.90$$

$$H_e = 0.6373\ V_{f1}^2 = 0.90 \times 60$$

$$V_{f1} = 9.205 \text{ m/s}$$

$$u_1 = 2.6514 \times 9.205 = \frac{\pi \times 0.60 \times N}{60}$$

$$\mathbf{N = 776.9 \text{ rpm}}$$

Discharge $$Q = (1 - 0.05) \times \pi D_1 b_1 V_{f1}$$

$$= 0.95 \times \pi \times 0.6 \times 0.06 \times 9.205$$

$$= 0.989 \text{ m}^3/\text{s}$$

Power delivered $$= \gamma Q H\ \eta_m$$

$$= (9.81 \times 998) \times 0.989 \times (0.90 \times 60) \times 0.95$$

$$= 497 \times 10^3 \text{ W}$$

$$= \mathbf{497 \text{ kW}}$$

16.7 Show that for a Francis turbine the hydraulic efficiency is given by

$$\eta_H = \left[1 - \frac{1}{1 + 2\cot \alpha_1 (\cot \alpha_1 + \cot \beta_1)}\right]$$

Assume radial discharge at the outlet and constant velocity of flow. In the equation α_1 = guide vane angle at the inlet and β_1 = blade angle at the inlet.

Solution: The velocity of flow is constant.
Hence $V_{f1} = V_{f2}$.
Since the outlet flow is radial $V_2 = V_{f2}$.
From the inlet velocity triangle (see Fig. 16.23)

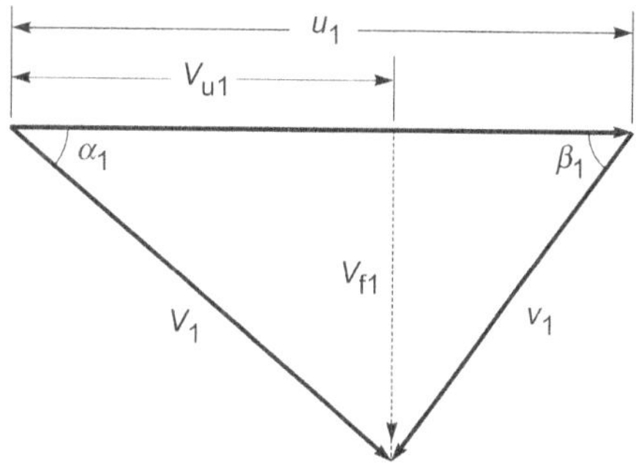

Fig. 16.23 *Inlet Velocity Triangle*

$$\frac{V_{f1}}{V_{u1}} = \tan \alpha, \quad \text{hence } V_{u1} = V_{f1} \cot \alpha_1$$

Also $$\frac{V_{f1}}{u_1 - V_{u1}} = \tan \beta_1, \quad \text{hence} \quad u_1 - V_{u1} = V_{f1} \cot \beta_1$$

$$u_1 = V_{u1} + V_{f1} \cot \beta_1$$

$$= V_{f1} (\cot \alpha_1 + \cot \beta_1)$$

Head extracted

$$H_e = \frac{u_1 V_{u1}}{g} = H - \frac{V_2^2}{2g} = H - \frac{V_{f1}^2}{2g}$$

where H = net available head.

$\therefore$ η_H = hydraulic efficiency

$$= \frac{u_1 V_{u1}}{gH} = \frac{u_1 V_{u1}/g}{\frac{u_1 V_{u1}}{g} + V_2^2/2g}$$

$$u_1 V_{u1} = V_{f1}^2 \cot \alpha_1 (\cot \alpha_1 + \cot \beta_1)$$

$$H = \frac{u_1 V_{u1}}{g} + \frac{V_{f1}^2}{2g}$$

$$gH = \frac{1}{2} [2\ V_{f1}^2 \cot \alpha_1 (\cot \alpha_1 + \cot \beta_1) + V_{f1}^2]$$

$$= \frac{1}{2} V_{f1}^2 [2 \cot \alpha_1 (\cot \alpha_1 + \cot \beta_1) + 1]$$

$$\eta_H = \frac{u_1 V_{u1}}{gH} = \frac{V_{f1}^2 (2 \cot \alpha_1)(\cot \alpha_1 + \cot \beta_1)}{V_{f1}^2 [2 \cot \alpha_1 (\cot \alpha_1 + \cot \beta_1) + 1]}$$

$$\eta_H = 1 - \left[\frac{1}{1 + 2\cot\alpha_1(\cot\alpha_1 + \cot\beta_1)}\right]$$

B. Kaplan Turbine

16.8 A Kaplan turbine develops 15000 kW power at a head of 30 m. The diameter of the boss is 0.35 times the diameter of the runner. Assuming a speed ratio of 2.0, a flow ratio of 0.65 and an overall efficiency of 90% calculate the (i) diameter of the runner, (ii) rotational speed and (iii) specific speed.

Solution:

Speed ratio $\phi = \dfrac{u_1}{\sqrt{2gH}} = 2.0$

Here $H = H_t = 30$ m

Henc $u_1 = 2.0 \times \sqrt{2 \times 9.81 \times 30} = 48.52$ m/s

Diameter of boss

$D_b = 0.35\ D$

Flow ratio $= V_{f1}/\sqrt{2gH} = 0.65$

$V_{f1} = 0.65\sqrt{2 \times 9.81 \times 30} = 15.77$ m/s

Power $\mathbf{P} = \gamma Q H\ \eta_0$

$15000 \times 10^3 = (9.81 \times 998) \times Q \times 30 \times 0.90$

$Q = 56.745$ m^3/s

But discharge $Q = \dfrac{\pi}{4}(D^2 - D_b^2) \times V_{f1}$

$$56.745 = \frac{\pi}{4}[D^2 - (0.35D)^2] \times 15.77$$

$$= 10.868\ D^2$$

$\boldsymbol{D = 2.285}$ **m**

$D_b = 0.35 \times 2.285 = 0.80$ m

If speed = N rpm,

$$u_1 = \frac{\pi DN}{60}$$

$$48.52 = \frac{\pi \times 2.285 \times N}{60}$$

$N =$ **405.5 rpm**

(iii) Specific speed

$$N_s = \frac{N\sqrt{\mathbf{P}}}{H^{5/4}}$$

Here $H = 30$ m, $\mathbf{P} = 15000$ kW, $N = 405.5$ rpm

$$N_s = \frac{405.5 \times \sqrt{15000}}{(30)^{5/4}} = \mathbf{707.4}$$

16.9 A Kaplan turbine has a runner diameter of 4.0 m and a hub diameter of 1.2 m. The discharge through the turbine is 70 m^3/s. The hydraulic and mechanical efficiencies can be assumed to be 0.9 and 0.93 respectively. Assuming an absence of whirl at the outlet and the discharge is free, estimate the net available head on the turbine and the power developed. If the speed ratio is 2.0, estimate the specific speed.

Solution:

Discharge $= Q = \dfrac{\pi}{4}(D^2 - D_b^2) \times V_{f1}$

$$70 = \frac{\pi}{4}[(4)^2 - (1.2)^2]V_{f1}$$

Velocity of flow = $V_{f1} = V_{f2} = 6.1213$ m/s

Let H = Net available head on the turbine.

Since at the outlet $V_2 = V_{f2} = 6.1213$ m/s, by energy equation

$$H - \frac{V_2^2}{2g} = \text{Head extracted} = H_e = \eta_H H$$

$$= 0.9\ \text{H}$$

$$\therefore \quad 0.1\ H = \frac{(6.1213)^2}{2 \times 9.81}$$

$\boldsymbol{H = 19.1}$ **m**

Power developed

$$\mathbf{P} = \gamma Q H \times \eta_H \times \eta_m$$

$$= 9.79 \times 70 \times 19.1 \times 0.9 \times 0.93$$

$$= \mathbf{10955\ kW}$$

Speed ratio $= \dfrac{u_1}{\sqrt{2gH}} = 2.0$

$$u_1 = 2.0\sqrt{2 \times 9.81 \times 19.1} = 38.72 \text{ m/s}$$

But $u_1 = \dfrac{\pi DN}{60} = \dfrac{\pi \times 4 \times N}{60}$

$$\therefore \quad N = \frac{60 \times 38.72}{\pi \times 4} = 184.9 \text{ rpm}$$

Specific speed

$$N_s = \frac{N\sqrt{\mathbf{P}}}{H^{5/4}} = \frac{184.9 \times \sqrt{10955}}{(19.1)^{5/4}} = \mathbf{485}$$

16.10 A Kaplan turbine develops 20,000 kW at a head of 35 m and at a rotational speed of 420 rpm. The outer diameter of the blades is 2.5 m and the hub diameter is 0.85 m. If the overall efficiency is 85% and the hydraulic efficiency is 88%, calculate the discharge, the inlet flow angle and the blade angle at the inlet.

Solution:

Power $\mathbf{P} = \eta_o\ \gamma Q H$

$20{,}000 = 0.85 \times 9.79 \times Q \times 35$

Discharge $Q = \dfrac{20.000}{0.85 \times 9.79 \times 35} = \mathbf{68.67\ m^3/s}$

$$Q = \frac{\pi}{4}(D^2 - D_b^2)\, V_{f1}$$

$$68.67 = \frac{\pi}{4}\{(2.5)^2 - (0.85)^2\}\, V_{f1} = 4.3413\, V_{f1}$$

Flow velocity at inlet

$$V_{f1} = \frac{68.67}{4.3413} = 15.82 \text{ m/s}$$

Peripheral velocity at inlet

$$u_1 = \frac{\pi DN}{60} = \frac{\pi \times 2.5 \times 420}{60} = 54.98 \text{ m/s}$$

Hydraulic efficiency

$$\eta_H = \frac{V_{u1} u_1}{gH}$$

$$0.88 = \frac{V_{u1}(54.98)}{9.81 \times 35}$$

Whirl velocity at inlet $V_{u1} = 5.495$ m/s

Since $V_{u1} < u_1$, the inlet velocity triangle is as in Fig. 16.24.

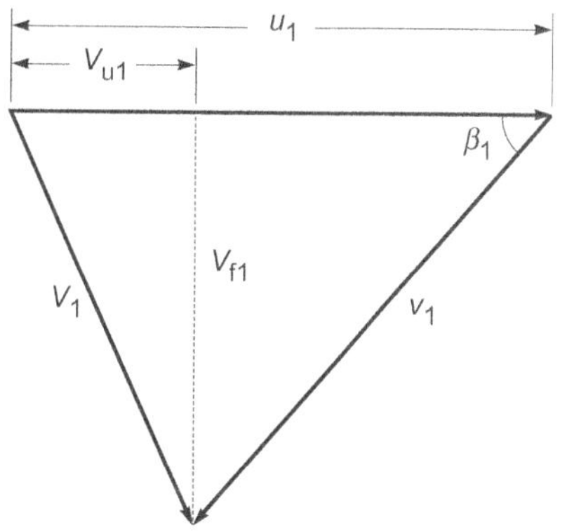

Fig. 16.24 *Inlet Velocity Triangle*

Let α_1 = Inlet flow angle

$$\tan \alpha_1 = \frac{V_{f1}}{V_{u1}} = \frac{15.82}{5.495} = 2.879$$

$$\boldsymbol{\alpha_1 = 70.84°}$$

Let β_1 = Inlet blade angle

$$\tan \beta_1 = V_{f1}/(u_1 - V_{u1}) = \frac{15.82}{(54.98 - 5.495)}$$

$$= 0.320$$

$$\boldsymbol{\beta_1 = 17.73°}$$

C. Pelton Turbine

16.11 A Pelton wheel 2.5 m diameter operates under the following conditions.

Net available head	= 300 m,
Speed	= 300 rpm
Coefficient of velocity of the jet	= 0.98
Blade friction coefficient	= 0.95
Blade angle	= 165°
Diameter of the jet	= 20 cm
Mechanical efficiency	= 0.95

Determine: (i) the power developed, (ii) hydraulic efficiency and (iii) specific speed

Solution:

$H = 300$ m, $D = 2.5$ m, $d = 0.20$ m

Here $C_v = 0.98$, $k = 0.95$, $\beta' = 180 - \beta = 15°$

V_1 = jet velocity = $C_v\sqrt{2gH}$

$$= 0.98\sqrt{2 \times 9.81 \times 300} = 75.186 \text{ m/s}$$

Discharge $Q = \frac{\pi}{4}(d)^2 V_1$

$$= \frac{\pi}{4} \times (0.2)^2\,(75.186) = 2.362 \text{ m}^3\text{/s}$$

$$u = \frac{\pi \times 2.5 \times 300}{60} = 39.27 \text{ m/s}$$

H_e = head extracted

$$= \frac{1}{g}\, u(V_1 - u)(1 + k\cos\beta')$$

$$= \frac{1}{9.81} \times 39.27 \times (75.186 - 39.27)\,(1 + 0.95\cos 15°)$$

$$= 275.7 \text{ m}$$

Hydraulic efficiency

$$= \eta_H = \frac{275.7}{300} = \mathbf{0.919}$$

Power developed

$$\mathbf{P} = \gamma Q H\, \eta_o = (\eta_H \cdot \eta_m)\, \gamma Q H$$

$$= 0.919 \times 0.95 \times 9.79 \times 2.362 \times 300 \text{ kW}$$

$$= \mathbf{6057\ kW}$$

Specific speed (per jet)

$$N_s = \frac{N\sqrt{\mathbf{P}}}{H^{5/4}} = \frac{300 \times \sqrt{6057}}{(300)^{5/4}} = \mathbf{18.7}$$

16.12 A Pelton wheel is working under a head of 45 m and the discharge is 0.8 m³/s. The mean bucket speed is 14 m/s. Find the overall efficiency and the power produced if the jet is deflected by the blades through an angle of 165°. Assume the coefficient of velocity = 0.985 and mechanical efficiency = 0.95.

Solution:

$u = 14.0$ m/s

$$V_1 = C_v\sqrt{2gH} = 0.985\sqrt{2 \times 9.81 \times 45} = 29.27 \text{ m}$$

$\beta = 165°$, $\beta' = 180 - 165 = 15°$, $k = 1.0$ (assumed)

Power

$$\mathbf{P} = \rho\, Q u\,(V_1 - u)(1 + \cos\beta')$$

$= 998 \times 0.8 \times 14.0 \times (29.27 - 14.0)(1 + \cos 15°)$

$= 335550 \text{ W} = 335.6 \text{ kW}$

Power delivered to shaft

$= 335.6 \times \eta_m = 335.6 \times 0.95 = 318.77 \text{ kW}$

Overall efficiency

$$\eta_0 = \frac{\text{power delivered to shaft}}{\gamma QH}$$

$$= \frac{318.77}{\dfrac{(9.81 \times 998)}{1000} \times 0.8 \times 45} = 0.905$$

16.13 A double jet Pelton wheel has a specific speed of 14 and is required to deliver 1000 kW. The turbine is supplied through a pipeline from a reservoir whose level is 400 m above the nozzles. Allowing 5% for friction loss in the pipe, calculate (a) Speed in rpm (b) diameter of jets (c) mean diameter of bucket circle. Take $C_v = 0.98$, Speed ratio = 0.46 and overal efficiency = 85%.

Solution:

Net available Head $H = 400(1 - 0.05) = 380$ m

Power per jet $= 500$ kW

Specific speed $= \dfrac{N\sqrt{\mathbf{P}}}{H^{5/4}}$

[*Note:* For multiple jet Pelton wheels the specific speed is based on brake power per jet.]

$$14 = \frac{N\sqrt{500}}{(380)^{5/4}} = 0.01333\,N$$

Rotational speed $N =$ **1050 rpm**

$$V_1 = C_v\sqrt{2gH} = 0.98\sqrt{2 \times 9.81 \times 380}$$

$$= 84.62 \text{ m/s}$$

Speed ratio $\phi = \dfrac{u}{\sqrt{2gH}} = 0.46$

$$u = 0.46 \times \sqrt{19.62 \times 380} = 39.72 \text{ m/s}$$

$$u = 39.72 = \frac{\pi DN}{60} = \frac{\pi \times D \times 1050}{60}$$

$D =$ **0.722 m**

= mean diameter of bucket circle

Power developed $= 1000 \text{ kW} = \gamma QH\,\eta_o$

$$1000 = 9.79 \times Q \times 380 \times 0.85 = 3162.17\,Q$$

$$Q = 0.3162 \text{ m}^3\text{/s}$$

As there are two jets of diameter d,

$$2 \times \frac{\pi}{4} \times d^2 \times 84.62 = 0.3162$$

$d = 0.04877$ m = **4.88 cm**

16.14 In a Pelton wheel the jet is deflected by 170°. The initial jet velocity is 96 m/s and the peripheral velocity of the wheel is 44 m/s. Calculate the kinetic energy of the jet leaving the bucket and the direction of the absolute velocity at the exit from the bucket.

Solution: From the outlet velocity triangle (see Fig. 16.25),

$$\beta' = 180 - \beta = 10°$$

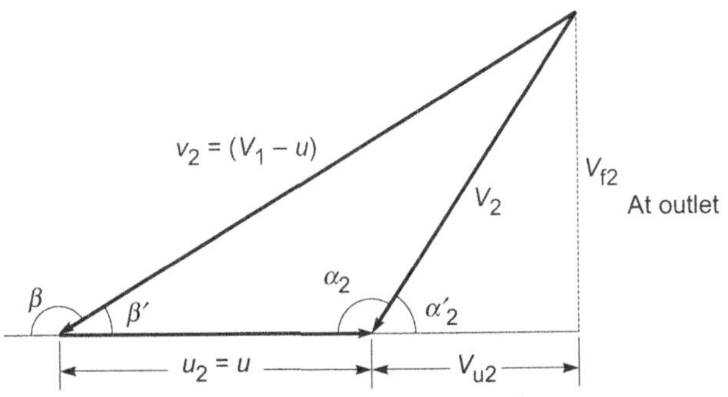

Fig. 16.25 *Outlet Velocity Triangle*

Let α'_2 be the direction of the absolute velocity V_2 with the peripheral velocity.

$$V_{f2} = V_2 \sin \alpha'_2$$

$$= (V_1 - u)\sin \beta' = (96 - 44)\sin 10° = 9.03$$

$$V_{u2} = V_2 \cos \alpha_2 = (V_1 - u)\cos \beta' - u$$

$$= (96 - 44)\cos 10° - 44 = 7.21$$

$$\tan \alpha'_2 = \frac{9.03}{7.21} = 1.252, \ \alpha'_2 = \mathbf{51.39°}$$

$$V_2 = \frac{9.03}{\sin 51.39°} = 11.555 \text{ m/s}$$

$$\frac{V_2^2}{2g} = \frac{(11.555)^2}{2 \times 9.81} = \mathbf{6.81 \ m}$$

16.15 Show that a Pelton turbine with coefficient of velocity = C_v, and blade friction coefficient k can have a maximum hydraulic efficiency of

$$\eta_H = \frac{1}{2}C_v^2(1 + k\cos\beta')$$

where $\beta' = (180 - \text{blade angle } \beta)$.

Solution: Let the net head at the base of the nozzle = H.

Velocity of jet $V_1 = C_v\sqrt{2gH}$

$$H = \left(\frac{V_1^2}{C_v^2 \cdot 2g}\right)$$

Head extracted in the turbine

$$H_e = \frac{1}{g}(V_1 - u) \cdot u(1 + k\cos\beta')$$

Hydraulic efficiency

$$\eta_H = \frac{H_e}{H} = 2C_v^2\frac{u}{V_1}\left(1 - \frac{u}{V_1}\right)(1 + k\cos\beta')$$

$$= 2C_v^2\, \varepsilon (1 - \varepsilon)(1 + k \cos \beta')$$

where $\varepsilon = u/V_1$

For maximum η_H, $\dfrac{d\eta_H}{d\varepsilon} = 0$

$$(1 - 2\varepsilon) = 0 \quad \text{or} \quad \varepsilon = 1/2$$

Maximum value of hydraulic efficiency

$$(\eta_H)_{Max} = 2C_v^2 \frac{1}{2}\left(1 - \frac{1}{2}\right)(1 + k \cos \beta')$$

$$= \mathbf{\frac{1}{2}\, C_v^2 (1 + k \cos \beta')}$$

16.16 A double overhung 1.5 m diameter impulse turbine installation is to develop 3000 kW at 400 rpm under a net head of 270 m. If the overall efficiency is 0.90, determine the (i) diameter of the nozzle (ii) speed ratio ϕ and (iii) specific speed. (Take $C_v = 0.95$).

Solution:

Power per wheel $= \dfrac{3000}{2} = 1500$ kW

Power $\mathbf{P} = \eta_o\, \gamma QH$

$$1500 = 0.90 \times 9.79 \times Q \times 270$$

$$Q = 0.6305 \text{ m}^3/\text{s}$$

Velocity of the jet $= V_1 = C_v \sqrt{2gH}$

$$= 0.95\sqrt{2 \times 9.81 \times 270} = 69.14 \text{ m/s}$$

For the nozzle: $Q = \dfrac{\pi}{4} \times d^2 \times V_1$

$$0.6305 = \frac{\pi}{4} \times d^2 \times 69.14$$

d = diameter of the nozzle = 0.1078 m

= **10.78 cm**

Peripheral velocity of the bucket $u = \dfrac{\pi DN}{60}$

$$u = \frac{\pi \times 1.5 \times 400}{60} = 31.42 \text{ m/s}$$

Specific ratio $\phi = \dfrac{u}{\sqrt{2gH}} = \dfrac{31.42}{\sqrt{2 \times 9.81 \times 270}} = \mathbf{0.432}$

Specific speed $N_s = \dfrac{N\sqrt{\mathbf{P}}}{H^{5/4}}$

$$= \frac{400 \times \sqrt{1500}}{(270)^{5/4}} = \mathbf{14.15}$$

16.17 A Pelton turbine is to operate under a net head of 500 m at 420 rpm. If a single jet with diameter 18 cm is used, find the specific speed of the machine. Take $C_v = 0.98$, $\phi = 0.45$ and overall efficiency $\eta_o = 0.85$.

Solution:

$$H = 500 \text{ m}$$

$$V_1 = C_v\sqrt{2gH} = 0.98\sqrt{2 \times 9.81 \times 500}$$

$$= 97.06 \text{ m/s}$$

Discharge $Q = \dfrac{\pi}{4} d^2 V_1 = \dfrac{\pi}{4} \times (0.18)^2 \times 97.06$

$$= 2.47 \text{ m}^3/\text{s}$$

Power developed

$$\mathbf{P} = \eta_o\, \gamma QH$$

$$= 0.85 \times 9.79 \times 2.47 \times 500 = 10{,}277 \text{ kW}$$

Specific speed

$$N_s = \frac{N\sqrt{\mathbf{P}}}{H^{5/4}} = \frac{420\sqrt{10{,}277}}{(500)^{5/4}} = \mathbf{18}$$

D. Similitude in Turbines

16.18 A Francis turbine of diameter 3.0 m develops 6750 kW at 300 rpm under a net head of 45 m. A geometrically similar model of scale ratio 1:8 is to be tested at a head of 9 m. Estimate the size, speed, discharge and power developed by the model. What is the specific speed of the model? (Assume overall efficiency of 0.82 for both the prototype and model).

Solution:

Power $\mathbf{P} = \eta_o\, \gamma QH$

For the prototype:

$$6750 = 0.82 \times \frac{(9.81 \times 998)}{1000} \times Q \times 45$$

$$= 361.265\, Q$$

Discharge $Q_p = 18.685 \text{ m}^3/\text{s}$

Using the suffixes *m* and *p* to denote model and prototype parameters respectively

$$\frac{D_m}{D_p} = \frac{1}{8} = \text{Scale ratio}$$

As $D_p = 3.0$ m, $D_m = \dfrac{3.0}{8} = \mathbf{0.375\ m}$

Also $H_p = 45$ m, $H_m = 9.0$ m, $\dfrac{H_m}{H_p} = \dfrac{9}{45} = \dfrac{1}{5}$

Speed: $\dfrac{N_m D_m}{\sqrt{H_m}} = \dfrac{N_p D_p}{\sqrt{H_p}}$

∴ N_m = speed of the model

$$= N_p \left(\frac{D_p}{D_m}\right)\left(\frac{H_m}{H_p}\right)^{1/2}$$

$$= 300\,(8)\left(\frac{1}{5}\right)^{1/2} = \mathbf{1073.3\ rpm}$$

Discharge:

$$\frac{Q_m}{N_m D_m^3} = \frac{Q_p}{N_p D_p^3}$$

$$Q_m = \text{Model discharge} = Q_p\left(\frac{N_m}{N_p}\right)\left(\frac{D_m}{D_p}\right)^3$$

$$= 18.685\left(\frac{1073.3}{300}\right)\left(\frac{1}{8}\right)^3$$

$$= \mathbf{0.13056\ m^3/s}$$

Power: $$\frac{\mathbf{P}_m}{N_m^3 D_m^3} = \frac{\mathbf{P}_p}{N_p^3 D_p^5}$$

$$\mathbf{P}_m = \text{Model power} = \mathbf{P}_p\left(\frac{N_m}{N_p}\right)^3\left(\frac{D_m}{D_p}\right)^5$$

$$= 6750\left(\frac{1073.3}{300}\right)^3\left(\frac{1}{8}\right)^5 = \mathbf{9.433\ kW}$$

Specific speed Since the model and the prototype are similar we expect them to have the same specific speed.

Hence $$N_{sm} = \frac{N_m\sqrt{\mathbf{P}_m}}{H_m^{5/4}} = \frac{1073.3\sqrt{9.433}}{(9)^{5/4}}$$

$$= \mathbf{211.5\ rpm}$$

$$N_{sp} = \frac{N_p\sqrt{\mathbf{P}_p}}{H_p^{5/4}} = \frac{300\sqrt{6760}}{(45)^{5/4}} = 211.5\text{ rpm}$$

It is seen that $N_{sm} = N_{sp}$, as expected. This is a check on the calculations.

16.19 A turbine develops 8,000 kW when running at 100 rpm. The head on the turbine is 30 m. If the head is reduced to 18 m, determine the speed and power developed by the turbine.

Solution: For geometrically similar turbines, the unit speed

$$N_u = \frac{N}{\sqrt{H}}$$

$$\therefore \quad \frac{N_1}{\sqrt{H_1}} = \frac{N_2}{\sqrt{H_2}}$$

$$N_2 = N_1\sqrt{H_2/H_1} = 100\sqrt{18/30}$$

$$= \mathbf{77.46\ rpm}$$

Unit power $$\mathbf{P}_u = \frac{\mathbf{P}}{H^{3/2}}$$

$$\frac{\mathbf{P}_1}{H_1^{3/2}} = \frac{\mathbf{P}_2}{H_2^{3/2}}$$

$$\mathbf{P}_2 = \mathbf{P}_1 \times \left(\frac{H_2}{H_1}\right)^{3/2}$$

$$= 8000 \times \left(\frac{18}{30}\right)^{3/2} = \mathbf{3718\ kW}$$

16.20 A Francis turbine produces 6750 kW at 300 rpm under a net head of 45 m with an overall efficiency of 85%. What would be revolutions per minute, discharge and brake power of the same turbine under a net head of 60 m under homologous conditions?

Solution: Power developed

$$\mathbf{P} = \eta_o\, \gamma Q H$$

$$6750 = 0.85 \times \frac{(9.81 \times 998)}{1000} \times Q \times 45$$

Discharge $Q = 18.03\ \text{m}^3/\text{s}$.

For calculating N, Q and $\mathbf{P}$ at $H = 60$ m the unit relationships are used.

$$N_u = \frac{N}{\sqrt{H}}$$

$$\therefore \quad \frac{N_1}{\sqrt{H_1}} = \frac{N_2}{\sqrt{H_2}}$$

or $$N_2 = N_1\sqrt{H_2/H_1} = 300\sqrt{60/45} = \mathbf{346.4\ rpm}$$

$$Q_u = \frac{Q}{\sqrt{H}}$$

$$\therefore \quad Q_2 = Q_1\sqrt{H_2/H_1} = 18.03\sqrt{60/45} = \mathbf{20.82\ m^3/s}$$

$$\mathbf{P}_u = \frac{\mathbf{P}}{H^{3/2}}$$

$$\therefore \quad \mathbf{P}_2 = \mathbf{P}_1\left(\frac{H_2}{H_1}\right)^{3/2} = 6750\left(\frac{60}{45}\right)^{3/2} = \mathbf{10{,}392\ kW}$$

E. Rotodynamic Pumps

16.21 A centrifugal pump has an impeller of outer diameter 30 cm. The vane tips are radial at the outlet. For a rotative speed of 1450 rpm calculate the net head developed. Assume manometric efficiency = 0.82.

Solution: Refer to Fig. 16.26.

From the outlet velocity triangle

$$V_{f2} = v_2 \quad \text{and} \quad V_2 \cos \alpha_2 = V_{u2} = u_2$$

$$u_2 = \pi D_2 N/60 = \pi \times 0.30 \times 1450/60 = 22.78\text{ m/s}$$

Hence, manometric efficiency

$$\eta_H = \frac{gH}{u_2 V_{u2}} = \frac{gH}{u_2^2}$$

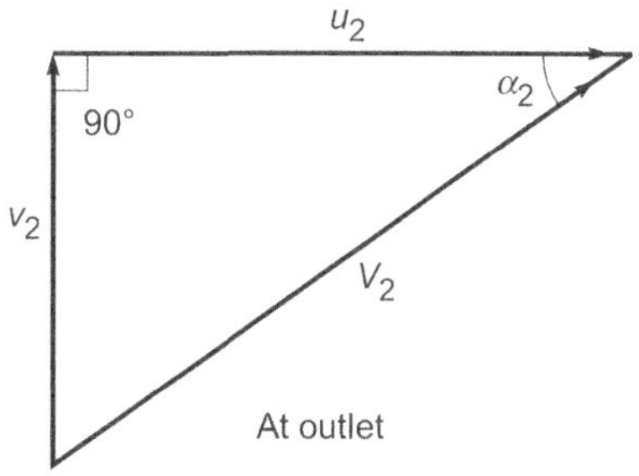

Fig. 16.26

$$0.82 = \frac{9.81 \times H}{(22.78)^2}$$

H = Net head developed = **43.36 m**

16.22 A centrifugal pump delivers water against a net head of 10.0 m at a design speed of 1000 rpm. The vanes are curved backwards and make an angle of 30° with the tangent at the outer periphery. The impeller diameter is 30 cm and has a width of 5 cm at the outlet. Determine the discharge of the pump if the manometric efficiency is 95%.

Solution:

Given $H = 10.0$ m; $N = 1000$ rpm

$\beta_2 = 30°$; $D_2 = 0.30$ m

$b = 0.05$ m; $\eta_H = 0.95$

u_2 = tangential velocity of the impeller at the outlet

$$= \frac{\pi D_2 N}{60} = \frac{\pi \times 0.30 \times 1000}{60} = 15.708 \text{ m/s}$$

Manometric efficiency

$$\eta_H = \frac{gH}{u_2 V_{u2}} \quad \text{(The oulflow is assumed to be radial.)}$$

$$0.95 = \frac{9.81 \times 10.0}{15.708 \times V_{u2}}$$

$V_{u2} = 6.574$ m/s

From the outlet velocity triangle (Fig. 16.27), since $\beta_2 < 90°$

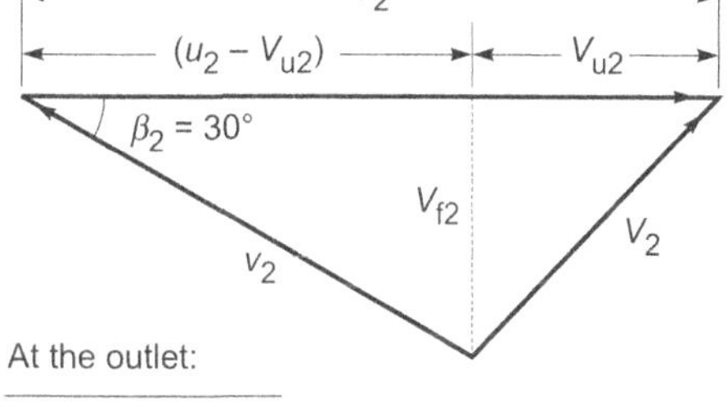

Fig. 16.27

$$\tan \beta_2 = \frac{V_{f2}}{u_2 - V_{u2}}$$

$$\tan 30° = \frac{V_{f2}}{15.708 - 6.574}$$

$V_{f2} = 5.274$ m/s

Discharge $Q = \pi D_2 b_2 V_{f2}$

$= \pi \times 0.30 \times 0.05 \times 5.274$

$= 0.249$ m^3/s = **249 L/s**

16.23 A centrifugal pump is to discharge 100 L/s at a speed of 1450 rpm against a head of 15 m. The impeller has an outer diameter of 25 cm with a width at the outlet of 6 cm. The manometric efficiency is 0.08. Estimate the blade angle at the outlet.

Solution: At the outlet

$$\text{Peripheral velocity } u_2 = \frac{\pi D_2 N}{60} = \frac{\pi \times 0.25 \times 1450}{60}$$

$= 18.98$ m/s

Assuming radial flow at the inlet, the manometric efficiency

$$\eta_H = \frac{gH}{u_2 V_{u2}}$$

$$0.80 = \frac{9.81 \times 15}{18.98 \times V_{u2}}$$

Swirl velocity at outlet $V_{u2} = 9.69$ m/s

The discharge $Q = \pi D_2 b_2 V_{f2}$

$0.100 = \pi \times 0.25 \times 0.06 \times V_{f2}$

Velocity of flow at outlet $V_{f2} = 2.122$ m/s

From the outlet velocity triangle (Fig. 16.28)

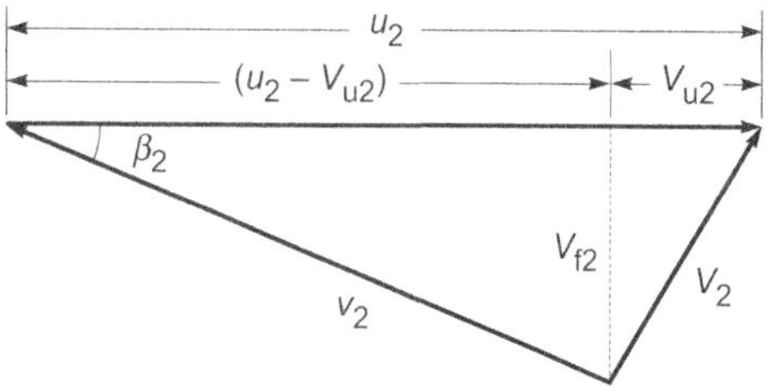

Fig. 16.28 *Volocity Triangle of Outlet*

$$\tan \beta_2 = \frac{V_{f2}}{u_2 - V_{u2}} = \frac{2.122}{(18.98 - 9.69)} = 0.2284$$

$\beta_2 = \tan^{-1} 0.2284 =$ **12.87°**

16.24 A centrifugal pump impeller has an outer diameter of 30 cm and an inner diameter of 15 cm. The pump runs at 1200 rpm. The impeller vanes are set at a blade angle of 30° at the outlet. If the velocity of flow constant at 2.0 m/s, calculate

(i) the velocity and direction of water at outlet,

(ii) the head developed, by assuming a manometric efficiency of 0.85, and

(iii) the blade angle at the inlet.

Solutioin: At the outlet

$$u_2 = \frac{\pi \times D_2 \times N}{60} = \frac{\pi \times 0.30 \times 1200}{60}$$

$$= 18.85 \text{ m/s}$$

$$V_{f2} = 2.0 \text{ m/s} \quad \text{and} \quad \beta_2 = 30°$$

From the outlet velocity triangle [Fig. 16.29(a)]

$$\tan \beta_2 = \frac{V_{f2}}{u_2 - V_{u2}}$$

$$\tan 30° = \frac{2.0}{18.85 - V_{u2}}$$

$18.85 - V_{u2} = 3.464$; and hence $V_{u2} = 15.386$ m/s

(i) V_2 = absolute velocity at outlet = $\sqrt{V_{u2}^2 + V_{f2}^2}$

$$= \sqrt{(15.386)^2 + (2.0)^2} = \mathbf{15.515 \text{ m/s}}$$

If α_2 = Inclination of V_2 to tangential direction at outlet

$$\tan \alpha_2 = \frac{V_{f2}}{V_{u2}} = \frac{2.0}{15.386} = 0.13$$

$$\alpha_2 = \mathbf{7.4°}$$

(ii) Manometric efficiency

$$\eta_H = \frac{gH}{u_2 V_{u2}}$$

Head developed

$$= H = \eta_H \frac{u_2 V_{u2}}{g} = \frac{0.85 \times 18.85 \times 15.386}{9.81}$$

$$= \mathbf{25.13 \text{ m}}$$

(iii) From the inlet velocity diagram [Fig. 16.29(b)]

$$V_1 = V_{f1} = V_{f2} = 2.0 \text{ m/s}$$

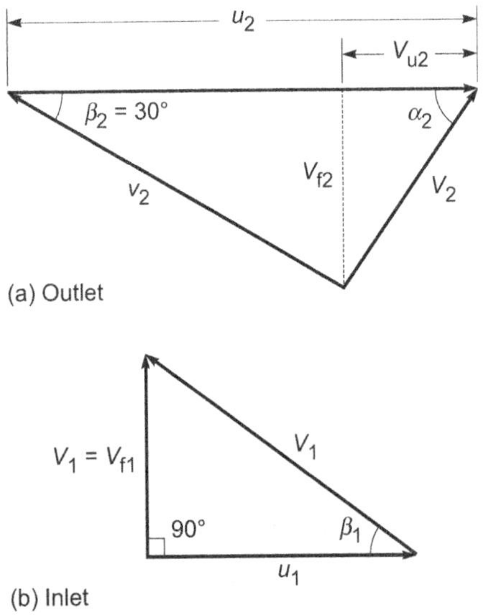

(a) Outlet

(b) Inlet

Fig. 16.29

$$u_1 = \text{peripheral velocity} = \frac{\pi \times 0.15 \times 1200}{60}$$

$$= 9.425 \text{ m/s}$$

The inlet blade angle β_1 is given by

$$\tan \beta_1 = \frac{V_{f1}}{u_1} = \frac{2.0}{9.425} = 0.2122$$

$$\beta_1 = \mathbf{11.98°}$$

16.25 A centrifugal pump lifts water under a static lift of 40 m of which 3 m is suction lift. The suction and delivery pipes are both of 35 cm diameter. The friction loss in suction pipe is 2.0 m and in delivery pipe it is 6.0 m. The impeller is 0.5 m in diameter and 3 cm wide at outlet and runs at a speed of 1200 rpm. The exit blade angle is 20°. If the manometric efficiency of the pump is 85% determine the pressures at the suction and delivery ends of the pump and the discharge.

Solution: Figure 16.30(a) shows the schematic layout.

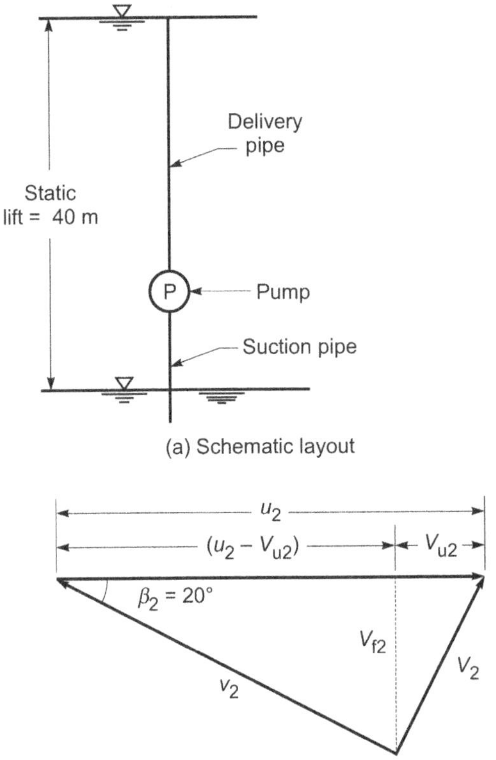

(a) Schematic layout

(b) Outlet velocity triangle

Fig. 16.30

Net head H = Static lift + friction loss

$$= 40.0 + 2.0 + 6.0 = 48 \text{ m}$$

u_2 = peripheral velocity at outlet

$$u_2 = \frac{\pi \times 0.5 \times 1200}{60} = 31.42 \text{ m/s}$$

Assume the flow to be radial at the inlet.

Manometric efficiency $\eta_H = \frac{gH}{u_2 V_{u2}}$

$$0.85 = \frac{9.81 \times 48.0}{31.42 \times V_{u2}}$$

$$V_{u2} = 17.63 \text{ m/s}$$

Blade angle $\beta_2 = 20°$

From the outlet velocity triangle [Fig. 16.25(b)]

$$\tan \beta_2 = \frac{V_{f2}}{u_2 - V_{u2}}$$

$$\tan 20° = \frac{V_{f2}}{31.42 - 17.63} = 0.3639$$

$$V_{f2} = 5.02 \text{ m/s}$$

Discharge $Q = \pi D_2 b_2 V_{f2}$

$$= \pi \times 0.5 \times 0.03 \times 5.02 = \mathbf{0.2366\ m^3/s}$$

V_d = Velocity in delivery pipe

= V_s = Velocity in suction pipe

$$= \frac{Q}{\pi \times D^2/4} = \frac{0.2366}{\frac{\pi}{4} \times (0.35)^2} = 2.459 \text{ m/s}$$

$$\frac{V_d^2}{2g} = \frac{V_s^2}{2g} = \frac{(2.459)^2}{2 \times 9.81} = 0.308 \text{ m}$$

Let the pressure on delivery side = p_d

$$\frac{p_d}{\gamma} + \frac{V_d^2}{2g} = H_d + H_{Ld}$$

$$\frac{p_d}{\gamma} = H_d + H_{Ld} - \frac{V_d^2}{2g} = 37 + 6 - 0.308$$

$$= 42.69 \text{ m}$$

$$P_d = 42.69 \times \frac{(9.81 \times 998)}{1000}$$

$$= 418 \text{ kN/m}^2 \text{ (gauge)}$$

Let the pressure on the suction side be p_s and the atmospheric pressure be p_{atm}.

Then $\frac{p_{atm}}{\gamma} = H_s + H_{Ls} + \frac{p_s}{\gamma} + \frac{V_s^2}{2g}$

$$= 3 + 2 + \frac{p_s}{\gamma} + 0.308$$

$$\frac{p_s}{\gamma} = \frac{p_{atm}}{\gamma} - 5.308 = 5.308 \text{ m vacuum}$$

Assuming $\frac{p_{atm}}{\gamma} = 10.35$ m

$$\frac{p_s}{\gamma} = 10.35 - 5.308$$

$$= 5.042 \text{ m of water (absolute)}$$

$$p_s = 5.042 \times 9.79 = \mathbf{49.36\ kN/m^2\ (abs)}$$

16.26 A centrifugal pump while running at 1000 rpm discharges 80 L/s against a net head of 16 m. The manometric efficiency of the pump is 0.85. If the vane angle at the outlet is 35° and the velocity of flow is 1.50 m/s, estimate the outer diameter of the impeller and its width at that diameter.

Solution:

Manometric efficiency, $\eta_H = \frac{gH}{u_2 V_{u2}}$

$$0.85 = \frac{9.81 \times 16}{u_2 V_{u2}}$$

$$u_2 V_{u2} = 184.66 \quad (1)$$

From the outlet velocity triangle (Fig. 16.31)

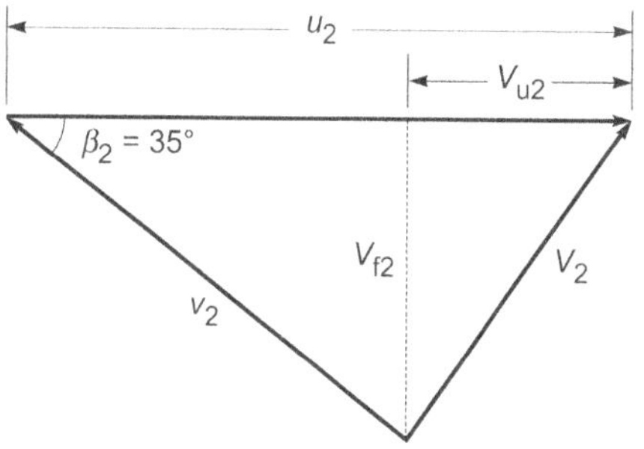

Fig. 16.31 *Outlet Velocity Triangle*

$$\tan 35° = \frac{V_{f2}}{u_2 - V_{u2}}$$

$$u_2 - V_{u2} = \frac{1.50}{\tan 35°} = 2.142$$

$$V_{u2} = (u_2 - 2.142) \quad (2)$$

Substituting for V_{u2} in Eq. 1, $(u_2 - 2.142)\ u_2 = 184.66$

$$u_2^2 - 2.142\ u_2 - 184.66 = 0$$

Taking the positive root

$$u_2 = 14.702$$

$$V_{u2} = 12.560 \text{ m/s}$$

Since $u_2 = \frac{\pi D_2 N}{60}$

$$14.702 = \frac{\pi \times D_2 \times 1000}{60}$$

$$D_2 = \mathbf{0.280\ m}$$

Discharge $Q = \pi D_2 b_2 V_{f2}$

$$0.080 = \pi \times 0.280 \times b_2 \times 1.50$$

$$b_2 = 0.061 \text{ m}$$

Width of impeller at outlet = **6.1 cm**

16.27 Assuming no loss of energy, show that the increase in piezometric head across the impeller of a centrifugal pump can be expressed as

$$\frac{(p_2 - p_1)}{\gamma} + (Z_2 - Z_1) = \frac{(u_2^2 - u_1^2)}{2g} - \frac{(v_2^2 - v_1^2)}{2g}$$

where u = peripheral velocity, v = relative velocity, Z = datum height, p = pressure, subscripts 1 and 2 refer to inlet and outlet parameters.

Solution: For a rotodynamic pump the net head developed H is given by the manometric efficiency

$$\eta_H = \frac{gH}{(u_2 V_{u2} - u_1 V_{u1})}$$

For no-loss situation $\eta_H = 1.0$. Hence

$$H = \frac{1}{g}(u_2 V_{u2} - u_1 V_{u1})$$

From the velocity vector diagram (Fig. 16.32), absolute velocity V, peripheral velocity u and relative velocity v are related as

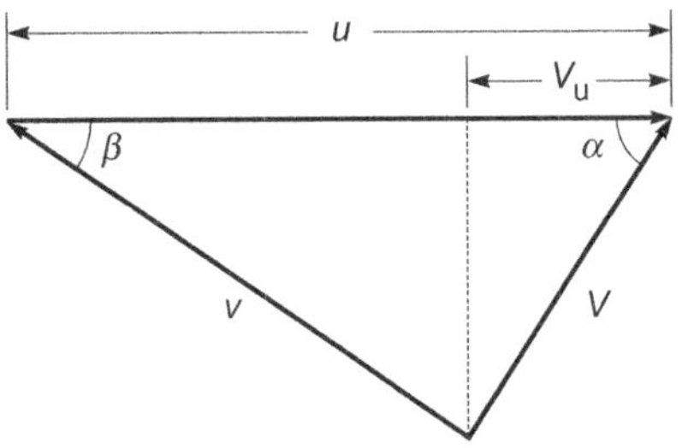

Fig. 16.32

$$V^2 + u^2 - 2uV \cos \alpha = v^2$$

$$uV \cos \alpha = uV_u = \frac{1}{2}(V^2 + u^2 - v^2)$$

Hence $$H = \frac{1}{2g}[(V_2^2 - V_1^2) + (u_2^2 - u_1^2) - (v_2^2 - v_1^2)]$$

By Bernoulli equation, between a point on the inlet and a point on the outlet,

$$H = \left(\frac{p_2}{\gamma} + Z_2 + \frac{V_2^2}{2g}\right) - \left(\frac{p_1}{\gamma} + Z_1 + \frac{V_1^2}{2g}\right)$$

Hence

$$\left(\frac{p_2}{\gamma} - \frac{p_1}{\gamma}\right) + (Z_2 - Z_1) = H - \frac{(V_2^2 - V_1^2)}{2g}$$

$$= \frac{(u_2^2 - u_1^2)}{2g} - \frac{(v_2^2 - v_1^2)}{2g}$$

= increase in piezometric head

16.28 A centrifugal pump of impeller diameters of 20 cm on the outer and 10 cm on the inner side is rotating at 1000 rpm. If the discharge pipe is shut off completely, calculate the difference in pressure between the outer and inner periphery. (Neglect all losses).

Solution: Increase in piezometric head

$$\frac{(p_2 - p_1)}{\gamma} + (Z_2 - Z_1) = \frac{(u_2^2 - u_1^2)}{2g} - \frac{(v_1^2 - v_2^2)}{2g}$$

When the outlet is shut off, the flow is zero and hence $v_1 = v_2 = 0$,

Assuming $Z_2 = Z_1$

$$\frac{p_2 - p_1}{\gamma} = \frac{u_2^2 - u_1^2}{2g}$$

$$u_2 = \frac{\pi D_1 N}{60} = \frac{\pi \times 0.2 \times 600}{60} = 6.283 \text{ m/s}$$

$$u_1 = \frac{\pi D_2 N}{60} = \frac{\pi \times 0.1 \times 600}{60} = 3.142 \text{ m/s}$$

$\frac{p_2 - p_1}{\gamma}$ = Difference in pressure head across the periphery.

$$= \frac{1}{2 \times 9.81}[(6.283)^2 - (3.142)^2] = \mathbf{1.51 \text{ m}}$$

16.29 In a centrifugal pump it is usual to make the outside diameter (D_1) of the impeller to be twice the inner diameter (D_2). For this condition, show that the minimum diameter of an impeller which will enable it to pump water to a head H metres at a speed of N rpm at a manometric efficiency of 0.70 is

$$D_1 = \frac{81.7\sqrt{H}}{N} \text{ metres}$$

Solution: For a pump to just start pumping, if there are no loses, the centrifugal head must be equal to the actual lift. Hence under ideal conditions

$$H_i = \frac{u_2^2 - u_1^2}{2g}$$

Allowing for manometric efficiency, if the net head delivered by the pump is H, then $H_i = H/\eta_H$

Hence $$H = \frac{1}{\eta_H}\left(\frac{u_2^2 - u_1^2}{2g}\right)$$

But $$u_2 = \frac{\pi D_2 N}{60} \text{ and } u_1 = \frac{\pi D_1 N}{60}$$

$$H = \frac{1}{\eta_H}\frac{\pi^2 D_1^2 N^2}{2g(60)^2}\left[1 - \left(\frac{D_2}{D_1}\right)^2\right]$$

Since $D_2/D_1 = 0.5$ and $\eta_H = 0.70$

$$H = \frac{1}{0.70} \times \frac{\pi^2 \times D_1^2 \times N^2}{3600 \times 2 \times 9.81}(1 - 0.25)$$

$$D_1^2 = 6679.4 \frac{H}{N^2}$$

or $$D_1 = 81.7 \frac{\sqrt{H}}{N}$$

Least diameter of the impeller $\boldsymbol{D_1 = 81.7\sqrt{H}/N}$

16.30 A centrifugal pump has an impeller 0.5 m outer diameter and when running at 600 rpm discharges 8000 litres/minute against a head of 8.5 m. The water enters the impeller without whirl and shock. The inner diameter is 0.25 m and the vanes are set back at the outlet at an angle of 45° to the periphery. The area of flow is constant from inlet to outlet of the impeller and is 0.06 m^2. Determine the (a) vane angle at the inlet (b) monomeric efficiency of the pump and (c) speed at which the pump commences to work.

Solution:

(a) Figure 16.33 show the velocity triangles at the inlet and oulet

$$V_{f1} = V_{f2} = \frac{8000 \times 10}{60 \times 0.06} = 2.22 \text{ m/s}$$

$$u_1 = \frac{\pi D_1 N}{60} = \frac{\pi \times 0.25 \times 600}{60} = 7.85 \text{ m/s}$$

$$u_2 = \frac{\pi D_2 N}{60} = \frac{\pi \times 0.5 \times 600}{60} = 15.71 \text{ m/s}$$

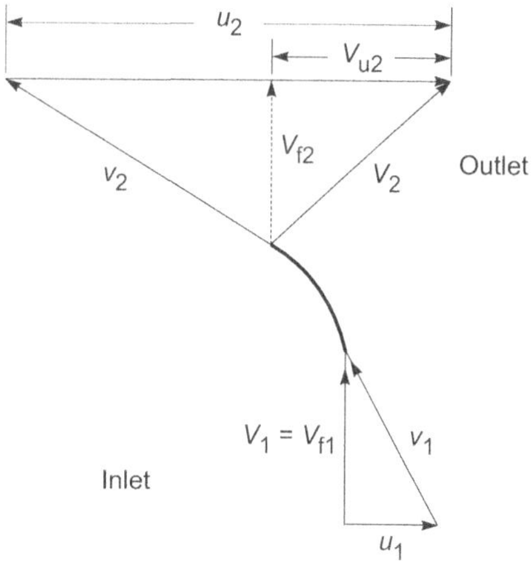

Fig 16.33 *Velocity Triangles*

From the inlet triangle,

$$\tan \beta_1 = \frac{V_{f1}}{u_1} = \frac{2.22}{7.85} = 0.2828$$

β_1 = vane angle at inlet = **15°47°**

(b) From outer velocity triangle,

$$\tan \beta_2 = \tan 45 = \frac{u_2 - V_{u2}}{V_{f2}}$$

$$15.71 - V_{u2} = 2.22$$

$$V_{u2} = 15.71 - 2.22 = 13.49 \text{ m/s}$$

Manometric efficiency

$$= \frac{gH}{u_2 V_{u2}} = \frac{9.81 \times 8.5}{15.71 \times 13.49} = 0.3935$$

$$= 39.35\ \%$$

(c) Minimum starting speed, N, in rpm is given by

$$\frac{\pi^2 N_m^2}{(60)^2} [D_2^2 - D_1^2] = 2gH$$

$$\frac{\pi^2 N_m^2}{(60)^2} [(0.5)^2 - (0.25)^2] = 2 \times 9.81 \times 8.5$$

$$\mathbf{N_m = 570 \text{ rpm}}$$

F. Similitude in Pumps

16.31 An axial flow pump delivers 350 L/s at a head of 8.0 m when rotating at 2000 rpm. It the overall efficiency is 0.70, calculate the power input to the pump. What is the specific speed of the pump? If the same pump were to operate at 2500 rpm, what would be the head, discharge and power required for homologous conditions? (Assume the same overall efficiency at both speeds).

Solution: $$\text{BP}_1 = \text{Power required} = \frac{\gamma Q H}{\eta_o}$$

$$= \frac{9.79 \times 0.350 \times 8.0}{0.70} = \mathbf{39.16 \text{ kW}}$$

Specific speed $$N_s = \frac{N\sqrt{Q}}{H^{3/4}}$$

$$= \frac{2000\sqrt{0.35}}{(8)^{3/4}} = \mathbf{248.7}$$

At higher speed: For homologous conditions

(i) $$\frac{Q_1}{N_1 D_1^3} = \frac{Q_2}{N_2 D_2^3}$$

$$Q_2 = Q_1 \left(\frac{N_2}{N_1}\right)\left(\frac{D_2}{D_1}\right)^3$$

Here $D_2 = D_1$

hence $$Q_2 = 0.350 \left(\frac{2500}{2000}\right)(1) = \mathbf{0.4375\ m^3/s}$$

(ii) $$\frac{H_1}{N_1^2 D_1^2} = \frac{H_2}{N_2^2 D_2^2}$$

$$H_2 = \left(\frac{N_2}{N_1}\right)^2 \left(\frac{D_2}{D_1}\right)^2 H_1$$

$$= \left(\frac{2500}{2000}\right)^2 (1)\,(8) = \mathbf{12.5\ m}$$

(iii) BP_2 = brake power required $= \dfrac{\gamma Q_2 H_2}{\eta_o}$

$$= \frac{9.79 \times 0.4375 \times 12.5}{0.70} = \mathbf{76.48\ kW}$$

16.32 Two homologous pumps A and B are to run at the same speed of 600 rpm. Pump A has an impeller of 50 cm diameter and discharges 0.4 m^3/s of water under a net head of 50 m. Determine the diameter of impeller of pump B and its net head if it is to discharge 0.3 m^3/s.

Solution: For homologous pumps the specific speed is the same.

Hence $$N_{sA} = \frac{N_A\sqrt{Q_A}}{H_A^{3/4}} = \frac{600\sqrt{0.4}}{50^{3/4}} = 20.18$$

For pump B, $N_{sB} = N_{sA} = 20.18$

$$20.18 = \frac{N_B\sqrt{Q_B}}{H_B^{3/4}} = \frac{600\sqrt{0.3}}{H_B^{3/4}}$$

$$= \frac{328.63}{H_B^{3/4}}$$

$$H_B = \mathbf{41.28\ m}$$

From the similarity relation

$$\left(\frac{Q_A}{N_A D_A^3}\right) = \left(\frac{Q_B}{N_B D_B^3}\right)$$

Hence $$D_B^3 = \frac{Q_B}{Q_A} \cdot \frac{N_A}{N_B} \cdot D_A^3$$

$$D_B = \left(\frac{0.3}{0.4}\right)^{1/3} \times (0.5) = \mathbf{0.454\ m}$$

16.33 A centrifugal pump has a rotational speed of 1200 rpm, A design discharge of 0.5 m^3/s and a head of 20 m. The corresponding head is 27 m. The shut off head is 30 m. For what speed will the discharge be 0.60 m^3/s? What will be corresponding head? For what speed will the shut off head be 22 m?

Solution: In this case, $D_1 = D_2$.

$$\frac{Q_1}{N_1 D_1^3} = \frac{Q_2}{N_2 D_2^3}$$

Hence $$N_2 = \left(\frac{Q_2}{Q_1}\right)\left(\frac{D_1}{D_2}\right)^3 N_1$$

$$= \left(\frac{0.60}{0.50}\right)(1)^3\ 1200 = \mathbf{1440\ rpm}$$

$$\frac{H_1}{D_1^2 N_1^2} = \frac{H_2}{D_2^2 N_2^2}$$

$\therefore$ $$H_2 = \left(\frac{N_2}{N_1}\right)^2 \left(\frac{D_2}{D_1}\right)^2 H_1$$

$$= \left(\frac{1440}{1200}\right)^2 \times 20 = \mathbf{28.8\ m}$$

At shut off head: $H_1^* = 30$ m, discharge $Q_1^* = 0$ and $N_1 = 1200$ rpm. When shut off head $H_2^* = 22$ m

$$\frac{H_2^*}{D_2^2 N_2^2} = \frac{H_1^*}{D_1^2 N_1^2}$$

$$N_2^2 = \left(\frac{H_2^*}{H_1^*}\right)\left(\frac{D_1}{D_2}\right)^2 N_1^2$$

$$N_2 = N_1\sqrt{H_2^*/H_1^*} = 1200\sqrt{30/22}$$

$$= \mathbf{1401\ rpm}$$

G. Pump Characteristics

16.34 A pump is to deliver water from a tank against a static head of 40 m. The suction pipe is 50 m long of 25 cm diameter of $f = 0.022$ and the delivery pipe is of 20 cm diameter, 1600 m long and has an $f = 0.022$. The pump characteristic can be defined as $H_p = 100 - 6000\ Q^2$ where H_p = pump head in metres and Q is the discharge in m^3/s. Calculate the head and discharge of the pump.

Solution: Pipe system—
(1) Static head $H_s = 40$ m
(2) Friction head $H_L = h_{L1} + h_{L2}$

$$H_L = \frac{f_1 L_1 V_1^2}{2gD_1} + \frac{f_2 L_2 V_2^2}{2gD_1}$$

$$= \frac{0.02 \times 50}{2 \times 9.81 \times 0.25}\left\{Q\Big/\frac{\pi}{4} \times (0.25)^2\right\}^2$$

$$+ \frac{0.022 \times 1600}{2 \times 9.81 \times 0.20} \times \left\{Q\Big/\frac{\pi}{4} \times (0.20)^2\right\}^2$$

$$= (84.6 + 9089)\ Q^2$$
$$= 9173.6\ Q^2$$

At equilibrium, system head = pump head.

Hence $$40 + 9173.6\ Q^2 = 100 - 6000\ Q^2$$

$$Q^2 = \frac{60}{15173.6} = 3.954 \times 10^{-3}$$

$$Q = 0.0629\ m^3/s = \mathbf{62.9\ L/s}$$

Net head of pump $H = H_p = 100 - 6000 \times (0.0629)^2$
$= \mathbf{76.27\ m}$

16.35 A centrifugal pump running at 1500 rpm has the following characteristics.

Q(Lps)	10	15.5	21.5	30.3	36.0	42.5	48.0
H(m)	35.0	34.5	33.3	30.0	26.8	22.8	18.5
η (per cent)	60.0	67.5	72.0	73.5	71.0	66.0	60.0

Draw the characteristics of the pump and find the best efficiency point (BEP). What is the specific speed of the pump? Estimate the shut off head.

Solution: Plots of Q vs H and Q vs η are plotted on the same graph sheet as shown in Fig. 16.34 from which the BEP is $\eta_{max} = 73.5$ with $Q = 30.3$ L/s and $H = 30$ m. The specific speed N_s is calculated for the Best Efficiency Point (Design Point).

$$N_s = \frac{N\sqrt{Q}}{H^{3/4}} = \frac{1500\sqrt{0.0303}}{(30)^{3/4}} = 20.37$$

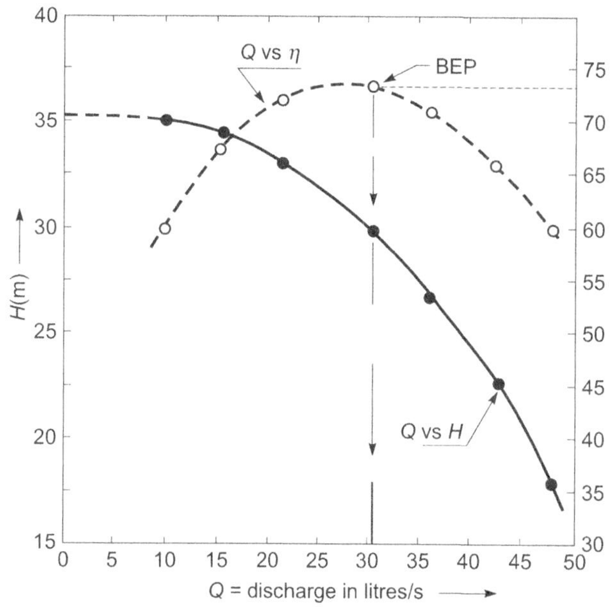

Fig. 16.34 *Pump Characteristics*

By extrapolating the Q vs H curve, the shut off head which corresponds to $Q = 0$ is found to be 35.5 m.

16.36 Two centrifugal pumps A and B are available for use in a pipe flow system and their characteristics are as follows:

Pump A		*Pump B*	
Discharge Q (m^3/s)	*Head* H (*m*)	*Discharge* Q (m^3/s)	*Head* H (*m*)
0	40.0	0	45
0.1	38.0	0.1	43
0.2	29.0	0.2	38
0.25	23.0	0.3	28
0.30	16.0	0.4	14

Determine the head-capacity curves for these pumps when both pumps are connected in (i) parallel and (ii) in series.

Solution:

(i) *Pumps in parallel:* The individual head-discharge curves for pump A and B are plotted in Fig. 16.35(a). When the pumps are connected in parallel, the discharge gets added up for a given head. The head-discharge data are obtained as in Table 16.2(a) and plotted in Fig. 16.35(a).

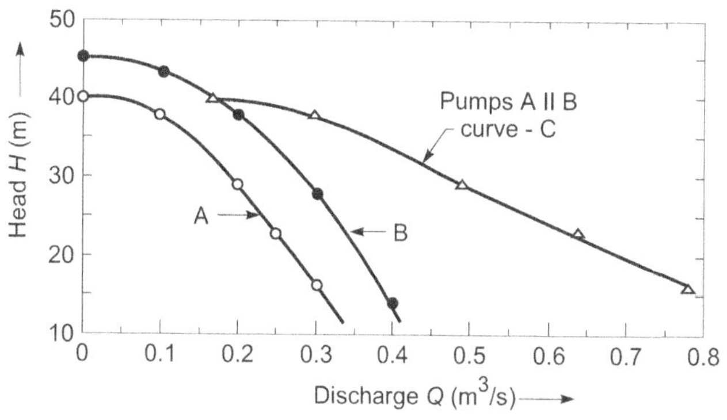

Fig. 16.35(a) *Pumps in Parallel*

Table 16.2(a) *Pumps A and B in Parallel*

Head	*Discharge from A alone* Q_a (m^3/s)	*Discharge from B alone* Q_b (m^3/s)	*combined discharge* $Q_l = Q_a + Q_b$ (m^3/s)
45	0	0	0
43	0	0.1	0.1
40	0	0.17*	0.17
38	0.1	0.20	0.30
29	0.2	0.29*	0.49
23	0.3	0.34*	0.64
16	0.4	0.38*	0.78

Values of Q_b indicated by an asterisk (*) are interpolated values of curve B from Fig. 16.35(a). Values of Q_t and corresponding H are plotted in Fig. 16.35 (a) to get curve C which is the head-discharge relationship for pumps A and B operating in parallel.

(ii) *Pumps in series:* When the pumps are connected in series the heads of individual pumps for a given discharge get added up. Thus the combined head $H_t = H_a + H_b$ for a given discharge. Table 16.2(b) is prepared from the individual H-Q relationship shown in Fig. 16.35(a).

Table 16.2(b) *Pumps A and B in Series*

Discharge Q (m^3/s)	*Head(m)* H_a (*for pump A alone*)	*Head (m)* H_b (*for pump B alone*)	*Combined head* H_t (*m*) ($H_t = H_a + H_b$)
0	40	45	85
0.1	38	43	81
0.2	29	38	67
0.25	23	34*	57
0.30	16	28	44
0.35	10*	22*	32

* Interpolated values from curves in Fig. 16.35(a).

The variation of H_t vs Q is shown plotted as curve D in Fig. 16.35(b) and represents the head-discharge curve for pumps A and B in series.

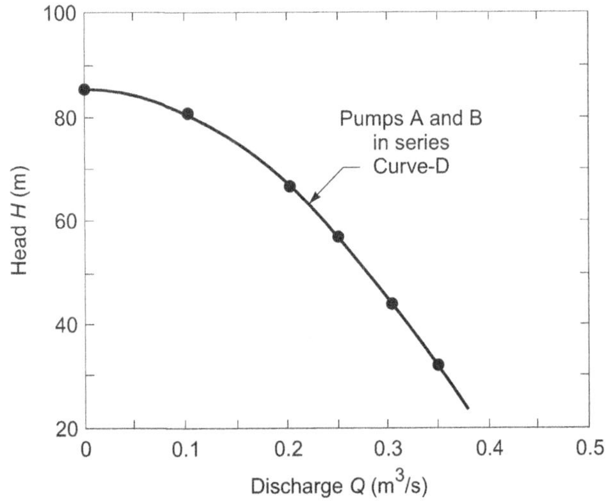

Fig. 16.35(b) *Pumps in Series*

16.37 A pump can delivery discharge of 0.10 m^3/s to a head of 30 m. The critical cavitation number σ_c for the pump is found to be 0.12. The pump is to be installed at a location where the barometric pressure is 96.0 kPa (abs) and the vapour pressure is 3.0 kPa (abs). Assuming an intake pipe friction of 0.3 m, determine the minimum value of NPSH. What would be the maximum allowable elevation above the sump water surface at which the pump can be located?

Solution:

(i) By Eq. 16.53 $\sigma_c = \dfrac{(\text{NPSH})_{min}}{H}$

$$0.12 = \frac{(\text{NPSH})_{min}}{30}$$

Minimum NPSH = **3.6 m**

(ii) $$\text{NPSH} = \frac{(p_{atm})_{abs}}{\gamma} - \frac{p_v}{\gamma} - Z_s - h_L$$

where Z_s = elevation of the pump above the sump water surface. $(Z_s)_{max}$ corresponds to σ_c. Hence

$$(Z_s)_{max} = \frac{(p_{atm})_{abs}}{\gamma} - \frac{p_v}{\gamma} - h_L - (\text{NPSH})_{min}$$

$$= \frac{96.0}{9.79} - \frac{3.0}{9.79} - 0.3 - 3.6 = \mathbf{5.6\ m}$$

H. Reciprocating Pumps

16.38 A single acting reciprocating pump has a 15 cm piston with a crank of radius 15 cm. The delivery pipe is 10 cm diameter. At a speed of 60 rpm, 310 litres/s of water is lifted to a total height of 15 m. Find the slip, coefficient of discharge and theoretical power in kW required to drive the pump.

Solution: Theoretical discharge $Q_t = \dfrac{ALN}{60}$

$$A = \frac{\pi}{4}(0.15)^2 = 0.01767 \text{ m}^2$$

$L = 2r = 2 \times 0.15 = 0.30$ m; $N = 60$ rpm

$$Q_t = \frac{0.01767 \times 0.30 \times 60}{60} = 0.0053 \text{ m}^3/\text{s} = 318\ l/\text{min}$$

Q_a = actual discharge = 310 l/min

$$\text{Slip} = \frac{(318 - 310) \times 100}{318} = \mathbf{2.52\ \%}$$

Coefficient of discharge = 310/318 = **0.975**

Total head = $H_t = (H_s + H_d) = 15$ m

Power = $\mathbf{P}_t = \rho Q_a H_t = 9800 \times 0.0053 \times 15 = 778$ W

= **0.778 kW**

16.39 A three-throw pump has a 20 cm diameter cylinder and a stroke of 40 cm. The pump is required to deliver 5000 litres/minute of water at a static head of 80 m. Friction losses can be taken as 2.0 m in suction pipe and 18 m in the delivery pipe. Assume a slip of 2% and pump efficiency of 90%. The velocity head in the delivery pipe can be neglected as too small. Determine the speed of the pump and the power required.

Solution:

$$A = \frac{\pi}{4}(0.2)^2 = 0.3142 \text{ m}^2, \quad L = 0.4 \text{ m},$$

$$Q_a = (50000/(60 \times 1000) = 0.0833 \text{ m}^3/\text{s}$$

For a three-throw pump

$$Q_t = \frac{3ALN}{60}$$

$$Q_t = \frac{0.3142 \times 0.40 \times N}{60} = 6.283 \times 10^{-4} \text{ N}$$

Since, Slip = 2%, $\dfrac{Q_t - Q_a}{Q_t} = 0.02$

$$\frac{(6.28 \times 10^{-4} N - 0.0833)}{6.28 \times 10^{-4} N} = 0.02$$

N = **135.4 rpm**

Total head $= H_t = (H_s + h_{fs}) + (H_d + h_{fd})$

$= 80 + 2 + 18 = 100$ m

Power = $\mathbf{P} = \gamma Q_a H_t/\eta = (9800 \times 0.00833 \times 100)/0.90$

= 778 W = **90730 W**

= **90.73 kW**

16.40 A single acting reciprocating pump has a 20 cm piston with a crank of radius 40 cm. The delivery pipe is 10 cm diameter and 45 m long. Water is lifted to a height of 40 m above the axis of the cylinder. Find the maximum speed at which the pump can be run without cavitation. Assume atmospheric pressure = 9.75 m water (abs) and cavitation occurs at 2.75 m water (abs).

Solution: At incipient cavitation in the delivery pipe

$$H_{atmo} + H_d - H_{ad} = H_v$$
$$9.75 + 40 - H_{ad} = 2.75$$
$$H_{ad} = 47 \text{ m}$$

At the end of thc delivery stroke

$$H_{ad} = \frac{L_d}{g}\left(\frac{A}{A_d}\right)\omega^2 r$$

Here $L_d = 45.0$ m, $(A/A_d) = (20/10)^2 = 4$,

and $r = 0.40$ m

$$\omega = \frac{2\pi N}{60} = 0.1047 \text{ N}$$

Thus $H_d = 47 = \dfrac{45}{9.81} \times 4 \times (0.1047 \text{ N})^2 \times 0.40$

$$= 0.0805 \text{ N}^2$$

$$\mathbf{N = 24.17 \text{ rpm}}$$

16.41 A single acting reciprocating pump has the following characteristics:
Cylinder diameter = 15 cm
Stroke length = 30 cm
Suction head = 2.5 m
Diameter of suction pipe = 5.0 cm
Suction pipe length = 5.0 m
Assuming atmospheric pressure = 10.0 m water (abs) and cavitation occurs at 2.0 m water (abs) determine the maximum speed at which the pump can be run without cavitation.

Solution: Maximum acceleration

$$H_{asm} = \frac{L_s}{g}\left(\frac{A}{A_s}\right)\omega^2 r$$

$$\omega = \frac{2\pi N}{60} = 0.1047 \text{ N}, r = 0.30/2 = 0.15 \text{ m}$$

$$(A/A_s) = (15/5)^2 = 9$$

$$H_{asm} = \frac{5.0}{9.81} \times 9 \times (0.1047 \text{ N})^2 \times 0.15$$
$$= 0.007543 \text{ N}^2$$

At limiting condition for a suction pipe

$$H_{asm} + H_v + H_s = H_{atmo}$$
$$H_{asm} + 2.0 + 2.5 = 10.0$$

Hence $H_{asm} = 5.5 = 0.007543 \text{ N}^2$

$$\mathbf{N = 27.0 \text{ rpm}}$$

16.42 A single acting reciprocating pump has a 20 cm cylinder with a stroke of 40 cm. The suction pipe is 10 cm diameter and 5 m long. If the speed of the pump is 30 rpm, determine the maximum suction lift if separation occurs at 2.5 m water (abs). Assume atmospheric pressure = 10.0 m water (abs).

Solution: Maximum acceleration head in suction pipe

$$H_{asm} = \frac{L_s}{g}\left(\frac{A}{A_s}\right)\omega^2 r$$

$$\omega = \frac{2\pi \times 30}{60} = 3.141 \text{ rad/s}, r = 0.40/2 = 0.20 \text{ m}$$

$$(A/A_s) = (20/10)^2 = 4, L_s = 5.0 \text{ m}$$

$$H_{asm} = \frac{5.0}{9.81} \times 4 \times (3.141)^2 \times 0.20 = 4.024 \text{ m}$$

At limiting condition for a suction pipe

$$H_{asm} + H_v + H_s = H_{atmo}$$
$$4.24 + 2.5 + H_s = 10.0$$

Hence H_s = suction lift = **3.476 m**

16.43 A single acting reciprocating pump has a 25 cm cylinder with a stroke of 40 cm. The diameters of suction and delivery pipes are 15 cm and 20 cm respectively. If the piston makes 40 strokes/minute estimate the maximum velocity and acceleration of water in the suction and delivery pipes.

Solution:

$$\omega = \frac{2\pi \times 40}{60} = 4.189 \text{ rad/s}, r = 0.40/2 = 0.20 \text{ m}$$

$$(A/A_s) = (25/15)^2 = 2.778, (A/A_d) = (25/20)^2 = 1.5625,$$

Item	*Suction Pipe*	*Delivery pipe*
Maximum velocity	$V_{sm} = \left(\frac{A}{A_s}\right)\omega r$ $= 2.778 \times 4.189 \times 0.20$ $= \mathbf{2.327 \text{ m/s}}$	$V_{dm} = \left(\frac{A}{A_d}\right)\omega r$ $= 1.5625 \times 4.189 \times 0.20$ $= \mathbf{1.309 \text{ m/s}}$
Maximum acceleration	$a_{sm} = \left(\frac{A}{A_s}\right)\omega^2 r$ $=2.778 \times (4.189)^2 \times 0.20$ $= \mathbf{9.749 \text{ m/s}^2}$	$a_{dm} = \left(\frac{A}{A_d}\right)\omega^2 r$ $= 1.5625 \times (4.189)^2 \times 0.20$ $= \mathbf{5.484 \text{ m/s}^2}$

16.44 A single acting reciprocating pump has the following data:
Cylinder diameter = 10 cm
Stroke = 25 cm
Static Suction head = 4.0 m
Diameter of suction pipe = 5.0 cm
Suction pipe length = 6.0 m
Crank speed = 30 rpm
Delivery pipe diameter = 5.0 cm
Length of delivery pipe = 25.0 m
Static delivery head = 16.0 m
Estimate the pressure head on the piston at the (1) beginning (2) mid and (3) end of the suction and delivery strokes. Assume atmospheric pressure = 10.0 m water (abs) and Darcy–Weisbach friction factor $f = 0.02$.

Solution:

$$\omega = \frac{2\pi \times 30}{60} = 3.1412 \text{ rad/s},\ r = 0.25/2 = 0.125 \text{ m}$$

$(A/A_s) = (10/5)^2 = 4$, $(A/A_d) = (10/5)^2 = 4$,

Suction side: Acceleration head:

$$H_{as} = \frac{L_s}{g}\left(\frac{A}{A_s}\right)\omega^2 r \cos\theta$$

$$= \frac{6.0}{9.81} \times 4 \times (3.1412)^2 \times 0.125 \times \cos\theta$$

$$= 3.0182 \cos\theta$$

Friction head:

$$h_{fs} = \frac{fL_s}{2gd_s}\left(\frac{A}{A_s}\omega r \sin\theta\right)^2$$

$$= \frac{0.02 \times 6}{2 \times 9.81 \times 0.05}(4 \times 3.1412 \times 0.125 \times \sin\theta)^2$$

$$= 0.302 \sin^2\theta$$

Pressure head on the piston

$$H_{ts} = H_{atmo} - (H_s + H_{as} + h_{fs})$$
$$= 10.0 - (4.0 + H_{as} + h_{fs})$$
$$= 6.0 - (H_{as} + h_{fs}) \text{ m (abs)}$$

Condition	θ	H_{as} *(m)*	h_{fs} *(m)*	*Pressure head H_{ts} (abs)*
Beginning	0°	3.0182	0	**2.982 m**
Mid	90°	0	0.302	**5.698 m**
End	180°	–3.0182	0	**9.018 m**

Delivery Side: Acceleration head:

$$H_{ad} = \frac{L_d}{g}\left(\frac{A}{A_d}\right)\omega^2 r \cos\theta$$

$$= \frac{25}{9.81} \times 4 \times (3.1412)^2 \times 0.125 \times \cos\theta$$

$$= 1.2576 \cos\theta$$

Friction head:

$$h_{fd} = \frac{fL_d}{2gd_d}\left(\frac{A}{A_d}\omega r \sin\theta\right)^2$$

$$= \frac{0.02 \times 25}{2 \times 9.81 \times 0.05}(4 \times 3.1412 \times 0.125 \times \sin\theta)^2$$

$$= 1.2576 \sin^2\theta$$

Pressure head on the piston

$$H_{td} = H_{atmo} + (H_d + H_{ad} + h_{fd})$$
$$= 10.0 + (16.0 + H_{ad} + h_{fd})$$
$$= 26.0 + (H_{ad} + h_{fd}) \text{ m (abs)}$$

Condition	θ	H_{ad} *(m)*	h_{fd} *(m)*	*Pressure head H_{td} (abs)*	*Pressure head H_{td} (gauge)*
Beginning	0°	12.576	0	38.576 m	**28.576 m**
Mid	90°	0	1.258	27.258 m	**17.258 m**
End	180°	–12.576	0	13.424 m	**3.424 m**

16.45 A double acting reciprocating pump has a 25 cm cylinder with a stroke of 40 cm. The suction pipe is 5.0 cm diameter and 3.0 m long. If the piston executes 50 strokes/minute, determine the minimum diameter of suction pipe to prevent cavitation. Separation occurs at 2.5 m water (abs). Assume atmospheric pressure = 10.0 m water (abs).

Solution:

Crank radius $r = L/2 = 0.4/2 = 0.20$ m

Since there are two strokes per revolution,

$$N = 50/2 = 25 \text{ rpm}$$

$$\omega = \frac{2\pi \times 25}{60} = 2.618 \text{ rad/s},\ (A/A_s) = (d/d_s)^2$$

Maximum acceleration head in suction pipe

$$H_{asm} = \frac{L_s}{g}\left(\frac{A}{A_s}\right)\omega^2 r$$

$$H_{asm} = \frac{5.0}{9.81} \times \left(\frac{d}{d_s}\right)^2 \times (2.618)^2 \times 0.20 = 0.699\ (d/d_s)^2$$

At limiting condition for a suction pipe

$$H_{asm} + H_v + H_s = H_{atmo}$$
$$0.699\ (d/d_s)^2 + 2.5 + 3.0 = 10.0$$
$$(d/d_s)^2 = 6.438 \text{ and as such } (d/d_s) = 2.537$$

Since $d = 25$ cm, $\boldsymbol{d_s = 9.86}$ **cm**

16.46 Estimate the percentage of work saved in one cycle of a single acting reciprocating pump when the air vessel is provided on the delivery side very near to the cylinder. What would be the value if the pump is double acting?

Solution:

(*a*) *Single Acting Pump*

With Air Vessel: When an air vessel is provided very near to the cylinder in the delivery pipe the flow in the pipe is steady. The average velocity in the delivery pipe.

$$V_d = Q_t/A_d = \left(\frac{A}{A_d}\right)\frac{LN}{60}.$$

h_{fd1} = head loss in the delivery pipe

$$= \frac{fL_dV_d^2}{2gd_d} = \frac{fL_d}{2gd_d}\left(\frac{A}{A_d}\frac{LN}{60}\right)^2$$

If W = weight of water pumped per stroke,
Work done per stroke $\mathbf{P}_1$ = W h_{fd1}

With No Air vessel:

$$h_{fd} = \frac{fL_d}{2gd_d}\left(\frac{A}{A_d}r\omega\sin\theta\right)^2$$

Maximum head loss occurs when $\theta = 0°$

h_{fdm} = maximum head loss

$$= \frac{fL_d}{2gd_d}\left(\frac{A}{A_d}r\omega\right)^2$$

$$= \frac{fL_d}{2gd_d}\left(\frac{A}{A_d}\frac{L}{2}\frac{2\pi N}{60}\right)^2$$

$$= \frac{fL_d}{2gd_d}\left(\frac{A}{A_d}\frac{LN}{60}\pi\right)^2 = \pi^2 h_{fd1}$$

Time averaged head loss

$$h_{fd2} = (2/3)\ h_{fdm} = \frac{2}{3}\pi^2 h_{fd1}$$

Work done per stroke

$$\mathbf{P}_2 = \text{W}\ h_{fd2} = \frac{2}{3}\pi^2\mathbf{P}_1$$

Saving in work due to air vessel

$$\frac{\mathbf{P}_2 - \mathbf{P}_1}{2} = \frac{\left(\frac{2}{3}\pi^2\right) - 1}{\left(\frac{2}{3}\pi^2\right)} = 0.848$$

Percentage savings in work done per stroke = **84.8 %**

(*b*) *Double Acting Pump*

$$V_d = Q_t/A_d = 2\left(\frac{A}{A_d}\right)\frac{LN}{60}.$$

h_{fd3} = head loss in the delivery pipe

$$= \frac{fL_dV_d^2}{2gd_d} = \frac{4fL_d}{2gd_d}\left(\frac{A}{A_d}\frac{LN}{60}\right)^2$$

If W = weight of water pumped per stroke,
Work done per stroke $\mathbf{P}_3$ = W h_{fd3}

With No Air vessel:

$$h_{fd} = \frac{fL_d}{2gd_d}\left(\frac{A}{A_d}r\omega\sin\theta\right)^2$$

Maximum head loss occurs when $\theta = 0°$

h_{fdm} = maximum head loss

$$= \frac{fL_d}{2gd_d}\left(\frac{A}{A_d}r\omega\right)^2$$

$$= \frac{fL_d}{2gd_d}\left(\frac{A}{A_d}\frac{L}{2}\frac{2\pi N}{60}\right)^2$$

$$= \frac{fL_d}{2gd_d}\left(\frac{A}{A_d}\frac{LN}{60}\pi\right)^2 = (\pi^2/4)\ h_{fd3}$$

Time averaged head loss

$$h_{fd4} = (2/3)\ h_{fdm} = \frac{2}{3}\frac{\pi^2}{4}h_{fd3}$$

Work done per stroke

$$\mathbf{P}_4 = \text{W}\ h_{fd3} = \frac{2}{3}\frac{\pi^2}{4}P_3$$

Savings in work due to air vessel

$$= \frac{\mathbf{P}_4 - \mathbf{P}_3}{\mathbf{P}_4} = \frac{\left(\frac{2}{3}\pi^2\frac{1}{4}\right) - 1}{\left(\frac{2}{3}\pi^2\frac{1}{4}\right)} = 0.392$$

Percentage savings in work done per stroke = **39.2%**

16.47 A single acting reciprocating pump has a stroke length of 45 cm and a cylinder diameter of 30 cm. The suction pipe is 6 m long and has a diameter of 15 cm. The water level in the sump is 3.0 m below the cylinder. (1) Calculate the maximum speed of the pump, if separation is known to occur at 2.5 m water (abs). (2) If an air vessel is fitted on the suction side at a length of 2.0 m from the cylinder, calculate the admissible maximum speed and the percentage change in the discharge. Take Darcy–Weisbach friction factor $f = 0.02$. Assume atmospheric pressure = 10.0 m water (abs).

Solution:

$$\omega = \frac{2\pi\times N}{60} = 0.1047\text{ N rad/s},\ r = 0.45/2 = 0.225\text{ m}$$

$(A/A_s) = (30/15)^2 = 4,$

Case-1*: No Air Vessel*

At $\theta = 0°$, $h_{fs} = 0$ and $H_{as} = H_{asm}$

$$H_{asm} = \frac{L_s}{g}\left(\frac{A}{A_s}\right)\omega^2 r$$

$$= \frac{6.0}{9.81} \times 4 \times (0.1047\ N)^2 \times 0.225 = 0.00634\ N^2$$

At limiting condition for a suction pipe

$$H_{asm} + H_v + H_s = H_{atmo}$$

$$H_{asm} + 3.0 + 2.5 = 10.0$$

Hence $\quad H_{asm} = 4.5 = 0.006034\ N^2$

$$N = 27.3 \text{ rpm}$$

Case-2 : With Air Vessel

When an air vessel is fitted in the suction pipe at 2.0 m from the cylinder.

Acceleration pressure head is confined to a 2.0 m length next to the cylinder.

Friction loss in remaining 4.0 m of suction pipe is constant over time, as the flow is steady.

$$H_{asm} = \frac{L_s}{g}\left(\frac{A}{A_s}\right)\omega^2 r$$

$$= \frac{2.0}{9.81} \times 4 \times (0.1047\ N)^2 \times 0.225 = 0.002011\ N^2$$

V_s = average steady velocity in pipe below air vessel

$$= \left(\frac{A}{A_s}\right)\frac{\omega r}{\pi} = \frac{1}{\pi} \times 0.1047\ N \times 0.225 = 0.03\ N$$

$$h_{fs} = \frac{fL_s}{2gd_s} V_s^2$$

$$= \frac{0.02 \times 40}{2 \times 9.81 \times 0.15} (0.03\ N)^2 = 2.446\ 10^{-5}\ N^2$$

At limiting condition for a suction pipe

$$H_{asm} + H_v + h_{fs} = H_{atmo}$$

$$0.002011\ N^2 + 2.5 + 3.0 + 2.446 \times 10^{-5}\ N^2 = 10.0$$

$$N^2 = 4.5/2.0355 = 2210.8$$

$$N = 47 \text{ rpm}$$

Discharge $Q_t = \dfrac{ALN}{60}$

Ratio of discharge $Q_2/Q_1 = N_2/N_1 = 47.0/27.3 = 1.722$

Percentage change in discharge = **72.2% increase after fitting the air vessel.**

16.48 A single acting reciprocating pump has a cylinder of diameter 15 cm and a stroke length of 30 cm. The delivery pipe is 20 m long and has a diameter of 7.5 cm. The speed of the pump is 40 rpm and discharges water to a height of 15 m. If an air vessel is fitted on the delivery side at a length of 1.0 m from the cylinder, calculate the pressure head at the beginning and middle of the delivery stroke. Take Darcy–Weisbach friction $f = 0.02$.

Solution:

$$A = \frac{\pi}{4}(0.15)^2 = 0.0177\ \text{m}^2, \quad L = 0.3\ \text{m},$$

For a single acting pump

$$Q_t = \frac{ALN}{60} = \frac{0.0177 \times 0.3 \times 40}{60} = 0.00354\ \text{m}^3/\text{s}$$

$$\omega = \frac{2\pi \times 40}{60} = 4.189 \text{ rad/s}, \ r = 0.30/2 = 0.150 \text{ m}$$

$$(A/A_d) = (15/7.5)^2 = 4,\ A_d = (0.01777/4) = 0.004425\ \text{m}^3/\text{s}$$

V_d = steady velocity after the air vessel

$= Q_t/A_d = 0.00354/0.004425 = 0.8$ m/s

L_{da} = length of delivery pipe before air vessel = 1.0 m

L'_d = length of delivery pipe after air vessel = 19.0 m

At the beginning of the stroke: Friction head is confined only to length L_{da}

Acceleration head:

$$H_{ad} = \frac{L_{da}}{g}\left(\frac{A}{A_d}\right)\omega^2 r$$

$$= \frac{1.0}{9.81} \times 4 \times (4.189)^2 \times 0.15 = 1.073 \text{ m}$$

Friction head:

$$h_{fdl} = \frac{fL'_d}{2gd_d} V_d^2$$

$$= \frac{0.02 \times 19}{2 \times 9.81 \times 0.075} \times (0.8)^2 = 0.165 \text{ m}$$

Pressure head on the piston

$$H_{td} = H_d + H_{ad} + h_{fdl}$$

$= 15.0 + 1.073 + 0.165 =$ **16.238 m (gauge)**

At the middle of the stroke:

Acceleration head $H_{ad} = 0$

Friction head $\quad h_{fd} = h_{fd1} + h_{fd2}$

h_{fdl} = same as at the beginning of the stroke = 0.165 m

h_{fd2} = additional friction head in pipe of length L_{da}

$$h_{fd2} = \frac{fL_{da}}{2gd_d}\left(\frac{A}{A_d}\omega r\right)^2$$

$$= \frac{0.02 \times 1}{2 \times 9.81 \times 0.075} \times (4 \times 4.189 \times 0.15)^2$$

$= 0.086$ m

Pressure head on the piston

$$H_{td} = H_d + h_{fd1} + h_{fd2}$$

$= 15.0 + 0.165 + 0.086 =$ **15.251 m (gauge)**

16.49 A single acting reciprocating pump has a cylinder diameter of 15 cm and a stroke length of 30 cm. The delivery pipe is 30 m long and has a diameter of 7.5 cm. The speed of the pump is 60 rpm. Determine the power saved in overcoming friction in the delivery side by fitting an air vessel very near the cylinder on the delivery pipe.

Take Darcy-Weisbach friction factor $f = 0.02$.

Solution:

Without Air Vessel

$$A = \frac{\pi}{4}(0.15)^2 = 0.0177 \text{ m}^2, \quad L = 0.3 \text{ m},$$

For a single acting pump

$$Q_t = \frac{ALN}{60} = \frac{0.0177 \times 0.3 \times 60}{60} = 0.00531 \text{ m}^3/\text{s}$$

$$\omega = \frac{2\pi \times 60}{60} = 6.283 \text{ rad/s}, \; r = 0.30/2 = 0.150 \text{ m}$$

Maximum friction head

$$h_{fdm} = \frac{fL_d}{2gd_d}\left(\frac{A}{A_d}\omega r\right)^2$$

$$= \frac{0.02 \times 30}{2 \times 9.81 \times 0.075} \times (4 \times 6.283 \times 0.15)^2$$

$$= 5.795 \text{ m}$$

Time averaged friction head

$$h_{fda} = (2/3)\, h_{fdm} = (2/3) \times 5.795 = 3.863 \text{ m}$$

With Air Vessel: The velocity is steady in the delivery pipe with an average value of

$$V_d = Q_t/A_d = 0.00531/0.004425 = 1.2 \text{ m/s}$$

$$\text{Friction head: } h_{fdl} = \frac{fL}{2gd_d} V_d^2$$

$$= \frac{0.02 \times 30}{2 \times 9.81 \times 0.075} \times (1.2)^2 = 0.587 \text{ m}$$

Friction power saved due to air vessel in delivery pipe, in kW,

$$\mathbf{P}_f = \frac{\gamma Q_t}{1000}(h_{fda} - h_{fdl})$$

$$= (9800/1000) \times 0.00531 \times (3.863 - 0.587)$$

$$= \mathbf{0.17 \text{ kW}}$$

16.50 A double acting reciprocating pump has the following data:

Cylinder diameter = 20 cm
Stroke = 40 cm
Diameter of suction pipe = 10.0 cm
Crank speed = 90 rpm

If an air vessel is fitted on the suction side, calculate the flow into or out of the air vessel when the crank angle is (i) 45° and (ii) 150°

Solution:

$$A = \frac{\pi}{4}(0.2)^2 = 0.0314 \text{ m}^2, \quad L = 0.4 \text{ m},$$

$$\omega = \frac{2\pi \times 90}{60} = 9.425 \text{ rad/s}, \; r = 0.40/2 = 0.20 \text{ m}$$

$$(A/A_s) = (20/10)^2 = 4, \; A_s = A/4 = 0.007854 \text{ m}^2$$

Since the pump is double acting

$$Q_t = \frac{2ALN}{60} = \frac{2 \times 0.0314 \times 0.4 \times 90}{60}$$

$$= 0.03768 \text{ m}^3/\text{s}$$

v_s = Instantaneous velocity of unsteady flow in suction pipe between air vessel and the cylinder

$$= \frac{A}{A_s}\omega r \sin\theta = 4 \times 9.425 \times 0.20 \sin\theta$$

$$= 7.54 \sin\theta$$

Discharge in this part of suction pipe at any instant

$$Q_s = v_s A_s = 7.54 \sin\theta \times 0.007854 = 0.05922 \sin\theta$$

Discharge into/out of air vessel:

If $Q_t < Q_s$, then flow goes out of the air vessel
If $Q_t > Q_s$, then flow goes into the air vessel

Crank angle θ	*Q_t m³/s*	*Q_s m³/s*	*Remarks*
45°	0.03768	0.04187	Flow going out of air vessel = **0.00419 m³/s**
150°	0.03768	0.02961	Flow going into air vessel = **0.00807 m³/s**

16.51 Following data pertain to a single acting reciprocating pump installation:

Cylinder diameter = 15 cm
Stroke length = 20 cm
Static suction head = 4.0 m
Diameter of suction pipe = 7.5 cm
Suction pipe length = 8.0 m
Crank speed = 60 rpm
Delivery pipe diameter = 7.5 cm
Length of delivery pipe = 30.0 m
Static delivery head = 25.0 m
Darcy–Weisbach friction factor $f = 0.025$.

Air vessels are provided very near to the cylinder both on the suction and the delivery sides. Assume Darcy-Weisbach friction factor $f = 0.025$. Estimate the power required to drive the pump by assuming a pump efficiency of 80%.

Solution:

$$A = \frac{\pi}{4}(0.15)^2 = 0.01767 \text{ m}^2, \quad L = 0.2 \text{ m},$$

$$\omega = \frac{2\pi \times 60}{60} = 6.283 \text{ rad/s}, \; r = 0.20/2 = 0.10 \text{ m}$$

$$(A/A_s) = (15/7.5)^2 = 4,$$

$$A_d = A_s = A/4 = 0.004179 \text{ m}^2$$

$$Q_t = \frac{ALN}{60} = \frac{0.01767 \times 0.2 \times 60}{60} = 0.003534 \text{ m}^3/\text{s}$$

Since the air vessels are very near to the cylinder, it is assumed that the entire suction an delivery pipes will have steady flow.

$V_s = V_d$ = velocity in suction and delivery pipes
= 0.003534/0.004179 = 0.8457 m/s

Hence

$$h_{fs} = \frac{fL_s}{2gd_s} V_s^2 = \frac{0.025 \times 8}{2 \times 9.81 \times 0.075} \times (0.8457)^2$$

= 0.097 m

$$h_{fd} = \frac{fL_d}{2gd_d} V_d^2 = \frac{0.025 \times 30}{2 \times 9.81 \times 0.075} \times (0.8457)^2$$

= 0.364 m

Power required in kW

$$\mathbf{P} = \frac{\gamma Q_t}{1000 \times \eta} (H_s + h_{fs} + H_d + h_{fd})$$

$$= \frac{9800 \times 0.003534}{1000 \times 0.8} (4.0 + 0.097 + 2.5 + 0.364)$$

= **1.275 kW**

OBJECTIVE QUESTIONS

Turbines

16.1 A hydraulic turbine has a discharge of 3 m^3/s when operating under a head of 15 m and a speed of 500 rpm. If it is to operate under 12 m of head, the rotational speed will be

(a) 600 rpm (b) 559 rpm
(c) 447 rpm (d) 400 rpm

16.2 A turbine has a discharge of 3 m^3/s when operating under a head of 14 m and a speed of 400 rpm. If it is to operate under a head of 18 m, the discharge, in m^3/s, will be

(a) 3.86 (b) 3.40
(c) 2.65 (d) 2.23

16.3 A turbine works at 30 m head and 400 rpm speed. Its 1:2 scale model to be tested at a head of 30 m should have a rotational speed of

(a) 800 rpm (b) 566 rpm
(c) 400 rpm (d) 200 rpm

16.4 The unit speed N_u of a turbine of rotational speed N and head H is equal to

(a) $N\sqrt{H}$ (b) $N/\sqrt{H}$
(c) $\sqrt{H}/N$ (d) $\sqrt{HN}$

16.5 The unit power $\mathbf{P}_u$ of a turbine developing a power **P** under a head H is equal to

(a) $\frac{\mathbf{P}}{H^{5/2}}$ (b) $\mathbf{P}/\sqrt{H}$
(c) $\mathbf{P}\,H^{3/2}$ (d) $\frac{\mathbf{P}}{H^{3/2}}$

16.6 For a 1 : m scale model of a turbine the specific speed of the model N_{sm} is related to the prototype specific speed N_{sp} as N_{sm} =

(a) N_{sp}/m (b) mN_{sp}
(c) N_{sp} (d) $(N_{sp})^{1/m}$

16.7 A turbine works under a head of 20 m, has a speed of 375 rpm and develops 400 kW of power. Its specific speed is

(a) 375 (b) 83
(c) 177 (d) 1677

16.8 The cavitation parameter σ for hydraulic machines is defined as σ =

(a) $\frac{p - p_{atm}}{\rho V^2/2}$ (b) $\frac{p - p_v}{\rho V^2/2}$
(c) $\frac{p_{atm} - p_v}{\rho V^2/2}$ (d) $\frac{p - p_v}{p_{atm} - p_v}$

where p = absolute pressure at the point of interest, p_v = vapour pressure of liquid and V = reference velocity, p_{atm} = atmospheric pressure.

16.9 The specific speed for a turbine has the dimensions of

(a) T^{-1} (b) dimensionless
(c) $F^{1/2} L^{-3/4} T^{-3/2}$ (d) $F^{1/2} L^{-5/2} T^{-3/2}$

16.10 In all reaction turbines, maximum efficiency is obtained, if

(a) the guide vane angle is 90°
(b) the blade angle of the runners is 90° at the inlet
(c) the blade angle of the runners is 90° at the outlet
(d) the angle of the absolute velocity vector at the outlet is 90°

[*Note*: All angles are measured with respect to the direction of the peripheral velocity.]

16.11 Two hydraulic turbines are similar and homologous when they are geometrically similar and have

(a) the same Thoma number
(b) the same Froude number
(c) the same specific speed
(d) the same rotational speed

16.12 A turbine develops 2515 kW at 240 rpm. The torque in the shaft is
(a) 400 kN·m (b) 3335 kN·m
(c) 1000 kN·m (d) 100 kN·m

16.13 The moment of momentum of water in a turbine is reduced by 15915 N.m in a turbine rotating at 600 rpm. The power developed, in kW, is
(a) 1000 (b) 1500
(c) 2000 (d) 5000

16.14 A turbine develops 500 kW power under a net head of 30 m. If the overall efficiency of the turbine is 0.83, the discharge through the turbine, in m^3/s, is
(a) 20.5 (b) 2.05
(c) 1.41 (d) 1.51

16.15 A reaction turbine has a discharge of 30 m^3/s passing through it under a net head of 10 m. If the overall efficiency is 0.851 the power developed, in kW, is
(a) 3400 (b) 3450
(c) 2940 (d) 2500

16.16 Pelton wheels are available in the following range of specific speeds (N_s)
(a) 8–30 (b) 40–420
(c) 380–950 (d) 4–70

16.17 Francis turbines are available in the following range of specific speeds (N_s)
(a) 8–30 (b) 40–420
(c) 380–950 (d) 10–300

16.18 The speed factor ϕ of a Pelton turbine for most efficient operation is in the range
(a) 1.4–2.0 (b) 0.70–0.85
(c) 0.43–0.47 (d) 0.1–0.3

16.19 In all reaction turbines, for maximum efficency
(a) the velocity of swirl at entrance must be zero
(b) the velocity of flow at oultet must be zero
(c) the velocity of swirl at outlet must be zero
(d) the velocity of flow at entrance must be zero

16.20 The net availabe head H in a Pelton turbine installation is the
(a) kinetic energy of the jet issuing from the nozzle
(b) difference in elevation between forebay water level and the nozzle outlet
(c) head at the base of the nozzle
(d) difference in level between water levels at the forebay and the tailwater level

16.21 In Pelton turbines the flow from the nozzle is controlled by a needle which works inside the nozzle. The typical value of the coefficent of velocity C_v for needle nozzles, for partly open conditions, is
(a) 0.61 (b) 0.95
(c) 0.75 (d) 0.10

16.22 In a general sense, water turbines may be put in the following decreasing order of specific speeds, as:
(a) propeller turbine, Francis turbine and Pelton wheel
(b) Pelton wheel, Francis turbine, Kaplan turbine
(c) Kaplan turbine, impulse turbine, Francis turbine
(d) Francis turbine, Kaplan turbine, Pelton wheel

16.23 To generate 8100 kW under a head of 81 m while working at a speed of 540 rpm, the turbine of choice will be
(a) Pelton (b) Kaplan
(c) Bulb (d) Francis

16.24 A turbine develops 7500 kW of power when rotating at 135 rpm under a head of 25 m. The normal speed under a head of 20 m would be
(a) 150.9 rpm (b) 108 rpm
(c) 120.7 rpm (d) 168.8 rpm

16.25 A Francis turbine has runner of outer and inner diameters 4 m and 2 m respectively. The breadth at inlet and also at outlet is 0.8 m and the velocity of flow is constant at 3.0 m/s. The discharge through the turbine, in m^3/s, is
(a) 28.28 (b) 37.70
(c) 30.16 (d) 15.08

16.26 In a Kaplan turbine the diameter of the runner is 2.25 m. The boss is having a diameter one third of the runner. If the velocity of flow and the peripheral velocity are 16 m/s and 48 m/s respectively, the discharge through the runner, in m^3/s, is
(a) 254.47 (b) 63.62
(c) 169.6 (d) 56.55

16.27 In a Kaplan turbine working under a head of 35 m, the speed ratio is 2.0. If the outer diameter of the runner is 2.0 m and the hub diameter is 0.6 m, the rotational speed, in rpm, is
(a) 500 (b) 125
(c) 150 (d) 250

16.28 In a Francis turbine the discharge leaves the runner radially at the exit. For this turbine
(a) the blade tip is radial at the outlet
(b) the blade tip is radial at the inlet
(c) the guide vane angle is 90°
(d) the absolute velocity is radial at the outlet

16.29 In a Francis turbine, the runner blades are radial at the entrance and the discharge is radial at the outlet. For this turbine
(a) the relative velocity is radial at the outlet
(b) the absolute velocity is radial at the outlet

(c) the guide vane angle is 90°
(d) the velocity of flow is constant

16.30 A turbine with a specific speed of 400 is connected to a generator on the same shaft. The generator has 10 pairs of poles and produces an AC current with a frequency of 50 Hz. The synchronous speed, in rpm, is
(a) 300 (b) 400
(c) 500 (d) 600

16.31 A turbine has a runner of outer diameter 0.56 m and works under a head of 30 m. If the rotational speed of the runner is 620 rpm the speed factor ϕ for this runner is
(a) 0.46 (b) 2.0
(c) 1.33 (d) 0.75

16.32 A turbine works at 20 m head and 500 rpm speed. Its 1 : 2 scale model is to be tested at a head of 20 m. The model should have a rotational speed of
(a) 1000 rpm (b) 250 rpm
(c) 500 rpm (d) 2000 rpm

Centrifugal Pumps

16.33 In a centrifugal pump the inlet angle will be designed to have
(a) relative velocity vector in the radial direction
(b) absolute velocity vector in the radial direction
(c) velocity of flow to be zero
(d) peripheral velocity to be zero

16.34 A fast centrifugal pump impeller will have
(a) forward facing blades
(b) radial blades
(c) backward facing blades
(d) propeller type blades

16.35 In a centrifugal pump with redial entry of liquid, the manometric efficiency η_H is given by η_H =

(a) $\dfrac{u_2 V_{u2}}{gH}$ (b) $\dfrac{gH}{u_2 V_{u2}}$

(c) $\dfrac{H}{g u_2 V_{u2}}$ (d) $\dfrac{2gH}{u_2 V_{u2}}$

where H = net head, u_2 = peripheral velocity at outlet and V_{u2} = velocity of swirl at the outlet.

16.36 In a centrifugal pump of 0.50 m diameter, running at 1500 rpm the absolute velocity at exit is 8.0 m/s and makes an angle of 45° with the tangential direction. With a hydraulic efficiency of 80% the net head produced is
(a) 18.1 m (b) 28.3 m
(c) 9.1 m (d) 14.2 m

16.37 A centrifugal pump has outer and inner diameters of 0.30 m and 0.15 m respectively. When running at 1450 rpm, with the discharge pipe completely shut off, the head produced is
(a) 26.4 m (b) 6.6 m
(c) 15.7 m (d) 19.8 m

16.38 The specific speed of a pump has the dimensions of
(a) $L^{3/4}\,T^{-3/2}$ (b) $M^0\,L^0\,T^0$
(c) $M^{-1/2}\,L^{1/2}\,T^{-1/4}$ (d) $L^{3/4}\,T^{-1/2}$

16.39 A pump is required to deliver 150 L/s at a head of 45 m when running at 1750 rpm. The specific speed of the pump is
(a) 17.4 (b) 39
(c) 89 (d) 1233

16.40 A pump running at 1400 rpm is required to deliver 360 L/S of water at a head of 16 m. The pump of choice will be
(a) normal speed, radial type
(b) low speed, radial type
(c) mixed flow type
(d) axial flow type

16.41 Typical range of specific speeds for axial flow pumps is
(a) 380–950 (b) 80–200
(c) 10–100 (d) 200–300

16.42 A pump delivers 100 L/s at a head of 25 m when running at 1450 rpm. A homologous pump to deliver 100 L/s at 12.5 m head will have a specific speed of
(a) 1025 (b) 69
(c) 41 (d) 1450

16.43 A similar model of a pump is built to a ratio of 1 : 2. If the model pumps the same fluid as the prototype at the same rotative speed, the ratio of model power to prototype power input is
(a) $1/2\sqrt{2}$ (b) 1/32
(c) 1/8 (d) $1/16\sqrt{2}$

16.44 If two pumps identical in all respects and each capable of delivering a discharge Q against a head H are connected in parallel, the resulting discharge is
(a) $2Q$ against a head $2H$
(b) $2Q$ against a head H
(c) Q against a head $2H$
(d) $2Q$ against a head $\sqrt{2}\,H$

16.45 If two pumps identical in all respects and each capable of delivering a discharge Q against a head H are connected in series, the resulting discharge is
(a) $2Q$ against a head $2H$
(b) $2Q$ against a head H
(c) Q against a head $2H$
(d) $\sqrt{Q}$ against a head $\sqrt{2}\,H$

16.46 The maximum permissible suction lift for centrifugal pump in prectice (at sea level and 30° C) is
(a) 12 m (b) 10 m
(c) 6 m (d) 3 m

16.47 Cavitation parameter σ defined in terms of NPSH (net positive suction head) and net head H as, $\sigma =$

(a) $\frac{\text{NPSH}}{\sqrt{H}}$ (b) $\frac{H}{\text{NPSH}}$

(c) $\frac{H}{\sqrt{\text{NPSH}}}$ (d) $\frac{\text{NPSH}}{H}$

16.48 In a pump installation the local atmospheric pressure is 9.8 m of water, vapour pressure head is 0.4 m (abs), height of the pump above sump water level is 5 m. For a head loss in the suction side of 0.6 m, the NPSH is
(a) 3.8 m (b) 5.0 m
(c) 4.0 m (d) 15.8 m

16.49 Water is to be lifted by a net head of 180 m. Identical pumps with specific speed of 30 and rotative speed of 1450 rpm capable of discharging 200 L/s are available. The number of pumps required is
(a) 1 (b) 2
(c) 3 (d) 4

16.50 Water is pumped through a pipeline to a height of 10 m at the rate of 0.1 m^3/s. Frictional and other minor losses are 5 m. Pumping power required, in kW, is
(a) 14.7 (b) 9.80
(c) 20.0 (d) 13.3

Reciprocating Pumps

16.51 In a reciprocating pump without air vessel, the friction head in the delivery pipe is maximum at the crank angle $\theta =$
(a) 0° (b) 90°
(c) 135° (d) 180°

16.52 In a reciprocating pump without air vessels, the acceleration head in the suction pipe is maximum at the crank angle $\theta =$
(a) 0° (b) 90°
(c) 135° (d) 180°

16.53 In a single acting reciprocating pump without air vessel, the average velocity in the delivery pipe is given by $V_s =$

(a) $\left(\frac{A}{A_s}\right)\omega r$ (b) $\left(\frac{A}{A_s}\right)\frac{\omega r}{\pi}$

(c) $\left(\frac{A}{A_s}\right)\pi\omega r$ (d) $\left(\frac{A}{A_s}\right)\frac{\pi\omega r}{2}$

where ω = angular velocity of the crank of radius = r

16.54 In a single reciprocating pump without air vessel, the ratio of the average friction head to the maximum fraction head in the delivery pipe is given by $h_{fda}/h_{fdm} =$
(a) 1/2 (b) 1/3
(c) 2/3 (d) 3/2

16.55 The indicator diagram of a reciprocating pump is a plot of
(a) Work done *vs* stroke length
(b) Acceleration head *vs* stroke length
(c) Angular displacement *vs* stroke length
(d) Pressure head *vs* stroke length

16.56 An air vessels in the delivery side of a reciprocating pump
(a) maintains steady discharge output
(b) prevents cavitation in the system
(c) enables suction head to be increased
(d) enables the pump to run at higher speed

PROBLEMS

A. Turbines

Velocity Triangle Relationships

16.1 An inward flow reaction turbine has a runner of outer diameter 1.2 m and inner diameter 0.6 m. The blades occupy 5% of the peripheral area and the widths of the blades are 25 cm and 30 cm at the inlet and outlet respectively. If a discharge of 3.0 m^3/s enters radially, determine the flow velocities at the inlet and outlet of the runner.
(*Ans.* V_{f1} = 3.35 m/s; V_{f2} = 5.58 m/s)

16.2 A Francis turbine has a wheel of outer diameter = 1.25 m and inner diameter = 0.6 m. The runner blades are radial at inlet and the discharge is radial at outlet. If the flow enters the vanes at 10° and the velocity of flow is 3.0 m/s calculate the speed of the runner and the vane angle at the outlet.
(*Ans.* N = 260 rpm; β_2 = 20.18°)

16.3 A Francis turbine has wheel diameters of 0.9 m at entrance and 0.45 m at exit. The runner blades are radial at entrance and the guide vanes are at 12°.

The head developed in the turbine is 25 m. Assuming that the flow leaves the turbine radially, calculate the speed of rotation and the blade angle at the exit.

(*Ans.* $N = 332.3$ rpm; $\beta_2 = 23.03°$)

16.4 An inward flow reaction turbine develops 1750 kW at 750 rpm under a net head of 100 m. The guide vanes are at an angle of 15° with tangent at the inlet. The breadth of the blade at inlet is 0.1 times the inlet diameter. The blade thickness blocks 5% of the inlet area. The hydraulic efficiency of the wheel is 88% and the overall efficiency is 84%. Determine the (a) wheel diameter at inlet and (b) blade angle at the inlet.

(*Ans:* $D_1 = 0.519$ m; $\beta_1 = 152.7°$)

16.5 In an outward flow reaction turbine rotating at 300 rpm the inner and outer diameters are 1.50 m and 1.85 m respectively. The wheel has 32 vanes 15 mm thick at inlet and 30 mm thick at the outlet. The breadth of the passage is 25 cm throughout. The net head available is 40 m. The discharge is 7.5 m^3/s and takes place to atmosphere. Determine the (i) blade angles at the inlet and outlet and (ii) power developed. (The flow leaves the outlet radially).

[***Hint:*** $V_2 = V_{f2}$ and head extracted $H_e = u_1 V_{u1}/g = H_t$

$$\frac{V_2^2}{2g} = H - \frac{V_{f2}^2}{2g}.]$$

(*Ans:* (i) $\beta_1 = 42.56°$; $\beta_2 = 12.01°$; (ii) **P** = 2794 kW)

16.6 An inward flow reaction turbine works under a total head of 25 m. The outer and inner diameters of the runners are 0.70 m and 0.35 m respectively. The vane tip is radial at the inlet and the flow leaves the turbine radially. The guide vane angle is 12°. Calculate the speed of the runner and the vane angle at exit, if the velocity of flow at the exit of the draft tube is 4.0 m/s. Assume the velocity of flow to be constant and the hydraulic efficiency to be 0.90.

[***Hint:*** Head extracted = Net available head-velocity head at the end of the draft tube.]

(*Ans.* $N = 420.2$ rmp; $\beta_2 = 23.03°$)

16.7 An inward flow reaction turbine has a guide vane angle of 20° and a vane angle at inlet of 30°. Determine the hydraulic efficiency of the turbine. Assume the outflow to be radial and the velocity of flow to be constant.

[***Hint:*** Refer to Example 16.7.]

(*Ans.* $\eta_H = 96$ %)

16.8 An inward flow radial turbine works under a head of 30 m and discharges 10 m^3/s. The speed of the runner is 300 rpm. At inlet tip of runner vane, the peripheral velocity of wheel is $0.9\sqrt{2gH}$ and radial velocity of flow is $0.3\sqrt{2gH}$ where H is the head on the turbine. If the overall efficiency and hydraulic efficiency of the turbine are 80% and 90% respectively, (i) determine the power developed in kW, (ii) diameter and width of runner at inlet, (iii) guide blade angle at inlet, (iv) inlet angle of runner vane and (v) diameter of runner at outlet. [Assume radial flow at outlet]

(*Ans*: 2352 kW, 1.391 m, 30° 58′ and 1.323. m)

16.9 An inward flow reaction turbine has a setting in the tailwater without a draft tube. Show that for radial inlet and outlet vanes and for constant velocity of flow, the hydraulic efficiency η_H is given by

$$\eta_H = \frac{2\cot^2\alpha_1}{1 + 2\cot^2\alpha_1}$$

where α_1 = guide vane angle.

[***Hint:*** Refer to Example 16.7.]

16.10 A Kaplan turbine runner has an outer diameter of 4.5 m and an inner diameter of 2.0 m and develops 20,000 kW of power while runner at 140 rpm. The head is known to be 20 m. Assuming a hydraulic efficiency of 85% find the discharge through the turbine, the blade angle at the inlet and guide vane angle at inlet.

(*Ans.* $Q = 120.17$ m^3/s: $\alpha_1 = 59.3°$, $\beta_1 = 161.03°$)

16.11 A Kaplan turbnine develops 2300 kW at an available head of 32 m. The speed ratio and flow ratios are respectively 2.1 and 0.62. The diameter of the boss is 1/3 the diameter of the runner. Assuming an overall efficiency of 88%, estimate the diameter, speed of rotation and specific speed of the runner.

(*Ans.* $D = 0.877$ m; $N = 1146$ rpm; $N_s = 722$)

16.12 A Kaplan turbine has speed ratio of 2.0 and a specific speed of 450. Determine the diameter of the propeller in order that it will develop 10,100 kW under a head of 20 m.

(*Ans.* $D = 4.0$ m)

16.13 A Kaplan turbine develops 1 MW under a head of 4.5 m. For speed ratio $\phi = 1.8$, flow ratio $\psi = 0.5$, boss diameter = 0.35 times the outer diameter and overall efficiency of 90%, find the diameter, speed of rotation and specific speed of the runner.

(*Ans.* $D = 2.79$ m, $N = 115.8$ rpm, $N_s = 558.7$)

16.14 A Pelton wheel has a bucket diameter of 90 cm and has one jet of diameter 8 cm with a coefficient of velocity of 0.97. The wheel has a speed ratio of 0.45 and a blade angle of 170°. The blade friction coefficient is 0.93. If the net head available on the wheel is 500 m, calculate (i) the power extracted and (ii) specific speed of the wheel.

(*Ans.* (i) **P** = 2120 kW; (ii) N_s = 18.4)

16.15 A Pelton wheel has a diameter of 1.25 m and operates under a net available head of 200 m. The other characteristics of the installation are: C_v = 0.98, N = 250 rpm, blade angle β = 162°, blade friction coefficient k = 0.96, diameter of the jet d = 10 cm and mechanical efficiency η_m = 0.96. Determine the (i) power delivered to the shaft, (ii) hydraulic efficiency and (iii) specific speed.

(*Ans.* (i) **P** = 650.8 kW; (ii) η_H = 71.83%; (iii) N_s = 8.48)

16.16 A Pelton wheel has a mean bucket speed of 10 m/s with a jet of water flowing at a rate of 0.7 m^3/s under a head of 30 m. The bucket deflects the jet through an angle of 160°. Assuming C_v = 0.98, calculate the power and overall efficiency of the turbine.

(*Ans.* **P** = 186.7 kW, η_o = 0.908)

16.17 Find the hydraulic efficiency of an impulse turbine fo which: coefficient of velocity C_v = 0.98, bucket angle β = 165°, speed factor ϕ = 0.46 and ratio of outlet relative velocity to inlet relative velocity in a bucket k = 0.99.

(*Ans.* η_H = 0.936)

16.18 In a Pelton wheel the overal efficiency is 0.85, C_v = 0.95, speed factor ϕ = 0.46 and the ratio of wheel diameter to jet diameter is 12. Calculate the specific speed of the wheel.

(*Ans.* N_s = 17.0)

16.19 Show that with usual notations the specific speed of an impulse turbine can be written as

$$N_s = 493.7\ \phi\left(\frac{d}{D}\right)\sqrt{\eta_o C_v}$$

16.20 The water jet in a Pelton wheel is 8 cm in diameter and has a velocity of 93 m/s. The rotational speed of the wheel is 600 rpm and the deflection angle of the jet is 170°. If the speed ratio ϕ = 0.47, determine the (i) diameter of the wheel and (ii) power developed.
(Assume C_v = 1.0).

(*Ans.* (i) D = 1.39 m; (ii) **P** = 1995 kW)

16.21 A Pelton turbine has two jets of diameter 15 cm each and develops 6500 kW of power. If the net available head on the turbine is 300 m determine the overall efficiency of the turbine. Also, if the rotatinoal speed is 375 rpm, calculate the specific speed. Take C_v = 0.98.

(*Ans.* N_s = 17.12; η_o = 0.833)

16.22 The water available to a power house is 3.0 m^3/s and the total head from the reservoir to the nozzle is 250 m. There are three pelton wheels of two jets each. All the six jets have the same diameter and are supplied with water from a single pipe of length 500 m. The efficiency of power transmission through the pipe is 90% and the overall efficiency of the turbie is 85%. The C_v for each nozzle is 0.95 and the Darcy–Weisbach friction factor f = 0.02. Determine (i) the power developed, (ii) diameter of the jet and (iii) diameter of the pipe.

(*Ans.* (i) **P** = 5617 kW; (ii) d = 10 cm; (iii) D_1 = 0.785 m)

16.23 A double overhung impulse turbine installation is to develop 15000 kW at 260 rpm under a net head of 350 m. (a) Determine the specific speed and wheel pitch diameter (b) What would these values be when (i) a single wheel with a single nozzle is used and (ii) a single wheel with four nozzles are used? (Assume a velocity ratio of 0.46 and C_v = 1.0).

(*Ans.* (a) N_s = 14.87; D = 2.8 m; (b) (i) N_s = 21.03; D = 2.8 m; (ii) N_s = 10.52; D = 2.8 m)

16.24 In a hydroelectric station, water is available at the rate of 175 m^3/s under a head of 18 m. The turbines run at a speed of 150 rpm whith overall efficicney of 82%. Find the number of turbines required if they have the maximum specific speed of 460 rpm.

(*Ans*: 2 Nos)

Similarity Relationships in Turbines

16.25 A turbine is to operate under a head of 25 m at a speed of 300 rpm. The discharge is 12 m^3/s. Assuming an eficiency of 0.85 calculate the power developed. What would be the specific speed, power, discharge and rotational speed at a head of 15 m?

(*Ans.* (i) N_s = 268.1; $\mathbf{P}_1$ = 2496.5 kW; (ii) N_s = 268.1; $\mathbf{P}_2$ = 1160 kW; Q_2 = 9.30 m^3/s; N_2 = 232.4 rpm)

16.26 A turbine develops 150 kW while running at 120 rpm under a head of 10.0 m. The diameter of the runner is 1.5 m. A 1 : 3 scale model of this turbine is tested under a head of 3.0 m. Determine the speed and power developed in the model. Assuming an overall efficiency of 0.90 for both the model and prototype calculate the discharges in the model and prototype.

(*Ans.* (i) N_m = 197.18 rpm; $\mathbf{P}_m$ = 2.739 kW; (ii) Q_p = 1.7024 m^3/s; Q_m = 0.1036 m^3/s)

16.27 A 1 : 5 model of a turbine develops 2.0 kW of power at 400 rpm under a head of 3.0 m. What is the specific speed of the runner? Assuming an overall efficiency of 0.85 for both the model and prototype, calculate the rotational speed, power and discharge of the prototype when working under a head of 20 m.

(*Ans.* N_s = 143.3; N_p = 206.6 rpm; $\mathbf{P}_p$ = 861.2 kW; Q_p = 5.165 m^3/s)

16.28 In a small hydro development a Kaplan turbine runs under a head of 2.1 m. It has a runner of 3.5 m diameter and develops 600 kW at 80 rpm. Assuming an overall efficiency of 80%, estimate the discharge and specific speed of the machine. If a 1.5 m diameter homologous turbine is to be tested at a head of 3.0 m, What are the rotational speed, discharge and power of that unit?

(*Ans.* Q_p = 36.48 m^3/s; N_s = 775; N_m = 156.2 rpm; Q_m = 5.607 m^3/s; $\mathbf{P}_m$ = 64.6 kW)

16.29 A turbine is to operate under a head of 25 m at 200 rpm. The discharge is 9.0 m^3/s. If the eficiency is 90%, determine the performance of the turbine under a head of 20 m.

(*Ans.* N_2 = 178.9 rpm; Q_2 = 8.05 m^3/s; $\mathbf{P}_2$ = 1418.6 kW)

16.30 A Kaplan turbine operates under a head of 15.2 m, has a speed of 75 rpm and develops 50 MW of power. The overall efficiency of the turbine is 82%. Calculate the specific speed, unit speed, unit discharge and unit power.

(*Ans.* N_s = 558.8; N_u = 19.24; Q_u = 105.10; $\mathbf{P}_u$ = 843.7)

B. Rotodynamic Pumps

Velocity Triangle Relationships

16.31 A centrifugal pump having an overall efficiency of 70% delivers 1500 L/min of water against a static head of 20 m. The suction and delivery pipes are of 20 cm diameter and has a combined total length of 1000 m. Assuming f = 0.02 estimate the power input required.

(*Ans.* $\mathbf{P}_s$ = 8.122 kW)

16.32 A centrifugal pump runs at 800 rpm and delivers 5000 L/min against a head of 7 m. The impeller has an outer diameter of 25 cm and a width of 5 cm at the outlet. If the backward curved vane at the outlet makes an angle of 45°, determine the manometric efficiency. What is the specific speed of the pump?

(*Ans.* η_H = 78.5%; N_s = 53.65)

16.33 A centrifugal pump has an impeller of diameter 30 cm whose width at exit is 6.0 cm. The velocity of flow through the impeller is constant at 3.0 m./s. The impeller vanes are radial at the outer periphery. If the rotational speed is 1000 rmp and the manometric efficiency is 80%, calculate (i) the head produced and (ii) discharge.

(*Ans.* (i) H = 20.12 m; (ii) Q = 169.6 L/s)

16.34 A centrifugal pump has vanes which are radial at the outer periphery. The impeller has an outer diameter of 20 cm and a width of 3 cm at that diameter. If the diacharge is 1800 L/min and the net head produced is 3.5 m, calculate the (i) rotational speed of the impeller and (ii) magnitude and direction of absolute velocity at exit. Manometric efficiency can be assumed as 0.85.

(*Ans.* N = 607 rpm; (ii) V_2 = 6.552 m/s, α_2 = 14.06°)

16.35 A centrifugal pump impeller is 40 cm in outer diameter and 2.5 cm wide at the exit, and its blade angle is 30°. When run at a speed of 2100 rpm, the flow rate through the pump is 80 L/s.

(i) Calculate the radial, relative and absolute velocities at the impeller exit.

(ii) If there is no inlet whirl, what would be the head added to the water by the impeller?

(*Ans.* (i) 2.55 m/s, 5.10 m/s and 39.62 m/s. (ii) 177.18 m)

16.36 A centrifugal pump running at 1200 rpm delivers water at a net head of 10.0 m. At the outlet of the impeller the vane angle is 30° with the peripheral velocity. The impeller has an outer diameter of 25 cm and a width of 5 cm at the outlet. Estimate the discharge (Assume manometric efficiency = 85%).

(*Ans.* Q = 190 L/s)

16.37 A centrifugal pump impeller rotates at a speed of 1000 rpm. The external and internal diameters of the impeller are 0.40 m and 0.20 m respectively. The vanes make an angle of 35° with the tangential direction of rotation at the outlet. If the radial velocity of flow through the impeller is constant at 1.75 m/s, find the (i) angle of vanes at the inlet, (ii) absolute velocity and its direction at outlet and (iii) net head of the pump. (Assume manometric efficiency = 80%)

(*Ans.* (i) β_1 = 9.5°; (ii) V_2 = 18.52 m/s, α_2 = 5.4°; (iii) H = 31.5 m)

16.38 A centrifugal pump with impeller diameter of 50 cm at outlet and 25 cm at inlet runs at a speed of 1000 rpm. The discharge is 150 L/s. The width is 8 cm at the inlet and at the outlet it is 6 cm. The vanes are curved back and make an angle of 25° with the peripheral velocity direction at the outlet. Assuming a manometric eficiency of 0.85 and mechanical eficiency of 0.80, calculate (i) the net head produced by the pump and (ii) brake power input to the pump. (iii) What is the specific speed of the pump?

(*Ans*. (i) $H = 51.65$ m; (ii) **BP** = 111.5 kW; (iii) $N_s = 20.1$)

16.39 A centrifugal pump has an impeller of outer diameter 35 cm and an inside diameter of 20 cm. The impeller vanes make angles of 35° and 30° with the direction of the peripheral velocity at the outlet and inlet respectively. The rotational speed of the impeller is 1000 rpm. Assuming the flow velocity to remain constant throghout the impeller, estimate the net head developed by the pump. Assume manometric efficiency as 0.80.

(*Ans*. $H = 21.06$ m)

General Characteristics

16.40 The impeller of a centrifugal pump has an outer diameter of 50 cm and inner diameter of 25 cm . If the discharge pipe is closed and the pump is full of water, what would be the difference in pressure between the outer and inner periphery when the impeller rotates at 600 rpm?

(*Ans*. $H = 9.43$)

16.41 A centrifugal pump with impeller of outer diameter 45 cm and inner diameter 25 cm, is required to develop a net head of 20 m. Find the lowest speed to start pumping.

(*Ans*. $N = 1011$ rpm)

16.42 A pump has a head-discharge characteristic given by $H_p = 35 - 2200\,Q^2$ where H_p = head developed by the pump in metres and Q = corresponding discharge in m^3/s. The pump is to deliver a discharge against a static head of 12 m. The suction pipe is 15 cm in diameter, 20 m long and has an f value of 0.018. The delivery pipe is 20 cm in diameter, 400 m long and has an f value of 0.0. Calculate the head and discharge delivered by the pump. If the overall efficiency of the pump is 0.70, calculate the driving power supplied to the pump.

(*Ans*. $H = 24.14$ m; $Q = 70$ L/s; **BP** = 23.63 kW)

16.43 (a) A pump delivers water at a net head of 45 m. The atmospheric pressure is 90 kPa (abs) and the vapour pressure is 4.24 kPa (abs). If the head lost in the intake pipe due to friction is 0.76 m, calculate the maximum allowable elevation above the sump water level at which the pump can be located. The critical cavitation number for the pump can be taken as 0.15.

(b) In another location, with all other factors remaining same except the net head of the pump which is now 80 m, estimate the maximum allowable elevation of the pump above the sump water level.

(*Ans*. (a) $(Z_c)_m = 1.25$ m; (b) $(Z_c)_m = -4.0$ m (i.e. 4 m below the sump water level.)

16.44 The head-discharge characteristic of a centrifugal pump is given in the following. The pump delivers water through a 20 cm diameter, 2000 m long pipe. The friction factor of the pipe $f = 0.02$. Neglecting minor losses, determine the discharge in the pipe for a static lift of 12 m.

Discharge (L/s)	0	10	20	30	40	50
Head (m)	26	25.5	24.5	22.5	18.5	12.5

(*Ans*. $H = 22.0$ m and $Q = 32$ L/s)

Similarity Relations for Pumps

16.45 A centrifugal pump with 30 cm diameter impeller was found to be most efficient when discharging 0.20 m^3/s of water at 1200 rpm against a head of 15 m of water. A similar pump is required to deliver 2.0 m^3/s at 1000 rpm. Calculate the dimater of the impeller and the head that could be developed by this pump.

(*Ans*. $D_2 = 0.687$ m; $H_2 = 54.6$ m)

16.46 A centrifugal pump of 25 cm diameter runs at 1450 rpm and delivers 0.3 m^3/s against a head of 12 m. Calculate the specific speed of the pump. A similar pump with half the size is to run at a head of 10 m. Find the working speed, discharge and power required. The efficiencies of the pumps can be assumed to be 75%.

(*Ans*. $N_s = 123$; $N_2 = 4833$ rpm; $Q_2 = 0.0205$ m^3/s; $\mathbf{BP}_2 = 2.68$ kW)

16.47 A centrifugal pump with an impeller diameter of 20 m discharges 120 L/s at 1200 rpm and 10 m head. What is its specific speed? If a homologous pump produces 240 L/s at 20 m head, determine its size and speed of rotation.

(*Ans*. $N_s = 73.9$; $D_2 = 25.94$ cm; $N_2 = 1427$ rpm)

16.48 A centrifugal pump running at 750 rpm discharges water at 0.1 m^3/s against a head of 10 m at its best efficiency. A second pump of the same homologous series, when working at 500 rpm, is to deliver water at 0.05 m^3/s at its best efficiency. What will be the design head of the second pump and what is the scale ratio between the first and the second?

(*Ans.* H_1 = 3.67 m, D_1/D_2 = 1.1)

C. Reciprocating Pumps
—Basic Relationships

16.49 A single acting reciprocating pump has the following characteristics:
Piston diameter = 30 cm
Stroke = 50 cm
Speed = 40 rpm
Total lift = 25 m
If the discharge delivered by the pump at the outlet is 1380 litres/minute, calculate the slip, coefficient of discharge and theoretical power in kW required to drive the pump.

(*Ans*: Slip = 2.34%, Coefficient of discharge = 0.9766, Power = 5.78 kW)

16.50 A plunger is fitted to a vertical pipe filled with water. The lower end of the pipe is submerged in a sump. If the plunger is drawn up with an acceleration of 5.0 m/s^2, find the maximum height above the sump level at which the plunger will work without separation. Assume atmospheric pressure = 10.0 m water (abs) and separation occurs at 2.0 m water (abs). Take acceleration due to gravity as 10 m/s^2.

(*Ans*: L_s = 5.33 m)

16.51 A single acting reciprocating pump has the following data:
Piston diameter = 10 cm
Stroke = 30 cm
Suction head = 4.0 m
Diameter of suction pipe = 7.5 cm
Suction pipe length = 4.0 m
Assuming atmospheric pressure = 10.0 m water (abs) and cavitation occurs at 2.5 m water (abs) determine the maximum speed at which the pump can be run without cavitation. Assume Frictional losses = 1.0 m.

(*Ans*: N = 45.8 rpm)

16.52 A single acting reciprocating pump has the following characteristics:
Cylinder diameter = 22.5 cm
Stroke length = 45 cm
Suction head = 4.5 m
Diameter of suction pipe = 22.5 cm
Suction pipe length = 20.0 m
Assuming atmospheric pressure = 10.0 m water (abs) and cavitation occurs at 2.0 m water (abs) determine the maximum speed at which the pump can be run without cavitation.

(*Ans*: N = 26.36 rpm)

16.53 If for the pump in Problem 16.52, all the data except the diameter of the suction pipe remain the same and the crank speed is 30 rpm, what is minimum size of suction pipe to be used to avoid cavitation?

(*Ans*: d_s = 25.6 cm)

16.54 A double acting reciprocating pump has a 20 cm cylinder with a stroke of 20 cm. The suction pipe is 20 cm diameter and 15 m long. If the speed of the pump is 60 rpm, determine the maximum suction lift if separation occurs at 2.5 m water (abs). Assume atmospheric pressure = 10.0 m water (abs). Neglect frictional losses.

(*Ans*: H_s = 1.457 m)

16.55 A double acting reciprocating pump has a cylinder of diameter 20 cm and stroke of 30 cm. The piston makes 30 strokes/minute. Estimate the maximum velocity and acceleration in the suction pipe of diameter 20 cm and delivery pipe of diameter 25 cm.

(*Ans*: V_{sm} = 0.471 m/s, a_{sm} = 1.480 m/s^2: V_{dm} = 0.301 m/s, a_{dm} = 0.947 m/s^2)

16.56 A single acting reciprocating pump has a stroke length of 15 cm. The suction pipe is 7 m long. The water level in the sump is 2.5 m below the cylinder. The diameters of suction pipe and the plunger are 7.5 cm and 10.0 cm respectively. If the speed of the pump is 75 rpm, determine the pressure head on the piston at the beginning, mid and end of the suction stroke. Take Darcy–Weisbach friction factor f = 0.02.

(*Ans*: H_{ts} = – 8.369 m (gauge), –2.604 m (gauge), and +3.868 m (gauge)

16.57 A single acting reciprocating pump has the following data:
Cylinder diameter = 35 cm
Stroke = 35 cm
Static suction head = 3.0 m
Diameter of suction pipe = 20.0 cm
Suction pipe length = 6.0 m
Crank speed = 20 rpm
Delivery pipe diameter = 20.0 cm
Length of delivery pipe = 25.0 m
Static delivery head = 20.0 m
Estimate the power required to drive the pump. Take Darcy–Wesbach friction factor f = 0.02 and pump efficiency = 0.9.

(*Ans*: **P** = 1.7 kW)

Air Vessels

16.58 A single acting reciprocating pump has a stroke length of 37.5 and a cylinder of diameter 22.5 cm. The suction pipe is 12 m long and has a diameter of 15 cm. The water level in the sump is 3.0 m below the level of the cylinder. If the speed of the pump is 20 rpm, determine the pressure head on the piston at the beginning of the suction stroke (1) when no air vessel is fitted and (2) when an air vessel is fitted to the suction pipe at the level of the cylinder and at a distance of 1.5 m from the cylinder. Take Darcy–Weisbach friction factor $f = 0.02$.

(*Ans* : (1) H_{ts} = 5.263 m water (vacuum);
(2) H_{ts} = 3.289 m water (vacuum))

16.59 A single acting reciprocating pump has an air vessel in the delivery side fitted very close to the cylinder. The cylinder has a diameter of 30 cm and a stroke length of 45 cm. The delivey pipe is 40 m long and has a diameter of 20 cm. The speed of the pump is 60 rpm. Determine the power saved by the air vessel in overcoming friction in the delivery pipe. Take Darcy–Weisbach friction factor $f = 0.03$.

(*Ans*: Power savings = 5.45 kW)

16.60 A single acting reciprocating pump has a stroke length of 40 cm and a cylinder diameter of 25 cm. The delivery pipe is 20 m long and has a diameter of 15 cm. A large diameter air vessel is fitted to the delivery pipe. For a crank speed of 40 rpm, determine the quantity of water going in or coming out of the air vessel when the crank angle is (i) 15° (ii) 90° and (iii) 120°. Also, determine the crank angle at which there is no flow into or out of the air vessel.

(*Ans*: (i) 0.0025 m^3/s into the air vessel,
(ii) 0.028 m^3/s goes out of the air vessel
(iii) 0.016 m^3/s goes out of the air vessel;
θ = 18, 58° or 161.41°)

16.61 A double acting reciprocating pump has the following data:

Cylinder diameter = 10 cm
Stroke length = 20 cm
Static suction head = 3.0 m
Diameter of suction pipe = 8.0 cm
Suction pipe length = 8.0 m
Crank speed = 40 rpm
Delivery pipe diameter = 8.0 m
Length of delivery pipe = 15.0 m
Static delivery head = 15.0 m
Darcy–Weisbach friction factor $f = 0.03$.

Estimate the power required to drive the pump by assuming a pump efficiency of 90%.

(*Ans*: **P** = 4.97 kW)

Appendix - A

Multiple Choice Objective Questions

Series–1

1. Match List I with List II and select the correct answer using the code given below the lists.

List I
(Description)

A. Property which explains the spherical shape of the drop of a liquid
B. Property which explains the phenomenon of cavitation in a fluid flow
C. Property which explains the rise of sap in a tree
D. Property which explains the flow of a jet of oil in an unbroken stream

List II
(Property of Fluid)

1. Viscosity
2. Surface Tension
3. Compressibility
4. Vapour pressure
5. Capillarity

Codes:

	A	B	C	D
(a)	1	2	4	5
(b)	2	4	5	1
(c)	4	2	5	1
(d)	1	2	3	4

2. Fig. M.02 shows the rheological behaviour of four types of viscous fluids. With reference to this figure, match the following two lists.

List I
(Curve in Fig. M 02)

A. Curve A
B. Curve B
C. Curve C
D. Curve D

List II
(Classification)

1. Dilatant
2. Ideal Bingham Plastic
3. Pseudo Plastic
4. Newtonian

Codes:

	A	B	C	D
(a)	2	4	3	1
(b)	1	4	3	2
(c)	3	2	1	4
(d)	3	1	4	2

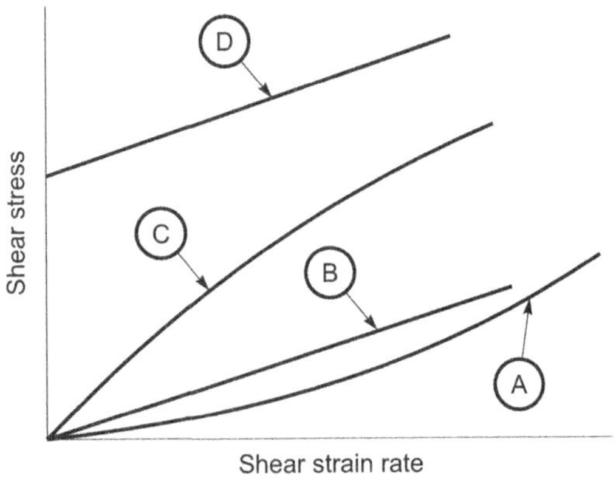

Fig. M.02 *Question No. 2*

3. Match List I with List II and select the correct answer using the code given below the lists.

List I

A. Submerged body
B. Floating body
C. Metacentric height
D. Buoyancy

List II

1. Force on a curved surface
2. Moment of inertia
3. Force acting vertically up
4. Metacentre
5. Centre of Buoyancy

Codes:

	A	B	C	D
(a)	5	4	2	1
(b)	1	2	4	3
(c)	5	4	2	3
(d)	1	2	5	3

4. Consider the following statements:

1. In a source, equipotential lines are circles.
2. Flownet is a representation of two-dimensional irrotational flow of incompressible fluid.
3. Boundary acts as limiting equipotential lines in a flownet.
4. In uniform flow region, streamlines will be parallel and equidistant.

Of these statements,

(a) 1, 2, and 3 are correct
(b) 1, 2, and 4 are correct
(c) 2, 3, and 4 are correct
(d) 1, 3, and 4 are correct

5. Match List I with List II and select the correct answer using the code given below the lists

List I	List II
A. Rotational flow	1. A fluid motion in which streamlines are concentric circles.
B. Vortex flow	2. The fluid particles moving in concentric circles may not rotate about their mass centres.
C. Free vortex	3. The fluid particles moving in concentric circles may rotate about their mass centres.
D. Forced vortex	4. Flow near a curved boundary

Codes:

	A	B	C	D
(a)	4	2	3	1
(b)	3	1	2	4
(c)	1	3	2	4
(d)	4	1	2	3

6. Match List I with List II and select the correct answer using the code given below the lists.

List I (*Phenomenon*)	List II (*Condition*)
A. Rotational flow	1. Velocity is zero or infinite.
B. Irrotational flow	2. Proportional to velocity
C. Singularities	3. Vorticity is zero
D. Streamline spacing	4. Vorticity exists

Codes:

	A	B	C	D
(a)	3	4	1	2
(b)	4	3	1	2
(c)	4	3	2	1
(d)	3	4	2	1

7. Match List I with List II and select the correct answer using the code given below the lists.

List I	List II
A. Stream lines	1. Tracing of motion of any one fluid particle
B. Streak lines	2. Tracing of motion of different fluid particles
C. Path lines	3. Identification of location of number of fluid particles
D. Equipotential lines	4. Orthogonal to streak lines
	5. Location of equal piezometric heads.

Codes:

	A	B	C	D
(a)	2	3	4	5
(b)	3	2	1	4
(c)	1	2	4	3
(d)	2	3	1	5

8. Which of the following pairs are correctly matched?

1. Piezometric head ... Sum of datum head and pressure head
2. Dynamic head ... Sum of datum head and velocity head

3. Stagnation head ... Sum of piezometric head and velocity head
4. Total head ... Sum of piezometric head and dynamic head

Select the correct answer using the codes given below:

Codes: (a) 1, 2 and 3 (b) 1, 3 and 4
(c) 1, 2 and 4 (d) 2, 3 and 4

9. Under which of the following conditions will the equation $\frac{p}{\gamma} + Z + \frac{V^2}{2g}$ = constant, will be valid in the whole flow field?

1. Flow is rotational
2. Flow is irrotational
3. Flow is incompressible
4. Flow is steady
5. Flow is laminar

Select the correct answer using the codes given below.

Codes: (a) 1, 3 and 5 (b) 2, 4 and 5
(c) 1, 3 and 4 (d) 2, 3 and 4

10. Match List I (phenomenon) with List II (equation/concept involved) and select the correct answer using the code given below the lists.

List I	List II
A. Force developed in a bend	1. Continuity equation
B. Pitot-static tube	2. Energy equation
C. Flow through smaller passage produces higher velocity	3. Momentum equation
D. Vortex flow	4. Moment of momentum equation

Codes:

	A	B	C	D
(a)	3	2	4	1
(b)	3	2	1	4
(c)	2	3	4	1
(d)	2	3	1	4

11. Match List I with List II and select the correct answer using the code given below the lists.

List I	List II
A. Moment of momentum equation	1. Equation to find energy loss in a pipeline having laminar flow
B. Bernoulli equation	2. Equation of motion for one-diamensional steady flow of ideal incompressible fluid.
C. Euler's equation	3. Equation based on conservation of momentum principle applicable to circulatory flow.
D. Hagen–Poiseuille equation	4. Three-dimensional equation of motion based on principle of conservation of momentum for ideal and incompressible fluid.

Codes:

	A	B	C	D
(a)	2	3	4	1
(b)	3	2	1	4
(c)	2	3	1	4
(d)	3	2	4	1

12. Match List I (units) with List II (dimensions) and select the correct answer using the code given below the lists.

List I	List II
A. Pressure	1. $M^0 L^0 T^0$
B. Power	2. $ML^{-2} T^{-2}$
C. Reynolds number	3. $ML^2 T^{-3}$
D. Specific weight	4. $ML^{-1} T^{-2}$

Codes:

	A	B	C	D
(a)	3	4	2	1
(b)	4	3	2	1
(c)	4	3	1	2
(d)	3	4	1	2

13. Match List I with List II and select the correct answer using the code given below the lists.

List I
(*Fluid property*)

A. Compressibility
B. Gravity
C. Viscosity
D. Vapour pressure

List II
(*Flow phenomenon*)

1. Flow of real fluid past a tiny sphere
2. Cavitation
3. Hydraulic jump
4. Flight of supersonic aircraft

Codes:

	A	B	C	D
(a)	4	3	2	1
(b)	4	3	1	2
(c)	3	4	1	2
(d)	3	4	2	1

14. Match List I (Phenomenon) with List II (Tool for dynamic similarity) and select the correct answer using the code given below the lists.

List I

A. Flow in open channel
B. Flow in pipes
C. Movement of bullet in air
D. Capillary flow

List II

1. Froude number
2. Reynolds number
3. Weber number
4. Mach number
5. Thoma number

Codes:

	A	B	C	D
(a)	1	2	3	4
(b)	1	2	3	5
(c)	2	3	4	5
(d)	1	2	4	3

15. Match List I with List II and select the correct answer using the code given below the lists.

List I
(*Non-Dimensional numbers*)

A. Mach number
B. Thoma number
C. Reynolds number
D. Weber number

List II
(*Application*)

1. Waves in an ocean
2. Launching of rockets
3. Cavitation phenomenon
4. Capillary flow in soil
5. Motion of a submarine

Codes:

	A	B	C	D
(a)	1	3	5	2
(b)	2	3	5	4
(c)	3	2	4	1
(d)	2	1	3	4

16. Match List I with List II and select the correct answer using the case given below the lists.

List I
(*Fluid problems*)

A. Flow over a spillway dam
B. Flow through a butterfly valve
C. Rise of moisture in the stem of a plant
D. Water hammer in a penstock

List II
(*Model laws*)

1. Euler number
2. Froude number
3. Mach number
4. Renynolds number
5. Weber number

Codes:

	A	B	C	D
(a)	2	4	5	3
(b)	4	2	3	1
(c)	2	3	1	4
(d)	5	1	4	2

17. In a distorted model of a free surface flow phenomenon L_r = horizontal scale and h_r = vertical scale. Match the following two lists pertaining to the model.

List I	List II
A. Slope ratio	1. $L_r^{-1/2} h_r^{3/2}$
B. Discharge ratio	2. $L_r h_r^{-1/2}$
C. Time scale	3. $L_r h_r^{3/2}$
D. Manning's roughness ratio	4. $L_r^{-1} h_r$

Codes:

	A	B	C	D
(a)	1	4	3	2
(b)	4	2	1	3
(c)	3	4	2	1
(d)	4	3	2	1

18. Consider the following statements regarding the laminar flow through a circular pipe:

1. The friction factor is constant
2. The friction factor depends upon the pipe roughness
3. The friction factor varies inversely with the Reynolds number
4. The velocity distribution is parabolic
5. The pressure drop varies directly with the square of mean velocity of these statements

(a) 1 and 4 are correct (b) 3 and 5 are correct
(c) 2 and 4 are correct (d) 3 and 4 are correct

19. Match List I with List II and select the correct answer using the code given below the lists.

List I	List II
A. Boundary layer thickness	1. Distance from the boundary where velocity is 99% of uniform velocity
B. Displacement thickness	2. Distance from the boundary by which the main flow can be assumed to be shifted
C. Laminar boundary layer thickness	3. Varies as $x^{4/5}$
D. Turbulent boundary layer thickness	4. Varies as $x^{(-1/2)}$
	5. Varies as $x^{(-1/2)}$

Codes:

	A	B	C	D
(a)	1	2	4	3
(b)	1	2	5	3
(c)	5	4	2	1
(d)	2	1	5	3

20. Match List I (Typical occurrence) with List II (Relevant flow condition) and select the correct answer using the code given below the lists.

List I *(Typical occurrence)*	List II *(Relevant flow condition)*
A. Cavitation	1. Absence of fluid velocity
B. Separation	2. Fluid pressure reduces to vapour pressure
C. Stagnation point	3. Bluff body in flow
D. Wake	4. Adverse pressure gradient in widening boundaries of flow

Codes:	A	B	C	D
(a)	4	2	3	1
(b)	2	4	3	1
(c)	4	2	1	3
(d)	2	4	1	3

21. Match List I with List II and select the correct answer using the code given below the lists.

List I (*Body*)	List II (*Drag coeficient*)
A. Aerofoil ($R_e > 10^7$)	1. 240
B. Sphere ($10^4 < R_e < 10^5$)	2. 1.90
C. Flat plate perpendicular to flow	3. 0.50
D. Droplet ($R_e \approx 0.1$)	4. 0.003

Codes:	A	B	C	D
(a)	1	2	3	4
(b)	4	3	1	2
(c)	4	3	2	1
(d)	4	1	3	2

22. Match List I with List II and select the correct answer using the code given below the lists.

List I (*Object in a fluid*)	List II (*Drag coefficient*)
A. Submarine	1. 1.33
B. Parachute	2. 0.10
C. Aircraft wing (air foil)	3. 0.15
D. Smoke stack (Chimney) (Reynolds number = 10^5)	4. 1.00

Codes:	A	B	C	D
(a)	1	3	2	4
(b)	3	1	4	2
(c)	1	3	4	2
(d)	3	1	2	4

23. Match List I with List II and select the correct answer using the code given below the lists.

List I (*Nature of flow*)	List II (*Friction factor equation*)
A. Turbulent flow with rough boundary	1. $f = 64/R_e$
B. Turbulent flow with smooth boundary	2. $f = \dfrac{0.316}{R_e^{0.25}}$
C. Turbulent flow with $R_e < 10^5$	3. $\dfrac{1}{\sqrt{f}} = 2 \log \dfrac{r_o}{\varepsilon} + 1.74$
D. Laminar flow	4. $\dfrac{1}{\sqrt{f}} = 2 \log R_e\sqrt{f} - 0.8$

Codes:	A	B	C	D
(a)	4	3	2	1
(b)	3	4	2	1
(c)	3	4	1	2
(d)	4	3	1	2

24. Consider the following statements:

1. Pipe network analysis is normally necessary in analyzing flow in pipes at city water systems.
2. Hardy–Cross method of solving pipe network is a method of successive approximations and is not a direct method.

3. The network must satisfy the momentum equation because the flow in each pipe satisfies the head loss equation.
4. Principle of continuity is satisfied in a pipe network.

Of these statements

(a) 1, 2 and 3 are correct
(b) 2, 3 and 4 are correct
(c) 1, 3 and 4 are correct
(d) 1, 2 and 4 are correct

25. Consider the following statements:

1. Flow is established in a pipe when the boundary layer thickness is equal to the radius of the pipe.
2. For laminar flow, the friction factor in Darcy–Weisbach equation varies inversely as the Reynolds number.
3. For turbulent flow, the friction factor in Darcy–Weisbach equation varies inversely as the square of Reynolds number.
4. When the boundary is rough, the friction factor varies with the relative roughness of the pipe.

Of these statements

(a) 1, 2 and 3 are correct
(b) 1, 2 and 4 are correct
(c) 2, 3 and 4 are correct
(d) 1, 3 and 4 are correct

26. The loss of head at various pipe fittings (h_L) is given by $h_L = K\dfrac{V^2}{2g}$. Match List I (Value of K) with List II (Pipe fitting) given below:

List I	List II
A. 0.40	1. Pump foot valve
B. 0.90	2. Standard 45° elbow
C. 1.50	3. Standard 90° elbow
D. 10.0	4. Globe valve (Fully open)

Codes:

	A	B	C	D
(a)	4	2	3	1
(b)	2	3	1	4
(c)	2	3	4	1
(d)	3	2	1	4

27. Match List I (Surface pofile) with List II (Description of the profile and select the correct answer using the code given below the lists.

List I		List II
A.	M_2	1. Convex upward; asymptotic to horizontal at downstream end; depth increasing in down-stream direction.
B.	S_3	2. Concave downwards; upstream asymptotic to normal depth with depth decreasing in downstream direction.
C.	C_1	3. Depth increasing downstream and meeting at an angle to CDL; a curve with an inflexion point
D.	A_3	4. Convex upward and depth increasing in flow direction; asymptotic to NDL at downstream.

Codes:

	A	B	C	D
(a)	2	4	1	3
(b)	2	1	4	3
(c)	3	4	1	2
(d)	3	1	4	2

28. Match List I with List II and select the correct answer using the code given below the lists.

List I (*Hydraulic structure*)	List II (*Water surface profile at the structure*)	
A. On a flat topped broad crested weir	1.	M_1
B. Immediately below a sluice gate on a level apron	2.	M_2
C. Behind the weir on an alluvial river	3.	S_1

D. In a chute spillway

4.	S_2
5.	H_2
6.	H_3

Codes:	A	B	C	D
(a)	5	6	1	4
(b)	4	6	2	1
(c)	2	5	6	4
(d)	5	4	1	3

29. Match List I with List II and select the correct answer using the code given below the lists. (y_o = Normal depth, y_c = critical depth and y = depth in gradually varied flow)

List I	List II
A. $y_c > y_o > y$	1. S_2
B. $y_o > y > y_c$	2. M_1
C. $y > y_c > y_o$	3. S_3
D. $y > y_o > y_c$	4. M_2
	5. S_1

Codes:	A	B	C	D
(a)	2	4	1	3
(b)	3	4	5	2
(c)	5	4	1	3
(d)	3	1	4	2

30. Match List I with List II and select the correct answer using the code given below the lists.

List I	List II
A. Specific force	1. Head loss due to friction
B. Specific energy	2. Rapidly varied flow
C. Hydraulic jump	3. Alternate depths
D. Darcy–Weisbach equation	4. Conjugate depths

Codes:	A	B	C	D
(a)	4	3	2	1
(b)	4	3	1	2
(c)	3	4	1	2
(d)	3	4	2	1

31. Match List I with List II and select the correct answer using the code given below the lists.

List I (*Devices*)	List II (*Use*)
A. Hydrometer	1. Measures kinematic viscosity
B. Pitot–Static tube	2. Measures flow velocity
C. Orifice meter	3. Measures specific gravity
D. Engler's apparatus	4. Measures volumetric flow rate

Codes:	A	B	C	D
(a)	1	4	2	3
(b)	3	1	2	4
(c)	3	2	1	4
(d)	3	2	4	1

32. Match List I with List II and select the correct answer using the code given below the lists.

List I	List II
A. Venturimeter	1. Flow rate
B. Current meter	2. Flow velocity
C. Piezometer	3. Gauge pressure
D. Barometer	4. Atmospheric Pressure

Codes:

	A	B	C	D
(a)	2	1	4	3
(b)	1	2	3	4
(c)	1	2	4	3
(d)	2	1	3	4

33. Match List I (Name of instrument) with List II (Variable measured) and select the correct answer using the code given below the lists.

List I	List II
A. Hot-wire anemometer	1. Boundary shear stress
B. Orifice meter	2. Discharge
C. Pitot tube	3. Velocity
D. Preston tube	4. Pressure
	5. Turbulence

Codes:

	A	B	C	D
(a)	2	3	4	1
(b)	5	2	3	4
(c)	2	5	1	3
(d)	5	2	3	1

34. Match List I with List II and select the correct answer using the code given below the lists.

List I *(Shape of the weir)*	List II *(Value of the exponent in the equation $Q = KH^n$)*	
A. Rectangular weir	1.	1.0
B. Triangular weir	2.	2.0
C. Sutro weir	3.	1.5
D. Parabolic weir	4.	2.5

Codes:

	A	B	C	D
(a)	4	2	1	3
(b)	2	3	4	1
(c)	3	4	2	1
(d)	3	4	1	2

35. Which of the following statements are true in relation to water hammer phenomenon?

1. It causes surface erosion in pipes.
2. The pressure rise is given by ρ CV for sudden closure of valve.
3. It is accompanied by serious cavitation.
4. The volume modulus of fluid is the relevant fluid property.
5. It is governed by the Reynolds number of the flow.

Select the correct answer using the codes given below.

Codes: (a) 3 and 5 (b) 2 and 4
(c) 2 and 5 (d) 1, 2, 3 and 4

36. Match List I with List II and select the correct answer using the code given below the lists.

List I	List II
A. Positive surge traveling upstream	1. Occurs on upstream of gate that is partly closed suddenly.
B. Positive surge traveling downstream	2. Occurs on downstream of gate that is partly closed suddenly.
C. Negative surge traveling upstream	3. Occurs on upstream of gate that is opened suddenly.
D. Negative surge traveling downstream	4. Occurs on the downstream that is opened suddenly.

Codes:	A	B	C	D
(a)	1	4	3	2
(b)	1	2	3	4
(c)	3	4	1	2
(d)	3	4	2	1

37. Match List I with List II and select the correct answer using the code given below the lists.

List I (*Type of turbines*)	List II (*Ranges of specific speeds in SI units*)
A. Francis	1. 8–30
B. Kaplan	2. 26–40
C. Pelton (one jet)	3. 40–420
D. Pelton (two jets)	4. 380–950

Codes:	A	B	C	D
(a)	3	4	2	1
(b)	4	3	2	1
(c)	3	4	1	2
(d)	4	3	1	2

Series–2

38. Which of the following will be satisfied by irrotational flow of an imcompressible fluid?

1. $\frac{\partial u}{\partial x}+\frac{\partial v}{\partial y}+\frac{\partial w}{\partial z}=0$
2. $\frac{\partial u}{\partial y}+\frac{\partial v}{\partial x}=\frac{\partial u}{\partial z}+\frac{\partial w}{\partial x}=\frac{\partial w}{\partial y}+\frac{\partial v}{\partial z}=0$
3. $\frac{\partial^2 u}{\partial x^2}+\frac{\partial^2 v}{\partial y^2}+\frac{\partial^2 w}{\partial z^2}=0$
4. $\frac{\partial v}{\partial x}=\frac{\partial u}{\partial y};\quad \frac{\partial u}{\partial z}=\frac{\partial w}{\partial x};\quad \frac{\partial v}{\partial z}=\frac{\partial w}{\partial y}$

Select the correct answer from the codes given below:

(a) 3 and 4 (b) 1 and 2 (c) 1 and 3 (d) 1 and 4

39. Which of the following conditions will be satisfied by two-dimensional steady, irrotational flow?

(a) $\frac{\partial u}{\partial y}+\frac{\partial v}{\partial x}=0$ (b) $\frac{\partial v}{\partial x}-\frac{\partial u}{\partial y}=0$ (c) $\frac{\partial u}{\partial x}+\frac{\partial v}{\partial y}=0$ (d) $\frac{\partial u}{\partial x}-\frac{\partial v}{\partial y}=0$

Select the correct answer using the codes given below:

Codes: (a) 1 and 2 (b) 1 and 3
(c) 2 and 4 (d) 2 and 3

40. Assuming the thrust T of a propeller depends upon the diameter D, speed of advance V, angular velocity ω, dynamic visocsity µ, and density ρ, which of the following dimensionless parameters can be derived by dimensional analysis?

(1) $\frac{T}{\rho D^2 V^2}$ (2) $\frac{VD\mu}{\rho}$ (3) $\frac{D\omega}{V}$ (4) $\frac{VD\rho}{\mu}$

Select the correct answer using the codes given below:

(a) 1, 2 and 3 (b) 2, 3 and 4 (c) 1, 3 and 4 (d) 1, 2 and 4

41. Which of the following statements are correct in respect of steady laminar flow through a circular pipe?

1. Shear stress is zero at the centre
2. Discharge varies directly with the viscosity of the fluid

3. Velocity is maximum at the centre
4. Hydraulic gradient varies directly with the velocity

Select the correct answer using the codes given below:

Codes: (a) 1, 2 and 4 (b) 1, 3 and 4
(c) 1 and 3 (d) 3 and 4

42. Given that, as the flow takes place between two parallel plates, the velocity midway between the plates is 2.0 m/s, the Reynolds number is 1200 and the distance between the plates is 10 cm, which of the following statements are true?

1. The velocity at the boundary is 1 m/s
2. The rate of flow is 0.13 m^3/s per metre width
3. The flow is turbulent
4. The energy correction factor α is 2.0.

Select the correct answer using the codes given below:

Codes: (a) 2 and 3 (b) only 2
(c) 1 and 3 (d) 1, 2 and 4

43. Consider the following statements:
In a laminar flow occurring in a circular conduit

1. The flow is rotational
2. Loss of head is proportional to the square of velocity
3. Loss of head is proportional to first power of viscosity
4. Other quantities remaining same, increase in diameter increases the Reynolds number.

Of these statements:

(a) 1 and 2 are correct (b) 1 and 3 are correct (c) 2, 3 and 4 are correct (d) 1, 2 and 4 are correct

44. Consider the following conditions for the pipe network shown is Fig. M-44

1. $Q_1 = Q_2$
2. $h_{f1} = h_{f3}$
3. $Q_2 = Q_1 + Q_3$
4. $h_{f1} = h_{f2} = h_{f3}$

Which of these conditions must be satisfied by the pipe system?

(a) 1 and 3 (b) 2 and 3
(c) 2 and 4 (d) 1 and 4

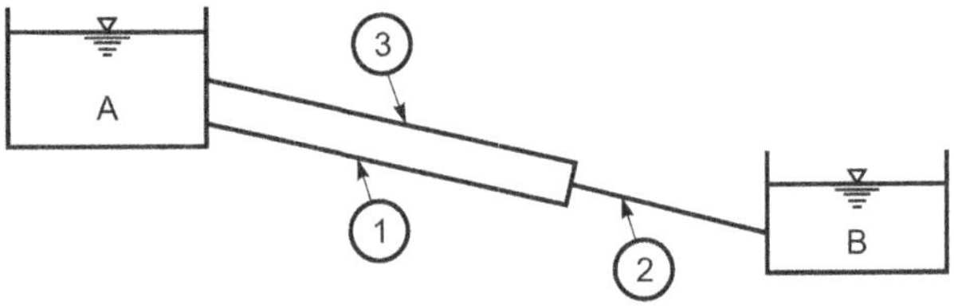

Fig. M-44 *Qusetion No. 44*

45. The following parametes relate to flow in a penstock:

1. water level in a reservoir
2. density of water
3. elasticity of water
4. roughness of pipe

Pressure rise due to water hammer is a penstock depends upon

(a) 1 and 2 (b) 2 and 3 (c) 3 and 4 (d) 1, 2, 3 and 4

46. Consider the follwing statements:
In a supersonic flow through a converging duct, along the flow the

1. velocity decreases
2. density increases
3. pressures decreases
4. temperature decreases

Of these, the correct statements are:

(a) 3 and 4 (b) 1 and 2 (c) 1, 3 and 4 (d) 3 and 4

47. Consider the following statements relating to compressible flow

1. In a contracting conduit, sonic velocity is reached in the minimum cross sectional area.
2. In a contracting conduit, supersonic velocities are reached only in the expanding section downstream from the minimum section.
3. An expanding shock wave is physically possible and can be exhibited.
4. In the Laval nozzle, it is possible to obtain a velocity greater than the velocity of sound in the expanding section of the nozzle.

Of these, statements

(a) 1, 2 and 3 are correct (b) 1, 2 and 4 are correct
(c) 2, 3 and 4 are correct (d) 1, 3 and 4 are correct

48. Consider the following statements:

1. Pumps in series operation allow the head to increase.
2. Pumps in series operation increase the flow rate.
3. Pumps is parallel operation increase the flow rate.
4. Pumps is parallel operation allow the head to increase.

Of these statements

(a) 1 and 3 are correct (b) 2 and 4 are correct
(c) 3 and 4 are correct (d) 1 and 2 are correct

Series–3

Directions: The following items consist of two statements labeled *Assertion A* and *Reason R*. You are to examine each statement carefully and decide if the Assertion A and Reason R are individually true and if so, whether the Reason R is a correct explanation of Assertion A. Select your answers to these items using the codes given below.
Codes:

(a) Both A and R are true and R is the correct explanation of A
(b) Both A and R are true but R is not a correct explanation of A
(c) A is true but R is false
(d) A is false but R is true

49. Assertion (A): The kinematic viscosity of both air and water decreases as the temperature increases.
Reason (R): The kinematic viscosity of liquids and gases at a given pressure is a function of temperature.

50. Assertion (A): The relative movement of two blocks of wood joined with hot glue requires greater and greater effort as the glue dries up.
Reason (R): Viscosity of liquids varies inversely with temperature.

51. Assertion (A): For submerged bodies, static forces in the lateral direction are balanced.
Reason (R): For submerged bodies static forces in the vertical directions are not balanced.

52. Assertion (A): The following potential function in two-dimensional flow field represents rotational flow

$$\varphi = 2x^2 - 3y^2$$

Reason (R): For the given function

$$\frac{\partial u}{\partial x} + \frac{\partial v}{\partial y} \neq 0$$

53. Assertion (A): In Raleigh's method of dimensional analysis, the dependent variable is written as the function of independent variables.
Reason (R): In Raleigh's method, when the number of independent variables exceeds three, the exponents of non-repeating variables are expressed as exponents of repeating variables.

54. Assertion (A): In dimensional analysis, dimensionless numbers can be expressed as ratios of forces acting on the system.
Reason (R): In dimensional analysis, Mach number is the ratio of inertia force to elastic force.

55. Assertion (A): If laminar flow of oil between two points of a given pipeline is doubled, then the power consumption is increased to four times the original power.
Reason (R): In laminar flow through circular pipes, head loss varies directly as the discharge.

56. Assertion (A): In the pipe flow, the upper critical Reynolds number is not well defined.
Reason (R): Flow changes from transition to full turbulence depending upon the external disturbance.

57. Assertion (A): Flow in boundary layer is always laminar.
Reason (R): In turbulent flow on a smooth boundary, a laminar sublayer still exists within the boundary layer.

58. Assertion (A): In the equation $h_f = f \frac{l}{d} \frac{V^2}{2g}$, for laminar flow through a pipe of diameter d, the term V (mean velocity) is given by

$$V = \frac{(p_1 - p_2)r^2}{8\mu l}$$

Reason (R): The term f (friction factor) in the above equation equals 64/R_e, where R_e is the Reynolds number.

59. Assertion (A): Water flows through a pipe connecting two reservoirs. The line joining the water surface levels in the reservoirs is the hydraulic grade line.
Reason (R): There will be no negative pressure anywhere in the pipeline, as long as the pipeline lies below the hydraulic gradient.

60. Assertion (A): Energy is lost in a sudden contraction in a pipeline.
Reason (R): If the flow is reversed, energy is gained at the transition which acts as an expansion.

61. Assertion (A): A loss of head at a sudden contraction in a pipe is smaller than that at a sudden expansion.
Reason (R): When the flow contracts, it tends to become irrotational.

62. Assertion (A): Loss of head at a sudden expansion in a pipe is greater than that at a sudden contraction.
Reason (R): Flow in a sudden expansion tends to be irrotational

63. Assertion (A): In an open channel flow, the maximum velocity filament does not occur on the free surface.
Reason (R): There is wind drag on the free surface of an open channel.

64. Assertion (A): A minor change in the specific energy at or close to critical state will cause a major change in depth.
Reason (R): Froude number being equal to unity characterizes a critical state of flow.

65. Assertion (A): At the critical state of flow, the specific force is a minimum for the given discharge.
Reason (R): For a minimum value of the specific force, the first derivative of force with respect to depth should be unity.

66. Assertion (A): A hydraulic jump cannot be expected in a long steep slope (fed by a large reservoir) when it is followed by a short stretch of adverse slope terminating in a deep and wide reservoir.
Reason (R): The terminal depth in such an adverse slope reach will be critical and the flow in the steep slope may be nearly at or at normal depth.

67. Assertion (A): In a running irrigation canal, a regulating gate is partially closed suddenly. Such movement gives rise to a positive surge travelling downstream and a negative surge travelling upstream of the gate.
Reason (R): The sudden change in the discharge gives rise to changes in water elevations both upstream and downstream of the gate and develops in the form of an unsteady, rapidly verying flow which propagates in the form of a surge wave moving on either side of the gate.

68. Assertion (A): The lower critical Reynolds number for pipe flow is well defined at 2000 while the upper critical Reynolds number has been found to exceed even 40,000.

Reason (R): When the Reynolds number increases above that of the lower critical value, the transition tends to depend on external disturbance.

69. Assertion (A): Surges can be positive or negative.
Reason (R): Negative surges may occur when the gate at the head of a channel is suddenly opened or when the gate at tail end of a channel is suddenly closed.

70. Assertion (A): The specific speed of a Pelton wheel is generally much less than that of a reaction turbine.
Reason (R): Pelton wheels generally use more than one nozzle and the specific speed is defined for power developed per nozzle.

71. Assertion (A): Inlet side of a pump is less susceptible to cavitation damage.
Reason (R): Cavitation occurs when the velocity is high and pressure is low.

72. Assertion (A): The inlet velocity triangle for a Pelton wheel is a straight line.
Reason (R): For a Pelton turbine, the vane angle at inlet is 180°.

73. Assertion (A): By providing air vessels on the suction and delivery sides of a reciprocating pump, it is possible to increase the delivery head of the pump.
Reason (R): The air vessel eliminates the acceleration head and makes discharge uniform.

74. Assertion: Reflux valve is provided on the suction side of a centrifugal pump.
Reason (R): Reflux valve prevents backflow from the line into the pump and consequent damage to the pump set.

Appendix B-1

Answers to Objective Questions

Chapter	Question No.	0	1	2	3	4	5	6	7	8	9
Chapter 1	1.00	_	d	b	b	a	a	c	b	b	a
	1.10	a	d	b	b	d	b	c	b	a	d
	1.20	c	a	c	d	d	d	a	c	b	d
	1.30	c	a	d	c	d	a	a	b	c	a
	1.40	d	b	b	d						
Chapter 2	2.00	_	c	c	d	c	a	b	c	b	d
	2.10	a	b	c	b	b	c	b	b	c	d
	2.20	a	a	b	b	c	d	c	a	d	c
	2.30	a	b	c	d	c	b	b	b	a	d
	2.40	a	c	b	c	a	c	d	b	d	d
	2.50	d	c	d	b	b	c	b	c	d	d
	2.60	c	b	d	a	c	b	b	d	a	b
	2.70	a									
Chapter 3	3.00	_	a	b	c	d	d	b	c	d	d
	3.10	a	c	b	d	b	c	d	a	c	d
	3.20	a	c	a	b	c	d	d	d	a	b
	3.30	c	b	c	c	c	d	c	d	a	a
	3.40	d	a	b	d	c	c	a	b		
Chapter 4	4.00	_	c	b	d	b	b	b	c	c	a
	4.10	a	d	c	d	c	a	c	b	a	d
	4.20	b	a	d	a	a	a	d	a	b	d
	4.30	d	c	d	a						
Chapter 5	5.00	_	c	b	d	c	a	c	a	d	c
	5.10	b	b	c	d	d	b	d	c	a	c
	5.20	a	c	b	a	c	d	b	d	b	a
Chapter 6	6.00	_	c	b	d	a	c	b	d	c	b
	6.10	d	b	c	b	c	a	b	b	a	b
	6.20	c	d	b	c	a	b	c	c		
Chpater 7	7.00	_	c	b	d	c	d	d	b	d	a
	7.10	b	c	a	c	d	a	d	c	c	d
	7.20	a	d	d	b	c	c	a	c	b	b
	7.30	d	c	a	d	d	b	b	c	d	
Chapter 8	8.00	_	d	c	a	c	b	c	d	a	c
	8.10	c	a	b	d	a	a	c	d	a	c
	8.20	a	d	c	b	a	d	b	c	d	c
	8.30	c	b	b	c	b	c	b	a	a	

Chapter	Question no	0	1	2	3	4	5	6	7	8	9
Chapter 9	9.00	_	c	d	b	a	c	b	b	b	d
	9.10	c	b	a	c	d	c	d	b	a	b
	9.20	c	d	c	b	c	c	b	d	a	d
	9.30	b	c	a	d						
Chapter 10	10.00	_	c	c	d	a	c	b	d	a	b
	10.10	c	b	d	a	b	b	c	b	d	c
	10.20	b	b	c	a	c	d	a	a	b	c
	10.30	d	c	b	a	b	a	c	b	d	d
Chapter 11	11.00	_	c	d	a	d	b	b	a	b	c
	11.10	c	a	b	d	c	d	d	c	a	c
	11.20	a	b	d	a	a	b	a	b	c	
Chapter 12	12.00	_	b	d	c	b	a	c	a	a	d
	12.10	d	a	b	d	c	a	d	d	a	c
	12.20	c	b	c	d	b	b	d	c	b	c
	12.30	a	b	d	c	a	d	d	c	b	c
	12.40	a	d	c	d	c	b	d	b	a	c
	12.50	b	a	b	a	c	a	c	a	b	c
	12.60	a	b	b	d	d	b	b	a	c	c
Chapter 13	13.00	_	b	c	a	c	d	b	d	d	a
	13.10	b	c	d	a	b	c	d	a	b	d
	13.20	c	b	c	a	d	c	b	a	d	c
	13.30	d	b	b	a	d	a	b	c	d	b
	13.40	c	d	c	b	c	a	b	d	a	d
	13.50	c	b	b	c	c	c	b	c		
Chapter 14	14.00	_	a	c	d	b	d	b	d	d	b
	14.10	d	a	a	b	c	d	b	b	b	a
	14.20	a	c	d	b	b	c	c	d	d	d
	14.30	b									
Chapter 15	15.00	_	b	a	c	d	b	b	a	c	b
	15.10	d	d	c	a	b	b	d	a	b	a
	15.20	d	c	a	c	b	d	b	c	b	b
	15.30	a	d	c	b	a	c	d	c		
Chapter 16	16.00	_	c	b	a	b	d	c	c	b	c
	16.10	d	c	d	a	b	d	a	b	c	c
	16.20	c	b	a	d	c	c	d	a	d	b
	16.30	a	d	a	b	c	b	a	d	a	b
	16.40	c	d	c	b	b	c	c	d	a	c
	16.50	a	b	a	b	c	d	a			

Appendix B-2

Answers to Multiple Choice Objective Questions of Appendix A

	Question No.	0	1	2	3	4	5	6	7	8	9
Series-1	00	_	b	b	c	b	b	b	d	b	d
	10	b	d	c	b	d	b	a	d	d	b
	20	d	c	d	b	b	b	b	a	a	b
	30	a	d	b	d	d	b	a	c		
Series-2	30									d	d
	40	c	b	b	d	b	d	b	b	a	
Series-3	40										d
	50	a	b	d	c	b	a	a	d	d	b
	60	c	b	c	b	b	a	a	d	a	c
	70	b	d	a	d	d					

Index